Fundamental Electron Interactions with Plasma Processing Gases

PHYSICS OF ATOMS AND MOLECULES

Recent volumes in this series:

COINCIDENCE STUDIES OF ELECTRON AND PHOTON IMPACT IONIZATION
Edited by Colm T. Whelan and H. R. J. Walters

COMPLETE SCATTERING EXPERIMENTS
Edited by Uwe Becker and Albert Crowe

DENSITY MATRIX THEORY AND APPLICATIONS, SECOND EDITION
Karl Blum

ELECTRON MOMENTUM SPECTROSCOPY
Erich Weigold and Ian McCarthy

FUNDAMENTAL ELECTRON INTERACTIONS WITH PLASMA PROCESSING GASES
Loucas G. Christophorou and James K. Olthoff

IMPACT SPECTROPOLARIMETRIC SENSING
S. A. Kazantsev, A. G. Petrashen, and N. M. Firstova

INTRODUCTION TO THE THEORY OF COLLISIONS OF ELECTRONS WITH ATOMS AND MOLECULES
S. P. Khare

NEW DIRECTIONS IN ATOMIC PHYSICS
Edited by Colm T. Whelan, R. M. Dreizler, J. H. Macek, and H. R. J. Walters

PHOTON AND ELECTRON COLLISION WITH ATOMS AND MOLECULES
Edited by Philip G. Burke and Charles J. Joachain

PRACTICAL SPECTROSCOPY OF HIGH-FREQUENCY DISCHARGES
Sergei A. Kazantsev, Vyacheslav I. Khutorshchikov, Günter H. Guthöhrlein, and Laurentius Windholz

RELATIVISTIC HEAVY-PARTICLE COLLISION THEORY
Derrick S. F. Crothers

A Chronological Listing of Volumes in this series appears at the back of this volume.

A Continuation Order Plan is available for this series. A continuation order will bring delivery of each new volume immediately upon publication. Volumes are billed only upon actual shipment. For further information please contact the publisher.

Fundamental Electron Interactions with Plasma Processing Gases

Loucas G. Christophorou

National Institute of Standards and Technology
Gaithersburg, Maryland

and

Academy of Athens
Athens, Greece

James K. Olthoff

National Institute of Standards and Technology
Gaithersburg, Maryland

Kluwer Academic / Plenum Publishers

New York, Boston, Dordrecht, London, Moscow

Library of Congress Cataloging-in-Publication Data

Christophorou, L. G.
Fundamental electron interactions with the plasma processing gases/Loucas G. Christophorou and James K. Olthoff.
p. cm. — (Physics of atoms and molecules)
Includes bibliographical references and index.
ISBN 0-306-48037-9
1. Electron-molecule collisions. 2. Plasma (Ionized gases) I. Olthoff, James Kenneth, 1958– II. Title. III. Series.

QC794.6.C6C48 2004
539.7′57—dc22

2003064029

ISBN 0-306-48037-9

233 Spring Street, New York, N.Y. 10013

http://www.wkap.nl/

10 9 8 7 6 5 4 3 2 1

A C.I.P. record for this book is available from the Library of Congress

Printed in the United States of America

Contents

Preface

This volume deals with the basic knowledge and understanding of fundamental interactions of low energy electrons with molecules. It provides an up-to-date and comprehensive account of the fundamental interactions of low-energy electrons with molecules of current interest in modern technology, especially the semiconductor industry.

The primary electron-molecule interaction processes of elastic and inelastic electron scattering, electron-impact ionization, electron-impact dissociation, and electron attachment are discussed, and state-of-the-art authoritative data on the cross sections of these processes as well as on rate and transport coefficients are provided. This fundamental knowledge has been obtained by us over the last eight years through a critical review and comprehensive assessment of "all" available data on low-energy electron collisions with plasma processing gases which we conducted at the National Institute of Standards and Technology (NIST). Data from this work were originally published in the *Journal of Physical and Chemical Reference Data*, and have been updated and expanded here.

The fundamental electron-molecule interaction processes are discussed in Chapter 1. The cross sections and rate coefficients most often used to describe these interactions are defined in Chapter 2, where some recent advances in the methods employed for their measurement or calculation are outlined. The methodology we adopted for the critical evaluation, synthesis, and assessment of the existing data is described in Chapter 3. The critically assessed data and recommended or suggested cross sections and rate and transport coefficients for ten plasma etching gases are presented and discussed in Chapters 4, 5, and 6.

It is hoped that the fundamental knowledge on electron-molecule interactions and the assessed data on the cross sections and rate and transport coefficients provided in this volume will aid the graduate student and the researcher in many fields of pure and applied science. It is also hoped that it will aid scientists and engineers in many industrial areas since this knowledge generically underpins advances in plasma, laser, lighting, discharge, environmental, and other technologies.

LOUCAS G. CHRISTOPHOROU

JAMES K. OLTHOFF

To our families

Acknowledgments

We would like to thank the many colleagues who freely made their unpublished data available to us, the scientists who redirected their research at our request to obtain much needed data, and the many reviewers who throughout the course of this work provided us with valuable suggestions and encouragement.

Chapter 1

FUNDAMENTAL ELECTRON-MOLECULE INTERACTIONS AND THEIR TECHNOLOGICAL SIGNIFICANCE

1. Introduction

The study of fundamental electron-molecule collision processes and the behavior of slow electrons in gases under an applied electric field traces back for about 100 years (see, for example, Refs. [1–21]). The systematic study of electron-molecule (or atom) interactions as a function of the electron energy for atoms in their ground electronic state and for molecules in their ground vibrational and electronic states, especially since the 1960's, has produced a wealth of knowledge. This knowledge has advanced our understanding of the structure of atoms and molecules, the phenomena which accompany the interaction of ionizing radiation with matter and the deposition of energy by radiation in matter, the transport of electricity in gases, the interactions of electrons in plasmas, and the behavior of electrons in the condensed phases of matter. Concomitantly, this fundamental knowledge has underpinned and is underpinning the development of many of today's advanced technologies, including the environmental and life sciences.

The material covered in this book is almost entirely on molecules. In fact, it deals primarily with up-to-date knowledge on electron collisions with molecules of current interest to modern technology, especially semiconductor fabrication. However, recently available information on atoms and molecules used for purposes other than plasma processing is also discussed when it furthers our understanding of fundamental electron-collision processes and their significance in plasma modeling and other applied areas.

The fundamental generality of this book derives from four factors: (i) the basic nature of electron-molecule collisions, (ii) the generic significance of electron-molecule interactions to modern technology, (iii) the definitive, state-of-the art, and authoritative source of information it provides via a comprehensive and thorough approach, and (iv) the generally applicable guidance it offers for the proper use of electron-molecule collision data and the deduction of otherwise unavailable information from critically assessed data. The material covered in this book is largely based upon the critical reviews and comprehensive assessments of knowledge on electron-molecule interactions with plasma processing gases conducted by the authors over the last eight years. The papers that have resulted from this work [22–32] constitute the main source of information presented in the book. However, the data and information presented here are updated, expanded, and modified in light of new work that has become available following the publication of the original papers. Throughout this book the word "data" is used only with reference to experimental measurements.

2. Low Energy Electron-Molecule Interaction Processes

In this section a brief description is given of the fundamental interactions between electrons and ground-state molecules at relatively low electron energies, mostly less than 100 eV. This low electron-energy range is the most significant for plasma processing. In addition, two important new directions of recent systematic studies of electron-molecule interactions are discussed. Namely, investigation of the effects on electron-molecule interactions of (i) the internal energy of the target molecule, and (ii) the nature, density, and state of the medium in which the electron-molecule interaction occurs. The former studies involve electron-"hot" (ro-vibrationally excited) molecule and electron-electronically excited molecule interactions, and the latter concern electron-molecule interactions in dense gases, clusters, liquids, and at surfaces.

There is a unique richness in the variety of reactions which occur when a low-energy electron encounters a molecule. The encounters themselves may be conveniently separated into two types: *direct* and *indirect*. By *direct* we mean *nonresonant* collisions occurring over a wide range of electron energies where the electron is scattered at a distance from the molecule and the electron-molecule collision time is very short ($\sim 10^{-16}$ s). By *indirect* collisions we mean *resonant* collisions, which occur only over limited energy ranges when the electron energy is low and the duration of the collision is long — comparable to or longer than the time taken by the bound electrons of the target molecule to complete their orbits. In such intimate collisions, the electron enters the empty orbitals of the target molecule and is temporarily retained (captured) by the molecule forming characteristic resonances, the so-called negative ion resonances. Both types of collisions are discussed below.

2.1 Interactions of Low-Energy Electrons with Ground-State Molecules

As mentioned in the preceding paragraph, the collisions of low-energy electrons with atoms and molecules are of two types: *direct* and *indirect*. The principal *direct* elastic and inelastic electron scattering processes are:

$$e(\varepsilon) + \mathrm{AX} \rightarrow \mathrm{AX} + e(\varepsilon) \tag{1.1a}$$

(Direct elastic electron scattering)

$$\Rightarrow \mathrm{AX}^* + e(\varepsilon') \tag{1.1b}$$

(Direct inelastic electron scattering)

$$\Rightarrow \mathrm{A} + \mathrm{X} + e(\varepsilon') \tag{1.1c}$$

(Dissociation into neutrals)

$$\Rightarrow \mathrm{A} + \mathrm{X}^* + e(\varepsilon') \tag{1.1d}$$

(Dissociative excitation)

$$\rightarrow \mathrm{AX}^+ + 2e \tag{1.1e}$$

(Molecular ionization)

$$\Rightarrow \mathrm{A} + \mathrm{X}^+ + 2e \tag{1.1f}$$

(Dissociative ionization)

$$\Rightarrow \mathrm{A}^{(*)} + \mathrm{X}^+ + 2e \tag{1.1g}$$

(Dissociative ionization with fragment excitation)

$$\rightarrow \mathrm{A}^+ + \mathrm{X}^- + e(\varepsilon') \tag{1.1h}$$

(Ion-pair formation)

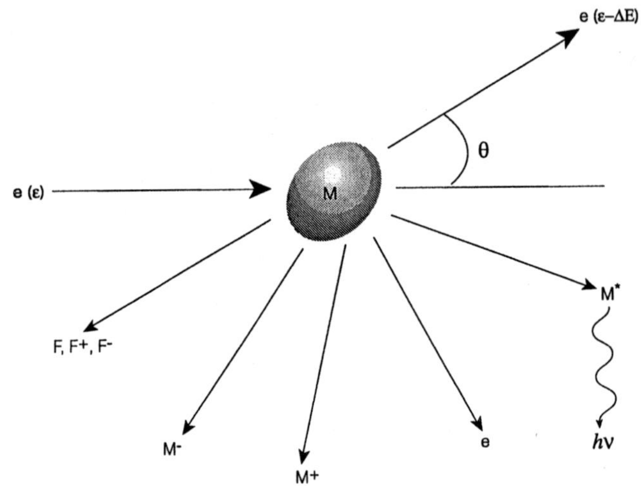

Figure 1.1. Schematic illustration of a generic electron-beam scattering experiment (based upon Fig. 1 in Trajmar and Cartwright [33]).

In the above reactions, $e(\varepsilon)$ and $e(\varepsilon')$ represent, respectively, the incident electron with energy ε and the scattered electron with energy ε'. The notation AX, AX^*, and AX^+ represents, respectively, ground-state, excited, and ionized molecules. Similarly, the notation A and X, X^*, X^+ and A^+, and X^- represents, respectively, ground-state fragments, excited fragments, positive ion fragments, and stable negative ion fragments. The asterisk denotes an excited species and the double arrow indicates multiple reaction pathways, which occur abundantly for polyatomic molecules. Reactions (1.1e) to (1.1g) can also lead to multiple ionization of AX or X, generating multiply charged positive ions AX^{n+} or $X^{n+} (n \geq 2)$.

Figure 1.1 shows schematically the general principle of electron-beam experiments. These experiments are conducted under single-collision conditions, that is, at sufficiently low pressures so that an incident electron has little chance of colliding with more than one molecule in the reaction chamber designed to investigate such processes. The incident electron $e(\varepsilon)$, upon collision with the molecule M, is scattered at an angle θ with respect to its incoming direction having undergone an energy loss ΔE. The result of such collision can be the formation of new species, such as excited molecules M^*, electrons, positive ions (M^+), negative ions (M^-), and neutral (F), positive (F^+), or negative (F^-) molecular fragments. The products of such reactions are studied as a function of the incident electron energy ε, and at times also as a function of the scattering angle (θ).

The probability of the various processes represented by the reaction channels (1.1) in an electron-molecule collision is described by the corresponding cross section, $\sigma(\varepsilon)$, for that reaction (see Ch. 2). The cross sections of these direct electron collision channels extend over a range of incident electron energies which depends on the reaction process. Figure 1.2 shows an example of the cross sections for some of these processes for the plasma processing gas CF_4 (see, also, Ch. 2 and Ch. 4). With the exception of elastic electron scattering, the cross section for each of the other processes has an energy threshold which depends on the physical process itself. The thresholds for molecular dissociation and ionization, for instance, are much higher than those for direct vibrational excitation. On the other hand, the cross section for elastic electron scattering is seen to be large over a wide range of electron energies.

In contrast with the direct electron collision processes just discussed for which the cross sections are substantial over a wide range of electron energies, in the low-energy region (usually $\lesssim$ 20 eV), electrons can be selectively captured by the molecule AX to form transient (metastable) negative ions AX^{-*}. These indirect, capturing collisions are resonant, and they occur over a narrow range of electron energies which are characteristic of the target molecule. The transient negative ions live for approximately 10^{-15} to 10^{-2} s [13, 34, 35]. They decay by autodetachment leaving the neutral molecule AX with or without excess internal energy, or via the processes of dissociative electron attachment. They can also form a stable parent negative ion AX^- if the molecule AX has a positive electron affinity and the excess energy in AX^{-*} is removed. These indirect electron-collision processes can be represented as

$$e(\varepsilon) + \mathrm{AX} \rightarrow \mathrm{AX}^{-*} \rightarrow \mathrm{AX} + e(\varepsilon) \tag{1.2a}$$

(Indirect elastic electron scattering)

$$\Rightarrow \mathrm{AX}^* + e(\varepsilon') \tag{1.2b}$$

(Indirect inelastic electron scattering)

$$\Rightarrow \mathrm{A} + \mathrm{X}^- \tag{1.2c}$$

(Dissociative electron attachment)

$$\rightarrow \mathrm{AX}^- + \text{energy} \tag{1.2d}$$

(Parent negative ion formation).

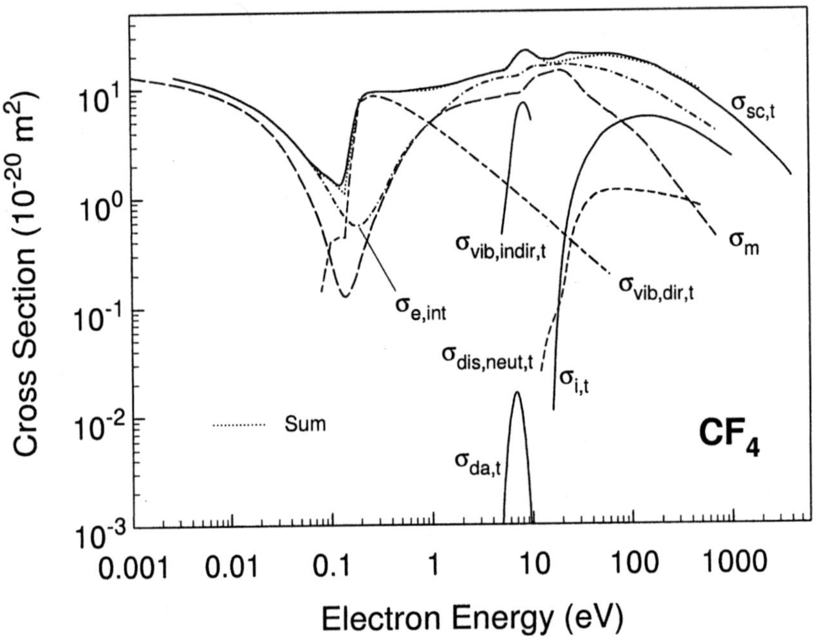

Figure 1.2. Cross sections for direct and indirect electron collision processes for the case of the plasma processing gas CF_4 (see Ch. 2 and Ch. 4). The symbols $\sigma_{vib,dir,t}$, $\sigma_{dis,neut,t}$, and $\sigma_{i,t}$ refer to the total cross sections for direct vibrational excitation, dissociation, and ionization of the molecule. The symbols $\sigma_{vib,indir,t}$, and $\sigma_{da,t}$ refer to the total cross sections for indirect vibrational excitation and dissociative electron attachment. The symbols σ_m, $\sigma_{e,int}$, and $\sigma_{sc,t}$ refer, respectively, to the cross sections for momentum transfer, integral elastic, and total electron scattering. The peaks in the last three cross sections reflect the contribution of resonant processes [22, 28].

To understand the energetics associated with Reactions (1.2) and the definitions of some of the physical quantities used to describe these reactions, we refer to Fig. 1.3 where schematic potential energy curves are depicted for the neutral molecule AX, the unstable negative ion AX^{-*}, and the stable negative ion AX^- (for a fuller discussion see Christophorou *et al.* [35]). In Fig. 1.3(a) the vertical arrow indicates the *vertical* transition from the initial state (neutral molecule AX in its lowest electronic and ro-vibrational states plus the electron at rest at infinity) to an attractive or a repulsive negative ion state of AX^{-*}. The transition energy is referred to as the *vertical attachment energy* (VAE) and is taken to be a positive quantity. Its negative gives the *vertical electron affinity* (VEA) of the molecule (VAE = –VEA). The negative sign of this quantity signifies that the potential energy curve of AX^{-*} in the Franck-Condon region lies energetically above that of the neutral molecule AX.

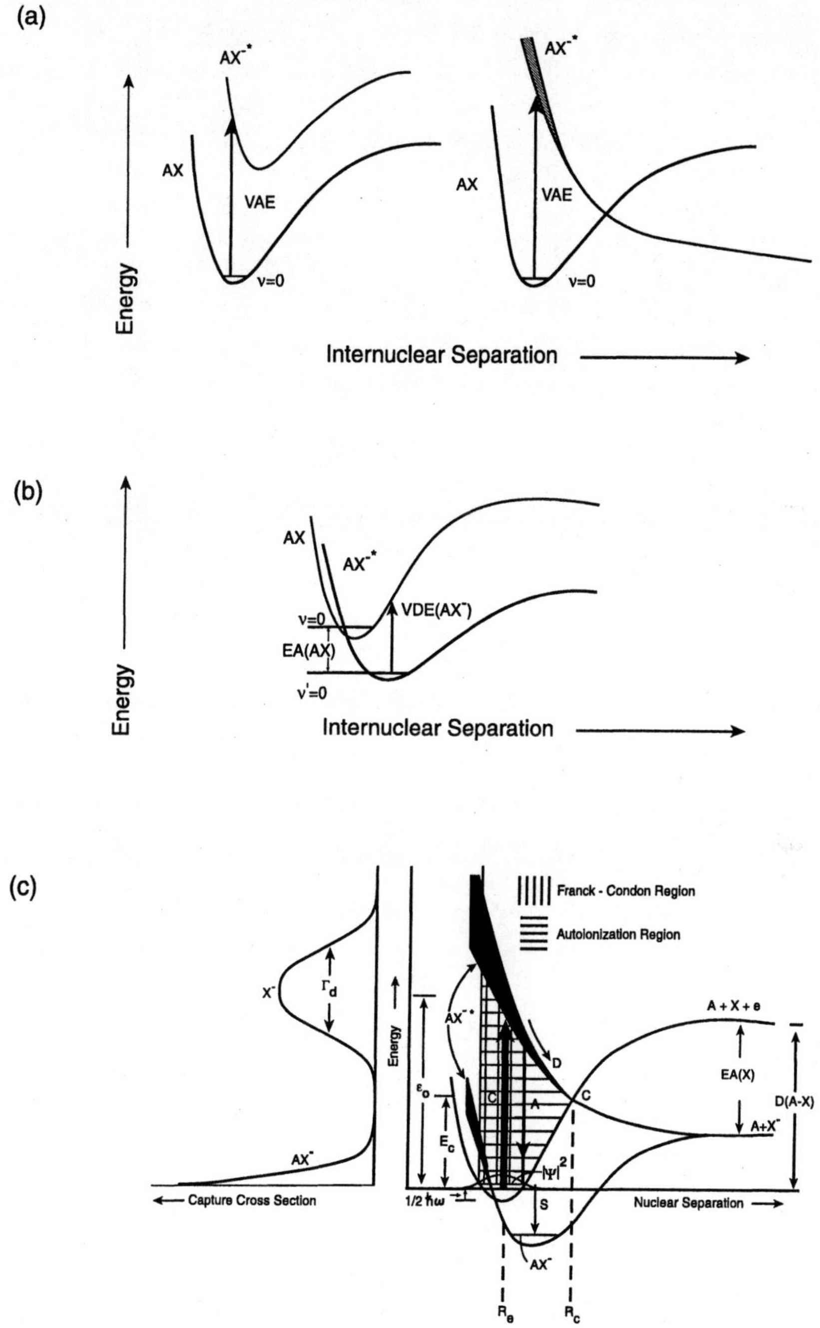

Figure 1.3. Schematic potential energy curves [35] illustrating the relations between (a) the vertical attachment energy VAE, (b) the vertical detachment energy VDE and the adiabatic electron affinity EA, and (c) the cross sections for the processes of dissociative and nondissociative electron attachment.

Figure 1.3(b) illustrates the case where the energy of the negative ion AX^- (in its lowest ro-vibrational and electronic states) lies lower than that of the neutral molecule (plus an electron at rest at infinity). The energy difference between the ground state of AX^- and the ground state of AX gives the *adiabatic electron affinity* (AEA $\equiv$ EA) of AX and this is taken to be a positive quantity. The vertical transition from AX^- (in its lowest ro-vibrational and electronic state) to AX + e is the most probable energy required to detach the electron from the stable negative ion AX^- and is called the *vertical detachment energy* (VDE) of AX^-. It is a positive quantity and is best measured using photodetachment experiments (e.g., see Refs. [35–38]). Because normally the minimum of the potential energy curve of AX^- is shifted to longer internuclear distances than the minimum of the potential energy curve of the corresponding neutral molecule AX, EA(AX) $\leq$ VDE(AX^-).

Furthermore, to aid our understanding of the processes of electron attachment to molecules, we reproduce in Fig. 1.3(c) the schematic diagrams used by Christophorou *et al.* [35] to illustrate the pertinent parameters in resonance dissociative and nondissociative electron attachment processes and the energy dependencies of the respective cross sections. The widths of the potential energy curves of AX^{-*} in Fig. 1.3(c) were drawn progressively larger as the internuclear separation R becomes smaller in order to illustrate the increased probability of autodetachment of AX^{-*} with decreasing R. The arrows identified by C, A, D, and S indicate, respectively, the processes of electron capture, autodetachment, dissociation, and stabilization of AX^{-*}.

The magnitude, energy position, and shape of the cross sections for dissociative and nondissociative electron attachment provide useful information about the electronic states of AX^-. Let us then discuss these processes a little further. Resonance nondissociative electron attachment processes leading to the formation of AX^- occur over a narrow electron energy range that is usually less than a few eV. In fact, in most known cases of parent negative-ion formation by electron impact in the field of the molecular ground state, the parent negative ions AX^- have a maximum probability of formation at $\sim$ 0.0 eV (e.g, see the cross section for the formation of SF_6^- in Fig. 6.112(a) of Ch. 6). To form AX^-, the adiabatic EA(AX) must be positive and the transient negative ion must be stabilized. On the other hand, resonance dissociative electron attachment processes leading to X^- occur over a wider energy range (0 to $\sim$ 20 eV) and do not require stabilization (the dissociation products take away the excess energy of AX^{-*}).

Resonance dissociative electron attachment is visualized [Fig. 1.3(c)] to proceed through the formation of a negative ion intermediate AX^{-*}

by the capture of the electron when the latter's energy lies in a restricted range defined by the Franck-Condon transition between the initial state (neutral molecule and electron at infinite separation) and the final state representing the transient negative ion, which dissociates into neutral and negative ion fragments. The cross section, $\sigma_{\mathrm{da}}(\varepsilon)$, for dissociative electron attachment as a function of the electron energy ε is taken to be equal to the product of the cross section, $\sigma_{\mathrm{c}}(\varepsilon)$, for capture of the electron by the molecule to form the transient negative ion AX^{-*} and the probability $p(\varepsilon)$ that AX^{-*} will decay via dissociative electron attachment, that is, $\sigma_{\mathrm{da}}(\varepsilon) = \sigma_{\mathrm{c}}(\varepsilon)p\,(\varepsilon)$. The quantity $\sigma_{\mathrm{c}}\,(\varepsilon)$ depends upon the symmetry properties and Franck-Condon overlap of the molecular and the negative ion states involved, while the quantity $p(\varepsilon)$ depends upon the mean lifetime of AX^{-*} and the separation time of the fragments A and X^- (see Ch. 2). Theoretical treatments have related $p(\varepsilon)$ to the properties of the potential energy curves of AX and AX^{-*} (e.g., see Ch. 2 and Refs. [39, 40]). Thus, $p(\varepsilon)$ is taken to represent the survival probability of AX^{-*}, which is itself approximated by $\exp(-\tau_{\mathrm{s}}/\tau_{\mathrm{a}})$, where τ_{a} is the mean autodetachment time of AX^{-*} and τ_{s} is the mean time taken by AX^{-*} to separate from the point of formation by capture of the electron with energy ε at an internuclear distance R_ε to the point C at an internuclear separation R_{c} where the potential-energy curves of AX and AX^{-*} cross. The AX^{-*} can be destroyed by autodetachment only when the potential energy curve of AX^{-*} lies energetically above that of AX. When $R > R_{\mathrm{c}}$, AX^{-*} dissociates with unit efficiency ($p = 1$) [see Fig. 1.3(c)]. This picture has proven useful in understanding dissociative electron attachment processes and in relating experimental observations and energetics to physical concepts.

The cross sections for the indirect electron scattering processes (1.2a) and (1.2b) show a different energy dependence than those of the corresponding direct processes (1.1a) and (1.1b). The thresholds of the indirect processes are determined by the position(s) of the corresponding negative ion states (resonances) rather than by the corresponding energetic thresholds. This can be seen from the example shown in Fig. 1.4 where the cross section, $\sigma_{\nu(0,1)}(\varepsilon)$, for direct excitation of the $\nu = 1$ vibrational level of N_2 (initial shoulder to the left of the main resonance in Fig. 1.4) and for indirect excitation of the $\nu = 1$ vibrational level of the N_2 molecule via the decay of the $N_2^{-*}\,(^2\Pi_{\mathrm{g}})$ shape resonance at 2.3 eV (main enhancement peaking near 2.3 eV in Fig. 1.4) is shown [41]. The energy thresholds for direct and indirect vibrational excitation are indicated in the figure by the vertical arrows. Such enhancements in the electron scattering cross section are responsible for the structure seen in the cross sections of the various elastic and inelastic

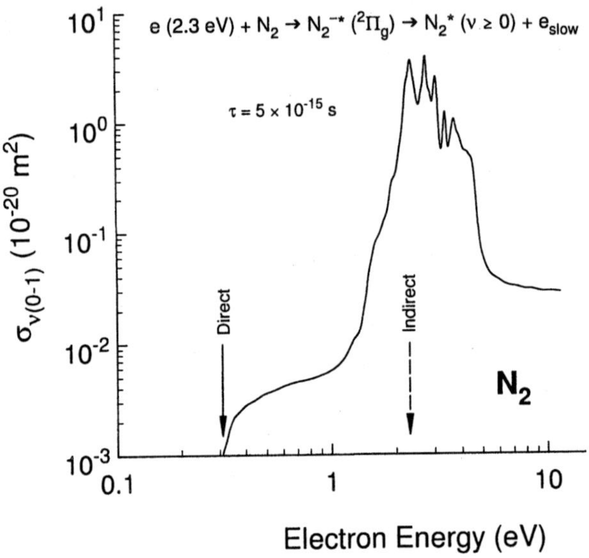

Figure 1.4. Cross section, $\sigma_{\nu(0,1)}(\varepsilon)$, for the $\nu = 0$–1 vibrational excitation of N_2 via the decay of the N_2^{-*} ($^2\Pi_g$) shape resonance at 2.3 eV [41]. The shoulder to the left of the main enhancement is due to direct vibrational excitation of the N_2 molecule.

processes in electron-molecule collisions. The large cross sections of resonant electron-molecule collisions make them efficient mechanisms for vibrational excitation of molecules, and, consequently, efficient mechanisms of influencing the energy distribution of electrons in gases under an applied electric field. Although vibrational excitation processes are low-energy processes, their effect on the electron energy distribution function can strongly influence high-energy processes such as dissociation and ionization.

In the electron capturing collisions (1.2c) and (1.2d), the incident electron disappears and its total energy is transferred to the transient anion AX^{-*}. Upon decay by electron attachment, the transient anion AX^{-*} generates either stable fragment anions (X^-) and free radicals (A), or a parent negative ion (AX^-), if the electron affinity of AX is positive (that is, if the molecule can bind permanently the extra electron) and the excess energy of the transient anion AX^{-*} is released through radiation or collisions with other molecules before autodetachment. While the cross sections for parent negative ions normally peak at thermal or near thermal energies and are very large ($\gtrsim 10^{-18}$ m^2) [13, 35], the cross sections for dissociative electron attachment depend strongly on the energy positions of the negative ion states involved. As can be seen from

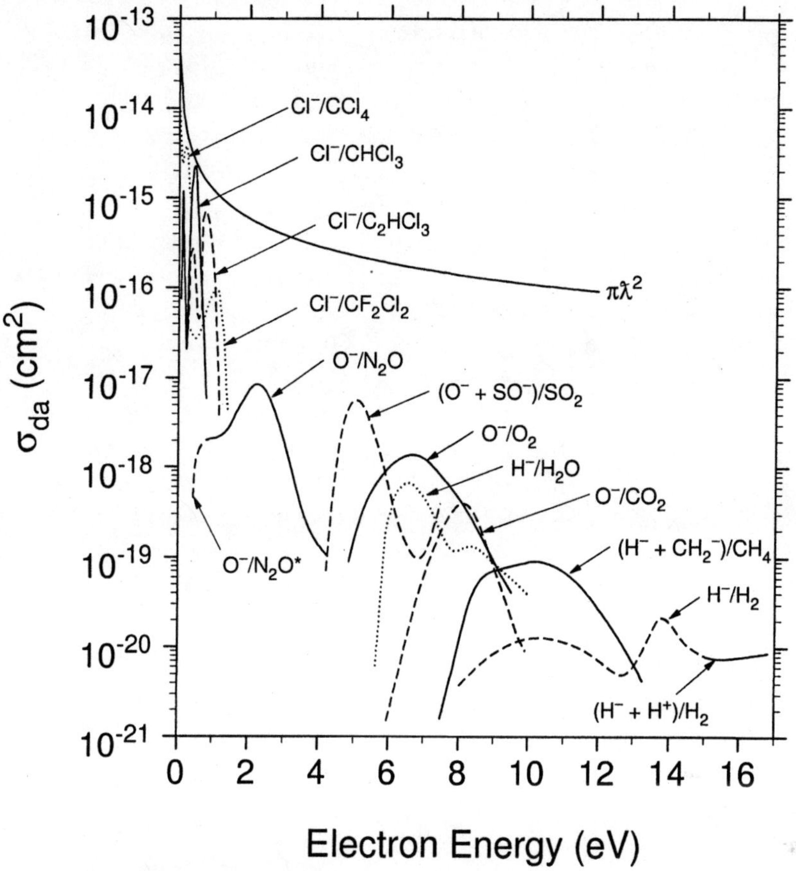

Figure 1.5. Dissociative electron attachment cross sections as a function of the electron energy for a number of molecules (from Christophorou *et al.* [35]). The line designated by $\pi\lambda\!\!\!^{-2}$ is the *s*-wave capture cross section.

the representative data in Fig. 1.5, in the energy range from thermal to $\sim$ 16 eV the cross sections for dissociative electron attachment to molecules vary by more than eight orders of magnitude. They generally increase sharply as the energy of the negative ion state decreases and approach maximum values at thermal energies. The data in Fig. 1.5 also show that low-lying dissociative electron attachment processes can be excellent sources of free radicals in plasmas.

Reactions (1.1a) to (1.1h) and (1.2a) to (1.2d) determine the energies and number densities of free electrons in electrically stressed gaseous matter. In such media, the energies of the free electrons are a function of the amount of energy they receive from the applied electric field and the energy they lose in elastic and inelastic collisions with the gas atoms and molecules. The main electron-energy-controlling reactions are the direct processes (1.1b) to (1.1h) and the indirect processes (1.2b) and (1.2c). The low-energy indirect electron scattering processes constitute the most effective way of transferring energy from the electrons to molecular vibrations, and thus such processes influence greatly the distribution of electron energies in electrically stressed gases and plasmas.

Similarly, the reactions controlling the electron-number-density consist of the electron-generating processes (1.1e) to (1.1g) and the electron-depleting processes (1.2c) and (1.2d). The former are direct (nonresonant) and latter are indirect (resonant). Other processes (see Sec. 3.3 of Ch. 1) such as electron detachment may also affect the electron number densities. In assessing the effect of electron attachment processes on free electron number densities in plasmas, it is important to recognize the strong dependence of the magnitude of the electron attachment cross section on the energy position of the negative ion states (Fig. 1.5). It is also important to recognize that the effectiveness of a given electron attachment process to control the free electron number density is a function of the electron scattering processes which determine the electron energy distribution function.

2.2 Interactions of Low-Energy Electrons With Excited Molecules

Interactions of low-energy electrons with atoms and molecules are not only a function of the kinetic energy of the electron, but also a function of the internal energy of the target atom or molecule [21, 42–44]. We shall designate the cross sections for the various reactions involving electron collisions with excited species with the same symbols as for the corresponding unexcited species, but with an asterisk, that is, as $\sigma^*(\varepsilon)$. The cross sections $\sigma^*(\varepsilon)$ often exceed the corresponding cross sections $\sigma(\varepsilon)$ by a large factor. This, and the overall effect of excited species on the electron energy distribution function, can significantly alter the nature of the particles in, and the properties of, plasmas even when the relative density of the excited species is small.

In spite of the basic and applied significance of electron interactions with excited molecules, studies of the interactions of slow electrons with excited atoms and molecules as a function of the internal energy of the latter are meager and relatively new. One of the reasons for this paucity of experimental data is the difficulty of producing sufficient numbers of excited atoms and molecules to study under experimentally controlled conditions. However, this has now been partially overcome with the availability of pulsed lasers which enable the production of much higher number densities and allow greater selectivity of excited states (e.g., see Refs. [21, 42–44] and Ch. 2).

Interactions of low-energy electrons with excited molecules can conveniently be separated into two groups: electron interactions with rovibrationally excited ("hot") molecules, and electron interactions with electronically excited molecules. Electron interactions with rotationally/vibrationally excited molecules can be symbolically represented as

$$e(\varepsilon) + \mathrm{AX}^*(G; \nu > 0; J > 0) \rightarrow \mathrm{AX}^{(*)} + e(\varepsilon') \tag{1.3}$$

(Direct elastic and inelastic electron scattering)

$$e(\varepsilon) + \mathrm{AX}^*(G; \nu > 0; J > 0) \rightarrow \mathrm{AX}^{*-} \rightarrow \mathrm{AX} + e(\varepsilon) \tag{1.4a}$$

(Indirect elastic electron scattering)

$$\rightarrow \mathrm{AX}^{*-} \Rightarrow \mathrm{AX}^* + e(\varepsilon') \tag{1.4b}$$

(Indirect inelastic electron scattering)

$$\rightarrow \mathrm{AX}^{*-} \Rightarrow \mathrm{A} + \mathrm{X}^- \tag{1.4c}$$

(Dissociative electron attachment)

$$\rightarrow \mathrm{AX}^{*-} \rightarrow \mathrm{AX}^- + \text{energy}, \tag{1.4d}$$

(Parent negative ion formation)

where the letters G, ν, and J refer to species, respectively, in their ground electronic state, in the ν vibrational state, and J rotational state.

The very limited data on Reactions (1.3) and (1.4b) have been discussed by Christophorou and Olthoff [21]. An example is shown in Fig. 1.6(a) where the total electron scattering cross sections for the unexcited CO_2 molecule and for the CO_2 molecule excited principally in the first vibrational bending mode are compared. The higher cross section values of the total electron scattering cross section observed for the vibrationally excited CO_2 molecule has been attributed to the dipole moment (~ 0.1 D) of the excited ($\nu = 1$) bending mode. Since, in addition to direct thermal excitation, such vibrationally excited molecules can be formed via the decay of negative ion states, such enhancements in the electron scattering cross section also can be seen in the energy regions where the negative ion states of the CO_2 molecule are located. This is

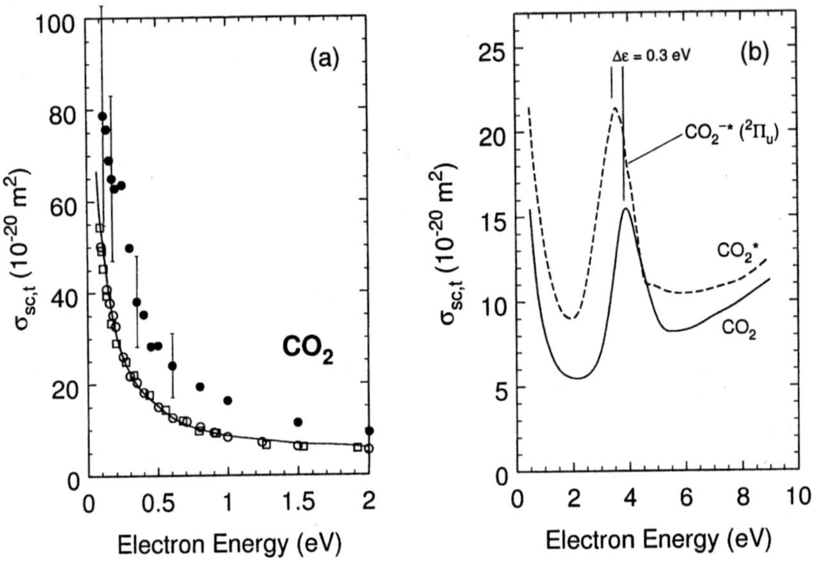

Figure 1.6. (a) Total cross section for electron scattering from the CO_2 molecule in the vibrational ground-state: (○) measurements of Ref. [45]; (□) measurements of Ref. [46]; (—) calculated values of Ref. [47]; and (•) from vibrationally-excited CO_2^* in the bending mode from measurements of Ref. [45]. (b) Total cross section for electron scattering from the CO_2 molecule in the vibrational ground-state (—) and from the vibrationally-excited CO_2^* molecule in the bending mode (- - -) covering the energy range of the CO_2^{-*} ($^2\Pi_u$) resonance [48].

illustrated by the data shown in Fig. 1.6(b). The effects of enhanced electron scattering from vibrationally excited molecules can significantly influence the behavior of free electrons in plasmas (for instance, their transport properties).

Of the Reactions (1.4a) to (1.4d), Reaction (1.4c) has been the one most studied [21, 35, 44]. Thus, it has been firmly established that the cross section for exothermic dissociative electron attachment processes generally increases with increasing total internal energy of the molecule. An example of this type of behavior is shown in Fig. 1.7 for the CH_3Br molecule. Such large increases in the dissociative electron attachment cross section are common for many halogenated molecules and indicate an efficient way to increase radical production by small increases in gas temperature. They find application in a wide range of fields, including environmental clean up where slow electrons would break up pollutants more efficiently if temperatures higher than ambient are employed.

In contrast with Reaction (1.4c), increases in the ro-vibrational energy of target molecules by increases in gas temperature often cause a decrease

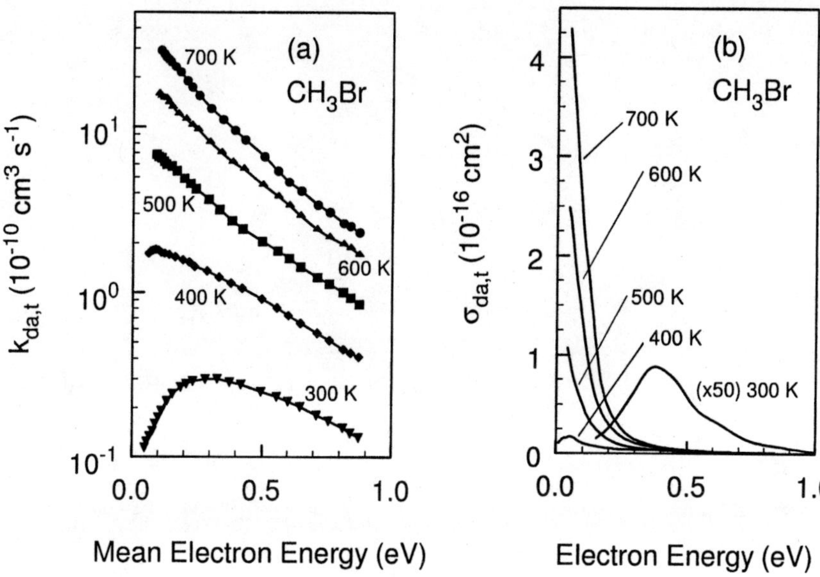

Figure 1.7. (a) Total dissociative electron attachment rate constant as a function of mean electron energy, $k_{\mathrm{da,t}}(\langle\varepsilon\rangle)$, for CH_3Br measured in N_2 buffer gas at temperatures of 300, 400, 500, 600, and 700 K. (b) Total dissociative electron attachment cross section as a function of the electron energy ε, $\sigma_{\mathrm{da,t}}(\varepsilon)$, for CH_3Br unfolded from the data in Fig. 1.7(a) [49].

in the overall production of parent negative ions, mainly because of enhanced autodetachment of AX^{-*} and thermally induced detachment of the stable parent anion AX^-, viz.

$$AX^- + \text{heat} \rightarrow AX^{-*} \rightarrow AX + e. \tag{1.5}$$

The effectiveness of Reaction (1.5) is a function of the electron affinity of the molecule AX. An example of the effect of temperature on Reactions (1.4d) and (1.5) is shown in Fig. 1.8 for *c*-C_4F_6 [21, 50]. The total electron attachment rate constant as a function of the mean electron energy $\langle\varepsilon\rangle$, $k_{\mathrm{a,t}}(\langle\varepsilon\rangle)$, for Reaction (1.4d) is seen not to be a function of temperature for this molecule for gas temperatures up to 600 K for all values of $\langle\varepsilon\rangle$ below 1 eV. In contrast, the electron detachment frequency, t_{d}^{-1}, for Reaction (1.5) increases with increasing mean internal energy, E_{int}, of the molecule, that is, with increasing gas temperature.

Thus the effect of ro-vibrational excitation of molecules on electron attachment depends on the mode (dissociative or nondissociative) of electron attachment and on the negative ion state involved. For dissociative electron attachment the cross section increases with increasing

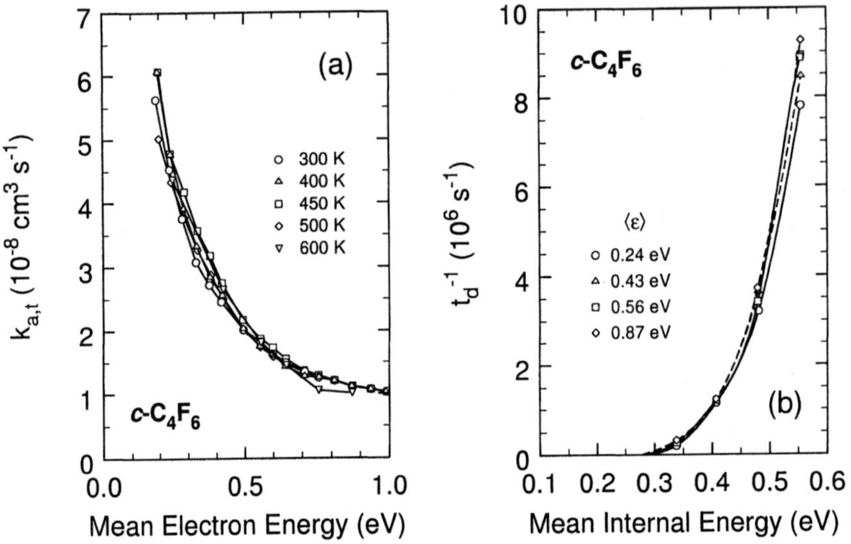

Figure 1.8. (a) Total electron attachment rate constant as a function of the mean electron energy, $k_{a,t}(\langle\varepsilon\rangle)$, for c-C_4F_6 measured in N_2 buffer gas at temperatures of 300, 400, 450, 500, and 600 K. (b) Thermally-induced electron detachment frequency, t_d^{-1}, for c-$C_4F_6^-$ as a function of the internal energy of the negative ion for a number of values of the mean electron energy $\langle\varepsilon\rangle$. The data points correspond to temperatures of 450, 500, 550, and 600 K (from Ref. [50]).

internal energy as long as the total internal energy of the molecule is less than the energy required to reach the dissociating negative ion state. For nondissociative electron attachment the cross section is little affected by increases in internal energy, but the detachment cross section is increased (see further discussion in Refs. [21, 44, 50, 51]).

Little quantitative work has been done on electron scattering from and electron attachment to electronically excited molecules. The processes involved can be represented as

$$e(\varepsilon) + \mathrm{AX}^*(E) \rightarrow \mathrm{AX} + e(\varepsilon') \tag{1.6}$$

(Direct elastic and inelastic electron scattering)

$$e(\varepsilon) + \mathrm{AX}^*(E) \rightarrow \mathrm{AX}^{*-}(E) \rightarrow \mathrm{AX} + e(\varepsilon) \tag{1.7a}$$

(Indirect elastic electron scattering)

$$\Rightarrow \mathrm{AX}^* + e(\varepsilon') \tag{1.7b}$$

(Indirect inelastic electron scattering)

$$\Rightarrow \mathrm{A} + \mathrm{X}^- \tag{1.7c}$$

(Dissociative electron attachment)

$$\rightarrow \mathrm{AX}^- + \text{energy} \tag{1.7d}$$

(Parent negative ion formation)

where $AX^*(E)$ refers to an electronically excited species, and AX^{*-} to a transient negative ion formed by electron attachment to the electronically excited molecule AX^*.

There have been only limited measurements (e.g., for O_2^* [21, 52] and N_2^* [21, 53]) and calculations (e.g., for N_2^* [21, 54, 55]) of the cross section for Reaction (1.6). Limited as these measurements and calculations are, they show that the cross section for excitation of a given electronic state is higher if the upward transition originates from an excited electronic state. This, for instance, is seen from the calculations of Huo [55] for N_2^* shown in Fig. 1.9.

To better illustrate the large effects of electronic excitation on the electron scattering cross sections, we refer to the data in Fig. 1.10 where the cross sections for electron scattering and electron-impact ionization from ground state Ar and from metastable Ar^* are compared. It is seen from Fig. 1.10(a) that the total electron scattering cross section for the excited argon atom $Ar^*(4\,^3P_2 + 4\,^3P_0)$ is much larger than the total electron scattering or the momentum transfer cross section for the ground state argon atom $Ar(3\,^1S_0)$. For this atom, the excited-state electron scattering cross section exceeds that for the ground state by a factor of more than 10^4 near 0.2 eV. This large enhancement and the disappearance of the Ramsauer-Townsend minimum in the excited-state cross section is largely due to the much higher polarizability of the excited atom (the polarizability of $Ar^*(4\,^3P_2)$ is $\sim 319.4\,a_0^3$) compared to that for the ground state atom (the polarizability of Ar (3^1S_0) is $\sim 11.07\,a_0^3$) (see Refs. [21, 66] and sources cited therein).

In contrast to ground-state atoms, the electron scattering cross sections for excited atoms have large contributions from higher angular momenta and are not significantly affected by electron exchange. These differences are also the direct consequence of the large dipole polarizabilities of excited states of atoms and the resultant dominance of the polarization potential in electron scattering. Similarly, as shown in Fig. 1.10(b), the cross section for ionization of metastable Ar^* far exceeds that from the ground state Ar atom. The lower ionization thresholds of the excited states of atoms and molecules cause a shift of their ionization cross

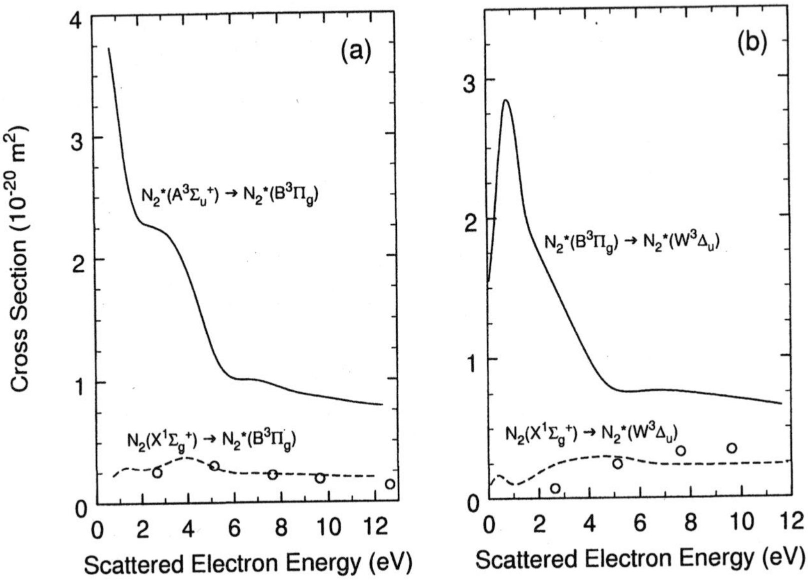

Figure 1.9. Calculated cross sections for excitation of the $B\,^3\Pi_g$ and $W\,^3\Delta_u$ excited states of N_2 from the ground state $N_2\,(X\,^1\Sigma_g^+)$ (- - -) and from the excited states (—) $N_2^*\,(A^3\Sigma_u^+)$ (Fig. 1.9(a)) and $N_2^*\,(B\,^3\Pi_g)$ (Fig. 1.9(b)) as a function of the scattered electron energy (from Ref. [55]). The open circles are the ground-state measurements of Ref. [56] as normalized in Ref. [57].

sections to lower energies, which, along with the higher ionization cross sections of the excited states, affect the rate coefficients of the various discharge processes in such systems.

There have been a few studies of Reaction (1.7c) [21]. An early study of this kind was concerned with the investigation of dissociative electron attachment to singlet oxygen $O_2^*(a\,^1\Delta_g)$ [67, 68], namely for the reaction,

$$O_2^*(a\,^1\Delta_g) + e \rightarrow O_2^{*-}(^2\Pi_u) \rightarrow O^- + O. \tag{1.8a}$$

Figure 1.11(a) compares the cross section for dissociative electron attachment to the excited $O_2^*(a\,^1\Delta_g)$ molecule via the negative ion state $^2\Pi_u$ (Reaction (1.8a)) with the corresponding cross section for the unexcited molecule $O_2(^3\Sigma_g^-)$, namely, reaction

$$O_2(^3\Sigma_g^-) + e \rightarrow O_2^{-*}(^2\Pi_u) \rightarrow O^- + O. \tag{1.8b}$$

The cross section for the production of O^- via the $O_2^{*-}(^2\Pi_u)$ negative ion state is seen to be approximately 3 to 4 times larger for the excited $a\,^1\Delta_g$ state compared to the $^3\Sigma^-_{g,\nu=0}$ ground state. These enhancements are

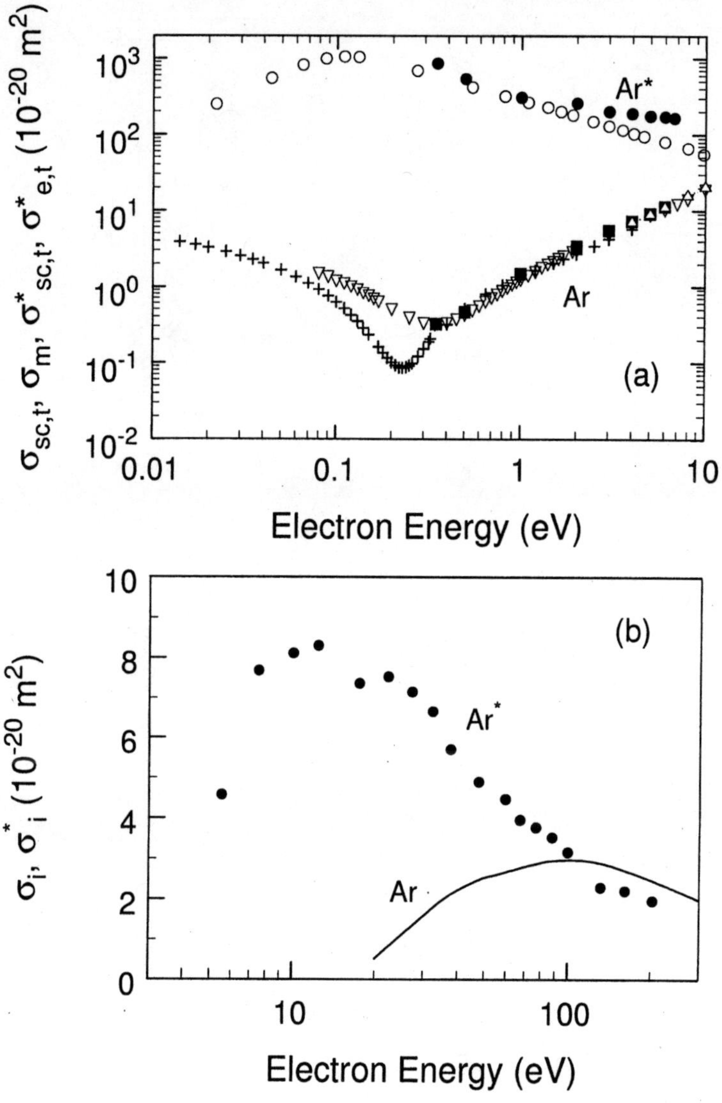

Figure 1.10. (a) Comparison of the electron scattering cross section from the excited Ar* $(4\,^3P_2 + 4\,^3P_0)$ argon atom, and from the unexcited Ar$(3\,^1S_0)$ argon atom: (•) experimental measurements of $\sigma^*_{\mathrm{sc,t}}(\varepsilon)\,(4^3P_2 + 4^3P_0)$ from Ref. [58] as reported in Ref. [43]; (○) calculated values of $\sigma^*_{\mathrm{e,t}}(\varepsilon)\,(4\,^3P_2)$ from Ref. [59]; (■, △, ▽) experimental measurements of $\sigma_{\mathrm{sc,t}}(\varepsilon)\,(3\,^1S_0)$ from Refs. [60], [61], and [62], respectively; (+) experimental measurements of $\sigma_{\mathrm{m}}(\varepsilon)\,(3\,^1S_0)$ from Ref. [63]. (b) Cross section $\sigma^*_{\mathrm{i}}(\varepsilon)$ for (•) electron-impact ionization of metastable Ar* from measurements of Ref. [64] in comparison with (—) the cross section $\sigma_{\mathrm{i}}(\varepsilon)$ for ground state Ar from Ref. [65].

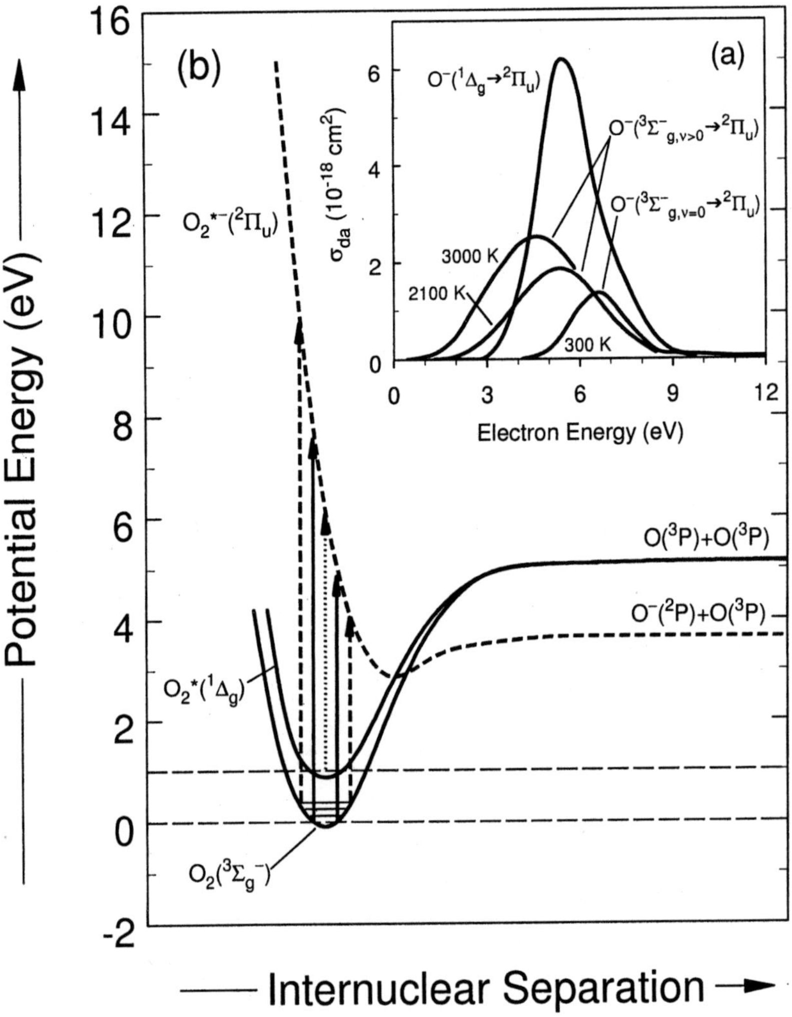

Figure 1.11. (a) Comparison of the measured dissociative electron attachment cross sections for ground-state $O_2(^3\Sigma_g^-)$ (data of Ref. [69]), electronically excited (singlet) $O_2^*(a^1\Delta_g)$ (data of Ref. [68]), and vibrationally excited $O_2^*(^3\Sigma^-_{g,\nu>0})$ oxygen molecules (from Ref. [21]). (b) Schematic potential energy curves for $O_2(^3\Sigma_g^-)$, $O_2^*(a^1\Delta_g)$, and $O_2^{*-}(^2\Pi_u)$, and transitions from ground state $O_2(^3\Sigma^-_{g,\nu=0})$, electronically excited $O_2^*(a^1\Delta_g)$, and vibrationally excited $O_2^*(^3\Sigma^-_{g,\nu>0})$ species to the $O_2^{*-}(^2\Pi_u)$ negative ion state illustrating dissociative electron attachment to the ground state, excited electronic state, and vibrationally excited states of O_2 (see text).

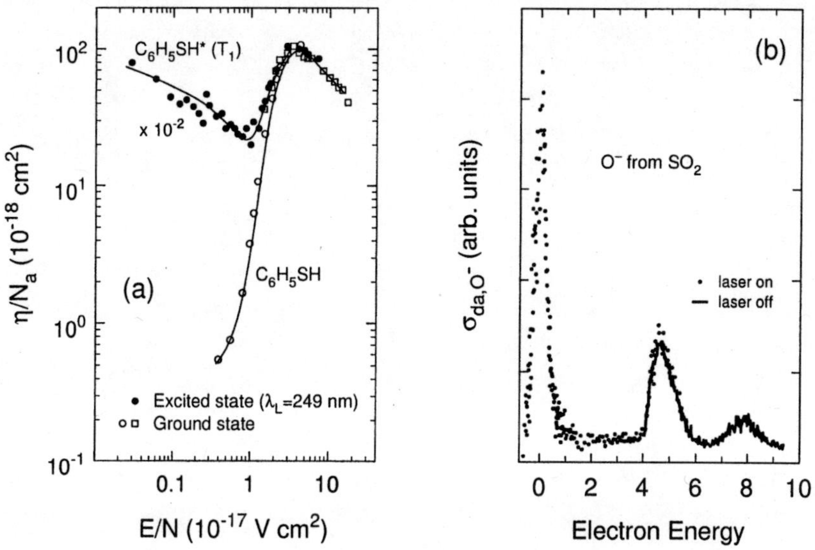

Figure 1.12. (a) Electron attachment coefficient η/N_a normalized to the unexcited attaching gas number density N_a as a function of E/N for C_6H_5SH in N_2 for the ground (○, □) and for the first-excited triplet state (•, data below $\sim 0.3\times10^{-17}$ Vcm^2) (from Refs. [44, 70]). (b) Relative cross section for the production of O^- from the ground state SO_2 ($X\,^1A_1$) (—) and from the electronically excited SO_2^* ($^1B_1$ or $^1A_2$) (···, below $\sim$ 0.5 eV) molecules as a function of the electron energy. The relative cross sections are normalized at 8 eV and the excited-state cross section is not corrected for the fraction of excited to unexcited gas number density (from Ref. [71]).

comparatively larger than those reported for "hot" $O_2^*(^3\Sigma^-_{g,\nu>0})$ ground-state oxygen molecules, namely, for the reaction,

$$O_2^*(^3\Sigma^-_{g,\nu>0}) + e \rightarrow O_2^{*-}(^2\Pi_u) \rightarrow O^- + O. \tag{1.8c}$$

Figure 1.11(b) schematically illustrates the transitions responsible for Reactions (1.8a) (the dotted arrow indicates the most probable vertical transition), (1.8b) (the solid arrows define the Franck-Condon region), and (1.8c) (the broken arrows show the broadening of the Franck-Condon region for the vibrationally excited oxygen molecules).

The most recent studies of electron attachment to electronically excited molecules utilized pulsed lasers to generate the excited states of molecules in electron swarm-type and in electron beam-type experiments [21, 44]. Figure 1.12(a) shows the results of the first observation of optically enhanced electron attachment to electronically excited molecular states made using an electron swarm-type experiment [70] (also see Ch. 2). The measured electron attachment coefficient normalized to the unexcited attaching gas number density N_a, $\eta/N_a(E/N)$, for the thiophenol molecule in its ground electronic state (C_6H_5SH) and in the first excited triplet state ($C_6H_5SH^*(T_1)$) are shown, respectively, by the symbols (□, ∘) and (•). The triplet state was populated indirectly by internal conversion from higher excited singlet states reached by the absorption of 249 nm laser radiation. At low E/N, the values of $\eta/N_a(E/N)$ for the excited state are more than 10^5 times larger than those for the ground state [note that in Fig. 1.12(a) η was normalized to the ground-state gas number density (rather than to the excited-state number density) which is about 100 times larger than the excited-state number density, thus the factor $\times 10^{-2}$ shown in the figure].

Similarly, in Fig. 1.12(b) are shown the results of the first observation of photoenhanced electron attachment in an electron beam/mass spectrometric study [71]. They refer to the relative cross section for the formation of O^- from the ground-state $SO_2(X\,^1A_1)$ molecule and from the excited-state $SO_2^*(^1B_1$ or $^1A_2)$ molecule, that is, respectively, for the reactions

$$e + SO_2(X\,^1A_1) \rightarrow SO_2^{-*} \rightarrow O^- + SO \qquad (1.9a)$$
$$(\sigma_{da} \text{ peaks at 4.5 eV and 7.3 eV})$$

and

$$h\nu\,(308\,\text{nm}) + SO_2(X\,^1A_1) \rightarrow SO_2^*\,(^1B_1\,\text{or}\,^1A_2)\,(+e < 0.5\,\text{eV})$$
$$\rightarrow SO_2^{*-} \qquad (1.9b)$$
$$\rightarrow O^- + SO$$
$$(\sigma_{da} \text{ peaks at } < 0.5 \text{ eV}).$$

In the above reactions, SO_2^{-*} represents a transient negative ion with excess energy formed by electron attachment to the ground state SO_2 and SO_2^{*-} represents a negative ion formed by electron attachment to the electronically excited SO_2^* molecule. The solid line in Fig. 1.12(b) is the relative cross section for the production of O^- from SO_2 via (1.9a) and the dots at electron energies below 1 eV represent the relative cross section for the production of O^- from SO_2 via (1.9b). The photoenhanced O^- signal was normalized to the 7.3 eV peak of Reaction (1.9a). According to Jaffke *et al.* [71], the dissociative attachment cross section σ_{da}^* for the photoenhanced O^- peak at energies $\lesssim 0.5$ eV exceeds the O^- cross section at the 4.5 eV peak from the ground state by more than a factor of 100, although another similar subsequent study [72] indicated smaller differences. A number of other studies using electron swarm methods (see Ch. 2) indicated that the cross section for dissociative electron attachment for excited electronic states of molecules can be as much as 10^6 times larger than for the corresponding ground state molecules [21]. The increases depend upon the relative positions of the excited electronic state and the negative ion state leading to dissociative electron attachment. The lower energy of the captured electron and the higher dipole polarizability of the excited states contribute to the observed enhancements.

The role of such knowledge in gas discharge physics and plasma-based technologies although expected to be significant is as yet unexplored. The effect of excited states on the practical properties of electrically stressed gases and plasmas may be large even when their percentage in the system is small because the cross sections of electron interactions with excited species are large and shifted to lower electron energies compared to the corresponding cross sections for the ground-state molecules. Indeed, as we pointed out earlier [21, 44], the field of electron interactions with excited molecules is young and is full of challenge and potential for applications. More work is needed on electron scattering from and ionization of excited molecules. Especially needed are reliable new methods for the quantitative study of electron attachment to electronically excited molecules. Such studies would offer unique opportunities to change and/or to modulate the electrical properties of matter using pulsed lasers, and would allow the proper consideration of "excited-state effects" in plasma modeling.

2.3 Interactions of Low-Energy Electrons with Molecules in High Pressure Gases, Liquids, Clusters, and with Molecules at Surfaces

Besides the dependence of the cross sections of electron-molecule collisions on the kinetic energy of the electron and the internal energy of the molecule, the cross sections and the energetics of electron-molecule interaction processes are a function of the nature, density, and state of the medium in which they occur. A number of investigators have summarized and discussed these effects in dense gaseous matter (e.g., Refs. [20, 73–76]), liquids (e.g., Refs. [20, 75–80]), at surfaces (e.g., Refs. [81–83], and in clusters (aggregates of 2 to $\sim 10^4$ atoms or molecules) (e.g., see Refs. [20, 84, 85]). Schematically, these environments are depicted in Fig. 1.13.

Under very-low gas pressures where single-collision conditions prevail, electron-molecule interaction processes are considered isolated, that is, the electrons interact with one molecule at a time and the collisions are not influenced by the presence of the other molecules. As the density of the gaseous medium in which these interactions occur is increased,

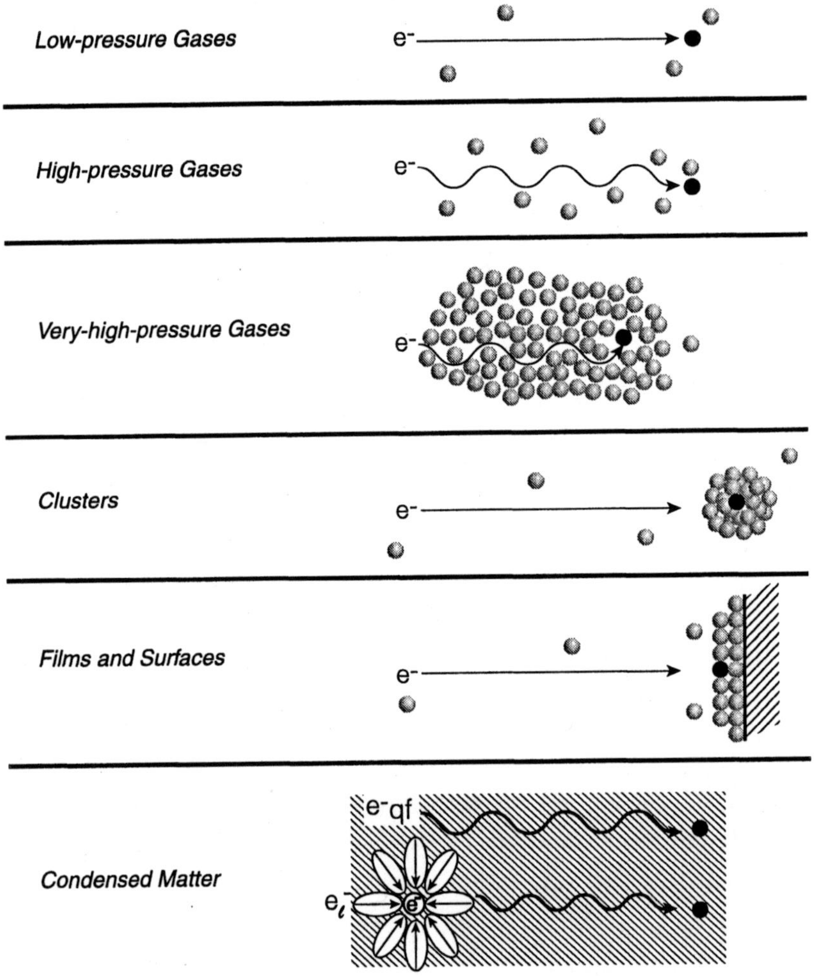

Figure 1.13. Schematic illustrations of electron-atom/molecule interactions in low-, high-, and very-high-pressure gases, in clusters, at surfaces, and in condensed matter.

the electron-molecule collisions are no longer isolated and they are affected by the nature, density, and state of the medium in which they are embedded. Indeed, interphase physics [73, 75, 86] has evolved as the systematic study of the effects of the nature and density of the medium on the elementary electron-molecule interactions as the density of the medium in which the interactions occur is increased from that of the very low pressure gas to the condensed phase. A somewhat complementary approach involves the study of electron-molecule collisions in clusters as a function of the size of the cluster (e.g., see Refs. [20, 84]). Both of these new directions provide valuable data that link the behavior of electron-molecule collisions under isolated conditions with that in dense gases, the condensed phases (liquid or solid) of matter, and at surfaces.

Charge-separated states (e.g., electrons, positive ions, negative ions), quasi-charge separated states (e.g., Rydberg states), and the physical quantities which describe their behavior and reactions are affected by the density and nature of the medium. Thus, among the physical quantities that are influenced, often profoundly, by the nature, density, and state of the medium are the electron state, electron transport, negative ion states, electron attachment, ionization, recombination, and their energetics. The electron energy ε, the quasi-free (excess) electron ground state energy V_0, the electron drift velocity w, the electron mobility μ, the cross sections for electron scattering, electron attachment and detachment, and ionization, and the associated physical quantities [e.g., electron affinity, vertical attachment energy, vertical detachment energy, ionization threshold energy, polarization energy of the positive (P^+) and the negative (P^-) ions] are all functions of the nature, density, and state of the medium in which these elementary physical processes occur. While these physical quantities have well-defined values in low-pressure gases, they assume a spectrum of values in dense media.

2.3.1 Changes in the Electron State and Electron Transport Coefficients

In a low pressure [$P \lesssim 100$ kPa (1 atm)] nonelectronegative gas, the electrons which are present in the gas are normally free and their collision (interaction) mean free path l is much longer than their de Broglie wavelength λ. When such a gaseous medium is under an applied uniform electric field E, the kinetic energies of the electrons in it can be well in excess of thermal (depending on the value of electric field to gas number density ratio, E/N) and their distribution, as well as their transport coefficients, can be determined using the Boltzmann transport equation. The steady-state electron energy distribution functions $f(\varepsilon, E/N)$ are non-Maxwellian (except at very low E/N) and — depending on the gas, temperature (T), and E/N — they peak at energies ranging from $3/2\,kT$ to nearly 10 eV [13, 18].

On the other hand, in a dense medium (high pressure gas, liquid, or solid) $l < \lambda$ and the excess electrons in such media are not free, but rather are either quasi-free or localized, depending on the medium, N, and T. Electrons are generally quasi-free in dense media for which the

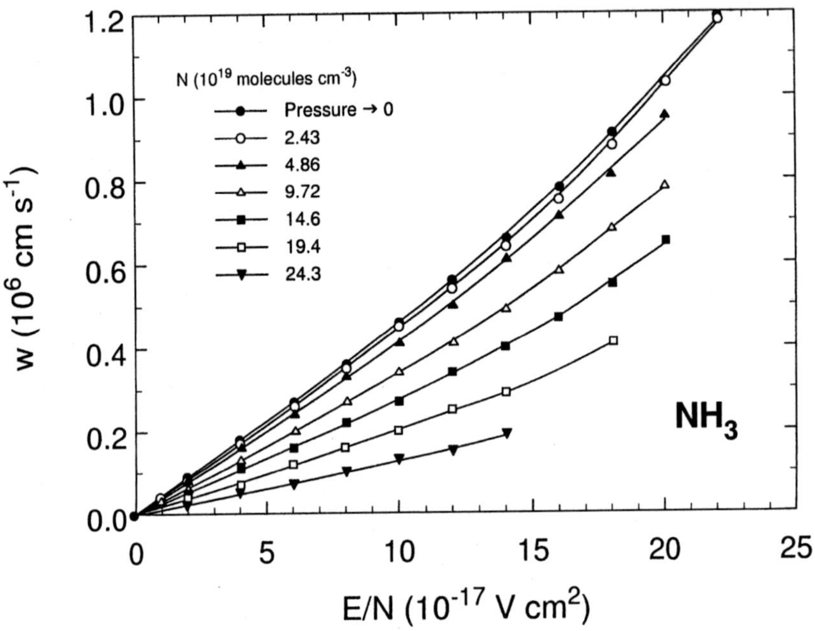

Figure 1.14. Electron drift velocity $w(E/N)$ in gaseous NH_3 at a number of values of the gas number density N ($T = 300$ K) (from Ref. [91]).

ground-state energy of the excess electron in the medium, V_0, is less than 0 eV, and electrons are localized in dense media whose $V_0 > 0$ eV. The V_0 itself can be a function of N (e.g., see Refs. [87–89]). The drift velocities w and the electron mobilities μ ($w = \mu E$) of localized electrons are low and their kinetic energies are thermal, while the drift velocities and mobilities of quasi-free electrons are high and their kinetic energies can be considerably in excess of thermal at high E/N (e.g., see Refs. [76, 90]). The connection between the electron behavior in the gaseous and the condensed phases of matter is more apparent — and more meaningful — for media with conduction bands ($V_0 < 0$ eV) where the excess electrons are quasi-free.

As examples of the effect of the medium and its density on electron motion, Fig. 1.14 shows the dependence of $w(E/N)$ on gas density for relatively low pressures of NH_3 gas, and Fig. 1.15 compares the $w(E/N)$ for gaseous and liquid Xe and gaseous and liquid tetramethylsilane ($Si(CH_3)_4$). The former medium (NH_3) is polar and electrons in it become quasi-localized as the gas density increases at relatively low densities. As the gas density approaches that of the liquid, the excess electrons become localized and their mobility ionic [95]. On the other hand,

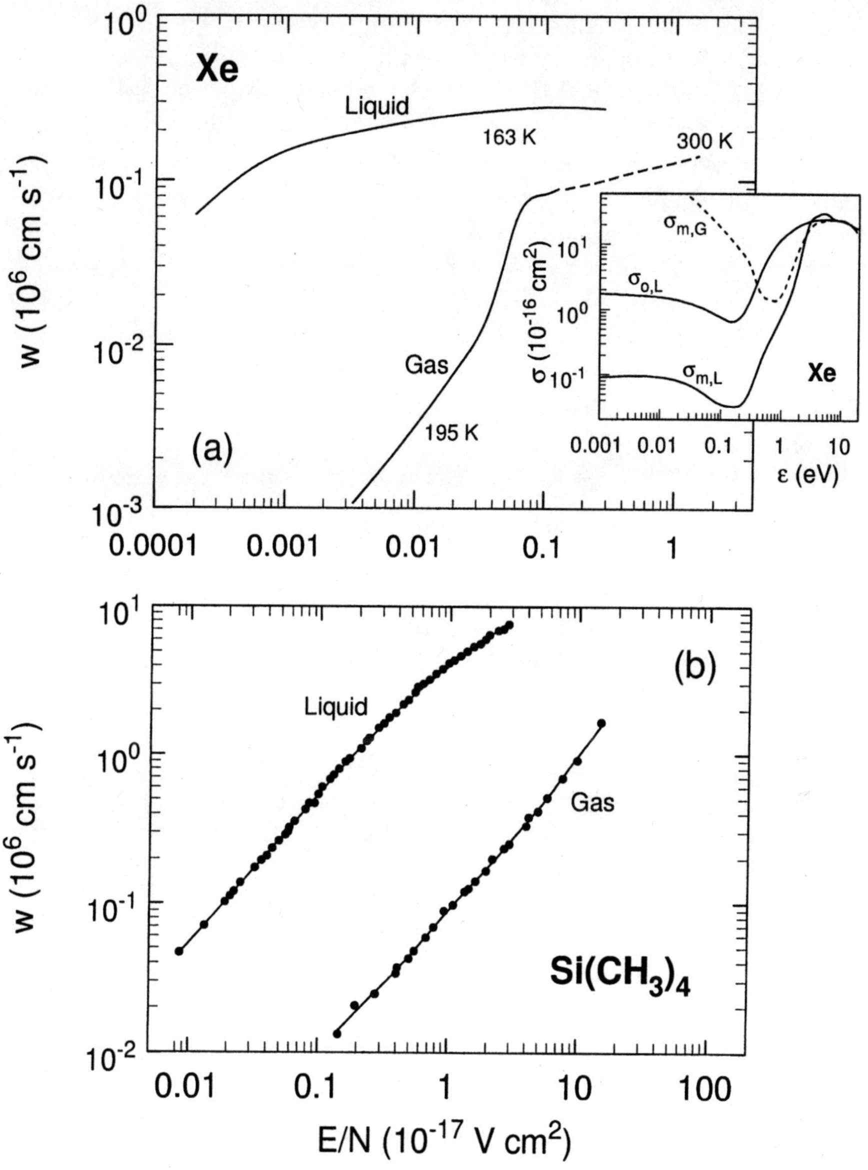

Figure 1.15. (a) Electron drift velocity $w(E/N)$ in liquid [92] and gaseous [93] xenon at the indicated temperatures (from Ref. [76]). The inset compares the elastic momentum transfer cross section in gaseous ($\sigma_{\mathrm{m,G}}(\varepsilon)$) and in liquid ($\sigma_{\mathrm{m,L}}(\varepsilon)$) xenon; $\sigma_{\mathrm{o,L}}(\varepsilon)$ is the cross section for elastic energy loss [94]. (b) Electron drift velocity $w(E/N)$ in liquid and gaseous tetramethylsilane ($Si(CH_3)_4$) [76, 90].

w (or μ) increases with increasing N for the heavier rare gases and the spherical hydrocarbon dielectric liquids (or solids) for which $V_0 < 0$ eV. This behavior is clearly shown by the comparisons in Fig. 1.15 between the electron drift velocity in gaseous and liquid Xe and between the electron drift velocity in gaseous and liquid tetramethylsilane ($Si(CH_3)_4$). Slow electrons are seen to drift much faster in these liquids than in the corresponding low-pressure gases. This has been attributed to changes in the electron scattering cross section in going from the gas to the liquid (or the solid). These are species with a Ramsauer-Townsend minimum in their electron scattering cross section, which is lowered in magnitude and is shifted to lower energies in the condensed medium compared to the low pressure gas as can be seen from the inset in Fig. 1.15(a) where the elastic momentum transfer cross section in liquid xenon [94], $\sigma_{\mathrm{m,L}}(\varepsilon)$, is compared with the gaseous momentum transfer cross section, $\sigma_{\mathrm{m,G}}(\varepsilon)$.

2.3.2 Effect of Medium on Negative Ion States (NISs)

While studies of negative ion states of molecules in low pressure gases are abundant, to our knowledge there has been no similar work for molecules embedded in liquids, and only a limited number of studies for molecules embedded in solids. To illustrate the effect of the medium on the negative ion states of molecules we focus on the 2.3 eV negative ion state of N_2, N_2^{-*} ($^2\Pi_g$). The N_2 molecule does not have a positive electron affinity and hence it does not attach electrons permanently (it also does not attach electrons dissociatively because the N atom cannot bind the extra electron either). However, the N_2 molecule, as we have seen earlier in this chapter, forms transient negative ions via a number of negative ion resonant states.

Figure 1.16 shows the relative cross section for indirect scattering of slow electrons via the N_2^{-*} ($^2\Pi_g$) negative ion state of N_2 in a solid film of N_2 leaving the N_2 molecule excited in the first ($\nu = 1$) vibrational level, as measured by Sanche and Michaud [96]. As in the isolated-molecule case, the cross section for excitation of the N_2 molecule in solid films of N_2 has distinct structure which originates from the vibrational levels of the transient N_2^{-*} ($^2\Pi_g$). The position of the first peak in Fig. 1.16 (first broken vertical line) lies $\sim$ 0.7 eV lower than the corresponding position in the low-pressure gas (first solid vertical line). This lowering of the energy position of the N_2^{-*} ($^2\Pi_g$) state is a general characteristic of the changes in the energetics of the NISs of atoms and molecules embedded in dense matter, and is due to the polarization of the medium which surrounds the temporarily localized electron on the molecule [97].

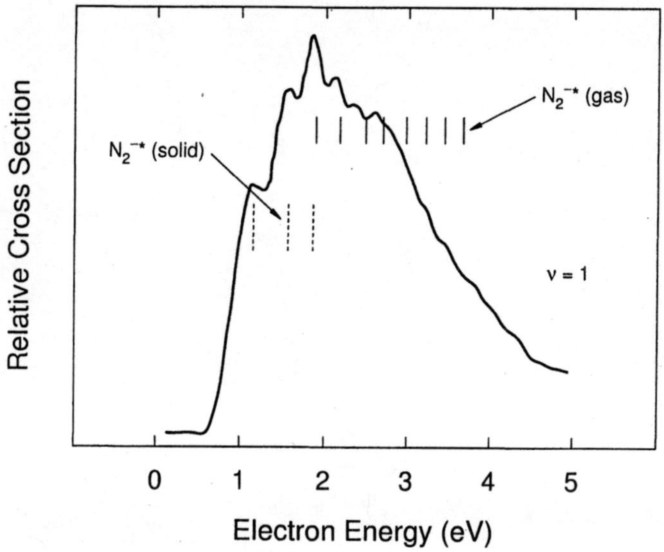

Figure 1.16. Relative cross section for indirect scattering of low energy electrons via the N_2^{-*} ($^2\Pi_g$) negative ion state of N_2 in a solid N_2 film leaving the N_2 molecule excited in the $\nu = 1$ vibrational level. The broken vertical lines indicate the positions of the peaks in the cross section as measured in the film, and the solid vertical lines the corresponding positions as measured in the gas (based on Fig. 13(b) of Ref. [99]).

The formation of the NIS in the condensed phase is localized in space and is dominated by short-range forces, as in the gaseous phase. For this reason, in the solid — as in the gas — the formation of a NIS leads to intra-molecular changes that enhance excitation of individual vibrational levels of molecules. The decay of the NIS in the condensed phase, however, differs from that in the gas phase because it is affected by the medium due to polarization. It has, for instance, been argued [98] that in the condensed phase inter-molecular changes lead to continuous acoustic excitations (phonons) which, in essence, replace the (low pressure) elastic scattering [99]. Medium interactions would also affect the autodetachment lifetime of the transient anion, as well as the cross section for its formation and decomposition by dissociative electron attachment. In making the transition from the gaseous to the condensed phase of matter, it is necessary to note that any such changes would depend on whether the electron affinity of a molecule in the gas phase $(EA)_G$ is negative (< 0 eV) or positive (> 0 eV) and on whether a negative $(EA)_G$ becomes positive in the condensed phase (see further discussion in [75]).

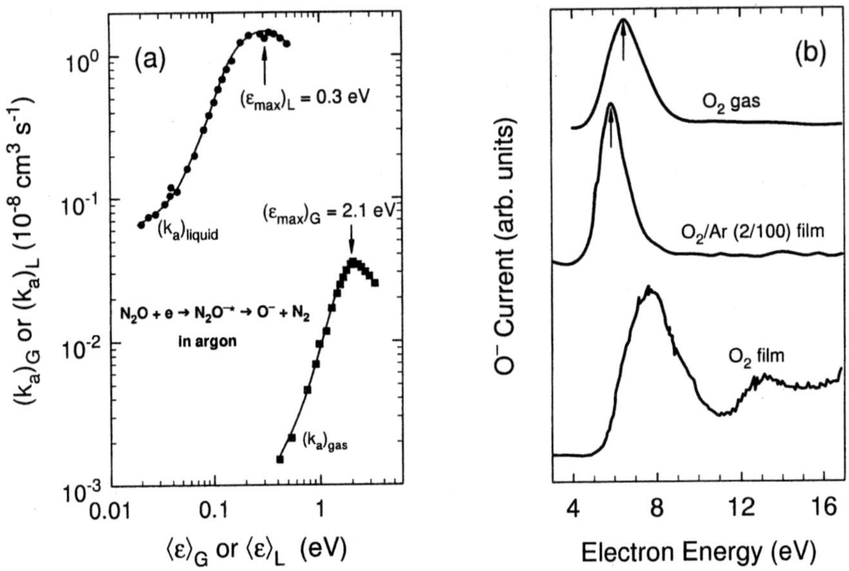

Figure 1.17. (a) Electron attachment rate constant for the production of O^- from N_2O in gaseous ($(k_a)_G$) and liquid ($(k_a)_L$) argon plotted as a function of the respective mean electron energies $\langle\varepsilon\rangle_G$ and $\langle\varepsilon\rangle_L$ [76, 100]. (b) Measured O^- current from O_2 in O_2 gas (top curve) and in solid films of O_2/Ar (middle curve) and O_2 (bottom curve) [99].

2.3.3 Effect of Medium on Dissociative and Nondissociative Electron Attachment Processes

The effect of the medium on electron attachment is a function of the mode of electron attachment (dissociative or nondissociative) and the anionic state(s) involved. It also depends on the state of the electron in the medium. For instance, the electron state crucially determines both the magnitude of the electron attachment cross section and its relation to the corresponding value in the gaseous phase. In the condensed phase of matter, the cross sections of attachment of quasi-free electrons to molecules far exceed those of localized electrons.

The effect of the medium on dissociative electron attachment is insignificant at low gas number densities because the fragment anions do not require collisional stabilization. However, at high gas number densities and in the condensed phase where the anionic state can be influenced by the medium and its energy position can be lowered, changes are expected and have indeed been observed. This is illustrated by the data in Fig. 1.17(a) where a comparison is made of the rate constant for dissociative electron attachment to N_2O (that is, for the reaction $e+N_2O \rightarrow N_2O^{-*} \rightarrow O^- + N_2$) in gaseous and in liquified argon [75, 76]. It is seen that while the reaction takes place in the liquid as it does in the gas, the energy position of the dissociative electron attachment resonance shifts to lower energy as a result of the polarization of the medium around the transient anion N_2O^{-*}, and the rate constant for the reaction is higher in the liquid than in the gas, in part because of this lowering of the energy of the negative ion state.

Studies of electron attachment to molecules in clusters [101] and in solid films [99] generally showed similar changes: the perturbation from the isolated-molecule case of the formation of the transient anion and its decay channels by the medium. An example of this is shown in Fig. 1.17(b) for the case of O^- from O_2 [99]. The negative ion resonance that produces O^- from O_2 and near 6.5 eV in the gas phase (Fig. 1.17(b), top), exists in solid films of mixtures of O_2 in Ar, but is shifted to lower energies by $\sim$ 0.7 eV (Fig. 1.17(b), middle). Furthermore, the yield of O^- from a solid O_2 film (Fig. 1.17(b), bottom) shows enhanced production of O^- near 8 and 14 eV that is attributed by Sanche and collaborators to transitions which are symmetry forbidden in the isolated e, O_2 system. However, these transitions become allowed when the e, O_2 system is embedded in the solid because of the perturbation by the medium (similar results were obtained for clusters [101]).

The effect of the medium and its density on nondissociative electron attachment is a function of the lifetime of the transient anion which is initially produced and can be large even at low gas number densities when the lifetime of the transient anion is short ($< 10^{-10}$ s). This is illustrated by the data in Fig. 1.18(a) where the rate constant for the production of O_2^- from O_2 in N_2 buffer gas is plotted as a function of the N_2 pressure. At low pressures, electrons with energies below $\sim$ 0.5 eV attach to O_2 producing O_2^- via the process

$$O_2\,(X\,^3\Sigma_g^-; \nu = 0) + e \rightarrow O_2^{-*}\,(X\,^2\Pi_g; \nu' \geq 4) \rightarrow O_2^{(*)} + e^{(\prime)} \quad (1.10a)$$

$$\rightarrow O_2^-\,(X\,^2\Pi_g; \nu' < 4) + \text{energy}. \quad (1.10b)$$

The stabilization of $O_2^{-*}\,(X\,^2\Pi_g;\ \nu' \geq 4)$ is usually through collision with other molecules. If the e, O_2 system is situated in a buffer gas (N_2 for the data in Fig. 1.18(a)), the magnitude of the electron attachment rate constant for the production of O_2^- increases and its energy dependence changes as the number density of N_2 is increased. The electron attachment process is then no longer described by Reaction (1.10), but it requires the involvement of more than one N_2 molecule [35, 102]. Such a dependence, however, is not to be expected for the formation of the par-

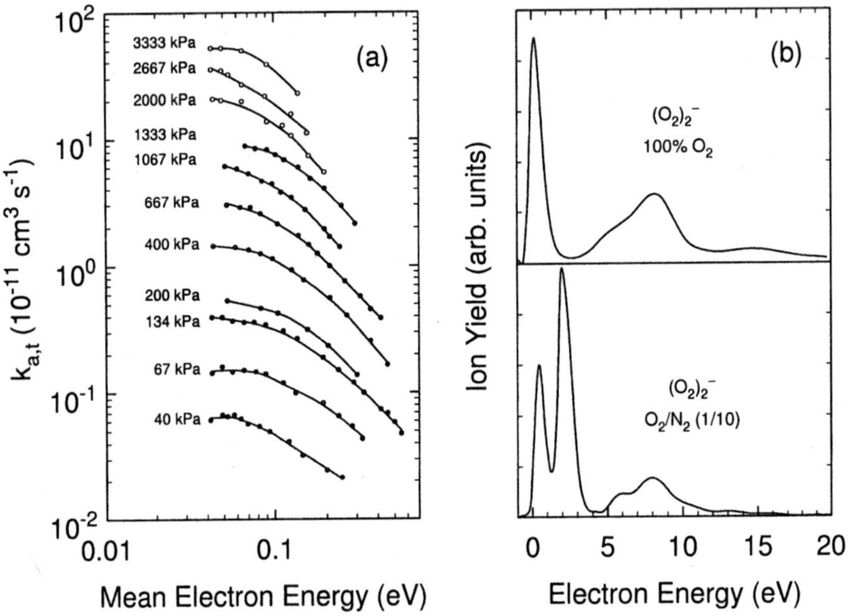

Figure 1.18. (a) Total electron attachment rate constant for the production of O_2^- from O_2 as a function of the mean electron energy, $k_{a,t}(\langle\varepsilon\rangle)$, measured in N_2 buffer gas at the indicated N_2 pressures (the percentage of O_2 in N_2 is negligibly small). The measurements were made at $T = 298$ K (data from Ref. [102]). (b) Yield of $(O_2)_2^-$ ions from pure O_2 clusters (upper curve) and from O_2/N_2 clusters (O_2/N_2 ratio 1/10, lower curve). The stagnation pressure was 400 kPa. The pronounced additional resonance in the $(O_2)_2^-$ production near 2.3 eV for the O_2/N_2 clusters is due to the formation of $(O_2)_2^-$ ions by capture of electrons which have been inelastically scattered to near-zero energy via the 2.3 eV N_2^{-*} $(^2\Pi_g)$ resonance of N_2 in the O_2/N_2 clusters (data from Ref. [103]).

ent negative ions of molecules such as SF_6, c-C_4F_8, and C_6F_6, since these molecules capture thermal and near-thermal-energy electrons with large cross sections forming long-lived ($\gtrsim 10^{-6}$ s) parent negative ions. Besides these changes, the proximity of collision partners within the condensed medium or the cluster may enhance the production of parent negative ions (e.g., O_2^- or cluster negative ions such as $(O_2)_2^-$). Furthermore, the negative ion properties of single molecules or clusters of molecules with strong "zero" or "near-zero" energy nondissociative electron attachment resonances are influenced by inelastic electron scattering processes by the cluster constituents within the cluster. This is illustrated in Fig. 1.18(b) where the production of $(O_2)_2^-$ ions in O_2/N_2 mixed clusters is influenced by the inelastic electron scattering by N_2, especially by inelastic electron scattering via the negative ion resonance $N_2^{-*}(^2\Pi_g)$ near 2.3 eV [103]. The yield of the $(O_2)_2^-$ ions exhibits a strong resonance near 2.3 eV in the case of the O_2/N_2 clusters (Fig. 1.18(b), lower part) which is absent in the homologous (pure O_2) clusters (Fig. 1.18(b), upper part).

Besides these changes, primary ions, such as O^- and O_2^- in the case of O_2, can be clustered and ions such as $O^-(O_2)_{n\geq 1}$ and $O_2^-(O_2)_{n\geq 1}$ can be formed [103, 104].

2.3.4 Changes in the Energetics of Electron-Impact Ionization and Electron Detachment

To our knowledge, there have been no measurements of the electron-impact ionization cross section (or coefficient) of molecules in high pressure gases or liquids. However, there have been a number of studies on the photoionization of molecules (e.g., Refs. [77, 80, 87, 89, 105–110]) and the photodetachment of negative ions [37, 111, 112] in dense media (and also of cluster negative ions [20, 113–115]), which have provided valuable information and insight into the effect of the medium on the energetics of the ionization and detachment processes. Photoionization studies [75, 76, 80, 89, 105, 108–110] have shown that the adiabatic ionization onset energy, $I(N)$, of a molecule in a dense medium having a number density N, is related to the low-gas-density adiabatic ionization onset energy, I_G, of the molecule by

$$I(N) = I_G + P^+(N) + V_0(N) + \Delta E(N), \tag{1.11}$$

where P^+ (a negative quantity) is the polarization energy of the positive ion produced by ionization, V_0 (a negative or a positive quantity) is the ground-state energy of the ejected electron, and ΔE is a small quantity associated with the broadening of the valence levels of the isolated molecule in the dense medium. The quantities V_0, P^+ and ΔE are all functions of N. Neglecting the small broadening ΔE of the valence electron levels in the dense medium, we see that $I(N)$ differs from I_G by an amount equal to $P^+(N) + V_0(N)$. For a number of nonpolar liquids, the measurements showed [75, 80, 105–109] that

$$I_L = I_G + P^+ + V_0. \tag{1.12}$$

In Eq. (1.12), P^+ is the polarization energy of the positive ion in the liquid and is normally represented by the Born expression

$$P^+ = -\frac{e^2}{2R}\left(1 - \frac{1}{\varepsilon_{\text{opt}}}\right), \tag{1.13}$$

where R is the effective radius of the positive ion cavity and ε_{opt} is the optical dielectric constant of the medium. For a number of organic molecules in nonpolar liquids, the energy difference $I_{\text{L}} - I_{\text{G}}$ $(= P^{+} + V_0)$ lies between -1 and -3 eV and $\langle R \rangle$ is ~ 2.7 Å.

Equations analogous to the energetic relations discussed above for photoionization can be envisioned for photodetachment. Indeed, such studies have been conducted (e.g., see Refs. [37, 38, 111, 112]) using lasers and transient photoconductivity techniques. The energetics for the reaction

$$\text{M}^{-} + h\nu \rightarrow \text{M} + e \tag{1.14}$$

under low-gas pressure conditions can be written as

$$E_{\text{th}} = \text{VDE} = \text{EA} + \Delta E, \tag{1.15}$$

where E_{th} is the threshold energy, EA is the adiabatic electron affinity of the molecule M, and VDE is the vertical detachment energy of M^{-}. The VDE exceeds the EA value of the molecule by ΔE, a quantity which depends on the relative positions of the potential energy curves (surfaces) of M and M^{-}.

Let us now consider the case where the photodetachment process (1.14) takes place in a dense medium, say a liquid L, and is represented as

$$\text{M}_{\text{L}}{}^{-} + h\nu \rightarrow \text{M}_{\text{L}} + e_{\text{L}}, \tag{1.16}$$

where M_{L}^{-}, M_{L}, and e_{L} are now species embedded in the liquid medium. The effect of the medium on the energetics of Reaction (1.16) would originate mainly from the effect of the medium on M_{L}^{-} and e_{L}. By analogy to Eq. (1.15), we can write

$$(E_{\text{th}})_{\text{L}} = \text{VDE}_{\text{L}} = \text{EA}_{\text{L}} + \Delta E_{\text{L}}. \tag{1.17}$$

If it is further assumed that $\text{EA}_{\text{L}} = \text{EA}_{\text{G}} + V_0 - P^{-}$ (e.g., see Refs. [37, 112]), we have

$$(E_{\text{th}})_{\text{L}} = \text{EA}_{\text{G}} + V_0 - P^{-} + \Delta E_{\text{L}}. \tag{1.18}$$

In these energetic relations P^{-} is the polarization energy of the negative ion in the medium, and EA_{L} and EA_{G} are, respectively, the values of the electron affinity of the molecule M in dense matter and in the gas phase.

The above energetic relations show the dominant role of the electric polarization of the medium on the energetics of electron attachment, detachment, and ionization, and can provide insight as to the energetics of such processes at surfaces.

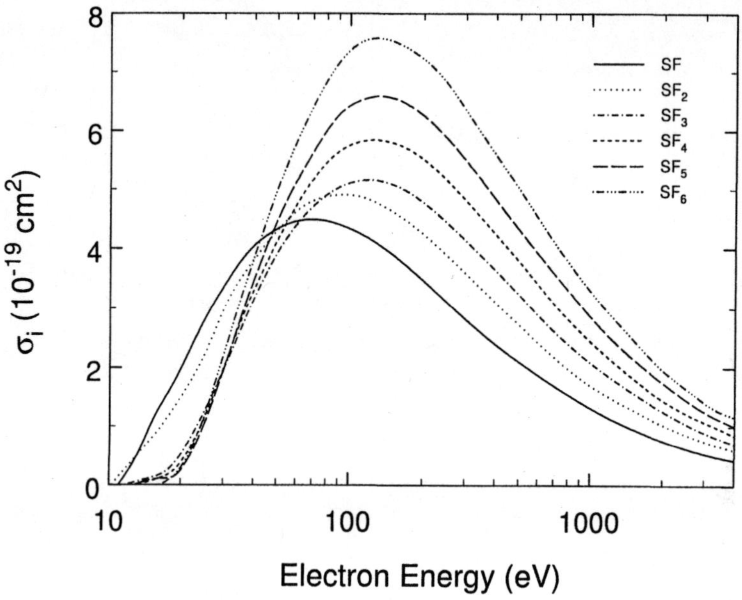

Figure 1.19. Electron-impact ionization cross sections for the radicals SF_x ($x = 1$–5) calculated by Ali *et al.* [122].

2.4 Interactions of Low-Energy Electrons With Transient Species (Radicals)

There have been only a few experimental measurements of the total or partial electron-impact ionization cross sections of radicals of interest to plasma processing (e.g., see the work of Tarnovsky and collaborators [116–120]). These are discussed in the appropriate subsequent chapters where the electron collision data for the parent molecules are presented. Similarly, there have been a number of computations of the total electron-impact ionization cross sections (e.g., see Refs. [120–125]) which are also discussed in later chapters. Figure 1.19 shows examples of the calculated electron-impact ionization cross sections for the radicals SF_x ($x = 1$–5) [122].

With regard to the cross sections for the other basic electron scattering processes for transient species in plasma processing gases, there exist only very few measurements or computations. Notable among these are the recent computations of elastic and inelastic electron scattering cross sections for a number of radicals by McKoy and collaborators (e.g., fluorocarbon radicals [126, 127] and fragments of BCl_3 and $SiCl_4$ [128]). Also, a number of experimental studies by Stoffels and collaborators [129–132] on plasma processing gases, while not providing cross sections for transient species, are nonetheless relevant to the present discussion and significant in that they have shown that in electronegative plasmas, negative ion densities (n_-) can be orders of magnitude higher than electron densities (n_e). For instance, the density ratio n_-/n_e was found to be of the order of 10 for O_2 and CF_4 plasmas, and as high as 10^3 for CCl_2F_2 plasmas [132]. Although it is difficult to account for electron attachment to radicals produced in the discharge by the dissociation of the parent gas molecules, these observations implicate the role of electron attachment to ground-state and/or excited fragments. It should be noted that very large electron attachment rate constants have been suggested [133] for dissociative electron attachment to CCl_3 radicals ($CCl_3 + e \rightarrow CCl_2 + Cl^-$) produced by dissociative electron attachment to CCl_4 ($CCl_4 + e \rightarrow CCl_3 + Cl^-$).

Not only can unstable molecules be responsible for increased electron attachment rates in a plasma, but stable electronegative molecules resulting from chemical conversion of the parent molecules can also contribute. Stoffels *et al.* [132] pointed out that plasma polymerization is such a case, and illustrated its significance for CF_4 and CHF_3 gases [132, 134]. They found that in CF_4 and CHF_3 plasmas, the rate constants for electron attachment (and hence the densities of the negative ions present) to both of these molecules are much higher than would be expected on the basis of the known electron attachment cross sections for the parent ground-state molecules [22, 23, 28].

In both cases, the experimental measurements of Stoffels *et al.* [132, 134] showed that this effect can be attributed to the conversion of the parent gases to polymers which have higher electron attachment cross sections than the parent gas. Stoffels *et al.* sampled electronegative species from the plasma with a mass spectrometer where the neutrals were converted into negative ions using an electron source with variable energy. This way, they measured the dependence of the effective cross section for various negative ion fragments on the plasma conditions. Figure 1.20 shows their results for F^- in CF_4 plasmas. The F^- signal is relatively weak in the absence of the plasma as is to be expected from the known [22] cross section for the production of F^- by dissociative electron attachment to CF_4. The F^- signal, however, shows large peaks near 4.5 and 3 eV in the presence of plasma. These peaks were attributed [132, 134] to dissociative electron attachment to the plasma-conversion products C_2F_6 and C_3F_8. These results indicate that despite the fact that CF_4 is the majority gas, dissociative electron attachment to CF_4 is only a minor contributor to the total negative ion formation in the plasma, and negative ion production in CF_4 plasmas is dominated by

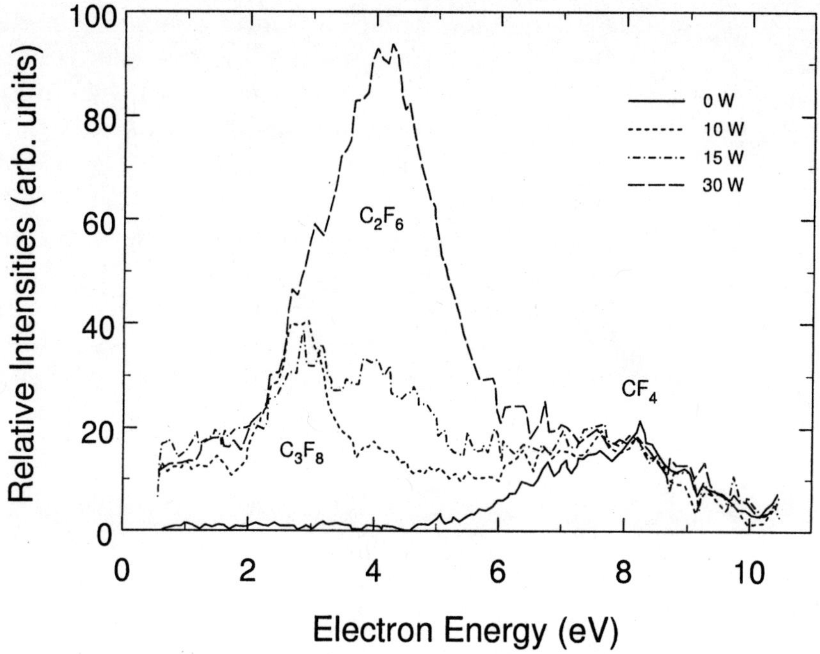

Figure 1.20. Relative cross sections measured by Stoffels *et al.* [132, 134] for the production of F^- in CF_4 gas at a pressure of 2 Pa and in CF_4 plasmas at power levels of: 0 W (—); 10 W (- - -); 15 W (– · –); and 30 W (— —) [132].

polymerization products which are more electronegative than the parent gas.

3. Significance of Electron-Molecule Interactions in Plasma Processing

Basic knowledge of the wide range of processes and novel reactions through which slow electrons and molecules embrace each other in rich and delicate ways has aided not only the advancement of many areas of science, but has also impacted significantly the advancement of many new technologies (e.g., see Refs. [135–139]). For instance, basic knowledge on electron-molecule interactions has contributed fundamentally to the advancement of electron and ion physics, gaseous electronics and dielectrics, and the physics of instrumentation, especially in the areas of radiation detection and dosimetry and particle accelerators. Additionally, it has led to new concepts of practical utility such as electron, ion, and chemical lasers. It also underpins a wide spectrum of areas in engineering and applied science that range from extinguishing of plasmas, arcs, and flames, to combustion and polymer construction; from radio, television, and other communications systems, to space- and defense-related applications; from the development of injection beams in fusion machines and intense negative ion beams, to high-power switching circuits and magnetohydrodynamics; from gas discharges and plasmas, to atmospheric, ionosphereric, and environmental and health sciences.

In most technological applications, the crucial knowledge is the combined effect of key electron-collision processes which collectively control the electron energies and electron number densities in the gaseous medium as a function of the applied electric field. For this reason, gases often are tailored (combined) to optimize the electron energies and electron number densities so as to enhance or to inhibit the critical reactions required for a particular application. To illustrate this important point, Christophorou and Hunter [135] discussed the physical basis of "the choice and tailoring of gases for new technologies" in five applied areas (gas lasers, electrical discharge lamps and plasma display panels, radiation detectors, gaseous dielectrics, and diffuse discharge opening switches) whose success crucially depends upon the properties of the gaseous medium they use. They indicated how basic knowledge of electron-molecule interactions permits the right choice of gas(es) for optimum performance. The basic knowledge on electron-molecule interactions covered in this book allows the proper choice and tailoring of gases for yet another technology, namely, plasma processing.

Low-temperature plasmas are used by the semiconductor industry to etch features, deposit materials, and clean reactor chambers. Development of these applications requires detailed understanding of the physical and chemical processes occurring in the plasmas themselves. For instance, as the push for smaller dimensions (down to a few hundred angstroms), higher quality devices, more complex patterns, and more application-specific properties have increased in the semiconductor industry, so has the need for sophisticated models with predictive capabilities to guide the technology and to serve as tools for design, control, and optimization of plasma etching reactors (e.g., see discussion in Refs. [140–146]). In addition, there is a need for advanced diagnostics to probe the details of the plasmas, and guidance is needed to more efficiently use global warming gases. Advancement in each of these areas requires knowledge of the basic collision processes taking place between the species existing in the plasma. The most fundamental of the discharge processes are collisions between electrons and atoms, radicals, or molecules. These collisions are the precursors of the ions and the radicals which drive the etching, cleaning, or deposition processes.

Hence, a quantitative understanding of the fundamental electron collision processes in terms of cross sections and rate coefficients is of utmost importance and the availability of reliable data on such quantities is a key to the further success of the technology.

To better understand the significance of electron interactions in underpinning this technology we briefly focus on low-temperature, low-density, non-equilibrium plasmas and the fundamental gas-phase processes occurring in the plasma volume/sheath.

3.1 Low-Temperature, Low-Density, Non-Equilibrium Plasmas

Among the many types of plasmas, those commonly employed for plasma processing are low-temperature, low-density, non-equilibrium, collision-dominated environments. By low temperature we mean "cold" plasmas with temperatures normally ranging from 300 and 600 K, by low density we mean plasmas with neutral gas number densities of approximately 10^{13} to 10^{16} molecules cm^{-3} (pressures between $\sim$ 0.1 to 10^3 Pa) which are weakly ionized (charge-carrier densities $\sim 10^8$ to 10^{11} cm^{-3} and degree of ionization between 10^{-6} to 10^{-1}), and by non-equilibrium we mean media for which the mean energies of the electrons are much higher than those of the neutral gas molecules. In such media, the electrons have a distribution of kinetic energies whose mean value is typically in the range of 0.5 to 5.0 eV, far in excess of the mean energy of the ions which is typically in the range 0.025 to 0.05 eV. While most of the electrons in such distributions have kinetic energies below the thresholds for electronic excitation, dissociation, or ionization of the gas molecules, a small fraction of the electrons have kinetic energies sufficiently high to induce such processes.

In contrast with previous practice, we refrain in this book from using the term "electron temperature" to describe electron kinetic energies, and rather use the term "temperature" for only the gas molecules. The main reason for this choice is the fact that electrons in plasmas do not have a single temperature since their energies are normally much higher than thermal and have a broad non-Maxwellian distribution. Also, while the term "gas temperature" for a molecular gas characterizes the translational energy distribution of the gas molecules, it also implies an amount of molecular internal energy, which has no equivalence for the electrons.

As the technology of microelectronic fabrication has advanced, the semiconductor industry has specifically tailored the chemistry of processing plasmas to produce favorable mixtures of reactive species and high energy ions. This combination is often referred to as reactive ion etching (RIE) and is the most commonly used process to achieve highly anisotropic etching while minimizing sputtering damage. The composition of the plasma depends strongly on both the gas-phase reactions of the feed gas and the reactions occurring at the surfaces in the reactors. Significantly more work needs to be done to obtain an understanding of the physical and chemical processes occurring at the surfaces. A more detailed description of the physical characteristics of the plasmas utilized in microelectronic fabrication may be found in Sec. 3.2 of this chapter and elsewhere [147–150].

3.2 Generation, Use, and Modeling of Low-Pressure Plasmas

Low-pressure plasmas employed by the semiconductor industry for the fabrication of microelectronics are often generated by applying radio frequency power between a pair of parallel-plate electrodes in a chamber filled with a low-pressure gas. This is shown schematically in Fig. 1.21(a). To produce a plasma with a higher density of charged species, the upper electrode may be replaced by an inductive coil that enables a more efficient transfer of power into the plasma. After ignition of the plasma, a capacitively coupled electrode acquires a negative charge because the electron mobility is much larger than that of the ions. The material on the electrode is thus exposed to energetic bombardment by positive ions, and the bulk of the plasma is exposed to energetic electrons produced by secondary electron emission from the surfaces. Both of these charged species obtain their energies by acceleration through the sheath that forms between the plasma glow and electrode surfaces. This sheath is usually a fraction of a millimeter to several millimeters in thickness, and visually manifests itself as the dark region between the surface and the glow. It is the nearly collisionless acceleration across this space that enables the anisotropic etching that is so important in many plasma processes.

Chemical reactions between the radicals and the neutral species and the material being etched occur at the surface and produce either volatile species or their precursors. The positive ion bombardment of the surface removes materials by sputtering and produces chemically active sites on the solid. The combination of the chemical activity of species and sputtering can result in much higher material erosion rates in the vertical than in the lateral direction, which makes reactive-ion etching an essential tool for replicating device patterns in the fabrication of very-large scale integrated circuits. As stated by Economou [143], the goals of any plasma etch process are to achieve high etch rate, uniformity (same etch

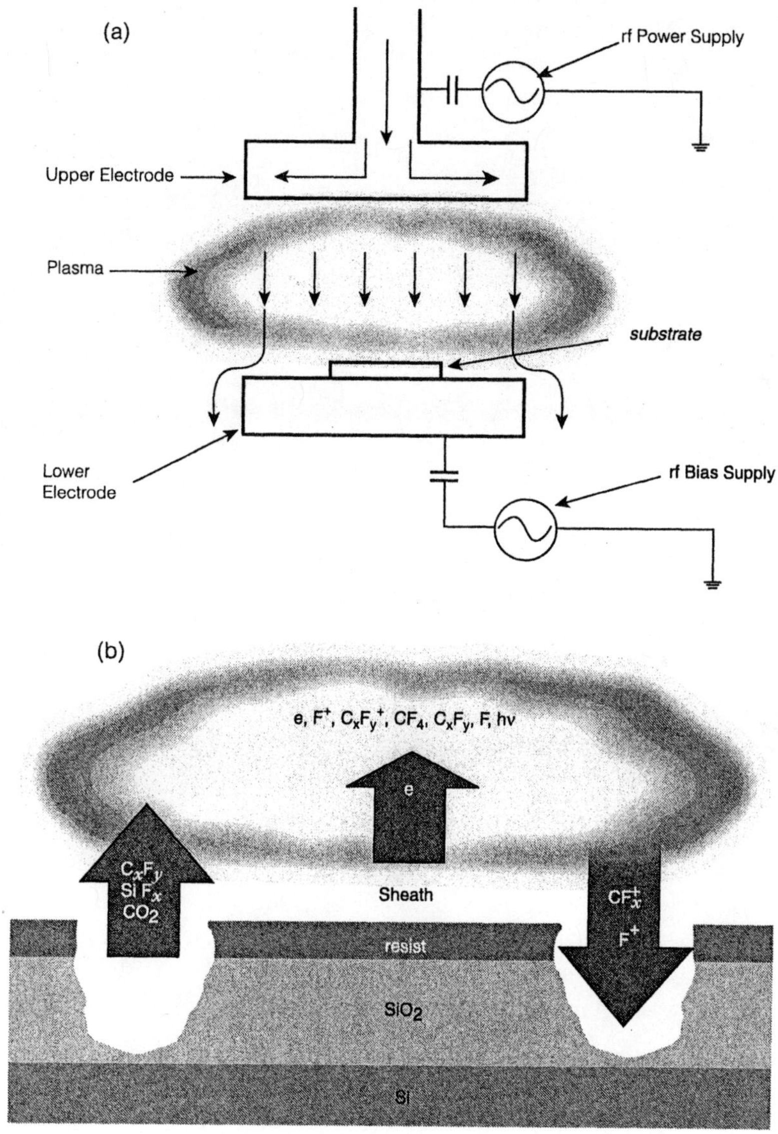

Figure 1.21. (a) Schematic of plasma reactor. The arrows indicate the common direction of the gas flow in industrial reactors. (b) Simple illustration of the chemistry of reactive ion etching in a CF_4 plasma.

characteristics), selectivity (relative rate of etching of one material with respect to another), controlled shape of the microscopic features etched into the film (anisotropy), and no radiation damage. These goals are realized by suitable control of key plasma properties such as the electron energy distribution function, the space and time variation of electron, ion, and neutral-species densities and energies, as well as by the control of the flux, energy, and angular distribution of ions and neutrals bombarding the electrodes and their uniformity across the electrodes.

The interactions of low-energy electrons in these plasmas occur under different conditions depending on the plasma region. Three regions can be identified: the bulk plasma, the plasma-surface interface (sheath), and the solid. From a plasma processing perspective, the most important processes are those occurring at the surfaces, since they actually drive the etching and deposition of material. However, those processes are all initiated by species generated in the gas phase (or glow) of the plasma. Thus, the discussions in this book primarily refer to the reactions occurring in the bulk plasma for which the generic primary processes are elastic and inelastic electron scattering, electron-impact ionization, electron-impact dissociation, and electron attachment. Lack of knowledge on these processes for the parent molecules (and their decomposition products) has often been cited as the main roadblock to the development of plasma models of proven importance to industry.

Let us, then, elaborate a little further on some of the important plasma parameters in low-temperature, low-density, non-equilibrium plasmas and the corresponding major electron interaction processes. These are listed in Table 1.1.

Electron-energy distributions in plasmas are determined by elastic and inelastic electron-scattering processes, even when the latter lie at low energies. Such is the case, for instance, of low energy inelastic processes due to vibrational excitation of molecules which affect the probability of high-energy processes via their effect on the electron energy distribution function. The determination of the electron energy distribution functions in the discharge is the initial calculation in models based upon Boltzmann or Monte Carlo techniques, and a successful calculation relies upon the availability of detailed and accurate electron scattering cross sections and electron transport coefficient data. Similarly, electron transport parameters and rate coefficients are often the first physical quantities which are calculated by plasma models. Accurate measurements of these values, then, are essential to validate model calculations. The transport parameters commonly calculated in such models are the electron drift velocity, transverse (and longitudinal) electron diffusion coefficient to electron mobility ratio, density-normalized electron-impact ion-

Table 1.1. Important plasma parameters in low-temperature, low-density, non-equilibrium plasmas and corresponding important major electron interaction processes.[a]

Plasma parameter	Major electron interaction process(es)
Electron-energy distributions	Momentum transfer Elastic scattering Inelastic scattering
Electron densities	Total ionization Total electron attachment
Positive ion densities	Partial ionization
Negative ion densities	Electron attachment Dissociative electron attachment
Radical densities	Neutral dissociation Dissociative electron attachment Partial ionization
Excited-species densities	Vibrational excitation Electronic excitation

[a]Besides information on the cross sections for these processes, cross section data are needed also for electron interactions with radicals and other decomposition products.

ization coefficient, density-normalized electron attachment coefficient, and density-normalized effective ionization coefficient.

Similarly, the *electron density in a plasma* is primarily determined by the electron-impact ionization processes (which produce free electrons), the electron attachment processes (which remove free electrons by creating negative ions), and by the secondary electron emission processes involving surfaces. The first two processes are described by the electron-impact ionization and electron attachment cross sections, respectively, and the third process while important and worthy of investigation is a surface process and is not discussed here. The ionization and attachment cross sections are of particular importance since they determine the ionization balance within the plasma and thus play an important role in determining the *positive and negative ion densities.* Another process of interest in this regard is ion-electron recombination, but since this is a secondary process, is not covered in the present discussion.

Additionally, plasma surface reactions induced by *positive ion bombardment* depend on the ion flux which is also a direct result of electron-impact ionization. Although the final identity and magnitude of the positive ion flux may be dependent upon secondary reactions, the initial ion-formation is driven by electron-impact processes. Partial ionization cross sections are then required to determine the identity and to quantity the number density of the initial ions created in the plasma. Moreover, since positive ion and negative ion recombination processes influence the positive ion densities, the latter are influenced by the production of negative ions by resonant electron attachment. Dissociative electron attachment is significant both as an efficient source of negative ions and as a source of free radicals.

Radical densities are determined by dissociative electron attachment, partial ionization, and dissociation into neutrals. Cross sections for all of these processes are considered here, although the available data for dissociation into neutrals is sparse. *Excited species densities* are influenced by electronic excitation and vibrational excitation cross sections. However, here too, rather less is known of these processes because of the experimental difficulties in performing relevant measurements.

Finally, the significance of fundamental electron collision processes for *plasma diagnostics* needs to be emphasized. The knowledge of such processes underpins the development of new diagnostic techniques such as negative-ion mass spectrometry and threshold-ion mass spectrometry used to determine the identity and to measure the number density of gas-phase products in industrial plasmas. Furthermore, *optical emission diagnostic* techniques often used to monitor plasma uniformity, excited-species densities, and electron-energy distributions, rely on light emission from a processing plasma and require electron-impact excitation cross sections for the feed gas and for the gas-phase products generated in the discharge.

Many models have been designed to emulate various aspects of reactive plasmas (see, for example, Ventzek *et al.* [151], Lymberopoulos and Economou [152], Bukowski *et al.* [153], Meyyappan and Govindan [154], and several comprehensive reviews of plasma reactor models and modeling techniques [143, 155–157]). From these and other reports, it is clear that models require data for both validation and prediction. Whatever the type of the model, it requires data on cross sections and rate coefficients, as well as thermodynamic data for electron-molecule and ion-molecule reactions of interest both in the gas phase and at the surface. Knowledge of the fundamental physics of electron collision processes enables the making of reasonable estimates of cross sections for probable reactions for which no data exist.

The energetics associated with the basic electron and ion interaction processes are also of significance. Such data are needed to characterize the plasma, because they determine the energy losses of the various processes and along with the corresponding cross sections constitute the critical input data in computational models. Additionally, they can provide information as to the nature of the particles that are to be expected depending on the energetics for their formation. While often the energy threshold of the cross section for a given process is a measure of the electron energy – and, thus, the energy loss – required for that process (e.g., the energy thresholds of direct vibrational excitation or electronic excitations are roughly the difference between the excited and ground-state energy levels), the energy thresholds for dissociation processes are generally significantly larger than their thermodynamic values. Moreover, attention should be given to the different electron energy losses associated with direct and indirect electron-collision processes. For instance, the thresholds for indirect vibrational excitation of molecules by electron impact are determined by the positions of the negative ion resonances rather than by the vibrational thresholds themselves. The former can be much larger than the latter. Also, proper consideration is necessary of the differences between the angular distributions of scattered electrons involving direct and indirect vibrational excitation processes.

The success of plasma modeling efforts have been mixed. Complex processing plasmas have been modeled, and the results of diagnostic measurements and surface conditions have been reproduced. However, such success is often obtained by the use of a significant number of adjustable parameters. The resulting model is therefore often not unique, and generally is not predictive. Predictive modeling is the end goal of most theoretical modeling efforts of the semiconductor industry, and its achievement will require continuing advancement in several technical areas, not the least of which is the accumulation and analysis of the fundamental data that feed the models.

3.3 Primary and Secondary Reactions

Tanaka and Inokuti [158] distinguished three temporal stages of the processes which occur in low temperature, low density, non-equilibrium plasmas. The *physical* or *initial* stage during which primary processes such as excitation, dissociation, and ionization of molecules take place; the *physicochemical or secondary* stage during which photons (emitted by excited species), positive and negative ions, excited atoms and molecules, and free radicals react with other molecules; and the *chemical* or *thermal* stage during which further reactions of the products of the second stage occur under nearly thermal-equilibrium conditions. The products of the third stage undergo diffusion and reactions with the surfaces of the reactor walls. Thus, the species that make up the plasma are created, transformed, or destroyed by a multiplicity of reactions which occur either in the discharge (volume plasma or the sheath) or at the surfaces.

For the purpose of our discussion in this book, we separate these important processes into three groups: primary electron interaction processes, secondary electron interaction processes, and secondary reactions between the products of the electron interactions.

Primary electron interaction processes. These are the primary interactions of electrons with the parent gas (the initial "feed" gas) which we outlined earlier in this chapter (Reactions (1.1) and (1.2)). They are the most significant plasma reactions in that they generate the key plasma particles: electrons, positive and negative ions, radicals, excited species, and photons that accompany the radiative decay of the excited species. These are the reactions and the products of the physical stage. They are the ones we are principally concerned with in this book. As we have indicated earlier in this chapter, these basic electron-molecule interaction processes largely control the energies and the number densities of free electrons in electrically stressed gases. Knowledge of the behavior of these reactions can be used to control the important species in the plasmas of many of today's technologies, foremost the microelectronics industry.

Secondary electron interaction processes. These are reactions the electrons undergo with the products (ions, excited species, radicals) of their primary interactions (see also a discussion by Hatano [159]). These processes consist of reactions such as (1.3), (1.4), (1.6) and (1.7), and reactions such as electron-positive ion recombination which remove electrons from the system, as well as electron interactions with metastable species and radicals. Much work is still needed regarding electron interactions with excited species and radicals especially since the cross sections of electron collisions with excited states oftentimes are much larger than with the ground-state species and since the number density of radicals can be substantial for certain plasma processing gases.

Secondary reactions among the products of electron interactions. The initial products of the primary and secondary electron collisions undergo subsequent collisions with the plasma constituents (neutral species, excited species, positive and negative ions, and photons). Via these secondary reactions the initial electron-generated products are transformed, often very efficiently, into other species. There is a multiplicity of such reactions which affect the identity and the number density of the "final" species in weakly ionized, low-temperature plasmas. In Table 1.2 are listed a number of photon-induced, ion-induced, and neutral-species-induced secondary reactions. Depending on the system, some of these reactions can be significant both in determining the densities of the desirable species in the plasma and also in plasma diagnostics. Because many molecular species readily react with other species, usually there is a large variety of chemically active species in the plasma which are often difficult to analyze, trace, or control. The secondary processes of electron-ion recombination can be significant in determining the overall electron number density, and the negative-ion destruction processes will deplete the negative ions. Ion-molecule reactions will change the identity of the ionic species in the discharge, and Penning ionization processes may be significant for mixtures containing, for instance, rare gas atoms. Similarly, charge-exchange reactions help determine the identity of the dominant ions in the discharge. These reactions also provide a route for energy exchanges between the charged particles and the neutral species in the plasma. Binary mixtures of rare gases with polyatomic molecules have frequently been used in reactive plasmas in which energy-transfer processes between excited rare gas atoms and molecules are considered important.

Thus, besides the need for data on the primary and secondary electron-collision processes, there is a need for data on the secondary processes, both in terms of cross sections and rate constants and in terms of energetics and thermochemical data. Although such data are sparse for many species of interest in plasma processing, when available they are provided in subsequent chapters. Also provided in subsequent chapters, are relevant spectroscopic data since these complement electron-collision data and constitute the basis of a number of sensitive radiative and laser-based techniques for species concentration and temperature measurements in processing plasmas.

4. The Present Work

This book deals mainly with the fundamental primary electron collision processes for ten gases commonly used for plasma etching. Namely, CF_4, SF_6, CCl_2F_2, Cl_2, BCl_3 which are used for the treatment of silicon; CHF_3, CF_4, C_2F_6, C_3F_8, c-C_4F_8 which are used for the treatment of silicon dioxide; and BCl_3, Cl_2, CCl_2F_2 which are used for the treatment of aluminum alloys. It draws heavily from the work we have conducted at the National Institute of Standards and Technology (NIST) over the last eight years under a project aimed at building a database for electronic processes in plasma processing gases, including cross sections and electron transport and rate coefficients.

Table 1.2. Photon-induced, ion-induced, and neutral-species-induced secondary reactions.

Process representation[a]	Process description
Photon-induced reactions	
$h\nu + AX \rightarrow AX^*$	Photoabsorption
$h\nu + AX \rightarrow AX^+ + e$	Photoionization
$\Rightarrow A + X^+ + e$	Dissociative photoionization
$h\nu + AX \Rightarrow A + X$	Photodissociation
$h\nu + AX^-(X^-) \rightarrow AX(X) + e$	Photodetachment
$h\nu + AX^- \rightarrow A + X^-$	Negative-ion photodissociation
Ion-induced reactions	
$AX^+ + e \rightarrow AX(AX^*)$	Electron recombination
$\Rightarrow A + X^{(*)}$	Dissociative recomb. with/without excitation
$\rightarrow AX + h\nu$	Radiative recombination
$A^+ + X^- \rightarrow AX(AX^*)$	Positive ion-negative ion recombination
$AX^+ + C^- \rightarrow$ products	Positive ion-negative ion recombination[b]
$AX^- + C \rightarrow AX + C + e$	Collisional detachment
$AX^- + C \rightarrow AXC + e$	Associative detachment
$AX^+(X^+) + CD(C) \Rightarrow$ products	Positive ion-molecule (atom) reactions[c]
$AX^-(X^-) + CD(C) \Rightarrow$ products	Negative ion-molecule (atom) reactions[c]
$AX^+ + nC \rightarrow AX^+nC \quad n \geq 1$	Cluster formation involving positive ions
$AX^- + nC \rightarrow AX^-nC \quad n \geq 1$	Cluster formation involving negative ions
Neutral-species-induced reactions	
$AX^* + C \rightarrow AX + C^*$	Energy transfer
$AX^*(X^*) + C \rightarrow AX(X) + C^+ + e$	Penning ionization
$A + BC(B) \Rightarrow$ products	Chemical reactions involving neutral species[d]

[a] AX represents an unexcited and AX^* represents an excited diatomic or polyatomic molecule. The double arrow indicates that the reaction can generate a multiplicity of products.
[b] Positive ion-negative ion recombination for molecular species. In this case, there may be multiple products.
[c] A multiplicity of reactions and products depending on the type and state of the reactants.
[d] These may include radical recombination ($2A \rightarrow A_2$), addition ($A + AX \rightarrow A_2X$), or abstraction ($A + AX \rightarrow A_2 + X$) reactions.

The fundamental electron-molecule interactions are first elaborated upon in Ch. 1. The quantities most often used to describe these interactions (cross sections, transport and rate coefficients) are defined in Ch. 2 where some recent advances in the methods employed for their measurement or calculation are outlined. The methodology we adopted for the critical evaluation, synthesis, and assessment of the existing data is described in Ch. 3. The critically assessed data and the recommended or suggested cross sections and transport coefficients for the ten plasma etching gases under consideration, are presented and discussed in the subsequent three chapters, Chapters 4, 5, and 6. For each molecule, the recommended or suggested data are given both in graphical and in tabular form and when possible they are used to derive needed data which at the present time are not otherwise available.

For all ten molecules under consideration in this book, the cross section for each electron-molecule collision is assessed independently. Therefore, each cross section is unique. However, the consistency of the cross sections is assessed (e.g., the sum of the cross sections for various electron collision processes is compared with the total electron scattering cross section to ensure that the former is not greater than the later). Also, in the present work, the rate and transport coefficients are similarly obtained by individual independent assessments of existing data under a set protocol (see Ch. 3).

The unique determination of the cross sections is in sharp contrast with the sets of electron collision cross sections which are determined from computational models and are then "validated" as complete sets through the ability of the computational code to reproduce the macroscopic rate coefficients of the molecule in question using such cross section sets. Such computationally-derived sets are not unique, and such "validation" only tests the cross sections sets when used as a whole. A cross section for a specific process may be incorrect, but the complete set may accurately reproduce the known rate coefficients because of a compensating error in another cross section.

While we realize that the acquisition of reliable data is a dynamic process, we nonetheless believe that at any given time the best possible values should be discerned and should be used as such — not as variables — in calculations, comparisons of codes, or models. Unless, that is, there is a scientific basis to question the data. We also believe that there is a need to understand the significance of the data, and the physical reality they represent and are a measure of. This is necessary for the proper use of the data. The assessment and validation of data is also essential to prevent unreliable data from entering the system and triggering a chain of erroneous results. Furthermore, the assessed synthesis and validation of data can provide bench marks for comparison of models and experiments. Of the ten molecules discussed in this book, the independently assessed cross sections for CF_4 shown in Fig. 1.2 (and discussed in Ch. 4), and possibly those for SF_6 (discussed in Ch. 6) may be considered for such purposes. Finally, it should be indicated that while the knowledge discussed in this book is tilted toward its use in plasma models for the semiconductor industry, its basic and applied significance transcends this technology. The knowledge of electron-interaction cross sections and transport coefficients is nearly universally needed.

A list of the basic electron collision processes considered in the book and their respective cross sections and coefficients is given in Table 1.3. Further definition of these processes is presented in Ch. 2.

Table 1.3. Electron collision processes considered, their respective cross sections and symbols, and electron-impact ionization, attachment, and transport coefficients.

Electron collision process	Cross section/Coefficient	Symbol
Electron scattering	Total electron scattering cross section	$\sigma_{\mathrm{sc,t}}(\varepsilon)$
	Differential electron scattering cross section(s)[a]	$\sigma_{\mathrm{sc,diff}}(\varepsilon)$
Rotational excitation	Total rotational electron scattering cross section	$\sigma_{\mathrm{rot,t}}(\varepsilon)$
Elastic electron scattering	Total elastic electron scattering cross section[b]	$\sigma_{\mathrm{e,t}}(\varepsilon)$
	Total elastic integral[b]	$\sigma_{\mathrm{e,int}}(\varepsilon)$
	Momentum transfer cross section (elastic)	$\sigma_{\mathrm{m}}(\varepsilon)$
	Elastic differential electron scattering cross section	$\sigma_{\mathrm{e,diff}}$
Vibrational excitation	Total vibrational excitation cross section	$\sigma_{\mathrm{vib,t}}(\varepsilon)$
	Total direct vibrational excitation cross section	$\sigma_{\mathrm{vib,dir,t}}(\varepsilon)$
	Total indirect vibrational excitation cross section	$\sigma_{\mathrm{vib,indir,t}}(\varepsilon)$
Electronic excitation	Electronic excitation cross section	$\sigma_{\mathrm{elec}}(\varepsilon)$
Dissociation	Total dissociation cross section	$\sigma_{\mathrm{dis,t}}(\varepsilon)$
	Total cross section for electron-impact dissociation into neutrals	$\sigma_{\mathrm{dis,neut,t}}(\varepsilon)$
	Partial dissociation cross section into neutrals	$\sigma_{\mathrm{dis,neut,part}}(\varepsilon)$

[a] For total or for a particular electron scattering process.
[b] The cross sections $\sigma_{\mathrm{e,int}}(\varepsilon)$ and $\sigma_{\mathrm{e,t}}(\varepsilon)$ refer to the same quantity (see Ch. 2).

Table 1.3 (continued)
Electron collision processes considered, their respective cross sections and symbols, and electron-impact ionization, attachment, and transport coefficients.

Electron collision process	Cross section/Coefficient	Symbol
Ionization	Total ionization cross section	$\sigma_{\mathrm{i,t}}(\varepsilon)$
	Partial ionization cross section	$\sigma_{\mathrm{i,part}}(\varepsilon)$
	Multiple ionization cross section	$\sigma_{\mathrm{i,mult}}(\varepsilon)$
	Density-normalized ionization coefficient	α/N
	Density-normalized effective ionization coefficient	$(\alpha - \eta)/N$
Attachment	Total electron attachment cross section	$\sigma_{\mathrm{a,t}}(\varepsilon)$
	Total dissociative electron attachment cross section	$\sigma_{\mathrm{da,t}}(\varepsilon)$
	Density-normalized electron attachment coefficient	η/N
	Total electron attachment rate constant	$k_{\mathrm{a,t}}$
Ion-pair formation	Cross section for ion-pair formation	$\sigma_{\mathrm{ip}}(\varepsilon)$
Electron drift	Electron drift velocity	w
Electron diffusion	Transverse electron diffusion coefficient to electron mobility ratio	D_{T}/μ
	Longitudinal electron diffusion coefficient to electron mobility ratio	D_{L}/μ

5. References

[1] J. J. Thomson and G. P. Thomson, *Conduction of Electricity Through Gases*, Vol. I, Cambridge, At the University Press, 1928.

[2] R. H. Healey and J. W. Reed, *The Behaviour of Slow Electrons in Gases*, Amalgamated Wireless (Australia) Limited, Sydney, 1941.

[3] J. Townsend, *Electrons in Gases*, Hutchinson's Scientific and Technical Publications, London, 1947.

[4] H. S. W. Massey and E. H. S. Burhop, *Electronic and Ionic Impact Phenomena*, Oxford, At the Clarendon Press, 1952.

[5] L. B. Loeb, *Basic Processes of Gaseous Electronics*, University of California Press, Berkeley and Los Angeles, 1955.

[6] A. von Engel, *Ionized Gases*, Oxford, At the Clarendon Press, 1955.

[7] W. P. Allis, in *Handbuch der Physik*, S. Flügge (Ed.), Springer-Verlag, Berlin, 1956, Vol. 21, p. 413.

[8] J. D. Craggs and H. S. W. Massey, in *Handbuch der Physik*, S. Flügge (Ed.), Springer-Verlag, Berlin, 1956, Vol. XXXVII, p. 314.

[9] J. B. Hasted, *Physics of Atomic Collisions*, Butterworths, London, 1964.

[10] E. W. McDaniel, *Collision Phenomena in Ionized Gases*, John Wiley & Sons, Inc., New York, 1964.

[11] H. S. W. Massey and E. H. S. Burhop, *Electronic and Ionic Impact Phenomena, Vol. I, Collisions of Electrons with Atoms*, Oxford, At the Clarendon Press, 1969.

[12] H. S. W. Massey, *Electronic and Ionic Impact Phenomena, Vol. II, Electron Collisions with Molecules and Photoionization*, Oxford, At the Clarendon Press, 1969.

[13] L. G. Christophorou, *Atomic and Molecular Radiation Physics*, Wiley-Interscience, London, 1971.

[14] G. J. Schultz, Rev. Mod. Phys. 45, 378 (1973) (also, *Resonances in Electron Impact on Atoms and Diatomic Molecules*, NSRDS-NBS 50, October 1973).

[15] L. G. H. Huxley and R. W. Crompton, *The Diffusion and Drift of Electrons in Gases*, Wiley-Interscience, New York, 1974.

[16] H. S. W. Massey, *Negative Ions*, 3rd Edition, Cambridge University Press, Cambridge, 1976.

[17] J. M. Meek and J. D Craggs (Eds.), *Electrical Breakdown of Gases*, John Wiley & Sons, Chichester, 1978.

[18] L. G. Christophorou (Ed.), *Electron-Molecule Interactions and Their Applications*, Academic, New York, Vols. 1 and 2, 1984.

[19] E. E. Kunhardt, L. G. Christophorou, and L. H. Luessen (Eds.), *The Liquid State and Its Electrical Properties*, Plenum, New York, 1988.

[20] L. G. Christophorou, E. Illenberger, and W. F. Schmidt (Eds.), *Linking the Gaseous and Condensed Phases of Matter: The Behavior of Slow Electrons*, Plenum, New York, 1994.

[21] L. G. Christophorou and J. K. Olthoff, Adv. At. Mol. Opt. Phys. **44**, 155 (2000).

[22] L. G. Christophorou, J. K. Olthoff, and M. V. V. S. Rao, J. Phys. Chem. Ref. Data **25**, 1341 (1996) [Electron Interactions with CF_4].

[23] L. G. Christophorou, J. K. Olthoff, and M. V. V. S. Rao, J. Phys. Chem. Ref. Data **26**, 1 (1997) [Electron Interactions with CHF_3].

[24] L. G. Christophorou, J. K. Olthoff, and Y. Wang, J. Phys. Chem. Ref. Data **26**, 1205 (1997) [Electron Interactions with CCl_2F_2].

[25] L. G. Christophorou and J. K. Olthoff, J. Phys. Chem. Ref. Data **27**, 1 (1998) [Electron Interactions with C_2F_6].

[26] L. G. Christophorou and J. K. Olthoff, J. Phys. Chem. Ref. Data **27**, 889 (1998) [Electron Interactions with C_3F_8].

[27] L. G. Christophorou and J. K. Olthoff, J. Phys. Chem. Ref. Data **28**, 131 (1999) [Electron Interactions with Cl_2].

[28] L. G. Christophorou and J. K. Olthoff, J. Phys. Chem. Ref. Data **28**, 967 (1999) [Electron Interactions with Plasma Processing Gases: An Update for CF_4, CHF_3, C_2F_6, and C_3F_8].

[29] L. G. Christophorou and J. K. Olthoff, J. Phys. Chem. Ref. Data **29**, 267 (2000) [Electron Interactions with SF_6].

[30] L. G. Christophorou and J. K. Olthoff, J. Phys. Chem. Ref. Data **29**, 553 (2000) [Electron Interactions with CF_3I].

[31] L. G. Christophorou and J. K. Olthoff, J. Phys. Chem. Ref. Data **30**, 449 (2001) [Electron Interactions with *c*-C_4F_8].

[32] L. G. Christophorou and J. K. Olthoff, J. Phys. Chem. Ref. Data **31** 971 (2002) [Electron Interactions with BCl_3].

[33] S. Trajmar and D. C. Cartwright, in *Electron-Molecule Interactions and Their Applications*, L. G. Christophorou (Ed.), Academic, New York, Vol. 1, Ch. 2, 1984.

[34] L. G. Christophorou, Adv. Electr. Electron Phys. **46**, 55 (1978).

[35] L. G. Christophorou, D. L. McCorkle, and A. A. Christodoulides, in *Electron-Molecule Interactions and Their Applications*, L. G. Christophorou (Ed.), Academic, New York, Vol. 1, Ch. 6, 1984.

[36] A. A. Christodoulides, D. L. McCorkle, and L. G. Christophorou, in *Electron-Molecule Interactions and Their Applications*, L. G. Christophorou (Ed.), Academic, New York, Vol. 2, Ch. 6, 1984.

[37] L. G. Christophorou, P. G. Datskos, and H. Faidas, J. Chem. Phys. **101**, 6728 (1994).

[38] P. G. Datskos, J. G. Carter, and L. G. Christophorou, Chem. Phys. Lett. **239**, 38 (1995).

[39] J. N. Bardsley, A. Herzenberg, and F. Mandl, Proc. Phys. Soc. (London) **89**, 321 (1966).

[40] T. F. O'Malley, Phys. Rev. **150**, 14 (1966).

[41] R. W. Crompton, in *Proc. XVI Int. Conf. on Phenomena in Ionized Gases, Invited Papers*, W. Bötticher, H. Wenk, and E. Schulz-Gulde (Eds.), Düsseldorf, August/September, 1983, p. 58.

[42] C. C. Lin and L. W. Anderson, Adv. At. Mol. Opt. Phys. **29**, 1 (1992).

[43] S. Trajmar and J. C. Nickel, Adv. At. Mol. Opt. Phys. **30**, 45 (1993).

[44] L. G. Christophorou, L. A. Pinnaduwage, and P. G. Datskos, in *Linking the Gaseous and Condensed Phases of Matter: The Behavior of Slow Electrons*, L. G. Christophorou, E. Illenberger, and W. F. Schmidt (Eds.), Plenum, New York, 1994, p. 415.

[45] S. J. Buckman, M. T. Elford, and D. S. Newman, J. Phys. B **20**, 5175 (1987).

[46] J. Ferch, C. Masche, and W. Raith, J. Phys. B **14**, L97 (1981).

[47] M. A. Morrison, N. F. Lane, and L. A. Collins, Phys. Rev. A **15**, 2186 (1977).

[48] J. Ferch, C. Masche, W. Raith, and L. Wiemann, Phys. Rev. A **40**, 5407 (1989).

[49] P. G. Datskos, L. G. Christophorou, and J. G. Carter, J. Chem. Phys. **97**, 9031 (1992).

[50] L. G. Christophorou and P. G. Datskos, Int. J. Mass Spectrom. Ion Physics **149/150**, 59 (1995).

[51] P. G. Datskos, L. G. Christophorou, and J. G. Carter, J. Chem. Phys. **99**, 8607 (1993).

[52] R. I. Hall and S. Trajmar, J. Phys. B **8**, L293 (1975).

[53] P. B. Armentrout, S. M. Tarr, A. Dori, and R. S. Freund, J. Chem. Phys. **75**, 2786 (1981).

[54] D. Ton-That and M. R. Flannery, Phys. Rev. A **15**, 517 (1977).

[55] W. M. Huo, in *Non-Equilibrium Processes in Partially Ionized Gases*, M. Capitelli and J. N. Bardsley (Eds.), Plenum, New York, 1990, p. 341.

[56] D. C. Cartwright, S. Trajmar, A, Chutjian, and W. Williams, Phys. Rev. A **16**, 1041 (1977).

[57] S. Trajmar, D. F. Register, and A. Chutjian, Phys. Repts. **97**, 219 (1983).

[58] R. Celotta, H. Brown, R. Molof, and B. Bederson, Phys. Rev A **3**, 1622 (1971).

[59] E. J. Robinson, Phys. Rev. **182**, 196 (1969).

[60] K. Jost, P. G. F. Bishing, F. Erchen, M. Felsmann, and J. Walther, in *XIIIth International Conference on the Physics of Electronic and Atomic Collisions*, Abstracts of Contributed Papers, Berlin, 1983, p. 91.

[61] J. C. Nickel, K. Imre, D. F. Register, and S. Trajmar, J. Phys. B **18**, 125 (1985).

[62] J. Ferch, B. Granitza, C. Masche, and W. Raith, J. Phys. B **18**, 967 (1985).

[63] H. B. Milloy, R. W. Crompton, J. A. Rees, and A. G. Robertson, Austr. J. Phys. **30**, 61 (1977).

[64] A. J. Dixon, M. F. A. Harrison, and A. C. H. Smith, in *Proceedings of the VIIIth International Conference on the Physics of Electronic and Atomic Collisions*, Book of Abstracts, Belgrade, 1973, p. 405.

[65] E. Krishnakumar and S. K. Srivastava, J. Phys. B **21**, 1055 (1988).

[66] L. G. Christophorou and E. Illenberger, Phys. Lett. A **173**, 78 (1993).

[67] P. D. Burrow, J. Chem. Phys. **59**, 4922 (1973).

[68] T. Jaffke, M. Meinke, R. Hashemi, L. G. Christophorou, and E. Illenberger, Chem. Phys. Lett. **193**, 62 (1992).

[69] D. Rapp and D. D. Briglia, J. Chem. Phys. **43**, 1480 (1965).

[70] L. G. Christophorou, S. R. Hunter, L. A. Pinnaduwage, J. G. Carter, A. A. Christodoulides, and S. M. Spyrou, Phys. Rev. Lett. **58**, 1316 (1987).

[71] T. Jaffke, R. Hashemi, L. G. Christophorou, E. Illenberger, H. Baumgärtel, and L. A. Pinnaduwage, Chem. Phys. Lett. **203**, 21 (1993).

[72] E. Krishnakumar, S. V. K. Kumar, S. A. Rangwala, and S. K. Mitra, Phys. Rev. A **56**, 1945 (1997).

[73] L. G. Christophorou, Chem. Revs. **76**, 409 (1976).

[74] S. R. Hunter and L. G. Christophorou, in *Electron-Molecule Interactions and Their Applications*, L. G. Christophorou (Ed.), Academic, New York, Vol. 2, Ch. 3, 1984.

[75] L. G. Christophorou, in *The Liquid State and its Electrical Properties*, E. E. Kunhardt, L. G. Christophorou, and L. H. Luessen (Eds.), Plenum, New York, 1988, p. 283.

[76] L. G. Christophorou, in *Linking the Gaseous and Condensed Phases of Matter: The Behavior of Slow Electrons*, L. G. Christophorou, E. Illenberger, and W. F. Schmidt (Eds.), Plenum, New York, 1994, p. 3.

[77] L. G. Christophorou and K. Siomos, in *Electron-Molecule Interactions and Their Applications*, L. G. Christophorou (Ed.), Academic, New York, Vol. 2, Ch. 4, 1984.

[78] C. Ferradini and J.-P. Jay-Gerin (Eds.), *Excess Electrons in Dielectric Media*, CRC Press, Boca Raton, 1991.

[79] R. A. Holroyd, in *Linking the Gaseous and Condensed Phases of Matter: The Behavior of Slow Electrons*, L. G. Christophorou, E. Illenberger, and W. F. Schmidt (Eds.), Plenum, New York, 1994, p. 443.

[80] W. F. Schmidt, *Liquid State Electronics of Insulating Liquids*, CRC Press, Boca Raton, 1997.

[81] L. Sanche, J. Phys. B **23**, 1597 (1990).

[82] L. Sanche, in *Linking the Gaseous and Condensed Phases of Matter: The Behavior of Slow Electrons*, L. G. Christophorou, E. Illenberger, and W. F. Schmidt (Eds.), Plenum, New York, 1994, p. 31.

[83] E. Illenberger, Chem. Revs. **92**, 1589 (1992).

[84] J. Jortner, Z. Phys. D **24**, 247 (1992).

[85] O. Ingólfsson, F. Weik, and E. Illenberger, Int. J. Mass Spectrom. Ion Physics **155**, 1 (1996).

[86] L. G. Christophorou, J. Phys. Chem. **76**, 3730 (1972).

[87] B. E. Springett, J. Jortner, and M. H. Cohen, J. Chem. Phys. **48**, 2720 (1968).

[88] Y. Yamaguchi, T. Nakajima, and M. Nishikawa, J. Chem. Phys. **71**, 550 (1979).

[89] H. Faidas, L. G. Christophorou, P. G. Datskos, and D. L. McCorkle, J. Chem. Phys. **90**, 6619 (1989).

[90] H. Faidas, L. G. Christophorou, D. L. McCorkle, and J. G. Carter, Nucl. Instr. Meth. Phys. Res. A **294**, 575 (1990).

[91] L. G. Christophorou, J. G. Carter, and D. V. Maxey, J. Chem. Phys. **76**, 2653 (1982).

[92] L. S. Miller, S. Howe, and W. E. Spear, Phys. Rev. **166**, 871 (1968).

[93] S. R. Hunter, J. G. Carter, and L. G. Christophorou, Phys. Rev. A **38**, 5539 (1988).

[94] Y. Sakai, S. Nakamura, and H. Tagashira, IEEE Trans. Electr. Insul. **EI-20**, 133 (1985).

[95] P. Krebs and M. Heintze, J. Chem. Phys. **76**, 5484 (1982).

[96] L. Sanche and M. Michaud, Phys. Rev. B **27**, 3856 (1983).

[97] L. G. Christophorou, Z. Phys. Chem. **195**, 195 (1996).

[98] U. Fano, J. A. Stevens, and M. Inokuti, J. Chem. Phys. **85**, 6239 (1986).

[99] L. Sanche, in *Excess Electrons in Dielectric Media*, C. Ferradini and J.-P. Jay-Gerin (Eds.), CRC Press, Boca Raton, 1991, p. 1.

[100] G. Bakale, U. Sawada, and W. F. Schmidt, J. Phys. Chem. **80**, 2556 (1976).

[101] T. Jaffke, R. Hashemi, L. G. Christophorou, and E. Illenberger, Z. Phys. D: **25**, 77 (1992).

[102] R. E. Goans and L. G. Christophorou, J. Chem. Phys. **60**, 1036 (1974).

[103] R. Hashemi, T. Jaffke, L. G. Christophorou, and E. Illenberger, J. Phys. Chem. **96**, 10605 (1992).

[104] E. Illenberger, in *Linking the Gaseous and Condensed Phases of Matter: The Behavior of Slow Electrons*, L. G. Christophorou, E. Illenberger, and W. F. Schmidt (Eds.), Plenum, New York, 1994, p. 49.

[105] B. Raz and J. Jortner, Chem. Phys. Lett. **4**, 155 (1969).

[106] R. A. Holroyd, J. Chem. Phys. **57**, 3007 (1972).

[107] R. A. Holroyd and R. L. Russell, J. Phys. Chem. **78**, 2128 (1974).

[108] W. F. Schmidt, IEEE Trans. Electr. Insul. **EI-19**, 389 (1984).

[109] H. Faidas and L. G. Christophorou, Rad. Phys. Chem. **32**, 433 (1988).

[110] I. T. Steinberger, in *The Liquid State and Its Electrical Properties*, E. E. Kunhardt, L. G. Christophorou, and L. H. Luessen (Eds.), Plenum, New York, 1988, p. 235.

[111] U. Sowada and R. A. Holroyd, J. Phys. Chem. **84**, 1150 (1980).

[112] H. Faidas, L. G. Christophorou, and D. L. McCorkle, Chem. Phys. Lett. **193**, 487 (1992).

[113] L.-S. Zheng, C. M. Karner, P. J. Brucat, S. H. Yang, C. L. Pettiette, M. J. Craycraft, and R. E. Smalley, J. Chem. Phys. **85**, 1681 (1986).

[114] D. G. Leopold, J. Ho, and W. C. Lineberger, J. Chem. Phys. **86**, 1715 (1987).

[115] M. J. DeLuca, C.-C. Han, and M. A. Johnson, J. Chem. Phys. **93**, 268 (1990).

[116] V. Tarnovsky, P. Kurunczi, D. Rogozhnikov, and K. Becker, Int. J. Mass Spectrom. Ion Processes **128**, 181 (1993).

[117] V. Tarnovsky, A. Levin, and K. Becker, J. Chem. Phys. **100**, 5626 (1994).

[118] V. Tarnovsky, A. Levin, K. Becker, R. Basner, and M. Schmidt, Int. J. Mass Spectrom. Ion Processes **133**, 175 (1994).

[119] V. Tarnovsky, A. Levin, H. Deutsch, and K. Becker, J. Phys. B **29**, 139 (1996).

[120] V. Tarnovsky, H. Deutsch, K. E. Martus, and K. Becker, J. Chem. Phys. **109**, 6596 (1998).

[121] Y.-K. Kim and K. K. Irikura, in *Atomic and Molecular Data and Their Applications*, K. A. Berrington and K. L. Bell (Eds.), American Institute of Physics Conference Proceedings, New York, Vol. 543, 2000, p. 220.

[122] M. A. Ali, K. K. Irikura, and Y.-K. Kim, Int. J. Mass Spectrom. **201**, 187 (2000).

[123] H. Deutsch, K. Becker, S. Matt, and T. D. Märk, Int. J. Mass Spectrom. **197**, 37 (2000).

[124] H. Deutsch, K. Becker, and T. D. Märk, Int. J. Mass Spectrom. Ion Processes **167/168**, 503 (1997).

[125] W. M. Huo, V. Tarnovsky, and K. H. Becker, Chem. Phys. Lett. **358**, 328 (2002).

[126] W. McKoy, C. Winstead, W. L. Morgan, and P. D. Haaland, *Data Compilation for Plasma Chemistries*, SEMATECH, Technology Transfer Report # 98063515A-TR, June 30, 1998.

[127] W. McKoy, W. L. Morgan, P. D. Haaland, C. Winstead, W. K. Trail, and C. Jiao, *Data Compilation for Plasma Chemistries*, SEMATECH, Technology Transfer Report # 00023909A-TR, March 31, 2000.

[128] W. McKoy, C. Winstead, and W. L. Morgan, *Data Compilation for Plasma Chemistries*, SEMATECH, Technology Transfer Report # 97043274A-TR, August 22, 1997.

[129] E. Stoffels, W. W. Stoffels, D. Vender, M. Haverlag, G. M. W. Kroesen, and F. J. de Hoog, Contrib. Plasma Phys. **4-5**, 331 (1995).

[130] E. Stoffels, W. W. Stoffels, D. Vender, M. Kando, G. M. W. Kroesen, and F. J. de Hoog, Phys. Rev. E **51**, 2425 (1995).

[131] E. Stoffels, W. W. Stoffels, D. Vender, G. M. W. Kroesen, and F. J. de Hoog J. Vac. Sci. Technol. A **13**, 2051 (1995).

[132] E. Stoffels, W. W. Stoffels, and G. M. W. Kroesen, Plasma Sources Sci. Technol. **10**, 311 (2001).

[133] N. G. Adams, D. Smith, and C. R. Herd, Int. J. Mass Spectrom. Ion Processes **84**, 243 (1988).

[134] E. Stoffels, W. W. Stoffels, and K. Tachibana, Rev. Sci. Instrum. **69**, 116 (1998).

[135] L. G. Christophorou and S. R. Hunter, in *Electron-Molecule Interactions and Their Applications*, L. G. Christophorou (Ed.), Academic, New York, Vol. 2, Ch. 5, 1984.

[136] H. S. W. Massey, E. W. McDaniel, and B. Bederson (Eds.), *Applied Atomic Collision Physics*, Vols. 1–5, Academic, New York, 1982-1984.

[137] M. A. Lieberman and A. J. Lichtenberg, *Principles of Plasma Discharges and Materials Processing*, Wiley, New York, 1994.

[138] M. Inokuti and K. H. Becker (Eds.), Adv. At. Mol. Opt. Phys. **43**, 2000.

[139] M. Kimura and Y. Itikawa (Eds.), Adv. At. Mol. Opt. Phys. **44**, 2000.

[140] J. N. Bardsley, in *Atomic and Molecular Data and Their Applications* (ICAMDATA-First International Conference, Gaithersburg, Maryland, October 1997), P. J. Mohr and W. L. Wiese (Eds.), The American Institute of Physics Conference Proceedings 434, Woodbury, New York, 1998, p. 333.

[141] C. Winstead and V. McKoy, Adv. At. Mol. Opt. Phys. **43**, 111 (2000).

[142] W. L. Morgan, Adv. At. Mol. Opt. Phys. **43**, 79 (2000).

[143] D. J. Economou, Thin Solid Films **365**, 348 (2000).

[144] E. Meeks and P. Ho, Thin Solid Films **365**, 334 (2000).

[145] IEEE Trans. Plasma Sci. **27** (5) 1999.

[146] K. Becker, H. Deutsch, and M. Inokuti, Adv. At. Mol. Opt. Phys. **43**, 399 (2000).

[147] S. M. Rossnagel, J. J. Cuomo, and W. D. Westwood (Eds.), *Handbook of Plasma Processing Technology — Fundamentals, Etching, Deposition, and Surface Interactions*, Noyes Publications, Park Ridge, NJ, 1990.

[148] Y. P. Raizer, M. N. Shneider, and N. A. Yatsenko, *Radio-Frequency Capacitive Discharges*, CRC Press, London, 1995.

[149] S. Wolf and R. N. Tauber, *Silicon Processing for the VLSI Era, Volume 1 – Process Technology*, Lattice Press, Sunset Beach, CA, 1986.

[150] T. Makabe (Ed.), *Advances in Low Temperature RF Plasmas*, Elsevier, Amsterdam, 2002.

[151] P. L. G. Ventzek, R. J. Hoekstra, and M. J. Kushner, J. Vac. Sci. Technol. B **12**, 461 (1994).

[152] D. P. Lymberopoulos and D. J. Economou, IEEE Trans. Plasma Science **23**, 573 (1995).

[153] J. D. Bukowski, D. B. Graves, and P. Vitello, J. Appl. Phys. **80**, 2614 (1996).

[154] M. Meyyappan and T. R. Govindan, J. Appl. Phys. **80**, 1345 (1996).

[155] L. E. Kline and M. J. Kushner, Crit. Rev. in Solid State and Materials Sciences **16**, 1 (1989).

[156] G. G. Lister, J. Appl. Phys. D **25**, 1649 (1992).

[157] D. B. Graves, IEEE Trans. Plasma Sci. **22**, 31 (1994).

[158] H. Tanaka and M. Inokuti, Adv. At. Mol. Opt. Phys. **43**, 1 (2000).

[159] Y. Hatano, Adv. At. Mol. Opt. Phys. **43**, 231 (2000).

Chapter 2

ELECTRON-MOLECULE INTERACTIONS IN THE GAS PHASE: CROSS SECTIONS AND COEFFICIENTS

1. Introduction

The quantity which characterizes an electron scattering process, n, is the electron scattering cross section, σ_{n}. The electron scattering cross section σ_{n} represents the probability for the occurrence of the particular process n in an electron-molecule collision. The cross sections for the various electron-molecule interaction processes n are a function of the electron energy ε, the type of the process n, and the scattering angle θ. As mentioned in Ch. 1, the electron scattering cross sections are also a function of the internal energy of the target molecule and the density and nature of the medium in which the electron-molecule collision occurs. Many books (e.g., Refs. [1–6]) have provided extensive discussions and definitions of the various cross sections for electron collision processes. The discussion below is based on these sources.

Electron-molecule collisions are normally divided into three groups: elastic, inelastic, and superelastic. *Elastic* collisions are those in which the electron loses only that amount of kinetic energy which is necessary for the conservation of momentum in the collision process. In such collisions only kinetic energy is exchanged between the colliding particles, the internal state of the target molecule remaining unchanged. *Inelastic* collisions are those in which the target gains internal energy at the expense of the kinetic energy of the impinging electron. The collision can result in excitation, dissociation, or ionization of the target. *Superelastic* collisions are those in which internal energy from the excited target is imparted to the impinging electron resulting in an increase of its kinetic energy. Superelastic collisions are important, especially at low electron energies (close to thermal where the fraction of the molecules in excited rotational and vibrational states can be appreciable) and in collisions of electrons with metastable atoms and molecules. In this book we deal with the first two types of collisions. The limited information on the third type of collision has been discussed in Ref. [7].

In this chapter the cross sections for the various electron collision processes discussed in Ch. 1 are defined. The symbols σ and σ^* are used, respectively, for the cross sections involving electron collisions with ground state molecules and excited molecules. Similarly, the coefficients w, D_{T} (or D_{T}/μ), D_{L} (or D_{L}/μ) which characterize the transport of electrons in gases, and the coefficients for electron attachment η, electron-impact ionization α, and effective ionization $\bar{\alpha}$, are defined and are related to the corresponding cross sections and the steady-state electron energy distribution function. Also, a brief discussion of the methods employed to measure or calculate the various electron-molecule collision cross sections and coefficients is given with emphasis on recent developments.

2. Collision Cross Sections

For a given molecule, the probability of an elastic or an inelastic process, n, in an electron-molecule collision is a function of the electron energy ε and the scattering angle θ. For each process, then, a differential electron scattering cross section, $\sigma_{\mathrm{n,diff}}(\varepsilon)$, and a total electron scattering cross section, $\sigma_{\mathrm{n,t}}(\varepsilon)$, are needed to describe the process. We shall define these cross sections by first considering elastic electron-molecule collisions.

2.1 Cross Sections for Elastic Electron Scattering

2.1.1 Differential Elastic Electron Scattering Cross Section, $\sigma_{\mathrm{e,diff}}(\varepsilon)$

Consider a parallel beam of N_{e} electrons per cm^2 per second striking the target gas containing N_{t} molecules in the volume from which scattered electrons are detected. Let dN_{s} be the number of these electrons which are scattered elastically per second at an angle θ into the solid angle $d\Omega$. Under single collision conditions, dN_{s} is given by

$$dN_{\mathrm{s}} = \sigma_{\mathrm{e,diff}}(\varepsilon) N_{\mathrm{e}} N_{\mathrm{t}} d\Omega \tag{2.1}$$

where $\sigma_{\mathrm{e,diff}}(\varepsilon)$ is the differential elastic electron scattering cross section. The cross section $\sigma_{\mathrm{e,diff}}(\varepsilon)$ is a function of the scattering angle θ and has the dimensions cm^2 molecule^{-1} steradian^{-1} (usually the units are written as cm^2 sr^{-1} or m^2 sr^{-1}); it defines the angular distribution of electrons which are elastically scattered by the molecule.

2.1.2 Total (Integral) Elastic Electron Scattering Cross Section, $\sigma_{\mathrm{e,t}}(\varepsilon)$

The probability of elastic electron scattering into the solid angle 4π-steradians is represented by the total elastic electron scattering cross section, $\sigma_{\mathrm{e,t}}(\varepsilon)$, defined by

$$\begin{aligned}\sigma_{\mathrm{e,t}}(\varepsilon) &= \int_0^{2\pi}\int_0^{\pi} \sigma_{\mathrm{e,diff}}(\varepsilon)\sin\theta d\theta d\phi \\ &= 2\pi\int_0^{\pi} \sigma_{\mathrm{e,diff}}(\varepsilon)\sin\theta d\theta.\end{aligned} \tag{2.2}$$

The cross section $\sigma_{\mathrm{e,t}}(\varepsilon)$ is expressed in m^2 or cm^2 or πa_0^2 ($= 0.88\times10^{-16}$ cm^2, a_0 is the Bohr radius), and at times is referred to as the elastic integral electron scattering cross section $\sigma_{\mathrm{e,int}}(\varepsilon)$.

2.1.3 Momentum Transfer Cross Section (Elastic), $\sigma_{\mathrm{m}}(\varepsilon)$

The cross section, $\sigma_{\mathrm{m}}(\varepsilon)$, for momentum transfer to a molecule by the colliding electron of energy ε is defined as

$$\begin{aligned}\sigma_{\mathrm{m}}(\varepsilon) &= \int_0^{2\pi}\int_0^{\pi} \sigma_{\mathrm{e,diff}}(\varepsilon)(1-\cos\theta)\sin\theta d\theta d\phi \\ &= 2\pi\int_0^{\pi} \sigma_{\mathrm{e,diff}}(\varepsilon)(1-\cos\theta)\sin\theta d\theta.\end{aligned} \tag{2.3}$$

2.2 Cross Sections for Inelastic Electron Scattering

Three cross sections are normally used to quantify the probability of inelastic electron-molecule collisions in which the molecule is excited from the initial state 0 to an excited state n'. These are the differential $\sigma_{\mathrm{n',diff}}(\varepsilon)$, total $\sigma_{\mathrm{n',t}}(\varepsilon)$, and momentum transfer $\sigma_{\mathrm{m,n'}}(\varepsilon)$, cross sections associated with the state n'. They can be defined in a way similar to the case for elastic electron scattering. Thus, $\sigma_{\mathrm{n',diff}}(\varepsilon)$ is defined by

$$dN_{\mathrm{s}} = \sigma_{\mathrm{n',diff}}(\varepsilon)N_{\mathrm{e}}N_{\mathrm{t}}d\Omega \tag{2.4}$$

where dN_{s} is now the number of electrons scattered per second at an angle θ into the solid angle $d\Omega$ having excited the target from the initial state 0 to a final state n'. The cross section $\sigma_{\mathrm{n}',\mathrm{diff}}(\varepsilon)$ now gives the probability of electron scattering into the solid angle $d\Omega$ for the particular inelastic process (n') as a function of scattering polar angles (θ, ϕ) and the electron energy ε.

Integration of $\sigma_{\mathrm{n}',\mathrm{diff}}(\varepsilon)$ over all scattering angles yields the total (integral) cross section $\sigma_{\mathrm{n}',\mathrm{t}}(\varepsilon)$ and the momentum transfer cross section $\sigma_{\mathrm{m},\mathrm{n}'}(\varepsilon)$ associated with the excitation of the process n', viz.,

$$\sigma_{\mathrm{n}',\mathrm{t}}(\varepsilon) = 2\pi \int_0^{\pi} \sigma_{\mathrm{n}',\mathrm{diff}}(\varepsilon) \sin\theta d\theta \tag{2.5}$$

and

$$\sigma_{\mathrm{m},\mathrm{n}'}(\varepsilon) = 2\pi \int_0^{\pi} \sigma_{\mathrm{n}',\mathrm{diff}}(\varepsilon)(1 - \cos\theta) \sin d\theta. \tag{2.6}$$

When the initial state is not the ground state, 0, but an excited state, the excitation cross sections for the transition to the state n' from the excited state are written as $\sigma^*_{\mathrm{n}'}(\varepsilon)$ (see, Ref. [7]).

If the inelastic collision leads to ionization, say to single ionization $(\mathrm{M} + e \rightarrow \mathrm{M}^+ + 2e)$, the cross section, $\sigma_{\mathrm{i}}(\varepsilon)$, for the process is written as

$$\sigma_{\mathrm{i}}(\varepsilon) = \int_{\varepsilon}\int_{\Omega} \sigma_{\mathrm{i},\mathrm{diff}}(\varepsilon) d\Omega d\varepsilon \tag{2.7}$$

where $\sigma_{\mathrm{i},\mathrm{diff}}(\varepsilon)$ is the differential cross section for an ionizing collision in which the energy of the ejected electron lies between ε and $\varepsilon + d\varepsilon$. The integration of Eq. (2.7) is carried out over all scattering angles and over all possible continuum states.

For collisions leading to multiple ionization, that is, for processes of the form $\mathrm{M} + e \rightarrow \mathrm{M}^{q+} + (q+1)e$ where $q \geq 1$, we shall designate the cross sections for single, double, and q-fold ionization (of the parent molecule or of its fragments) respectively by $\sigma_{\mathrm{i}}(\varepsilon)$, $\sigma_{2\mathrm{i}}(\varepsilon)$ . . . , and $\sigma_{q\mathrm{i}}(\varepsilon)$. The total ionization cross section, $\sigma_{\mathrm{i},\mathrm{t}}(\varepsilon)$, is then the sum of the cross sections for the production of each ion, parent or fragment (single and multiple), measured individually, viz.,

$$\sigma_{\mathrm{i},\mathrm{t}}(\varepsilon) = \sum_{\mathrm{species}} \sum_{q} \sigma_{q\mathrm{i}}(\varepsilon) = \sum_{\mathrm{species}} [\sigma_{\mathrm{i}}(\varepsilon) + \sigma_{2\mathrm{i}}(\varepsilon) + \ldots + \sigma_{q\mathrm{i}}(\varepsilon)]. \tag{2.8}$$

The cross section $\sigma_{\mathrm{i},\mathrm{t}}(\varepsilon)$ determined this way may differ from the cross section which is determined from total ion current measurements where multiply-charged ions (parent and/or fragment) are weighted by their net charge q, namely,

$$\sigma_{\mathrm{i},\mathrm{t}}(\varepsilon) = \sum_{\mathrm{species}} \sum_{q} q\sigma_{q\mathrm{i}}(\varepsilon). \tag{2.9}$$

It is only at low energies where the probability of forming multiply charged ions is negligibly small that the values of $\sigma_{\mathrm{i,t}}(\varepsilon)$ determined in these two ways agree. It should be noted, however, that even at high electron energies, the difference between the values of $\sigma_{\mathrm{i,t}}(\varepsilon)$ determined in these two ways is not large since the cross section for single ionization is normally much greater than the cross sections for multiple ionization. The comparison of the two types of cross sections may become complicated when doubly charged molecular ions break up into two singly charged fragments prior to their detection and are thus counted as two ionization events. Often, the differences between the two values is smaller than the stated uncertainties.

2.3 Partial Cross Sections

Many electron scattering processes involve multichannel mechanisms. For instance, the process of electron-impact ionization can result in ionization of the parent molecule and/or ionization of one of the molecular fragments (dissociative ionization). Similarly, electron-impact dissociation can lead to a multiplicity of neutral fragments, and dissociative electron attachment can likewise lead to a number of negative-ion fragments. In such cases, besides the total cross section $\sigma_{\mathrm{n,t}}(\varepsilon)$ for the process n, the partial cross sections for the production of the various individual products are measured. In such cases, then, the partial differential cross sections, $\sigma_{\mathrm{n,part,diff}}(\varepsilon)$, for the process n, are defined as the fraction dN_{n} of the initial electron flux N_0 which has been scattered into a solid angle $d\Omega$ after a collision implying process n, and the partial integral cross sections, $\sigma_{\mathrm{n,part,int}}(\varepsilon)$, for the process n are obtained by integration of the partial differential cross section, $\sigma_{\mathrm{n,part,diff}}(\varepsilon)$, over all scattering angles. The cross section $\sigma_{\mathrm{n,part,int}}(\varepsilon)$ is thus defined as the total flux of particles scattered in a 4π solid angle, after a collision that resulted in process n.

The sum, $\sigma_{\mathrm{n,t}}(\varepsilon)$, of $\sigma_{\mathrm{n,part,int}}(\varepsilon)$ at a given electron energy ε gives the total cross section at ε.

2.4 Total Electron Scattering Cross Section, $\sigma_{\mathrm{sc,t}}(\varepsilon)$

The total probability of scattering of an electron of energy ε by a molecule is the sum of the probabilities of all possible electron scattering mechanisms and is measured by the total electron scattering cross section, $\sigma_{\mathrm{sc,t}}(\varepsilon)$, expressed as

$$\begin{aligned}\sigma_{\mathrm{sc,t}}(\varepsilon) &= \sum_{n} \sigma_{\mathrm{n,t}}(\varepsilon) \\ &= \sigma_{\mathrm{e,t}}(\varepsilon) + \sigma_{\mathrm{exc,t}}(\varepsilon) + \sigma_{\mathrm{dis,t}}(\varepsilon) + \sigma_{\mathrm{i,t}}(\varepsilon) + \sigma_{\mathrm{a,t}}(\varepsilon) + \sigma_{\mathrm{oth,t}}(\varepsilon)\end{aligned} \tag{2.10}$$

where $\sigma_{\mathrm{e,t}}(\varepsilon)$, $\sigma_{\mathrm{exc,t}}(\varepsilon)$, $\sigma_{\mathrm{dis,t}}(\varepsilon)$, $\sigma_{\mathrm{i,t}}(\varepsilon)$, $\sigma_{\mathrm{a,t}}(\varepsilon)$, and $\sigma_{\mathrm{oth,t}}(\varepsilon)$ are, respectively, the total cross sections for elastic electron scattering, and inelastic electron scattering leading to molecular excitation, dissociation, ionization, electron attachment, and other processes. The cross section $\sigma_{\mathrm{sc,t}}(\varepsilon)$ is usually determined in electron beam experiments from a measurement of the attenuation of the electron beam intensity as it passes through a sample cell containing the gas. If I_0 and I are, respectively, the unscattered and attenuated electron beam intensities (currents), then $\sigma_{\mathrm{sc,t}}(\varepsilon)$ is related to I_0 and I by

$$I = I_0 \exp(-\sigma_{\mathrm{sc,t}} N l), \tag{2.11}$$

where N is the gas number density, and l is the effective path length of the electrons through the gas. Thus,

$$\sigma_{\mathrm{sc,t}}(\varepsilon) = \frac{1}{Nl} \ln \frac{I_0}{I}. \tag{2.12}$$

The experimental determination of $\sigma_{\mathrm{sc,t}}(\varepsilon)$ requires a careful consideration of all quantities in Eq. (2.12). Since these quantities are often difficult to measure accurately, the cross section $\sigma_{\mathrm{sc,t}}(\varepsilon)$ is usually determined from relative measurements with reference to well-known cross sections of a reference gas (such as Ar or Kr) and requires the simultaneous measurement of the known and the unknown cross sections under identical experimental conditions.

The total electron scattering cross section $\sigma_{\mathrm{sc,t}}(\varepsilon)$ is a useful quantity in that it represents the total probability of electron interaction with a molecule as a function of the kinetic energy of the electron. It is also a useful quantity for checking the consistency of the total cross sections for the other electron scattering processes. At any value of the electron energy, the sum of the total cross sections for the individual electron scattering processes cannot exceed, and should equal, the value of $\sigma_{\mathrm{sc,t}}(\varepsilon)$ for that energy.

2.5 Methods of Measurement

The experimental study of electron-molecule collisions has traditionally dealt with ground-state molecules and has been conducted using a variety of techniques which fall into two broad categories: electron beam methods and electron swarm methods. The former are more direct than the latter. There are many texts and review articles which fully describe both of these methods and for this reason we shall not attempt to do so here. For instance, excellent descriptions of electron beam methods can be found in Refs. [1–4, 6, 8–10]. Similarly, excellent descriptions of electron swarm methods can be found in Refs. [1, 4–6, 11]. Recent advances and improvements in both types of methods, which have expanded our capability of data acquisition and have broadened the database for molecules of basic and applied interest, are highlighted briefly below (see also Ref. [12]).

2.5.1 Electron-Beam Methods

Electron-beam methods are used to measure the cross sections for the various electron-molecule interaction processes as a function of the electron energy ε, and also as a function of the electron scattering angle θ. In these experiments, a well collimated monoenergetic beam of electrons is directed into a collision chamber containing the gas under study. A small fraction of these electrons collides with the molecules M. These electrons are scattered at various angles with respect to the initial direction of the electron beam, each having lost an energy ΔE to the target molecule as a result of the collision. When the collision is elastic, the electron energy loss ΔE is small and the internal energy of the molecule M remains unchanged. When the collision is inelastic, the energy imparted to the molecule can lead to rather profound changes in the molecule and can create new electrons and new neutral and/or charged particles. As discussed in Ch. 1, the various types of primary electron-molecule reactions are either direct or indirect via negative ion resonances. High-resolution electron beam experiments, especially high-resolution crossed-beam experiments, provide the most detailed information about the individual electron scattering collision processes. Unfortunately, it is generally difficult to obtain accurate absolute cross section data with the electron beam method below ~ 1 eV. For this reason, the relative data obtained by the electron beam methods are often converted to absolute cross section values by normalization to well established values of reference gases studied under identical experimental conditions. Up until recently (see later this section), electron beam methods did not permit measurement of differential cross sections in the forward and backward directions. Thus, the conversion of differential cross sections to integral cross sections (total or momentum) required extrapolation of the data to 0° and to 180°.

Among the recent developments in the electron-beam method, four areas are of particular interest to us in this book: (a) the techniques which have been developed to measure differential electron scattering cross sections to 0° and to 180° scattering angles [13–19]; (b) the methods developed to measure electron-impact ionization of radicals [20–32]; (c) the methods developed to measure electron-impact dissociation of molecules into neutral fragments [33–47]; and (d) the techniques developed for the measurement of electron attachment cross sections at very low electron energies [48–64].

The magnetic-angle-changer technique for the measurement of differential electron scattering cross sections to 0° and to 180° scattering angles. The measurement of differential electron scattering cross sections has traditionally been limited to a range of angles that typically varies from 10° to 135° due to mechanical constrains imposed by the physical size of the electron source and the analyzer/detector in such experiments. Although several schemes of extending the angular range have been attempted, such as the techniques that have been devised to study differential cross sections for elastic and inelastic back scattering and scattering at angles larger than 135° [13], none have been universally applicable. Recently, a novel magnetic angle-changing technique has been developed by Read and collaborators [13, 14, 16] that allows measurement of elastic and inelastic electron scattering cross sections continuously from nearly 0° to 180° (at scattering angles close to 0° the unscattered electron beam and the elastically scattered electrons cannot be differentiated).

In this technique, the scattering geometry of an apparatus could effectively be rotated with respect to the position of the monochromator/analyzer by surrounding the interaction volume by two or more pairs of magnetic field coils producing opposed, but coaxial magnetic fields. Careful positioning of the coils and balancing of the coil currents can generate a highly localized magnetic field that is symmetric about the axis of the molecular beam allowing deflection of the incident beam through an arbitrary large angle in the scattering plane. The scattered electrons are also deflected as they emerge from the interaction volume and, if there is no energy loss during the collision, they suffer the same angular rotation as the incident beam. For example, deflection of the incident beam by an angle of 45° results in the unscattered or elastically scattered electrons undergoing a similar additional rotation as they exit the field. Thus, the incident beam emerges from the region of the magnetic field at an angle of 90° to the original direction and the scattering geometry is effectively rotated by 90°. As a result, an electron analyzer placed

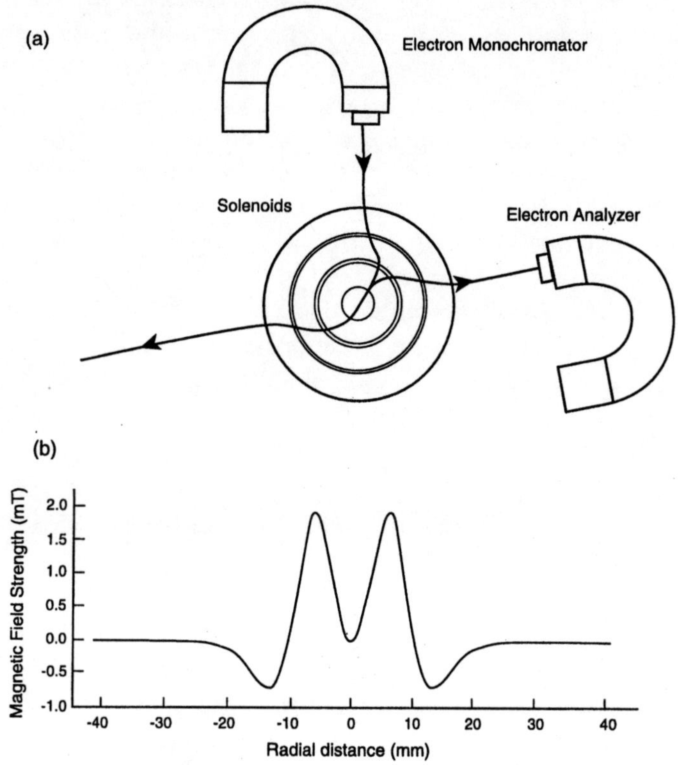

Figure 2.1. (a) Schematic diagram of the electron spectrometer used by Zubek *et al.* [16] with a localized, static magnetic field in the interaction region of the spectrometer generated by three pairs of coaxial solenoids, showing the incident electron beam, the unscattered electron beam (deflected away from the electron analyzer), and the trajectory of the electrons which have been scattered through 180°. (b) Variation of the magnetic field strength with radial distance from the scattering center in the symmetry plane between the solenoids [16].

at the conventional 90° position will record elastic electron scattering events occurring at a real angle of 180°.

Figure 2.1 shows the arrangement employed by Zubek *et al.* [16]. The schematic diagram of their electron spectrometer is shown in Fig. 2.1(a). Three pairs of short, coaxial solenoids produce a localized, static magnetic field at the interaction region of the spectrometer. The solenoids in each pair are separated by a gap for the entry of the incident electron beam and the exit of the scattered electron beam. The incident electron beam, the unscattered electron beam, and the trajectory of the electrons that have been scattered at 180° are also shown in Fig. 2.1(a). The magnetic field is symmetric about the axis of the target-particle beam and the currents in the three pairs of solenoids are set to produce a zero magnetic field at the scattering center and a zero resultant magnetic dipole moment. This ensures that the electron-atom/molecule scattering occurs in a field-free region and that the magnetic field is negligible at the positions of the electron beam source and the analyzer of the scattered electrons. The variation of the magnetic field strength with radial distance from the interaction region in the symmetry plane between the solenoids is shown in Fig. 2.1(b). The incident electron beam is deflected by the magnetic field and the symmetry of the field ensures that an electron beam directed initially towards the interaction center will cross the center after the angular deflection. Similarly, the scattered electrons are deflected by the magnetic field and at a sufficient distance from the interaction center they move in a radial direction.

The magnetic-angle-changer technique has been applied successfully in a number of studies (e.g., see [14, 16–19]). Further details of its general application can be found in these papers.

Methods for measurement of electron-impact ionization cross sections of molecules and radicals. The measurement of electron-impact ionization cross sections of molecules has been significantly improved recently and has been extended to radicals. Both of these developments have generated new information on the electron-impact ionization cross sections of the plasma processing gases of interest in this book. Systematic studies have been made of the effects of the kinetic energy of the fragment ions on their detection efficiency and the corresponding partial ionization cross sections. Early experiments using mass spectrometers tended to miss ions with appreciable kinetic energies (about a few eV) and this may be the reason why the magnitude of the total ionization cross section as determined experimentally oftentimes is lower than its value as calculated by semiempirical methods. The reliability of mass spectrometric techniques for absolute measurements of parent and fragment ionization cross sections was demonstrated by Märk *et al.* [65–68] who emphasized the proper correction for discrimination effects and the loss of energetic fragment ions in order to obtain reliable cross sections (see, also, Refs. [32, 69]). Practical solutions of this ion discrimination problem involve extensive ion trajectory modeling in conjunction with *in situ* experimental studies. Recent reliable total and partial electron-impact ionization cross sections for the molecules of interest in this book have been provided by a number of groups foremost those of Bonham and coworkers [70–73], Stebbings and coworkers [74, 75], Nishimura *et al.* [76], Märk *et al.* [65–68], and Jiao *et al.* [77–81]. The last group of investigators used Fourier transform mass spectrometry and provided data on partial and total ionization cross sections and also at times data on rate constants for ion-molecule reactions of molecules of interest in this work (see later chapters). More recent measurements by Torres *et al.* [82], using a time-of-flight mass spectrometer designed to minimize the effect of energetic fragment ions, agree well with the results of Jiao *et al.* [77–81]. It should be noted that the early measurements of electron-impact ionization cross sections of atoms and molecules made by Rapp and Englander-Golden [83] are still some of the most reliable data available.

A fast-neutral-beam apparatus that was originally developed for the measurement of ionization cross section of atoms and molecules, has subsequently been used to measure electron-impact ionization cross sections of radicals of plasma processing gases [20–32]. The method, its performance, potential sources of systematic uncertainties, and an estimate of the overall accuracy of the measured cross sections have been discussed in Refs. [21, 25, 26, 28, 32]. For the measurement of electron-impact ionization cross sections of radicals, the method uses a dc discharge through the gas as the primary ion source. The primary ions are extracted into a beam and mass selected. A fraction of them is then neutralized by near-resonant charge transfer in a charge-transfer cell filled with an appropriate gas for resonant or near resonant charge transfer. The residual ions are removed from the target gas beam by electrostatic deflection and most species in Rydberg states are quenched in a region of high electric field. The resulting neutral beam is subsequently crossed at right angles by a well-characterized electron beam. The product ions are focused at the entrance plane of an electrostatic hemispherical analyzer, which separates ions of different charge-to-mass ratio. The ions leaving the analyzer are detected by a channel electron multiplier. A schematic diagram of the fast-beam apparatus is shown in Fig. 2.2. The neutral beam density in the interaction region is determined from a measurement of the energy deposited by the fast neutral beam into a pyroelectric crystal whose response is first calibrated by a well-characterized ion beam [21], or alternatively, the well-established Kr or Ar absolute ionization cross sections (known to better than $\pm 5\%$) are used to calibrate the pyroelectric crystal. This method has provided valuable data on the electron-impact ionization cross sections of radicals of plasma processing gases, which are presented and discussed in subsequent chapters.

Methods for measurement of the cross section for electron-impact dissociation of molecules into neutral fragments. Of the many electron-molecule collision processes, electron-impact disso-

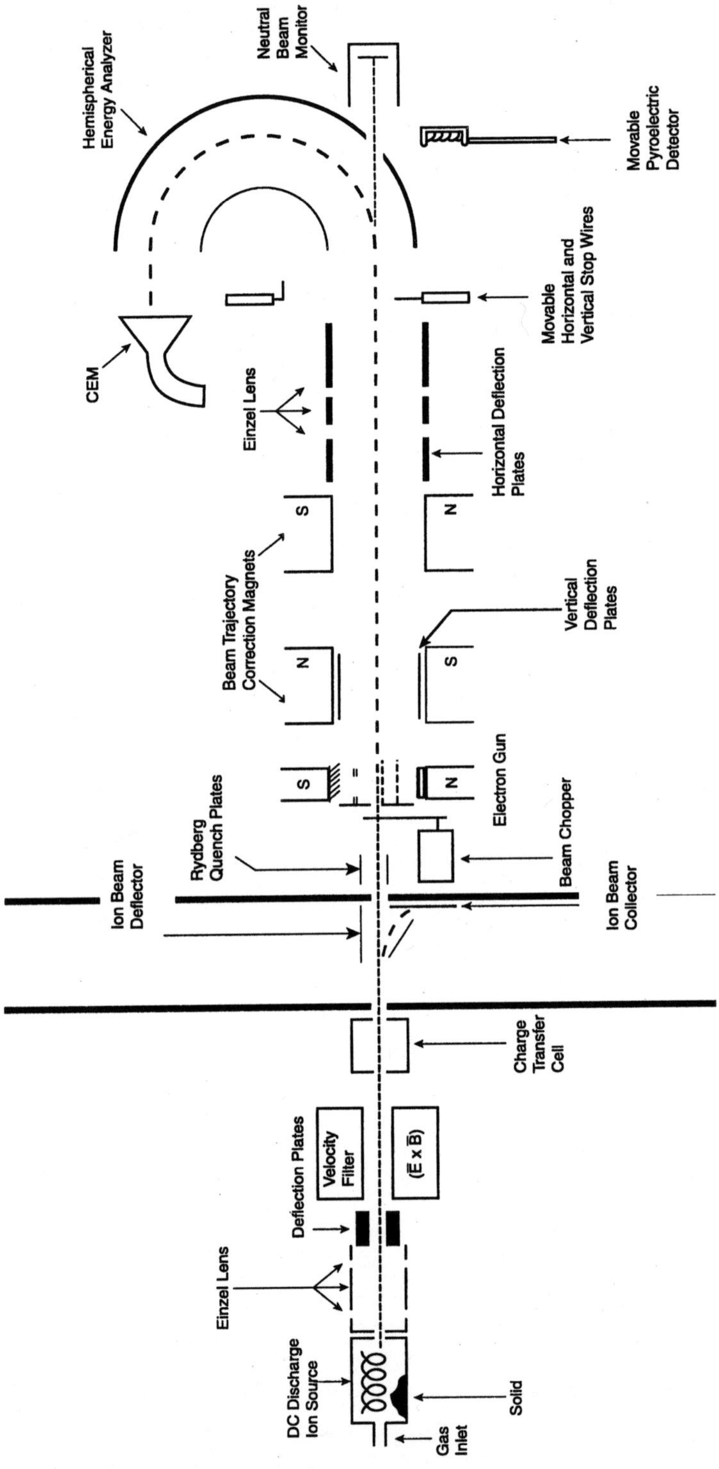

Figure 2.2. Schematic diagram of the fast-neutral-beam apparatus of Tarnovsky *et al.* [28].

ciation of molecules into neutral fragments is experimentally the least studied. Apart from destroying the original molecule, the dissociation process provides an efficient mechanism for the conversion of the kinetic energy of the electron into translational energy of heavier species. Earlier measurements (see, for instance, Ref. [84]) have generally been on partial cross sections for the production of highly excited products that can be detected by spontaneous fluorescence, field ionization, or secondary emission at a surface. Ground-state atomic or molecular fragments are not accessible with these techniques. A number of techniques have been developed for the detection of ground-state neutral fragments and measurement of the dissociation cross sections for their formation, which generally fall into three categories: those using getters for the detection of the radicals, those using double-electron beam/mass spectrometer arrangements, and those employing special electron beam apparatus.

Of the first type, Corrigan [85] derived the cross section for the dissociation of H_2 from observations of the pressure decrease in a closed system whereby H_2 gas was subjected to electron bombardment and the dissociation fragments (H atoms) were trapped (with an efficiency of ~ 1) on molybdenum trioxide. Similar type experiments by Winters [44–46] allowed determination of the total electron-impact dissociation cross sections for a number of molecules, including some of direct interest to plasma processing, by measuring the pumping speed produced by the gettering of their dissociation fragments by metal surfaces.

More recently, Motlagh and Moore [47] mass spectrometrically detected molecular and atomic radicals from electron-impact dissociation of methane and a variety of fluoroalkanes as organotellurides produced by the reaction of the radicals at the surface of a tellurium mirror. They measured relative cross sections for the various radicals which they placed on an absolute scale by comparison with other related measurements. For instance, they measured the relative cross section for the production of CF_3 by electron-impact on CF_4 by monitoring the intensity of the $Te(CF_3)_2^+$ peak in the mass spectrum and the cross section for the production of F from CF_4 by monitoring the intensity of the TeF_2^+ peak. The two processes contributing to CF_3 production are

$$e + CF_4 \rightarrow CF_3 + F + e \quad \text{(neutral dissociation)} \tag{2.13a}$$

and

$$e + CF_4 \rightarrow CF_3 + F^+ + 2e \quad \text{(dissociative ionization).} \tag{2.13b}$$

The measured relative cross section for the production of CF_3 via Processes (2.13a) and (2.13b) was normalized to the total cross section for neutral dissociation of Mi and Bonham [43]. The cross section for the production of CF_3 by dissociative ionization (Reaction (2.13b)) was taken equal to the cross section for the production of F^+ by dissociative ionization and the cross section for the production of CF_3 by neutral dissociation (Reaction (2.13a)) was taken equal to the difference between the normalized cross section for the production of CF_3 by neutral dissociation and dissociative ionization, and the cross section for the production of CF_3 by dissociative ionization.

Of the second type, Niehaus [86] first obtained relative cross sections for electron-impact dissociation of several molecules using a dual filament mass spectrometer. More recent similar techniques include those of Cosby [33–35] and Sugai *et al.* [36–42].

The technique of Cosby consists of four steps: production of a molecular ion beam, charge-transfer neutralization of this beam to produce the fast molecular beam, intersection of the fast neutral beam by an electron beam, and detection of the fragments using a position-sensitive detector for correlated fragments. A diagram of Cosby's apparatus is shown in Fig. 2.3. Because of the high beam velocity, the dissociation products are constrained to a narrow cone and can be detected through secondary emission at an electron multiplier. At keV energies, the secondary emission efficiency of the multiplier is relatively insensitive to electronic excitation in the fragments, allowing near-uniform detection sensitivity of both ground and excited-state products.

Cosby has shown that position and time-sensitive detection of the two correlated products from a single dissociating molecule permits accurate definition of the center-of-mass translational energy release in the dissociation and the angular distribution of the fragments. Charged particles produced by electron interaction are collected by an electric field such that dissociative ionization cannot contribute to the correlated product signal. Cosby determined absolute cross sections for electron-impact dissociation of a number of diatomic molecules (CO, NO, O_2, and Cl_2) into neutral atoms. He produced the fast neutral beams of these molecules by neutralization of the respective positive ions by resonant charge transfer in the corresponding parent gas. The data of Cosby for Cl_2 are presented and discussed in Ch. 6. Significant as the Cosby method is, it is limited in its applicability in that when multiple fragments are formed (as is usually the case for polyatomic molecules), there is no way to achieve unambiguous correlation.

Sugai and coworkers [36–42] have measured partial and total electron-impact dissociation cross sections for a number of molecules, including some which are under consideration in this book, using a dual-electron beam device combined with a quadrupole mass spectrometer (QMS).

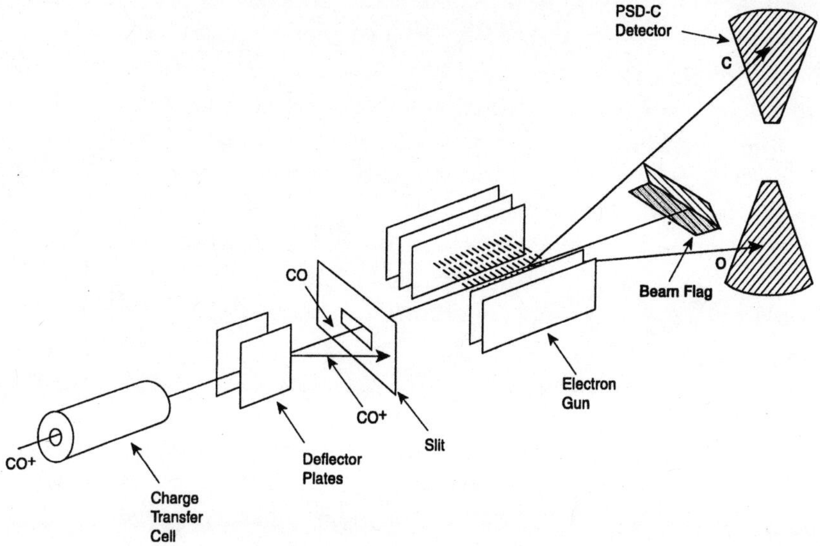

Figure 2.3. Schematic diagram of the electron-impact dissociation apparatus of Cosby [34], as used for the measurement of the cross section for electron-impact dissociation of CO into C and O atoms. A fast (keV) CO beam is first produced by charge-transfer neutralization of CO^+ and is subsequently intersected by an electron beam. The correlated fragments C and O from the electron-impact dissociation of CO are spatially and temporally detected at the position sensitive detector for correlated fragments (PSD-C).

This system has three compartments which are differentially pumped. The first compartment is a dissociation cell where a primary electron beam dissociates the molecules. The second compartment is a detection cell where a probing electron beam selectively ionizes neutral radicals effusing from the dissociation cell (radical selectivity is accomplished using a threshold ionization technique based on the finding that the ionization threshold of the parent gas differs by ~ 5 eV from that of its radicals). The ionized radicals are mass-separated by the QMS and the output pulses from a secondary electron multiplier in the QMS system are counted and processed by a computer. Putting the relative cross section measurements on an absolute scale involves a number of variables which are difficult to determine. The fact that the results obtained using this technique have been at variance with other measurements and analyses (see Ch. 4) indicates that not all possible sources of error have as yet been identified.

Finally, an electron-ion coincidence technique has been described by Mi and Bonham [43] which was employed for the measurement of the sum of the integrated cross sections for neutral dissociation and excitation of N_2 at two values (24.5 eV and 33.4 eV) of the electron energy.

In spite of the aforementioned efforts, there still remain formidable difficulties in the accurate measurement of partial and total cross sections for dissociation of molecules into neutral fragments by electron impact, which are mostly connected with the detection of low-energy neutral particles. Consideration may thus be given to obtaining values of the total cross section for electron-impact dissociation of molecules into neutrals via reliable measurements of the cross sections for total electron scattering and total ionization.

Techniques for the measurement of electron attachment cross sections with very-high electron energy resolution at very low electron energies. Two similar techniques of this type have been developed, one by Chutjian and collaborators [48–55] and the other by Hotop and collaborators [56–64]. These techniques allow sub-meV energy resolution and measurement of electron attachment cross sections down to $\sim 20\ \mu$eV.

The technique of Chutjian and coworkers is referred to as the threshold photoelectron spectroscopy for attachment (TPSA) technique. It involves the *in situ* production, in a mixture of a rare gas (usually Ar, Kr or Xe) and the molecular gas under investigation, of photoelectrons of well defined and variable energy by VUV photoionization of the rare-gas atom at and above its ${}^2P_{1/2}$ ionization limit. The technique provides relative cross sections at electron energies from $\sim$ 0 meV to 150 meV with an energy resolution of $\sim$ 4 meV to 6 meV [55]. The energy dependence of the cross section is obtained using convolution techniques, and its absolute value is determined by normalization of the relative measurements to electron swarm data.

Recently, Chutjian and coworkers [55] generated the ultra-low-energy electrons at high resolution using single-step rare-gas ionization by tunable 92-nm laser radiation produced from a 1064-nm Nd:YAG laser by doubling, by mixing with the frequency-shifted 573.6-nm output of a dye laser, or by frequency tripling in vacuum to obtain 92 nm. Using this approach, they measured the electron attachment cross section of SF_6 over the range from $\sim$ 0 meV to 127 meV. They found that below 5 meV the electron attachment cross section they measured for SF_6 can be described by the Wigner *s*-wave law [87].

The technique of Hotop and collaborators is referred to as the laser-photoelectron attachment (LPA) method and is based on the controlled production of variable-energy photoelectrons using lasers to photoionize a well-collimated beam of metastable Ar* ($4s$, ${}^3P_2$) atoms from a differentially pumped dc discharge source. The technique can extend the electron energy scale down to 20 μeV at energy widths as low as 20 μeV [62]. In order to reach sub-meV energy resolution, stray electric and magnetic fields must be minimized in the electron collision region (see discussion in Ref. [55]). The experimental setup and excitation scheme for the LPA technique of Hotop *et al.* [64] are shown in Fig. 2.4. The generation of free electrons with variable energy and narrow energy spread was accomplished by resonant two-step photoionization of metastable Ar* atoms. A collimated beam of metastable argon atoms Ar* ($4s$, $^3P_2$) is excited to Rydberg states (ns, nd), or to the ionization continuum by two lasers via the Ar* ($4p$, $^3D_3$) intermediate level. A continuous wave (CW) single mode dye laser ($\lambda_1 = 811.75$ nm) transversely excites the transition Ar* ($4s$, $^3P_2 \rightarrow 4p$, $^3D_3$) and a small fraction of the Ar* ($4p$, $^3D_3$) atoms is excited further by a tunable intracavity laser. Both lasers are linearly polarized with electric field vectors parallel to each other and to the direction of ion extraction. By tuning the intracavity laser over the wavelength range 473 nm to 425 nm, bound Ar** (nl) Rydberg levels ($n \gtrsim 15$, $\varepsilon < 0$) or free electrons ($\varepsilon < 230$ meV) are created. The photoelectrons react with the molecules of an ambient diffuse target under conditions of low residual electric field (≤ 0.5 V/m) and negligible magnetic field ($< 2 \times 10^{-6}$ T). Negative ions resulting from electron attachment reactions are extracted by a pulsed electric field and are analyzed by a quadrupole mass spectrometer. The experiment is pulsed at a repetition rate of 140 kHz. A time interval of essentially field-free conditions for electron production and attachment (phase I), is followed by a period for ion extraction with a delayed pulsed field of 23 V/cm (phase II). During phase II, the infrared laser and thereby the electron production are switched off by an acousto-optical modulator. The accelerated negative ions are imaged into a quadrupole mass spectrometer, and the mass-selected ions are detected by a dual channel plate following a 90° deflection by an electric field. The measured relative electron attachment cross section after having been corrected for broadening is converted to an absolute electron attachment cross section using the relation

$$k(\langle\varepsilon\rangle) = \left(\frac{2}{m}\right)^{1/2} \int \sigma_{\rm a}(\varepsilon)\varepsilon^{1/2} f_{\rm M}(\varepsilon)d\varepsilon, \qquad (2.14)$$

where $k(\langle\varepsilon\rangle)$ is the thermal electron attachment rate constant at a mean electron energy $\langle\varepsilon\rangle$ measured in swarm experiments and $f_{\rm M}(\varepsilon)$ is the Maxwellian electron energy distribution function. Naturally, the uncertainty in the measurement of $k(\langle\varepsilon\rangle)$ enters the overall uncertainty of the cross section via this method. Absolute energies are claimed to be ac-

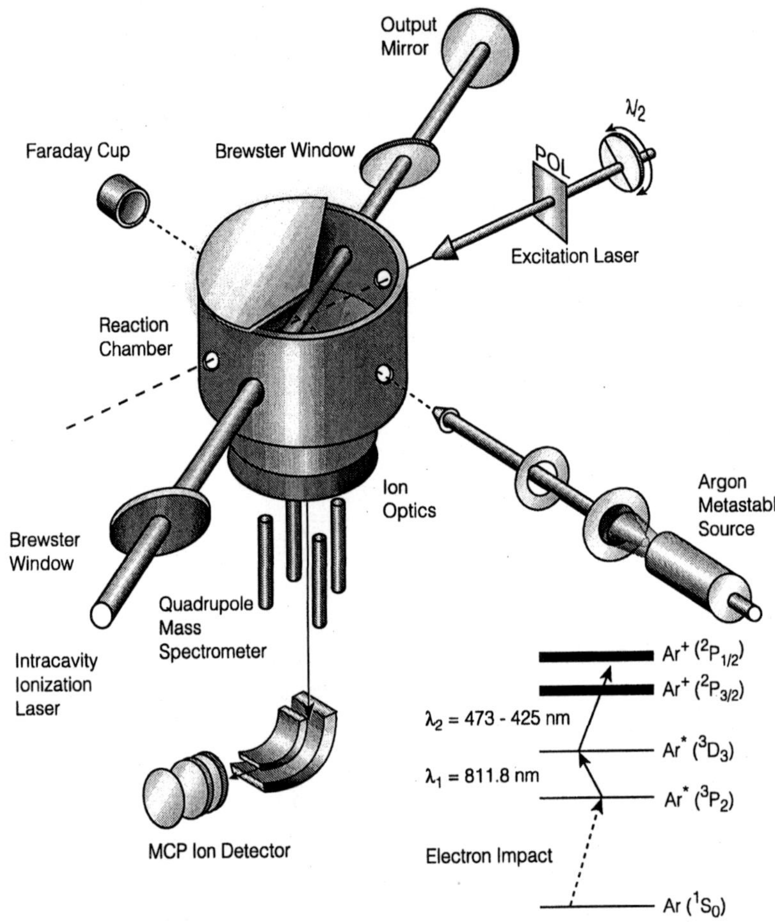

Figure 2.4. Schematic of the laser photoelectron attachment arrangement of Hotop and coworkers. The metastable $Ar^*(4s\ ^3P_{0,2})$ beam is intersected in the center of the reaction chamber by the excitation laser and the intracavity ionization laser. Photoelectrons of variable energy react with the molecules of the diffuse gas in the reaction chamber. The negative ions produced are extracted from the reaction chamber into the ion optics by a pulsed electric field, are mass analyzed by a quadrupole mass spectrometer, and after a 90° deflection are detected by a dual microchannel plate detector [64].

curate to within ± 0.4 meV or less [64]. The technique has been used to measure electron attachment cross sections for a number of molecules below ~ 200 meV (see Ch. 6 for data on SF_6).

Another method that has been applied to investigate electron attachment to molecules XY at energies down to $\sim 4\ \mu$eV [88] involves the transfer of an electron to a molecule XY from a highly excited Rydberg atom $A^{**}(nl)$ [57–60, 62, 64, 88–93], viz.,

$$A^{**}(nl) + XY \xrightarrow{k_{a,be}(nl)} [A^+ XY^{-*}] \tag{2.15a}$$

$$\longrightarrow A^+ + X + Y^-. \tag{2.15b}$$

Reactions (2.15a) and (2.15b) have been referred to earlier as "bound-electron capture" reactions [94].

In almost all such studies, $A^{**}(nl)$ is a rare-gas or an alkali-metal atom in a Rydberg state with principal quantum number n and angular momentum quantum number l. The rate constant $k_{a,be}$ for the formation of XY^- or Y^- is measured as a function of the effective principal quantum number n^* ($= n - \delta_l$, where δ_l is the l-dependent quantum defect). From these rate-constant data, the cross section for bound-electron capture, $\sigma_{a,be}(v)$, is determined from

$$\sigma_{a,be}(v) = \frac{k_{a,be}(n^*)}{v_{rms}} \tag{2.16}$$

where v_{rms} is the root-mean-square velocity of the Rydberg electron. While a number of authors (e.g., Refs. [59, 88]) have shown Eq. (2.16) to be appropriate for Rydberg states with l small compared to n, and whereas the post attachment electrostatic attraction between the products (A^+ and XY^{-*}) may have significant effects on the size of the measured cross section (e.g., see Ref. [88]), a number of studies [88, 95–97] have shown that the cross sections for bound-electron capture are consistent with the free electron model for sufficiently large values (> 30) of the principal quantum number n. In such cases, then, the Rydberg electron and the core can be considered as independent particles with the Rydberg electron being essentially equivalent to a free electron with kinetic energy equal to the binding energy of the Rydberg electron.

In the framework of the quasifree Rydberg-electron model [59, 88, 89, 98], the rate constants $k_{a,be}(nl)$ for Rydberg electron attachment and $k_{a,fe}(v)$ for free electron attachment are related by

$$k_{a,be}(nl) = \int k_{a,fe}(v) f_{nl}(v) dv, \tag{2.17}$$

where $f_{nl}(v)$ is the normalized Rydberg-electron velocity distribution.

2.5.2 Electron Swarm Methods

The physical quantities normally measured in electron swarm experiments are the electron transport coefficients [the electron drift velocity

$w(E/N)$, the transverse electron diffusion coefficient $D_T(E/N)$, and the longitudinal electron diffusion coefficient $D_L(E/N)$], and the electron attachment coefficient $\eta\,(E/N)$ and electron-impact ionization coefficient $\alpha\,(E/N)$ (see Sec. 3 of this chapter).

The electron swarm transport and other coefficients measured as a function of E/N can be used to determine absolute cross sections for electron scattering at low electron energies. The link between the microscopic description of the swarm in terms of cross sections and the macroscopic description of the swarm in terms of electron transport and other coefficients is provided by the Boltzmann transport equation.

The derivation of electron scattering cross sections from measured electron transport coefficients is a complex and convoluted process. Cross sections are derived from electron transport, electron attachment, and electron-impact ionization coefficients measured in either the pure (unitary) gas or in mixtures of the gas under study with simple gases (e.g., He, Ar, or N_2), by an iterative procedure in which a trial set of cross sections is continually refined until the transport coefficients calculated with this assumed set of cross sections matches the experimental data for all values of E/N at which measurements are made. The swarm-based cross sections are absolute and cover the most difficult but interesting energy region below $\sim$ 10 eV. However, swarm-based cross sections are indirectly determined quantities based on complex analyses (from the measured coefficients to cross sections, and from cross sections to calculated coefficients) and lack uniqueness. The lack of uniqueness is a major shortcoming that limits this method, especially for polyatomic molecules which have multiple electron-molecule reaction channels even at very low energies.

Iterative numerical techniques whereby calculated transport coefficients are matched to experimental data are now universally used to give cross section sets for various molecules of industrial interest. Besides the initial work by Phelps and collaborators [99, 100], and Crompton and collaborators [5, 11, 101], many other authors (see Sec. 4 of this chapter, subsequent chapters where data on individual molecules are presented, and Refs. [102–104] and sources cited therein) have generated cross section sets for molecules using the Boltzmann code with various approximations.

2.6 Calculations

The various theoretical methods that have been employed to calculate electron-molecule collision cross sections have been discussed by many authors (e.g., see Refs. [105, 106]) and will not be discussed further.

However, advances in the following three areas which are of direct interest to the subject of this book will be elaborated upon, namely, *ab initio* calculations of electron scattering cross sections, semiempirical calculations of total electron-impact ionization cross sections, and theories of electron attachment.

2.6.1 *Ab initio* Calculations

Ab initio quantum mechanical calculations provide electron collision cross sections which can be used in modeling. Recent theoretical calculations of elastic and momentum transfer cross sections include those by Lima *et al.* [107–109] for CF_4 and CHF_3. These researchers used the Schwinger multichannel method, pseudopotentials to represent the core electrons of the heavy atoms, and the static exchange approximation. Due to the nature of the approximations made, the results of these calculations are not expected to be valid at energies below $\sim$ 10 eV. Similar calculations were conducted by Winstead *et al.* for CF_4 [110] and c-C_4F_8 [111], by McKoy *et al.* [112, 113] for C_2F_6, C_3F_8, c-C_4F_8, CHF_3, NF_3, and BCl_3, by Varella *et al.* [114] for CH_3F, CH_2F_2, CHF_3, and CF_4, and by Morgan *et al.* [115] for CHF_3. The nature of the approximation employed in these computations makes the low energy data less reliable.

Valuable information has also been provided by the calculations of Rescigno and collaborators [116–118] who used the complex Kohn method to determine the elastic differential and/or momentum transfer cross sections for CF_4, BCl_3 and Cl_2. Their calculations accounted for the long range interaction potential and also included both target polarization and electron correlation effects. This approach should provide an adequate description of the cross section at low energies.

Two other theoretical studies which provided electron scattering cross sections for CF_4 are those of Huo [119] who used the Schwinger multichannel method with polarization, and of Gianturco *et al.* [120, 121] who used the one-center method with density functional theory polarization potential.

In closing this section, it is noted that oftentimes effective range theory [122–125] is used to extrapolate the elastic, momentum transfer, and total electron scattering cross sections to close-to-zero energy using low-energy electron beam measurements of such cross sections. The modified effective range theory of O'Malley [123–125] develops the near-to-zero energy total cross section for electron-atom scattering (defined in terms of the scattering length A_0 [with $\sigma(\varepsilon = 0) = 4\pi A_0^2$] and the electron wavenumber k) into a series of k, viz.

$$\begin{aligned}\sigma_{\mathrm{sc,t}}(\varepsilon) = 4\pi[A_0^2 + \left(\frac{2\pi}{3a_0}\right)\alpha A_0 k \\ + \left(\frac{8\pi}{3a_0}\right)\alpha A_0^2 k^2 \ln(ka_0) + \mathrm{B}k^2 + \ldots,\end{aligned} \tag{2.18}$$

where α is the electric polarizability of the atom and a_0 is the Bohr radius. The parametrization of the energy dependence of the scattering phase-shifts in terms of the dipole polarizability, effective range, and scattering length, has been shown to be useful for rare gas atoms and some simple, spherically symmetric molecules such as CF_4. However, it is generally not applicable to electron-molecule scattering, and thus not applicable to the molecules under consideration here, except CF_4. The accuracy of its implementation depends both on the energy range of the experimental data used and the number of terms in Eq. (2.18) [126–129].

2.6.2 Semiempirical Calculations of Electron-Impact Total Ionization Cross Sections

Of the various semiclassical, semiempirical, and *ab initio* treatments of electron-impact ionization cross sections, two recent semiempirical additivity-rule-based theories have provided extensive results on the total ionization cross sections of plasma processing gases: the Binary-Encounter-Bethe (BEB) theory of Kim and Rudd [130, 131] and the formalism developed by Märk and Deutsch and their collaborators [65, 132–134]. Each has been shown to generally reproduce with reasonable accuracy the ionization cross sections determined experimentally for a number of molecules. They are useful from our point-of-view in that they have provided needed results for molecules of interest to plasma processing, and in that they can provide electron-impact ionization cross sections for radicals for which experimental measurements are difficult, scarce, or absent.

The Binary-Encounter-Bethe theory developed by Kim and Rudd [130, 131] calculates total ionization cross sections of atoms and molecules as the sum of the ionization cross sections for ejecting one electron from each of the atomic or molecular orbitals. It combines the Mott cross section [135] as modified by the binary-encounter theory [136] for low electron energies, with the Bethe [137] theory for high electron energy, and gives the singly differential ionization cross section for electron-impact ionization of atoms or molecules, or the energy distribution of ejected electrons $d\sigma/dW$ with the ejected electron energy W for each atomic or molecular orbital. For each orbital, the calculation requires the electron binding energy B, the average kinetic energy $U = \langle \mathbf{p}^2/2m \rangle$ ($\mathbf{p}$ is the bound-electron momentum and m is the electron mass), the electron occupation number N, and the continuum dipole oscillator strength df/dW. Because df/dW is difficult to determine for each orbital, Kim and collaborators [130, 131, 138, 139] assumed for this parameter the simple form

$$\frac{df}{dw} = \frac{N}{(\mathrm{w}+1)^2}, \tag{2.19}$$

where $\mathrm{w} = W/B$, based on the approximate shape of df/dW for the ionization of the ground state of the hydrogen atom.

With this approximation, they arrived at a simple analytic formula for the integrated (total) single ionization cross section, $\sigma_{\mathrm{i}}^{\mathrm{BEB}}$, for the molecular orbital, referred to as the Binary-Encounter-Bethe (BEB) ionization cross section, given by

$$\sigma_{\mathrm{i}}^{\mathrm{BEB}} = \frac{S}{t+u+1}\left[\frac{\ln t}{2}\left(1-\frac{1}{t^2}\right)+1-\frac{1}{t}-\frac{\ln t}{t+1}\right]. \tag{2.20}$$

In Eq. (2.20), $S = 4\pi a_0^2 N R^2/B^2$, $t = \varepsilon/B$ is the reduced incident electron energy, $u = U/B$ is the reduced orbital kinetic energy, ε is the kinetic energy of the incident electron, $a_0 = 0.5292$ Å is the Bohr radius, and $R = 13.61$ eV is the binding energy of the H_{1s} orbital.

The quantities N, B, U can be calculated for a number of molecules using standard molecular wave function codes, and the total ionization cross section $\sigma_{\mathrm{i,t}}^{BEB}$ can, thus, be obtained by summing Eq. (2.20) over all occupied orbitals of the molecule. The applicability of this theory is limited by the availability of a molecular structure code that can provide basic information on molecular orbitals. As discussed by Kim and collaborators, the cross sections for tightly bound inner-shell orbitals (those with $B \geq 400$ eV) contribute very little to $\sigma_{\mathrm{i,t}}^{BEB}$ at $\varepsilon < 1$ keV. They also noted that at low electron energies $\sigma_{\mathrm{i,t}}^{BEB}$ is sensitive to the lowest binding energy and thus to the vertical ionization threshold energy. Furthermore, implicit in the calculation of $\sigma_{\mathrm{i,t}}^{BEB}$ is the assumption that all energy transfers above a given orbital binding energy result in the ejection of an electron from either that orbital or other orbitals with lower binding energies [76]. This assumption does not hold well for the ionization of molecules, since an energy transfer higher than the lowest ionization threshold energy of the molecule may result in dissociation of the molecule into neutral fragments and the BEB theory counts neutral dissociation as ionization as long as the required energy for dissociation is higher than the lowest ionization threshold energy. Thus, the $\sigma_{\mathrm{i,t}}^{BEB}$ values should be upper limits to the total ionization cross section.

Cross sections $\sigma_{\mathrm{i,t}}^{BEB}$ have been determined for a number of molecules and radicals and have been compared with existing experimental data (see Chapters 1, 4, 5, and 6, and Refs. [76, 131, 138–144]) with reasonable success in terms of reproducing the magnitude and energy dependence of the ionization cross section. Especially significant is the determination of ionization cross sections for radicals and neutral fragments for which experimental data are scarce.

The BEB formula has recently been slightly modified to allow calculation of ionization cross sections of singly charged molecular ions [145]. Hopefully, it will also be applied to the electron-impact ionization of negative ions.

The Deutsch-Märk (DM) additivity rule method [65, 132–134, 146, 147] uses empirically determined cross sections for the atomic constituents of a molecule and sums these atomic cross sections up often after scaling them according to the Mulliken population analysis which gives the atomic populations in a molecular orbital. Thus, the method constructs a total single ionization cross section $\sigma_{\mathrm{i,t}}^{DM}$ by adding the ionization cross sections for the constituent atoms, viz.,

$$\sigma_{\mathrm{i,t}}^{DM} = \sum_{\mathrm{i},nl} \pi r_{\mathrm{i},nl}^{2} \sum_{j} \left(g_{\mathrm{i},nl}^{j} N_{\mathrm{i},nl}^{j} \right) f(u_j) \tag{2.21}$$

where $r_{\mathrm{i},nl}^{2}$ is the mean square radius of the n,l subshell, $N_{\mathrm{i},nl}^{j}$ is the occupation number of the atomic orbital n,l, and $g_{\mathrm{i},nl}^{j}$ are weighting factors. The quantity

$$f(u_j) = \frac{1}{u_j}\left(\frac{u_j - 1}{u_j + 1}\right)^{3/2} \left\{1 + \frac{2}{3}\left(1 - \frac{1}{2u_j}\right) \ln\left[2.7 + (u_j - 1)^{1/2}\right]\right\} \tag{2.22}$$

with $u_j = \varepsilon / E_j$, where ε is the incident electron energy and E_j is the ionization energy of the nth sub-shell. The method of calculating the weighting factors, binding energies, and occupation number is described by Margreiter *et al.* [132, 133]. The DM method was originally devised for the calculation of the total cross section for the single ionization of ground-sate atoms, but has subsequently been successfully applied to excited atoms, molecules, radicals, and clusters. According to Huo and Kim [142], for molecules for which the necessary empirical/scaled parameters are known, the DM method has had a fair share of success. A drawback of this type of empirical model is that the necessary parameters often need to be readjusted when new sets of experimental data become available.

2.6.3 Theories of Electron Attachment to Molecules

As we have indicated in Ch. 1, resonance dissociative electron attachment can be visualized to proceed through a negative ion intermediate which dissociates into neutral and negative ion fragments. The cross section, $\sigma_{\rm da}(\varepsilon)$, for dissociative electron attachment as a function of the electron energy ε is equal to the product of the cross section $\sigma_{\rm c}(\varepsilon)$ for capture of the electron by the molecule to form the negative-ion intermediate, and the probability $p(\varepsilon)$ that the negative-ion intermediate will decay via dissociative electron attachment, viz.,

$$\sigma_{\rm da}(\varepsilon) = \sigma_{\rm c}(\varepsilon)p(\varepsilon). \tag{2.23}$$

Expressions for $\sigma_{\rm c}(\varepsilon)$ and $p(\varepsilon)$ have been given for diatomic molecules within the resonance scattering theory and the Born-Oppenheimer approximation.

Following O'Malley [148], we write the $\sigma_{\rm da}(\varepsilon)$ for a diatomic molecule AX initially in its lowest vibrational level ($\nu = 0$) and with a purely repulsive negative-ion state AX^{-*} (upper AX^{-*} curve in Fig. 1.3(c) of Ch. 1) as

$$\sigma_{\rm da}(\varepsilon)_{\nu=0} = \frac{4\pi^{3/2}\bar{g}}{(2m/\hbar^2)\varepsilon}\frac{\Gamma_{\bar{a}}}{\Gamma_d}\exp\left(\frac{\Gamma_a^2 - 4(\bar{\varepsilon}_0 - \varepsilon)^2}{\Gamma_d^2}\right)\exp[-\rho(\varepsilon)]. \tag{2.24}$$

In Eq. (2.24), m is the electron mass, $\hbar$ is Planck's constant divided by 2π, $\bar{g}$ is a statistical factor, Γ_a is the total autodetachment width, $\Gamma_{\bar{a}}$ is the partial autodetachment width, Γ_d is the dissociative attachment resonance width, $\bar{\varepsilon}_0 = \varepsilon_0 + \frac{1}{2}\hbar\omega$, ε_0 is the true electron energy at the peak of $\sigma_{\rm da}(\varepsilon)_{\nu=0}$, and $\frac{1}{2}\hbar\omega$ is the zero-point energy. The quantity $\exp[-\rho(\varepsilon)]$ is defined explicitly [148, 149] by

$$\exp[-\rho(\varepsilon)] = \exp\left(-\int_{R_\varepsilon}^{R_C}\frac{\Gamma_a(R)}{\hbar}\frac{dR}{v(R)}\right) \approx \exp\left(-\frac{\tau_s}{\tau_a}\right), \tag{2.25}$$

and is the survival probability of AX^{-*} (i.e., the probability that AX^{-*} will not autoionize while separating from the point of formation by capture of an electron with energy ε at an internuclear separation R_ε to the point C (Fig. 1.3(c), Ch. 1) at an internuclear separation R_C where the potential-energy curves of AX and AX^{-*} cross). For $R > R_C$, AX^{-*} will dissociate with unit efficiency [$p(\varepsilon) = 1$]. The time τ_s needed for A-X^- to separate from R_ε to R_C is

$$\tau_s = \int_{R_\varepsilon}^{R_C}\frac{dR}{v(R)} \tag{2.26}$$

where $v(R)$ is the relative velocity of separation of A and X^- and $\bar{\tau}_a$ (= $\hbar/\bar{\Gamma}_a$) is the mean autodetachment lifetime of AX^{-*} ($\bar{\Gamma}_a$ is the average autodetachment width in the Franck-Condon region).

Strictly speaking, expression (2.24) is limited to situations in which the potential energy curve of AX^{-*} in the Franck-Condon region is purely repulsive. Negative-ion potential energy curves (or surfaces) have a variety of shapes, and $\sigma_{\rm da}(\varepsilon)$ is affected by these and the competitive decay channels. Since at low energies, the energy dependence of the Franck-Condon factor (second term on the right-hand-side of Eq. (2.24)) and the survival factor (third term on the right-hand-side of Eq. (2.24)) are relatively weak, the threshold behavior of $\sigma_{\rm da}(\varepsilon)_{\nu=0}$, is predicted to be $\sim \varepsilon^{l-1/2}$ where l is the lowest angular momentum allowed by the symmetry of the intermediate negative-ion state in accord with the Bethe-Wigner law (Refs. [87, 150]; see, also, relevant theoretical work in Refs. [63, 151, 152]).

Electron attachment in the very-low energy range ($\lesssim 0.1$ eV) is of special interest in connection with the test of quantum mechanical threshold laws [54, 55, 57, 58, 87, 88, 150] and the investigation of cross section shapes at the thresholds of vibrational excitation of molecules [57, 58, 61, 62, 151, 153–157]. Theory predicts that for $\varepsilon \to 0$ the s-wave ($l = 0$) electron attachment cross section diverges as

$$\sigma_{\rm a}(\varepsilon \to 0, l = 0) \propto \varepsilon^{l-1/2} \propto v^{-1}. \tag{2.27}$$

For molecules without permanent electric dipole or quadrupole moments for which the leading long-range interaction between the electron and the molecule is the polarization potential

$$V(r) = -\frac{\alpha e^2}{2r^4}, \tag{2.28}$$

it is also predicted [158] that the capture cross section $\sigma_{\rm c}(\varepsilon \to 0, l = 0)$ is

$$\sigma_{\rm c}(\varepsilon \to 0, l = 0) = 4\pi a_0^2 \left(\frac{\alpha}{2\varepsilon}\right)^{1/2}, \tag{2.29}$$

where a_0 is the Bohr radius, and α and ε are, respectively, the polarizability of the molecule and the electron energy both in atomic units. An interpolation between the Vogt and Wannier formula (2.29) for $\varepsilon \to 0$, and the s-wave attachment (reaction) cross section $\pi\lambda\!\!\!^{-2} = \pi a_0^2/(2\varepsilon)$ was suggested [159], viz.,

$$\sigma_{\rm a''} = \left(\frac{\pi a_0^2}{2\varepsilon}\right)\left(1 - \exp\left[-4(2\alpha\varepsilon)^{1/2}\right]\right). \tag{2.30}$$

The semiempirical formula (2.30) has been found to be in reasonable agreement with experimental measurements at very low electron energies (e.g., see Refs. [54, 57, 58, 88]).

3. Coefficients and Rate Constants

3.1 Electron Transport Coefficients

In electron swarm experiments, electrons are injected by various means most notably by shining short-duration laser pulses on a metallic electrode surface into the gas which is normally at atmospheric or sub-atmospheric pressure. Quickly upon their injection into the gaseous medium, the electrons attain an equilibrium (steady-state) kinetic energy distribution $f(\varepsilon, E/N, T)$ which is characteristic of the gas, the applied uniform electric field E divided by the gas number density N, and the temperature T, and drift toward the anode. The gas pressure and the drift distance are such that the time taken by the injected electrons to attain their steady-state energy distribution is very much shorter than their drift time. The experimental methods are normally designed to measure specific coefficients over a range of E/N values that vary depending on the experimental arrangement and the coefficient to be measured (see Refs. [1, 3–6]).

For an electron swarm in thermal equilibrium with the gas, the diffusion is symmetric and is characterized by a coefficient of diffusion which is a constant independent of direction which will be designated as the thermal equilibrium electron diffusion coefficient D_{th}. When equilibrium is disturbed by the application of an electric field E, the diffusion is asymmetric and the diffusion coefficient becomes a tensor, having components D_{T} and D_{L}, respectively in directions perpendicular to and along the direction of the applied electric field. As $E/N \to 0$, both D_{T} and $D_{\text{L}} \to D_{\text{th}}$. Experimentally, one normally measures directly $D_{\text{T}}/\mu(E/N)$ and $D_{\text{L}}/\mu(E/N)$, where μ is the electron mobility $[\mu N = w/(E/N)]$. At atmospheric and sub-atmospheric pressures, the electron transport coefficients are pressure independent and the experimental data are at times conveniently presented as $D_{\text{T}}N(E/N)$ and $D_{\text{L}}N(E/N)$. However, at high gas pressures they can become pressure dependent (see Ref. [102] and references cited therein). The quantity eD_{T}/μ has the dimensions of energy (eV) and is often called the characteristic energy of the electrons in the swarm. As $E/N \to 0$, $eD_{\text{T}}/\mu \to kT$.

The electron transport coefficients are related to the electron energy distribution function and the cross sections for the various elastic and inelastic electron scattering processes via the Boltzmann transport equation [5, 99, 100, 102–104]. To the first approximation, the following simplified relations hold for w, $D_{\rm T}$, and $D_{\rm L}$ (e.g., see [4, 5, 11, 102, 160]):

$$w = -\frac{4\pi eE}{3mN}\int_0^\infty \frac{v^2}{\sigma_m(v)}\frac{df_0(v)}{dv}dv \tag{2.31}$$

$$D_{\rm T} = \frac{4\pi}{3N}\int_0^\infty \frac{v^3}{\sigma_m(v)}f_0(v)dv \tag{2.32}$$

$$D_{\rm L} = D_{\rm T} - \frac{4\pi eE}{3m}\int_0^\infty b_1(v)f_0(v)\frac{d}{dv}\frac{v^3}{\nu}dv \tag{2.33}$$

In the equations above, N is the gas number density, $\sigma_{\rm m}(v)$ is the elastic momentum transfer cross section, $f_0(v)$ is the spherically symmetric term in the expansion of the electron velocity distribution function, $b_1(v)$ is a velocity-dependent term which also depends on the functional dependence of $\sigma_{\rm m}(v)$ [102, 161–163], $\nu = Nv\sigma_{\rm m}(v)$, and e and m are the electron charge and mass, respectively. The electron velocity distribution function $f_0(v)$ used in the above equations to define the transport coefficients and later on in this chapter to define the rate coefficients is normalized by

$$4\pi\int_0^\infty f_0(v)v^2dv = 1.$$

Expressions (2.31), (2.32), and (2.33) can be written in terms of the electron energy by noting that the energy distribution function $f_0(\varepsilon)$ is related to the velocity distribution function $f_0(v)$ by $\varepsilon^{1/2}f_0(\varepsilon)d\varepsilon = 4\pi f_0(v)v^2dv$.

3.2 Coefficients for Electron Attachment, and Electron-Impact Ionization and Excitation

Besides the electron transport coefficients, a number of other coefficients can be defined to describe the rate at which particular inelastic processes occur, such as the commonly measured coefficients η, α, and β, respectively for electron attachment, electron-impact ionization, and electron-impact excitation. These coefficients (in units of $\rm cm^{-1}$) are related to the respective cross sections $\sigma_{\rm a}(v)$, $\sigma_{\rm i}(v)$, and $\sigma_{\rm E}(v)$ (in units of $\rm cm^2$) by

$$\eta = \frac{4\pi}{w}\int_0^\infty Nv^3\sigma_{\rm a}(v)f_0(v)dv \tag{2.34}$$

$$\alpha = \frac{4\pi}{w}\int_0^\infty Nv^3\sigma_{\rm i}(v)f_0(v)dv \tag{2.35}$$

$$\beta = \frac{4\pi}{w} \int_0^\infty N v^3 \sigma_{\mathrm{E}}(v) f_0(v) dv. \tag{2.36}$$

In addition, another commonly used coefficient is the so-called effective ionization coefficient, $\overline{\alpha} = \alpha - \eta$. These coefficients are measured as a function of E/N and at times also as a function of N and T either in the pure gas or in mixtures with a buffer gas. When the measurements are made in mixtures with a buffer gas for which the electron energy distribution function is known at each E/N, the coefficient (or the rate constant, see below) can be plotted as a function of the mean electron energy $\langle \varepsilon \rangle$ and unfolded to obtain the respective total cross sections [e.g., see Ref. [94] for data on η; see also Tables 6.81 and 6.82 in Ch. 6 for values of $\langle \varepsilon \rangle$ as a function of E/N for two buffer gases (Ar and N_2) commonly used in electron attachment studies].

In the case of mixtures, the total gas number density, N, which is used to determine the value of E/N is equal to the sum of the number density of the gas under study and the number density of the buffer gas. Thus, if we designate the number density of, say, an electron attaching gas by N_{a}, the number density of the buffer gas by N_{b}, and the total gas number density N $(= N_{\mathrm{a}} + N_{\mathrm{b}})$, the attaching-gas-number-density-normalized electron attachment coefficient η/N_{a} (in units of cm^2) can be expressed as

$$\eta/N_{\mathrm{a}}(E/N) = \frac{4\pi}{w} \int_0^\infty v^3 \sigma_{\mathrm{a}}(v) f_0(v, E/N) dv. \tag{2.37}$$

3.3 Rate Constants

The rate constants (in units of $\mathrm{cm}^3\ \mathrm{s}^{-1}$) for electron attachment k_{a}, electron-impact ionization k_{i}, and electron-impact excitation k_{E}, are given, respectively, by

$$k_{\mathrm{a}} = \eta w/N, \qquad k_{\mathrm{i}} = \alpha w/N, \qquad \text{and} \qquad k_{\mathrm{E}} = \beta w/N \tag{2.38}$$

where η, α, and β are, respectively, the coefficients of electron attachment, electron-impact ionization, and electron-impact excitation. Thus, the rate constant for electron attachment (and similarly for ionization and excitation) can be written as

$$\begin{aligned} k_{\mathrm{a}}(E/N) &= \eta/N_{\mathrm{a}}(E/N) \times w(E/N) \\ &= \left(\frac{2}{m}\right)^{1/2} \int_0^\infty \sigma_{\mathrm{a}}(\varepsilon) \varepsilon^{1/2} f(\varepsilon, E/N) d\varepsilon. \end{aligned} \tag{2.39}$$

When, as in many experiments, the measurements are made only at very low E/N where the electrons are in thermal equilibrium with the gas molecules, the electron energy distribution function is Maxwellian $f_{\mathrm{M}}(\varepsilon, T)$, characteristic of only the gas temperature T. In these cases, the value, $(k_{\mathrm{a,t}})_{\mathrm{th}}(T)$, of the total electron attachment rate constant at thermal energies can be expressed as

$$(k_{\mathrm{a,t}})_{\mathrm{th}}(T) = \left(\frac{2}{m}\right)^{1/2} \int_0^\infty \sigma_{\mathrm{a,t}}(\varepsilon)\varepsilon^{1/2} f_{\mathrm{M}}(\varepsilon, T) d\varepsilon \qquad (2.40)$$

$$= \frac{2^{3/2}}{(m\pi)^{1/2}(kT)^{3/2}} \int_0^\infty \varepsilon \sigma_{\mathrm{a,t}}(\varepsilon, T) \exp\left(-\frac{\varepsilon}{kT}\right) d\varepsilon.$$

3.4 Methods of Measurement

The methods for measuring the electron transport coefficients and the coefficients for electron attachment, electron-impact ionization, and electron-impact excitation have been fully described elsewhere (e.g., see, Refs. [1, 3–6]). Among the recent developments in the electron-swarm method of direct interest to the subject of this book are: (a) the various improvements to the pulsed-Townsend technique (e.g., see, Refs. [102, 164–171, 174]); (b) the various approaches to analyze charges formed in electron (and ion) swarms using mass analysis [172–178]; (c) the various techniques developed to study electron-hot molecule interactions [7, 164, 179], notably the time-resolved electron swarm technique [180–184]; (d) the swarm techniques for the study of electron attachment to electronically excited molecules [7, 164, 185–187]; and (e) the swarm techniques for measurement of photodetachment cross sections in low and high-pressure gases and in liquids [188–190].

We discuss some of these advances here.

3.4.1 Time-Resolved Electron Swarm Technique

This technique [180–184, 191] allows information on electron attachment and detachment processes to be obtained simultaneously from an analysis of transient electron current waveforms. This knowledge is especially important for the quantitative understanding of the effect of temperature on the formation ($\mathrm{M} + e \rightarrow \mathrm{M}^{-*} \rightarrow \mathrm{M}^-$) and thermal autodestruction ($\mathrm{M}^- + \text{heat} \rightarrow \mathrm{M} + e$) of long-lived parent negative ions such as SF_6^-, c-$C_4F_6^-$, and $C_6F_6^-$ [94, 180–182]. The principle of the time-resolved electron swarm technique [180–182] is shown in Fig. 2.5.

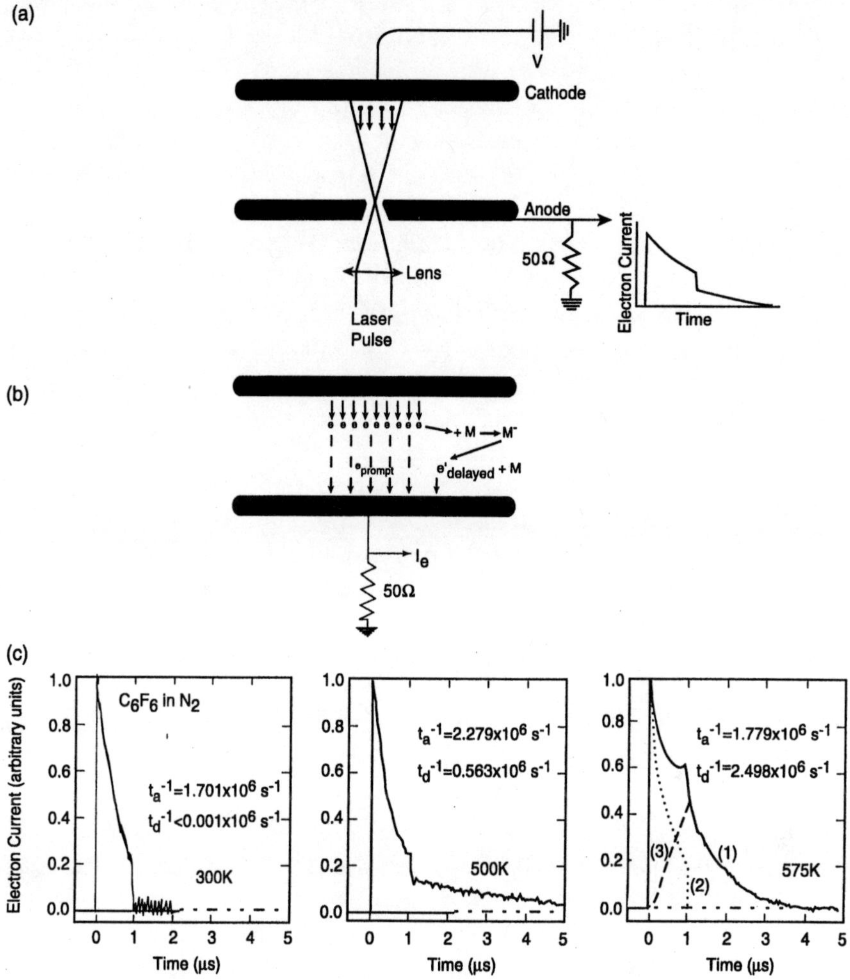

Figure 2.5. (a) and (b). Schematic diagram of the experimental arrangement and principle of the time-resolved electron swarm technique. (c) Electron-current waveforms for C_6F_6 in N_2 at 300, 500, and 575 K at an E/N value of 1.09×10^{-17} V cm^2. The solid curves (—) are the measured electron current waveforms, the dotted (···) curve represents the contribution to the total electron current of the initial (prompt) electrons when only electron attachment is present, and the broken (- - -) curve represents the contribution to the total electron current of the autodetached (delayed) electrons. The calculated waveforms using the t_a^{-1} and t_d^{-1} values obtained from a nonlinear least squares fit and shown in the figure, reproduce well the experimental waveforms (solid curves) [180–182].

The electron swarm is produced by a narrow nitrogen laser pulse (pulse full width at half maximum (FWHM) $\sim 6 \times 10^{-10}$ s), which strikes the cathode through a hole in the anode electrode. The interelectrode space is filled with minute amounts of the electron attaching gas molecules M (for instance SF_6) in a buffer gas (for instance N_2) at such densities that the initially produced (prompt) electrons quickly attain (through collisions with the gas molecules) a steady-state equilibrium energy distribution $f(\varepsilon, E/N, T)$ and drift toward the anode under the influence of an externally applied uniform electric field. As they drift, a fraction of them is removed by attachment, forming unstable parent negative ions which are rapidly stabilized via collisions with the buffer gas molecules to form stable negative ions (M^- in Fig. 2.5). These stabilized negative ions can subsequently thermally autodetach giving rise to delayed electrons.

The motion of the electrons which reach the anode without ever having been attached (prompt electrons) and the electrons which have been attached and then thermally released (delayed electrons) induces a current in the anode circuit which is observed through a resistor to ground. The time and space evolution of the electron swarm when both electron attachment and detachment (but no ionization) processes are present can be described [180–184, 191] by

$$\frac{\partial \rho_{\mathrm{e}}(x,t)}{\partial t} + w_{\mathrm{e}} \frac{\partial \rho_{\mathrm{e}}(x,t)}{\partial x} = -\frac{1}{t_{\mathrm{a}}} \rho_{\mathrm{e}}(x,t) + \frac{1}{t_{\mathrm{d}}} \rho_{\mathrm{i}}(x,t) \tag{2.41}$$

$$\frac{\partial \rho_{\mathrm{i}}(x,t)}{\partial t} + w_{\mathrm{i}} \frac{\partial \rho_{\mathrm{i}}(x,t)}{\partial x} = \frac{1}{t_{\mathrm{a}}} \rho_{\mathrm{e}}(x,t) - \frac{1}{t_{\mathrm{d}}} \rho_{\mathrm{i}}(x,t) \tag{2.42}$$

where $\rho_{\mathrm{e}}(x,t)$ and $\rho_{\mathrm{i}}(x,t)$, are, respectively, the number densities of electrons and negative ions, w_{e} and w_{i} are their respective drift velocities, and t_{a}^{-1} and t_{d}^{-1} are, respectively, the electron attachment and detachment frequencies. The solutions to Eqs. (2.41) and (2.42) when the initial electron swarm is produced by a short-duration laser pulse, that is, for the initial conditions $\rho_{\mathrm{e}}(x,0) = n_0\delta(x)$ and $\rho_{\mathrm{i}}(x,0) = 0$, are given in Refs. [183, 184, 191]. For the microsecond time scales of interest in these swarm experiments, the current is solely due to the motion of the electrons because the contribution of the slowly moving negative ions is negligible and can be neglected. The electron current is given by [182–184]

$$i_{\mathrm{e}}(t) = \frac{e w_{\mathrm{e}}}{d} \int_{w_{\mathrm{i}} t}^{\min(w_{\mathrm{e}} t, d)} \rho_{\mathrm{e}}(x,t) dx \tag{2.43}$$

where d is the drift distance, and the electron number density $\rho_e(x,t)$ is

$$\rho_e(x,t) = \frac{n_0}{w_e - w_i} \exp\left[-\frac{1}{t_a}\left(\frac{x - w_i t}{w_e - w_i}\right) + \frac{1}{t_d}\left(\frac{x - w_e t}{w_e - w_i}\right)\right] \times \left\{\delta\left(\frac{w_e t - x}{w_e - w_i}\right) + \left(\frac{1}{t_a t_d}\frac{x - w_i t}{w_e t - x}\right)^{1/2} \times I_1\left[\frac{2}{w_e - w_i}\left(\frac{1}{t_a t_d}(w_e t - x)(x - w_i t)\right)^{1/2}\right]\right\}. \tag{2.44}$$

In Eq. (2.44) n_0 is the initial electron number density (at $x = 0$), t_a^{-1} and t_d^{-1} are respectively the electron attachment and detachment frequencies, $\delta(w_e t - x)/(w_e - w_i)$ is a delta function, and I_1 is the first order modified Bessel function. Figure 2.5(c) shows an example of the recorded electron waveforms for the case of $C_6F_6^-$ [180, 181]. The evolution of the autodetachment process as the temperature is increased is clearly evident. The solid curves are the measured waveforms, and the dotted curve (curve 2 in Fig. 2.5(c)) is the contribution to the total electron current when no autodetachment occurs, that is, when all electrons reaching the anode are prompt. The difference between the total electron current (solid lines) and the prompt electron current (dotted curves) represents the contribution to the total electron current of the autodetached (delayed) electrons, which is represented by the dashed curve (curve 3 in Fig. 2.5(c)). As the temperature is increased, this contribution increases due to the faster autodetachment of the parent ions. From such recorded electron waveforms, the t_a^{-1} and t_d^{-1} are obtained at each value of T and E/N (or $\langle\varepsilon\rangle$) using a nonlinear least squares fit to the experimental waveforms. The electron attachment rate constant is then determined from $k_a(T) = (t_a N_a)^{-1}$ where N_a is the attaching gas number density.

3.4.2 Electron Swarm Techniques for the Study of Electron Attachment to Electronically Excited Molecules

The study of electron attachment to electronically excited molecules requires production of an appreciable number of excited molecules under conditions allowing electron attachment to the excited molecules within their lifetimes. Since the excited electronic states of most molecules lie energetically a few electron volts above the ground state, the excimer laser lines are ideal for studying electron attachment to electronically excited molecules. With the availability of pulse energies of $\sim$ 100 mJ in 10–20 ns pulses at the XeF (355 nm, 3.5 eV), XeCl (308 nm, 4.0 eV),

KrF (248 nm, 5.0 eV), and ArF (193 nm, 6.4 eV) lines, appreciable number densities of low-lying excited states of molecules can be generated with monophotonic excitation. Similarly, appreciable number of high-lying excited electronic states of molecules can be generated with 2- or 3-photon excitation processes. Furthermore, useful wavelength tunability can be obtained using dye lasers pumped by excimer-laser lines. Depending on the lifetime of the excited electronic state under study, a number of swarm techniques have been developed by Christophorou and collaborators [164, 185–187] for electron attachment studies using pulsed lasers (for electron beam techniques using pulsed lasers see Refs. [192–195]. Figure 2.6(a) shows the principle of their technique for the study of electron attachment to long-lived (lifetimes, $\tau > 10^{-5}$ s) electronically excited molecules. Depending on the absorption coefficient of the molecule under study at the particular laser wavelength, 0.1 Pa to 100 Pa of the gas is mixed with a suitable buffer gas (usually N_2 or Ar) at high pressure (1–100 kPa). A laser pulse enters the interaction region through the gridded bottom electrode, excites the molecules in the interaction region, and produces a pulse of electrons at the top electrode. The energies of the electrons in this pulse very quickly (within $< 10^{-8}$ s) reach a steady-state distribution and the electrons drift as an electron swarm through the partially excited gas to the anode. The drift time taken by the electron swarm to reach the bottom electrode is $\lesssim 10^{-5}$ s, and thus the excited molecules must live longer than 10^{-5} s. The experimental arrangement in Fig. 2.6(b) is an improvement over that in Fig. 2.6(a) in that the production of the excited species is decoupled from that of the attaching electrons by using two lasers, allowing the time delay between the production of the excited species and the arrival of the attaching electrons in the interaction region to be varied. Also, the use of three electrodes in the arrangement of Fig. 2.6(b) allows the detection of negatively charged particles unambiguously (see below).

In Fig. 2.6(c) is shown the principle of an experimental arrangement which has been developed [187] to measure electron attachment to short-lived electronically excited species ($\tau < 10^{-8}$ s) in a high-pressure environment (1 to 50 kPa). The basic idea behind this technique is to produce the excited species and the electrons that will attach to them by a single laser pulse. Since the excited species and the electrons are produced concomitantly and in close proximity, electron attachment may occur in spite of the short lifetime of the excited species. As with the arrangement in Fig. 2.6(b), the positive ions generated by laser photoionization are separated from the negative charges (electrons and negative ions) unambiguously by separating the detection region (located between electrodes 2 and 3) from the interaction region (located between elec-

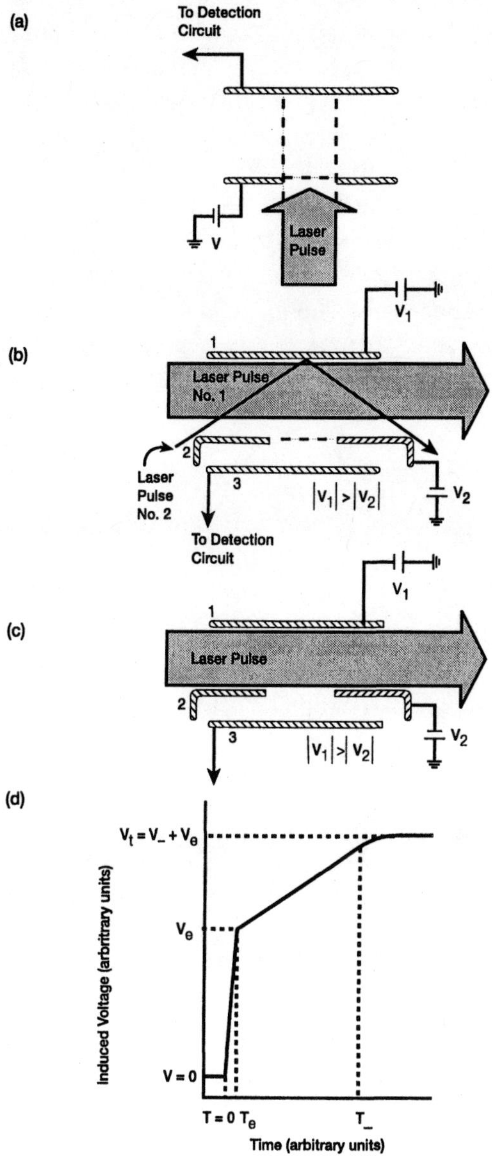

Figure 2.6. (a) and (b). Schematic electrode-laser pulse arrangements used in the study of photoenhanced electron attachment to long-lived ($\tau > 10^{-5}$ s) excited electronic states of molecules (see text and Refs. [164, 185, 186]). (c) Schematic diagram showing the principle of the technique used for the measurement of electron attachment to short-lived ($\tau < 10^{-8}$ s) excited electronic states of molecules (see text and Refs. [164, 187]). (d) Schematic drawing showing the components of the voltage induced in the anode circuit due the motion of the electrons and the negative ions (see text).

trodes 1 and 2) in a three-electrode arrangement. By applying suitable electric fields in the two regions, either positive or negative charges can be extracted to the detection region. In the case of negative charges, the signal components due to electrons and negative ions are distinguished due to the different drift velocities as shown schematically in Fig. 2.6(d). In the presence of electron attachment, the voltage induced in the anode circuit as a function of time ($t = 0$ being the time at which the laser pulse is fired) consists of an initial fast rise due to the unattached electrons and a subsequent slow rise due to the slow-moving negative ions. If N_e and N_i are, respectively, the number of electrons and negative ions reaching the anode, and V_e and V_i are the saturated values of the induced voltage due to the motion of the electrons and the negative ions respectively, then

$$R(E/N) = \frac{N_\mathrm{e} + N_\mathrm{i}}{N_\mathrm{e}} = \frac{V_\mathrm{e} + V_\mathrm{i}}{V_\mathrm{e}} = \frac{\eta d}{1 - \exp(-\eta d)}, \tag{2.45}$$

where d is the drift distance and $\eta(E/N)$ is the electron attachment coefficient. From the measured values of $R(E/N)$, the coefficient $\eta(E/N)$ is determined by an iterative procedure, and the total electron attachment rate constant is calculated from $k_\mathrm{a}(E/N) = \eta(E/N)w(E/N)/N_\mathrm{a}$, where $w(E/N)$ is the electron drift velocity and N_a is the electron attaching gas number density.

Data obtained by these techniques have indicated large increases in the electron attachment cross sections for a number of molecules compared to their ground electronic states. These results have been summarized and discussed in Refs. [7, 164].

3.4.3 Electron Swarm Techniques for the Measurement of the Photodetachment Cross Section in Low and High Pressure Gases and in Liquids

Electron detachment processes are secondary reactions (see Ch. 1). They play an important role in determining the nature of the species present in the plasma, especially the processes of collisional detachment and photodetachment. Collisional detachment has been investigated for a number of years [196], foremost by Champion and collaborators (e.g., see Refs. [197–200]). Electron detachment processes, also, play a crucial role in plasma diagnostics. For instance, photodetachment has been used by many authors as a tool to identify the negative ions and their number densities in plasmas. For this reason, we describe in this section a recent technique developed by Christophorou and collaborators [188–190] for the measurement of the photodetachment cross section of negative ions in low and high pressure gases and in liquids. A schematic of their technique is shown in Fig. 2.7(a).

The cell consists of a six-way stainless steel cube with two windows and electrical feedthroughs at opposite sides. The two parallel electrodes have a gridded circular area and are held at distance d (= 0.7 cm) apart. The gas under study is mixed with a buffer gas (N_2 or other nonelectron attaching gas with high-electron mobility such as methane (CH_4) or tetramethylsilane (TMSi)). An excimer laser beam (308 nm, FWHM $\approx 15 \times 10^{-9}$ s) enters the cell through a circular aperture having a diameter $2a$ (= 0.508 cm) and produces an electron swarm by injecting into the gas photoelectrons from the gridded part of the cathode electrode. The electrons in the swarm move toward the anode under the influence of an externally applied electric field. For the molecules investigated by Christophorou *et al.* (SF_6 and C_6F_6) and the experimental conditions they employed, the electrons are totally depleted at distances $z < d/2$ by forming parent negative ions M^- (SF_6^- or $C_6F_6^-$). A second counter propagating laser beam (FWHM $\approx 0.6 \times 10^{-9}$ s) from a tunable dye laser enters the cell coaxially with a time delay of $\sim$ 3500 ns from the first laser pulse. Photons from this second tunable dye-laser pulse with energies in excess of the photodetachment threshold, $E_{\rm th}$, detach the electron from the parent negative ion, giving rise to a transient current due to photodetached electrons. As the photodetached electrons move along the z-axis towards the anode, they are depleted by again forming M^- anions, which are finally collected at the anode electrode.

The circular aperture with diameter $2a$ allows a practically constant (flat) intensity profile of the excimer laser beam to strike the gridded cathode producing an electron swarm with a number density which is constant across r (the radial direction along a plane parallel to that of the aperture) and proportional in magnitude to the excimer laser intensity, that is, $n_{\rm e}(r, z = 0) = AI_{\rm e}(r)$, where A is a proportionality constant. These electrons are depleted by attachment, forming M^-. The electron number density $n_{\rm e}(r, z)$ and the negative ion number density $n_{\rm i}(r, z)$ for times t less than the electron drift time $t_{\rm e}$, are described by

$$n_{\rm e}(r,z) = AI_{\rm e}(r)e^{-\eta z} \quad \text{and} \quad n_{\rm i}(r,z) = AI_{\rm e}(r)(1 - e^{-\eta z}), \tag{2.46}$$

where η is the electron attachment coefficient.

The dye-laser beam with a Gaussian profile $I_{\rm d}(r)$ and a spatial opening less than $2a$ enters the reaction region between the two electrodes $\sim$ 3500 ns after the excimer laser and well after the electrons have been attached. For photon energies $h\nu = E > E_{\rm th}$, this gives rise to photodetached electrons. Under the conditions that the photodetached electron number density $n_{\rm ed}(r, z)$ is much smaller than the negative ion density

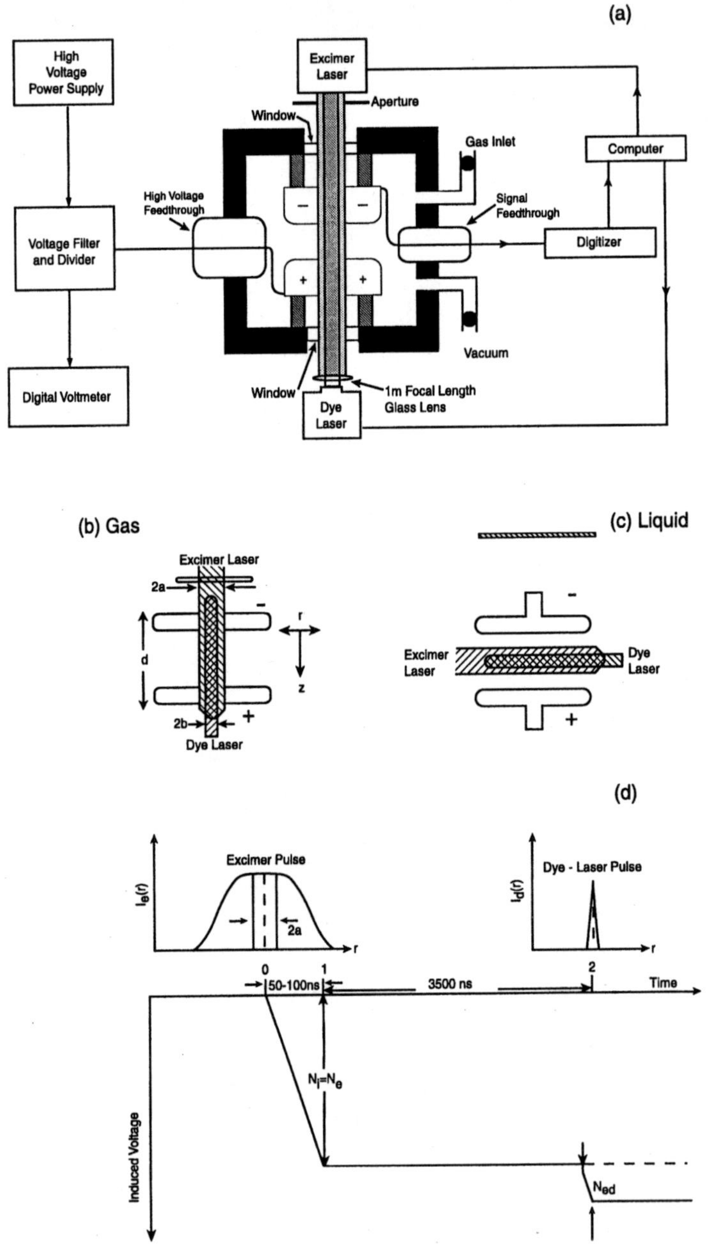

Figure 2.7. Schematic arrangements of the two-photon photodetachment technique [188–190] illustrating (a) the overall arrangement, (b) the geometry used to measure the photodetachment cross section in gases, (c) the geometry used to measure the photodetachment cross section in liquids, and (d) the spatial profiles of the excimer and tunable dye-laser pulses and typical induced voltage waveforms.

$n_i(r, z)$, $n_{ed}(r, z)$ is given by

$$n_{ed}(r, z) = \sigma_{pd}(E) n_i(r, z) I_d(r) \tag{2.47}$$

where $\sigma_{pd}(E)$ is the photodetachment cross section and $I_d(r)$ is the number of photons per unit area per dye-laser pulse. The total number N_{ed} of photodetached electrons is then obtained by integrating over the interaction volume between the two circular electrodes which gives [189]

$$N_{ed} = \frac{\sigma_{pd}(E) N_i I_t}{\pi a^2}. \tag{2.48}$$

In Eq. (2.48), N_i is the initial number of negative ions (in the experiments of Christophorou *et al.* it is equal to the initial number of photoelectrons N_e) and I_t is the total number of electron-detaching photons that enter the volume defined by the gridded part of the two circular electrodes. From Eq. (2.48) the photodetachment cross section as a function of the photon energy E, $\sigma_{pd}(E)$, is determined from

$$\sigma_{pd}(E) = \frac{N_{ed}}{N_i} \frac{\pi a^2}{I_t} \tag{2.49}$$

by measuring the ratio N_{ed}/N_i (Fig. 2.7(d)), the total number of photons I_t in the dye-laser pulse, and the area πa^2.

The measurement of the ratio N_{ed}/N_i is accomplished from data as depicted schematically in Fig. 2.7(d). As the electrons produced by the excimer laser pulse from the cathode electrode drift toward the anode, their motion induces a transient voltage in the anode electrode which is detected as a fast initial drop in the measured voltage waveform. This drop stops when all of the electrons have been attached to the molecules M forming M^-. The magnitude of the voltage drop is proportional to the number of electrons N_e in the gap, which is the same as the total number of negative ions N_i formed. The motion of the slower drifting M^- causes an additional small gradual drop in the waveform. When photons from the tunable dye-laser pulse strike the negative ions present in the gap, electrons can be detached whenever $E > E_{th}$, inducing a second fast drop in the voltage waveform. This drop is smaller in magnitude than the initial drop ($N_{ed}/N_i < 0.04$). The detached electrons are again captured causing an additional small and slow decline of the total voltage. From such waveforms, the ratio N_{ed}/N_i is measured and used to determine $\sigma_{pd}(E)$. Data obtained this way for SF_6^- can be found in Ch. 6.

4. Boltzmann-Code-Generated Collision Cross-Section Sets

The derivation of cross sections from electron transport coefficients requires solving the Boltzmann transport equation and taking into account all possible elastic and inelastic processes. Boltzmann (and Monte Carlo) codes have provided useful information on electron collision cross sections for a number of gases including those discussed in this book (e.g., Refs. [201–213]). In these investigations a cross-section set is assumed and then modified iteratively until the electron-transport parameters calculated using the Boltzmann equation best agree with the measured values. Such computations rely heavily on electron swarm transport coefficients measured over wide ranges of E/N, and on knowledge of electron collision cross sections from other sources used as input to the procedure. A serious difficulty of this procedure is that the derived electron-collision cross section set is not a unique solution, and if little is known about the cross sections for the molecule under study, the Boltzmann-code-generated cross sections for individual processes may be physically unrealistic (compare, for instance, the momentum transfer cross section $\sigma_{\rm m}(\varepsilon)$ in Fig. 3.1 of Ch. 3 as computed by Bordage *et al.* [210] and as deduced from experiments), or, in some cases, fictitious (e.g., the cross section for electronic excitation for CF_4 by Hayashi and Nakamura [211]).

A solution to this problem was suggested by Christophorou *et al.* [212, 214] who proposed to use the assessed data on electron-collision cross sections as essentially invariant (within the experimental uncertainties) inputs to such computations, and to use the assessed electron transport coefficients as the reliable set of reference data for comparison with the computed values. Constraining the computational procedure this way, the codes can serve as a means of generating cross sections and coefficients which are not available, and/or as a means of checking the validity of questionable data. The assessed cross sections for CF_4 have indeed been used in such a manner by Bordage *et al.* [212] (see also Bordage and Ségur [213] for similar work on CHF_3).

In addition to invariant cross section measurements, computational methods also often suffer from a lack of accurate measurements of the electron transport coefficients over wide ranges of E/N, angular electron scattering distribution measurements which extend from 0° to 180° scattering angles, and accurate energy losses assigned to each particular electron-collision process. Additionally, for applications in plasma models, these computations need knowledge of the cross sections for electron scattering from and electron attachment to excited species.

For plasma processing gases, it seems that a reasonably complete set of independently assessed electron collision cross sections and coefficients exists only for CF_4. The assessed knowledge for this gas can serve as a benchmark for experimental, theoretical, and model-calculation studies. The assessed data on electron collision cross sections and coefficients presented later in the book for other gases have significant gaps. While the data needs for ground-state molecules vary from one molecule to another, there seems to be a universal need for certain types of data, foremost for measurement of vibrational excitation cross sections, electronic excitation cross sections, differential scattering cross sections at small and large scattering angles, as well as for cross sections for electron-impact dissociation of molecules into neutral fragments. Cross sections for ion-pair and for multiple ionization are also scarce.

5. References

[1] E. W. McDaniel, *Collision Phenomena in Ionized Gases*, John Wiley & Sons, Inc., New York, 1964.

[2] H. S. W. Massey and E. H. S. Burhop, *Electronic and Ionic Impact Phenomena, Vol. I, Collisions of Electrons with Atoms*, Oxford, At the Clarendon Press, 1969.

[3] H. S. W. Massey, *Electronic and Ionic Impact Phenomena, Vol. II, Electron Collisions with Molecules and Photoionization*, Oxford, At the Clarendon Press, 1969.

[4] L. G. Christophorou, *Atomic and Molecular Radiation Physics*, Wiley-Interscience, London, 1971.

[5] L. G. H. Huxley and R. W. Crompton, *The Diffusion and Drift of Electrons in Gases*, Wiley-Interscience, New York, 1974.

[6] L. G. Christophorou (Ed.), *Electron-Molecule Interactions and Their Applications*, Academic, New York, Vols. 1 and 2, 1984.

[7] L. G. Christophorou and J. K. Olthoff, Adv. Atom. Molec. Opt. Phys. **44**, 155 (2000).

[8] S. Trajmar, D. F. Register, and A. Chutjian, Phys. Rep. **97**, 219 (1983).

[9] J. C. Nickel, P. W. Zetner, G. Shen, and S. Trajmar, J. Phys. E **22**, 730 (1989).

[10] S. Trajmar and J. W. McConkey, Adv. At. Mol. Opt. Phys. **33**, 63 (1994).

[11] R. W. Crompton, Adv. At. Mol. Opt. Phys. **33**, 97 (1994).

[12] M. J. Brunger and S. J. Buckman, Phys. Rep. **357**, 215 (2002).

[13] F. H. Read and J. M. Channing, Rev. Sci. Instrum. **67**, 2372 (1996).

[14] M. Zubek, N. Gulley, G. C. King, and F. H. Read, J. Phys. B **29**, L239 (1996).

[15] C. J. Sweeney and T. W. Shyn, Phys. Rev. A **53**, 1576 (1996).

[16] M. Zubek, B. Mielewska, J. Cahnning, G. C. King, and F. H. Read, J. Phys. B **32**, 1351 (1999).

[17] M. Allan, J. Phys. B **33**, L215 (2000).

[18] H. Cho, R. J. Gulley, and S. J. Buckman, J. Phys. B **33**, L309 (2000).

[19] H. Cho, R. J. Gulley, K. W. Trantham, L. J. Uhlmann, C. J. Dedman, and S. J. Buckman, J. Phys. B. **33**, 3531 (2000).

[20] F. A. Baiocchi, R. C. Wetzel, and R. S. Freund, Phys. Rev. Lett. **53**, 771 (1984).

[21] R. C. Wetzel, F. A. Baiocchi, T. R. Hayes, and R. S. Freund, Phys. Rev. A **35**, 559 (1987).

[22] T. R. Hayes, R. J. Shul, F. A. Baiocchi, R. C. Wetzel, and R. S. Freund, J. Chem. Phys. **89**, 4035 (1988).

[23] T. R. Hayes, R. C. Wetzel, F. A. Baiocchi, and R. S. Freund, J. Chem. Phys. **88**, 823(1988).

[24] R. J. Shul,T. R. Hayes, R. C. Wetzel, F. A. Baiocchi, and R. S. Freund, J. Chem. Phys. **89**, 4042(1988).

[25] R. S. Freund, R. C. Wetzel, R. J. Shul, and T. R. Hayes, Phys. Rev. A **41**, 3575 (1990).

[26] V. Tarnovsky and K. Becker, Z. Phys. D **22**, 603 (1992).

[27] V. Tarnovsky and K. Becker, J. Chem. Phys. **98**, 7868 (1993).

[28] V. Tarnovsky, P. Kurunczi, D. Rogozhnikov, and K. Becker, Int. J. Mass Spectrom. Ion Processes **128**, 181 (1993).

[29] V. Tarnovsky, A. Levin, and K. Becker, J. Chem. Phys. **100**, 5626 (1994).

[30] V. Tarnovsky, A. Levin, H. Deutch, and K. Becker, J. Phys. B **29**, 139 (1996).

[31] V. Tarnovsky, H. Deutch, K. E. Martus, and K. Becker, J. Chem. Phys. **109**, 6596 (1998).

[32] R. Basner, M. Schmidt, K. Becker, and H. Deutsch, Adv. At. Mol. Opt. Phys. **43**, 147 (2000).

[33] P. C. Cosby, *Dissociation Rates of Diatomic Molecules*, Wright Laboratory Report #WL-TR-93-2004, December 1992.

[34] P. C. Cosby, J. Chem. Phys. **98**, 7804 (1993).

[35] P. C. Cosby, J. Chem. Phys. **98**, 9560 (1993).

[36] T. Nakano, H. Toyoda, and H. Sugai, Jpn. J. Appl. Phys. **30**, 2908 (1991).

[37] T. Nakano, H. Toyoda, and H. Sugai, Jpn. J. Appl. Phys. **30**, 2912 (1991).

[38] T. Nakano and H. Sugai, Japn. J. Appl. Phys. **31**, 2919 (1992).

[39] T. Nakano and H. Sugai, J. Phys. D **26**, 1909 (1993).

[40] M. Goto, K. Nakamura, H. Toyoda, and H. Sugai, Jpn. J. Appl. Phys. **33**, 3602 (1994).

[41] H. Sugai, H. Toyoda, T. Nakano, and M. Goto, Contrib. Plasma Phys. **35**, 415 (1995).

[42] H. Toyoda, M. Iio, and H. Sugai, Jpn. J. Appl. Phys. A **36**, 3730 (1997).

[43] L. Mi and R. A. Bonham, J. Chem. Phys. **108**, 1904 (1998).

[44] H. F. Winters, J. Chem. Phys. **44**, 1472 (1966).

[45] H. F. Winters, J. Chem. Phys. **63**, 3462 (1975).

[46] H. F. Winters and M. Inokuti, Phys. Rev. A **25**, 1420 (1982).

[47] S. Motlagh and J. H. Moore, J. Chem. Phys. **109**, 432 (1998).

[48] J. M. Ajello and A. Chutjian, J. Chem. Phys. **65**, 5524 (1976).

[49] A. Chutjian, Phys. Rev. Lett. **46**, 1511 (1981); **48**, 289E (1982).

[50] A. Chutjian and S. H. Alajajian, Phys. Rev. A **31**, 2885 (1985).

[51] A. Chutjian and S. H. Alajajian, Phys. Rev. A **35**, 4512 (1987).

[52] A. Chutjian, S. H. Alajajian, and K.-F. Man, Phys. Rev. A **41**, 1311 (1990).

[53] A Chutjian, in *Electronic and Atomic Collisions*, W. R. MacGillivray, I. E. McCarthy, and M. C. Standage (Eds.), Adam Hilger, Bristol, 1992, pp. 127-138.

[54] A. Chutjian, A. Garscadden, and J. M. Wadehra, Phys. Rep. **264**, 393 (1996).

[55] P.-T. Howe, A. Kortyna, M. Darrach, and A. Chutjian, Phys. Rev. A **64**, 042706 (2001).

[56] K. Harth, M.-W. Ruf, and H. Hotop, Z. Phys. D **14**, 149 (1989).

[57] D. Klar, M.-W. Ruf, and H. Hotop, Chem. Phys. Lett. **189**, 448 (1992).

[58] D. Klar, M.-W. Ruf, and H. Hotop, Austr. J. Phys. **45**, 263 (1992).

[59] D. Klar, B. Mirbach, H. J. Korsch, M.-W. Ruf, and H. Hotop, Z. Phys. D **31**, 235 (1994).

[60] D. Klar, M.-W. Ruf, and H. Hotop, Meas. Sci. Technol. **5**, 1248 (1994).

[61] H. Hotop, D. Klar, J. Kreil, M.-W. Ruf, A. Schramm, and J. M. Weber, in *The Physics of Electronic and Atomic Collisions*, L. J. Dubé, J. B. A. Mitchell, J. W. McConkey, and C. E. Brion (Eds.), AIP Conf. Proc. **360**, 267 (1995).

[62] A. Schramm, J. M. Weber, J. Kreil, D. Klar, M.-W. Ruf, and H. Hotop, Phys. Rev. Lett. **81**, 778 (1998).

[63] A. Schramm, I. I. Fabrikant, J. M. Weber, E. Leber, M.-W. Ruf, and H. Hotop, J. Phys. B **32**, 2153 (1999).

[64] D. Klar, M.-W. Ruf, and H. Hotop, Int. J. Mass Spectrom. **205**, 93 (2001).

[65] D. Margreiter, G. Waldner, H. Deutsch, H. U. Poll, C. Winkler, K. Stephan, and T. D. Märk, Int. J. Mass Spectrom. Ion Processes **100**, 143 (1990).

[66] K. Stephan, H. Deutsch, and T. D. Märk, J. Chem. Phys. **83**, 5712 (1985).

[67] K. Leiter, P. Scheier, G. Walder, and T. D. Märk, Int. J. Mass Spectrom. Ion Processes **87**, 209 (1989).

[68] H. U. Poll, C. Winkler, V. Margreiter, V. Grill, and T. D. Märk, Int. J. Mass Spectrom. Ion Processes **112**, 1 (1992).

[69] T. Fiegele, V. Grill, S. Matt, M. Lezius, G. Hanel, M. Probst, P. Scheier, K. Becker, H. Deutsch, O. Echt, A. Stamatovic, and T. D. Märk, Vacuum **63**, 561 (2001).

[70] C. Ma, M. R. Bruce, and R. A. Bonham, Phys. Rev. A **44**, 2921 (1991) (see also Erratum **45**, 6932 (1992).

[71] M. R. Bruce, C. Ma, and R. A. Bonham, Chem. Phys. Lett. **190**, 285 (1992).

[72] M. R. Bruce and R. A. Bonham, Int. J. Mass Spectrom. Ion Processes **123**, 97 (1993).

[73] R. A. Bonham, Jpn. J. Appl. Phys. **33**, 4157 (1994).

[74] D. R. Sieglaff, R. Rejoub, B. G. Lindsay, and R. F. Stebbings, J. Phys. B **34**, 799 (2001).

[75] R. Rejoub, D. R. Sieglaff, B. G. Lindsay, and R. F. Stebbings, J. Phys. B **34**, 1289 (2001).

[76] H. Nishimura, W. M. Huo, M. A. Ali, and Y.-K. Kim, J. Chem. Phys. **110**, 3811 (1999).

[77] C. Q. Jiao, R. Nagpal, and P. D. Haaland, Chem. Phys. Lett. **269**, 117 (1997).

[78] C. Q. Jiao, A. Garscadden, and P. D. Haaland, Chem. Phys. Lett. **297**, 121 (1998).

[79] C. Q. Jiao, A. Garscadden, and P. D. Haaland, Chem. Phys. Lett. **310**, 52 (1999).

[80] C. Q. Jiao, A. Garscadden, and P. D. Haaland, Chem. Phys. Lett. **325**, 203 (2000).

[81] C. Q. Jiao, B. Ganguly, C. A. DeJoeph Jr., and A. Garscadden, Int. J. Mass Spectrom. **208**, 127 (2001).

[82] I. Torres, R. Martínez, and R. Castaño, J. Phys. B: At. Mol. Opt. Phys. **35**, 2423 (2002).

[83] D. Rapp and P. Englander-Golden, J. Chem. Phys. **43**, 1464 (1965).

[84] E. C. Zipf, in *Electron-Molecule Interactions and Their Applications*, L. G. Christophorou (Ed.), Academic, New York, Vol. 1, Ch. 4, 1984.

[85] S. J. B. Corrigan, J. Chem. Phys. **43**, 4381 (1965).

[86] A. Niehaus, Z. Naturforschung **22a**, 690 (1967).

[87] E. P. Wigner, Phys. Rev. **73**, 1002 (1948).

[88] F. B. Dunning, J. Phys. B **28**, 1645 (1995).

[89] F. B. Dunning, J. Phys. Chem. **91**, 2244 (1987).

[90] T. Kondow, J. Phys. Chem. **91**, 1307 (1987).

[91] T. Kraft, M.-W. Ruf, and H. Hotop, Z. Phys. D **14**, 179 (1989).

[92] H. S. Carman Jr., C. E. Klots, and R. N. Compton, J. Chem. Phys. **99**, 1734 (1993).

[93] C. Desfrancois, H. Abdoul-Carime, and J.-P. Schermann, Int. J. Mod. Phys. B **12**, 1339 (1996).

[94] L. G. Christophorou, D. L. McCorkle, and A. A. Christodoulides, in *Electron-Molecule Interactions and Their Applications*, L. G. Christophorou (Ed.), Academic, New York, Vol. 1, Ch. 6, 1984.

[95] B. G. Zollars, C. Higgs, F. Lu, C. W. Walter, L. G. Gray, K. A. Smith, F. B. Dunning, and R. F. Stebbings, Phys. Rev. A **32**, 3330 (1985).

[96] X. Ling, B. G. Lindsay, K. A. Smith, and F. B. Dunning, Phys. Rev. A **45**, 242 (1992).

[97] M. T. Frey, S. B. Hill, K. A. Smith, F. B. Dunning, and I. I. Fabrikant, Phys. Rev. Lett. **75**, 810 (1995).

[98] M. Matsuzawa, J. Phys. Soc. Jpn. **33**, 1108 (1973).

[99] L. S. Frost and A. V. Phelps, Phys. Rev. **127**, 1621 (1962).

[100] A. V. Phelps, Rev. Mod. Phys. **40**, 399 (1968).

[101] R. W. Crompton, M. T. Elford, and R. L. Jory, Austr. J. Phys. **20**, 369 (1967).

[102] S. R. Hunter and L. G. Christophorou, in *Electron-Molecule Interactions and Their Applications*, L. G. Christophorou (Ed.), Academic, New York, Vol. 2, Ch. 3, 1984.

[103] R. Winkler, Adv. Atom. Molec. Opt. Phys. **43**, 19 (2000).

[104] W. L. Morgan, Adv. Atom. Molec. Opt. Phys. **43**, 79 (2000).

[105] G. Csanak, D. C. Cartwright, S. K. Srivastava, and S. Trajmar, in *Electron-Molecule Interactions and Their Applications*, L. G. Christophorou (Ed.), Academic, New York, Vol. 1, Ch. 1, 1984.

[106] C. Winstead and V. McKoy, Adv. Atom. Molec. Opt. Phys. **43**, 111 (2000).

[107] M. H. F. Bettega, L. G. Ferreira, and M. A. P. Lima, Phys. Rev. A **47**, 1111 (1993).

[108] A. P. P. Natalense, M. H. F. Bettega, L. G. Ferreira, and M. A. P. Lima, Phys. Rev. A **52**, p. R1, 1995.

[109] A. P. P. Natalense, M. H. F. Bettega, L. G. Ferreira, and M. A. P. Lima, Phys. Rev. A **59**, 879 (1999).

[110] C. Winstead, Q. Sun, and V. McKoy, J. Chem. Phys. **98**, 1105 (1993).

[111] C. Winstead and V. McKoy, J. Chem. Phys. **114**, 7407 (2001).

[112] V. McKoy, C. Winstead, and W. L. Morgan, *Data Compilation for Plasma Chemistries*, SEMATECH Report #97043274A-TR, August 22, 1997.

[113] V. McKoy, C. Winstead, W. L. Morgan, and P. D. Haaland, *Data Compilation for Plasma Chemistries*, SEMATECH Report #98063515A-TR, June 30, 1998.

[114] M. T. do N. Varella, C. Winstead, V. McKoy, M. Kitajima, and H. Tanaka, Phys. Rev. A **65**, 022702 (2002).

[115] W. L. Morgan, C. Winstead, and V. McKoy, J. Appl. Phys. **90**, 2009 (2001).

[116] W. A. Issacs, C. W. McCurdy, and T. N. Rescigno, Phys. Rev. **58**, 309 (1998).

[117] W. A. Issacs, C. W. McCurdy, and T. N. Rescigno, Phys. Rev. **58**, 2881 (1998).

[118] T. N. Rescigno, Phys. Rev. A **50**, 1382 (1994).

[119] W. H. Huo, in *Computational Methods for Electron-Molecule Collisions*, W. H. Huo and F. A. Gianturco (Eds.), Plenum, New York, 1995, pp. 341-342.

[120] F. A. Gianturco, R. R. Lucchese, and N. Sanna, J. Chem. Phys. **100**, 6464 (1994).

[121] F. A. Gianturco, R. R. Lucchese, and N. Sanna, J. Chem. Phys. **104**, 6482 (1996).

[122] J. M. Blatt and J. D. Jackson, Phys. Rev. **26**, 18 (1949).

[123] T. F. O'Malley, L. Spruch, and J. Rosenberg, Math. Phys. **2**, 491 (1961).

[124] T. F. O'Malley, L. Rosenberg, and L. Spruch, Phys. Rev. **125**, 1300 (1962).

[125] T. F. O'Malley, Phys. Rev. **130**, 1020 (1963).

[126] S. R. Hunter, J. G. Carter, and L. G. Christophorou, Phys. Rev. A **38**, 5539 (1988).

[127] S. J. Buckman and J. Mitroy, J. Phys. B **22**, 1365 (1989).

[128] S. J. Buckman and M. J. Brunger, Aust. J. Phys. **50**, 483 (1997).

[129] A. Zecca, G. P. Karwasz, and R. S. Brusa, La Rivista del Nuovo Cimento **19** (3), 1 (1996).

[130] Y.-K. Kim and M. E. Rudd, Phys. Rev. A **50**, 3954 (1994).

[131] W. Hwang, Y.-K. Kim, and M. E. Rudd, J. Chem. Phys. **104**, 2956 (1996).

[132] D. Margreiter, H. Deutsch, M. Schmidt, and T. D. Märk, Int. J. Mass Spectrom. Ion Processes **100**, 157 (1990).

[133] D. Margreiter, H. Deutsch, and T. D. Märk, Int. J. Mass Spectrom. Ion Processes **139**, 127 (1994).

[134] H. Deutsch, K. Becker, S. Matt, and T. D. Märk, Int. J. Mass Spectrom. **197**, 37 (2000).

[135] N. F. Mott, Proc Roy. Soc. (London) **A126**, 259 (1930).

[136] L. Vriens, in *Case Studies in Atomic Physics*, E. W. McDaniel and M. R. C. McDowell (Eds.), North Holland, Amsterdam, 1969, Vol. 1, p. 335.

[137] H. A. Bethe, Ann. Phys. (Leipzig) **5**, 325 (1930).

[138] Y.-K. Kim, W. Hwang, N. M. Weinberger, M. A. Ali, and M. E. Rudd, J. Chem. Phys. **106**, 1026 (1997).

[139] M. A. Ali, Y.-K. Kim, W. Hwang, N. M. Weinberger, and M. E. Rudd, J. Chem. Phys. **106**, 9602(1997).

[140] Y.-K. Kim, M. A. Ali, and M. E. Rudd, J. Res. NIST, **102**, 693 (1997).

[141] Y.-K. Kim and M. E. Rudd, Comments At. Mol. Phys. **34**, 309 (1999).

[142] W.M. Huo and Y.-K. Kim, IEEE Trans. Plasma Sci. **27**, 1225 (1999).

[143] Y.-K. Kim and K. K. Irikura, *Atomic and Molecular Data and Their Applications*, K. A. Berrington and B. L. Bell (Eds.), American Institute of Physics Conference Proceedings, Vol. 543, 2000, p. 220.

[144] M. A. Ali, K.K. Irikura, and Y.-K. Kim, Int. J. Mass Spectrom. **201**, 187 (2000).

[145] Y.-K. Kim, K. K. Irikura, and M. A. Ali, J. Res. NIST, **105**, 285 (2000).

[146] H. Deutsch, C. Cornelissen, L. Cespiva, V. Bonacic-Koutecky, D. Margreiter, and T. D. Märk, Int. J. Mass Spectrom. Ion Processes **129**, 43 (1993).

[147] H. Deutsch, T. D, Märk, V. Tarnovsky, K. Becker, C. Cornelissen, L. Cespiva, and V. Bonacic-Koutecky, Int. J. Mass Spectrom. Ion Processes **137**, 77 (1994).

[148] T. F. O'Malley, Phys. Rev. **150**, 14 (1966).

[149] J. N. Bardsley, A. Herzenberg, and F. Mandl, Proc. Phys. Soc. (London) **89**, 321 (1966).

[150] H. A. Bethe, Phys. Rev. **47**, 747 (1935).

[151] J. P. Gauyacq and A. Herzenberg, J. Phys. B **17**, 1155 (1984).

[152] I. I. Fabrikant, Phys. Rev. A **43**, 3478 (1991).

[153] R. Abouaf and D. Teillet-Billy, J. Phys. B **10**, 2261 (1977).

[154] A. I. Baz', Sov. Phys. JETP **6**, 709 (1958).

[155] J. P. Gauyacq and A. Herzenberg, Phys. Rev. A **25**, 2959 (1982).

[156] J. Horáček and W. Domcke, Phys. Rev. A **53**, 2262 (1996).

[157] J. Horáček, W. Domcke, and H. Nakamura, Z. Phys. D **42**, 181 (1997).

[158] E. Vogt and G. H. Wannier, Phys. Rev. **95**, 1190 (1954).

[159] C. E. Klots, Chem. Phys. Lett. **38**, 61(1976).

[160] W. P. Allis, in *Handbuch der Physik*, S. Flügge (Ed.), Springer-Verlag, Berlin, Vol. 21, p. 383.

[161] J. H. Parker Jr. and J. J. Lowke, Phys. Rev. **181**, 290 (1969).

[162] H. R. Skullerud, J. Phys. B **2**, 696 (1969).

[163] H. R. Skullerud, Aust. J. Phys. **27**, 195 (1974).

[164] L. G. Christophorou, L. A. Pinnaduwage, and P. G. Datskos, in *Linking the Gaseous and the Condensed Phases of Matter: The Behavior of Slow Electrons*, L. G. Christophorou, E. Illenberger, and W. Schmidt (Eds.) Plenum, New York, 1994, p. 415.

[165] S. R. Hunter, J. G. Carter, and L. G. Christophorou, J. Appl. Phys. **60**, 24 (1986).

[166] J. de Urquijo, C. A. Arriaga, C. Cisneros, and I. Alvarez, J. Phys. D **32**, 41 (1999).

[167] H. Faidas, L. G. Christophorou, D. L. McCorkle, and J. G. Carter, Nucl. Instr. Meth. Phys. Res. A **294**, 575 (1990).

[168] J. de Urquijo, C. A. Arriaga, C. Cisneros, and I. Alvarez, J. Phys. D **32**, 41 (1999).

[169] J. de Urquijo, I. Alvarez, E. Basurto, and C. Cisneros, J. Phys. D **32**, 1646 (1999).

[170] J. de Urquijo and E. Basurto, J. Phys. D **34**, 1352 (2001).

[171] J. de Urquijo, E. Basurto, and J. K. Hernández-Ávita, J. Phys. D **34**, 2151 (2001).

[172] R. A. Kennedy and C. A. Mayhew, Int. J. Mass Spectrom. **206**, *i* (2001).

[173] Y. Liu, C. A. Mayhew, and R. Peverall, Int. J. Mass Spectrom. and Ion processes **152**, 225 (1996).

[174] G. K. Jarvis, C. A. Mayhew, L. Singleton, and S. M. Spyrou, Int. J. Mass Spectrom. and Ion Processes **164**, 207 (1997).

[175] G. K. Jarvis, R. A. Kennedy, and C. A. Mayhew, Int. J. Mass Spectrom. **205**, 253 (2001).

[176] E. Stoffels, W. W. Stoffels, D. Vender, G. M. W. Kroesen, and F. J. de Hoog, J. Vac. Sci Technol. A **13**, 2051 (1995).

[177] W. W. Stoffels, E. Stoffels, M. Haverlag, G. M. W. Kroesen, and F. J. de Hoog, J. Vac. Sci Technol. A **13**, 2058 (1995).

[178] E. Stoffels, W. W. Stoffels, and G. M. W. Kroesen, Plasma Sources Sci. Technol. **10**, 311 (2001).

[179] D. Smith and P. Spaněl, Adv. At. Mol. Opt. Phys. **32**, 307 (1994).

[180] L. G. Christophorou and P. G. Daskos, Int. J. Mass Spectrom. Ion Processes **149/150**, 59 (1995).

[181] P. G. Datskos, L. G. Christophorou, and J. G. Carter, J. Chem. Phys. **98**, 7875 (1993).

[182] P. G. Datskos, L. G. Christophorou, and J. G. Carter, J. Chem. Phys. **99**, 8607 (1993).

[183] C. Wen, Ph.D. dissertation, Eindhoven University of Technology, The Netherlands, 1989.

[184] C. Wen and J. M. Wetzer, IEEE Trans. Electr. Insul. **23**, 999 (1988).

[185] L. G. Christophorou, S. R. Hunter, L. A. Pinnaduwage, J. G. Carter, A. A. Christodoulides, and S. M. Spyrou, Phys. Rev. Lett. **58**, 1316 (1987).

[186] L. A. Pinnaduwage, L. G. Christophorou, and S. R. Hunter, J. Chem. Phys. **90**, 6275 (1989).

[187] L. A. Pinnaduwage, L. G. Christophorou, and A. P. Bitouni, J. Chem. Phys. **95**, 274 (1991).

[188] H. Faidas, L. G. Christophorou, and D. L. McCorkle, Chem. Phys. Lett. **193**, 487(1992).

[189] L. G. Christophorou, P. G. Datskos, and H. Faidas, J. Chem. Phys. **101**, 6728 (1994).

[190] P. G. Datskos, J. G. Carter, and L. G. Christophorou, Chem. Phys. Lett. **239**, 38 (1995).

[191] L. Frommhold, Fortschr. Phys. **12**, 597 (1964).

[192] T. Jaffke, R. Hashemi, L. G. Christophorou, E. Illenberger, E. Baumgärtel, and L. A. Pinnaduwage, Chem. Phys. Lett. **203**, 21 (1993).

[193] E. Krishnakumar, S. V. K. Kumar, S. A. Rangwala, and S. K. Mitra, J. Phys. B **29**, L657 (1996).

[194] E. Krishnakumar, S. V. K. Kumar, S. A. Rangwala, and S. K. Mitra, Phys. Rev. A **56**, 1945 (1997).

[195] S. A. Rangwala, S. V. K. Kumar, and E. Krishnakumar, Phys. Rev. A **64**, 012707 (2001).

[196] L. G. Christophorou, Contrib. Plasma Phys. **27**, 237 (1987).

[197] R. L. Champion and L. D. Doverspike, in *Electron-Molecule Interactions and Their Applications*, L. G. Christophorou (Ed.), Academic, New York, Vol. 1, Ch. 7, 1984.

[198] J. K. Olthoff, R. J. Van Brunt, Y. Wang, R. L. Champion, and L. D. Doverspike, J. Chem. Phys. **91**, 2261 (1989).

[199] Y. Wang, R. L. Champion, L. D. Doverspike, J. K. Olthoff, and R. J. Van Brunt, J. Chem. Phys. **91**, 2254 (1989).

[200] R. L. Champion, in *Gaseous Dielectrics VI*, L. G. Christophorou and I. Sauers (Eds.), Plenum, New York, 1991, p. 1.

[201] M. Hayashi and T. Nimura, J. Phys. D **17**, 2215 (1984).

[202] M. Hayashi, in *Swarm Studies and Inelastic Electron-Molecule Collisions*, L. C. Pitchford, B. V. McKoy, A. Chutjian, and S. Trajmar (Eds.), New York, Springer, 1987, p. 167.

[203] M. Hayashi and A. Niwa, in *Gaseous Dielectrics V*, L. G. Christophorou and D. W. Bouldin (Eds.), Pergamon, New York, 1987, p. 27.

[204] M. Hayashi, in *Gaseous Electronics and Its Applications*, R. W. Crompton, M. Hayashi, D. E. Boyd, and T. Makabe (Eds.), KTK Scientific Publishers, Tokyo, Japan, 1991, p. 9.

[205] A. V. Phelps and R. J. Van Brunt, J. Appl. Phys. **64**, 4269 (1988).

[206] H. Itoh, M. Kawaguchi, K. Satoh, Y. Miura, Y. Nakao, and H. Tagashira, J. Phys. D **23**, 229 (1990).

[207] Y. Nakamura, in *Gaseous Electronics and Their Applications*, R. W. Crompton, M. Hayashi, D. E. Boyd, and T. Makabe (Eds.), KTK Scientific Publishers, Tokyo, Japan, 1991, p. 178.

[208] H. Itoh, T. Miyachi, M. Kawaguchi, Y. Nakao, and H. Tagashira, J. Phys. D **24**, 277 (1991).

[209] H. Itoh, T. Matsumura, K. Satoh, H. Date, Y. Makao, and H. Tagashira, J. Phys. D **26**, 1975 (1993).

[210] M.-C. Bordage, P. Ségur, and A. Chouki, J. Appl. Phys. **80**, 1325 (1996).

[211] Y. Hayashi and Y. Nakamura, in *International Conference on Atomic and Molecular Data and their Applications*, W. L. Wiese and P. J. Mohr (Eds.), NIST Special Publication 926 (National Institute of Standards and Technology, Gaithersburg, MD), 1998, p. 248.

[212] M.-C. Bordage, P. Ségur, L. G. Christophorou, and J. K. Olthoff, J. Appl. Phys. **86**, 3558(1999).

[213] M.-C. Bordage and P. Ségur, in *Proceedings of the XXVth International Conference on Phenomena in Ionized Gases*, T. Goto (Ed.), Nagoya, Japan, July 2001, Vol. 3, p. 253.

[214] L. G. Christophorou and J. K. Olthoff, Adv. At. Mol. Opt. Phys. **44**, 59 (2000).

Chapter 3

SYNTHESIS AND ASSESSMENT OF ELECTRON COLLISION DATA

1. Introduction

In this chapter we describe the three major components of our effort to provide electron-impact reference data relevant to plasma processing, namely, (i) the synthesis, assessment, and recommendation of electron collision data, (ii) the deduction of unavailable data and understanding from the assessed data, including identification of new measurements and data needs, and (3) the dissemination and updating of the database.

2. Synthesis, Assessment, and Recommendation of Data

The task begins with thorough literature searches and contacts with researchers in the field who generate the most recent data. It is followed by a critical review and assessment of the available data as obtained from the literature and the workers in the field. It ends with a recommendation of data. The critical review and assessment requires understanding of the physical processes themselves, and oftentimes is aided by complementary information on the physical and chemical properties of the molecules under consideration.

In general, there exist three main sources of electron collision data: experimental measurements (obtained principally by using electron beam and electron swarm techniques), calculations (of varied levels of sophistication), and Boltzmann- and Monte Carlo-based computations. The first source provides directly — and at times also indirectly — cross sections or coefficients for individual electron collision processes. The second source gives, in principle, cross sections for any electron collision process, but in practice is limited in its utility by the complexity the calculations themselves entail for polyatomic molecules. The third source relies on electron transport data and other inputs and gives only self-consistent sets of cross sections, not independent and unique cross sections for each individual collision process (see discussion in Sec. 4 of Ch. 2). A joint assessment of cross sections from all of these sources is essential since the values determined from each are often inconsistent. The assessment of the cross sections from these diverse sources is complex, involving consideration of the uncertainty of the measurements, the restrictions of the calculation, and the physical reasonableness of the data obtained as sets rather than for each individual electron-molecule reaction.

Any assessment of data relies upon some established protocol for determining which data are the most reliable. In the present work, "recommended" or "suggested" values of cross sections and transport coefficients are determined, where possible, for each type of cross section and coefficient for which data exist. These values are derived from fits to the most reliable data, as determined by the following criteria: (i) the data are published in peer reviewed literature; (ii) there is no evidence of unaddressed errors; (iii) the data are absolute determinations; (iv) multiple data sets exist and are consistent with one another within combined stated uncertainties over common energy ranges; and (v) in regions where both experimental and theoretical results exist, the experimental data are preferred. Cross sections and coefficients for the various processes that meet these criteria are designated as "recommended" and fits to these data represent the best current estimates for the cross sections and coefficients for each electron collision process. For cross sections and coefficients for which the only data that are available do not meet all of the above criteria, the best available data may be used to designate a "suggested" cross section or coefficient. In cases where no reasonable data exist, or where two or more measurements are in an unresolved contradiction, the raw data are presented for information and no recommendation is made.

To most clearly identify the reliable data for electron interactions with a given gas, all published data for each cross section are considered in our assessment, even those which have been subsequently superseded. This is done in order to aid in the understanding of the evolution of the data, assist in the determination of the reliability of the data, and draw attention to these changes for researchers who might have used the earlier data in their work. When possible, data are obtained from published tables. For literature data which are available only in graphical form, the published figures are scanned and the data digitized for use in this work. Depending on the quality of the original figure, the values of the data obtained in this manner are within 1–3% of the values in the original figure.

No uncertainties are assigned to the recommended data since no means exist of confirming the experimental uncertainties reported in some of the original papers (see discussions in the papers by Christophorou and co-workers [1–11]). Some measure of uncertainty can be obtained from an analysis of the combined relative uncertainties of the original data that are fitted to derive the recommended values. Although for many cross sections there exist published values that differ by as much as two orders of magnitude (see the next section, for example), for many cases critical analysis of all the available data allows the determination of recommended cross sections whose uncertainties are expected to be between ±10% and ±20%.

The assessment process is illustrated below by four examples. The first example shows how reasonable values for a given cross section can be extracted from conflicting data. The second example shows how a recommended cross section for a particular electron collision process can be determined by considering its relationship to known cross sections for other electron collision processes. The third example shows how the magnitude of an electron-beam determined cross section can be adjusted based on knowledge provided by the electron-swarm method. The fourth example shows how the total electron scattering cross section can be employed to check the required consistency between the cross sections of the various electron collision processes.

2.1 Determination of the Cross Section for Momentum Transfer, $\sigma_m(\varepsilon)$, of CF_4

Figure 3.1 shows an example of how a recommended cross section is extracted from multiple sets of experimental data and theoretical results. Shown in the figure are the many determinations of the momentum transfer cross section, $\sigma_m(\varepsilon)$, of CF_4 for which some values differ by more than two orders of magnitude. The data in Fig. 3.1 cover the energy range from 0.001 to 1000 eV and come from three sources: direct measurements, calculations, and swarm-based Boltzmann codes. The $\sigma_m(\varepsilon)$ derived from direct measurements are the high-energy data of Sakae *et al.* [12] and Boesten *et al.* [13], which are in agreement within their combined uncertainty ($\sim$ ±10% for Sakae *et al.* [12] and ±15% to ±20% for Boesten *et al.* [13]), and the low-energy values of Mann and Linder [14] which were determined from their elastic differential cross section measurements (uncertainty of ±20% to ±30%) and modi-

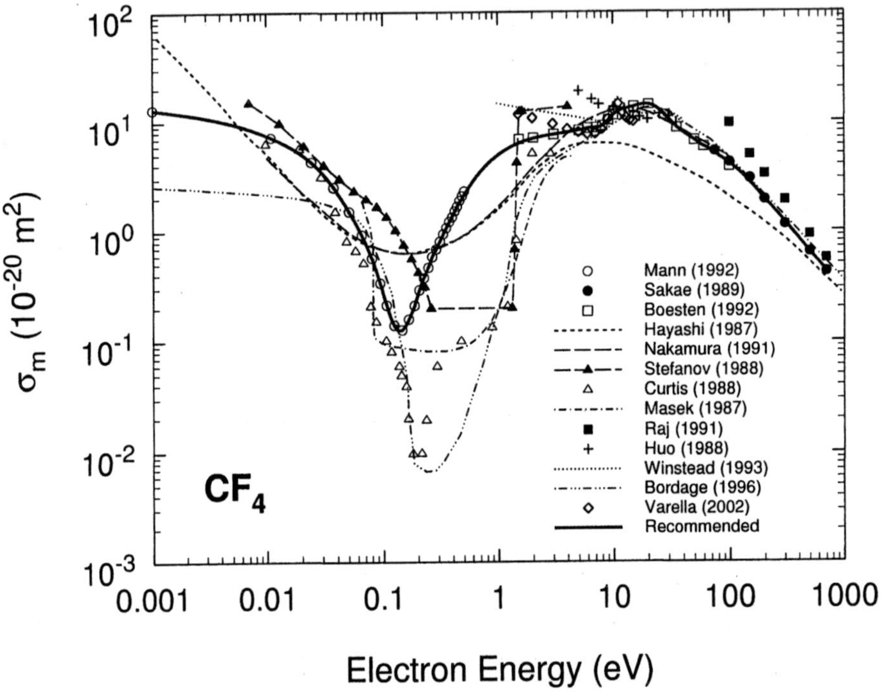

Figure 3.1. Extraction of a recommended cross section for the elastic momentum transfer cross section, $\sigma_m(\varepsilon)$, of CF_4 from widely varied experimental and calculated results from various sources (see text and Christophorou *et al.* [1]).

fied effective range theory. The calculated $\sigma_m(\varepsilon)$ are the results of three investigations, one using an independent-atom model with partial waves [15] and the other two using the static exchange approximation [16, 17], and they agree only partially with the experimental measurements of Sakae *et al.* and Boesten *et al.* The results of a more recent calculation in the energy range 1.5 to 15 eV by Varella *et al.* [18] using the Schwinger multichannel method are in better agreement with the experimental data. The swarm-based Boltzmann-code cross sections [19–24] are model dependent and have different degrees of uncertainty depending upon their input data and on the electron transport data they used. None agree well with the measurements. In spite of these disparities, a recommended cross section can be obtained [1] over the energy range 0.001 to 700 eV (solid line in Fig. 3.1) using the criteria discussed earlier in Sec. 2 of this chapter. This recommended cross section is derived from a least squares fit to the experimental cross sections of Mann and

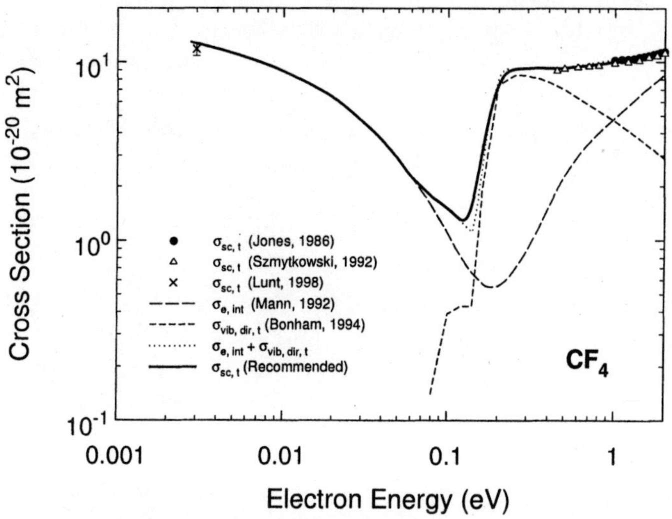

Figure 3.2. Determination of the recommended total electron scattering cross section, $\sigma_{sc,t}(\varepsilon)$, of CF_4 below 1 eV (see text).

Linder below 0.5 eV, the experimental cross sections of Sakae *et al.* and Boesten *et al.* above 1.5 eV, and an interpolation between 0.5 to 1.5 eV (see also Sec. 2.2.4 of Ch. 4).

2.2 Determination of the Total Electron Scattering Cross Section, $\sigma_{sc,t}(\varepsilon)$, of CF_4 Below 1 eV

A recommended value for the total electron scattering cross section $\sigma_{sc,t}(\varepsilon)$ of the CF_4 molecule was derived with confidence above 1 eV because reliable experimental measurements exist over wide energy ranges which agree within specified uncertainties (see Ref. [1] and Sec. 2.2.1 of Ch. 4). However, below 1 eV this is not the case. Nonetheless, recommended values for the $\sigma_{sc,t}(\varepsilon)$ of the CF_4 molecule down to 0.003 eV were determined by considering the relationship of $\sigma_{sc,t}(\varepsilon)$ and the cross sections for other electron collision processes as shown in Fig. 3.2. Firstly, the measurements of $\sigma_{sc,t}(\varepsilon)$ above ~ 0.5 eV by Jones [25] (solid circles) and by Szmytkowski *et al.* [26] (open triangles) were accepted as accurate higher-energy reference data points. Secondly, the measurements of Mann and Linder [14] for the elastic integral cross section, $\sigma_{e,int}(\varepsilon)$, below 2 eV (long-dash curve in Fig. 3.2) were accepted since they were determined from their differential electron scattering cross section measurements and a sound analysis. Thirdly, below the lowest vibrational threshold of CF_4 at 0.054 eV, $\sigma_{sc,t}(\varepsilon)$ was taken to be equal to $\sigma_{e,int}(\varepsilon)$, i.e., $\sigma_{sc,t}(\varepsilon) = \sigma_{e,int}(\varepsilon)$. Fourthly, since there are no known negative ion states of CF_4 below 2 eV (see Ref. [1], and Table 4.2 of Ch. 4), it was assumed that all vibrational excitation below $\sim$ 2 eV is due to the direct excitation by electron impact of the two infrared active modes, ν_3 and ν_4, of CF_4. Furthermore, since in this energy range electronic excitation is absent and rotational excitation is negligible (the CF_4 molecule has neither a dipole nor a quadrupole moment), it was assumed that between approximately 0.06 and 2 eV, $\sigma_{sc,t}(\varepsilon) = \sigma_{e,int}(\varepsilon) + \sigma_{vib,dir,t}(\varepsilon)$, where $\sigma_{vib,dir,t}(\varepsilon)$ is the sum of the cross sections for the direct excitation of the ν_3 and ν_4 infrared active modes of CF_4 as calculated by Bonham [27] in the Born-dipole approximation (short-dash curve in Fig. 3.2).

The dotted curve in Fig. 3.2 represents $\sigma_{sc,t}(\varepsilon) = \sigma_{e,int}(\varepsilon) + \sigma_{vib,dir,t}(\varepsilon)$ and was obtained by using the recommended values of Christophorou *et al.* [1] for $\sigma_{e,int}(\varepsilon)$ and $\sigma_{vib,dir,t}(\varepsilon)$. Note the excellent agreement between the derived values (dotted line) and the directly measured values at 1 eV. The cross section $\sigma_{sc,t}(\varepsilon)$ estimated in this manner was therefore used to obtain a best estimate of $\sigma_{sc,t}(\varepsilon)$ from 0.003 to $\sim$ 1 eV. This, then, with the experimental values available at higher energies, allowed recommended values to be determined for $\sigma_{sc,t}(\varepsilon)$ from 0.003 to 4000 eV (see Ref. [1] and Sec. 2.2.1 of Ch. 4). A subsequent measurement by Lunt *et al.* [28], shown in Fig. 3.2 by the symbol ($\times$), is in excellent agreement with the assessed cross section at the lowest energies. The results of a recent *ab initio* calculation by Isaacs *et al.* [29] are also in general agreement with the recommended cross section.

2.3 Determination of the Total Electron Attachment Cross Section, $\sigma_{a,t}(\varepsilon)$, of Cl_2

The total electron attachment cross section of Cl_2 has been measured by Kurepa and Belić [30] using an electron beam experiment. From a critical assessment of the available swarm data, Christophorou and Olthoff [6] adjusted the cross section of Kurepa and Belić upward by 30% (see Sec. 2.5 of Ch. 6). It is instructive to see how this adjustment was made, for it shows an example of the assessment process itself. It is also an excellent example of the value of absolute electron-swarm measurements for adjusting (normalizing) the often difficult to establish magnitude of electron-beam-determined cross section data. The process essentially utilizes the strength of each experimental procedure: the determination of well resolved relative cross sections by electron beam experiments, and the determination of the absolute magnitude of the cross section by electron swarm experiments.

Let us then first refer to the measurements of McCorkle *et al.* [31] of the rate constant, $k_{a,t}$ of electron attachment to Cl_2 in the buffer gas N_2 over a wide range of values of the density-reduced electric field E/N. Since for these measurements the content of Cl_2 in N_2 was kept very small, the electron energy distribution function in the mixture is virtually the same as that in the pure buffer gas N_2. Furthermore, since the electron energy distribution functions in N_2 can be reliably calculated at the E/N values for which the $k_{a,t}$ measurements were made, the $k_{a,t}(E/N)$ data can be plotted as a function of the mean electron energy $\langle \varepsilon \rangle$, i.e., the quantity $k_{a,t}(\langle \varepsilon \rangle)$ can be accurately determined. The $k_{a,t}(\langle \varepsilon \rangle)$ data determined this way for Cl_2 at room temperature ($\sim$ 298 K) are shown in Fig. 3.3(a).

In Fig. 3.3(a) are also plotted three sets of calculated values for $k_{a,t}(\langle \varepsilon \rangle)$. One was calculated by McCorkle *et al.* [31] using the electron energy distribution functions $f(\varepsilon, E/N)$ in N_2 and the total electron attachment cross section of Kurepa and Belić [30], and the other two were calculated by Kurepa *et al.* [32] and by Chantry [33] using the Kurepa and Belić [30] cross section and a Maxwellian electron energy distribution function for the electron energies. Clearly, the assumption of a Maxwellian distribution function for the electron energies is unrealistic at high E/N as is shown by the large difference in energy dependence between the calculated rates of the last two groups and the measurements at energies above $\sim$ 0.2 eV. In the low-energy region (around the $k_{a,t}(\langle \varepsilon \rangle)$ maximum) all three calculated values of $k_{a,t}(\langle \varepsilon \rangle)$ using the Kurepa and Belić [30] total electron attachment cross section have an energy dependence similar to the directly measured rate constants of McCorkle *et al.* [31]. However, each of the calculated values is lower in magnitude by $\sim$ 30%, suggesting that the Kurepa and Belić electron attachment cross section is lower than its true value by this amount. Hence the swarm-based adjustment of the electron-beam dissociative electron attachment cross section for Cl_2 shown in Fig. 3.3(b). Recent calculations by Fabrikant *et al.* [35] are in agreement with this adjustment.

2.4 Consistency Between the Assessed Cross Sections

It is important to stress the significance of the total electron scattering cross section $\sigma_{sc,t}(\varepsilon)$ in efforts to establish the consistency of the independently assessed cross sections for the various electron collision

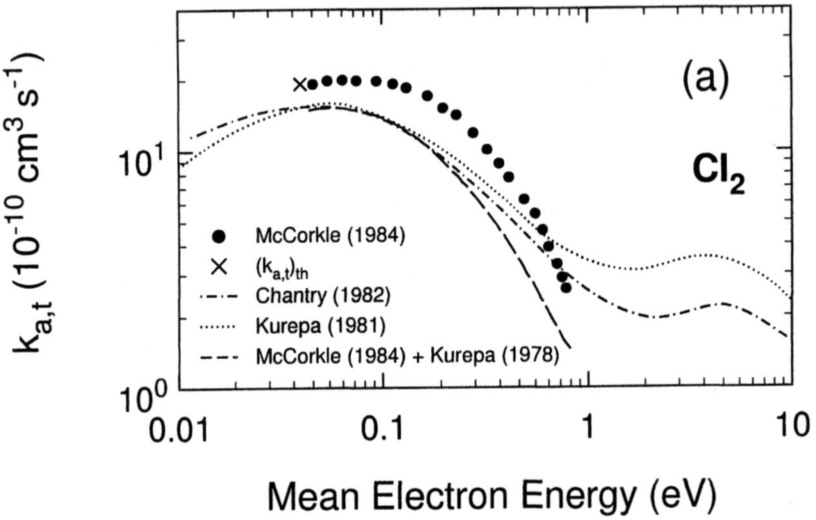

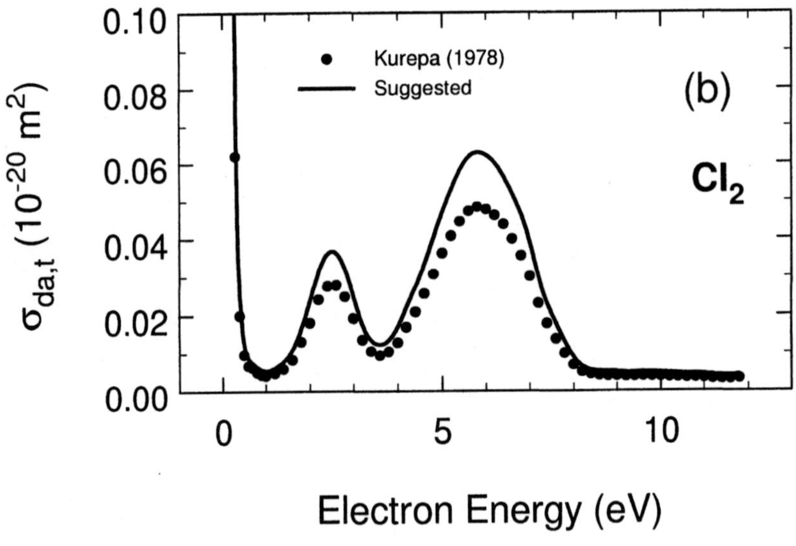

Figure 3.3. (a) Total electron attachment rate constant as a function of the mean electron energy, $k_{a,t}(\langle\varepsilon\rangle)$, of Cl_2 ($T \sim 298$ K): (•) Ref. [31]; (×) $(k_{a,t})_{th}$ [average of the two most recent values [31, 34] of the thermal ($T \sim 298$ K) electron attachment rate constant]; (– · –) Ref. [33]; (· · ·) Ref. [32]; (– –) Ref. [31] (see text and Ref. [6]). (b) Total dissociative electron attachment cross section, $\sigma_{da,t}(\varepsilon)$, of Cl_2 ($T \sim 298$ K): (•) Ref. [30]; (—) cross section of Ref. [30] adjusted upwards by 30% (see text and Ref. [6]).

processes for a given gas. The cross section $\sigma_{\mathrm{sc,t}}(\varepsilon)$ is usually measured with lower uncertainty compared to other cross sections. While the total electron scattering cross section is rarely used directly in plasma models, it provides a way to normalize or to validate the other cross sections: for a given molecule, the sum of the independently assessed cross sections for all possible electron-collision processes should add up to and not exceed the total electron scattering cross section.

This has been shown nicely by Christophorou *et al.* [1] for the CF_4 molecule for which the sum of the independently assessed cross sections nearly equals the independently assessed total electron scattering cross section at all electron energies (see Fig. 4.35 in Ch. 4.) The small dip around 20 eV in the sum of the cross sections may be due to an uncertainty in the recommended values of $\sigma_{\mathrm{e,int}}(\varepsilon)$ for CF_4. In this energy range the two experimental measurements of $\sigma_{\mathrm{e,int}}(\varepsilon)$ [13, 14] differ appreciably. Indeed, if instead of taking the average of the two sets of measurements of $\sigma_{\mathrm{e,int}}(\varepsilon)$ in this energy range, one considers only the values of Boesten *et al.* [13], the dip in the sum of the independently assessed cross sections disappears and the sum agrees well with the assessed values of $\sigma_{\mathrm{sc,t}}(\varepsilon)$ in this energy region as well. Conversely, this suggests that the Boesten *et al.* cross section may be preferred to that of Mann and Linder in this energy range.

3. Deduction of Unavailable Data and Understanding from Assessed Data, New Measurements, and Data Needs

A thoroughand critical assessment of the available data on electron-molecule collisions and related physical and chemical properties for each plasma processing gas often helps deduce needed data, which at the time are not otherwise available. It also enhances our understanding of the dependence of the cross sections of the various electron collision processes on the structural and electronic properties of molecules, suggests the need for new measurements, and identifies needed basic data. Examples of these benefits are given below.

3.1 Deduction of Unavailable Data from Assessed Data

Two examples of cross sections deduced from critically assessed data are given in this section (for other examples see Christophorou *et al.* [1–11], and subsequent chapters).

The first example is the deduced cross section for indirect (resonance enhanced) vibrational excitation of CF_4. Figure 3.4(a) shows the sum of the cross sections for direct vibrational excitation of the two infrared active modes ν_3 and ν_4 of CF_4 as calculated in the Born-dipole approximation [27]. The sum is taken to give the cross section, $\sigma_{\mathrm{vib,dir,t}}(\varepsilon)$, for total direct vibrational excitation of CF_4. This cross section $(- \cdot -)$ is compared in Fig. 3.4(a) with the total inelastic cross section $(\times)$ [which is approximately equal to $\sigma_{\mathrm{vib,dir,t}}(\varepsilon)$ in the energy range below ~ 3 eV] measured by Mann and Linder [14], and with the values (•, ∘) of the total inelastic cross section $\sigma_{\mathrm{inel,t}}(\varepsilon) = [(\sigma_{\mathrm{sc,t}}(\varepsilon) - \sigma_{\mathrm{e,int}}(\varepsilon)]$ deduced from the assessed values of Christophorou *et al.* [1] for $\sigma_{\mathrm{sc,t}}(\varepsilon)$ and $\sigma_{\mathrm{e,int}}(\varepsilon)$. Since electronic excitation is not energetically possible below the electronic excitation threshold at 12.5 eV, the difference between $\sigma_{\mathrm{inel,t}}(\varepsilon)$ and the Born $\sigma_{\mathrm{vib,dir,t}}(\varepsilon)$ gives the cross section, $\sigma_{\mathrm{inel,indir,t}}(\varepsilon)$, for indirect inelastic electron scattering for CF_4. The indirect vibrational excitation cross section $\sigma_{\mathrm{vib,indir,t}}(\varepsilon)$ can then be obtained by subtracting the assessed total dissociative electron attachment cross section $\sigma_{\mathrm{da,t}}(\varepsilon)$ from $\sigma_{\mathrm{inel,indir,t}}(\varepsilon)$ below 12.5 eV, i.e.,

$$\sigma_{\mathrm{vib,indir,t}}(\varepsilon) = [\sigma_{\mathrm{sc,t}}(\varepsilon) - \sigma_{\mathrm{e,int}}(\varepsilon)] - [\sigma_{\mathrm{vib,dir,t}}(\varepsilon) + \sigma_{\mathrm{da,t}}(\varepsilon)]. \qquad (3.1)$$

This is shown by the solid line in Fig. 3.4(a). The data shown in the figure by the open circles were not considered in this process since the recommended values of $\sigma_{\mathrm{e,int}}(\varepsilon)$ are less reliable in the 10 to 20 eV energy range due to a large difference between the two experimental measurements [13, 14] of $\sigma_{\mathrm{e,int}}(\varepsilon)$ for CF_4. The work of Boesten *et al.* [13] does not indicate any contribution to the electron scattering cross section in this energy range due to the existence of negative ion resonances.

The deduced cross section clearly shows that for this molecule indirect vibrational excitation is a dominant inelastic electron scattering process in the energy range from near 6 to nearly 15 eV. It plays a crucial role in the yield of the various discharge products by virtue of its effect on the electron energy distribution function. This effect results not only from the large value of $\sigma_{\mathrm{vib,indir,t}}(\varepsilon)$ near its peak, but also from the large electron energy loss associated with indirect vibrational excitation (as compared to direct vibrational excitation) via the negative ion states of CF_4 in this energy range (for further discussion see Christophorou *et al.* [1], and Bordage *et al.* [36]).

The second example, shown in Figs. 3.4(b) and 3.4(c), refers to the deduced total vibrational excitation cross section, $\sigma_{\mathrm{vib,t}}(\varepsilon)$, and total dissociation cross section into neutral fragments, $\sigma_{\mathrm{dis,neut,t}}(\varepsilon)$, of SF_6. The cross section $\sigma_{\mathrm{vib,t}}(\varepsilon)$ for SF_6 was deduced by Christophorou and

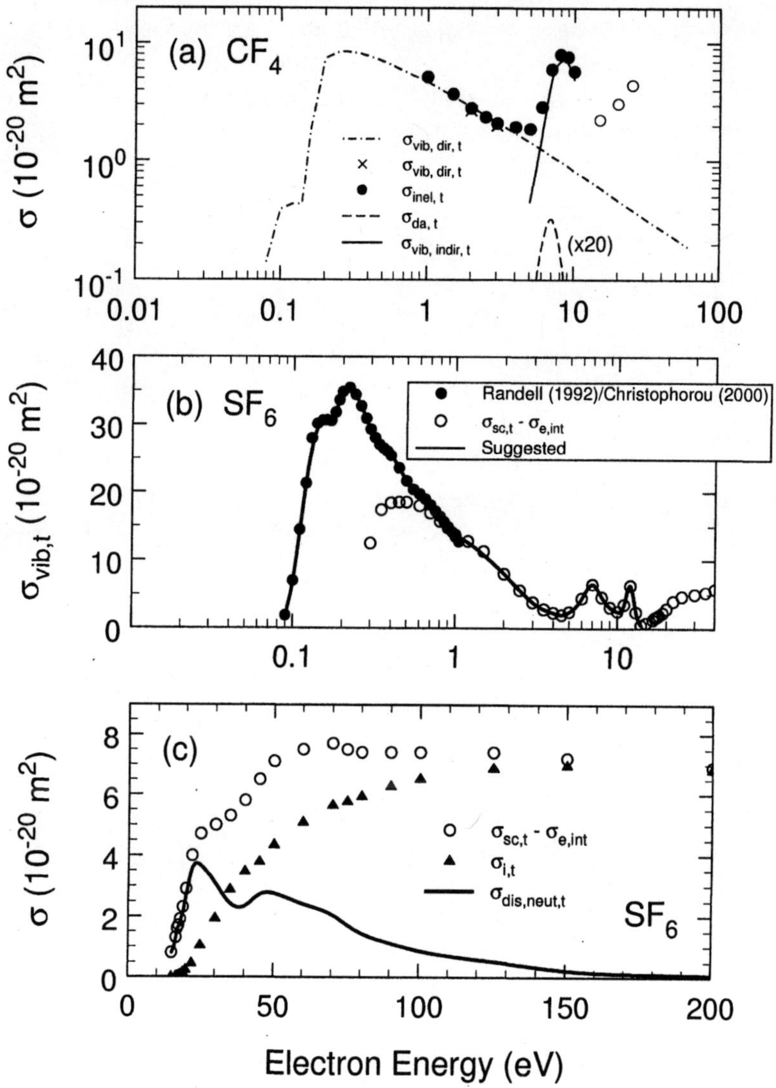

Figure 3.4. (a) Total deduced indirect vibrational excitation cross section $\sigma_{\text{vib,indir,t}}(\varepsilon)$ for CF_4: $(- \cdot -)$ total direct vibrational excitation cross section $\sigma_{\text{vib,dir,t}}(\varepsilon)$ for CF_4 (sum of the Born-approximation calculation [27] for the two infrared active modes of CF_4); $(\times)$ measurements of $\sigma_{\text{vib,dir,t}}(\varepsilon) \approx \sigma_{\text{inel,t}}(\varepsilon)$ of Ref. [14]; $(\bullet, \circ)$ $\sigma_{\text{sc,t}}(\varepsilon) - \sigma_{\text{e,int}}(\varepsilon)$, where the values of these two cross sections are those assessed in Ref. [1]; (- - -) assessed total dissociative electron attachment cross section $\sigma_{\text{da,t}}(\varepsilon)$ in Ref. [1]; (—) deduced $\sigma_{\text{vib,indir,t}}(\varepsilon)$ (see text). (b) Deduced total vibrational excitation cross section, $\sigma_{\text{vib,t}}(\varepsilon)$ for SF_6: $(\bullet)$ as determined in Ref. [8] based on the measurements of Refs. [37–39]; $(\circ)$ difference $\sigma_{\text{sc,t}}(\varepsilon) - \sigma_{\text{e,int}}(\varepsilon)$ of the recommended values [8] of $\sigma_{\text{sc,t}}(\varepsilon)$ and $\sigma_{\text{e,int}}(\varepsilon)$; (—) suggested values. (c) Deduced cross section for $\sigma_{\text{dis,neut,t}}(\varepsilon)$ of SF_6 from other recommended cross sections (see text, Ref. [8], and Ch. 6).

Olthoff [8] over the electron energy range from ~ 0.009 to 14.0 eV using two sources of related assessed cross sections. Between 1 and 14 eV (that is, in the electron energy range where the total electron attachment cross section of SF_6 is very much smaller than the total electron scattering cross section, and electronic excitation is not significant), $\sigma_{\text{vib,t}}(\varepsilon)$ was deduced from

$$\sigma_{\text{vib,t}}(\varepsilon) = \sigma_{\text{sc,t}}(\varepsilon) - \sigma_{\text{e,int}}(\varepsilon) - \sigma_{\text{a,t}}(\varepsilon) \approx \sigma_{\text{sc,t}}(\varepsilon) - \sigma_{\text{e,int}}(\varepsilon) \tag{3.2}$$

using the assessed cross sections for $\sigma_{\text{sc,t}}(\varepsilon)$, $\sigma_{\text{e,int}}(\varepsilon)$, and $\sigma_{\text{a,t}}(\varepsilon)$ for this molecule [8]. These deduced values are shown in Fig. 3.4(b) by the open circles ($\circ$). Below 1 eV, the values of $\sigma_{\text{vib,t}}(\varepsilon)$ for SF_6, shown by the solid points in Fig. 3.4(b), were determined from the relative vibrational excitation cross sections of Randell *et al.* [37] for ν_1, ν_3, $2\nu_1$ and $\nu_1 + \nu_3$. Christophorou and Olthoff [8] argued that since the vibrational energy-loss measurements by Rohr [38] and Randell *et al.* [37] showed that the major contribution to the vibrational excitation of SF_6 by electrons below ~ 3 eV comes from ν_1 (indirect excitation of SF_6 via the SF_6^{-*} resonances) and ν_3 (direct excitation of SF_6 via the dipole interaction) vibrations and their combined excitations, the sum of the relative vibrational excitation cross sections would represent a reasonable estimate of $\sigma_{\text{vib,t}}(\varepsilon)$, if put on an absolute scale. This relative total vibrational excitation cross section determined from the measurements of Randell *et al.* by Christophorou and Olthoff was put on an absolute scale by comparison to the Randell *et al.* elastic electron scattering cross section, which itself was put on an absolute scale by normalizing its relative value at 1 eV to the absolute value of Rohr [39] for $\sigma_{\text{e,int}}(\varepsilon)$ at this energy. Although the two sets of values differ significantly at energies below ~ 0.5 eV, indicating potential errors in the various cross sections employed, in the absence of any direct measurements of $\sigma_{\text{vib,t}}(\varepsilon)$, the values of $\sigma_{\text{vib,t}}(\varepsilon) = \sigma_{\text{sc,t}}(\varepsilon) - \sigma_{\text{e,int}}(\varepsilon)$ between 1 and 14 eV and the normalized data of Randell *et al.* [37] below 1 eV are considered to be a good estimate of $\sigma_{\text{vib,t}}(\varepsilon)$. The solid line in Fig. 3.4(b) represents the suggested values for $\sigma_{\text{vib,t}}(\varepsilon)$ [8].

Above ~ 15 eV, the $\sigma_{\text{vib,t}}(\varepsilon)$ for SF_6 becomes small compared to the other inelastic processes. Indeed, the difference $\sigma_{\text{sc,t}}(\varepsilon) - \sigma_{\text{e,int}}(\varepsilon)$ can be identified with the sum of the cross sections for electron-impact ionization and dissociation of SF_6 and may be used in combination with the assessed total ionization cross section $\sigma_{\text{i,t}}(\varepsilon)$ to deduce values for the total cross section for dissociation of SF_6 into neutral fragments, $\sigma_{\text{dis,neut,t}}(\varepsilon)$, from

$$\sigma_{\text{dis,neut,t}}(\varepsilon) = [\sigma_{\text{sc,t}}(\varepsilon) - \sigma_{\text{e,int}}(\varepsilon)] - \sigma_{\text{i,t}}(\varepsilon) \tag{3.3}$$

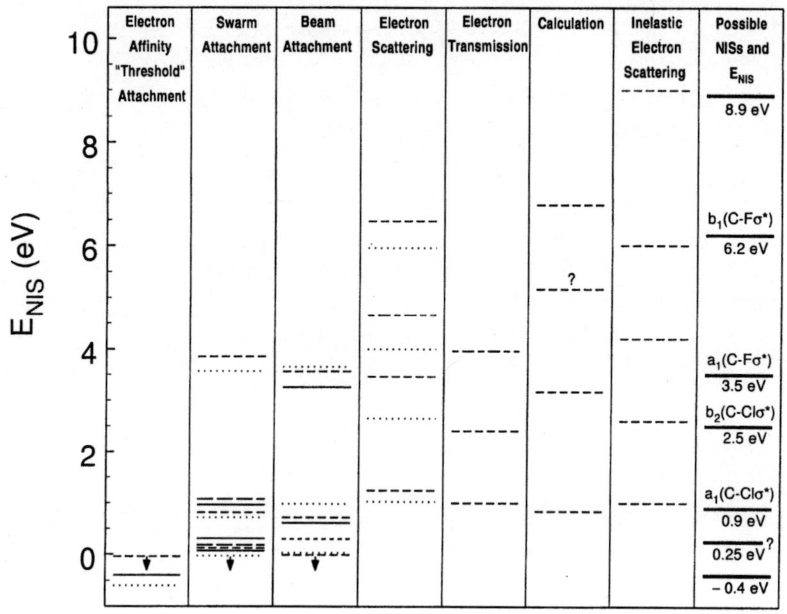

Figure 3.5. Energy positions of the negative ion states of CCl_2F_2 below 10 eV obtained from various experimental and theoretical sources [3]. The last column gives the assessed energies of the negative ion states and their assignments.

(see Fig. 3.4(c), Ref. [8], and Ch. 6). While the uncertainty of the deduced cross section may be large, it represents an acceptable estimate in the absence of direct measurements.

3.2 A Better Understanding from Assessed Data

The critical review and assessment of existing data on electron collisions with the ten plasma processing gases covered in this book has enhanced our understanding of electron-molecule interactions in a number of ways. This is illustrated by the following two examples.

The first example pertains to the enhancement of our knowledge of the negative ion states (NISs) of the molecules under consideration and their effects on the various types of electron collision processes and the respective cross sections. Let us then look at a specific molecule, namely, CCl_2F_2, and the number and energy positions of its negative ion states below $\sim$ 10 eV. The available data on the number and energy positions of the negative ion states of CCl_2F_2 are summarized in Fig. 3.5. The last column in the figure lists the assessed energies of the negative ion states of the CCl_2F_2 molecule. This assessment is based on published data on the electron affinity, electron attachment using the electron swarm method, electron attachment using the electron beam method, electron scattering, electron transmission, information on indirect electron scattering deduced from assessed cross sections, and related calculations [3]. Thus, the lowest negative ion states of CCl_2F_2 have been identified with the average positions as follows: a_1(C–Clσ^*) at −0.4 eV [= −EA(CCl_2F_2)] and 0.9 eV, b_2(C–Clσ^*) at 2.5 eV, a_1(C–Fσ^*) at 3.5 eV, and b_1(C–Fσ^*) at 6.2 eV. The lowest negative ion state, a_1(C–Clσ^*), accounts for both the production of Cl_2^- at −0.4 eV and the production of Cl^- at 0.9 eV. Similar information has been obtained for the other molecules discussed in this book and can be found in the subsequent chapters (see also Refs. [1–11]).

The second example pertains to an increased understanding of the interactions of low-energy electrons with the perfluoroalkane molecules. For instance, the thorough review of the literature used to determine the assessed data for CF_4 reveals a rather simple picture of the collisional behavior of the CF_4 molecule with low-energy electrons:

- Vibrational excitation is the dominant inelastic process below 12.5 eV, i.e., below the threshold for electronic excitation. It is dominated by the excitation of the infrared active modes ν_3 and ν_4 via direct dipole scattering at electron energies below the negative ion resonance region, and via indirect electron scattering in the resonance region (approximately 6–20 eV).

- All electronic excitations of CF_4 lead to dissociation.

- Dissociation of CF_4 into neutral fragments begins near 12.5 eV, dominates the inelastic electron scattering process until ionization sets in, and then progressively yields to dissociative ionization.

- Cross sections for positive ion-negative ion pair formation, and multiple ionization, are generally smaller than those for single ionization in the low-energy range of interest to plasma reactor modeling.

Similarly, a systematic review of the assessed data for all three perfluoroalkane molecules (CF_4, C_2F_6, C_3F_8) reveals:

- Large direct vibrational excitation cross sections at low electron energies, and very large indirect vibrational excitation cross sections in the energy regions of the negative ion resonances.

- Maxima in the various cross sections at the location of negative ion states.

- Increase of the electron scattering cross sections with molecular polarizability.
- Existence of a Ramsauer-Townsend (R-T) minimum in the total scattering, elastic scattering, and momentum transfer cross sections below ~ 0.2 eV.
- Dissociation of all electronic states into charged and/or neutral fragments (a property largely shared also by the other molecules discussed in this work).

3.3 Determination of Data Needs and New Related Measurements and Calculations

The data assessment for each molecule naturally identifies gaps in the database. In general, there are two types of data needs: (i) new data to replace existing data judged to be incorrect, and (ii) data that are needed, but are not available.

In connection with the first kind of data needs, we present here as an example the cross section for electron-impact dissociation of molecules into neutral fragments. At the time the first CF_4 review was performed [1], there was only one direct measurement [40] (revised by Sugai *et al.* [41]) of the cross section for the production of CF_3, CF_2, and CF radicals by electron impact on CF_4. In the absence of any other data at the time, the sum of the partial cross sections for these three radicals was designated [1] as the recommended value of $\sigma_{\text{dis,neut,t}}(\varepsilon)$. This cross section is plotted (short dashed curve line) in Fig. 3.6. For comparison, the total electron-impact dissociation cross section, $\sigma_{\text{dis,t}}(\varepsilon)$, [42] and the total electron-impact ionization cross section, $\sigma_{\text{i,t}}(\varepsilon)$, [1] are also plotted in the figure. An estimated value (dotted line) of $\sigma_{\text{dis,neut,t}}(\varepsilon)$ deduced from $\sigma_{\text{dis,t}}(\varepsilon) - \sigma_{\text{i,t}}(\varepsilon)$ for energies < 70 eV [using the presently recommended values of $\sigma_{\text{i,t}}(\varepsilon)$ derived in Sec. 2.3 of Ch. 4], is also shown in Fig. 3.6. Clearly, the measurements of Sugai *et al.* [41] are inconsistent with the recommended values of $\sigma_{\text{dis,t}}(\varepsilon)$ and $\sigma_{\text{i,t}}(\varepsilon)$. The need for more accurate measurements of the cross section for this important process was pointed out in Ref. [1] and led to new measurements by Mi and Bonham [43] and Motlagh and Moore [44]. The results of both of these groups are also shown in Fig. 3.6, and confirm the conclusions of the initial assessment, namely that the cross section values of Sugai *et al.* for $\sigma_{\text{dis,t}}(\varepsilon)$ are much too small. The new suggested cross section $\sigma_{\text{dis,neut,t}}(\varepsilon)$ is shown in the figure by the solid line. The fairly large discrepancy remaining between the values of Motlagh and Moore and the values deduced from $\sigma_{\text{dis,t}}(\varepsilon) - \sigma_{\text{i,t}}(\varepsilon)$ requires further investigation.

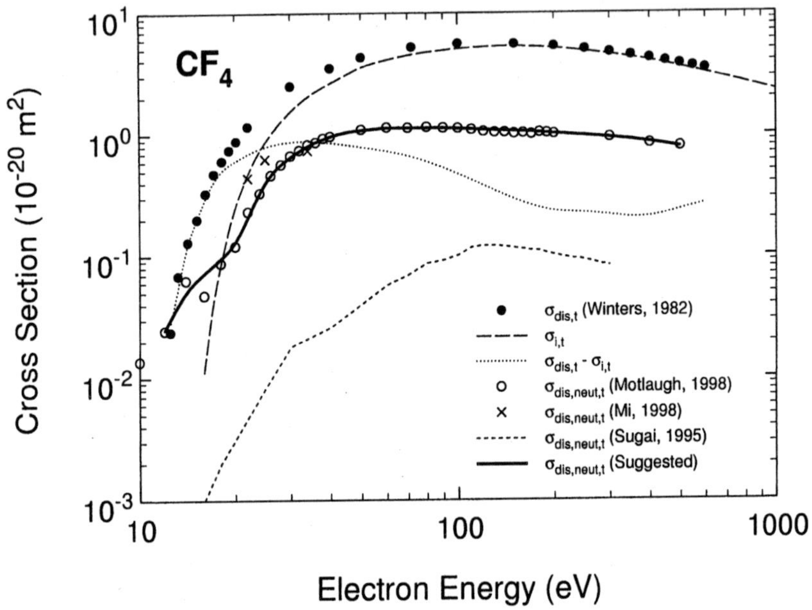

Figure 3.6. Measured cross sections for electron-impact dissociation $\sigma_{\mathrm{dis,neut,t}}(\varepsilon)$ of CF_4 into neutrals: (- - -) Ref. [41]; (o) Ref. [44]; (×) Ref. [43]; (—) suggested $\sigma_{\mathrm{dis,neut,t}}(\varepsilon)$. For comparison, the total dissociation cross section $\sigma_{\mathrm{dis,t}}(\varepsilon)$ (•) from Ref. [42], the assessed $\sigma_{\mathrm{i,t}}(\varepsilon)$ (– –) from Sec. 2.3 of Ch. 4, and $\sigma_{\mathrm{dis,t}}(\varepsilon) - \sigma_{\mathrm{i,t}}(\varepsilon)$ (· · ·) are also plotted in the figure.

In connection with the second kind of data needs, we point to the situation with CHF_3, a plasma processing gas at times used in place of CF_4. In this instance, when the review and assessment work for CHF_3 was begun about six years ago by Christophorou *et al.* [2], there were no measurements of most of the main electron scattering cross sections or electron transport coefficients. The cross section of Sugai *et al.* [41] for electron-impact dissociation into neutral fragments was judged to be incorrect, and there were no absolute cross section measurements for dissociative electron attachment. Partly as a consequence of the discussions during the review and assessment process and the review paper itself [2], measurements have since been made of the cross section for total electron scattering [45–48], partial and total ionization cross sections [49, 50], dissociation into CHF_2 and CF_3 neutrals [44], elastic differential electron scattering cross section [51, 52], electron drift velocity [53–56], electron attachment coefficient [53, 54, 57], electron-impact ionization coefficient [55], and electron diffusion coefficients [56]. In addition, a study has been made of the ion chemistry in CHF_3 using

Fourier-transform mass spectrometry [58], and cross sections have been calculated for a number of electron collision processes using various types of calculations [59–62]. This impressive amount of new knowledge underlines the significance of the assessment process in identifying basic data needs and triggering new investigations. These new electron collision data for this important molecule are presented and discussed in Ch. 5.

In terms of determining remaining data needs, the state of our knowledge regarding electron-collision data for the ten gases under consideration in this work is summarized in Table 3.1. With the sole exception of CF_4, and possibly also SF_6, the database for the other eight gases needs improvement. In general, the cross sections for total electron scattering, elastic integral, total ionization, total dissociation, and total electron attachment are better known than the cross sections for momentum transfer, vibrational excitation, partial ionization, multiple ionization, ion-pair formation, and dissociation into neutrals. Existing data for electron transport and electron attachment and ionization rate coefficients are reliably known for CF_4, C_2F_6, C_3F_8, SF_6, and to some extent also for CCl_2F_2, CHF_3, and c-C_4F_8. However, such knowledge is meager for Cl_2, CF_3I, and BCl_3. For some of the molecules, the coefficients although accurately known in a restricted E/N range, are not known or are poorly known in other or wider E/N ranges. The reader is referred to subsequent chapters where the available knowledge on each of these molecules is assessed and discussed. Also, see Christophorou and Olthoff [64] for information and needs for cross sections/coefficients on electron interactions with excited molecules.

4. Dissemination and Updating of the Database

For the review and assessment process that had been performed by us at NIST, the end product has been a comprehensive article for each gas published in the Journal of Physical and Chemical Reference Data. These critical reviews contain the best efforts of the authors to provide a complete, yet concise, review of data relevant to electron collision cross sections and coefficients for these gases [1–11].

A summary of relevant data, containing primarily the recommended and suggested cross section data and coefficients (in downloadable form) for the ten plasma processing gases studied, is available via the Internet at `http://www.eeel.nist.gov/811/refdata`.

The data on CF_4, C_2F_6, C_3F_8, and CHF_3 were updated previously by Christophorou and Olthoff [7]. The data for all ten molecules are further updated in this book.

Table 3.1. The state of present knowledge of electron collision data for CF_4, C_2F_6, C_3F_8, CHF_3, CCl_2F_2, Cl_2, SF_6, CF_3I, c-C_4F_8, and BCl_3.

Cross section/ Coefficient	CF_4	C_2F_6	C_3F_8	CHF_3	CCl_2F_2	Cl_2	SF_6	CF_3I	c-C_4F_8	BCl_3
$\sigma_{sc,t}(\varepsilon)$	R[a]	R	R	R	S[b]	S	R	S	S	none
$\sigma_{m}(\varepsilon)$	R	S	S	C[c]	C	C	S	C	none	C
$\sigma_{e,diff}(\varepsilon)$	M[d]/C	M/C	M	M/C	M/C	none	M/C	M	M/C	C
$\sigma_{e,int}(\varepsilon)$	R	S	S	C	S	S	S	none	none	C
$\sigma_{vib,indir,t}(\varepsilon)$	D[e]	–[f]	–[f]	none	–[f]	D	D	none	–[f]	none
$\sigma_{vib,dir,t}(\varepsilon)$	M/C	–[f]	–[f]	none	C	–[f]	D	none	–[f]	C
$\sigma_{i,t}(\varepsilon)$	R	R	R	R	S	S	R	S	S	S
$\sigma_{i,part}(\varepsilon)$	S	S	S	S	S	M	M	S	S	S
$\sigma_{i,pair}(\varepsilon)$	M	none	none	M	none	S	none	none	none	none
$\sigma_{i,mult}(\varepsilon)$	S	none	none	none	S	none	none	none	none	none
$\sigma_{dis,t}(\varepsilon)$	R	R	R	R	none	–[f]	–[f]	none	none	C
$\sigma_{dis,neut,t}(\varepsilon)$	S	C	M	M	none	M/C	D	none	M[g]	none
$\sigma_{a,t}(\varepsilon)$	S	R	S	none	S	S	R	S	S	M
$\alpha/N(E/N)$	R	R	R	–[f]	R	S	R	none	S	none
$\eta/N(E/N)$	R	R	R	–[f]	R	S	R	none	M	none
$(\alpha-\eta)/N(E/N)$	R	R	S[h]	S	S	S	R	none	M	none
$k_{a,t}(\langle\varepsilon\rangle)$	R	R	R	–[f]	R	S	R	S	R	M
$w(E/N)$	R	S	S	R	S	M	R/S	none	S	none
$D_T/\mu(E/N)$	S	R	S	none	S	M	S	none	S	none

[a]R = Recommended.
[b]S = Suggested.
[c]C = Calculated.
[d]M = Measured.
[e]D = Deduced.
[f]See discussion in appropriate book chapter.
[g]Sum of the reported partial dissociation cross sections [10].
[h]Deduced from the recommended values of $\alpha/N(E/N)$ and the density-independent values of $\eta/N(E/N)$ [63].

5. References

[1] L. G. Christophorou, J. K. Olthoff, and M. V. V. S. Rao, J. Phys. Chem. Ref. Data **25**, 1341 (1996).

[2] L. G. Christophorou, J. K. Olthoff, and M. V. V. S. Rao, J. Phys. Chem. Ref. Data **26**, 1 (1997).

[3] L. G. Christophorou, J. K. Olthoff, and Y. Wang, J. Phys. Chem. Ref. Data **26**, 1205 (1997).

[4] L. G. Christophorou and J. K. Olthoff, J. Phys. Chem. Ref. Data **27**, 1 (1998).

[5] L. G. Christophorou and J. K. Olthoff, J. Phys. Chem. Ref. Data **27**, 889 (1998).

[6] L. G. Christophorou and J. K. Olthoff, J. Phys. Chem. Ref. Data **28**, 131 (1999).

[7] L. G. Christophorou and J. K. Olthoff, J. Phys. Chem. Ref. Data **28**, 967 (1999).

[8] L. G. Christophorou and J. K. Olthoff, J. Phys. Chem. Ref. Data **29**, 267 (2000).

[9] L. G. Christophorou and J. K. Olthoff, J. Phys. Chem. Ref. Data **29**, 553 (2000).

[10] L. G. Christophorou and J. K. Olthoff, J. Phys. Chem. Ref. Data **30**, 449 (2001).

[11] L. G. Christophorou and J. K. Olthoff, J. Phys. Chem. Ref. Data **31**, 971 (2002).

[12] T. Sakae, S. Sumiyoshi, E. Murakami, Y. Matsumoto, K. Ishibashi, and A. Katase, J. Phys. B **22**, 1385 (1989)

[13] L. Boesten, H. Tanaka, A. Kobayashi, M. A. Dillon, and M. Kimura, J. Phys. B **25**, 1607 (1992).

[14] A. Mann and F. Linder, J. Phys. B **25**, 533 (1992).

[15] D. Raj, J. Phys. B **24**, L431 (1991).

[16] W. M. Huo, Phys. Rev. A **38**, 3303 (1988).

[17] C. Winstead, Q. Sun, and V. McKoy, J. Chem. Phys. **98**, 1105, (1993).

[18] M. T. do N. Varella, C. Winstead, V. McKoy, M. Kitajima, and H. Tanaka, Phys. Rev. A **65**, 022702 (2002).

[19] M. Hayashi, in *Swarm Studies and Inelastic Electron-Molecule Collisions*, L. C. Pitchford, B. V. McKoy, A. Chutjian, and S. Trajmar (Eds.), New York: Springer, 1987, p. 167.

[20] K. Masek, L. Laska, R. D'Agostino, and F. Cramarossa, Contrib. Plasma Phys. **27**, 15 (1987).

[21] Y. Nakamura, in *Gaseous Electronics and Their Applications*, R. W. Crompton, M. Hayashi, D. E. Boyd, and T. Makabe (Eds.), KTK Scientific, Tokyo, Japan, 1991, p. 178.

[22] B. Stefanov, N. Popkirova, and L. Zarkova, J. Phys. B **21**, 3989 (1988).

[23] M. G. Curtis, I. C. Walker, and K. J. Mathieson, J. Phys. D **21**, 1271 (1988).

[24] M.-C. Bordage, P. Ségur, and A. Chouki, J. Appl. Phys. **80**, 1325 (1996).

[25] R. K. Jones, J. Chem. Phys. **84**, 813 (1986).

[26] C. Szmytkowski, A. M. Krzysztofowicz, P. Janicki, and L. Rosenthal, Chem. Phys. Lett. **199**, 191 (1992).

[27] R. A. Bonham, Jpn. J. Appl. Phys. **33**, 4157 (1994).

[28] S. L. Lunt, J. Randell, J.-P. Ziesel, G. Mrotzek, and D. Field, J. Phys. B **31**, 4225 (1998).

[29] W. A. Isaacs, C. W. McCurdy, and T. N. Rescigno, Phys. Rev. A **58**, 693 (1987).

[30] M. V. Kurepa and D. S. Belić , J. Phys. B **11**, 3719 (1978).

[31] D. L. McCorkle, A. A. Christodoulides, and L. G. Christophorou, Chem. Phys. Lett. **109**, 276 (1984).

[32] M. V. Kurepa, D. S. Babić , and D. S. Belić , Chem. Phys. **59**, 125 (1981).

[33] P. J. Chantry, in *Applied Atomic Collision Physics, Vol. 3, Gas Lasers*, H. S. W. Massey, E. W. McDaniel, and B. Bederson (Eds.), Academic, New York, 1982, p. 35.

[34] D. Smith, N. G. Adams, and E. Alge, J. Phys. B **17**, 461 (1984).

[35] I. I. Fabrikant, T. Leininger, and F. X. Gadéa, J. Phys. B **33**, 4575 (2000).

[36] M.-C. Bordage, P. Ségur, L. G. Christophorou, and J. K. Olthoff, J. Appl. Phys. **86**, 3558 (1999).

[37] J. Randell, D. Field, S. L. Lunt, G. Mrotzek, and J. P. Ziesel, J. Phys. B **25**, 2899 (1992).

[38] K. Rohr, J. Phys. B **10**, 1175 (1977).

[39] K. Rohr, J. Phys. B **12**, L185 (1979).

[40] T. Nakano and H. Sugai, Jpn. J. Appl. Phys. **31**, 2919 (1992).

[41] H. Sugai, H. Toyoda, T. Nakano, and M. Goto, Contrib. Plasma Phys. **35**, 415 (1995).

[42] H. F. Winters and M. Inokuti, Phys. Rev. A **25**, 1420 (1982).

[43] L. Mi and R. A. Bonham, J. Chem. Phys. **108**, 1910 (1998):

[44] S. Motlagh and J. H. Moore, J. Chem. Phys. **109**, 432 (1998).

[45] J. E. Sanabia, G. D. Cooper, J. A. Tossell, and J. H. Moore, J. Chem. Phys. **108**, 389 (1998).

[46] O. Sueoka, H. Takaki, A. Hamada, H. Sato, and M. Kimura, Chem. Phys. Lett. **288**, 124 (1998).

[47] I. Iga, P. R. Pinto, and M. G. P. Homem, in *Proceedings International Symposium on Electron-Molecule Collisions and Swarms*, Y. Hatano, H. Tanaka, and N. Kouchi (Eds.), Tokyo, Japan, July 1999, p. 123.

[48] H. Nishimura, K. Okuda, and Y. Nakamura, in *Proceedings International Symposium on Electron-Molecule Collisions and Swarms*, I. Fabrikant, G. Gallup, and P. Burrow (Eds.), Lincoln, NE, July 2001, p. 105.

[49] I. Iga, I. P. Sanches, S. K. Srivastava, and M. Mangan, Int. J. Mass Spectrom. **208**, 159 (2001).

[50] I. Torres, R. Martinez, and F. Castaño, J. Phys. B **35**, 2423 (2002).

[51] H. Tanaka, T. Masai, M. Kimura, T. Nishimura, and Y. Itikawa, Phys. Rev. A **56**, R3338 (1997).

[52] H. Tanaka, private communication, 1998.

[53] Y. Wang, L. G. Christophorou, J. K. Olthoff, and J. K. Vergrugge, Chem. Phys. Lett. **302**, 303 (1999).

[54] J. D. Clark, B. W. Wright, J. D. Wrbanek, and A. Garscadden, in *Gaseous Dielectrics IX*, L. G. Christophorou and J. K. Olthoff (Eds.), Plenum, New York, 1998, p. 23.

[55] J. de Urquijo, E. Basurto, C. Cisneros, and I. Alvarez, in *Proceedings XXIV International Conference on Phenomena in Ionized Gases*, P. Pisarczyk, T. Pisarczyl, and J. Wolowski (Eds.), Warsaw, Poland, July 1999, Contributed Papers, Vol. 1, p. 73.

[56] M. Mitani and Y. Nakamura, in *Proceedings XXV International Conference on Phenomena in Ionized Gases*, T. Goto (Ed.), Nagoya, Japan, July 2001, Vol. 3, p. 19.

[57] G. K. Jarvis, C. A. Mayhew, L. Singleton, and S. M. Spyrou, Int. J. Mass Spectrom. and Ion Processes **164**, 207 (1997).

[58] C. Q. Jiao, R. Nagpal, and P. D. Haaland, Chem. Phys. Lett. **269**, 117 (1997).

[59] R. B. Diniz, M. A. P. Lima, and F. J. da Paixão, J. Phys. B **32**, L539 (1999).

[60] M. J. Kushner and D. Zhang, J. Appl. Phys. **88**, 3231 (2000).

[61] W. L. Morgan, C. Winstead, and V. McKoy, J. Appl. Phys. **90**, 2009 (2001).

[62] M. C. Bordage and P. Ségur, in *Proceedings of the XXV International Conference on Phenomena in Ionized Gases*, T. Goto (Ed.), Nagoya, Japan, July 2001, Vol. 3, p. 33.

[63] L. G. Christophorou and J. K. Olthoff, Adv. At. Mol. Opt. Phys. **44**, 59 (2000).

[64] L. G. Christophorou and J. K. Olthoff, Adv. At. Mol. Opt. Phys. **44**, 155 (2000).

Chapter 4

ELECTRON INTERACTIONS WITH CF_4, C_2F_6, AND C_3F_8

1. Introduction

In this chapter the current state of our knowledge on the cross sections and the rate coefficients for collisional interactions of the plasma processing gases CF_4 [1], C_2F_6 [2], and C_3F_8 [3] with electrons is reviewed, assessed, and discussed. As we have indicated previously, such knowledge is necessary to assess the behavior of these gases in the atmosphere and in their many applications, especially in the semiconductor industry. For instance, the testing of theoretical models for plasma reactors requires knowledge of collision cross sections over a wide energy range below approximately 100 eV. These are crucial for investigating, understanding, and characterizing the gas-phase reactions in the plasma, and estimating the fluxes of species that are ultimately responsible for the multitude of surface interactions.

The collisional interactions of electrons with CF_4, C_2F_6, and C_3F_8 under controlled single and multiple collision conditions have been studied by many groups and in this chapter we assess, synthesize, and present this knowledge comprehensively. We also refer to interactions of these gases with photons which are relevant to the present discussion and of interest to applications and to the environment. The symbols and units of the collision cross sections, coefficients, and rate constants used to quantify the various processes which result from the collisions of low-energy electrons with the CF_4, C_2F_6, and C_3F_8 molecules are defined in Table 1.3 of Ch. 1.

One goal of this work is to reach a conclusion as to the most reliable available data for the various electron collision processes in CF_4, C_2F_6, and C_3F_8, and to recommend or suggest values for each type of cross section and coefficient when possible. The recommended or suggested values are derived from fits to the most reliable data that are available, based on the criteria discussed in Chapter 3. It is important to note that the recommended or suggested cross sections are based predominantly on experimental measurements and are, thus, different from model-dependent cross section sets such as those of Hayashi [4], Nakamura [5], and Bordage *et al.* [6].

2. Electron Interactions with CF_4

Carbon tetrafluoride (CF_4) is a man-made gas with wide technological applications: plasma etching in the semiconductor industry [7–11], pulse power switching [12–14], gaseous dielectrics [15–17], particle detectors [18–21], and a host of other areas in plasma and space sciences, gas discharges, and atmospheric physics and chemistry. In non-equilibrium plasmas used for plasma assisted material processing applications, CF_4 is one of the most widely used components of feed gas mixtures [22–25]. It serves as a source of the reactive species (ions, neutrals, radicals) that are largely responsible for surface reactions in various etching and deposition applications. The CF_4 molecule is attractive as a feed gas component because it is relatively inert in its electronic ground state and because it has no stable excited electronic states. The CF_4^+ parent ion is also unstable or metastable both in its ground and excited electronic states. As a consequence of these properties, in a plasma environment, the CF_4 molecule is an ideal source for a variety of reactive neutral and ionic fragments formed in either ground or excited states, and especially neutral F atoms which is a desirable active species in etching processes.

Carbon tetrafluoride is, unfortunately, a greenhouse gas with a high global warming potential [26, 27]. Its half-life in the atmosphere is greater than 50 000 years [26] and its global warming potential over a one-hundred year period is 6 300 with reference to the absolute global warming potential for CO_2 [26]. The CF_4 molecule is not expected to cause ozone depletion in the stratosphere because the catalytic destruction of stratospheric ozone by free fluorine atoms formed in the photodissociation of CF_4 is negligible [26, 28]. Recently, electron and ion reactions have been shown to reduce the atmospheric lifetimes of fully fluorinated compounds [29, 30].

In the aforementioned areas and uses, it is necessary to have accurate information about the fundamental properties and reactions of CF_4, particularly its interactions with low energy electrons ($\lesssim$ 100 eV). Assessed information is presented in this section on: (i) cross sections for electron scattering (total, momentum, elastic differential, elastic integral, inelastic), electron-impact ionization (total, partial, multiple, dissociative), electron-impact dissociation (total, and for dissociative excitation), and electron attachment (total, and for specific anions); (ii) coefficients for electron transport (electron drift velocity, transverse and longitudinal electron diffusion coefficients), electron attachment, and electron impact-ionization; and (iii) cross section sets derived from analyses of electron transport data. The limited ionization data on CF_x (x = 1–3) radicals are also presented, and references are made to measurements of electron transport properties of CF_4 gas mixtures. Based upon the assessment of published experimental data, recommended or suggested values for various cross sections and coefficients are generated which are presented in graphical and tabular form.

2.1 Electronic and Molecular Structure of CF_4

The CF_4 molecule is a tetrahedral and has spherical top structure. It does not have a dipole or a quadrupole moment [31], and its electric dipole polarizability is small. Beran and Kevan [32] listed 27.3×10^{-25} cm^3 and 29.3×10^{-25} cm^3, and Au *et al.* [33] reported 28.24×10^{-25} and 28.42×10^{-25} cm^3 for the electric dipole polarizability of CF_4. It belongs to the T_d point group and the ground-state configuration [34, 35] of its outer valence shell is $[\ldots(4a_1)^2\,(3t_2)^6\,(1e)^4\,(4t_2)^6\,(1t_1)^6]\,{}^1A_1$.

There have been many studies of the electronic and molecular structure of CF_4 (see, for example, Kuroki *et al.* [34] and Robin [36] and literature cited therein; see, also, a discussion by Boesten *et al.* [37] concerning the symmetries of the various unoccupied orbitals of the CF_4 molecule). A number of energy-loss studies have been published [34, 38, 39], and in Fig. 4.1 is shown the electron energy-loss spectrum from a recent study [34] obtained using incident electrons of 200 eV energy at scattering angles of 3° and 9°. All excitations from the outer valence shell appear to follow a Rydberg pattern and have been so classified [34].

The energy-loss regime shown in Fig. 4.1 covers the entire range of transitions from the outer valence shell. Excitation energies, type of orbital transitions, and ionization potential values [34, 38–40] are given in Table 4.1 (see also later in this section). In Fig. 4.1 the angular dependence exhibited by the peaks labelled 1, 4, and 11 in the electron energy-loss spectrum at 12.56, 14.84, and 21.63 eV, respectively, correspond to optically forbidden transitions [34] (see also Iga *et al.* [44]). The excitation energies determined in the electron energy-loss spectroscopy (EELS) study [34] are compared in Table 4.1 with those obtained from two earlier EELS measurements [38, 39]. The energy-loss results are

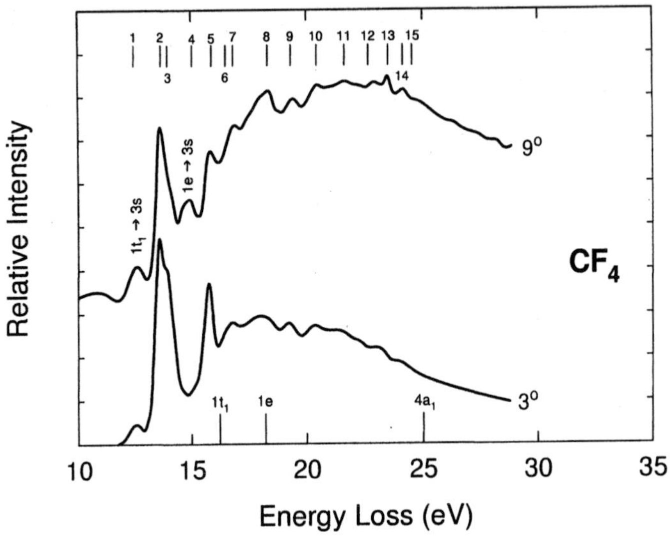

Figure 4.1. Electron energy-loss spectrum of CF_4 obtained by Kuroki *et al.* [34] using 200 eV incident energy electrons and scattering angles equal to 3° and 9°. See the text and Table 4.1 for discussion and explanation of symbols.

consistent with each other, and with photoabsorption data [41, 42, 45, 46].

Figure 4.2 shows the photoabsorption cross section data and also the photoabsorption cross sections $\sigma_{\text{pa}}(\varepsilon)$ as a function of the photon energy ε obtained from measurements of the differential oscillator strength, $df/d\varepsilon$, for the valence-shell electronic excitation [33, 47]. The $df/d\varepsilon$ data were converted to $\sigma_{\text{pa}}(\varepsilon)$ via the relationship [33]

$$\sigma_{\text{pa}}(10^{-18}\,\text{cm}^2) = 109.75\,df/d\varepsilon\ (\text{eV}^{-1}). \tag{4.1}$$

The $\sigma_{\text{pa}}(\varepsilon)$ obtained from the $df/d\varepsilon$ measurements are more detailed and cover a much broader energy range than the photoabsorption measurements. All measurements are consistent in showing the large photoabsorption maximum around 20 eV. The significance of the data in Fig. 4.2 within the purpose of the present work is that $\sigma_{\text{pa}}(\varepsilon)$ represents the total electron-impact excitation cross section for optically-allowed transitions of CF_4 and reflects the distribution of electron-energy losses for sufficiently energetic electrons.

All electronically excited states of CF_4 dissociate or predissociate with high probability [49, 50] and this is consistent with the absence of optical emission from the CF_4 molecule [36] itself. The parent molecular ion CF_4^+ also must be unstable or metastable (lifetime $< 10\ \mu$s) since

Table 4.1. Excitation energies, types of transition, and ionization potentials (IP) of CF_4 [34].

	Excitation energy (eV)[b,c]				
No.[a]	EELS Ref. [34]	EELS Ref. [38]	EELS Ref. [39]	Orbital transition[d]	IP (eV) Refs. [39, 40]
1	12.56	12.69	12.51	$1t_1 \to 3s$	16.20
2	13.60	13.67	13.59	$1t_1 \to 3p$	
3	13.94	13.96	13.89	$4t_2 \to 3s$	17.40
4	14.84	14.71		$1e \to 3s$	18.50
5	15.82	15.86	15.81	$4t_2 \to 4s, 3d$ $1e \to 3p$	
6	16.56	16.53		$4t_2 \to 5s, 4d$ $1e \to 3d$	
7	16.87	16.88	16.86	$1e \to 4s$, 3d	
8	18.39	18.0	18.01	$3t_2 \to 3s$	22.12
9	19.43	19.45	19.42	$3t_2 \to 3p$	
10	20.48	20.45	20.53	$3t_2 \to 4s, 3d$	
11	21.63	(21.55)		$4a_1 \to 3s$	25.1
12	22.68	22.78		$4a_1 \to 3p$	
13	23.55			$4a_1 \to 3d$	
14	24.21			$4a_1 \to 4d$	
15	24.7			$4a_1 \to 5d$	

[a]Numbers in Column 1 identify the transitions in this table and in Fig. 4.1.
[b]See Refs. [41, 42] for relevant information obtained from photoabsorption and photoionization studies.
[c]The excitation maxima in the threshold electron spectra of Ref. [43] are consistent with the EELS values.
[d]Orbital transition assignments are from Refs. [34, 38, 39].

it has not been observed [51–53]. Weak signals of metastable CF_4^+ ions have however been reported in some mass spectrometric studies [54–56] (metastable CF_4^+ ions have also been detected in collisions of electrons with CF_4 clusters [57]). It can be seen from Table 4.1 that the threshold for electronic excitation is rather high — near 12.6 eV. Consequently, below this energy, collisions of electrons with the CF_4 molecule lead to elastic scattering, vibrational excitation, and dissociative electron attachment. Above 12.6 eV, electronic excitation becomes energetically possible, and the dissociation of the CF_4 molecule into neutral and/or charged fragments becomes significant (see Secs. 2.3 and 2.4 of Ch. 4).

The CF_4 molecule has four fundamental vibrational modes [58]: the symmetric stretch $h\nu_1$ (0.112 eV) [this mode is singly degenerate], the

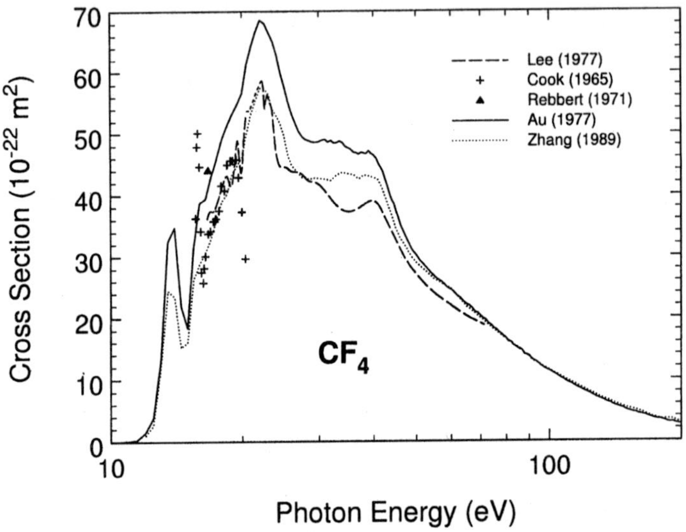

Figure 4.2. Comparison of photoabsorption cross sections of CF_4 [48]. Photoabsorption data: (– –) Ref. [45]; (+) Ref. [41]; (▲) Ref. [46]. Differential oscillator strength data: (—) Ref. [33]; (···) Ref. [47].

symmetric bend $h\nu_2$ (0.054 eV) [this mode is doubly degenerate], the asymmetric stretch $h\nu_3$ (0.157 eV) [this mode is triply degenerate], and the asymmetric bend $h\nu_4$ (0.078 eV) [this mode is triply degenerate]. Most electron collision studies cannot resolve the ν_2 and ν_4 levels and often two vibrational cross sections are obtained for the stretching modes ($\sigma_{\nu(1,3)}$) and for the bending modes ($\sigma_{\nu(2,4)}$) with the former having a statistical weight of 4 and the latter having a statistical weight of 5 [4, 5, 59] (see Sec. 2.2.5 of Ch. 4). The excitation energy of the ν_3 mode almost coincides with the deep Ramsauer-Townsend minimum in the momentum and total electron scattering cross sections at 0.16 eV [60]. Threshold-electron excitation studies have indicated strong vibrational excitation by electron impact below 2.0 eV [61] (see Sec. 2.2.5 of Ch. 4). Rotational excitation cross sections are expected to be very small for the CF_4 molecule due to the absence of a dipole and a quadrupole moment.

Carbon tetrafluoride is a weak electronegative gas. Electron attachment to the CF_4 molecule occurs mainly in the 6 to 8 eV range via two negative ion resonances; one, at 6.8 eV, associated with the ground state of CF_4^- producing F^- and CF_3^-, and another, at 7.6 eV, associated with the first electronically excited CF_4^{*-} producing only F^- (see Refs. [62–64] and Sec. 2.5 of Ch. 4). Besides the electron attachment studies, electron scattering experiments and calculations confirm the locations of these two negative ion resonances and assign the T_2 symmetry to the 6.8 eV resonance [65–76] (see Table 4.2). A number of other negative ion resonances have been identified with various degrees of certainty at higher energies (see Table 4.2 and subsequent sections). These negative ion resonances play a crucial role in electron-impact induced (indirect) vibrational excitation of CF_4 (Sec. 2.2.5 of Ch. 4).

The parent negative ion CF_4^- has not been observed in the gas phase. Two theoretical calculations [77, 78] give a value of −0.7 eV for the electron affinity (EA) of the CF_4 molecule. (See Ref. [79] for EA values of the F atom and radicals formed by electron impact on CF_4). There have been, however, studies reporting the observation of the CF_4^- parent anion in van der Waals aggregates (clusters) of CF_4 [80, 81]. This does not mean that the EA of CF_4 is positive, but it rather indicates that in CF_4 clusters the CF_4^{-*} transient anion can be in a potential minimum where, due to small Franck-Condon overlap, autodetachment is sufficiently slow (autodetachment lifetime $> 1\ \mu s$) to allow its detection with mass spectrometric techniques. According to Lotter and Illenberger [81] "it is likely that CF_4^- represents a weakly bound $F^- \cdot CF_3$ adduct with one bond significantly weakened, rather than a tetrahedral CF_4^-."

The structure of the CF_4 molecule accounts for its high ionization threshold energy. Bonham and Bruce [82] reported that the average value of the ionization threshold energy of CF_4 obtained from a number of electron-impact studies is 15.9 eV. Consistent with this high value are the results of electron impact [39, 83], photoelectron [40], and photoabsorption [47] studies which give a value of 16.20 eV. Bonham and Bruce [82] gave the value of 14.7 eV for the zero translational energy threshold of the reaction $CF_4 + e \rightarrow CF_3^+ + F + 2e$. Two recent studies [84, 85] on the heat of formation of CF_3^+ put the threshold energy (adiabatic) for the process $CF_4 + \text{energy} \rightarrow CF_3^+ + F + e$ at 14.28 eV [85] and 14.67 ± 0.04 eV ($T = 0$ K) [84]. This is similar to earlier values of 14.2 ± 0.1 eV ($T = 298$ K) [86], 14.24 ± 0.07 eV ($T = 298$ K) [87], and $\leq 14.7 \pm 0.3$ eV ($T = 0$ K) [88]. These values are clearly consistent with the zero translational energy threshold of 14.7 eV, and since the threshold energy for the formation of CF_3^+ may be taken as the lower energy for ionization of CF_4, at room temperature the adiabatic and vertical ionization energies of CF_4 may be identified, respectively, with the values of 14.7 and 16.2 eV. It should be mentioned also that Jarvis and Tuckett [89] put the adiabatic ionization energy of the CF_3 radical at $\leq 8.8 \pm 0.2$ eV.

The dissociation process generating neutral fragments via electron impact has an energy threshold near 12.5 eV [50]. Since all excited electronic states of CF_4 and CF_4^+ are unstable (see discussion earlier

Table 4.2. Negative ion resonance states of CF_4.

Energy (eV)[a]	Type of resonance	Symmetry	Method of observation
6.8	Shape	T_2	Electron impact
~ 8.0	Feshbach	T_1	spectroscopy [65]
8.0		T_2	Electron
~ 9.0		A_1	scattering [66, 67]
3.6[b]			Time-of-flight electron
~ 8.9			transmission [68]
~ 9.0			Electron [69] transmission
7	Shape	T_2	Electron scattering [37]
6.7 (F^- yield)			Dissociative
7.1 (CF_3^- yield)			attachment [70]
6.15 (F^- yield)			Dissociative
6.9 (CF_3^- yield)			attachment [71]
~ 7.5 (F^- yield)			
7.3 (total)[c]			Electron swarm [72]
7–8			Dissociative
6.9 (CF_3^- yield)			attachment [73]
~ 7.0 (F^- yield)			
12.0	Core-excited Feshbach		Threshold-electron
13.0	Core-excited Feshbach		excitation [43][d]
6.8 (F^- yield)			Dissociative
6.8 (CF_3^- yield)			attachment [62–64][e]
7.6 (F^- yield)			

[a]Energy at the cross section maximum.
[b]It was argued in Ref. [66] that this cross section maximum cannot be due to a resonance process.
[c]For the production of both F^- and CF_3^-.
[d]See also Ref. [61].
[e]Electron capture in the 6 to 8 eV range occurs via two negative ion states, the ground state of CF_4^- at 6.8 eV producing both F^- and CF_3^- and an electronically excited state CF_4^{*-} at 7.6 eV producing only F^-.

Table 4.2 (continued)
Negative ion resonance states of CF_4.

Energy (eV)[a]	Type of resonance	Symmetry	Method of observation
6.6	Shape	T_2	Static-exchange
11.7	Shape	A_1	approximation
27.5	Shape	E	calculation [74][f]
29.1	Shape	T_2	
9.9		T_2	Schwinger multichannel
11.0		A_1	calculation [76]
3.2		T_2	CMS-Xα
5.2		A_1	calculation [75][g]

[a]Energy at the cross section maximum.
[f]Mann and Linder [67] do not agree with the findings of Huo [74]. However, the calculated values at 6.6 eV and at 11.7 eV are consistent with the experiment.
[g]Mann and Linder [66] argued that the energy positions calculated by Tossell and Davenport [75] for the T_2 and A_1 resonances are located at energies that are too low. In the calculation of Ref. [75] the energy position of the T_2 resonance was rather sensitive to the C–F distance.

in this section), the total electronic excitation cross section of CF_4 is effectively equal to the total dissociation cross section. In view of the energetic thresholds for electron-impact dissociation and ionization, the total dissociation cross section is dominated by the dissociation processes producing neutral fragments near the threshold, while the dissociative ionization process (i.e., production of neutral-ion dissociation products) progressively takes over as the electron energy increases above the ionization threshold and dominates above about 30 eV. Furthermore, since no stable parent CF_4^+ ions have been observed, the total ionization cross section of CF_4 is equal to the total dissociative ionization cross section (see Secs. 2.3 and 2.4 of Ch. 4).

From the preceding discussion and the results summarized in subsequent sections, it becomes apparent that the CF_4 structure leads to a rather simple picture of the collisional behavior of this molecule with low energy electrons: Vibrational excitation is the dominant inelastic process below 12.5 eV, i.e., below the threshold for electronic excitation, and is dominated by the excitation of the infrared active modes ν_3 and ν_4 via direct dipole scattering below the negative ion resonance region 6–8 eV and via indirect scattering in the resonance region. All electronic excitations of CF_4 lead to dissociation. Dissociation of CF_4 into neutrals begins near 12.5 eV, dominates until the ionization sets in, and progressively yields to dissociative ionization which takes over and accounts, at sufficiently high electron impact energies (> 35 eV), for the total electronic excitation cross section.

Cross sections for positive ion pair formation, multiple ionization, and positive ion-negative ion pair formation generally are significantly smaller than those for single ionization in the low energy range of interest in this work.

2.2 Electron Scattering from CF_4

In this section information is presented and discussed on the following cross sections used to describe the various electron scattering processes: total electron scattering cross section, elastic differential electron scattering cross section, elastic integral electron scattering cross section, momentum transfer (elastic) electron scattering cross section, and inelastic electron scattering cross section for total vibrational excitation and total electronic excitation. The data are first presented in ways that facilitate their comparison and usefulness and they are subsequently assessed and discussed. Recommended or suggested cross section values are given when possible. (See also reviews by Morgan [90, 91], Bonham [92] and Karwasz *et al.* [93], and the model-based cross section sets of Hayashi [4], Nakamura [5], and Bordage *et al.* [6]).

2.2.1 Total Electron Scattering Cross Section, $\sigma_{sc,t}(\varepsilon)$, for CF_4

In Fig. 4.3 are summarized the measured and the calculated values of the total electron scattering cross section as a function of electron energy for CF_4. The experimental measurements of the total electron scattering cross section are: those of Jones [68] made using a time-of-flight electron transmission spectrometer in the electron energy range 1.0 to 50 eV; those of Szmytkowski *et al.* [69] made using the linear transmission technique in the electron energy range 0.5 to 200 eV; those of Zecca *et al.* [94] made using a Ramsauer-type apparatus in the electron energy range 75 to 4000 eV; those of Sueoka *et al.* [95] (see also Refs. [97, 101–104]) made using a time-of-flight apparatus in the electron energy range 1 to 400 eV; and those of Nishimura *et al.* [98] made using a linear electron transmission apparatus in the electron energy range 1.5 to 3000 eV. Jones [68] reported that the most probable uncertainty in his measurements was ±2.3% below 4.0 eV, +3.3% and −3.0% between 4.2 and 15.0 eV, +3.4% and −2.1% between 16.0 and 25.0 eV, and +7.5% and −2.8% between 26 and 50 eV. The direct sum of all potential individual systematic uncertainties in the experiment of Szmytkowski *et*

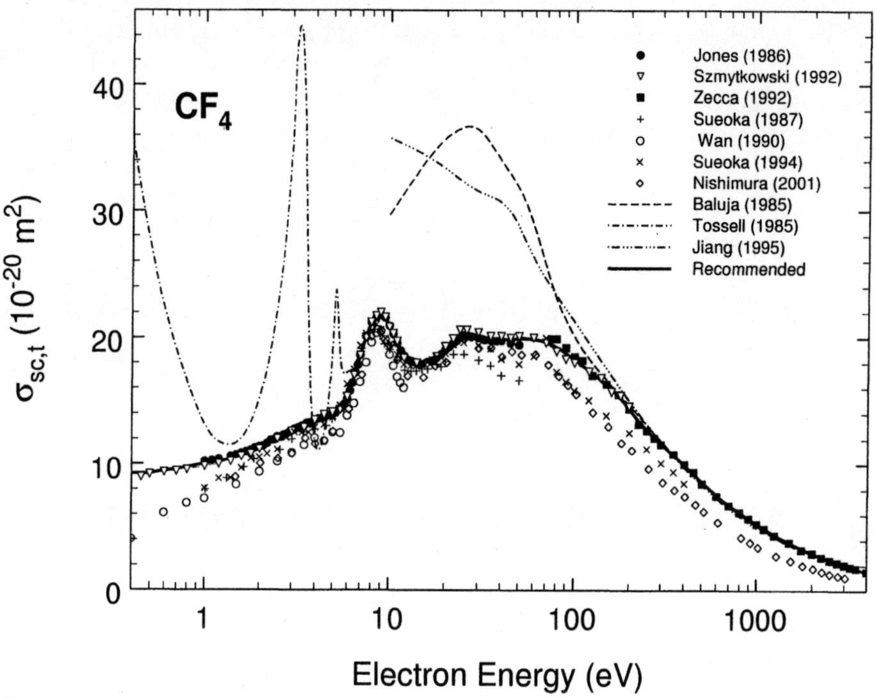

Figure 4.3. Total electron scattering cross section $\sigma_{sc,t}(\varepsilon)$ as a function of electron energy for CF_4: (•) Ref. [68]; (▽) Ref. [69]; (■) Ref. [94]; (+) Ref. [95]; (○) Ref. [96]; (×) Ref. [97]; (◊) Ref. [98]; (- - -) Ref. [99]; (— · —) Ref. [75]; (— · · · —) Ref. [100]; (—) recommended.

al. [69] was estimated to be ±4% below 1 eV, gradually decreasing to ±3% near 20 eV, and increasing to ±4% at higher energies. Zecca *et al.* [94] estimated their systematic errors to be $< \pm 3\%$ at all energies, and the uncertainty of the Sueoka *et al.* [97] measurements is usually $< 4\%$. These cross section measurements are generally in good quantitative agreement in the energy ranges over which they overlap although the measurements of Sueoka *et al.* give smaller values at the extremes of their energy range. Szmytkowski *et al.* [69] pointed out that a correction might be needed for the data of Sueoka *et al.* [95] related to the normalization procedure [102, 103] they applied. Another absolute measurement [96] of $\sigma_{sc,t}(\varepsilon)$ in the energy range 0.2 to 12 eV gave much lower values than the other absolute measurements in this energy range (see Fig. 4.3).

The results of two calculations for $\varepsilon > 10$ eV, one based on the additivity rule [100] and the other based on a parameter-free spherical complex potential optical potential [99], are also shown in Fig. 4.3. The calculated total electron scattering cross sections for electron energies below ~ 200 eV are seen to be substantially larger than the experimental values, with agreement at higher energies where the Born approximation is valid. The total cross sections obtained by a multiple scattering Xα calculation [75] in the low energy regime are in disagreement with the measured cross sections both in magnitude and shape.

To arrive at recommended values for $\sigma_{sc,t}(\varepsilon)$, we considered the experimental measurements of Jones [68], Szmytkowski *et al.* [69], and Zecca *et al.* [94], which are in agreement over their common energy ranges. These data were fitted with equal weights. The resultant average cross section values are indicated by the solid line in Fig. 4.3, and are listed in Table 4.3 as our recommended values for $\sigma_{sc,t}(\varepsilon)$ for energies above 0.5 eV (recommended values at lower energies are determined from the data presented in Fig. 4.4).

The overall energy dependence of $\sigma_{sc,t}(\varepsilon)$ is rather interesting. Above ~ 100 eV the usual systematic fall-off of the cross section with increasing electron energy sets in. In this high-energy range the Born-based calculations generally agree with the experimental measurements and can be used to normalize or to check the latter. In the energy range of approximately 6 to 50 eV there are two broad structureless enhancements in $\sigma_{sc,t}(\varepsilon)$. The peak at 9 eV is due to indirect electron scattering via short-lived negative ion states (see Table 4.2). The second broad maximum in the experimental total electron scattering cross section centered near 24 eV may be due to both direct electron scattering and indirect electron scattering via negative ion resonances. A very broad resonance has been reported [37] around 21 eV, and the theoretical work of Huo [74] indicated an E-type resonance at 27.5 eV (see Table 4.2). The predicted [75] resonant effects near 3 eV are not clearly and unambiguously reflected in the measurements.

Below ~ 6 eV the cross section $\sigma_{sc,t}(\varepsilon)$ decreases smoothly with electron energy. Analysis of electron beam [37, 43, 61] and electron swarm [4–6, 106–110] data indicates that in the lower part of this energy range (≤ 2 eV) vibrational excitation of the CF_4 molecule (mainly the ν_3 symmetric mode) contributes appreciably to the scattering.

In order to determine $\sigma_{sc,t}(\varepsilon)$ below ~ 2 eV, it is necessary to initially consider data on other types of cross sections which are discussed later in this section. In Fig. 4.4 are shown measurements of $\sigma_{sc,t}(\varepsilon)$, $\sigma_{m}(\varepsilon)$, and $\sigma_{e,int}(\varepsilon)$ below ~ 2 eV. The total electron scattering cross section [106], the momentum transfer cross section [67], and the elastic integral cross section [67] continuously decrease with decreasing electron energy to ~ 0.2 eV where a deep Ramsauer-Townsend minimum appears [1, 4, 5,

Table 4.3. Recommended total electron scattering cross section, $\sigma_{sc,t}(\varepsilon)$, for CF_4.

ε (eV)	$\sigma_{sc,t}(\varepsilon)$ (10^{-20} m^2)	ε (eV)	$\sigma_{sc,t}(\varepsilon)$ (10^{-20} m^2)	ε (eV)	$\sigma_{sc,t}(\varepsilon)$ (10^{-20} m^2)
0.0030	12.7	0.25	9.12	25	20.4
0.0035	12.2	0.30	9.26	30	20.2
0.0040	11.9	0.35	9.28	35	19.9
0.0045	11.5	0.40	9.25	40	19.9
0.0050	11.2	0.45	9.23	45	19.9
0.0060	10.6	0.50	9.27	50	19.9
0.0070	10.1	0.60	9.45	60	19.9
0.0080	9.69	0.70	9.60	70	19.6
0.0090	9.26	0.80	9.75	80	19.2
0.010	8.89	0.90	9.89	90	18.8
0.015	7.40	1.0	10.0	100	18.3
0.020	6.35	1.5	10.6	150	16.2
0.025	5.41	2.0	11.3	200	14.4
0.030	4.67	2.5	12.0	250	12.8
0.035	4.12	3.0	12.6	300	11.6
0.040	3.63	3.5	13.0	350	10.8
0.045	3.21	4.0	13.5	400	9.95
0.050	2.86	4.5	13.8	450	9.24
0.060	2.30	5.0	14.0	500	8.60
0.070	1.98	6.0	15.3	600	7.57
0.080	1.76	7.0	18.6	700	6.80
0.090	1.62	8.0	21.1	800	6.21
0.10	1.50	9.0	21.8	900	5.70
0.14	1.60	10.0	20.8	1000	5.28
0.15	2.17	12.0	18.5	2000	2.95
0.16	3.05	15.0	17.9	3000	2.05
0.20	7.35	20.0	19.2	4000	1.49

66, 67, 106, 107], and increase again as the electron energy decreases to the left of the minimum [66, 67, 111]. We have arrived (see also Ch. 3) at a recommended set of values for $\sigma_{sc,t}(\varepsilon)$ below $\sim$ 2 eV by considering the available data as follows: (i) We accepted the values of $\sigma_{sc,t}(\varepsilon)$ of Jones [68] and Szmytkowski *et al.* [69] (• and Δ in Fig. 4.4), but the data of Curtis *et al.* [106] were not considered because they were indirectly determined. (ii) We accepted the $\sigma_m(\varepsilon)$ and $\sigma_{e,int}(\varepsilon)$ values of Mann and Linder [67] in the energy range below 1 eV because they were deduced from more direct measurements. (iii) Below the lowest vibrational threshold of 0.054 eV, $\sigma_{sc,t}(\varepsilon) = \sigma_{e,int}(\varepsilon)$. (iv) In the energy range between about 0.08 and 1 eV, a value of $\sigma_{sc,t}(\varepsilon)$ can be obtained by

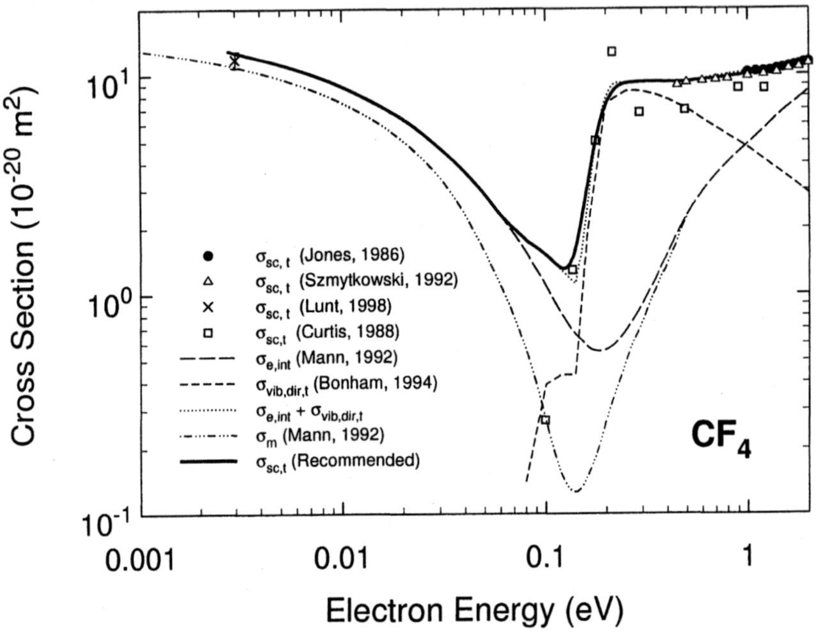

Figure 4.4. Electron scattering cross sections for CF_4 in the low energy range (electron-impact energies less than ~ 2 eV). $\sigma_{sc,t}(\varepsilon)$: (•) Ref. [68]; (Δ) Ref. [69]; (×) Ref. [105]; (□) Ref. [106]. $\sigma_m(\varepsilon)$: (– · · · –) Ref. [67]. $\sigma_{e,int}(\varepsilon)$: (– –) Ref. [67]. $\sigma_{vib,dir,t}(\varepsilon)$: Ref. [92]. The sum of $\sigma_{e,int}(\varepsilon)$ and $\sigma_{vib,dir,t}(\varepsilon)$ is shown by (· · ·), and the recommended value of $\sigma_{sc,t}(\varepsilon)$ is shown by the solid line (see Ref. [1] for further explanation).

adding $\sigma_{e,int}(\varepsilon)$ and $\sigma_{vib,dir,t}(\varepsilon)$, the latter being the total cross section for vibrational excitation (which is discussed in detail later; see Table 4.8), since in this energy range electronic excitation is absent and rotational excitation is negligible (due to the absence of a dipole and a quadrupole moment). Therefore, we assume that in the energy range 0.08 to 1 eV $\sigma_{sc,t}(\varepsilon) = \sigma_{e,int}(\varepsilon) + \sigma_{vib,dir,t}(\varepsilon)$, and determine the $\sigma_{sc,t}(\varepsilon)$ using the $\sigma_{e,int}(\varepsilon)$ of Mann and Linder [67] and the $\sigma_{vib,dir,t}(\varepsilon)$ listed in column 4 of Table 4.8. The values so obtained are plotted in Fig. 4.4 (dotted line from 0.08 to 2 eV).

The values of $\sigma_{sc,t}(\varepsilon)$ we determined agree well with the measured $\sigma_{sc,t}(\varepsilon)$ for energies near 1 eV and with the $\sigma_{e,int}(\varepsilon)$ for energies below ~ 0.08 eV. More recently, a direct measurement of $\sigma_{sc,t}(\varepsilon)$ has been reported by Lunt *et al.* [105] [$(11.8 \pm 1)\ 10^{-20}$ m^2 at 0.003 eV], which confirmed the value of 12.7×10^{-20} m^2 at 0.003 eV deduced as described above [1]. Figure 4.4 provides a direct comparison of $\sigma_{sc,t}(\varepsilon)$, $\sigma_{e,int}(\varepsilon)$, and $\sigma_m(\varepsilon)$ in the region of the Ramsauer-Townsend cross section minimum. It appears that the minimum is the deepest for $\sigma_m(\varepsilon)$. Below ~ 0.5 eV, $\sigma_m(\varepsilon) < \sigma_{e,int}(\varepsilon)$ indicating small angle scattering.

The cross section $\sigma_{sc,t}(\varepsilon)$ estimated in the manner outlined above from 0.08 to 2 eV was used along with the measurements of Mann and Linder [67] for $\sigma_{e,int}(\varepsilon)$ below 0.08 eV, and the measurements of Jones [68] and Szmytkowski [69] for $\sigma_{sc,t}(\varepsilon)$ above ~ 0.5 eV to obtain a best estimate of $\sigma_{sc,t}(\varepsilon)$ below ~ 1 eV. The values of this best estimate are shown by the solid line in Fig. 4.4 and are listed in Table 4.3 (along with the data for energies above 1 eV from Fig. 4.3). The data in Table 4.3 are our recommended $\sigma_{sc,t}(\varepsilon)$ from 0.003 to 4000 eV.

2.2.2 Elastic Differential Electron Scattering Cross Section, $\sigma_{e,diff}(\varepsilon)$, of CF_4

There are three measurements of the elastic differential electron scattering cross section, $\sigma_{e,diff}(\varepsilon)$, of CF_4 covering various ranges of incident electron energies and scattering angles. The first set of measurements of $\sigma_{e,diff}(\varepsilon)$ for CF_4 was made by Sakae *et al.* [112] for scattering angles between 5° to 135°, and incident electron energies of 75, 100, 150, 200, 300, 500, and 700 eV. The experimental cross sections were extrapolated to 0° and 180° scattering angles by fitting the square of the Legendre polynomials to the measured values [112]. The uncertainty in their data was estimated to be about ±10%. These results are shown in Fig. 4.5 for 100, 200, 300, 500, and 700 eV incident electron energies. The cross section $\sigma_{e,diff}(\varepsilon)$ is seen to increase steeply at the forward angles.

The second set of measurements of $\sigma_{e,diff}(\varepsilon)$ was made by Boesten *et al.* [37] in the energy range 1.5–100 eV and for scattering angles from 15° to 130° using a crossed-beam apparatus. The relative cross section measurements were put on an absolute scale by normalization to the cross section data for He. The measured cross sections were extrapolated to 0° and 180° by phase shift fitting [37]. These cross sections are given in Table 4.4, and are shown in Fig. 4.5 for 1.5, 5, 8, 10, 15, 35, and 100 eV incident electron energies.

The third measurement of $\sigma_{e,diff}(\varepsilon)$ is that of Mann and Linder [67]. These measurements were made at scattering angles between 10° to 105° and for electron energies from 0.5 to 20 eV. The absolute values of $\sigma_{e,diff}(\varepsilon)$ were obtained by integration and by normalization of the sum of the elastic integral cross section and the total cross section for inelastic electron scattering they measured at 3 eV to the total cross section of Jones [68] at this energy. The absolute uncertainty of the measurements was estimated to be between ±20 and ±30%. These data are given in

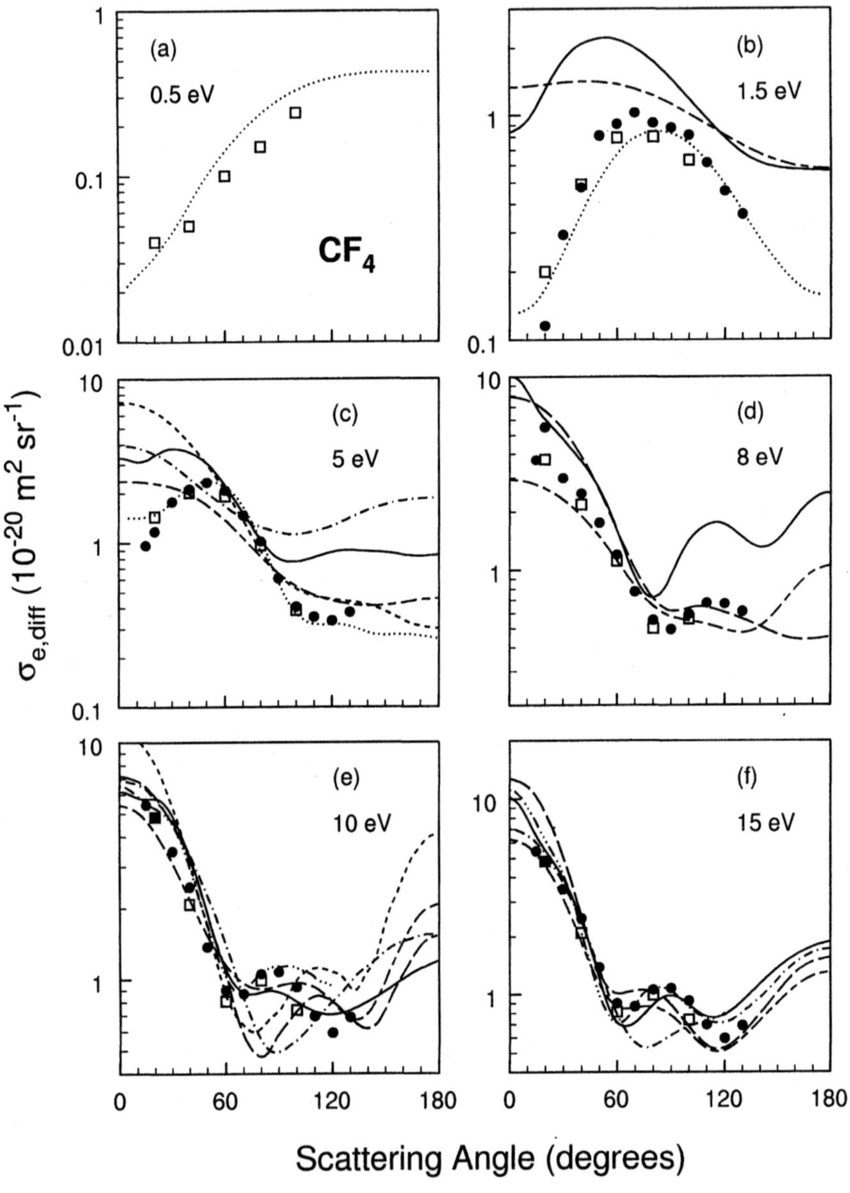

Figure 4.5. Comparison of elastic differential electron scattering cross sections $\sigma_{e,diff}(\varepsilon)$ at representative incident electron energies. Experimental data: (•) Boesten *et al.* [37]; (□) Mann and Linder [67]; (∘) Sakae *et al.* [112]. Calculations: (···) Isaacs *et al.* [113]; (- - -) Varella *et al.* [76]; (– · –) Huo *et al.* [74]; (- - -) Natalense *et al.* [114]; (— —) Winstead *et al.* [115]; (– · · · –) Gianturco *et al.* [116]; (– – –) Raj [118]; (—) Curik *et al.* [119] (HFEGE exchange results).

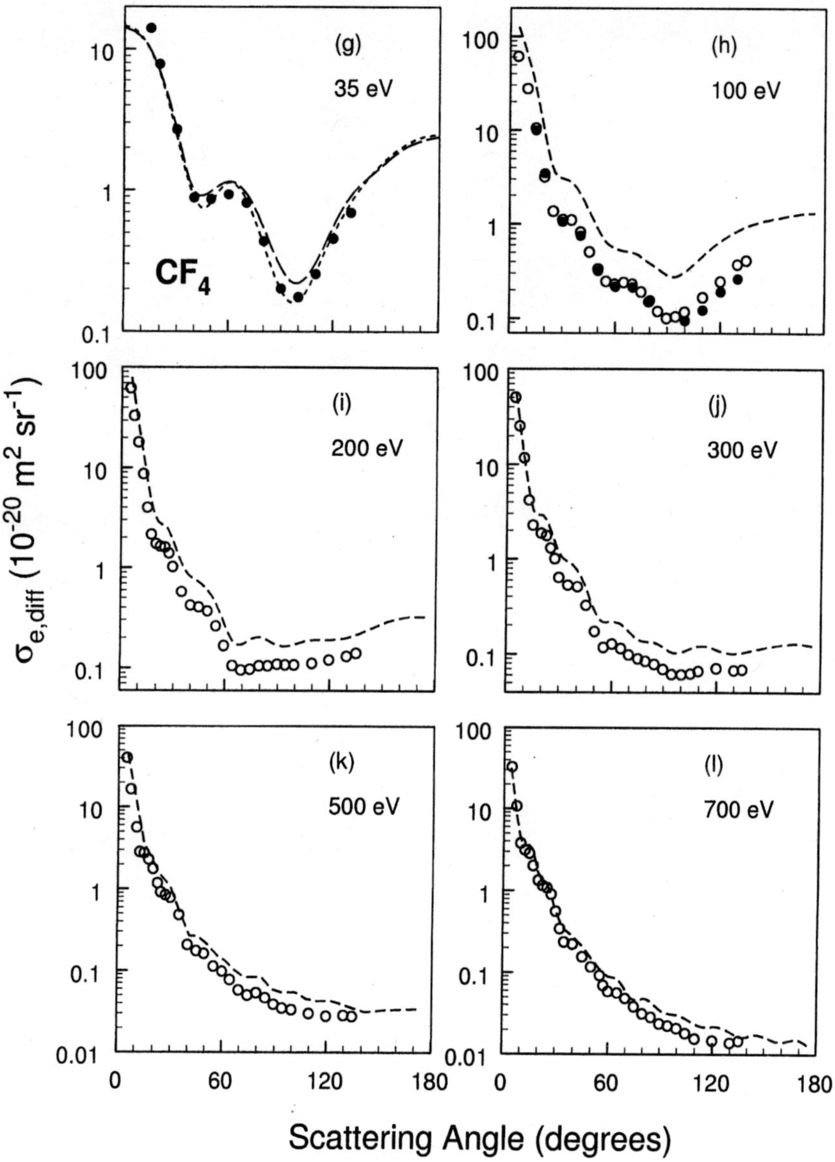

Figure 4.5 (continued)
Comparison of elastic differential electron scattering cross sections $\sigma_{e,diff}(\varepsilon)$ at representative incident electron energies. Experimental data: (•) Boesten *et al.* [37]; (□) Mann and Linder [67]; (○) Sakae *et al.* [112]. Calculations: (···) Isaacs *et al.* [113]; (– - –) Varella *et al.* [76]; (– · –) Huo *et al.* [74]; (- - -) Natalense *et al.* [114]; (— —) Winstead *et al.* [115]; (–···· –) Gianturco *et al.* [116]; (– – –) Raj [118]; (—) Curik *et al.* [119] (HFEGE exchange results).

Table 4.4. Elastic differential electron scattering cross section, $\sigma_{e,diff}(\varepsilon)$, measurements of Boesten *et al.* [37] for CF_4 in units of 10^{-20} m^2 sr^{-1}.

Angle	1.5 eV	2 eV	3 eV	5 eV	6 eV
15°	—	0.119	0.341	0.965	1.553
20°	0.116	0.211	0.544	1.178	1.674
30°	0.293	0.517	0.957	1.778	2.041
40°	0.476	0.753	1.256	2.131	2.363
50°	0.811	1.118	1.591	2.334	2.418
60°	0.915	1.399	1.603	2.081	1.938
70°	1.026	1.258	1.513	1.472	1.484
80°	0.923	1.044	1.179	1.023	0.903
90°	0.878	0.807	0.891	0.607	0.543
100°	0.815	0.726	0.538	0.408	0.377
110°	0.615	0.486	0.440	0.355	0.435
120°	0.458	0.403	0.317	0.336	0.447
130°	0.362	0.298	0.264	0.378	0.450

Angle	7 eV	8 eV	9 eV	10 eV	15 eV
15°	2.557	3.704	4.715	4.401	5.433
20°	2.472	3.506	4.097	4.768	4.822
30°	2.378	2.987	3.513	4.116	3.473
40°	2.422	2.489	2.656	2.884	2.464
50°	2.052	1.757	1.714	1.685	1.383
60°	1.780	1.197	1.115	0.999	0.901
70°	1.110	0.775	0.720	0.730	0.869
80°	0.657	0.552	0.605	0.782	1.058
90°	0.435	0.495	0.681	0.800	1.076
100°	0.468	0.592	0.767	0.794	0.931
110°	0.539	0.678	0.754	0.727	0.698
120°	0.593	0.670	0.731	0.622	0.595
130°	0.574	0.612	0.658	0.651	0.691

Table 4.5, and are shown in Fig. 4.5 for 0.5, 1.5, 5, 8, 10, and 15 eV incident electron energies.

The two sets of experimental measurements at low energies [37, 67] are in reasonably good agreement for all common values of the incident electron energy. At higher energies, only at one energy value (100 eV) can the experimental data be compared. At this energy, the data of Boesten *et al.* [37] and Sakae *et al.* [112] are in good agreement.

A number of computational methods have been employed for the calculation of the $\sigma_{e,diff}(\varepsilon)$ of CF_4. The results of these calculations are

Table 4.4 (continued)
Elastic differential electron scattering cross section, $\sigma_{e,diff}(\varepsilon)$, measurements of Boesten *et al.* [37] for CF_4 in units of 10^{-20} m^2 sr^{-1}.

Angle	20 eV	35 eV	50 eV	60 eV	100 eV
15°	6.757	14.10	13.32	12.19	9.926
20°	5.146	7.827	6.938	5.889	3.461
30°	3.167	2.691	1.409	1.015	1.056
40°	1.720	0.878	0.738	0.746	0.753
50°	0.912	0.861	0.857	0.759	0.319
60°	0.795	0.927	0.672	0.429	0.217
70°	1.004	0.809	0.360	0.219	0.215
80°	1.095	0.435	0.170	0.129	0.157
90°	0.988	0.201	0.136	0.128	0.010
100°	0.690	0.176	0.133	0.124	0.095
110°	0.530	0.255	0.200	0.185	0.122
120°	0.554	0.455	0.409	0.322	0.192
130°	0.823	0.691	0.657	0.433	0.262

compared with the experimental data in Fig. 4.5. It is rather difficult to compare in detail the results of the various computations with the measurements, mainly because there is only a limited overlap in the two key variables: incident electron energy and scattering angle. Such comparisons are however possible for some values of the incident electron energy and are presented in Fig. 4.5.

At low energies, the calculated values of the $\sigma_{e,diff}(\varepsilon)$ of CF_4 by Isaacs *et al.* [113], who used the complex Kohn method, are in excellent agreement with the experimental data at 0.5, 1.5 and 5 eV. The calculated values of $\sigma_{e,diff}(\varepsilon)$ by Varella *et al.* [76], who used the Schwinger multichannel method, are in less satisfactory agreement with the experimental data. (The measurements reported by Varella *et al.* [76] are identical to those reported by Boesten *et al.* [37]). The calculated results of Huo [74], who employed the fixed-nuclei static-exchange approximation, for 10 and 15 eV agree reasonably well with the experimental data, but at 5 eV her results show large deviations from the experimental measurements, which may be attributed to neglect of the polarization effects in the calculations. The results of four other calculations (the static exchange approximation calculation of Winstead *et al.* [115], the static exchange calculation of Gianturco *et al.* [116, 117], the pseudopotential calculation of Natalense *et al.* [114], and the HFEGE exchange interaction and semiclassical exchange (SMCE) calculation of Curik *et al.* [119]) at low incident electron energies (5, 8, 10, and 15 eV) are in various de-

Table 4.5. Elastic differential electron scattering cross sections, $\sigma_{e,diff}(\varepsilon)$, for CF_4 measured by Mann and Linder [67] in units of 10^{-20} m^2 sr^{-1}.

	Scattering angle				
ε (eV)	20°	40°	60°	80°	100°
0.5	0.04	0.05	0.10	0.15	0.24
1.0	0.08	0.20	0.40	0.50	0.52
1.5	0.20	0.49	0.79	0.80	0.63
2.0	0.38	0.82	1.14	1.00	0.64
2.5	0.58	1.09	1.42	1.09	0.61
3.0	0.78	1.34	1.59	1.11	0.55
3.5	0.97	1.57	1.73	1.10	0.50
4.0	1.16	1.77	1.83	1.06	0.46
4.5	1.30	1.92	1.90	1.02	0.42
5.0	1.44	2.02	1.92	0.96	0.39
5.5	1.54	2.05	1.90	0.87	0.37
6.0	1.72	2.04	1.81	0.77	0.36
6.5	2.09	2.01	1.64	0.67	0.38
7.0	2.62	2.05	1.44	0.56	0.43
7.5	3.20	2.10	1.25	0.51	0.49
8.0	3.73	2.18	1.11	0.50	0.56
8.5	4.18	2.29	1.01	0.52	0.62
9.0	4.54	2.37	0.95	0.57	0.66
9.5	4.74	2.40	0.92	0.64	0.69
10.0	4.81	2.44	0.93	0.71	0.70
10.5	4.79	2.46	0.95	0.78	0.73
11.0	4.73	2.47	0.95	0.81	0.74
11.5	4.68	2.44	0.96	0.86	0.74
12.0	4.62	2.41	0.93	0.88	0.75
12.5	4.64	2.36	0.93	0.89	0.75
13.0	4.62	2.32	0.90	0.92	0.77
13.5	4.68	2.26	0.87	0.94	0.76
14.0	4.71	2.21	0.84	0.96	0.75
14.5	4.76	2.15	0.82	0.96	0.75
15.0	4.83	2.08	0.81	0.99	0.74
15.5	4.91	2.02	0.77	1.01	0.74
16.0	4.91	1.99	0.77	1.03	0.72
16.5	5.03	1.96	0.76	1.02	0.70
17.0	5.12	1.92	0.73	1.04	0.69
17.5	5.21	1.87	0.74	1.05	0.67
18.0	5.35	1.83	0.73	1.06	0.67
18.5	5.46	1.80	0.73	1.06	0.65

grees of agreement with the experimental data. The agreement with the measurements is poor at the lowest energy (5 eV) for which comparisons can be made. At the higher energies (10 and 15 eV), the agreement is good for scattering angles between 30° and 130°, but outside this angle range the agreement between the calculated and the experimental data varies depending on the electron energy, the scattering angle, and the type of calculation.

At high energies, Raj [118] used an independent-atom model along with partial waves and calculated the $\sigma_{e,diff}(\varepsilon)$ of CF_4 for energies between 100 and 700 eV. The results of this calculation are compared in Fig. 4.5 with the experimental data of Boesten *et al.* [37] and Sakae *et al.* [112] for 100 eV, and with the experimental data of Sakae *et al.* [112] for 200, 300, 500, and 700 eV. It is seen that the calculations reproduce qualitatively the overall behavior of the experimental data, but the calculated values are higher than the experimental data. The agreement between the experimental data and the calculated results improves, as expected, with increasing incident electron energy. At low energies, the values of $\sigma_{e,diff}(\varepsilon)$ as calculated within the independent-atom model are larger than the measured values possibly because such factors as orbital overlap in molecules decrease the atomic contributions compared with those in the free atomic state.

2.2.3 Elastic Integral Electron Scattering Cross Section, $\sigma_{e,int}(\varepsilon)$, of CF_4

From the elastic differential electron scattering cross sections obtained from the experiments discussed in the preceding section, Sakae *et al.* [112], Boesten *et al.* [37], and Mann and Linder [67] obtained values of the elastic integral cross section $\sigma_{e,int}(\varepsilon)$ for electron energies above 1 eV. These cross sections are plotted in Fig. 4.6.

Mann and Linder [67] obtained $\sigma_{e,int}(\varepsilon)$ by a simple integration procedure having weighted the cross section values they measured between 10° and 105° by $\sin\theta$ and then linearly extrapolating towards the integration limits of 0° and 180°. According to the authors, the total uncertainty in their data is expected to be ±20%. In order to put their data on an absolute scale, the sum of the elastic integral cross section and the total cross section for inelastic scattering was normalized to the total cross section value measured by Jones [68] at 3 eV. The observed broad enhancements in $\sigma_{e,int}(\varepsilon)$ below $\sim$ 20 eV are due to resonances, especially the peak at 5 eV, which agrees well with the position of resonances which have been well-established by other studies (e.g., on dissociative electron attachment; Table 4.2). The $\sigma_{e,int}(\varepsilon)$ data in Fig. 4.6 below 0.5 eV were

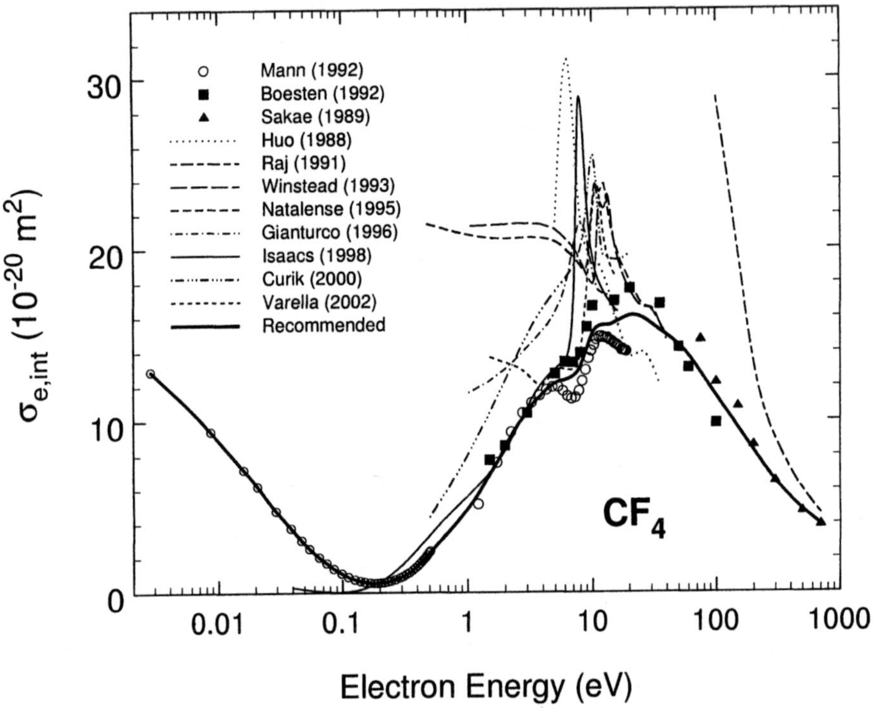

Figure 4.6. Elastic integral electron scattering cross section $\sigma_{e,int}(\varepsilon)$ for CF_4. Experimental values: (o) Ref. [67]; (■) Ref. [37]; (▲) Ref. [112]. Calculated values: (···) Ref. [74]; (– - –) Ref. [118]; (— —) Ref. [115]; (– –) Ref. [114]; (– · –) Ref. [116]; (—) Ref. [113]; (– ··· –) Ref. [119]; (- - -) Ref. [76]. The recommended data for $\sigma_{e,int}(\varepsilon)$ for CF_4 are shown by the heavy solid line.

obtained by Mann and Linder [67] using modified effective range theory (MERT) analysis.

The experimental cross sections [37, 67, 112] in Fig. 4.6 are compared with the results of eight calculations [74, 76, 113–116, 118, 119]. The low energy results of Huo [74] indicate a structure near 6 eV, but her cross section is not in general agreement with the experimental data. Similarly, the results of the more recent calculations [76, 114–116, 119] exhibit structure the location of which varies from calculation to calculation. The calculated values exceed the experimental ones and the overall agreement between the two is generally poor. The calculation of Raj [118], gives results which are in satisfactory agreement with the measurements of Sakae *et al.* [112] only for energies above ~ 500 eV.

Table 4.6. Recommended elastic integral cross section, $\sigma_{e,int}(\varepsilon)$, for CF_4.

ε (eV)	$\sigma_{e,int}(\varepsilon)$ (10^{-20} m^2)	ε (eV)	$\sigma_{e,int}(\varepsilon)$ (10^{-20} m^2)	ε (eV)	$\sigma_{e,int}(\varepsilon)$ (10^{-20} m^2)
0.0030	12.7	0.20	0.56	10.0	15.1
0.0035	12.2	0.25	0.68	12.5	15.5
0.0040	11.9	0.30	0.89	15.0	15.7
0.0045	11.5	0.35	1.18	17.5	15.9
0.0050	11.2	0.40	1.53	20.0	16.1
0.0060	10.6	0.45	1.91	25.0	15.9
0.0070	10.1	0.50	2.29	30.0	15.6
0.0080	9.68	0.60	2.96	35.0	15.2
0.0090	9.27	0.70	3.52	40.0	14.9
0.010	8.88	0.80	4.02	45.0	14.6
0.015	7.39	0.90	4.46	50.0	14.4
0.020	6.35	1.0	4.86	60.0	13.7
0.025	5.42	1.5	6.87	70.0	13.1
0.030	4.68	2.0	8.48	80.0	12.4
0.035	4.11	2.5	9.68	90.0	11.9
0.040	3.62	3.0	10.5	100.0	11.4
0.045	3.21	3.5	11.1	200.0	8.22
0.050	2.85	4.0	11.6	300.0	6.47
0.060	2.29	4.5	11.9	400.0	5.46
0.070	1.87	5.0	12.1	500.0	4.78
0.080	1.54	6.0	12.4	600.0	4.29
0.090	1.29	7.0	12.6	700.0	3.91
0.10	1.09	8.0	13.2		
0.15	0.62	9.0	14.2		

We have fitted a line to the three sets of experimental data giving equal weight to each data set. This is shown by the solid line in Fig. 4.6 and represents our recommended values of $\sigma_{e,int}(\varepsilon)$ as listed in Table 4.6.

2.2.4 Momentum Transfer (Elastic) Electron Scattering Cross Section, $\sigma_m(\varepsilon)$, of CF_4

There exist three types of data on $\sigma_m(\varepsilon)$: experimental determinations [37, 67, 112], swarm-unfolded cross sections [4–6, 106, 107, 109], and calculated cross sections [74, 76, 113, 115, 118]. These are compared in Fig. 4.7 from 0.001 to 1000 eV. There is a large uncertainty in these cross section values due to the indirect determination of some of the cross section values in this region and to the fact that various methods (cal-

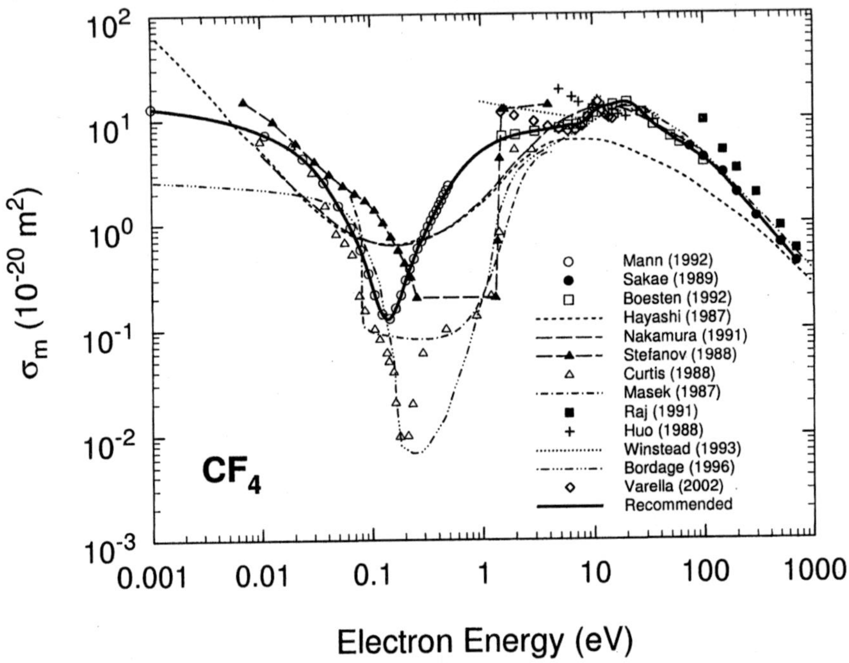

Figure 4.7. Momentum transfer (elastic) cross sections $\sigma_m(\varepsilon)$ as a function of electron energy for CF_4. Experimentally derived: (○) Ref. [67]; (•) Ref. [112]; (□) Ref. [37]. Swarm-unfolded (effective): (- - -) Ref. [4]; (— —) Ref. [5]; (▲) Ref. [107]; (△) Ref. [106]; (– · –) Ref. [109]; (– · · · –) Ref. [6]. Calculated: (■) Ref. [118]; (+) Ref. [74]; (· · ·) Ref. [115]; (◊) Ref. [76]. (—) Recommended.

culations and experiments) yield different types of cross sections. The experimental and the calculated $\sigma_m(\varepsilon)$ are for elastic electron scattering. The swarm-based determinations [4–6, 106, 107, 109] have employed the Boltzmann code [4–6, 107, 109] or the Monte Carlo method [106]. It should be noted that swarm-based cross sections are model-specific and are thus of limited value.

The measurements of Sakae *et al.* [112] were made using a crossed-beam method with an estimated uncertainty of about ±10%, and are in agreement with the data of Boesten *et al.* [37], which were also obtained in a crossed-beam experiment. Boesten *et al.* determined their momentum transfer cross sections from their elastic differential cross section measurements which had an uncertainty of ±15% to ±20%. At lower energies, Mann and Linder [67] determined the momentum transfer cross section using their measurements of the elastic differential cross sections and MERT analysis. Their elastic differential cross sections were measured using a crossed-beam apparatus with a quoted uncertainty of ±20% to ±30%.

The swarm-unfolded cross sections have different degrees of uncertainty and rely heavily on the accuracy of the electron transport measurements and energy range over which they were made. Although none of the cross sections of the six studies [4–6, 106, 107, 109] of this kind agree well with the low energy data from the beam measurements [67], the data of Nakamura *et al.* [5] are in reasonable agreement with the higher energy experimental values. Nakamura's cross sections are based on measurements of electron drift velocities and longitudinal electron diffusion coefficients in CF_4/Ar mixtures which were especially designed to reduce the uncertainty of the derived cross sections.

Hayashi [4] determined a set of cross sections for momentum transfer and vibrational excitation such that the values of $w(E/N)$ for CF_4 and for mixtures of CF_4 in rare gases, and the values of $D_T/\mu\,(E/N)$ calculated using the derived cross sections, best agreed with the experimental data of $w(E/N)$ [18, 19, 120] for pure CF_4, $w(E/N)$ [18, 19] for mixtures of CF_4 with the rare gases, and $D_T/\mu\,(E/N)$ [120, 121] for pure CF_4. Hayashi used the conventional two-term expansion approximation to the solution of the Boltzmann equation and did not consider the effects of superelastic electron scattering. Morgan [90, 91] pointed out that the cross section set determined by Hayashi might have been influenced by the effect of vibrational excitation because at 300 K, 17% of the CF_4 molecules are in the ν_2 and 10% in the ν_4 vibrational states. Superelastic collisions between electrons and vibrationally excited CF_4 molecules should, therefore, be significant. Experimental measurements have indicated a rather large effect of vibrational excitation on the electron drift velocity in similar systems, for example, C_2F_6 [122]. The effect of superelastic collisions on the cross sections that are derived in this manner for CF_4 (and similar molecules) needs exploration. The effect of vibrational excitation on the cross section set derived for CF_4 by Nakamura *et al.* [5] also needs to be investigated.

A Boltzmann equation analysis was also performed by Stefanov *et al.* [107] who derived a cross section set based on the measurements of $w(E/N)$ by Hunter *et al.* [13] and on the measurements of Lakshminarasimha *et al.* [121] for the $D_T/\mu\,(E/N)$. In the Boltzmann equation calculation of Bordage *et al.* [6] the derived set of electron collision cross sections for CF_4 was evaluated by fitting the calculated swarm parameters to experimental electron transport and other coefficients for the pure gas and for Ar-CF_4 mixtures. Curtis *et al.* [106] reported cross sections derived from measurements they made of the characteristic energy in CF_4 using a Monte Carlo method. The results of these investigations differ substantially from those of Hayashi and Nakamura, and from the more direct measurements (see Fig. 4.7).

In Fig. 4.7 are also plotted the results of four calculations [74, 76, 115, 118]. The results of these calculations agree only partially with the experimental measurements of Sakae *et al.* [112] and Boesten *et al.* [37]. However, the results of another recent calculation [113] are in agreement with the energy position of the Ramsauer-Townsend minimum present in the $\sigma_{\mathrm{e,int}}(\varepsilon)$ recommended by Christophorou *et al.* [1], and also in reasonable agreement with the recommended values of $\sigma_{\mathrm{e,int}}(\varepsilon)$ for electron energies ranging from 0.5 to 5 eV.

The recommended values for the elastic momentum transfer cross section $\sigma_{\mathrm{m}}(\varepsilon)$ are determined from the experimental cross sections of Mann and Linder [67] below 0.5 eV (see discussion earlier in this section), and those of Sakae *et al.* [112] and Boesten *et al.* [37] above 1.5 eV. We have fitted a line through the data of the three experimental measurements and interpolated between these data (0.5 to 1.5 eV) as shown in Fig. 4.7. These recommended values for the elastic momentum transfer cross section $\sigma_{\mathrm{m}}(\varepsilon)$ are listed in Table 4.7.

2.2.5 Inelastic Electron Scattering Cross Sections, $\sigma_{\mathrm{inel}}(\varepsilon)$, of CF_4

It is convenient to divide the inelastic electron scattering cross sections for CF_4 into two groups: those for vibrational excitation and those for electronic excitation.

Vibrational Excitation. Information on the vibrational excitation of the CF_4 molecule comes from direct measurements using electron beam methods [37, 66], from indirect determinations via electron swarm-data analyses [4–6, 106, 107, 109, 110] and from theory [66, 92, 99].

As discussed in Sec. 2.1 of this chapter, the CF_4 molecule has four fundamental frequencies: the symmetric stretch ν_1, the symmetric bend ν_2, the asymmetric stretch ν_3, and the asymmetric bend ν_4. Vibrational excitation of the CF_4 molecule is a mixture of direct excitation and indirect excitation via resonances. In Fig. 4.8(a) is shown the energy dependence of the cross section for excitation of the asymmetric ν_3 mode obtained by Boesten *et al.* [37] at a scattering angle of 90°. The excitation cross section increases sharply for energies approaching the vibrational excitation threshold (0.157 eV) due to direct excitation of the ν_3 mode which has a large infrared (IR) activity [37]. At higher energies, the data in Fig. 4.8(a) show a strong enhancement near 8 eV due to the T_2 resonance ($\ell = 1$ partial wave) near 6.6 eV (see Table 4.2), and a weak broad

Table 4.7. Recommended momentum transfer (elastic) cross section, $\sigma_m(\varepsilon)$, for CF_4.

ε (eV)	$\sigma_m(\varepsilon)$ (10^{-20} m^2)	ε (eV)	$\sigma_m(\varepsilon)$ (10^{-20} m^2)	ε (eV)	$\sigma_m(\varepsilon)$ (10^{-20} m^2)
0.0010	13.0	0.08	0.55	6	8.55
0.0015	12.3	0.09	0.39	7	8.68
0.0020	11.8	0.10	0.26	8	8.96
0.0025	11.3	0.15	0.13	9	10.1
0.0030	10.9	0.20	0.27	10	11.2
0.0035	10.6	0.25	0.48	15	13.4
0.0040	10.2	0.30	0.76	20	14.1
0.0045	9.93	0.35	1.05	25	12.5
0.0050	9.65	0.40	1.39	30	10.4
0.0060	9.14	0.45	1.76	35	8.80
0.0070	8.67	0.50	2.13	40	7.80
0.0080	8.25	0.60	2.82	45	7.24
0.0090	7.85	0.70	3.45	50	6.66
0.010	7.52	0.80	4.01	60	5.80
0.015	6.15	0.90	4.48	70	5.28
0.020	5.06	1.0	4.92	80	4.77
0.025	4.16	1.5	6.26	90	4.37
0.030	3.44	2.0	6.92	100	4.03
0.035	2.82	2.5	7.30	200	1.92
0.040	2.29	3.0	7.53	300	1.17
0.045	1.90	3.5	7.72	400	0.82
0.050	1.54	4.0	7.89	500	0.62
0.060	1.10	4.5	8.04	600	0.50
0.070	0.78	5.0	8.21	700	0.41

enhancement around 21 eV possibly due to indirect excitation via the decay of the E and T_2 resonances located in the vicinity of this energy (see Table 4.2). The strong direct excitation of the ν_3 mode is clearly seen in Fig. 4.8(b) where typical energy-loss spectra are shown [37] for the vibrational excitation of CF_4 at 2 and 8 eV. At small scattering angles [20° data in Fig. 4.8(b)], excitation of the ν_3 mode becomes stronger than elastic electron scattering. The Ramsauer-Townsend minimum occurs at (162 ± 25) meV [60], i.e., it almost coincides with the threshold for excitation of the ν_3 mode. The relative weakness of the ν_4 excitation is consistent with its weak IR absorption intensity as compared to that of ν_3 [123].

Another significant electron beam study of CF_4 has been carried out by Mann and Linder [66] for electron energies ranging from $\sim$ 0.5 to 12 eV. The results of these investigators are similar to those of Boesten

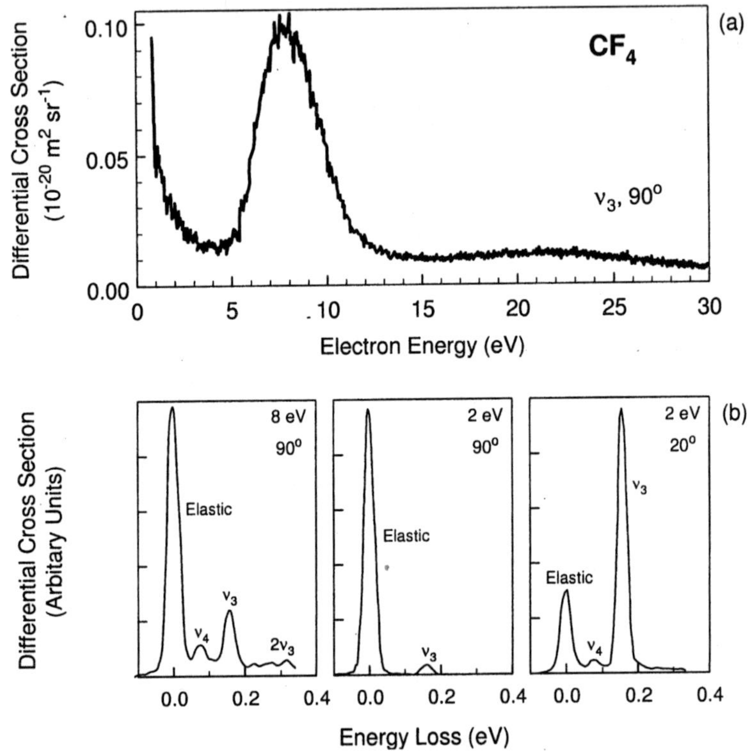

Figure 4.8. (a) Differential cross section for excitation of the asymmetric stretch ν_3 mode of CF_4 as a function of electron energy for a scattering angle of 90° [37]. (b) Electron energy-loss spectra of Boesten *et al.* [37] for vibrational excitation of CF_4 at the indicated incident electron energies and scattering angles.

et al. [37], namely, strong inelastic electron scattering is observed in two distinct energy regions: one in a resonance region between 6 and 11 eV (due to the T_2 negative ion resonance), and the other in a region of direct excitation below $\sim$ 2 eV via the dipole moment associated with ν_3. For low scattering angle and/or low incident electron energy, the cross section for direct excitation can be much larger than the resonant cross section. Excitation of the ν_3 asymmetric stretch mode is the dominant energy loss process over the entire range of collision energies below the threshold for electronic excitation. The excitation of the ν_4 and $2\nu_3$ modes is much smaller than elastic electron scattering under all conditions. For the excitation behavior of other modes (including overtones and combination modes) see Refs. [37, 66].

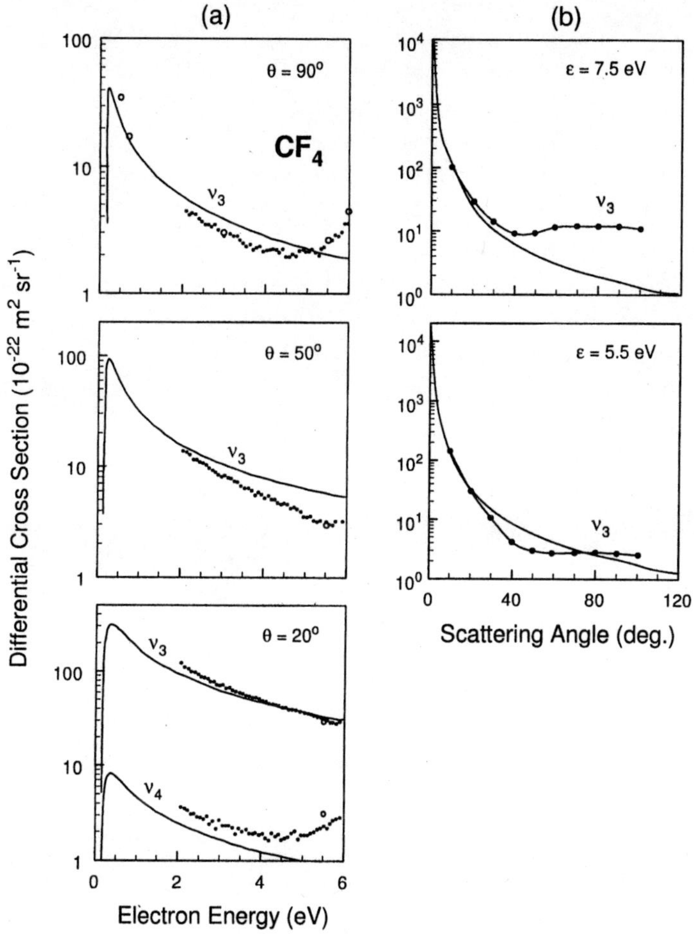

Figure 4.9. (a) Differential cross sections for the ν_3 and ν_4 vibrational modes of CF_4 as a function of electron energy at 20°, 50°, and 90° scattering angles. Comparison of the experimental results of Mann and Linder [66] with the Born dipole approximation (solid lines). (b) Differential cross section for the ν_3 vibrational mode for 5.5 and 7.5 eV incident electron energies; comparison of the experimental results of Mann and Linder [66] with the Born approximation (solid lines).

A comparison of measured cross sections with values calculated using the Born dipole approximation has been made by Mann and Linder [66] and is shown in Fig. 4.9(a). For energies below 5 eV, the experimental data are roughly described by the Born dipole approximation. At 90°, the agreement is good up to ~ 5.5 eV where the resonant part of the cross section becomes visible. At 50°, the measured cross section for ν_3 between 5 and 6 eV is half of the calculated one by the Born dipole approximation. At 20°, the agreement is good for ν_3, but the measured cross section for ν_4 is larger than that obtained from the Born dipole approximation. This may be due to a contribution to the scattering cross section by higher multiple moments and/or polarization as has been observed [66, 124] for modes with weak infrared (IR) activity [37, 124]. In Fig. 4.9(b) is shown the large cross section which is obtained with the Born dipole model in the forward direction. For a scattering angle up to 30° the calculated values agree with the measurements, but for larger scattering angles the two sets of data deviate. However, for the 7.5 eV data the difference at larger scattering angles may be ascribed to a resonant contribution to the measured cross section.

Hayashi [4] and Nakamura [5] obtained a set of cross sections for various processes using the two-term expansion approximation to the solution of the Boltzmann transport equation but with different sets of electron transport data for pure CF_4 and for CF_4 mixtures (see original references and Sec. 2.2.4 of Ch. 4). Hayashi used the electron drift velocity data of Christophorou *et al.* [18, 19] and Naidu and Prasad [120] which are 10–15% higher than the more recent values of Hunter *et al.* [108]. The vibrational cross sections $\sigma_{\nu 1,3}(\varepsilon)$ [$\approx \sigma_{\nu 3}(\varepsilon)$], and $\sigma_{\nu 2,4}(\varepsilon)$ [$\approx \sigma_{\nu 4}(\varepsilon)$] obtained by Hayashi for the ν_3 and ν_4 excitations, respectively, are shown by the solid lines in Figs. 4.10(a) and 4.10(b). These and the momentum transfer cross section were chosen so as to give $w(E/N)$ and $D_T/\mu\,(E/N)$ values that best agree with the measured $w(E/N)$ [18, 19, 120] and $D_T/\mu\,(E/N)$ [121] values.

The vibrational cross sections of Nakamura for the excitation of the same modes are also shown in Figs. 4.10(a) and 4.10(b) by the short dashed lines. Nakamura obtained these cross sections using the electron drift velocities and the longitudinal electron diffusion coefficients he measured in CF_4-Ar mixtures. The data of Hayashi and Nakamura differ in the threshold region and the resonant peak near 8 eV in the Nakamura cross section is absent from the Hayashi cross section function. The swarm-based cross sections [4, 5, 110] and the swarm studies [106] at low E/N, indicated the effect of direct vibrational excitation of infrared active modes and the effect of indirect vibrational excitation via resonances at high E/N. Nakamura reported that the measured ND_L at high E/N necessitates a large inelastic process around 7 eV. He obtained a reasonable agreement between the measured and the calculated electron transport coefficients by assuming a resonance peak in the vibrational excitation cross section. The magnitude of this cross section is consistent with the measurements of Jones [68], the beam results described earlier in this section, and the positions of the negative ion resonances obtained by other methods (see Table 4.2).

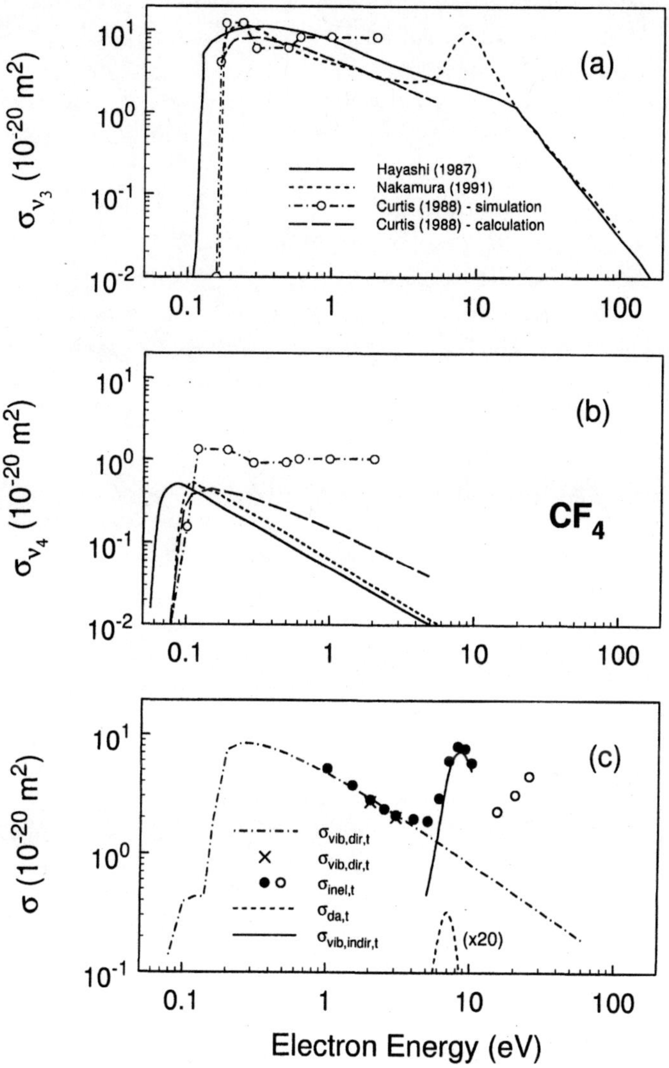

Figure 4.10. (a) Vibrational excitation cross section functions σ_{ν_3} for the ν_3 vibrational mode deduced from electron swarm data using a Boltzmann code (—, Ref. [4]; - - -, Ref. [5]) and a Monte Carlo simulation method (– o –, Ref. [106]), and calculations using the Born approximation (– –, Ref. [106]). (b) The same as above, but for σ_{ν_4} of the ν_4 mode. (c) Determination of indirect vibrational excitation cross section: (– · –) total direct vibrational excitation cross section obtained from a Born approximation calculation of $\sigma_{\text{vib,dir,t}}(\varepsilon) = \sigma_{\nu_3} + \sigma_{\nu_4}$ by Bonham [92]; (×) measurements of $\sigma_{\text{vib,dir,t}}(\varepsilon)$ by Mann and Linder [66]; (•) measurements of $\sigma_{\text{inel,t}}$ by Boesten *et al.* [37]; (—) indirect vibrational contribution due to resonances (see text).

The results of another swarm-based analysis which used a Monte Carlo simulation method [106] instead of Boltzmann equation analysis are compared in Figs. 4.10(a) and 4.10(b) with the results of the other two swarm-based studies [4, 5]. The Monte Carlo studies also show the strong direct excitation of the IR active modes ν_3 and ν_4 at near-threshold energies, but the energy dependence of these cross sections is difficult to rationalize physically. Curtis *et al.* [106] also calculated the cross section for direct excitation of ν_3 and ν_4 using the Born approximation and measured IR absorption intensities; these cross sections are also shown in Figs. 4.10(a) and 4.10(b).

Based on the reasonable agreement between the measured cross section for direct vibrational excitation and the cross sections predicted by the Born approximation, Bonham [92] calculated total vibrational excitation cross sections σ_{ν_3} and σ_{ν_4}, respectively, for the ν_3 and the ν_4 modes using the Born dipole formula. These are given in Table 4.8. Furthermore, Bonham added σ_{ν_3} and σ_{ν_4} for the two vibrational modes, which represents the total direct vibrational excitation cross section, and compared it with the experimental data on $\sigma_{\mathrm{inel,t}}(\varepsilon)$ of Boesten *et al.* [37] and Mann and Linder [66]. The agreement is seen (Fig. 4.10(c), Table 4.8) to be satisfactory only below $\sim$ 5 eV. Above this energy, indirect electron scattering through the negative ion resonances in the 6–8 eV range [solid line in Fig. 4.10(c)] makes a large contribution to vibrational excitation and accounts for the much larger measured cross sections compared to the values predicted by the Born approximation in this energy range.

Above $\sim$ 12.5 eV, the $\sigma_{\mathrm{inel,t}}(\varepsilon)$ data of Boesten *et al.* contain a contribution from electronic excitation. The difference between the measurements of Boesten *et al.* and the Born $\sigma_{\mathrm{vib,dir,t}}(\varepsilon)$ gives the cross section for indirect inelastic electron scattering $\sigma_{\mathrm{inel,indir,t}}(\varepsilon)$ from the CF_4 molecule. The indirect vibrational excitation cross section $\sigma_{\mathrm{vib,indir}}(\varepsilon)$ can then be obtained by subtracting the total electron attachment cross section $\sigma_{\mathrm{a,t}}(\varepsilon)$ (Sec. 2.5 of Ch. 4) from $\sigma_{\mathrm{inel,indir,t}}(\varepsilon)$ below 12.5 eV. This has been done in the last column of Table 4.8, and is shown as the solid line in Fig. 4.10(c). Clearly, indirect vibrational excitation is the predominant inelastic electron scattering process in the energy range from approximately 7 to 13 eV.

It should be noted that the cross section recommended in Ref. [1] for indirect vibrational excitation, $\sigma_{\mathrm{vib,indir,t}}(\varepsilon)$, was based on a deducement of the total inelastic electron scattering cross section by Boesten *et al.* [37] who subtracted their values for the elastic integral cross section $\sigma_{\mathrm{e,int}}(\varepsilon)$ from the values of Jones [68] for the total scattering cross section $\sigma_{\mathrm{sc,t}}(\varepsilon)$. While this technique is valid, a reassessment of this

Table 4.8. Vibrational cross sections for CF_4 in units of 10^{-20} m^2.

ε (eV)	$\sigma_{\nu 3}$ [a]	$\sigma_{\nu 4}$ [a]	$\sigma_{vib,dir,t} = \sigma_{\nu 3} + \sigma_{\nu 4}$	$\sigma_{inel,t} = \sigma_{sc,t} - \sigma_{e,int}$	$\sigma_{inel,t} - \sigma_{vib,dir,t} = \sigma_{inel,indir,t}$	$\sigma_{vib,indir,t} = \sigma_{inel,indir,t} - \sigma_{a,t}$
0.08		0.14	0.14			
0.10		0.39	0.39			
0.12		0.43	0.43			
0.14		0.43	0.43			
0.16	1.43	0.43	1.86			
0.20	7.03	0.40	7.43			
0.25	8.04	0.36	8.40			
0.275	8.11	0.35	8.46			
0.30	8.06	0.33	8.39			
0.35	7.80	0.30	8.10			
0.40	7.46	0.28	7.74			
0.50	6.77	0.24	7.01			
1.0	4.52	0.15	4.67			
1.5	3.43	0.11	3.54	3.70[b]		
2.0	2.79	0.09	2.88	2.81[b];2.7[c]		
3.0	2.06	0.06	2.12	2.09[b];2.0[c]		
5.0	1.39	0.04	1.43	1.88[b]	0.45	0.45
6.0	1.20	0.04	1.24	2.89[b]	1.65	1.64
7.0	1.06	0.03	1.09	6.02[b]	4.93	4.91
8.0	0.95	0.03	0.98	7.98[b]	7.00	6.99
9.0	0.87	0.03	0.90	7.63[b]	6.73	6.73
10.0	0.79	0.02	0.81	5.76[b]	4.95	4.95
15.0	0.57	0.02	0.59	2.25[b]	1.66[d]	1.66
20.0	0.45	0.01	0.46	3.09[b]	2.63[d]	
35.0	0.28	0.01	0.29	4.72[b]	4.43[d]	
50.0	0.21	0.01	0.22	5.56[b]	5.34[d]	
60.0	0.18	0.01	0.19			

[a] From Ref. [92] .
[b] Difference between the assessed $\sigma_{sc,t}(\varepsilon)$ (Table 4.3) and the assessed $\sigma_{e,int}(\varepsilon)$ (Table 4.6).
[c] From Ref. [66].
[d] Contains a contribution from inelastic scattering due to electronic excitation which increases with increasing energy above 12.5 eV.

calculation indicated that a more reliable value of $\sigma_{vib,indir,t}(\varepsilon)$ could be derived if the values of $\sigma_{e,int}(\varepsilon)$ and $\sigma_{sc,t}(\varepsilon)$ used in the calculation were those recommended in this work and in Ref. [1] rather than the ones used by Boesten *et al.* [37]. The values listed in the last three columns of Table 4.8 and the solid line in Fig. 4.10(c) were determined using the assessed values of $\sigma_{e,int}(\varepsilon)$ and $\sigma_{sc,t}(\varepsilon)$ (Tables 4.6 and 4.3, respectively). The peak value of $\sigma_{vib,indir,t}(\varepsilon)$ at 8 eV is ~ 11% higher than for the pre-

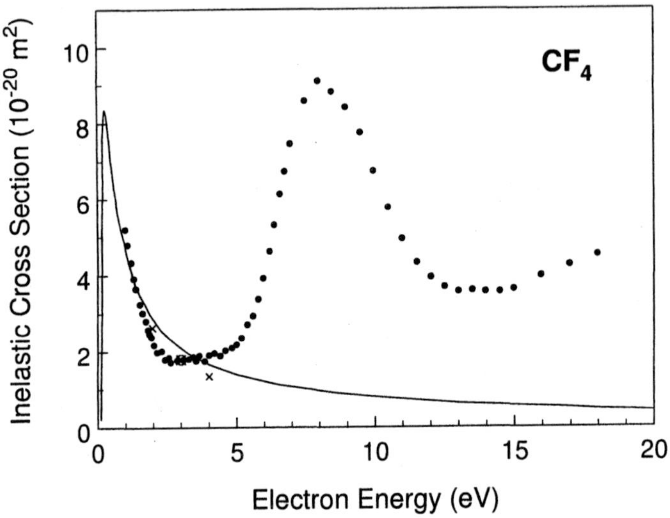

Figure 4.11. Inelastic integral cross section as a function of electron energy, $\sigma_{\mathrm{inel,int}}(\varepsilon)$, for CF_4: (•) Difference between the total cross section of Jones [68] and the elastic integral cross section of Mann and Linder [66]; (×) measured total vibrational excitation cross section of Mann and Linder [67]; (—) Born dipole approximation for vibrational excitation [67]; (⊗) normalization point.

vious data [1]. The difference is primarily due to the uncertainty in the measurements of $\sigma_{\mathrm{e,int}}(\varepsilon)$ in this energy range. Additional measurements of $\sigma_{\mathrm{vib,indir,t}}(\varepsilon)$ to reduce the uncertainty of this cross section are indicated due to its large influence on the calculated swarm parameters [110]. These measurements could either take the form of new measurements of $\sigma_{\mathrm{e,int}}(\varepsilon)$ to reduce the uncertainty of the deduced value of $\sigma_{\mathrm{vib,indir,t}}(\varepsilon)$, or direct measurements of $\sigma_{\mathrm{vib,indir,t}}(\varepsilon)$.

Electronic Excitation. There are no direct measurements of the cross sections for electronic excitation of the CF_4 molecule. Mann and Linder [67] determined the inelastic integral cross section for electron scattering from CF_4 by taking the difference between the total cross section of Jones [68] and their elastic integral cross section. This cross section is presented in Fig. 4.11 and is in essential agreement with the data of Boesten *et al.* [37] shown in Fig. 4.10(c). Below the onset of electronic excitation near 12.5 eV, this cross section is due to vibrational excitation with a small contribution from dissociative electron attachment (see Sec. 2.5 of Ch. 4). Above this energy, progressively the cross section has a larger contribution from electronic excitation [see Table 4.8 and Fig. 4.10(c)]. It has been argued [50], and is supported by mea-

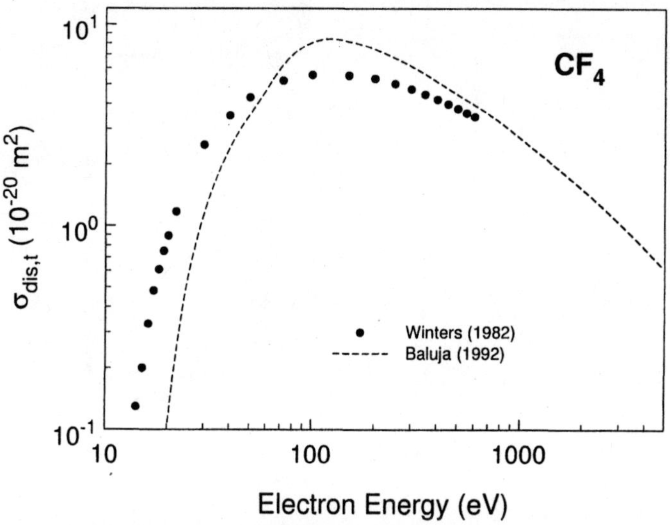

Figure 4.12. Total dissociation cross section $\sigma_{\mathrm{dis,t}}(\varepsilon)$ as a function of electron energy for CF_4: (•) Experimental measurements [50]; (- - -) calculated absorption cross section [99].

surements, that the total dissociation cross section for CF_4 is equal to the total electronic excitation cross section. The total electron-impact dissociation cross section for CF_4 has been measured by Winters and Inokuti [50] from threshold up to 600 eV with an overall uncertainty of ±20%. This cross section is shown in Fig. 4.12 and is listed in Table 4.9. The cross section rises from threshold to a maximum value of 5.55×10^{-20} m^2 near 100 eV, and decreases monotonically at higher energies. The results of a recent spherical complex potential calculation [99] of the absorption cross section of CF_4 are also shown in Fig. 4.12, and are seen to be in reasonable agreement with the experiment over much of the energy range.

An extrapolation of the experimental cross section data [50] to lower energies gave an apparent threshold for dissociation of $\sim$ 12.5 eV which coincides with the onset for electronic excitation. Dissociative electron attachment processes occur at lower energies than this energy (see Sec. 2.5 of Ch. 4). Dissociation into neutral fragments dominates near 12.5 eV, but the dissociative ionization process progressively (see Sec. 2.3 of Ch. 4) takes over above $\sim$ 30 eV. All ionization processes apparently lead to dissociation since the CF_4^+ ion has not been observed except in a few mass spectrometric studies, and even then with small abundances. Even if some excited electronic states of CF_4^+ are stable against disso-

Table 4.9. Total dissociation cross section, $\sigma_{\rm dis,t}(\varepsilon)$, measured by Winters and Inokuti [50] for CF_4.

ε (eV)	$\sigma_{\rm dis,t}(\varepsilon)$ (10^{-20} m^2)	ε (eV)	$\sigma_{\rm dis,t}(\varepsilon)$ (10^{-20} m^2)	ε (eV)	$\sigma_{\rm dis,t}(\varepsilon)$ (10^{-20} m^2)
12.5	0.024[a]	22	1.17	300	4.72
13.2	0.069[a]	30	2.50	350	4.45
14.2	0.13[a]	40	3.50	400	4.20
15.2	0.20[a]	50	4.30	450	3.98
16.2	0.33[a]	72	5.20	500	3.78
17.2	0.48[a]	100	5.55	550	3.60
18.2	0.61[a]	150	5.51	600	3.45
19.2	0.75[a]	200	5.32		
20.2	0.89[a]	250	5.02		

[a]Obtained from Fig. 5 in Ref. [50].

ciation [31, 55], they would dissociate upon deexcitation to the ground ionic state which is unstable. The cross sections for dissociative ionization are discussed in Sec. 2.3 of Ch. 4, and those for dissociation into neutral fragments in Sec. 2.4 of Ch. 4.

If all electronically excited states of CF_4 dissociate or predissociate, the total electronic excitation cross section should be equal to the total dissociation cross section. Since, moreover, at higher energies dissociative ionization dominates, the total ionization cross section should also be about equal to the total electronic excitation cross section (See Secs. 2.3, 2.4, and 2.8 of Ch. 4):

$$\sigma_{\rm exc,elec,t}(\varepsilon) = \sigma_{\rm dis,neut,t}(\varepsilon) + \sigma_{\rm i,t}(\varepsilon).$$

Model-based total electronic excitation cross sections have been reported [4, 5], but they are suspect because they are not consistent with the above relationship.

2.3 Electron-Impact Ionization of CF_4

2.3.1 Total Ionization Cross Section, $\sigma_{\rm i,t}(\varepsilon)$, of CF_4

The total ionization cross section is defined as

$$\sigma_{\rm i,t}(\varepsilon) = \sum_{i}^{\rm all} \left[q_i \sigma'_{\rm i,part}(\varepsilon)\right] \tag{4.2}$$

where $\sigma'_{\mathrm{i,part}}(\varepsilon)$ is the partial ionization cross section including contributions from positive ion pair formation, q_i is the number of charges on the corresponding ionic fragment, and the sum is over "all" ions produced (or, more correctly, over all ions "detected"). Many ionic fragments contribute to $\sigma_{\mathrm{i,t}}(\varepsilon)$ as can be seen from Table 4.10, where the thresholds and the translational energies for various positive ions observed in electron impact studies of CF_4 are given. A major difficulty in obtaining accurate values for the thresholds of the various fragment positive ions produced by electron impact on CF_4 is the high excess kinetic energy (1–15 eV) of many of the fragment ions (see Table 4.10 and Refs. [1, 53, 130]). Besides the energy threshold values listed in Table 4.10, a recent study [53] of the ionization threshold behavior of positive ion fragments from CF_4 gave values for the "appearance energies" of CF_3^+, CF_2^+, CF^+, and C^+, of (15.69±0.05), (21.47±0.1), (29.14±0.2), and (34.77±0.2) eV, respectively.

Dissociative ionization is the dominant process in CF_4 at electron impact energies above 30 eV, mainly due to the reaction [125]

$$e + \mathrm{CF_4} \rightarrow \mathrm{CF_3^+} + \mathrm{F} + 2e. \tag{4.3}$$

Since all excited electronic states of CF_4 and CF_4^+ are unstable or metastable, the total ionization cross section may serve as a lower limit to the total inelastic cross section.

An early discussion of electron-impact ionization of CF_4 was given by Märk [131]. Measurements until 1995 were reviewed, assessed, and discussed by Christophorou *et al.* [1]. We shall first follow the analysis of Christophorou and collaborators and then update their results. Christophorou *et al.* considered three sets of measurements of $\sigma_{\mathrm{i,t}}(\varepsilon)$. The first set of measurements of absolute partial and total electron impact ionization cross sections from threshold to 180 eV was that made by Stephan *et al.* [51]. These measurements were subsequently shown [130] to be in error due to strong discrimination in the ion extraction characteristics of their apparatus. The errors resulted in smaller cross section values and have been corrected by Poll *et al.* [130]. The original measurements of Stephan *et al.* [51] and the corrected measurements as reported by Poll *et al.* [130] are shown in Fig. 4.13(a).

The second set of measurements was by Ma *et al.* [125] who measured the partial electron impact ionization cross sections of CF_4 from threshold to 500 eV using a pulsed electron beam and a time-of-flight apparatus. They obtained the absolute value of the total ionization cross section by charge-weighted summing of all of the observed partial ionization cross sections. The uncertainty of their partial cross sections was estimated to be about ±15%. These cross sections are also plotted in

Table 4.10. Energy thresholds and excess kinetic energies of products from dissociative ionization of CF_4 by electron impact (from Ref. [82] unless otherwise stated).

Reaction products	Zero translation energy threshold (eV)	Energy threshold (eV)	Excess energy at threshold (eV)	Observed translation energy (eV)
A. Positive ion-neutral fragments				
$CF_3^+ + F$	14.7	15.9[a,b];16.2[c]	1.2	1.2[d]
$CF_2^+ + F_2$	19.3	22	2.7	
$CF_2^+ + 2F$	20.9		1.1	
$CF^+ + F + F_2$	22.1	27	4.9	
$CF^+ + 3F$	23.7		3.3	
$C^+ + 2F_2$	28.2	34.5	5.8	
$C^+ + F_2 + 2F$	29.8		4.7	
$C^+ + 4F$	31.4		3.1	
$F^+ + CF_3$	27.9	34.5	6.6	
$F^+ + CF_2 + F$	27.0		7.5	
$F^+ + CF + F_2$	30.3		4.2	
$F^+ + CF + 2F$	31.9		2.6	
$F^+ + C + F + F_2$	36.0			
$F^+ + C + 3F$	37.6			
$F_2^+ + CF_2$	23.6	35	11.4	
$F_2^+ + CF + F$	28.6		6.4	
$F_2^+ + CF_2$	32.6		2.4	
$F_3^+ + C + 2F$	34.2		0.8	
B. Doubly ionized fragments				
$CF_3^{++} + F_2$		41[e]		
$CF_2^{++} + F_2$		42[f]		
$CF_2^{++} + 2F$				
$CF^{++} + F + F_2$		52[g]		
$CF^{++} + 3F$				

[a] Average of a number of values in Refs. [47, 51, 125] (see also Ref. [82]).
[b] Ref. [83].
[c] From Table 4.1.
[d] Ref. [125].
[e] A synchrotron radiation study of dissociative photoionization of CF_4 [126] gives 40.7 eV for the production of CF_3^{++}.
[f] A synchrotron radiation study of dissociative photoionization of CF_4 [126] gives 41.7 eV for the production of CF_2^{++}.
[g] CF^{++} has been reported in Ref. [51] to have an energy threshold of (52.1 ± 0.5) eV.

Fig. 4.13(a). A subsequent paper by Bruce *et al.* [127] reported values for the "gross" ionization cross section [which is identical to what we

Table 4.10 (continued)
Energy thresholds and excess kinetic energies of products from dissociative ionization of CF_4 by electron impact (from Ref. [82] unless otherwise stated).

Reaction products	Zero translation energy threshold (eV)	Energy threshold (eV)	Excess energy at threshold (eV)	Observed translation energy (eV)
C. Positive ion pair formation				
$CF_3^+ + F^+$	32.4	36^h;37.6^i		$3.9–5.0^i$
$CF_2^+ + F^+ + F$	38.1	40^h;42.4^i	1.9	$3–4.7^i$
CF^+ $F^+ + F_2$	39.3	42^h	2.7	$3.6^{i,j}$
$CF^+ + F^+ + 2F$	40.9		1.1	
$C^+ + F^+ + F + F_2$	47.2	63^h	15.8	$15^{i,k}$
$C^+ + F^+ + 3F$	48.8		14.2	

[h] Ref. [127].
[i] Ref. [129].
[j] Ref. [129] gives 47.5 eV for the threshold energy for the production of $CF^+ + F^+$.
[k] Ref. [129] gives 62.0 eV for the threshold energy for the production of $C^+ + F^+$.

call here the total ionization cross section $\sigma_{\rm i,t}(\varepsilon)$], defined as

$$\sigma_{\rm i,gross}(\varepsilon) = \sum_{i}^{\rm all} [q_i \sigma_{\rm i,part}(\varepsilon) + 2\sigma_{\rm coinc,t}(\varepsilon)] , \quad (4.4)$$

where

$$\sigma_{\rm coinc,t}(\varepsilon) = \sum_{n=0}^{3} \left[\sigma_{{\rm CF}_n^+ + {\rm F}^+}(\varepsilon)\right] . \quad (4.5)$$

The $\sigma_{{\rm CF}_n^+ + {\rm F}^+}(\varepsilon)$ are the cross sections for the positive ion pair formation $CF_n^+ + F^+$ (for $n = 0$–3). The $\sigma_{\rm i,part}(\varepsilon)$ of Bruce *et al.* [127] differ from those, $\sigma'_{\rm i,part}(\varepsilon)$, of Ma *et al.* [125] by the fact that the contribution of positive ion pair formation (see Sec. 2.3.3 of Ch. 4) was not considered by Ma and others. However, the total ionization cross section is not affected by positive ion pair formation and would be the same in the two measurements as can be seen by the overlapping symbols in Fig. 4.13(a). Subsequently, the data of Ma and others underwent a number of further corrections. First, Bruce and Bonham [132] corrected the earlier data [125] upward by 5% to 15% to account for detection efficiency errors connected with insufficient ion impact energy on the front surface of the detector used in their equipment in the earlier measurements. The sum of the cross sections given in Table 1 of Ref. [132] can be used to determine the total ionization cross section. Second, Bonham [92, 133]

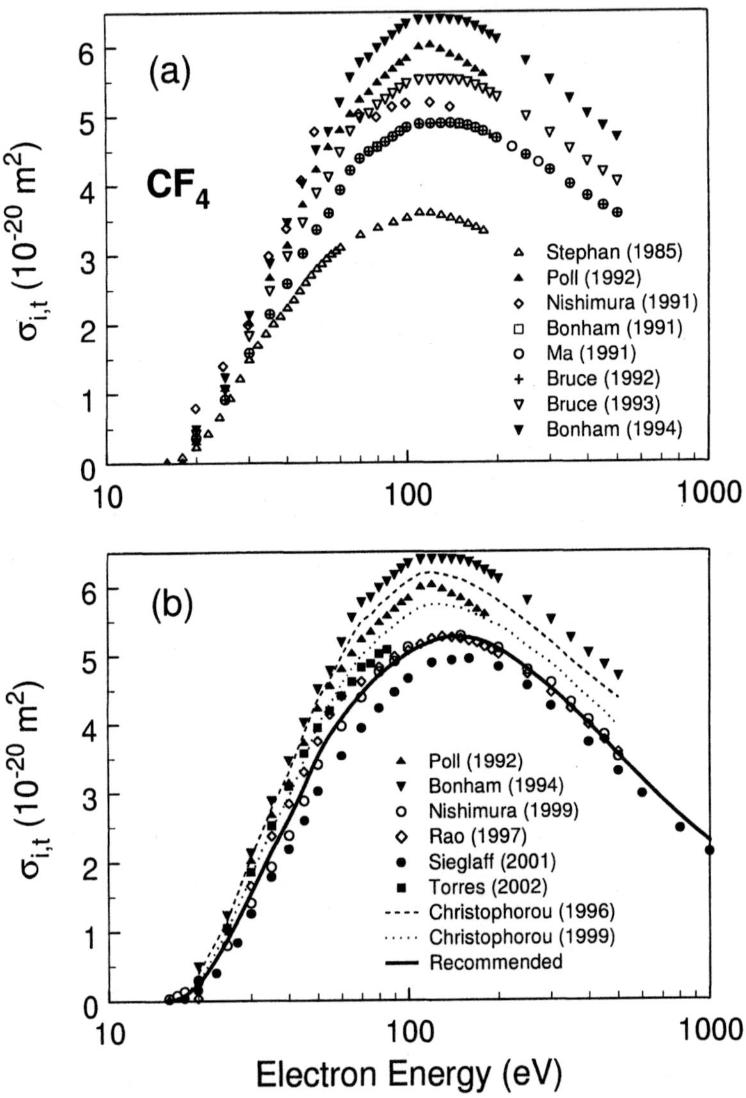

Figure 4.13. (a) Total ionization cross section $\sigma_{i,t}(\varepsilon)$ as a function of electron energy for CF_4. Measured values: (△) Ref. [51]; (▲) Ref. [130]; (□) Ref. [82]; (◇) Ref. [134]; (○) Ref. [125]; (+) Ref. [127]; (▽) Ref. [132]; (▼) data of Ref. [132] multiplied by 1.16 (per Ref. [92]. (b) Total ionization cross section $\sigma_{i,t}(\varepsilon)$ as a function of electron energy for CF_4 updated with inclusion of the data of (◇) Rao and Srivastava [136]; (○) Nishimura *et al.* [137]; (•) Sieglaff *et al.* [138]; and (■) Torres *et al.* [139]. Also shown in the figure are the experimental data, (▲) Ref. [130] and (▼) Refs. [92, 132], previously used to determine the recommended values in Ref. [1]. The previously recommended values are shown by the dashed line [1] and the dotted line [48]. The new recommended $\sigma_{i,t}(\varepsilon)$ (—) is based on the four newest sets of experimental data shown in Fig. 4.13(b) (see text).

reported that the data of Bruce and Bonham [132] have to be revised upward by 16% due to an additional instrumental detection efficiency correction [133]. Christophorou *et al.* [1] introduced such a correction to the Bruce and Bonham data [132]. The agreement between the corrected [92] cross sections of Ma and others and the corrected [130] cross sections of Stephan and others for the total ionization of CF_4 is within the ±15% uncertainty assigned to each set of the experimental data. All of these data, and the values of $\sigma_{i,t}(\varepsilon)$ for two incident electron energies obtained from Bonham and Bruce [82], are shown in Fig. 4.13(a). These final sets of data [92, 130, 132] were used by Christophorou *et al.* [1] to derive the original recommended values of $\sigma_{i,t}(\varepsilon)$ of CF_4 that are shown by the broken line in Fig. 4.13(b).

A third measurement that had not appeared in the archival literature at the time of the first review by Christophorou and collaborators was that of Nishimura [134] which fell below the final set of values from the other two groups. Similarly, at that time, unpublished results by Rao and Srivastava [135] also indicated lower values than the final set of values from the other two groups. In a subsequent assessment study by Christophorou and Olthoff [48], the final values of $\sigma_{i,t}(\varepsilon)$ by Bonham and others [▼, Fig. 4.13(a)] and Märk and others [▲, Fig.4.13(a)], and the more recent measurements of Rao and Srivastava [136] and Nishimura *et al.* [137] [see Fig. 4.13(b)] were considered to be reliable, and from a fit of these four sets of measurements they obtained the updated recommended values of $\sigma_{i,t}(\varepsilon)$ for CF_4 shown by the dotted line in Fig. 4.13(b).

Two additional measurements of $\sigma_{i,t}(\varepsilon)$ by Sieglaff *et al.* [138] and by Torres *et al.* [139] have recently become available, which also exhibit magnitudes below the two earliest assessments. The solid line in Fig. 4.13(b) is the presently recommended $\sigma_{i,t}(\varepsilon)$ for CF_4 and is based upon the four most recent sets of experimental data [136–139]. The cross sections of Refs. [136, 137] are total ionization measurements, while those of Refs. [138, 139] are the sum of the measured partial ionization cross sections. The latter have quoted uncertainties of ±5% to ±8% [138] and less than ±10% [139]. Values obtained from the solid line are listed in Table 4.11.

The recommended cross section $\sigma_{i,t}(\varepsilon)$ [Fig. 4.13(b), Table 4.11] can now be used to assess the results of the various calculations. A comparison between the recommended experimental values of $\sigma_{i,t}(\varepsilon)$ and the results of calculations is given in Fig. 4.14. Margreiter *et al.* [140] calculated the dissociative ionization cross section for singly charged species from CF_4 from threshold up to 200 eV using a semiclassical formula which is a combination of the classical binary encounter approximation, the Born-Bethe approximation, and the additivity rule. According to

Table 4.11. Recommended total ionization cross section, $\sigma_{i,t}(\varepsilon)$, for CF_4.

ε (eV)	$\sigma_{i,t}(\varepsilon)$ (10^{-20} m^2)	ε (eV)	$\sigma_{i,t}(\varepsilon)$ (10^{-20} m^2)	ε (eV)	$\sigma_{i,t}(\varepsilon)$ (10^{-20} m^2)
16	0.011	50	3.53	250	4.79
17	0.038	60	4.08	300	4.51
18	0.084	70	4.45	350	4.25
19	0.155	80	4.71	400	4.00
20	0.247	90	4.89	450	3.76
22	0.476	100	5.03	500	3.54
25	0.872	120	5.20	600	3.18
30	1.55	140	5.27	700	2.88
35	2.15	160	5.25	800	2.64
40	2.63	180	5.19	900	2.44
45	3.09	200	5.09	1000	2.28

Margreiter *et al.*, this calculation has the advantage that the ionization cross section can be described by a simple analytical formula which gives results in better agreement with the experimental data than the classical binary encounter approximation (i.e., the "Gryzinski formula" [141]). The results obtained using the Gryzinski formula are much higher and are in poor agreement with the measured values [140]. The results of the calculations by Margreiter *et al.* [140] are (Fig. 4.14) lower than the recommended values. In somewhat better agreement with the recommended cross sections are the results of the semiclassical calculation of Poll *et al.* [130], and the recent semiempirical results of Deutsch *et al.* [142, 143]. The results of the BEB calculation (see Ch. 2) of Kim and coworkers [137, 144] are high. In best overall agreement with the recommended values are the results of a recent "amplified binary encounter dipole model" calculation of Huo *et al.* [145].

Another source of total ionization cross sections is the "swarm-based" computations [4, 5]. In these analyses, however, the ionization cross sections come from other sources, but they might be adjusted to yield electron transport, ionization, and attachment coefficients consistent with the experiment. The total ionization cross sections from analysis of swarm data reported by Hayashi [4] and by Nakamura [5] shown in Fig. 4.14 do not agree well with the experimental values. It is worth noting that Hayashi started with the cross section of Leiter *et al.* [146], which is presumably the data of Stephan *et al.* [51] that were found later to be in error. It is interesting to observe that Hayashi found it necessary to multiply the cross section of Leiter *et al.* by 1.1 in order to fit

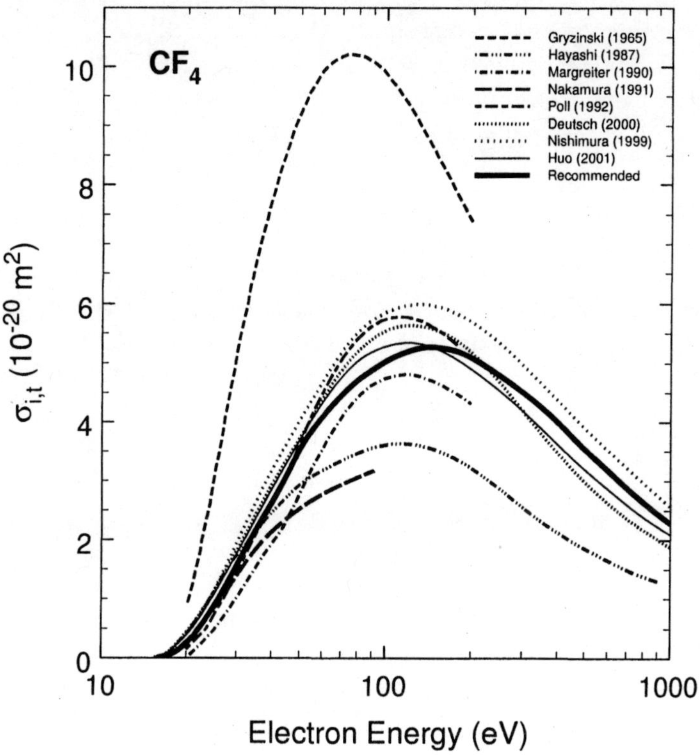

Figure 4.14. Comparison of the recommended $\sigma_{\mathrm{i,t}}(\varepsilon)$ with the results of various calculations: (—) Recommended from Fig. 13(b); (- - -) Ref. [141]; (– - –) Ref. [130]; (–·–) Ref. [140]; (– · · · –) Ref. [4]; (——) Ref. [5]; (· · ·) Ref. [137]; (· · ·) Ref. [143]; (—) Ref. [145].

the experimental data on $\alpha/N\,(E/N)$. No information has been given by Nakamura [5] about the source of the ionization cross section used as input in his analysis, but the results of his calculation are also in poor agreement with the experimental cross section data.

2.3.2 Partial Dissociative Ionization Cross Sections, $\sigma_{\mathrm{i,part}}(\varepsilon)$, of CF_4

There are four sets of measurements of the CF_4 partial dissociative ionization cross sections. These include the early measurements from the groups at the Universities of Innsbruck and Indiana for the production of the fragment ions CF_3^+, CF_2^+, CF^+, C^+, and F^+, and the most recent measurements by Sieglaff *et al.* [138] and Torres *et al.* [139]. The first set is that of Ma *et al.* [125] who quoted an uncertainty of ±15% in their measurements, and the second is that of Stephan *et al.* [51] who quoted an estimated uncertainty of about ±10%. Both sets of data have subsequently been revised and are discussed below.

The original data of Ma *et al.* [125] [$\sigma'_{\text{i,part}}(\varepsilon)$, plotted as open circles in Fig. 4.15] have been revised [converted to $\sigma_{\text{i,part}}(\varepsilon)$] by Bruce *et al.* [127] to consider the effect of positive ion pair formation on these cross sections. The revised data were further revised upward by Bruce and Bonham [132] to correct for detection errors due mainly to the effect of too low ion impact energy on the front surface of their detector in the earlier measurements as follows: 16% for CF_3^+, 13% for CF_2^+, 9% for CF^+, 5% for C^+, and 5% for F^+. Finally, another correction was made yet again by Bonham [92, 132], increasing their cross section values upward by 16% to account "for changes in instrumental detection efficiency." All of these cross sections are plotted in Fig. 4.15.

Similarly, we show in Fig. 4.15 the cross sections, $\sigma'_{\text{i,part}}(\varepsilon)$, of Stephan *et al.* [51] as corrected by Poll *et al.* [130]. These cross sections were not corrected for positive ion pair formation. Christophorou *et al.* [1] made such a correction as follows: they subtracted from the Poll *et al.* cross sections the corresponding positive ion pair cross sections of Bruce *et al.* [127] which they adjusted upward by 16% as suggested by Bonham [92]. The resultant cross sections are plotted in Fig. 4.15 and they represent the final values of $\sigma_{\text{i,part}}(\varepsilon)$ from the University of Innsbruck group.

Also shown in Fig. 4.15 are the recent measurements of $\sigma_{\text{i,part}}(\varepsilon)$ by Sieglaff *et al.* [138] and Torres *et al.* [139], which have, respectively, total quoted uncertainties of ±5% to ±8% and less than ±10%. The agreement between the two most recent sets of data is poor and it depends on the positive ion fragment.

To obtain suggested values of the $\sigma_{\text{i,part}}(\varepsilon)$ for the various positive ion fragments, we considered only the final data (represented by the symbols ▲ and ▼ in Fig. 4.15) from the University of Innsbruck and the University of Indiana groups (these are within the quoted combined error of the experiments) and the two most recent sets of measurements [138, 139]. A fit was made to the four sets of data for each ion fragment. These fits were then summed and normalized at each energy to the recommended total ionization cross section derived in the previous section. This is recognized to be an approximation, since the sum of these cross sections does not include contributions due to ion-pair formation and multiple ionization. The solid lines in the figure represent our suggested values, although it should be recognized that these values are uncertain especially for the CF^+, C^+, and F^+ ions. Values from the solid lines in Fig. 4.15 are listed in Table 4.12.

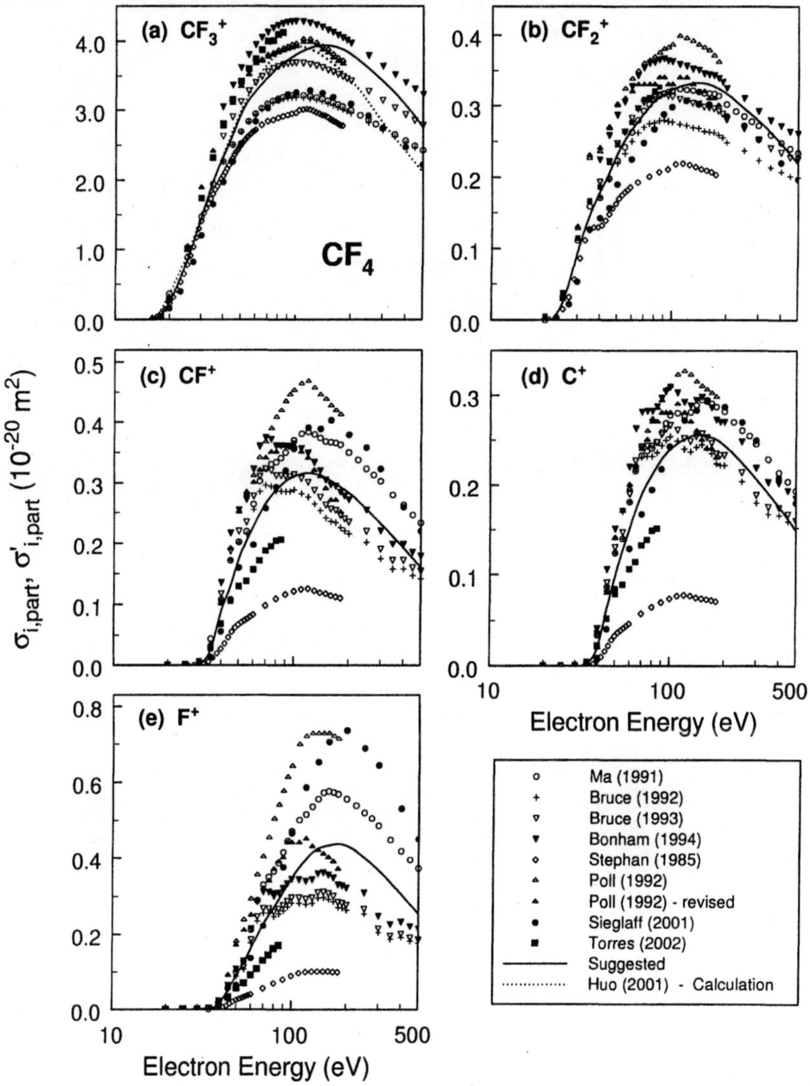

Figure 4.15. Partial ionization cross section for the production of (a) CF_3^+, (b) CF_2^+, (c) CF^+, (d) C^+, and (e) F^+ by electron collision on CF_4. $\sigma'_{\mathrm{i,part}}(\varepsilon)$: (◇) Ref. [51]; (△) Ref. [130]; (○) Ref. [125]. $\sigma_{\mathrm{i,part}}(\varepsilon)$: (+) Ref. [127]; (▽) Ref. [132]; (▼) Ref. [92]; (▲) revised data of Ref. [130] (see text); (●) Ref. [138]; (■) Ref. [139]. (—) Suggested data based upon a fit to the 4 data sets represented by solid symbols (see text). The data from the solid lines are presented in Table 4.12. Also shown (···) is a calculation of the cross section for CF_3^+ production by electron impact on CF_4 [147].

Table 4.12. Suggested values for partial ionization cross sections, $\sigma_{i,part}(\varepsilon)$, in units of 10^{-20} m^2, for the production of CF_3^+, CF_2^+, CF^+, C^+, and F^+ by electron impact on CF_4.

ε (eV)	CF_3^+	CF_2^+	CF^+	C^+	F^+
16	0.011				
18	0.084				
20	0.247				
22	0.474	0.002			
25	0.847	0.025			
30	1.45	0.093	0.004		0.001
35	1.97	0.145	0.026	0.003	0.003
40	2.33	0.179	0.088	0.021	0.014
45	2.64	0.205	0.130	0.063	0.052
50	2.94	0.232	0.173	0.099	0.090
60	3.28	0.266	0.227	0.151	0.161
70	3.48	0.289	0.265	0.189	0.227
80	3.63	0.306	0.288	0.212	0.272
90	3.73	0.316	0.303	0.228	0.311
100	3.81	0.323	0.311	0.239	0.347
120	3.91	0.330	0.316	0.251	0.397
140	3.94	0.333	0.312	0.256	0.425
160	3.93	0.330	0.303	0.254	0.434
180	3.88	0.324	0.293	0.251	0.438
200	3.81	0.317	0.283	0.244	0.433
250	3.61	0.296	0.255	0.225	0.401
300	3.42	0.279	0.232	0.207	0.368
400	3.08	0.247	0.192	0.176	0.306
500	2.76	0.219	0.159	0.150	0.255

2.3.3 Positive Ion Pair Formation Cross Section, $\sigma_{i,pair}(\varepsilon)$, of CF_4

As we have discussed in the preceding section, Bruce *et al.* [127] corrected their previous data on the partial ionization cross sections by taking into account the contributions to the reported cross section coming from the production of two-fragment positive ions in a single electron CF_4 collision. Bruce and others used a coincidence technique that enabled simultaneous detection of two positive ions. They measured cross sections for the production of the positive pair fragment cations C^+ + F^+, CF^+ + F^+, CF_2^+ + F^+, and CF_3^+ + F^+ from threshold to 500 eV with a reported uncertainty of ±20%. These cross sections are comparable to or larger than the double-ionization cross sections (see Ref. [127] and Sec. 2.3.5 of Ch. 4) with the $CF^+ + F^+$ production having the largest cross section.

According to Bonham [92], the data of Bruce *et al.* [127] must be adjusted upward by 16% to account for a change in instrumental efficiency [133] from 0.36 to 0.31. Presumably, this correction is independent of the electron energy and ionic mass. Thus the positive ion pair formation cross section data of Bruce *et al.* [127] were raised [1] by 16 % and the resultant cross sections are plotted in Fig. 4.16. There is still substantial disagreement between these data and those reported elsewhere by Bruce *et al.* [128] (see Fig. 4.16). However, the data of Bruce *et al.* [127] for the ion pairs $CF^+ + F^+$ and $CF_2^+ + F^+$ as adjusted by Christophorou *et al.* [1], are in good agreement with the recent measurements for these ion pairs by Sieglaff *et al.* [138]. The cross sections of Sieglaff *et al.* have a quoted uncertainty of ±20% and ±30%, respectively, for $CF^+ + F^+$ and $CF_2^+ + F^+$.

2.3.4 Total Counting Ionization Cross Section, $\sigma_{i,count,t}(\varepsilon)$, of CF_4

The total counting ionization cross section has been defined [125] as

$$\sigma_{i,count,t}(\varepsilon) = \sum_{i}^{all} \sigma_{i,part}(\varepsilon), \tag{4.6}$$

where $\sigma_{i,part}(\varepsilon)$ are the partial cross sections for single ionization and single ion formation. The values of $\sigma_{i,count,t}(\varepsilon)$ reported for CF_4 by Ma *et al.* [125], however, were not corrected for positive ion pair formation. Unlike the total ionization cross section, the counting cross section is affected by the positive ion pair formation process. As discussed in Sec. 2.3.3 of Ch. 4, Bruce *et al.* [127] measured positive ion pair formation cross sections by a coincidence method and were able to correct the partial ionization cross sections of CF_4 for this process. Consequently, they were able to determine "true" counting cross sections for ionization using the expression

$$\sigma_{i,count,t}(\varepsilon) = \sum_{i}^{all} [\sigma_{i,part}(\varepsilon) + \sigma_{coinc,t}(\varepsilon)] \tag{4.7}$$

where

$$\sigma_{coinc,t}(\varepsilon) = \sum_{n=0}^{3} \left[\sigma_{CF_n^+ + F^+}(\varepsilon)\right].$$

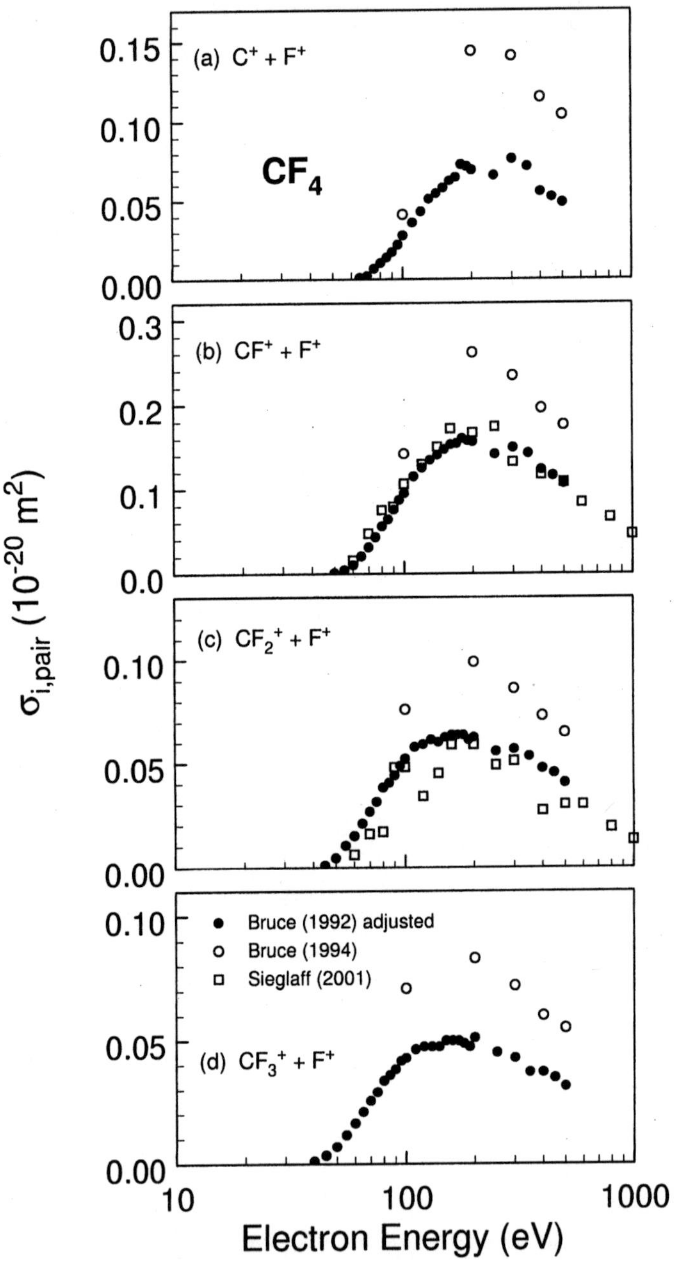

Figure 4.16. Positive ion pair formation cross sections as a function of electron energy for CF_4: (•) data of Bruce *et al.* [127] multiplied by 1.16 as suggested by Bonham [92]; (○) Bruce *et al.* [128]; (□) Sieglaff *et al.* [138].

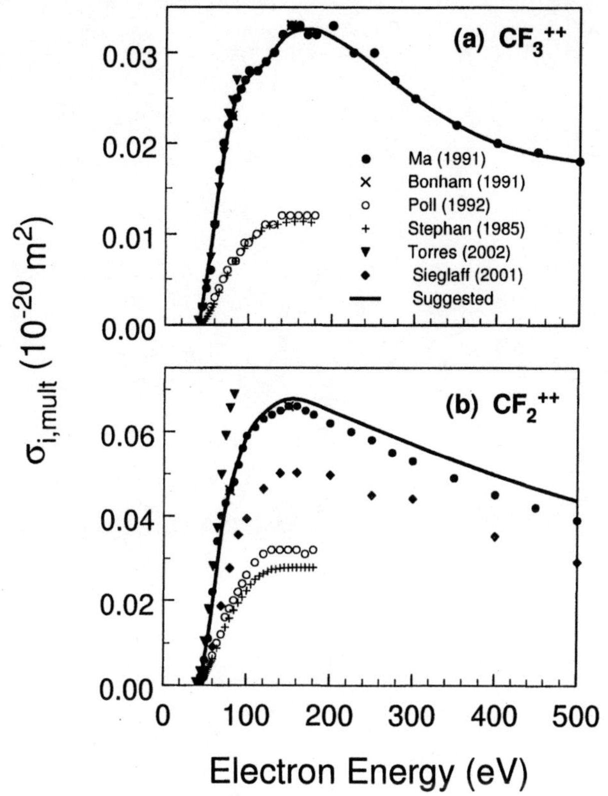

Figure 4.17. Multiple ionization cross sections $\sigma_{i,mult}(\varepsilon)$ as a function of electron energy for CF_4: (+) Ref. [51]; (○) Ref. [130]; (●) Ref. [125]; (×) Ref. [82]; (▼) Ref. [139]; (♦) Ref. [138]; (—) suggested.

2.3.5 Multiple Ionization Cross Sections, $\sigma_{i,mult}(\varepsilon)$, of CF_4

Multiple ionization cross sections have been measured for only CF_3^{++} and CF_2^{++} from CF_4 [51, 82, 92, 101, 125, 127, 130, 132, 135, 138, 139]. No multiply-charged parent ions of CF_4 have been reported. The data of Refs. [51, 82, 125, 130, 138, 139] for CF_3^{++} and CF_2^{++} are compared in Figs. 4.17(a) and 4.17(b), respectively. Ma *et al.* [125] quoted an uncertainty of ±15% in their values and Stephan *et al.* [51] of ±10%. Similarly, Sieglaff *et al.* [138] quoted an uncertainty for the production of CF_2^{++} equal to ±7% and Torres *et al.* [139] quoted less than ±10%. The cross section measurements of these four groups differ by more than the combined quoted experimental error. No explanation of this discrepancy has been advanced in the literature, but differences in the flight times of

Table 4.13. Suggested multiple ionization cross sections, $\sigma_{i,mult}(\varepsilon)$, for CF_4 in units of 10^{-20} m^2.

ε (eV)	CF_3^{++}	CF_2^{++}
40		0.0003
42	0.0005	
45	0.0022	0.0018
50	0.0050	0.0065
60	0.0115	0.0197
70	0.0185	0.0361
80	0.0235	0.0464
90	0.0261	0.0534
100	0.0274	0.0588
120	0.0290	0.0643
140	0.0315	0.0671
160	0.0325	0.0677
180	0.0325	0.0666
200	0.0317	0.0650
300	0.0250	0.0570
400	0.0200	0.0497
500	0.0180	0.0437

the ions in the various instruments employed may affect their respective detection efficiencies for the multiply-charged ions. It is worth noting, however, that the unpublished results of Rao and Srivastava [135] are in better agreement with the measurements of Ma and others than with the measurements of Stephan and others. For instance, at 100 eV the measured cross sections of Rao and Srivastava for CF_3^{++} and CF_2^{++} are, respectively, 0.024×10^{-20} m^2 and 0.075×10^{-20} m^2.

If we accept the data of Refs. [51, 130] as too low, the rest of the measurements are in good agreement for CF_3^{++} and in reasonable agreement for CF_2^{++}. The solid line in Fig. 4.17(a) is a least squares fit to the data of Refs. [82, 125, 139] for CF_3^{++}, and in Fig. 4.17(b) the solid line is a least squares fit to the data of Refs. [82, 125, 139, 138] for CF_2^{++}. Values taken off the solid lines are listed in Table 4.13 as our suggested data for the $\sigma_{i,mult}(\varepsilon)$ of the two doubly charged positive ions formed by electron impact on CF_4.

As expected, the multiple ionization cross sections are much smaller and the energy thresholds for them are much higher than those for single ionization. The cross sections for multiple ionization are, however, comparable to those for positive pair ion fragment formation.

2.3.6 Dissociative Ionization Cross Section, $\sigma_{i,dis}(\varepsilon)$, of CF_4

Since no parent CF_4^+ ion has been observed, the total ionization cross section $\sigma_{i,t}(\varepsilon)$ for CF_4 is equal to the total dissociative ionization cross section $\sigma_{i,dis}(\varepsilon)$ of CF_4.

2.3.7 Ionization Coefficients of CF_4

Density-Normalized Electron-Impact Ionization Coefficient, $\alpha/N\,(E/N)$. The coefficient $\alpha/N\,(E/N)$ is related to the normalized electron energy distribution function $f(\varepsilon, E/N)$ and the total ionization cross section $\sigma_{i,t}(\varepsilon)$ by

$$\alpha/N\,(E/N) = \left(\frac{2}{m}\right)^{1/2} w^{-1} \int_I^\infty f(\varepsilon, E/N)\varepsilon^{1/2}\sigma_{i,t}(\varepsilon)d\varepsilon, \qquad (4.8)$$

where I is the ionization threshold energy of the CF_4 molecule and m is the electron mass. There have been a number of room temperature measurements of $\alpha/N\,(E/N)$ [120, 121, 148–152] for CF_4 which have been reviewed by a number of authors (e.g., Refs. [1, 152, 153]). The more recent measurements by Shimozuma *et al.* [151] using a steady-state Townsend technique and by Hunter *et al.* [152] using a pulsed Townsend technique are considered to be the most accurate. The data of Hunter *et al.* [152] were corrected for diffusion and are deemed to be the most reliable. They extend up to E/N values of 200×10^{-17} V cm^2 while the measurements by Shimozuma *et al.* [151] extend up to 600×10^{-17} V cm^2.

Figure 4.18 shows the data for $\alpha/N\,(E/N)$ from the seven sources of measurements [120, 121, 148–152]. The measurements of Hunter *et al.* [152] and Shimozuma *et al.* [151] are shown by solid symbols. The data from these two measurements were used [48] to derive recommended values of $\alpha/N\,(E/N)$ because of the improved experimental techniques employed as compared to the earlier measurements. The solid line in Fig. 4.18 shows a fit to these two sets of measurements from which the recommended values of $\alpha/N\,(E/N)$ have been derived and are listed in Table 4.14.

Effective Ionization Coefficient, $\bar{\alpha}/N(E/N)$. The effective ionization coefficient is equivalent to the difference between the density-normalized electron-impact ionization coefficient, $\alpha/N\,(E/N)$, and the density-normalized electron attachment coefficient, $\eta/N\,(E/N)$, (see Sec. 2.5.3 of Ch. 4). The recent measurements of Hunter *et al.* [152] and Datskos *et al.* [154] on $(\alpha - \eta)/N\,(E/N)$ for pure CF_4 are presented in

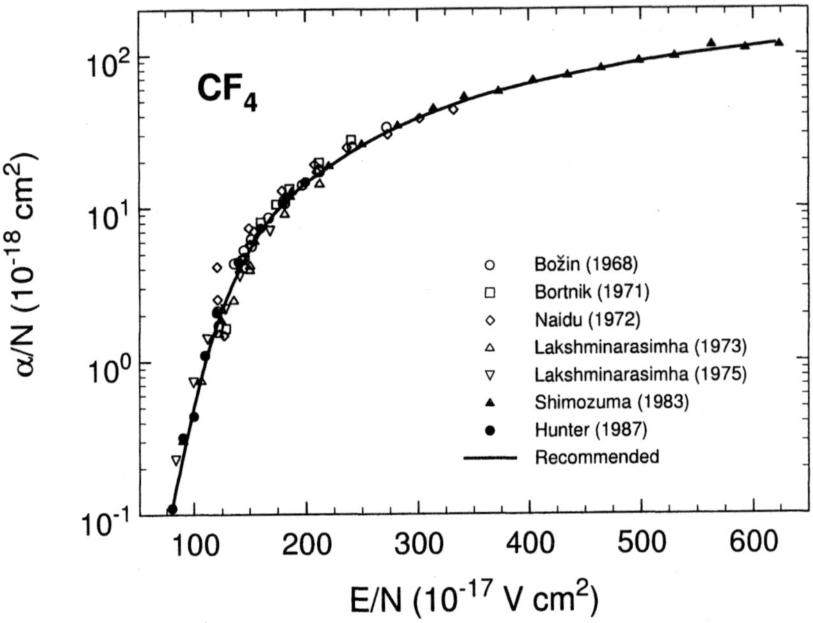

Figure 4.18. Measurements of the density-normalized electron-impact ionization coefficient as a function of E/N, $\alpha/N\,(E/N)$, of CF_4: (▲) Ref. [151]; (•) Ref. [152]; (○) Ref. [148]; (□) Ref. [149]; (◊) Ref. [120]; (△) Ref. [121]; (▽) Ref. [150]. The solid line is a fit to the data of Hunter *et al.* [152] and Shimozuma *et al.* [151], and represents the recommended values for the $\alpha/N\,(E/N)$ of CF_4.

Fig. 4.19. There is a general agreement between these data and the earlier measurements [120, 151]. The fit to all the data in Fig. 4.19 is represented in the figure by the solid line. Values from this line are listed in Table 4.15 as our recommended values for the effective ionization coefficient of CF_4. All data were equally weighted in the fitting process.

Measurements of effective ionization coefficients of mixtures of CF_4 with Ar, CO_2, H_2O, and CH_4 have also been made and can be found in Refs. [154, 155].

Average Energy to Produce an Electron-Ion Pair, *W*. The average energy to produce an electron-ion pair, W, for α particles (initial energy $\sim$ 5.1 MeV) has been measured for pure CF_4 and found to be 34.3 eV per ion pair [156]. The large value of this quantity for CF_4 is consistent with the high ionization onset and the large cross section for electron-impact dissociation of CF_4 into neutral species. W values for

Table 4.14. Recommended values of the density-normalized ionization coefficient, $\alpha/N\,(E/N)$, for CF_4.

E/N (10^{-17} V cm^2)	α/N (10^{-18} cm^2)	E/N (10^{-17} V cm^2)	α/N (10^{-18} cm^2)
80	0.11	360	54.2
100	0.50	380	59.3
120	1.71	400	64.2
140	3.82	420	69.1
160	6.97	440	74.0
180	10.5	460	79.1
200	14.4	480	84.3
220	18.3	500	89.7
240	22.8	520	94.8
260	27.8	540	99.3
280	32.9	560	103.5
300	38.1	580	107.9
320	43.4	600	112.4
340	48.8	620	116.9

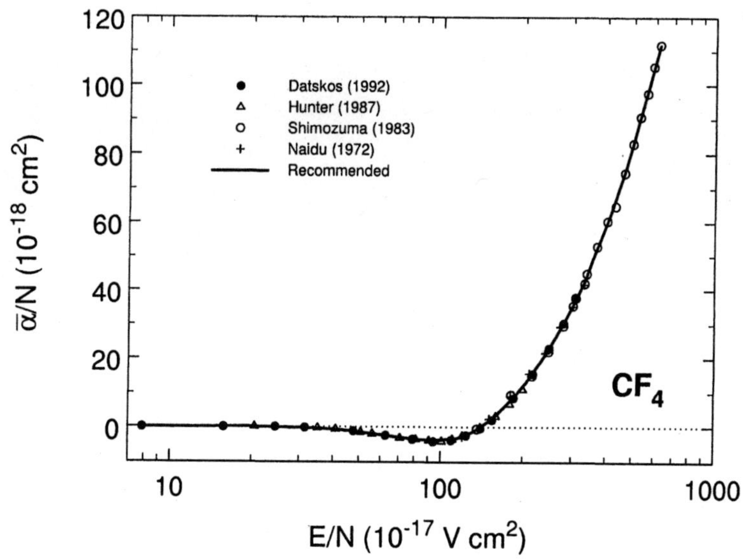

Figure 4.19. Effective ionization coefficient $\bar{\alpha}/N(E/N) = (\alpha - \eta)/N\,(E/N)$ for CF_4: (•) Ref. [154]; (△) Ref. [152]; (○) Ref. [151]; (+) Ref. [120]; (—) recommended.

Table 4.15. Recommended values of the effective ionization coefficient, $\bar{\alpha}/N(E/N) = (\alpha - \eta)/N\,(E/N)$, for CF_4.

E/N (10^{-17} V cm^2)	$\bar{\alpha}/N$ (10^{-18} cm^2)	E/N (10^{-17} V cm^2)	$\bar{\alpha}/N$ (10^{-18} cm^2)
8	−0.052	100	−3.66
9	−0.056	110	−3.43
10	−0.058	120	−2.68
15	−0.051	130	−1.54
20	−0.043	140	−0.033
25	−0.143	150	1.61
30	−0.281	200	12.0
35	−0.438	250	23.5
40	−0.730	300	35.0
45	−1.09	350	47.2
50	−1.47	400	59.4
60	−2.24	450	71.4
70	−2.92	500	83.6
80	−3.47	600	107.4
90	−3.73	620	111.7

CF_4/Ar and CF_4/CH_4 mixtures for α particles have also been made and can be found in Ref. [156].

2.4 Electron-Impact Dissociation of CF_4 into Neutral Fragments

2.4.1 Total Dissociation Cross Section for Neutral Species, $\sigma_{\mathrm{dis,neut,t}}(\varepsilon)$, of CF_4

Direct measurement of the total cross section for processes producing only neutral fragments by electron impact with CF_4 is difficult because of the problems connected with the detection of low-energy neutrals. One such study by Nakano and Sugai [83] accomplished direct measurement of the electron energy dependence of the partial cross sections for dissociation of CF_4 into neutral CF_3, CF_2, and CF radicals, by using threshold-ionization mass spectrometry in a dual-electron beam system. Their original measurements [83] have subsequently been revised by Sugai *et al.* [157] (their data have been renormalized using the absolute cross sections for parent and dissociative ionization of CF_x radicals of Tarnovsky and co-workers [158, 159]. See also Huo *et al.* [160]). The

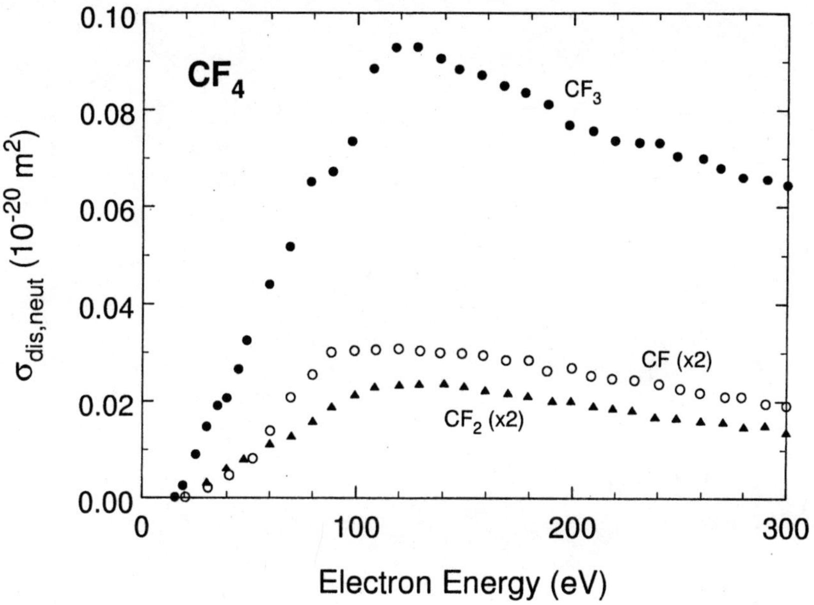

Figure 4.20. Cross sections for electron impact dissociation of CF_4 into CF_3 (•), CF_2 (▲), and CF (○). Data of Nakano and Sugai [83] as revised by Sugai *et al.* [157].

revised data differ substantially from the initially reported values. In the energy range 100–130 eV, the revised cross sections are lower than their initially reported values by factors of approximately 4, 13, and 16 for CF_3, CF_2, and CF, respectively. The revised data [157] are shown in Fig. 4.20. These data are too low, as can be seen from the discussion below (see Fig. 4.21).

The total neutral dissociation cross sections for CF_4 were calculated by summing the partial cross sections obtained by direct measurement [83, 157]. These values represent an estimate of the total cross section for electron impact dissociation of CF_4 into the neutral fragments CF_3, CF_2, and CF, and are presented in Fig. 4.21 for both the original [83] (open circles) and the revised [157] (closed circles) data. These are compared in Fig. 4.21 with other total neutral dissociation cross sections obtained by indirect measurements [92, 125, 127] and by swarm-based calculations [4, 5]. It is obvious that little agreement exists among these data.

The first indirect determination of the total dissociation cross section into neutral-neutral fragment pairs (regardless of the nature of the internal degree of excitation of the fragments) was obtained by Ma *et al.* [125] (open triangles) who subtracted their total ionization cross section from the total dissociation cross section of Winters and Inokuti [50].

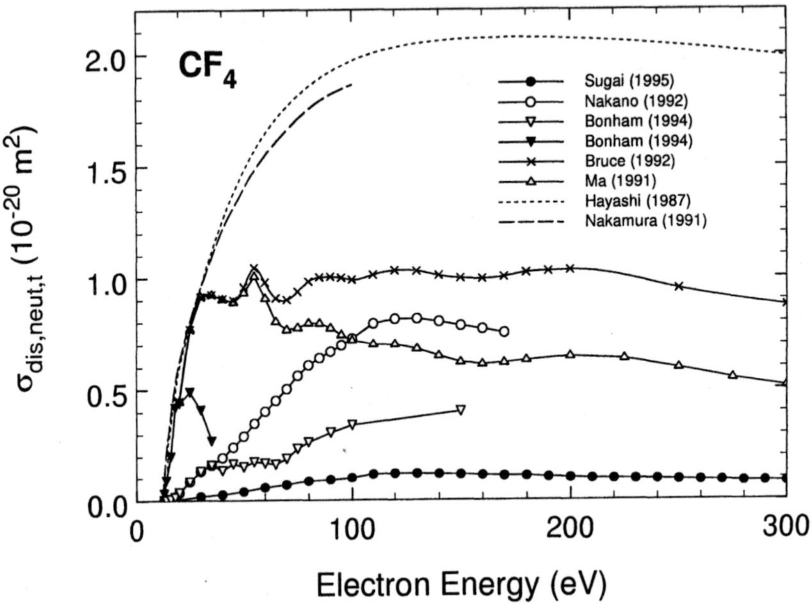

Figure 4.21. Total cross section for electron-impact dissociation of CF_4 into neutral-neutral pairs [(△) Ref. [125]; (×) Ref. [127]; (- - -) Ref. [4]; (– –) Ref. [5]; (▼) Ref. [92]], and into CF_n radicals [(○) Ref. [83]; (▽) measurements of Nakano and Sugai [83] as revised by Bonham [92]; (•) measurements of Nakano and Sugai [83] as revised by Sugai *et al.* [157]; see text].

Since in the determination of the total neutral dissociation cross section Ma and others ought to have used the total counting cross section rather than the total ionization cross section, these values were revised by Bruce *et al.* [127] to correct for double positive ion production (× symbols). However, the ionization cross sections used to derive both of these cross section values have been subsequently revised [92, 132] upward so these values for $\sigma_{\text{dis,neut,t}}(\varepsilon)$ are no longer valid. In fact, the magnitude of the total counting cross section [92] now approaches (or exceeds) the magnitude of the total dissociation cross section [50], so that the error in the difference of the two cross sections is too large for this method to be used to determine $\sigma_{\text{dis,neut,t}}(\varepsilon)$ for electron energies above 30 eV. Below 30 eV, an estimate can be made for $\sigma_{\text{dis,neut,t}}(\varepsilon)$ using this method, since the magnitude of the measured total dissociation cross section [50] is significantly greater than the total ionization cross section in this energy range. Bonham [92] performed this calculation, and the results are presented (solid inverted triangles) in Fig. 4.21. Interestingly, these data are very much greater than those measured by

Sugai *et al.* [157]. Cross section values for $\sigma_{\rm dis,neut,t}(\varepsilon)$ were derived by Bonham [92] from the original data of Nakano and Sugai [83] and are also shown in Fig. 4.21 (open inverted triangles), but these are no longer valid since the original data of Sugai and co-workers have been superseded by their more recent analysis [157]. Also shown in Fig. 4.21 are the swarm-based $\sigma_{\rm dis,neut,t}(\varepsilon)$ cross sections derived by Hayashi [4] and Nakamura [5], which significantly exceed all of the experimentally derived values.

Clearly, the available data at the time of the review by Christophorou *et al.* [1] were inconsistent. Indeed, although the cross sections measured by Sugai *et al.* [157] were designated by Christophorou *et al.* as recommended because they were the only measurements of $\sigma_{\rm dis,neut,t}(\varepsilon)$ available at the time, Christophorou *et al.* emphasized that the Sugai *et al.* cross sections were too low and inconsistent with the accepted values of $\sigma_{\rm i,t}(\varepsilon)$ and $\sigma_{\rm dis,t}(\varepsilon)$ cross sections, and they called for further measurements of this important cross section. Indeed, two recent measurements [161, 162] of $\sigma_{\rm dis,neut,t}(\varepsilon)$ confirm that earlier values by Sugai *et al.* [157] are too small. Figure 4.22 shows the absolute measurements of Mi and Bonham [161] and the relative measurements of Motlagh and Moore [162] normalized to the data of Mi and Bonham [161], along with the measurements by Sugai *et al.* [157]. The solid line in Fig. 4.22 are the values of $\sigma_{\rm dis,neut,t}(\varepsilon)$ suggested by Christophorou and Olthoff [48] (see Table 4.16). As can be seen from Fig. 4.22, the data still exhibit significant differences when compared to the difference between the recommended values of $\sigma_{\rm dis,t}(\varepsilon)$ and $\sigma_{\rm i,t}(\varepsilon)$. This suggests the need for additional measurements of $\sigma_{\rm dis,neut,t}(\varepsilon)$.

Finally, Bruce *et al.* [127] were able to determine the total cross section $\sigma_{\rm F,t}(\varepsilon)$ for production of neutral fluorine by electron impact on CF_4. This cross section is shown in Fig. 4.23. As noted by Bruce *et al.* [127], the cross section for the production of neutral fluorine is large (even larger than the total ionization cross section) due to the many processes that lead to the production of one or more fluorine atoms. Above approximately 40 eV most of the neutral fluorine results from ionization processes.

2.4.2 Dissociative Excitation Cross Section, $\sigma_{\rm dis,exc}(\varepsilon)$, of CF_4

Becker [24] has discussed the formation of radiating and energetic metastable fragments following single electron impact on CF_4. A number of different processes contribute to the production of such excited and energetic fragments [163]. The study of Van der Burgt and Mc-

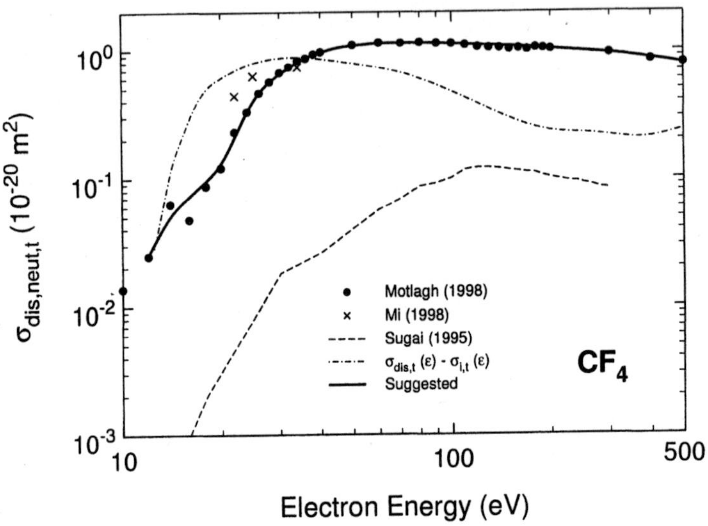

Figure 4.22. Total cross section for electron-impact dissociation into neutrals, $\sigma_{\rm dis,neut,t}(\varepsilon)$, for CF_4 based upon recent measurements from Ref. [161] (×) and Ref. [162] (•). The solid line represents the suggested values [48] for this cross section. For comparison, the previously recommended values [1] (based on the measurements of Ref. [157]) are shown as a dashed line. Also shown (– · –) is the value of $\sigma_{\rm dis,t}(\varepsilon) - \sigma_{\rm i,t}(\varepsilon)$ determined from the suggested values derived in Secs. 2.2.5 and 2.3.1 of this chapter.

Table 4.16. Suggested total cross section for electron-impact dissociation of CF_4 into neutrals, $\sigma_{\rm dis,neut,t}(\varepsilon)$.

ε (eV)	$\sigma_{\rm dis,neut,t}(\varepsilon)$ (10^{-20} m^2)	ε (eV)	$\sigma_{\rm dis,neut,t}(\varepsilon)$ (10^{-20} m^2)	ε (eV)	$\sigma_{\rm dis,neut,t}(\varepsilon)$ (10^{-20} m^2)
12	0.025	65	1.13	160	1.06
14	0.050	70	1.14	170	1.05
16	0.072	75	1.14	180	1.04
18	0.095	80	1.13	190	1.03
20	0.13	85	1.13	200	1.02
25	0.40	90	1.13	250	0.97
30	0.65	95	1.12	300	0.94
35	0.84	100	1.12	350	0.90
40	0.96	110	1.11	400	0.86
45	1.05	120	1.09	450	0.82
50	1.09	130	1.08	500	0.78
55	1.11	140	1.07		
60	1.13	150	1.07		

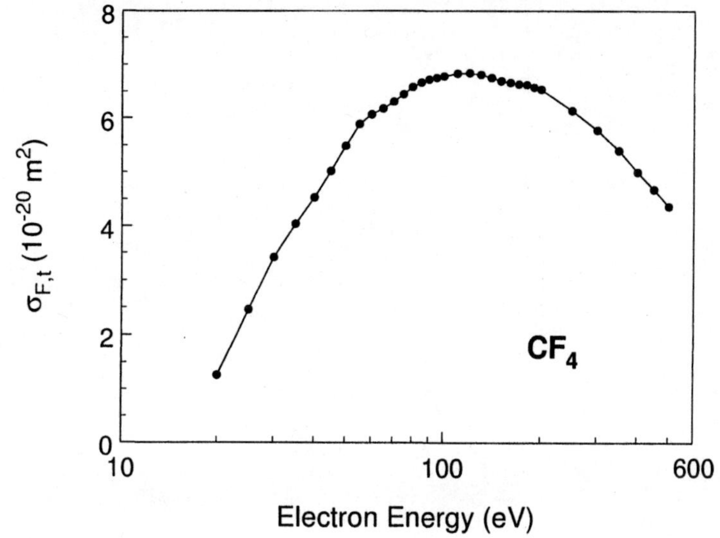

Figure 4.23. Total cross section for the production of neutral atomic fluorine by electron impact on CF_4 from Bruce *et al.* [127].

Conkey [163] indicated the importance of two-fragment dissociation generating simultaneously excited F* and excited CF_3^{+*} from the process

$$e + CF_4 \rightarrow CF_4^{**} \rightarrow CF_3^{+*} + F^*. \tag{4.9}$$

Electron impact-induced light emission studies [24, 163–165] have shown that by far the most prominent emission produced by electron impact on CF_4 is the intense continuous emission from 200 to 500 nm. This emission (Fig. 4.24) has a maximum near 285 nm and, with the exception of a series of weak discrete emission bands in the 360–420 nm range and a shoulder at 245 nm, is essentially structureless. The source of this emission is still in question [164] although Becker *et al.* [24] assigned it to the CF_3^{+*} species.

Cross sections for the most intense $3p \rightarrow 3s$ line emissions from atomic fluorine between 620 and 780 nm following electron impact on CF_4 have been reported by Blanks *et al.* [166, 167], Van Sprang *et al.* [164], and more recently by Torres *et al.* [168]. These lines are emitted by atomic fluorine formed in the various excited states associated with the $1s^22s^22p^43p$ electronic configuration [166–169]. The measured emission cross sections for the fluorine resonance lines are listed in Table 4.17 for 100 eV electron impact. The cross section values of Van Sprang *et al.* [164] are generally higher than those of Blanks *et al.* [167] and Torres *et*

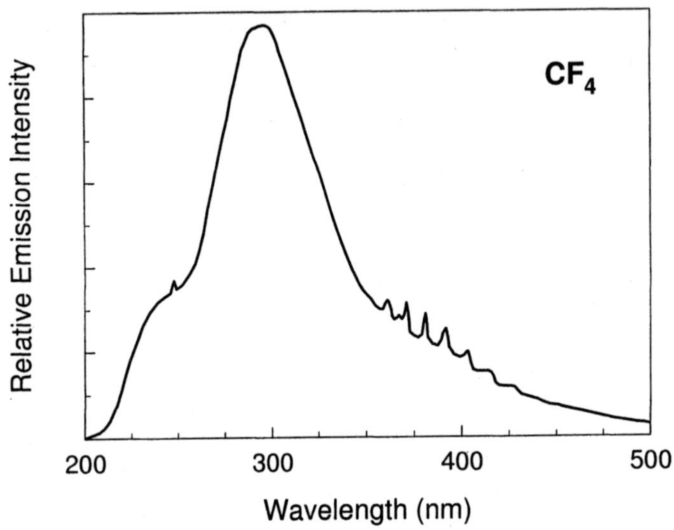

Figure 4.24. Optical emission from CF_4 in the wavelength range 200–500 nm produced by collisions of 100 eV electrons with CF_4. Measurement by Becker and coworkers [24, 165].

al. [168]. Similarly, the cross section values of Torres *et al.* [168] at 624.1, 635.0, and 677.6 nm are substantially higher than those of Blanks *et al.* [167]. This difference has been attributed by Torres *et al.* [168] to the contribution of the CF_3^{+*} continuum at these wavelengths in their experiments. (See also Ref. [170] for emission spectra in the range 50–130 nm via dissociative excitation of CF_4 by electron impact.)

It is worth noting that a number of photoabsorption studies of CF_4 have been made using synchrotron radiation which provided complementary information on the decomposition of CF_4 and the nature of the emitting species. Thus photolytic studies have been made on the decay pathways of the excited electronic states of CF_4 [171], the emission spectra of CF_3 radicals [172], the fluorescence cross sections, the nature of emitting species, and the possible emitting states [173]. Interestingly, Lee *et al.* [173] measured the photoabsorption and fluorescence cross sections for CF_4 and CF_3X ($X = H$, Cl, Br) at 50–106 nm using synchrotron light and identified the emitting species to be mainly excited CF_2^*, CF_3^*, and CF_3X^{+*}, but not CF_3^{+*}. (See also Becker [174]).

Laser-induced fluorescence has been used (e.g., see Refs. [175–179]) effectively to identify radicals and to follow their space- and time-distribution in high density CF_4 plasmas.

Table 4.17. Absolute emission cross sections, σ_{em}, at 100 eV, for various visible FI line emissions for CF_4.

		σ_{em} (10^{-19} cm^2)		
Transition	Line (nm)	Ref. [167][a]	Ref. [164][b]	Ref. [168][c]
	624.0 (624.1)[d]	1.2	—	6.6
$^4S^0 \rightarrow {}^4P$	634.9 (635.0)	0.8	—	4.2
	641.3	0.6	—	—
	677.4 (677.6)	0.8	1.6	1.9
	679.5	0.2	—	—
	683.4 (683.6)	1.2	—	2.0
$^4D^0 \rightarrow {}^4P$	685.6 (685.8)	4.7	6.1	4.7
	687.0	0.9	—	—
	690.2 (690.4)	2.8	4.7	3.4
	691.0	0.9	—	—
	703.7 (703.9)	1.2	2.1	0.83
$^2P^0 \rightarrow {}^2P$	712.8	0.7	1.3	—
	720.2	0.4	—	—
$^2S^0 \rightarrow {}^2P$	731.1 (731.1)	3.3	2.0	1.2
	733.2 (733.4)	—	2.3	1.2
	739.9 (740.1)	3.9	4.1	1.1
$^4P^0 \rightarrow {}^4P$	746.5 (742.8)	0.6	1.2	0.6
	755.2	—	—	—
	757.3	1.9	—	—
	760.7	—	1.0	—
$^2D^0 \rightarrow {}^2P$	775.5 (775.7)	3.5	4.9	2.6
	780.0 (780.2)	2.1	2.0	1.8

[a] The authors quoted an uncertainty of ±20% for cross section values larger than 1×10^{-19} cm^2, and about ±25% for cross sections smaller than 1×10^{-19} cm^2. Cross sections for emissions at wavelengths longer than 750 nm have an uncertainty of ±20% [167].
[b] Quoted uncertainty ±10%.
[c] Quoted uncertainty ±10% – ±15%.
[d] Values in parentheses are those of Ref. [168].

2.5 Electron Attachment to CF_4

In Sec. 2.1 of this chapter we summarized and discussed information on the negative ion states of the CF_4 molecule as obtained from electron scattering, electron attachment, and theoretical studies. In this section we assess and discuss (i) electron beam data on the cross sections for the production of specific ions by resonance and nonresonance electron attachment to CF_4, and (ii) electron swarm data on the total electron attachment cross section as a function of electron energy, electron attachment rate constants as a function of E/N and $\langle \varepsilon \rangle$, and electron attachment coefficients as a function of E/N.

The negative ions that have been observed in the majority of the single-collision electron beam studies of CF_4 are the complementary ions F^- and CF_3^- (see, however, Iga *et al.* [180] and later in this section). These fragment negative ions are produced via two broad and overlapping resonances located between 4.5 and 10 eV [62, 63, 70–73, 81, 181]. The energetics and the cross sections for their production have been studied by a number of electron swarm [72, 182] and electron beam measurements [63, 70–72, 81, 175, 180, 181, 183].

Electron beam data, especially ion kinetic energy measurements [62, 63, 73, 183], have shown that the resonant electron attachment to CF_4 occurs mainly in the 6 to 8 eV energy range via two negative ion states: (i) Via the ground state of CF_4^- at 6.8 eV producing F^- and CF_3^- through the complementary channels

$$CF_4^- \rightarrow F^- + CF_3, \tag{4.10}$$

$$CF_4^- \rightarrow F + CF_3^-. \tag{4.11}$$

Time-of-flight measurements [62, 64, 73] reveal that the decomposition of CF_4^- at 6.8 eV is associated with remarkably high translational energy imparted to the products. (ii) Via the first electronically excited CF_4^{*-} state at 7.6 eV producing only F^-. This electronic state is likely a core excited resonance and may correlate with the formation of an electronically excited CF_3^* radical, viz.

$$CF_4^{*-} \rightarrow F^- + CF_3^*. \tag{4.12}$$

Since F^- is produced from this reaction with only thermal kinetic energy, the CF_3^* radical must contain an amount of excess energy ranging from 4 eV at the onset of the reaction to $\sim$ 8 eV at the higher energy side of the F^- resonance (the first excited electronic state [184] of CF_3 is at 5.9 eV). Thus, the final channel of Reaction (4.12) may consist of three fragments such as $F^- + CF_2 + F$ or $F^- + CF + F_2$. These products will be formed with little kinetic energy compared to the kinetic energy of the products formed via the ground state of CF_4^- at 6.8 eV [62–64, 73]. Interestingly, studies of dissociative electron attachment to CF_4 in CF_4 films and to CF_4 condensed onto multilayers of Kr [185–187] have shown that, in contrast to the gas phase, the kinetic energy release

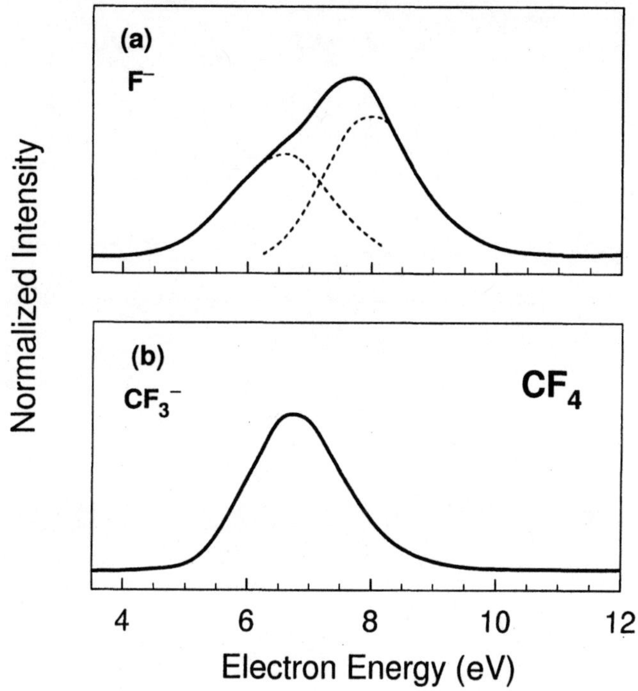

Figure 4.25. Relative intensities of F^- and CF_3^- from CF_4 as a function of electron energy measured by Illenberger and coworkers [62–64]. The production of F^- from the two resonances at 6.8 and 7.6 eV is illustrated by the dashed lines.

of the desorbed F^- ions formed via the core excited resonance $^2T_1$ is appreciable.

The energy dependence of the relative cross section for the formation of F^- and CF_3^- as given in the electron beam experiments of Illenberger and coworkers is shown in Fig. 4.25. The energy-dependent yield of the two complementary ions as measured by various investigators is shown in Fig. 4.26. The agreement is rather poor due partly to the different electron energy resolutions of the various experimental methods. For example, for F^- the beam experiments with lower energy resolution [70, 71, 181] agree among themselves and those with higher energy resolution [63, 180] also tend to agree between themselves. However, this does not seem to be the case for the CF_3^- ion where the data of Iga *et al.* [180] and Lifshitz and Grajower [181] differ substantially from the rest.

There are very limited absolute measurements of the cross sections for these negative ions. Hunter and Christophorou [72] reported that the cross section values for the complementary anions at their respective

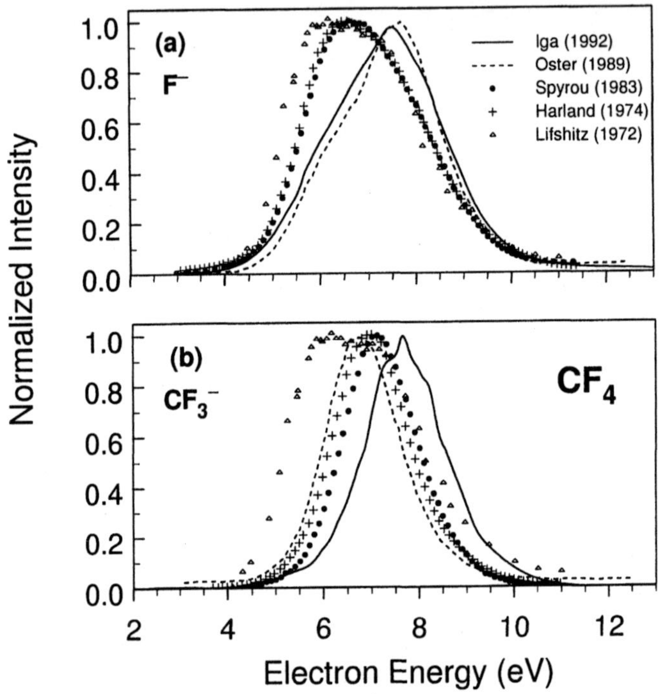

Figure 4.26. Normalized intensities of CF_3^- and F^- from electron impact on CF_4 as a function of electron energy: (—) Ref. [180]; (- - -) Ref. [63]; (•) Ref. [70]; (+) Ref. [71]; (△) Ref. [181].

maxima are about the same and equal to $\sim 0.8 \times 10^{-22}$ m^2 (see below).

2.5.1 Total Electron Attachment Cross Section, $\sigma_{a,t}(\varepsilon)$, of CF_4

The total electron attachment cross section for the CF_4 molecule is small compared to the cross sections for electron impact dissociation and ionization of CF_4. Total electron attachment cross sections have been unfolded from electron swarm measurements [72] and are presented in Fig. 4.27 along with the total electron attachment cross section obtained by normalizing the total relative cross section measured in a single collision beam experiment to the swarm-unfolded absolute total electron attachment cross section. Values from a fit to the latter cross section (solid line in Fig. 4.27) are listed in Table 4.18. The uncertainty of these cross section values were quoted to be ±20%. At the position

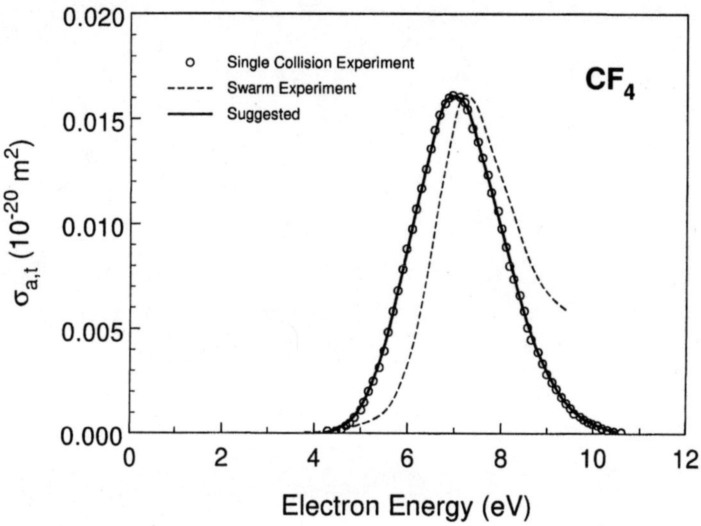

Figure 4.27. Swarm-unfolded and swarm-normalized electron beam total electron attachment cross sections for CF_4 of Hunter *et al.* [72]: (o) Normalized total electron attachment cross section from single collision experiment; (- - -) swarm-unfolded electron attachment cross section; (—) suggested data from fit to the normalized single collision experiment data.

of the maximum in the swarm-unfolded cross section, near 7.3 eV, the value of the total cross section is 1.58×10^{-22} m^2 [72], which can be compared with the mass spectrometric value of 1.04×10^{-22} m^2 [71].

2.5.2 Dissociative Electron Attachment Cross Section for the Production of F^- and CF_3^- for CF_4

The yields of F^- and CF_3^- in Fig. 4.26 can be put on an absolute scale by normalizing their respective peaks to the peak cross section value of 0.8×10^{-22} m^2 [72]. Besides these data, Iga *et al.* [180] have reported dissociative electron attachment cross sections for F^-, CF_3^-, and F_2^- for CF_4 over a wide electron energy range that covered the resonance region below 10 eV to energies of 50 eV. At energies above $\sim$ 20 eV, F^- and F_2^- production was observed due to nonresonant processes, probably via negative ion-positive ion pair formation. These results are presented in Fig. 4.28. The yield of F_2^- is much lower than that of either F^- or CF_3^-. Interestingly, no CF_3^- ions were detected outside the resonance region. The sum of the peak values of the cross sections for F^- and CF_3^- in

Table 4.18. Total electron attachment cross section, $\sigma_{a,t}(\varepsilon)$, for CF_4, obtained by normalizing the total relative cross section measured in a single-collision beam experiment to the swarm-unfolded total cross section [72].

ε (eV)	$\sigma_{a,t}(\varepsilon)$ (10^{-20} m^2)	ε (eV)	$\sigma_{a,t}(\varepsilon)$ (10^{-20} m^2)
4.3	0.00004	7.25	0.0155
4.5	0.00019	7.5	0.0137
5.0	0.00127	8.0	0.00969
5.5	0.00420	8.5	0.00567
6.0	0.00898	9.0	0.00291
6.5	0.0138	9.5	0.00127
6.75	0.0157	10.0	0.00046
7.0	0.0161	10.5	0.00007

Fig. 4.28, is $\sim 1.6 \times 10^{-22}$ m^2, which is in excellent agreement with the value of Hunter and Christophorou [72]. However, in contrast to the other studies [70, 71] which show the peak cross section values for F^- and CF_3^- to be about the same, the peak cross section value of Iga *et al.* [180] for the F^- ion is ~ 3 times larger than that for the CF_3^- ion (Fig. 4.28).

Production of negative ions by electron impact on CF_4 via nonresonant electron attachment processes occurs at energies above ~ 19 eV and is rather small [180, 188]. The results of the two recent studies on the cross sections for positive ion-negative ion pair formation differ substantially. The sum of the cross sections of Iga *et al.* [180] for nonresonant production of F^- and F_2^- ranges from 3×10^{-24} m^2 at 18.7 eV to 24×10^{-24} m^2 at 43.7 eV, while the sum of the cross sections measured by Mi *et al.* [188] for the production of F^- via a number of positive ion-negative ion pair processes [$F^- + CF_n^+$ ($n = 0$–3) and $F^+ + F^-$] ranges [188] from 0.09×10^{-24} m^2 at 18.7 eV to 6.9×10^{-24} m^2 at 43.7 eV. Clearly, more work is needed to quantify the cross sections for these processes.

It should be noted that a negative ion mass spectrometric study of positive ion-negative ion pair formation using synchrotron radiation [189] puts the threshold for F^- production at (13.25±0.03) eV, which is ~ 2 eV higher than the thermodynamic threshold of 11.3 eV for the formation of the ground state ions $F^-\,(^1S_g) + CF_3^+\,(X\,^1A_1')$. Mitsuke *et al.* [189] concluded that the onset for the F^- production from CF_4 occurs at the adiabatic excitation energy for the Rydberg state near 13.6 eV.

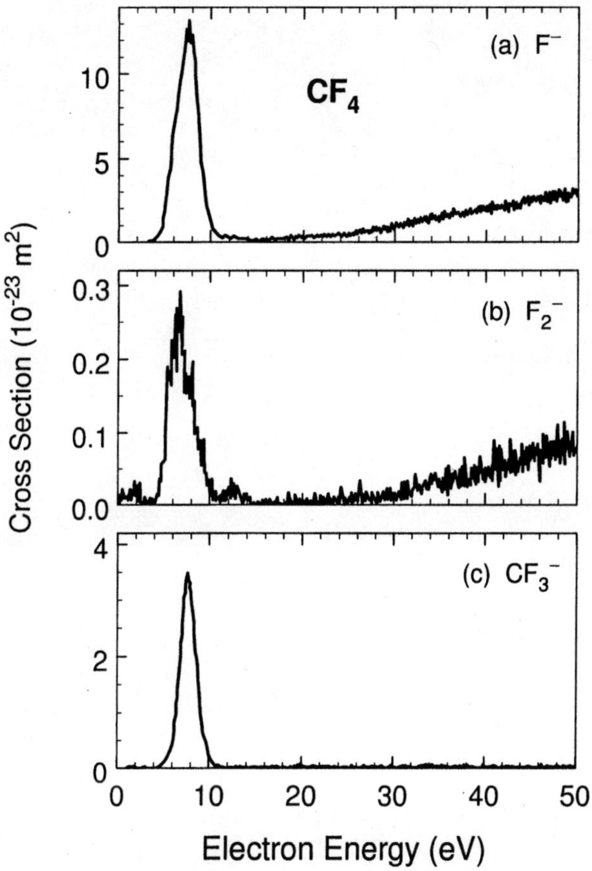

Figure 4.28. Cross sections for F^-, F_2^-, and CF_3^- production by electron impact on CF_4 as a function of electron energy as measured by Iga *et al.* [180].

2.5.3 Electron Attachment Coefficients and Electron Attachment Rate Constants for CF_4

Density-Normalized Electron Attachment Coefficient, η/N (E/N). The density-normalized electron attachment coefficient η/N is measured as a function of E/N. It is related to the total electron attachment cross section $\sigma_{a,t}(\varepsilon)$ and $f(\varepsilon, E/N)$ by

$$\eta/N_a\,(E/N) = \left(\frac{2}{m}\right)^{1/2} w^{-1} \int_0^\infty f(\varepsilon, E/N)\varepsilon^{1/2}\sigma_{a,t}(\varepsilon)d\varepsilon, \qquad (4.13)$$

where N_a is the number density of the electron attaching gas and w is the electron drift velocity. For a unitary gas, the total number density

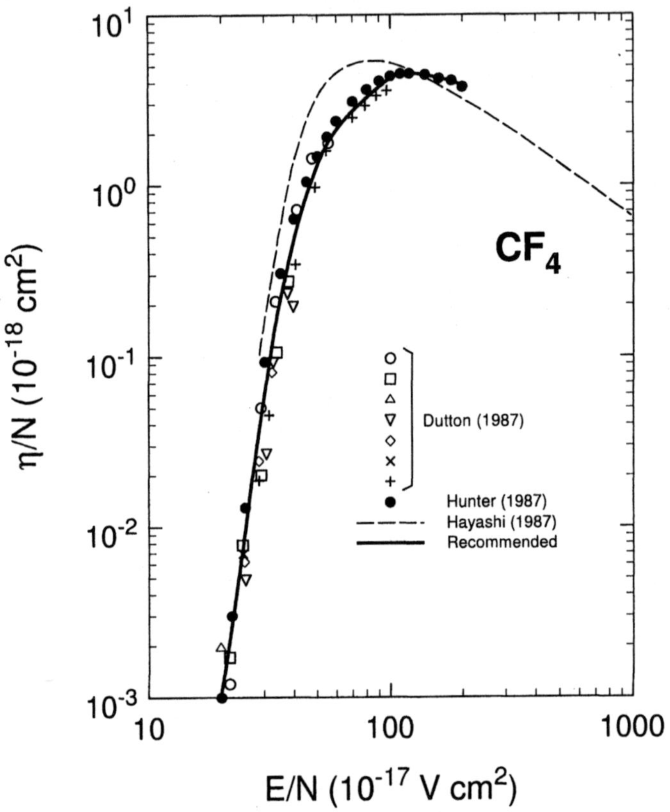

Figure 4.29. Density-normalized electron attachment coefficient as a function of E/N, $\eta/N\,(E/N)$, for CF_4: (•) Ref. [152]; (– –) Ref. [4]; (○, □, △, ▽, ◇, ×, +) measurements of Dutton *et al.* [182] at various CF_4 pressures; (—) recommended.

$N = N_a$; for mixtures of an electron attaching gas in a buffer nonelectron attaching gas of density N, N is normally much larger than N_a. The density-normalized electron attachment coefficient for CF_4 has been measured by a number of investigators [120, 121, 148–152, 182]. With the exception of the data by Dutton *et al.* [182], the rest of these measurements have been summarized by Hunter *et al.* [152]. In Fig. 4.29 are compared the results of Dutton *et al.* [182] with those of Hunter *et al.* [152] and the predicted values of Hayashi [4]. The two latest sets of experimental data are in good agreement, while the calculated values are generally higher than the experimental data. We have fitted the results of Dutton *et al.* and Hunter *et al.*, and the resultant values of $\eta/N\,(E/N)$ are listed in Table 4.19 and represent our recommended values.

Table 4.19. Recommended values of the density-normalized electron attachment coefficient, $\eta/N\,(E/N)$, for CF_4.

E/N (10^{-17} V cm^2)	η/N (10^{-18} cm^2)	E/N (10^{-17} V cm^2)	η/N (10^{-18} cm^2)
20	0.0011	70	2.73
22	0.0026	75	3.00
24	0.0063	80	3.26
26	0.015	85	3.52
28	0.031	90	3.76
30	0.059	95	4.02
32	0.109	100	4.24
34	0.178	110	4.46
36	0.259	120	4.51
38	0.361	130	4.49
42	0.629	140	4.42
44	0.783	150	4.33
46	0.949	160	4.25
48	1.13	170	4.16
50	1.31	180	4.05
55	1.75	190	3.95
60	2.11	200	3.86
65	2.43		

Total Electron Attachment Rate Constant, $k_{a,t}(E/N)$. The density-normalized electron attachment coefficient $\eta/N_a(E/N)$ is related to the total electron attachment rate constant $k_{a,t}(E/N)$ by

$$k_{a,t}(E/N) = \eta/N_a\,(E/N) \times w(E/N). \qquad (4.14)$$

The total electron attachment rate constant $k_{a,t}(E/N)$ for CF_4 has been measured [72, 152] in mixtures with Ar as a function of E/N and $\langle\varepsilon\rangle$. The measurements of $k_{a,t}$ of Hunter and Christophorou [72] are plotted in Fig. 4.30 as a function of the mean electron energy $\langle\varepsilon\rangle$, and are listed in Table 4.20 as our recommended values.

Thermal Value, $(k_{a,t})_{th}$, of the Total Electron Attachment Rate Constant. The value of $k_{a,t}(E/N)$, when the electron energy distribution function $f(\varepsilon, E/N)$ is Maxwellian, $f_M(\varepsilon, T)$, and characteristic of only the gas temperature T, i.e., when $E/N \to 0$, is referred to as the

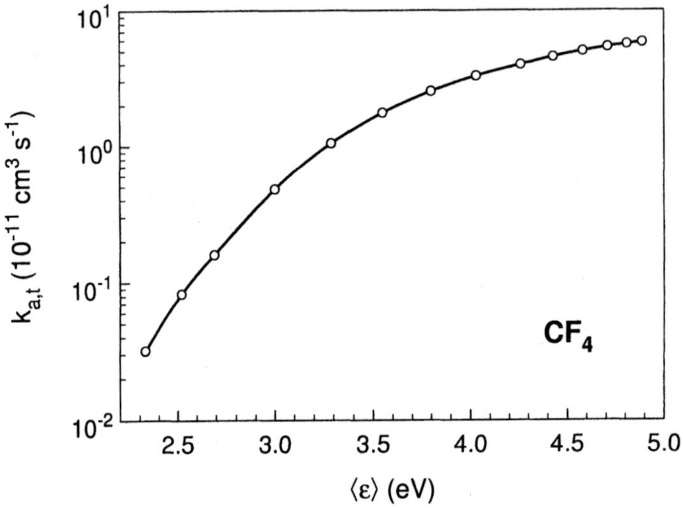

Figure 4.30. Total electron attachment rate constant as a function of mean electron energy, $k_{a,t}(\langle\varepsilon\rangle)$, measured by Hunter and Christophorou [72] in mixtures of CF_4 with Ar.

Table 4.20. Total electron attachment rate constant, $k_{a,t}$, for CF_4 in a buffer gas of Ar as a function of E/N and $\langle\varepsilon\rangle$ as measured by Hunter and Christophorou [72].

E/N (10^{-18} V cm^2)	$\langle\varepsilon\rangle$ (eV)	$k_{a,t}$ (10^{-11} cm^3 s^{-1})
9.32	2.33	0.032
10.97	2.52	0.083
12.4	2.69	0.16
15.5	3.00	0.48
18.6	3.29	1.05
21.7	3.55	1.76
24.9	3.80	2.53
27.9	4.03	3.28
31.1	4.26	3.97
34.2	4.43	4.54
37.3	4.58	5.00
40.4	4.71	5.35
43.5	4.81	5.62
46.6	4.89	5.80

total thermal electron attachment rate constant $(k_{a,t})_{th}$ and is given by

$$(k_{a,t})_{th} = \left(\frac{2}{m}\right)^{1/2} \int_0^\infty f_M(\varepsilon, T)\varepsilon^{1/2}\sigma_{a,t}(\varepsilon)d\varepsilon. \tag{4.15}$$

The reported values of $(k_{a,t})_{th}$ for CF_4 are very small ($< 1\times10^{-13}$ cm^3 s^{-1} [72], $< 3.1 \times 10^{-13}$ cm^3 s^{-1} [190], $< 1 \times 10^{-16}$ cm^3 s^{-1} [191, 192]), and vary significantly. These experimentally derived values may have been affected by trace amounts of electronegative impurities in the gases.

Electron Detachment in CF_4 Plasmas. It is worth pointing out that a number of interesting studies [193–195] have been initiated to probe the role of electron detachment in radio-frequency plasmas in CF_4 using laser photodetachment.

2.6 Electron Transport in CF_4

2.6.1 Electron Drift Velocity, $w(E/N)$, in CF_4

Measurements of the electron drift velocity as a function of E/N, $w(E/N)$, in pure CF_4 have been made at room temperature and over a range of pressures and E/N [18, 19, 108, 120, 122, 196–201]. Some of these measurements have been summarized and discussed earlier in the literature [153, 202]. The measurements of Hunter *et al.* [108] were made using a pulsed Townsend method in the E/N range from 0.03×10^{-17} to 300×10^{-17} V cm^2, and were corrected for the effects of electron attachment and ionization. Their estimated maximum uncertainty above $(E/N)_{lim}$ (i.e., above the value of $E/N = 140 \times 10^{-17}$ V cm^2 at which $\alpha = \eta$; see Table 4.24 later in this section) was ±5% and below $(E/N)_{lim}$ decreased to ±2%. These are compared in Fig. 4.31 with other measurements [18, 120, 199, 200]. The $w(E/N)$ data of Hunter *et al.* [108] are reproduced in Table 4.21 and are our recommended values.

Electron drift velocities for a number of CF_4 gas mixtures have also been measured: CF_4 in He [18, 20]; CF_4 in Ne [18]; CF_4 in Ar [5, 18, 21, 198, 199]; CF_4 in Kr [18]; CF_4 in Xe [19]; CF_4 in CO_2 [122, 198]; CF_4 in CH_4 [21, 122, 197]; CF_4 in C_2H_2 [18]; CF_4 in C_2H_6 and C_3H_8 [21]; CF_4 in i-C_4H_{10} [21, 197]; CF_4 in Ar + CO_2 [122, 198]; CF_4 in Ar + C_2H_2 [18]; CF_4 in Xe + C_2H_2 [19]; CF_4 in Ar + NH_3 [122]; and CF_4 in Ar + H_2O [122]. Most of the measurements on CF_4 mixtures have been made for the purpose of identifying fast gases (i.e., gases with very large w for E/N values employed in practice) for particle detectors. An example of these data for mixtures is reproduced in Fig. 4.32 for CF_4 in Ar [18].

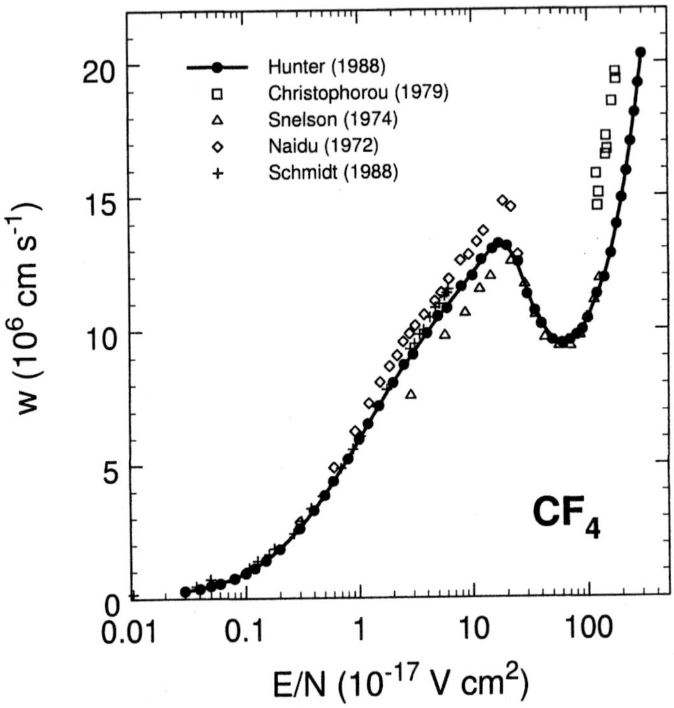

Figure 4.31. Electron drift velocity as a function of E/N in CF_4: (•) Ref. [108]; (□) Ref. [18]; (△) Ref. [200]; (◇) Ref. [120]; (+) Ref. [199]; (—) recommended.

2.6.2 Ratio of the Transverse, D_T/μ, and Longitudinal, D_L/μ, Electron Diffusion Coefficients to Electron Mobility of CF_4

Measurements of $D_T/\mu\,(E/N)$ and $D_L/\mu(E/N)$ in CF_4 are limited. The available data on $D_T/\mu\,(E/N)$ [106, 120, 121, 199] were taken near 293 K and are summarized in Fig. 4.33(a). Curtis *et al.* [106] quoted an uncertainty of about ±5% below 17×10^{-17} V cm^2 and ±8% above this E/N value. Naidu and Prasad [120] reported that the uncertainty in their measurements varied from about ±5% at low E/N to about ±3% at high E/N. The data of Lakshminarasimha *et al.* [121] agree with the data of Naidu and Prasad at high E/N and Curtis *et al.* [106] at low E/N. The measurements of Schmidt and Polenz [199] differ substantially from the other measurements. They are also at variance with the results [$-\cdot-$ in Fig. 4.33(a)] of a recent Monte Carlo calculation [203] in which use was made of the total elastic and inelastic electron scattering cross sections recommended by Christophorou *et al.* [1]. The

Table 4.21. Electron drift velocity, $w(E/N)$, for CF_4 as a function of E/N as measured by Hunter *et al.* [108].

E/N (10^{-17} V cm^2)	w (10^6 cm s^{-1})	E/N (10^{-17} V cm^2)	w (10^6 cm s^{-1})
0.04	0.36	15	13.0
0.05	0.46	17	13.2
0.06	0.55	20	13.1
0.08	0.74	25	12.5
0.10	0.93	30	11.3
0.12	1.11	35	10.7
0.15	1.40	40	10.2
0.2	1.83	50	9.6
0.3	2.61	60	9.5
0.4	3.28	70	9.6
0.5	3.85	80	9.8
0.6	4.38	90	10.0
0.8	5.22	100	10.4
1.0	5.95	120	11.3
1.2	6.53	140	11.9
1.5	7.20	160	12.8
2.0	8.05	180	13.9
2.5	8.72	200	14.9
3.0	9.10	220	15.9
4.0	9.88	240	17.0
5.0	10.5	260	18.1
6.0	10.8	280	19.2
8.0	11.6	300	20.3
10.0	12.0		

solid line in Fig. 4.33(a) is a fit to the measurements of Refs. [106, 120, 121]. Values taken off the solid line are given in Table 4.22 as the presently suggested data for $D_T/\mu\,(E/N)$.

In Fig. 4.33(b) are plotted the measurements of Schmidt and Polenz [199] for the longitudinal electron diffusion coefficient to electron mobility ratio, $D_L/\mu(E/N)$. Also plotted (solid points) is a set of values of $D_L/\mu(E/N)$ which was obtained by Christophorou and Olthoff [48] using the measurements of Hayashi and Nakamura [204] for $ND_L(E/N)$ (Table 4.23), the recommended values of $w(E/N)$, and the relation

$$\frac{D_L}{\mu} = \frac{D_L N}{w} \times \frac{E}{N}. \tag{4.16}$$

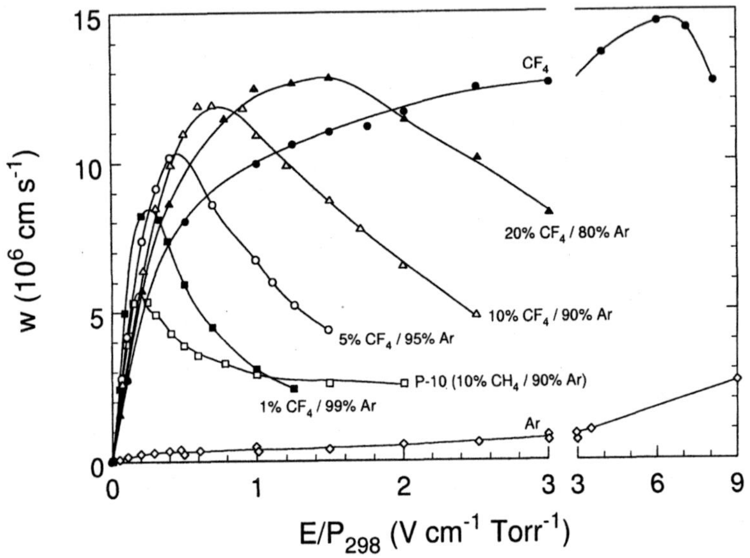

Figure 4.32. Electron drift velocity as a function of E/P ($T = 298$ K), $w(E/P)$, for CF_4/Ar mixtures measured by Christophorou *et al.* [18].

Table 4.22. Suggested $D_T/\mu\,(E/N)$ for CF_4.

E/N (10^{-17} V cm^2)	D_T/μ (V)	E/N (10^{-17} V cm^2)	D_T/μ (V)	E/N (10^{-17} V cm^2)	D_T/μ (V)
0.14	0.025	3	0.036	40	1.01
0.17	0.026	4	0.037	45	1.29
0.20	0.027	5	0.039	50	1.58
0.25	0.028	6	0.041	60	2.16
0.30	0.029	8	0.046	70	2.68
0.35	0.030	10	0.052	80	3.12
0.40	0.031	12	0.062	90	3.49
0.50	0.032	15	0.084	100	3.81
0.60	0.033	20	0.155	150	4.78
0.80	0.034	25	0.293	200	5.16
1.00	0.035	30	0.492	250	5.29
2.00	0.035	35	0.736	300	5.39

Although these data and the measurements of Schmidt and Polenz [199] for D_L/μ converge to their asymptotic value as $E/N \to 0$, they disagree considerably at higher E/N, stressing the need for further measure-

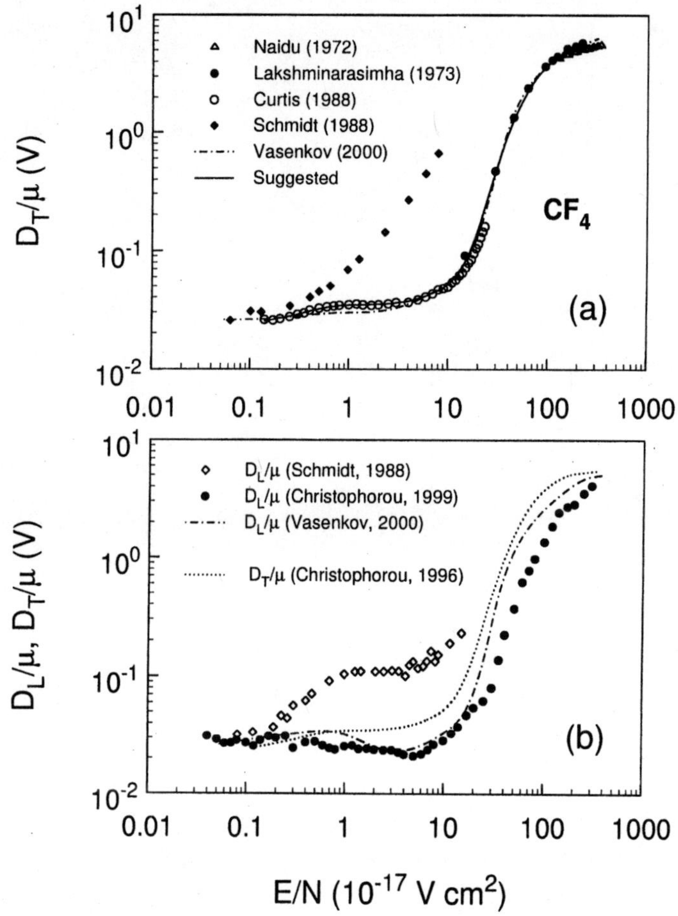

Figure 4.33. (a) Ratio of the lateral electron diffusion coefficient to electron mobility as a function of E/N, $D_T/\mu\,(E/N)$, for CF_4. Measured values: (•) Ref. [121]; (△) Ref. [120]; (○) Ref. [106]; (♦) Ref. [199]. Calculated values: (– · –) Ref. [203]. Suggested values: (—) based on a fit to the measurements of Refs. [106, 120, 121]. (b) Ratio of the longitudinal electron diffusion coefficient to electron mobility as a function of E/N, $D_L/\mu(E/N)$, for CF_4. Measured values: (◊) Ref. [199]. Calculated values (see text): (•) Ref. [48] based on the measurements of $ND_L\,(E/N)$ of Ref. [204]. Computed values: (– · –) Ref. [203]. Also shown for comparison are the suggested values of $D_T/\mu\,(E/N)$ (· · ·) from Fig. 4.33(a).

ments. It is noted that the measurements of Schmidt and Polenz [199] also disagree with the Monte Carlo calculation results of Vasenkov [203].

The dotted line in the Fig. 4.33(b) shows the recommended values of $D_T/\mu\,(E/N)$ from Fig. 4.33(a). It is interesting to observe that the values of $D_T/\mu\,(E/N)$ lie above the corresponding values of $D_L/\mu(E/N)$ derived from the Hayashi and Nakamura data but fall below the Schmidt and Polenz data. Since $D_T/\mu\,(E/N)$ is expected [205] to be greater or equal to $D_L/\mu(E/N)$, the data of Hayashi and Nakamura are considered to be more reliable. The values of $ND_L(E/N)$ from Hayashi and Nakamura [204] and the values of $D_L/\mu(E/N)$ derived using their data and Eq. 4.16 are listed in Table 4.23 as our suggested values for CF_4.

2.6.3 Mean Electron Energy, $\langle\varepsilon\rangle\,(E/N)$, for CF_4

Mean electron energies $\langle\varepsilon\rangle$ as a function of E/N have been computed using a Boltzmann code for pure CF_4 [122, 154] and for mixtures of CF_4 in CO_2 and CF_4 in Ar [122]. These data are approximate and should be used for guiding purposes only. The $D_T/\mu\,(E/N)$ data in Table 4.22 can be used to determine $\langle\varepsilon\rangle$ [$= (3/2)(eD_T/\mu)$] as a function of E/N.

2.6.4 $(E/N)_{\mathrm{lim}}$ for CF_4

$(E/N)_{\mathrm{lim}}$ is an interesting and useful quantity which comes naturally from the measurements of the ionization and attachment coefficients made as a function of E/N. It is the value of the density-reduced electric field at which $(\alpha - \eta)/N = 0$. This value should also be equal to the breakdown voltage of CF_4 as measured under uniform field conditions. Measured values of this quantity are given in Table 4.24. If we exclude the two largest values, the average of the rest of the data gives a suggested value of $(E/N)_{\mathrm{lim}}$ for CF_4 equal to 140×10^{-17} V cm^2.

2.7 Electron Interactions with CF_4 Neutral Fragments

The studies of the interactions of low energy electrons with radicals of the CF_4 molecule are very limited. Very little experimental work, for example, has been done on the electron-impact ionization of CF_4 fragments, CF_x ($x = 1$–3), and nothing has been reported on their electron scattering and electron attachment properties. Such studies, however, are significant in many applications, especially in plasma processing applications of CF_4 gas [24, 158, 175, 206, 207]. The CF_3, CF_2, and CF radicals and their ions are the most abundant and reactive species that result from electron impact dissociation of CF_4 and are therefore important in modeling CF_4 gas discharges. Fortunately, there recently have been some theoretical studies on electron-impact excitation [208, 209], and some experimental and theoretical determinations of total ioniza-

Table 4.23. Suggested values of $ND_L(E/N)$ and $D_L/\mu(E/N)$ for CF_4.

E/N (10^{-17} V cm^2)	ND_L [a] (10^{21} cm^{-1} s^{-1})	D_L/μ [b] (V)
0.04	29.3	0.031
0.05	27.1	0.029
0.06	24.9	0.027
0.07	25.0	0.027
0.08	26.3	0.028
0.10	25.1	0.027
0.12	23.4	0.025
0.14	26.3	0.028
0.17	28.5	0.031
0.20	27.6	0.030
0.25	28.1	0.031
0.30	24.7	0.024
0.40	23.1	0.027
0.50	22.0	0.028
0.60	19.2	0.026
0.70	17.3	0.024
0.80	16.0	0.024
1.0	15.4	0.025
1.2	14.4	0.026
1.4	12.3	0.024
1.7	11.1	0.024
2.0	9.87	0.024
2.5	8.48	0.024
3.0	7.48	0.023
3.5	6.41	0.023
4.0	5.57	0.022
5.0	4.57	0.021
6.0	4.10	0.022
7.0	3.96	0.024
8.0	3.98	0.026
10.0	3.62	0.028
12.0	3.62	0.033
14.0	3.64	0.037
17.0	3.92	0.047
20.0	3.90	0.055
25.0	3.41	0.062
30.0	3.40	0.081
35.0	4.56	0.138

[a] Data of Hayashi and Nakamura [204].
[b] Determined by Christophorou and Olthoff [48] from the measurements of Hayashi and Nakamura [204] of $ND_L(E/N)$, the recommended values of $w(E/N)$, and Eq. 4.16 (see text).

Table 4.23 (continued)
Suggested values of $ND_L(E/N)$ and $D_L/\mu(E/N)$ for CF_4.

E/N (10^{-17} V cm^2)	ND_L [a] (10^{21} cm^{-1} s^{-1})	D_L/μ [b] (V)
40	6.05	0.224
50	7.38	0.368
60	10.3	0.621
70	11.2	0.778
80	12.6	0.976
100	15.0	1.36
120	18.5	1.85
140	22.3	2.41
170	23.1	2.70
200	23.2	2.84
250	27.4	3.51
300	30.5	4.07

[a] Data of Hayashi and Nakamura [204].
[b] Determined by Christophorou and Olthoff [48] from the measurements of Hayashi and Nakamura [204] of $ND_L(E/N)$, the recommended values of $w(E/N)$, and Eq. 4.16 (see text).

Table 4.24. Published $(E/N)_{lim}$ values for CF_4.[a]

$(E/N)_{lim}$ (10^{-17} V cm^2)	Reference
137	[120]
138	[148]
140	[151]
141	[16][b]
142	[154]
143	[150]
149	[149]
151	[15][b]

[a] If we exclude the highest two values in the table, the average of the rest of the data is 140×10^{-17} V cm^2.
[b] Uniform field breakdown measurements; the rest of the data are the values of E/N at which $\alpha = \eta$.

tion cross sections [143, 160, 207, 210, 211] of the CF_3, CF_2, and CF radicals. These are discussed below.

2.7.1 Electron-Impact Ionization Cross Sections, $\sigma_{\mathrm{i,fragment}}(\varepsilon)$

Significant experimental results have been reported [158–160, 207] on the electron-impact ionization of the free radicals CF_3, CF_2, and CF prepared by near-resonance charge transfer reactions of CF_3^+, CF_2^+, and CF^+ with various species. In Fig. 4.34 are shown the electron-impact ionization cross sections for the formation of the parent CF_x^+ ions from the CF_x radicals (x = 1–3) measured by Tarnovsky and Becker [159] from threshold to 200 eV by fast neutral beam techniques. These data are listed in Table 4.25, and will be discussed further in the next section. The reported overall uncertainties are: ±15% for CF, ±16% for CF_2, and ±18% for CF_3.

2.7.2 Electron-Impact Dissociative Ionization Cross Sections, $\sigma_{\mathrm{i,dis,fragment}}(\varepsilon)$

Tarnovsky *et al.* [158] also measured the cross sections for dissociative electron-impact ionization of the CF_x (x = 1–3) free radicals of CF_4.

CF_3. The measurements of Tarnovsky *et al.* [158] for the production of CF_2^+ and CF^+ from the CF_3 fragment are shown in Fig. 4.34(a), and are listed in Table 4.25. The assigned overall uncertainty to the CF_2^+ and CF^+ cross sections is ±20%. These data agree well with the cross section values reported by Wetzel *et al.* [212] [$(0.7 \pm 0.2) \times 10^{-20}$ m^2 for CF_2^+ and $(0.6 \pm 0.2) \times 10^{-20}$ m^2 for CF^+ at 70 eV]. Tarnovsky *et al.* [158] found that the processes of positive double ionization CF_2^+ + F^+ and ion pair CF_2^+ + F^- play an insignificant role in the dissociative ionization of CF_3 and the fragment ions are produced with little kinetic energy [158].

Also shown in Fig. 4.34(a) is the cross section for the formation of F^+ (at 70 eV the magnitude of the cross section for this ion is 0.35×10^{-20} m^2 with an estimated [158] uncertainty of ±30%). No significant production of C^+ was observed (an upper limit of 0.1×10^{-20} m^2 was estimated [158] for the cross section of this ion at 70 eV.

Interestingly, the cross sections for the molecular fragments (CF_2^+ and CF^+ from CF_3) exceed the parent ionization cross section (CF_3^+ from CF_3). This is similar to the case of CF_4 where dissociative ionization is the dominant process. See Tarnovsky *et al.* [158] for information on threshold energies and other energetics of electron-impact processes for CF_3.

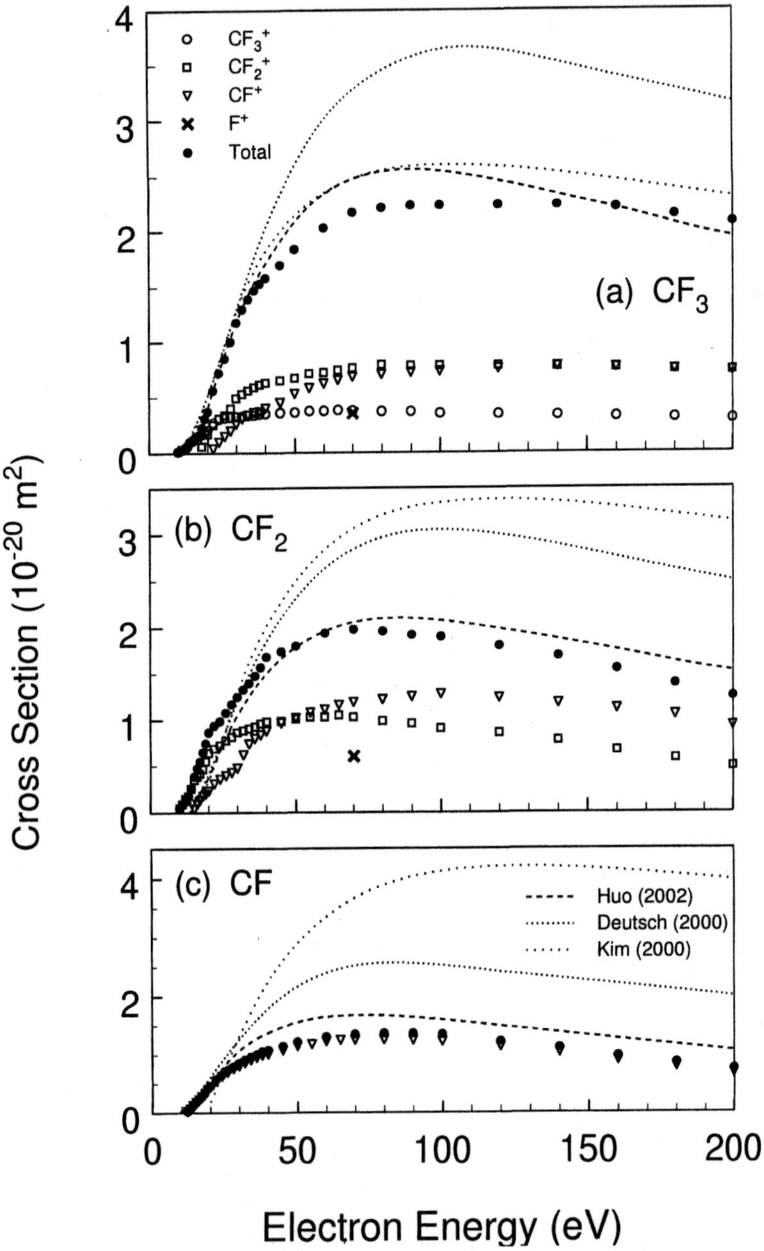

Figure 4.34. Electron-impact ionization cross section measurements of Becker and coworkers for the fragments (a) CF_3, (b) CF_2, and (c) CF: (•) total electron-impact ionization cross section, $\sigma_{i,t,fragment}(\varepsilon)$ [160]; (○) CF_3^+ production [158, 159]; (□) CF_2^+ production [158, 159]; (▽) CF^+ production [158, 159]; (×) F^+ production [158]. Also shown are calculations of $\sigma_{i,t,fragment}(\varepsilon)$: (- - -) Ref. [160]; (···) Ref. [143]; (· · ·) Ref. [210].

Table 4.25. Cross sections for total [160] and partial [158, 159] electron-impact ionization of CF_x (x = 1–3) fragments measured by Becker and coworkers in units of 10^{-20} m^2.

	CF_3					CF_2 [a]				CF	
Energy (eV)	CF_3^+	CF_2^+	CF^+	F^+	Total	CF_2^+	CF^+	F^+	Total[a]	CF^+	Total
10	0.02				0.02	0.05			0.05		
11	0.03				0.03	0.09			0.09		
12	0.04				0.04	0.15			0.15	0.03	0.03
13	0.06				0.06	0.18			0.18	0.07	0.07
14	0.10				0.10	0.26			0.26	0.13	0.13
15	0.11				0.11	0.35	0.04		0.39	0.18	0.18
16	0.15				0.15	0.39	0.09		0.48	0.23	0.23
17	0.16				0.16	0.42	0.13		0.55	0.28	0.28
18	0.17	0.06			0.23	0.47	0.18		0.65	0.33	0.34
19	0.19	0.12			0.31	0.55	0.20		0.75	0.40	0.42
20	0.20	0.17			0.37	0.64	0.23		0.87	0.45	0.47
22	0.27	0.25	0.04		0.56	0.69	0.31		0.94	0.55	0.58
24	0.30	0.31	0.10		0.72	0.73	0.36		0.99	0.63	0.66
26	0.32	0.34	0.15		0.85	0.78	0.40		1.08	0.70	0.74
28	0.32	0.40	0.20		1.00	0.82	0.43		1.17	0.76	0.80
30	0.33	0.49	0.26		1.18	0.87	0.48		1.25	0.81	0.86
32	0.33	0.53	0.31		1.30	0.89	0.62		1.33	0.86	0.92
34	0.34	0.56	0.34		1.39	0.91	0.74		1.40	0.91	0.97
36	0.34	0.59	0.36		1.47	0.93	0.80		1.48	0.95	1.01
38	0.35	0.61	0.37		1.53	0.96	0.84		1.57	0.99	1.06
40	0.35	0.63	0.40		1.58	0.98	0.88		1.68	1.01	1.08
45	0.36	0.65	0.45		1.70	0.99	0.97		1.74	1.08	1.15

[a] There is a discrepancy between the reported [158, 159] partial ionization cross sections for CF_2 and the reported [160] total ionization cross section for CF_2. For electron energies above ~ 20 eV, the sum of the partial cross sections exceeds the reported corresponding value of the total cross section. The reported values of the partial and total cross sections for CF_3 and CF appear to be consistent.

Table 4.25 (continued)
Cross sections for total [160] and partial [158, 159] electron-impact ionization of CF_x (x = 1–3) fragments measured by Becker and coworkers in units of 10^{-20} m^2.

	CF_3					CF_2 [a]				CF	
Energy (eV)	CF_3^+	CF_2^+	CF^+	F^+	Total	CF_2^+	CF^+	F^+	Total[a]	CF^+	Total
50	0.36	0.67	0.53		1.84	1.01	1.02		1.80	1.15	1.22
60	0.37	0.72	0.62		2.03	1.03	1.11		1.94	1.23	1.31
70	0.38	0.76	0.68	0.35	2.17	1.03	1.19	0.60	1.98	1.25	1.35
80	0.37	0.79	0.70		2.21	0.99	1.22		1.96	1.26	1.37
90	0.37	0.78	0.72		2.23	0.96	1.25		1.92	1.25	1.37
100	0.35	0.78	0.73		2.23	0.91	1.28		1.90	1.23	1.35
120	0.34	0.78	0.75		2.23	0.86	1.24		1.80	1.14	1.23
140	0.33	0.77	0.77		2.23	0.78	1.18		1.69	1.04	1.12
160	0.32	0.76	0.76		2.21	0.67	1.12		1.55	0.90	0.98
180	0.31	0.74	0.74		2.14	0.58	1.05		1.39	0.79	0.86
200	0.29	0.73	0.72		2.07	0.49	0.93		1.25	0.67	0.75

[a]There is a discrepancy between the reported [158, 159] partial ionization cross sections for CF_2 and the reported [160] total ionization cross section for CF_2. For electron energies above ~ 20 eV, the sum of the partial cross sections exceeds the reported corresponding value of the total cross section. The reported values of the partial and total cross sections for CF_3 and CF appear to be consistent.

CF_2. In Fig. 4.34(b) the cross section for the production of CF^+ by electron impact on CF_2 as measured by Tarnovsky *et al.* [158] is presented, and is seen to exceed that for the parent ion CF_2^+ above $\sim$ 50 eV. This cross section is listed in Table 4.25 and has an estimated [158] overall uncertainty of $\pm 16\%$, i.e., somewhat lower than that ($\pm 20\%$) for the parent ion CF_2^+. Also included in Fig. 4.34(b) is the cross section for F^+ production (0.6×10^{-20} m^2) at 70 eV. Tarnovsky *et al.* estimated that the F^+ cross section has a contribution of 0.4×10^{-20} m^2 from the process $CF^+ + F^+$ and a contribution of 0.2×10^{-20} m^2 from the F^+ single positive ion formation. The production of C^+ was reported [158] to be small (cross section $< 0.1 \times 10^{-20}$ m^2 at 70 eV).

It is clear from Fig. 4.34(b) that the cross section for CF^+ from CF_2 displayed two prominent onsets, a low energy onset corresponding to the formation of $CF^+ + F$ and a higher energy onset representing the double positive ion formation $CF_2 \rightarrow CF^+ + F^+$. The fragment ions are produced with little kinetic energy [158].

CF. The cross section for CF^+ production from the CF radical by electron is shown in Fig. 4.34(c). The dissociative ionization of this radical produces weak C^+ and F^+ fragment ions. The cross sections for these two ions at 70 eV have been measured [158] to be $(0.25 \pm 0.1) \times 10^{-20}$ m^2 for F^+ and less than 0.1×10^{-20} m^2 for C^+.

2.7.3 Total Ionization Cross Section, $\sigma_{i,t,fragment}(\varepsilon)$

From the partial ionization cross sections of each fragment in Fig. 4.34 (Table 4.25) the total ionization cross sections, $\sigma_{i,t,fragment}(\varepsilon)$, have been determined for the CF_3, CF_2, and CF radicals. The data designated by the solid points (•) in the figure are the recent determinations of $\sigma_{i,t,fragment}(\varepsilon)$ by Huo *et al.* [160] who re-analyzed the partial ionization cross sections for possible effects due to incomplete collection of fragment ions and ion pair formation.

Also in Fig. 4.34 are plotted the results of recent calculations [143, 160, 210]. For all three radicals, the results of the Binary-Encounter-Bethe (BEB) calculation of Kim and Irikura [210] are higher than the experimental data by about a factor of two, although for many molecules the predictions of this model are in good agreement with the measurements. According to Huo *et al.* [211], this difference occurs for fluorine-containing molecules in which the non-fluorine atom has non-bonding valence electrons. The theory needs to consider the effect of the non-bonding electron(s) and the building up of charges in the inner region of the molecule due to the fluorine polar bonds. In an attempt to consider

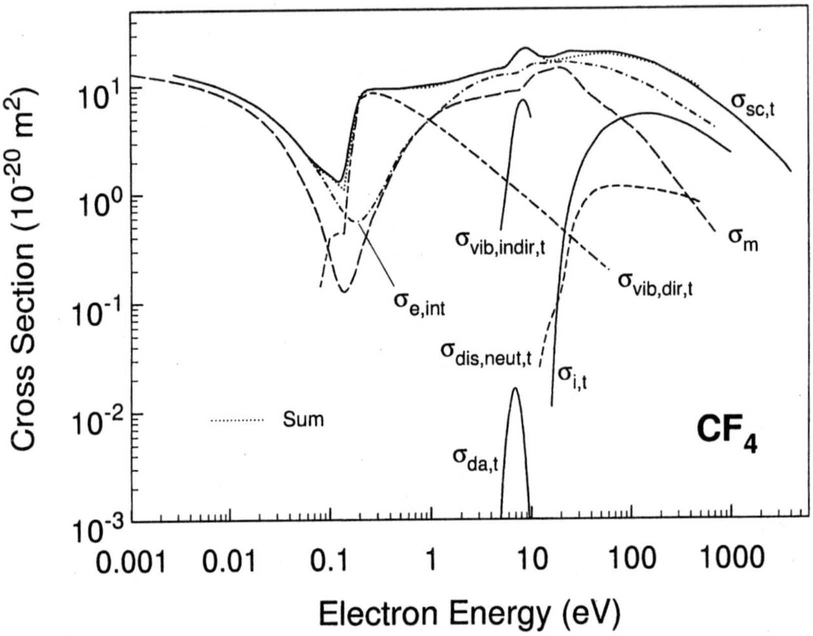

Figure 4.35. Recommended and suggested electron-impact cross sections for CF_4 (see the text).

these effects, Huo [145] developed a "simple-Binary-Encounter-Dipole" (sBED) model which gave results [160] in close agreement with the measurements (see Fig. 4.34). The results of the semiclassical Deutsch-Märk formalism [143] are also somewhat higher than the experimental data for all three radicals.

2.8 Summary of Recommended and Suggested Electron Collision Cross Sections and Electron Transport Coefficients for CF_4

In Fig. 4.35 are plotted the cross sections for CF_4 that were derived from several sets of data, and were designated as recommended or suggested. These are:

- $\sigma_{sc,t}(\varepsilon)$ — Table 4.3, Figs. 4.3 and 4.4;
- $\sigma_{e,int}(\varepsilon)$ — Table 4.6, Fig. 4.6;
- $\sigma_m(\varepsilon)$ — Table 4.7, Fig. 4.7;
- $\sigma_{i,t}(\varepsilon)$ — Table 4.11, Fig. 4.13(b);

- $\sigma_{\rm vib,dir,t}(\varepsilon)$ — Column 4 of Table 4.8, Fig. 4.10(c);
- $\sigma_{\rm vib,indir,t}(\varepsilon)$ — Column 7 of Table 4.8, Fig. 4.10(c);
- $\sigma_{\rm dis,neut,t}(\varepsilon)$ — Table 4.16, Fig. 4.22; and
- $\sigma_{\rm a,t}(\varepsilon)$ — Table 4.18, Fig. 4.27.

One may relate these cross sections [with the exception of $\sigma_{\rm m}(\varepsilon)$] by the simple equation:

$$\sigma_{\rm sc,t}(\varepsilon) = \sigma_{\rm e,int}(\varepsilon) + \sigma_{\rm inel,t}(\varepsilon) \tag{4.17}$$

Based on Eq. (4.17) one would expect

$$\begin{aligned}\sigma_{\rm sc,t}(\varepsilon) \approx \sigma_{\rm e,int}(\varepsilon) + \sigma_{\rm i,t}(\varepsilon) + \sigma_{\rm dis,neut,t}(\varepsilon) + \sigma_{\rm vib,dir,t}(\varepsilon) \\ + \sigma_{\rm vib,indir,t}(\varepsilon) + \sigma_{\rm a,t}(\varepsilon).\end{aligned} \tag{4.18}$$

Indeed, we have evaluated the right hand side of Eq. (4.18) using the recommended values listed above, and have plotted the results in Fig. 4.35 (dotted line). The sum reproduces $\sigma_{\rm sc,t}(\varepsilon)$ rather well over the entire energy range from 0.001 to 1000 eV. The overall observed agreement is gratifying and indicates the consistency and validity of the recommended/suggested cross sections. The cross sections in Fig. 4.35 were used by Bordage *et al.* [110] in a Boltzmann code and were found to give calculated transport coefficients in reasonable agreement with the recommended experimental values. Similar findings were reported by Vasenkov [203] who used the recommended cross sections [1, 48] in a Monte Carlo calculation. Vasenkov [203, 213] found that anisotropic electron scattering is important for molecules such as CF_4 with a Ramsauer-Townsend minimum in their elastic cross section, in the energy region of the minimum.

In addition to the cross sections presented in Fig. 4.35, the suggested partial ionization cross sections for CF_4 are given in Fig. 4.15 and Table 4.12.

The recommended or suggested data for the electron transport coefficients and the density-normalized electron attachment and ionization coefficients are as follows:

- $\eta/N\,(E/N)$ — Table 4.19, Fig. 4.29;
- $\alpha/N\,(E/N)$ — Table 4.14, Fig. 4.18;
- $(\alpha-\eta)/N\,(E/N)$ — Table 4.15, Fig. 4.19;
- $w\,(E/N)$ — Table 4.21, Fig. 4.31;

- $D_{\mathrm{T}}/\mu\,(E/N)$ — Table 4.22, Fig. 4.33(a); and
- $k_{\mathrm{a,t}}\,(E/N)$ — Table 4.20, Fig. 4.30.

* * *

3. Electron Interactions with C_2F_6

Perfluoroethane (C_2F_6, hexafluoroethane) is a man-made gas with many applications. It is used in the aluminum industry, in the semiconductor industry, in plasma chemistry and etching applications [7, 214–219], in pulsed power switching gas mixtures [12–14, 220–227], and in carbon-13 separation [228]. It is of concern to the environment because it is a greenhouse gas with a long residence time in the environment [229]. Its half-life in the environment is reported as 10 000 years [26, 230], and its global warming potential for 100-year period is 12 500 [26], with reference to the global warming potential of CO_2 taken equal to one (see also Maroulis *et al.* [231]). As is the case for other perfluorocarbon gases, it is difficult to remove C_2F_6 from the environment because of its extremely low decomposition rate and reactivity, although recently Morris *et al.* [29] presented evidence that C_2F_6 is destroyed by reacting with O^+, and thus have identified an atmospheric loss mechanism for this gas.

In the aforementioned and other uses, knowledge of the interactions of slow electrons is fundamental in optimizing the performance parameters involved in the particular application. A critical evaluation and synthesis of existing knowledge on electron interactions with C_2F_6 is therefore important. The following cross sections for C_2F_6 are presented and discussed: total electron scattering, elastic differential, elastic integral, momentum transfer, differential vibrational, vibrational inelastic, total ionization, partial ionization, total dissociation, and electron attachment. Information is presented also on the coefficients for electron-impact ionization, effective ionization, electron attachment, and electron transport (lateral diffusion coefficient and drift velocity), as well as on the rate constant for electron attachment as a function of the mean electron energy and gas temperature. While some information is available for these cross sections, additional measurements are needed for each of them, especially for inelastic electron scattering, momentum transfer, and dissociation of C_2F_6 into neutral fragments. The coefficients are generally better known than the cross sections although further measurements on electron diffusion coefficients and electron attachment at high E/N values are indicated.

3.1 Electronic and Molecular Structure of C_2F_6

The C_2F_6 molecule has D_{3d} symmetry. It has no permanent electric dipole moment [232] and has reported values [32] of static polarizability, α, ranging from 46.0×10^{-25} cm^3 to 65.0×10^{-25} cm^3 depending on the method of calculation used. More recently, McKoy *et al.* [233] report a calculated value of α equal to 67.5×10^{-25} cm^3 and an experimental value of α equal to 68.2×10^{-25} cm^3. The absence of electron-electric dipole scattering has a rather profound effect on the electron scattering cross sections at low energies (< 1 eV) as compared to polar gases, which can be seen from the data presented in the next section.

Photoelectron spectra [234] of C_2F_6 show that the highest filled molecular orbital (MO) in the ground state is mainly populated in the C–C bond, i.e., the uppermost molecular orbital is the C–C σ_g MO [36, 234]. Sauvageau *et al.* [234] interpreted the ultraviolet (UV) absorption spectrum of C_2F_6 in terms of Rydberg bands, some of which are superimposed on the ionization continuum. They ascribed the first UV band, observed at 12.10 eV, as a $3p$ type which is consistent with the assignment of Robin [36] who attributed the first allowed Rydberg excitation in C_2F_6 to the $\sigma_g \rightarrow 3p$ transition near 12.03 eV. The absorption spectrum of Sauvageau *et al.* [234] shown in Fig. 4.36 exhibits two peaks at 13.40 eV ($\lambda = 92.7$ nm) and 14.25 eV ($\lambda = 87$ nm).

The first ionization threshold energy has been found by Sauvageau *et al.* [234] to be at 14.6 eV and by Robin [36] at 14.48 eV. A mass spectrometric study of the photoionization of C_2F_6 conducted by Noutary [235] found no production of parent molecular ions. This study reported the energy thresholds for the photoionization processes

$$C_2F_6 + h\nu \rightarrow C_2F_5^+ + F + e, \tag{4.19}$$

$$C_2F_6 + h\nu \rightarrow CF_3^+ + CF_3 + e, \tag{4.20}$$

to be (15.46 ± 0.02) eV and (13.62 ± 0.015) eV, respectively. The photoionization threshold for the production of the CF^+ ion was found to be 16.75 eV. These threshold values may not be adiabatic since the ions may have excess energy. A photoelectron-photoion coincidence spectrometric study by Simm *et al.* [236, 237] showed that while the decomposition of ground state $C_2F_6^+$ ions gives entirely CF_3^+, and while the CF_3^+ ion is the predominant ion from highly excited states of $C_2F_6^+$, the $C_2F_6^+$ ions in their first excited state ($\tilde{A}$) decompose preferentially to $C_2F_5^+$ + F prior to internally converting to the cationic ground state ($\tilde{X}$).

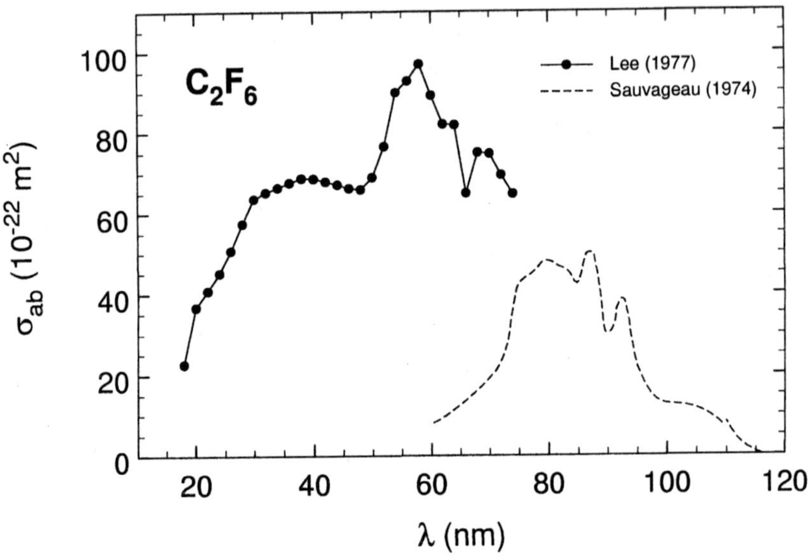

Figure 4.36. Photoabsorption cross section, $\sigma_{ab}(\lambda)$, as a function of photon wavelength, λ, for C_2F_6: (•) Ref. [45]; (– –) Ref. [234].

The photoabsorption cross section of C_2F_6 has been measured by Lee *et al.* [45] in the range 17.5–77.0 nm, and is shown in Fig. 4.36, along with the data of Sauvageau *et al.* [234]. The data of Sauvageau *et al.* are much lower, and Lee *et al.* suggested that this difference may be due to the effect of stray light in the measurements of Sauvageau *et al.* An electron-impact energy-loss spectrum was published by Robin [36] and is reproduced in Fig. 4.37. Its shape is compared with the absorption spectra from Fig. 4.36 which have been replotted in Fig. 4.37 as a function of energy. All spectra were normalized to one at their maximum intensity to facilitate the comparison. There is substantial disagreement between the optical and the electron energy-loss measurements, stressing the need for further measurements.

Absolute oscillator strength spectra in the C 1s (280–340 eV) and F 1s (680–740 eV) regions have been determined by Ishii *et al.* [238] from inner-shell electron energy-loss spectra using 2.5 keV energy electrons and scattering angles less than 2°. These investigators also measured the electron transmission spectrum of C_2F_6 and found negative ion resonances at 4.60 and 8.86 eV which they attributed to σ^* molecular orbitals. These values are in reasonable agreement with electron attachment, electron scattering, and vibrational excitation cross section data (see Table 4.26).

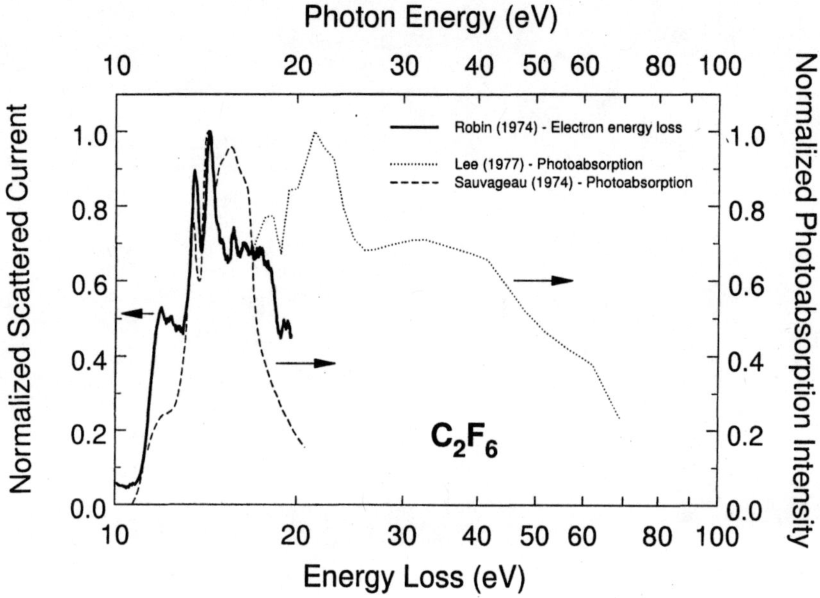

Figure 4.37. Electron energy-loss spectrum (—) of C_2F_6 taken at a scattering angle $\theta = 0°$ and at an incident electron energy of 100 eV [36]. For comparison the absorption data in Fig. 4.36 are also plotted as a function of photon energy: (- - -) Ref. [234]; (· · ·) Ref. [45]. The maximum intensity of each spectrum has been normalized to one.

Perfluoroethane is not a strongly electron attaching gas. However, it attaches electrons more efficiently than CF_4 and at relatively lower energies (see Sec. 3.5 of Ch. 4). No parent negative ion has been observed for perfluoroethane which is consistent with a negative electron affinity for this molecule (see Refs. [238, 244] and Table 4.26). The detection of $C_2F_6^-$ reported [245] in a study of C_2F_6 clusters is likely to be a metastable species associated with the lowest negative ion state of C_2F_6 at 4.0 eV. Clearly the data in Table 4.26 show that there are at least two negative ion states near 4 and 9 eV [238–241, 244]. Dissociative electron attachment studies indicate two other negative ion states at 4.8 and 12.5 eV; the latter may be due to an electron-excited Feshbach resonance. Electrons attach to C_2F_6 dissociatively mainly via the negative ion states at 4 eV and weakly via the negative ion states at 4.8, 9, and 12.5 eV. The predominant fragment negative ions are F^- and CF_3^- (see Sec. 3.5 of Ch. 4).

The energies of the 12 fundamental frequencies $\nu_1, \nu_2, \ldots \nu_{12}$ of C_2F_6 as listed by Shimanouchi [246] are: 0.1522 eV, 0.1001 eV, 0.0431 eV,

Table 4.26. Negative ion states of C_2F_6.[a]

Energy position	Type of study	Reference	Symmetry[a]
4.0 eV[b]	Dissociative attachment producing F^- and CF_3^-	Table 4.42	$3a_{2u}(\sigma^*_{CC})$ [239]
4.3 eV	Vibrational excitation cross section function	[241]	a_{2u} [241]
4.6 eV	Electron transmission	[238]	
~ 5.0 eV	Total electron scattering	[242, 243]	
4.8 eV[c]	Dissociative attachment producing $C_2F_5^-$	Table 4.42	$3a_{2u}(\sigma^*_{CF})$ [239]
9.3 eV	Maximum in the production of F^-	[240]	
8.6 eV	Maximum in the vibrational excitation cross section function	[241]	e_u [241]
8.86 eV	Electron transmission	[238]	
~ 9.0 eV	Total electron scattering	[242, 243]	
12.5 eV	Weak maximum in the production of F^-	[240]	

[a]Ishii *et al.* [238] calculated −3.73 eV, −4.24 eV, and −6.65 eV for the energies of the virtual orbitals e_u, a_{2u}, and a_{2u}. Lindholm and Li [244] calculated the electron affinity of C_2F_6 to be −3.1 eV and the orbital to be σ^*C–C antibonding.
[b]Average position of the F^- and CF_3^- as determined in studies of dissociative electron attachment to C_2F_6 (see Table 4.42).
[c]Average position of $C_2F_5^-$ as determined in studies of dissociative electron attachment to C_2F_6 (see Table 4.42).

0.0084 eV, 0.1385 eV, 0.0885 eV, 0.1550 eV, 0.0767 eV, 0.0461 eV, 0.1551 eV, 0.0645 eV, and 0.0273 eV, respectively.

A set of electron-C_2F_6 collision cross sections has been derived by Hayashi and Niwa [247] from a Boltzmann code analysis (two-term approximation). In their analysis Hayashi and Niwa used various cross sections from the literature as input, some of which they adjusted to optimize their final set of cross sections shown in Fig. 4.38. Their derived cross sections for individual processes and the ability of their cross section set to generate electron transport coefficients can be seen from the comparisons in subsequent sections. A more recent cross section set for C_2F_6 has been obtained by Okumo and Nakamura [248] using multiterm Boltzmann analysis and new measurements of $w(E/N)$ and $ND_L(E/N)$ they made in 0.524% and 5.4% C_2F_6–Ar mixtures over an E/N range of 0.04×10^{-17} to 100×10^{-17} V cm^2. The cross section set they reported

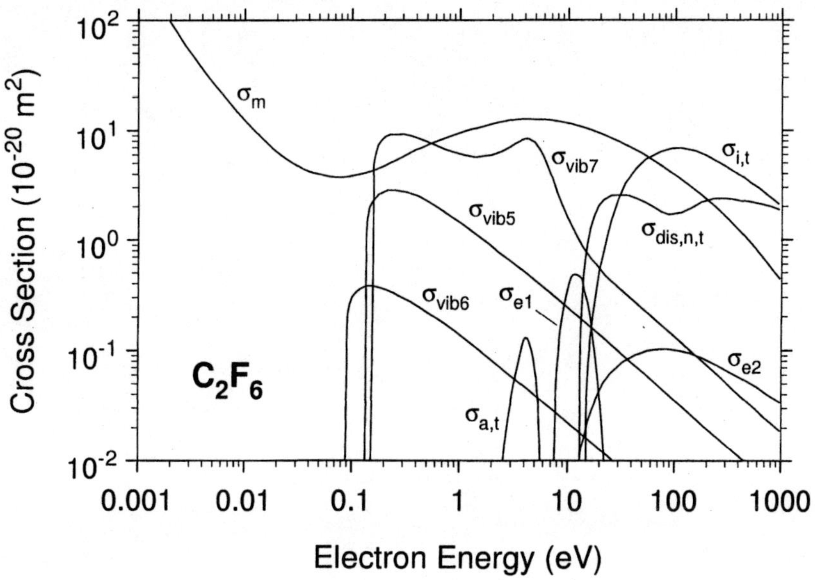

Figure 4.38. A Boltzmann code analysis cross section set for electron collision processes with C_2F_6 by Hayashi and Niwa [247]. The symbols σ_m, $\sigma_{i,t}$, $\sigma_{dis,n,t}$, σ_{e1}, σ_{e2}, $\sigma_{a,t}$, σ_{vib5}, σ_{vib6}, and σ_{vib7} refer, respectively, to the cross section for momentum transfer, total ionization, total dissociation into neutral fragments, excitation of the first electronic state, excitation of the second electronic state, total electron attachment, excitation of the ν_5 vibration, excitation of the ν_6 vibration, and excitation of the ν_7 vibration. It should be noted that σ_{e1} and σ_{e2} are *apparent* overall *non-dissociative* electronic excitation cross sections, and have no physical equivalent in regards to the measured cross sections discussed in this book.

[248] is similar to that of Hayashi and Niwa [247] except that the peak values of their σ_{vib6} and σ_{vib5} are larger, and the dip in their σ_{vib7} is more pronounced.

3.2 Electron Scattering from C_2F_6

In this section information is presented and discussed on the following cross sections: total electron scattering cross section $\sigma_{sc,t}(\varepsilon)$, elastic differential electron scattering cross section $\sigma_{e,diff}(\varepsilon)$, elastic integral electron scattering cross section $\sigma_{e,int}(\varepsilon)$, momentum transfer cross section (elastic) $\sigma_m(\varepsilon)$, and inelastic electron scattering cross section $\sigma_{inel}(\varepsilon)$.

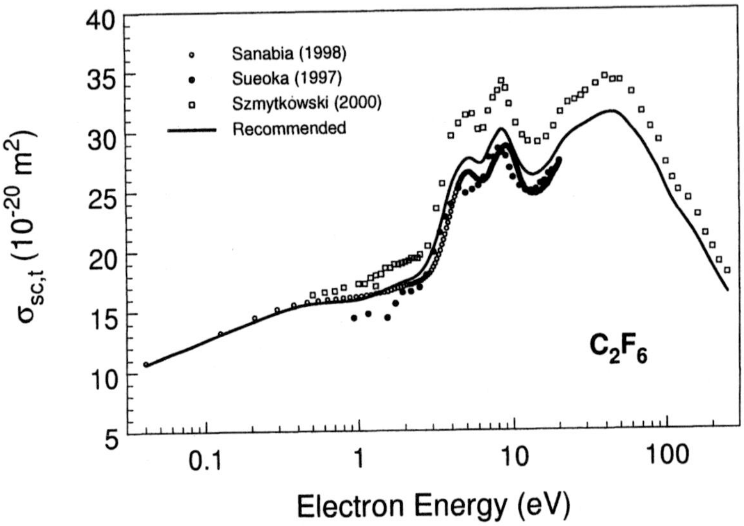

Figure 4.39. Total electron scattering cross section, $\sigma_{sc,t}(\varepsilon)$, for C_2F_6: (○) Ref. [242]; (•) Ref. [249]; (□) Ref. [243]; (—) recommended.

3.2.1 Total Electron Scattering Cross Section, $\sigma_{sc,t}(\varepsilon)$, for C_2F_6

There have been three recent measurements [242, 243, 249] of $\sigma_{sc,t}(\varepsilon)$ for C_2F_6. These are presented in Fig. 4.39. The measurements of Sanabia *et al.* [242] and Sueoka *et al.* [249] have estimated uncertainties of approximately ±20%. The estimated total uncertainty of the measurement of Szmytkowski *et al.* [243] is ±6% below 4 eV and about ±4% above this energy. The three sets of measurements exhibit structure with maxima at about 5.0 and 9.0 eV due to indirect electron scattering via the negative ion states located at these energies. The measurements of Szmytkowski *et al.* [243] lie systematically higher than the two earlier studies, but are still within the combined uncertainties of the measurements.

Below ~ 1 eV, the cross section decreases as the energy approaches zero due to the presence of a Ramsauer-Townsend minimum. However, due to the low-energy position of this minimum the experiment of Sanabia *et al.* [242] is unable to detect the low energy rise in $\sigma_{sc,t}(\varepsilon)$. A Ramsauer-Townsend minimum around 0.07 eV is clearly evident in the backward electron scattering cross section of Lunt *et al.* [105] (see also Sec. 3.2.3 of Ch. 4). Also, from their backward electron scattering experiments, Lunt *et al.* [105] determined a zero-energy total electron scattering cross section of $(90 \pm 4) \times 10^{-20}$ m^2. No calculated values of this cross section have been reported.

The recommended cross section was determined by considering all three sets of experimental data, and is shown in Fig. 4.39 by the solid line. Values from this line are listed in Table 4.27 as our recommended data for the $\sigma_{sc,t}(\varepsilon)$ of C_2F_6.

3.2.2 Elastic Differential Electron Scattering Cross Section, $\sigma_{e,diff}(\varepsilon)$, of C_2F_6

Takagi *et al.* [241] reported measurements of the elastic differential electron scattering cross section, $\sigma_{e,diff}(\varepsilon)$, for C_2F_6 at electron-impact energies between 2 and 100 eV and scattering angles between 10° and 130°. These are plotted in Fig. 4.40, and are listed in Table 4.28. The uncertainty in the values has been estimated by the authors to be between ±15% and ±20%. The data were fitted under certain assumptions and extrapolated to 0° and 180° scattering angles [241]. These fits are represented by the solid lines in the figure. They help provide data for scattering angles towards 0° and towards 180°, which the authors used to determine the elastic integral electron scattering cross section $\sigma_{e,int}(\varepsilon)$, and the momentum transfer cross section (elastic) $\sigma_m(\varepsilon)$ (see next two

Table 4.27. Recommended total electron scattering cross section, $\sigma_{sc,t}(\varepsilon)$, for C_2F_6.

ε (eV)	$\sigma_{sc,t}(\varepsilon)$ (10^{-20} m^2)	ε (eV)	$\sigma_{sc,t}(\varepsilon)$ (10^{-20} m^2)	ε (eV)	$\sigma_{sc,t}(\varepsilon)$ (10^{-20} m^2)
0.04	10.6	4.5	26.8	14	26.2
0.06	11.5	5.0	27.5	16	26.7
0.08	12.1	5.5	27.5	18	27.5
0.10	12.6	6.0	27.2	20	28.5
0.20	14.1	6.5	27.4	30	30.4
0.40	15.4	7.0	28.3	40	31.3
0.60	15.7	7.5	29.1	50	31.1
0.80	15.8	8.0	29.7	60	30.0
1.0	15.9	8.5	30.0	70	28.9
1.5	16.7	9.0	29.7	80	27.6
2.0	17.4	9.5	29.1	90	26.4
2.5	18.1	10.0	28.4	100	25.0
3.0	19.9	11.0	27.2	150	21.3
3.5	22.7	12.0	26.5	200	18.4
4.0	25.4	13.0	26.2	250	16.3

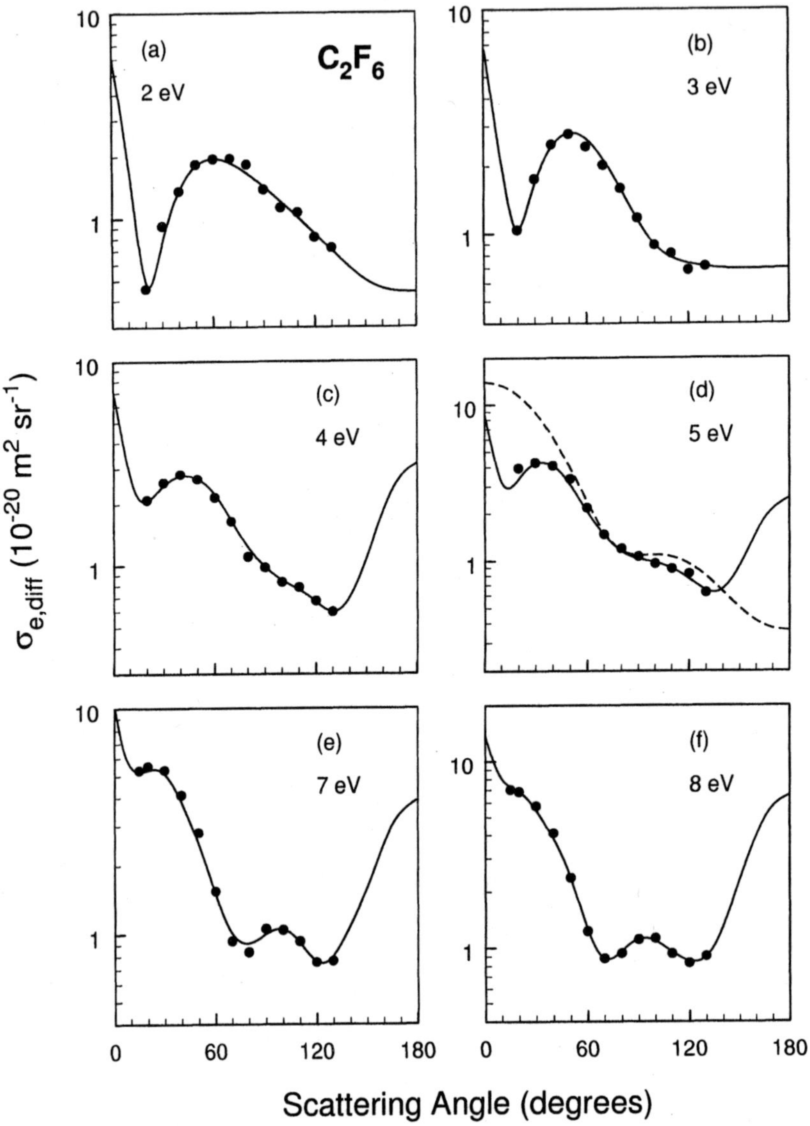

Figure 4.40. Elastic differential electron scattering cross sections, $\sigma_{e,\text{diff}}(\varepsilon)$, for C_2F_6: (•) measurements of Takagi *et al.* [241]; (—) fitting to the experimental data [241]; (---) calculated values of McKoy *et al.* [250].

sections). McKoy *et al.* [250] calculated the elastic differential electron scattering cross sections of C_2F_6 at electron-impact energies of 5, 7.5, 10, 12.5, 15, 20 and 30 eV using the Schwinger multichannel variational method. Their results for 5, 10, 15, 20, and 30 eV are compared in

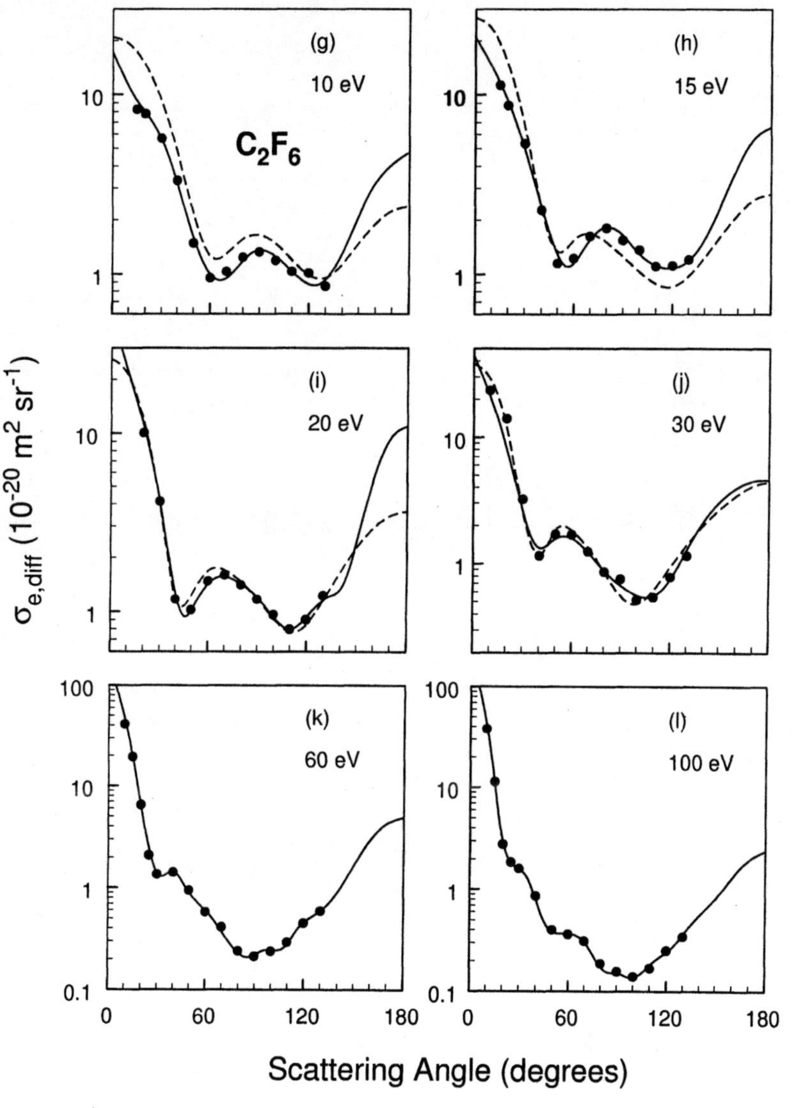

Figure 4.40 (continued)
Elastic differential electron scattering cross sections, $\sigma_{e,diff}(\varepsilon)$, for C_2F_6: (•) measurements of Takagi *et al.* [241]; (—) fitting to the experimental data [241]; (- - -) calculated values of McKoy *et al.* [250].

Fig. 4.40 with the experimental measurements of Takagi *et al.* [241] at these energies. Agreement between the calculations and the experimental results is good at energies above 10 eV, and poor for lower energies.

Table 4.28. Elastic differential electron scattering cross section, $\sigma_{e,diff}(\varepsilon)$, measurements of Takagi *et al.* [241] for C_2F_6 in units of 10^{-20} m^2 sr^{-1}.

Angle	2 eV	3 eV	4 eV	5 eV	7 eV	8 eV
10°						
15°					5.30	7.05
20°	0.458	1.04	2.11	3.96	5.54	6.88
25°						
30°	0.927	1.76	2.56	4.27	5.33	5.77
40°	1.37	2.50	2.81	4.09	4.13	4.09
50°	1.84	2.77	2.66	3.37	2.80	2.37
60°	1.95	2.44	2.16	2.19	1.55	1.23
70°	1.96	2.02	1.65	1.47	0.94	0.88
80°	1.84	1.59	1.11	1.19	0.84	0.94
90°	1.39	1.17	0.981	1.06	1.06	1.11
100°	1.13	0.890	0.834	0.954	1.05	1.13
110°	1.07	0.816	0.784	0.886	0.934	0.933
120°	0.814	0.688	0.675	0.825	0.757	0.833
130°	0.722	0.715	0.598	0.625	0.767	0.904

Angle	10 eV	15 eV	20 eV	30 eV	60 eV	100 eV
10°				23.5	41.1	37.9
15°	8.26	11.3			19.3	11.3
20°	7.80	8.70	10.1	14.1	6.47	2.77
25°					2.09	1.86
30°	5.68	5.33	4.13	3.22	1.37	1.61
40°	3.32	2.27	1.17	1.15	1.43	0.860
50°	1.49	1.15	1.02	1.69	0.948	0.398
60°	0.950	1.23	1.48	1.68	0.577	0.362
70°	1.03	1.63	1.60	1.24	0.409	0.311
80°	1.24	1.81	1.41	0.863	0.240	0.188
90°	1.33	1.55	1.17	0.758	0.213	0.157
100°	1.19	1.37	0.962	0.513	0.239	0.139
110°	1.03	1.11	0.797	0.541	0.292	0.169
120°	1.01	1.12	0.901	0.784	0.448	0.253
130°	0.852	1.21	1.23	1.15	0.592	0.344

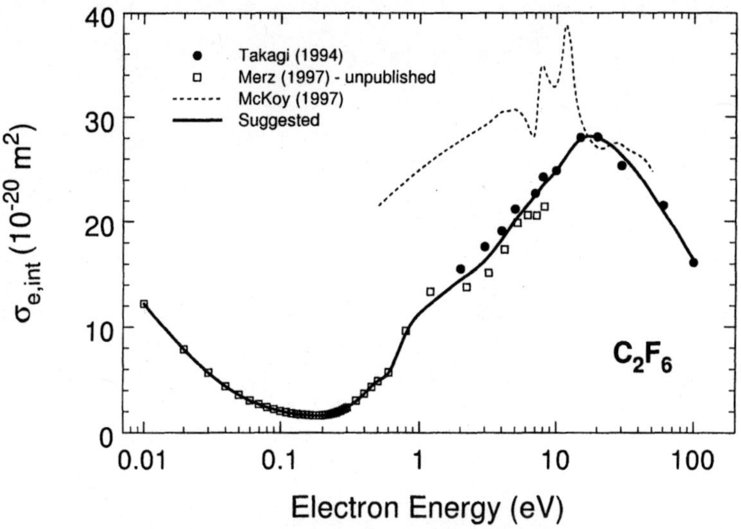

Figure 4.41. Elastic integral electron scattering cross section, $\sigma_{e,int}(\varepsilon)$, for C_2F_6: (•) Ref. [241]; (□) Ref. [251]); (- - -) Ref. [250]; (—) suggested.

3.2.3 Elastic Integral Electron Scattering Cross Section, $\sigma_{e,int}(\varepsilon)$, of C_2F_6

Takagi *et al.* [241] extrapolated their measured elastic differential electron scattering cross sections (Fig. 4.40) to 0° and 180° scattering angles, weighted them by $\sin\theta$, and obtained $\sigma_{e,int}(\varepsilon)$ by integration. Their values are plotted in Fig. 4.41. The quoted uncertainty is approximately ±25%. Also plotted in Fig. 4.41 are values of $\sigma_{e,int}(\varepsilon)$ that have been determined by Merz and Linder [2, 251] from their unpublished elastic differential electron scattering cross section measurements using MERT and phase-shift analysis for energies below 0.5 eV and phase-shift analysis for energies above 0.5 eV. These data are in good agreement with the lower energy measurements of Takagi *et al.* [241]. They extend the energy range of $\sigma_{e,int}(\varepsilon)$ down to 0.01 eV and clearly show the presence of the Ramsauer-Townsend minimum (near 0.18 eV). Suggested values of $\sigma_{e,int}(\varepsilon)$ are determined by fitting the two sets of data, and are shown by the solid line in Fig. 4.41. Values from this fit are listed in Table 4.29.

McKoy *et al.* [250] calculated the $\sigma_{e,int}(\varepsilon)$ for C_2F_6 using the Schwinger multichannel variational method. Their results are shown by the bro-

Table 4.29. Suggested elastic integral electron scattering cross section, $\sigma_{e,int}(\varepsilon)$, for C_2F_6.

ε (eV)	$\sigma_{e,int}(\varepsilon)$ (10^{-20} m^2)	ε (eV)	$\sigma_{e,int}(\varepsilon)$ (10^{-20} m^2)	ε (eV)	$\sigma_{e,int}(\varepsilon)$ (10^{-20} m^2)
0.01	12.2	0.5	4.91	8	23.5
0.02	7.86	0.6	5.71	9	24.2
0.03	5.68	0.7	7.54	10	24.9
0.04	4.41	0.8	9.39	15	27.9
0.05	3.61	0.9	10.6	20	28.0
0.06	3.07	1.0	11.3	25	27.3
0.07	2.70	1.5	13.2	30	26.3
0.08	2.43	2.0	14.5	40	24.4
0.09	2.23	2.5	15.4	50	22.5
0.10	2.07	3.0	16.3	60	20.9
0.15	1.69	4.0	18.3	70	19.6
0.20	1.66	5.0	20.1	80	18.4
0.25	1.91	6.0	21.4	90	17.3
0.30	2.39	7.0	22.5	100	16.4
0.40	3.70				

ken line in Fig. 4.41, and are seen to be in nominal agreement with the experimental data only for a small energy range around 20 eV.

3.2.4 Momentum Transfer Cross Section (Elastic), $\sigma_m(\varepsilon)$, of C_2F_6

There are two experimental determinations [241, 251] of the $\sigma_m(\varepsilon)$ for C_2F_6 which are shown in Fig. 4.42. Takagi *et al.* [241] determined $\sigma_m(\varepsilon)$ by extrapolation and integration of their measurements of the elastic differential electron scattering cross sections for C_2F_6. The latter measurements were made at electron energies of 2–100 eV and scattering angles of 10° and 130° and were extrapolated to 0° and 180°. The elastic momentum transfer cross section was determined by numerical integration of the fits to the measured differential cross sections weighted by the factor $\sin\theta(1-\cos\theta)$. The uncertainty of these cross sections was quoted by the authors to be between ±30% and ±35%. Merz and Linder [251] (see also Ref. [2]) deduced values of $\sigma_m(\varepsilon)$ from their unpublished measurements of the elastic differential electron scattering cross sections using a modified MERT analysis [67]. Their values cover the range of energies between 0.01 eV and 0.6 eV. They show a

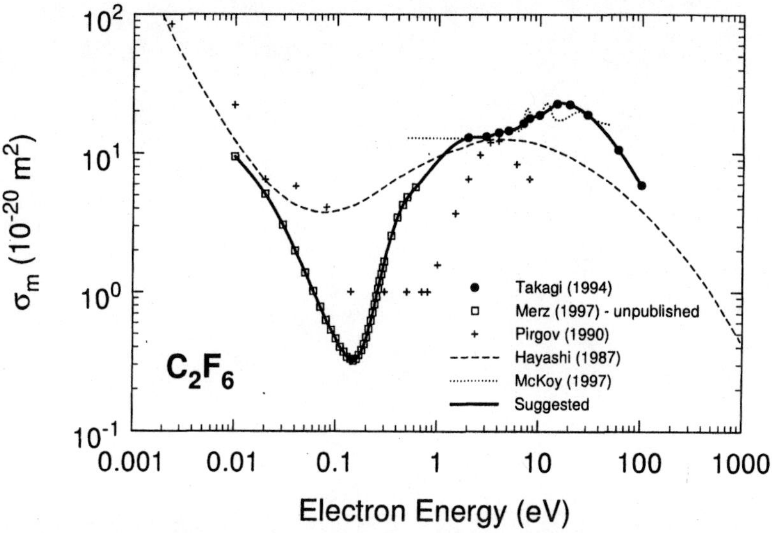

Figure 4.42. Momentum transfer cross section (elastic), $\sigma_m(\varepsilon)$, for C_2F_6. Measured values: (•) Ref. [241]; (□) [251]. Calculated values: (+) Ref. [252]; (- -) Ref. [247]; (···) Ref. [250]. Suggested values (—).

pronounced Ramsauer-Townsend minimum near 0.15 eV. The two sets of data [241, 251] are in general agreement.

Also plotted in Fig. 4.42 are the results of a theoretical calculation of the $\sigma_m(\varepsilon)$ of C_2F_6 by McKoy *et al.* [250] using the Schwinger multichannel variational method, and the computed values of $\sigma_m(\varepsilon)$ from two Boltzmann-type calculations [247, 252] based on measured values of $w(E/N)$ and $D_T/\mu\,(E/N)$ in the pure gas and in its mixtures with Ar. Pirgov and Stefanov [252] used the $w(E/N)$ and $D_T/\mu\,(E/N)$ data of Naidu and Prasad [120] and Hunter *et al.* [13] for pure C_2F_6 and the data of Hunter *et al.* [13] on $w(E/N)$ and $D_T/\mu\,(E/N)$ for mixtures of C_2F_6 with Ar. Also, they used the measurements of Hunter *et al.* [13] on the dissociative electron attachment cross section of C_2F_6 and the data of Milloy *et al.* [253] on the elastic momentum transfer cross section of argon. Hayashi and Niwa [247] used the same experimental data on $w(E/N)$, $D_T/\mu\,(E/N)$, and electron attachment as Pirgov and Stefanov [252], but in addition they employed input data for vibrational excitation, dissociation, and ionization [247]. The computed values of $\sigma_m(\varepsilon)$ indicate a maximum in $\sigma_m(\varepsilon)$ near 4 eV and a Ramsauer-Townsend minimum below 1 eV. The agreement amongst the computed results is rather poor both in regard to the magnitude and energy dependence of $\sigma_m(\varepsilon)$, and in regard to the position of the Ramsauer-Townsend min-

Table 4.30. Suggested values for the momentum transfer cross section, $\sigma_m(\varepsilon)$, of C_2F_6.

ε (eV)	$\sigma_m(\varepsilon)$ (10^{-20} m^2)	ε (eV)	$\sigma_m(\varepsilon)$ (10^{-20} m^2)	ε (eV)	$\sigma_m(\varepsilon)$ (10^{-20} m^2)
0.01	9.47	0.5	4.82	9	18.4
0.02	5.08	0.6	5.71	10	18.8
0.03	3.06	0.7	6.55	15	22.7
0.04	1.99	0.8	7.39	20	22.5
0.05	1.38	0.9	8.20	25	20.8
0.06	1.01	1.0	8.96	30	18.9
0.07	0.78	1.5	11.8	40	15.5
0.08	0.63	2.0	13.0	50	12.8
0.09	0.53	2.5	13.1	60	10.6
0.10	0.46	3.0	13.2	70	8.96
0.15	0.32	4.0	14.1	80	7.67
0.20	0.47	5.0	14.6	90	6.66
0.25	0.93	6.0	15.2	100	5.86
0.30	1.66	7.0	16.4		
0.40	3.45	8.0	17.9		

imum ($\sim$ 0.08 eV [247], $\sim$ 0.3 eV [252]). Also, compared to the two sets of experimental determinations [241, 251] of $\sigma_m(\varepsilon)$, the results of the two Boltzmann-code-based computations [247, 252] are in agreement with the measurements only near 3 eV.

In an effort to obtain values of $\sigma_m(\varepsilon)$ that can be suggested for possible use in modeling until reliable direct measurements are made over a wider energy range, a least squares fit was made to the data of Merz and Linder [251] and Takagi *et al.* [241]. These values are shown in Fig. 4.42 by the solid line, and are listed in Table 4.30 as our presently suggested values.

3.2.5 Inelastic Electron Scattering Cross Sections, $\sigma_{inel}(\varepsilon)$, of C_2F_6

The only experimental information on inelastic electron scattering processes in C_2F_6 is the work of Takagi *et al.* [241] on the vibrational excitation of this molecule. Takagi *et al.* studied the vibrational energy-loss spectra of C_2F_6 using various incident electron energies and scattering angles. Figure 4.43 shows the vibrational energy-loss spectra they obtained at incident electron energies of 2, 4, 7 and 8.5 eV and a scattering

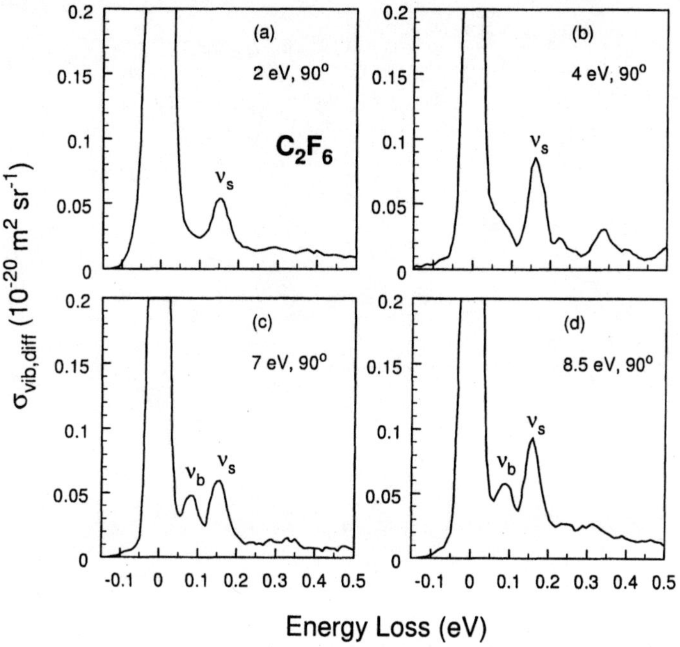

Figure 4.43. Vibrational differential cross section, $\sigma_{\text{vib,diff}}(\varepsilon)$, as a function of electron energy for various vibrational energy losses at a scattering angle of 90° and incident electron energies of 2, 4, 7, and 8.5 eV. The symbol ν_s refers to a stretching and ν_b to a bending mode. The ν_s peak consists mainly of the vibrational mode ν_1 [241].

angle of 90°. In this figure ν_s refers to excitation of a stretching mode and ν_b to a bending mode. The large energy-loss peak, ν_s, at 0.16 eV is mainly due to the excitation of the stretching vibrational mode, ν_1. The excitation spectrum of the ν_s energy-loss stretching mode at 0.16 eV in the energy range 1.5–16.5 eV is shown in Fig. 4.44 for four scattering angles. At small scattering angles direct electron scattering is evident while at the larger scattering angles two distinct maxima are prominent near 4.3 and 8.5 eV which are attributed to shape resonances at these energies [241].

Two Boltzmann-code calculations [247, 252] gave inelastic vibrational excitation cross sections as a function of electron energy which are shown in Fig. 4.45. Hayashi and Niwa [247] reported three such cross sections, σ_{vib5}, σ_{vib6}, and σ_{vib7}, while Pirgov and Stefanov [252] reported an overall inelastic vibrational excitation cross section. The sum of the three inelastic vibrational excitation cross sections [2] of Hayashi and Niwa is shown in Fig. 4.45 by the solid line. It is in reasonable agreement

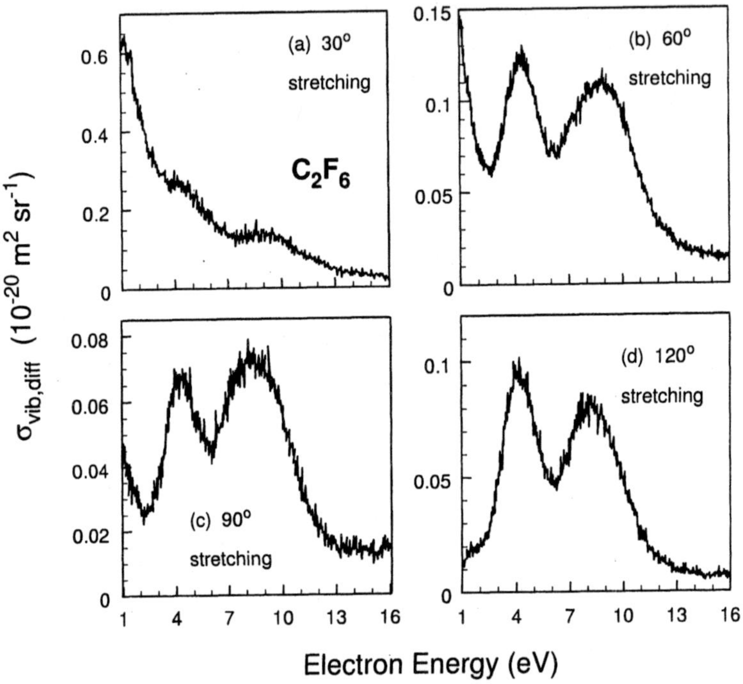

Figure 4.44. Excitation functions for the vibrational energy-loss peak ν_s (0.16 eV) at the indicated scattering angles [241].

with the overall cross section of Pirgov and Stefanov [252]. No experimental measurements exist to confirm these calculated cross sections. It might be noted that a number of studies have indicated enhanced direct electron scattering from vibrationally-excited C_2F_6 molecules [122, 254, 255]. It is possible that this enhancement is due to the dipole moments that some of the excited vibrational modes may possess.

For calculated excitation cross sections for a number of electronic transitions of C_2F_6 see McKoy *et al.* [233].

3.3 Electron-Impact Ionization of C_2F_6

3.3.1 Partial Ionization Cross Sections, $\sigma_{i,part}(\varepsilon)$, of C_2F_6

There have been three measurements [256–258] of the partial ionization cross sections, $\sigma_{i,part}(\varepsilon)$, of the C_2F_6 molecule. In these experimental studies the observed fragment ions are CF^+, CF_2^+, CF_3^+, $C_2F_4^+$,

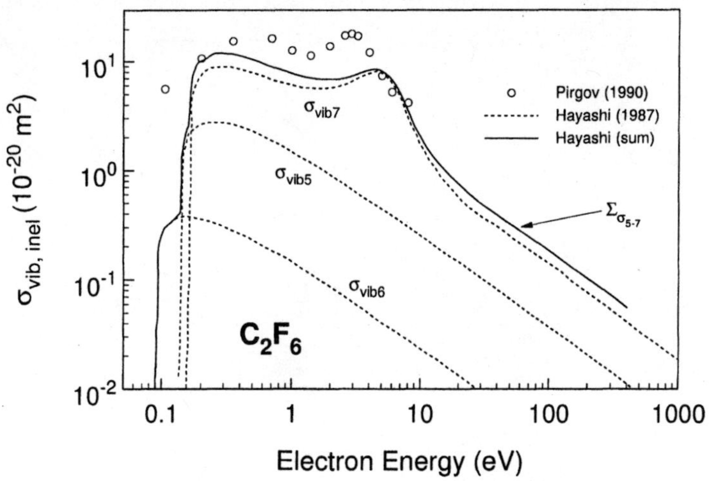

Figure 4.45. Inelastic vibrational excitation cross sections as a function of electron energy, $\sigma_{\rm vib,inel}(\varepsilon)$, for C_2F_6: (- - -) calculated cross sections for three vibrational modes by Hayashi and Niwa [247]; (—) sum of the three cross sections of Hayashi and Niwa [247]; (○) overall calculated inelastic vibrational excitation cross section of Pirgov and Stefanov [252].

and $C_2F_5^+$. In none of these studies was the parent positive ion $C_2F_6^+$ observed.

Poll and Meichsner [256] measured the $\sigma_{\rm i,part}(\varepsilon)$ for CF^+, CF_2^+, CF_3^+, and $C_2F_5^+$ produced by electron impact on C_2F_6 from threshold energies to approximately 130 eV. The CF_3^+ ion has the largest cross section of all four fragment ions. Their data are shown (dashed lines) in Fig. 4.46. Earlier analyses [2] indicate possible problems with the data of Poll and Meichsner [256] at low energies near threshold. The partial ionization cross section for CF_3^+ (and possibly $C_2F_5^+$) as determined from the figure of Poll and Meichsner has finite values below the accepted values of the appearance potentials (AP) for these ions (discussed below in Table 4.31). These "long tails" were noted by Poll and Meichsner who advanced two possible explanations: ionization of vibrationally excited molecules and/or the effect of the broad ("some eV") electron energy distribution in their experiments. The latter cause seems more likely. This uncertainty is the largest for the most abundant ion, CF_3^+. It is unclear why a similar effect is not observed for the less abundant ions, CF^+ and CF_2^+.

The available data on appearance potentials for ions from C_2F_6 are presented in Table 4.31. The suggested appearance potentials for the four fragment cations CF_3^+, $C_2F_5^+$, CF^+, and CF_2^+ are derived from the

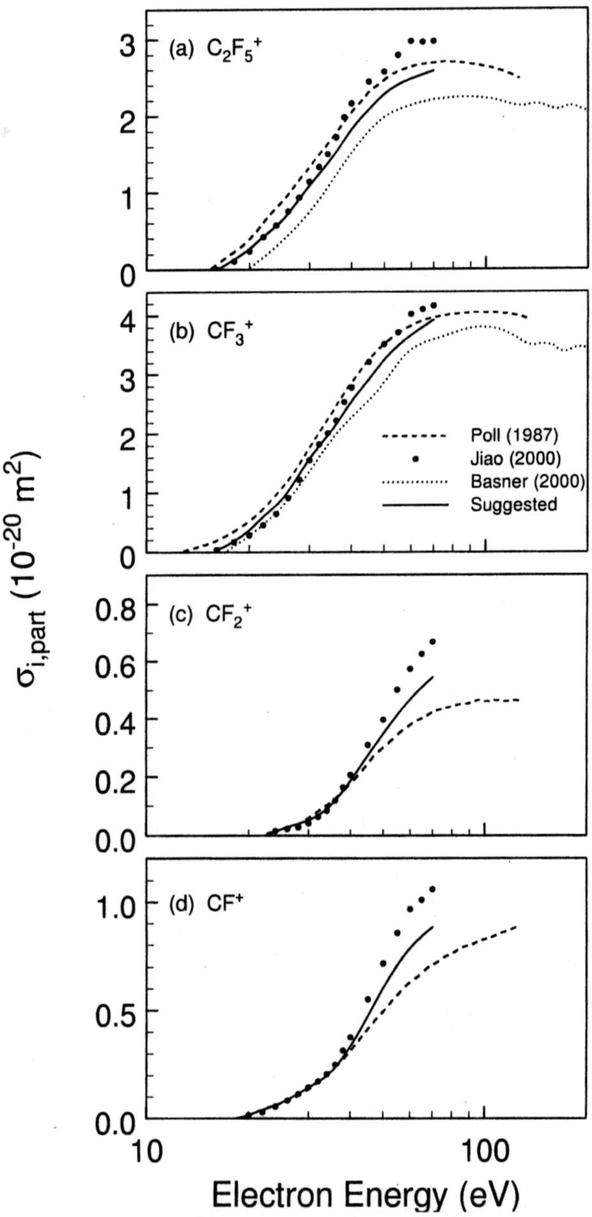

Figure 4.46. Partial electron-impact ionization cross sections, $\sigma_{i,part}(\varepsilon)$, as a function of electron energy for C_2F_6. (- - -) Ref. [256]; (•) Ref. [257]; (···) Ref. [258]; (—) suggested.

Table 4.31. Appearance potentials or ionization threshold energies[a,b] for C_2F_6.

Energy (eV)	Method	Reference
14.6	Photoabsorption spectra	[234]
14.48 (vertical)	Photoelectron spectra	[36]
13.62 ± 0.015 (CF_3^+) 15.46 ± 0.02 ($C_2F_5^+$) 16.75 (CF^+)	Photoionization	[235]
15.4 (CF_3^+) 16.05 ($C_2F_5^+$) 22.6 (F^+)	Electron impact	[259]
16.5 (CF_3^+) 15.5 ($C_2F_5^+$) 17.5 (CF_2^+) 18.0 (CF^+)	Electron impact	[256]

[a]The photoionization threshold energies are lower than the corresponding electron impact threshold energies. This reflects the large differences between the energies of the vertical and the adiabatic transitions leading to dissociative ionization for this molecule.
[b]See Ref. [250] for threshold energies for various neutral dissociation channels and for dissociative ionization channels.

available electron-impact data, and are, respectively, 16.0 eV (average of the two electron impact values), 15.8 eV (average of the two electron impact values), 18.0 eV, and 17.5 eV.

Jiao *et al.* [257] measured the electron-impact partial ionization cross sections in C_2F_6 using Fourier transform mass spectrometry. They observed four positive ion fragments: CF_3^+, $C_2F_5^+$, CF^+, and CF_2^+. The CF_3^+ is formed with the highest cross section at all energies up to 70 eV. They estimated the uncertainty in their cross section measurements to be ±16%. Their $\sigma_{\mathrm{i,part}}(\varepsilon)$ for all four fragment ions are shown (•) in Fig. 4.46. Basner *et al.* [258] reported measurements of $\sigma_{\mathrm{i,part}}(\varepsilon)$ for CF_3^+, $C_2F_5^+$, and $C_2F_4^+$ by electron impact on C_2F_6 from threshold to 1000 eV. Their results for CF_3^+ and $C_2F_5^+$ are also shown in Fig. 4.46 (dotted lines). The values for $C_2F_4^+$ production are very small ($< 10^{-22}$ m^2). There has also been an early measurement of $\sigma_{\mathrm{i,part}}(\varepsilon)$ of C_2F_6 by Bibby and Carter [259] for only one value of incident electron energy. Bibby and Carter reported the following cross section values at 35 eV: 0.022×10^{-20} m^2 for CF^+, 0.316×10^{-20} m^2 for CF_3^+, and 0.257×10^{-20} m^2 for $C_2F_5^+$. The sum of these partial ionization cross

Table 4.32. Suggested partial ionization cross sections, $\sigma_{i,part}(\varepsilon)$, for C_2F_6 in units of 10^{-20} m^2.

ε (eV)	$C_2F_5^+$	CF_3^+	CF_2^+	CF^+
16		0.04		
17	0.07	0.10		
18	0.13	0.19		
20	0.27	0.37		0.02
22	0.43	0.59		0.04
25	0.64	0.89	0.02	0.07
30	1.09	1.58	0.05	0.14
35	1.46	2.07	0.11	0.22
40	1.82	2.55	0.19	0.34
45	2.08	2.91	0.27	0.47
50	2.28	3.24	0.35	0.60
60	2.48	3.66	0.47	0.79
70	2.58	3.92	0.54	0.88

sections at 35 eV is 0.595×10^{-20} m^2, which is much smaller than the corresponding values discussed in the next section.

The solid lines in Fig. 4.46 are the result of a least squares fit to the experimental data for each cation up to 70 eV. The sum of these fits was then normalized at each energy to the recommended values of $\sigma_{i,t}(\varepsilon)$ derived in the next section. Values from these normalized fits are listed in Table 4.32 as our suggested values for the $\sigma_{i,part}(\varepsilon)$ for the production of CF_3^+, $C_2F_5^+$, CF^+, and CF_2^+ by electron impact on C_2F_6.

3.3.2 Total Ionization Cross Section, $\sigma_{i,t}(\varepsilon)$, of C_2F_6

There are four measurements of the total ionization cross section, $\sigma_{i,t}(\varepsilon)$, of the C_2F_6 molecule which were available at the time of the initial review of C_2F_6 by Christophorou and Olthoff [2]. The first was by Kurepa [260] for incident electron energies up to 100 eV. No uncertainty was assigned to these measurements. They are plotted in Fig. 4.47 as open squares. The second measurement of $\sigma_{i,t}(\varepsilon)$ was made by Beran and Kevan [261] for only three values of the incident electron energy. These results are shown in Fig. 4.47 by the $\times$ symbols. It should be noted that the data of Beran and Kevan [261] for a number of species are consistently higher than those of Rapp and Englander-Golden [262] which are generally accepted to be more accurate. The third set of values are those obtained by summation of the partial ionization cross sections

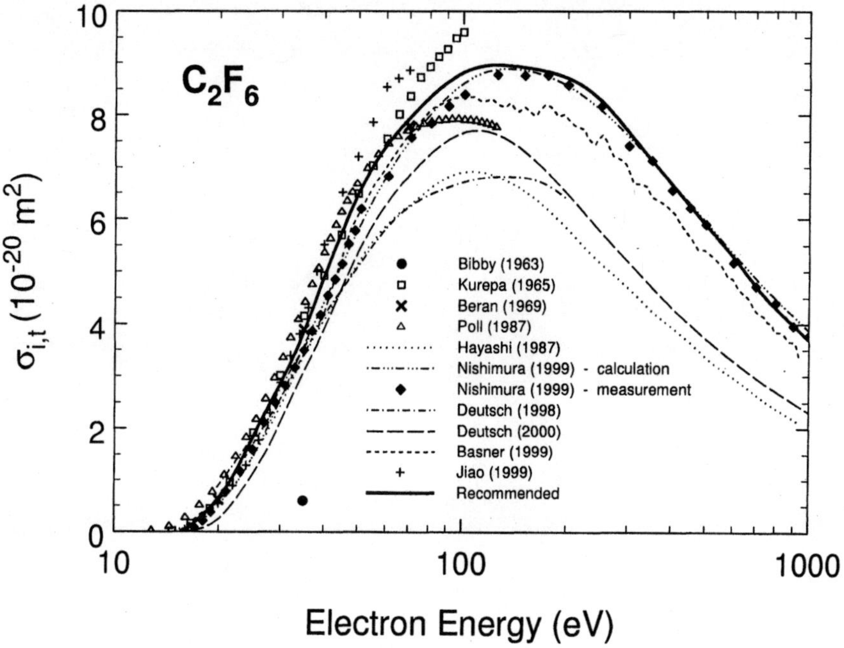

Figure 4.47. Total ionization cross section, $\sigma_{i,t}(\varepsilon)$, as a function of electron energy for C_2F_6. Experimental data: (•) Ref. [259]; (□) Ref. [260]; (×) Ref. [261]; (△) Ref. [256]; (♦) Ref. [137]; (+) Ref. [257]; (- - -) Ref. [258]. Calculated values: (— · · · —) Ref. [137]; (— · —) Ref. [263]; (– –) Ref. [143]. Computed values: (· · ·) Ref. [247]. (—) Recommended.

measured by Poll and Meichsner [256] as discussed in the previous section. These are shown in Fig. 4.47 by the open triangles. The fourth measurement, as mentioned in the preceding section, is an early report [259] of a measurement of the ionization cross section for the production of the CF^+, CF_3^+ and $C_2F_5^+$ ions from C_2F_6 at one value of the electron energy, namely 35 eV. At this electron energy the sum of the cross sections of Bibby and Carter [259] for the three ions is approximately 0.6×10^{-20} m^2. It is shown in Fig. 4.47 by the solid circle, and is obviously well below the other experimental data.

Since the original work of Christophorou and Olthoff [2] there have been a number of measurements and computations of the $\sigma_{i,t}(\varepsilon)$ of C_2F_6 which also are included in Fig. 4.47 (see also Ref. [48]). Nishimura *et al.* [137] measured the $\sigma_{i,t}(\varepsilon)$ of C_2F_6 up to 1000 eV with a reported uncertainty of ±6% for energies $<$ 900 eV and ±6.5% for energies $\geq$ 900 eV. Also, the measurements of Jiao *et al.* [257] and Basner *et al.* [258] on the partial ionization cross sections presented in the previous section

Table 4.33. Recommended total ionization cross section, $\sigma_{i,t}(\varepsilon)$, for C_2F_6.

ε (eV)	$\sigma_{i,t}(\varepsilon)$ (10^{-20} m^2)	ε (eV)	$\sigma_{i,t}(\varepsilon)$ (10^{-20} m^2)	ε (eV)	$\sigma_{i,t}(\varepsilon)$ (10^{-20} m^2)
16	0.04	45	5.74	250	8.28
17	0.17	50	6.47	300	7.68
18	0.32	60	7.39	400	6.67
19	0.47	70	7.93	500	5.89
20	0.65	80	8.33	600	5.29
22	1.06	90	8.61	700	4.73
25	1.82	100	8.80	800	4.31
30	2.86	120	8.95	900	3.98
35	3.85	150	8.91	1000	3.70
40	4.90	200	8.71		

have been summed up by the respective authors, and provide additional experimental data on the total ionization cross section of C_2F_6. The total cross section of Basner *et al.* [258] reaches a maximum of $\sim 8 \times 10^{-20}$ m^2 at 100 eV and that of Jiao *et al.* [257] rises to a maximum value of $(8.9 \pm 1.3) \times 10^{-20}$ m^2 at 70 eV.

The bold solid line in Fig. 4.47 represents our recommended values for the $\sigma_{i,t}(\varepsilon)$ of C_2F_6 derived by fitting a curve to the ionization threshold at 15.9 eV and to the four most recent experimental sets of measurements [137, 256–258] up to 70 eV. The recommended values are extended to 1000 eV by fitting the data of Nishimura *et al.* [137] and Basner *et al.* [258] above 70 eV, and normalized to the recommended value at 70 eV. Values taken off this solid line are listed in Table 4.33 as our recommended values for the $\sigma_{i,t}(\varepsilon)$ of C_2F_6.

Besides the experimental values of $\sigma_{i,t}(\varepsilon)$ in Fig. 4.47, a number of theoretical results are also plotted in the figure, namely, the BEB-model results of Nishimura *et al.* [137], and the modified additivity rule-model results of Deutsch *et al.* [143, 263]. Also shown in Fig. 4.47 is the $\sigma_{i,t}(\varepsilon)$ deduced using a Boltzmann code analysis by Hayashi and Niwa [247]. Only the calculations of Nishimura *et al.* [137] agree reasonably well with the recommended values.

3.3.3 Total Dissociation Cross Section, $\sigma_{dis,t}(\varepsilon)$, of C_2F_6

The only measurement of the total dissociation cross section, $\sigma_{dis,t}(\varepsilon)$, of C_2F_6 is that by Winters and Inokuti [50]. This cross section is pre-

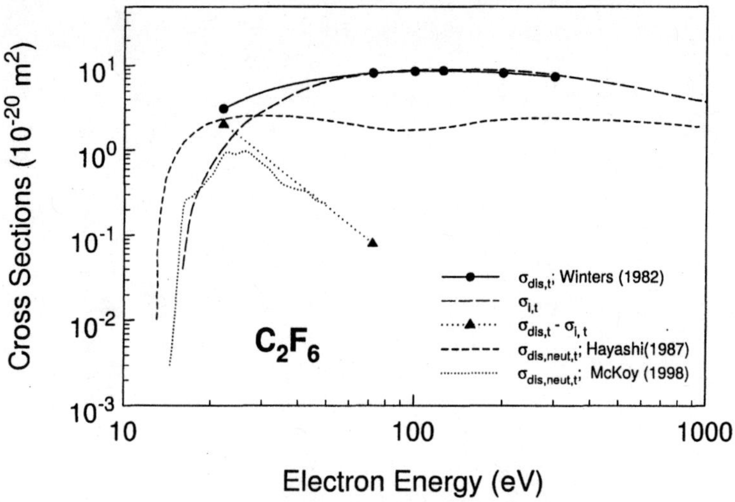

Figure 4.48. Total dissociation cross section as a function of electron energy, $\sigma_{dis,t}(\varepsilon)$, for C_2F_6: (•) Ref. [50]. Also plotted in the figure is the recommended total ionization cross section $\sigma_{i,t}(\varepsilon)$ (– –, Table 4.33), and the difference $\sigma_{dis,t}(\varepsilon)$ - $\sigma_{i,t}(\varepsilon)$ (▲). The short dashed curve (- - -) is the computed $\sigma_{dis,neut,t}(\varepsilon)$ of Ref. [247], and the dotted line (···) is the sum of the cross sections for three excitations in C_2F_6 as calculated in Ref. [233].

Table 4.34. Total electron-impact dissociation cross section, $\sigma_{dis,t}(\varepsilon)$, measurements of Winters and Inokuti [50] for C_2F_6.

ε (eV)	$\sigma_{dis,t}(\varepsilon)$ (10^{-20} m^2)	ε (eV)	$\sigma_{dis,t}(\varepsilon)$ (10^{-20} m^2)
22	3.1	125	8.6
72	8.1	200	8.1
100	8.5	300	7.3

sented in Fig. 4.48 and in Table 4.34, and represents the sum of the cross sections for dissociative ionization and the cross sections for electron-impact dissociation into neutral fragments with a reported uncertainty of ±20%.

An estimate may be obtained of the total cross section for dissociation into neutral fragments, $\sigma_{dis,neut,t}(\varepsilon)$, by subtracting the recommended total ionization cross section $\sigma_{i,t}(\varepsilon)$, which is exclusively due to dissociative ionization, from the total dissociation cross section of Winters and

Inokuti. This difference is shown in Fig. 4.48 by the closed triangles. The values of $\sigma_{\mathrm{dis,neut,t}}(\varepsilon)$ determined this way are reasonably consistent with the calculations of McKoy *et al.* [233] discussed below, and also with the limited measurements by Motlagh and Moore [162] (not shown). The measurements of Motlagh and Moore indicate a neutral dissociation cross section of $\sim 1 \times 10^{-20}$ m^2 at the maximum. Finally, in Fig. 4.48 is shown the sum of the cross sections for the $6a_{1g} \rightarrow 7a_{1g}$, $6a_{1g} \rightarrow 6e_u$, and $6a_{1g} \rightarrow 6a_{2u}$ excitations in C_2F_6 as determined in the *ab initio* calculations of McKoy *et al.* [233]. Since all excitations in C_2F_6 are dissociative, this sum represents a lower limit to the $\sigma_{\mathrm{dis,neut,t}}(\varepsilon)$ for this molecule.

3.3.4 Ionization Coefficients for C_2F_6

Density-Normalized Ionization Coefficient, $\alpha/N\,(E/N)$. There have been a number of measurements [120, 148, 149, 152] of $\alpha/N\,(E/N)$ for C_2F_6 at temperatures ranging from 293 to 298 K. These measurements were performed at pressures below $\sim$ 13 kPa and generally have an uncertainty of $\pm$10%, with the exception of the data of Naidu and Prasad [120] which have a reported overall uncertainty between $\pm$10% for E/N values lower than the $(E/N)_{\mathrm{lim}}$ and $\pm$20% for higher values of E/N. All measurements were made using the steady-state Townsend method except those of Hunter *et al.* [152] who used a pulsed-Townsend method. These measurements are compared in Fig. 4.49. The ionization coefficient is not expected to be sensitive to small variations in gas temperature and indeed the measurements of Hunter *et al.* [152] found no noticeable change in the values of α/N at 300 and at 500 K.

The data between 210×10^{-17} V cm^2 and 400×10^{-17} V cm^2 shown in Fig. 4.49 (the range in which all the data overlap) were least-squares fitted and their average is shown in the figure by the solid line. Below 210×10^{-17} V cm^2 the data of Hunter *et al.* were normalized to the average value at 210×10^{-17} V cm^2. Above 400×10^{-17} V cm^2, the data of Naidu and Prasad were normalized to the mean value at 400×10^{-17} V cm^2. The resultant values are shown in Fig. 4.49 by the solid line and are listed in Table 4.35 as our recommended data for the $\alpha/N\,(E/N)$ of C_2F_6. The broken line in Fig. 4.49 shows the calculated values of Hayashi and Niwa [247]. As far as we know there have been no reported measurements of $\alpha/N\,(E/N)$ for C_2F_6 in gas mixtures.

Effective Ionization Coefficient, $(\alpha - \eta)/N(E/N)$. Figure 4.50 shows the values of the effective ionization coefficient $(\alpha-\eta)/N(\equiv \bar{\alpha}/N)$ for C_2F_6 measured as a function of (E/N) by various groups at two

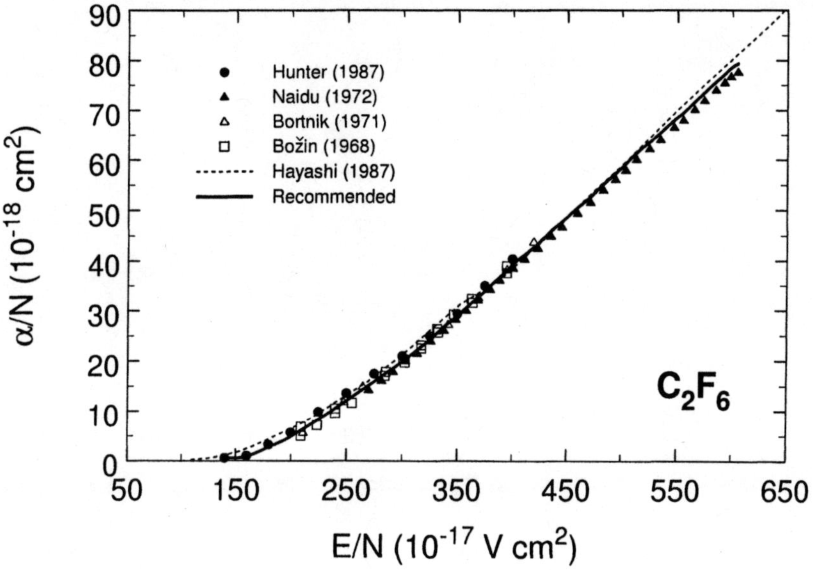

Figure 4.49. Density-normalized ionization coefficient α/N as a function of E/N for C_2F_6: (●) Ref. [152]; (▲) Ref. [120]; (△) Ref. [149]; (□) Ref. [148]; (- - -) Ref. [247]; (—) recommended.

Table 4.35. Recommended density-normalized ionization coefficients, $\alpha/N\,(E/N)$, for C_2F_6.

E/N (10^{-17} V cm^2)	α/N (10^{-18} cm^2)	E/N (10^{-17} V cm^2)	α/N (10^{-18} cm^2)
140	0.51	360	31.0
160	0.85	380	35.0
180	2.71	400	39.2
200	4.83	420	42.9
220	7.65	440	46.8
240	10.6	460	50.6
260	13.6	480	54.5
280	16.7	500	58.5
300	20.0	550	68.4
320	23.4	600	78.6
340	27.2		

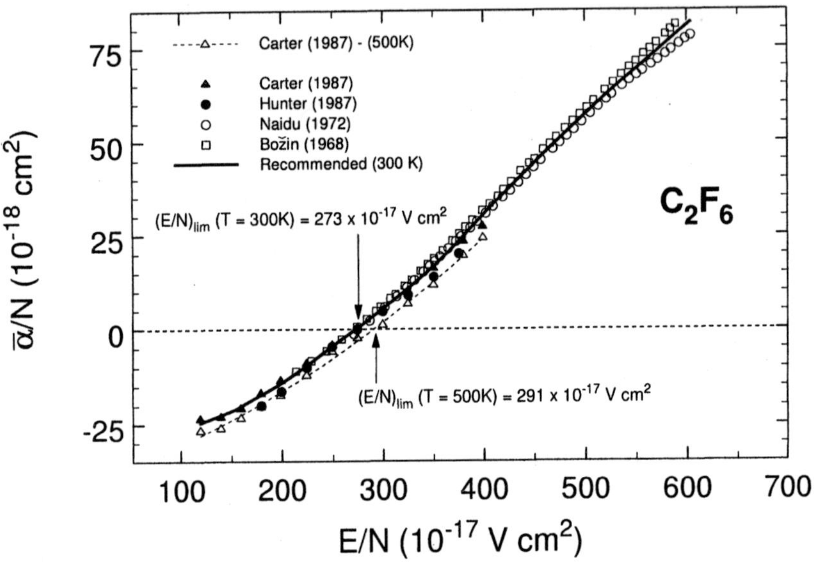

Figure 4.50. Density-normalized effective ionization coefficient $(\alpha - \eta)/N$ as a function of E/N for C_2F_6. Room temperature data: (▲) Ref. [264] (300 K); (•) Ref. [152] (300 K); (○) Ref. [120] (293 K); (□) Ref. [148] (293 K); (—) recommended (300 K). Data at 500 K: (△) difference between the values of $\alpha/N\,(E/N)$ and $\eta/N\,(E/N)$ reported by Carter *et al.* [264] at this temperature. Also indicated in the figure are the limiting values of E/N at 300 and 500 K.

temperatures ~ 300 and 500 K. The room temperature data are those of Carter *et al.* [264], Hunter *et al.* [152], Naidu and Prasad [120], and Božin and Goodyear [148]. The only data at a temperature other than ambient are those of Carter *et al.* [264] at 500 K. The reported uncertainties are all about ±10%, except for Hunter *et al.* (~ ±5%) and Naidu and Prasad (between ±10% and ±20% at high E/N).

Interestingly, the values of $(\alpha - \eta)/N(E/N)$ at 500 K are lower than at 300 K although at these two temperatures the values of $\alpha/N\,(E/N)$ are virtually the same. Clearly, the temperature dependence of $(\alpha - \eta)/N(E/N)$ is due to the temperature dependence of the electron attachment coefficient $\eta/N\,(E/N)$ (Sec. 3.5.1 of Ch. 4). A least-squares fit to the room-temperature data is represented in Fig. 4.50 by the solid line. Points taken off this line are listed in Table 4.36 as our recommended values for the $(\alpha - \eta)/N(E/N)$ of C_2F_6. Finally, it is noted that Byszewski *et al.* reported effective ionization rates for pure C_2F_6 [265, 266] and mixtures of C_2F_6 in Ar [266].

Table 4.36. Recommended values of the effective ionization coefficient, $(\alpha - \eta)/N$ ($\equiv \bar{\alpha}/N$), for C_2F_6 at 300 K.

E/N (10^{-17} V cm^2)	$(\alpha - \eta)/N$ (10^{-18} cm^2)	E/N (10^{-17} V cm^2)	$(\alpha - \eta)/N$ (10^{-18} cm^2)
120	−24.5	340	14.3
130	−23.5	360	19.3
140	−22.4	380	24.7
160	−20.0	400	30.4
180	−17.1	420	35.9
200	−14.0	440	41.3
220	−10.4	460	46.7
240	−6.34	480	51.9
260	−2.45	500	57.0
280	1.43	550	69.1
300	5.47	600	80.7
320	9.75		

$(E/N)_{\rm lim}$. The value of E/N, $(E/N)_{\rm lim}$, at which $(\alpha - \eta)/N = 0$, comes naturally from the measurements of $\eta/N\,(E/N)$ and $\alpha/N\,(E/N)$. The $(E/N)_{\rm lim}$ value for C_2F_6 should coincide with the breakdown voltage of C_2F_6 as measured under a uniform electric field. In Table 4.37 are listed values of $(E/N)_{\rm lim}$ measured at room temperature. Their average is 275×10^{-17} V cm^2. Interestingly, Christophorou *et al.* [267, 268] found that $(E/N)_{\rm lim}$ increases with increasing temperature (see Table 4.37), and this increase has been attributed to an increase in the dissociative electron attachment cross section with increasing temperature for this molecule (see Sec. 3.5.6 of Ch. 4 and Refs. [267, 268]). The least-squares fits to the data in Fig. 4.50 for 300 and 500 K give values of $(E/N)_{\rm lim}$ equal to 273×10^{-17} V cm^2 at 300 K and 291×10^{-17} V cm^2 at 500 K. Furthermore, measurements have been made of the $(E/N)_{\rm lim}$ of binary mixtures of C_2F_6 with Ar or CH_4 for pulse power applications [220]. Figure 4.51 shows these measurements.

Average Energy to Produce an Electron-Ion Pair, W, for C_2F_6. The average energy to produce an electron-ion pair (i.p.), W, for α particles (initial energy $\sim$ 5.1 MeV) has been measured by Nakanishi *et al.* [272] for pure C_2F_6 and found to be 34.7 eV per ion pair. A similar measurement by Reinking *et al.* [156] gave a value of 34.5 eV per ion pair. These values are almost identical with those measured by Reinking *et al.* [156] for CF_4 and C_3F_8. They are large compared

Table 4.37. Values[a,b] of $(E/N)_{\rm lim}$ for C_2F_6.

$(E/N)_{\rm lim}$ (10^{-17} V cm^2)	Reference
274.2 (298 K) 274.6 (373 K) 275.1 (423 K) 275.9 (473 K) 276.3 (523 K) 277.1 (573 K)	[267, 268]
275	[152, 220]
277	[149]
277	[120]
271	[148]
271[c]	[269]
278[c]	[270]
285[c]	[15]
289[c]	[16]
322[c]	[271]

[a] All values are for room temperature unless otherwise noted.
[b] From the least-squares fit to the 300 K data in Fig. 4.50 we obtained a value for $(E/N)_{\rm lim}$ equal to 273×10^{-17} V cm^2 which is recommended.
[c] Uniform field breakdown voltage measurements.

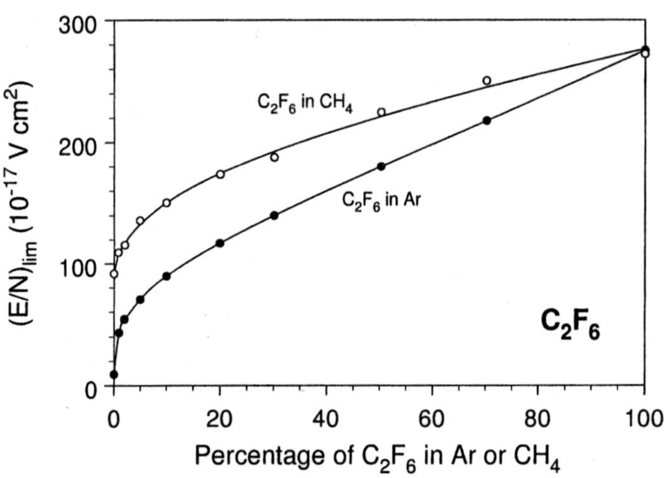

Figure 4.51. $(E/N)_{\rm lim}$ as a function of percentage of C_2F_6 in Ar or CH_4 [220].

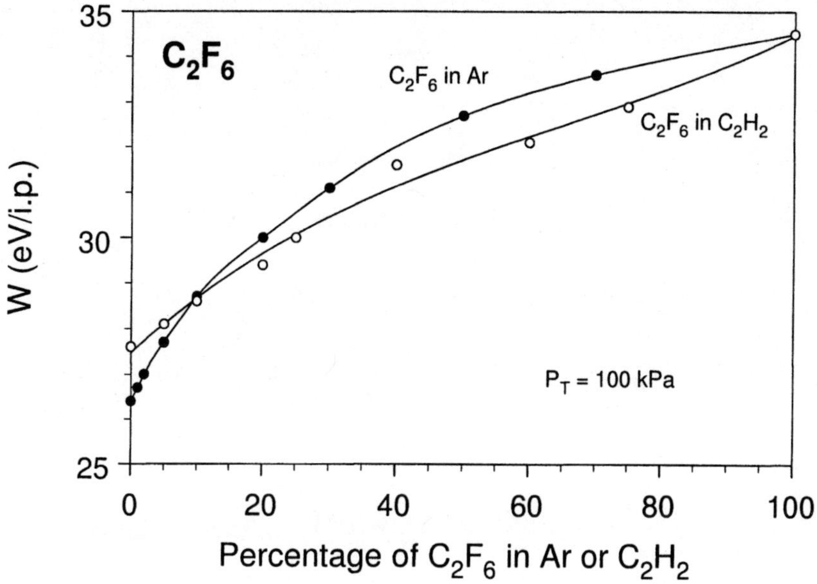

Figure 4.52. Energy, W, needed to produce an electron-ion pair by α particles (initial energy 5.1 MeV) in mixtures of C_2F_6 with Ar or C_2H_2 measured by Reinking *et al.* [156]. The total pressure of the mixtures was 100 kPa.

to the W values of other polyatomic molecules [273] reflecting the high ionization threshold energies for these perfluorocarbon molecules and the considerable amount of energy going into transnational and/or internal energy of the fragments that accompanies the dissociative ionization processes in these molecules. Nagra and Armstrong [274] measured the W of C_2F_6 using ^{60}Co γ-rays and found it to be 32.7 eV per ion pair. This value is lower than that for α particles, as expected [273].

Nakanishi *et al.* [272] and Reinking *et al.* [156] measured the W values for binary mixtures of C_2F_6 with Ar or C_2H_2. Figure 4.52 shows the measurements Reinking *et al.* Nakanishi *et al.* [272] also made measurements of W for binary mixtures of C_2F_6 with 2-C_4H_8.

3.4 Electron-Impact Dissociation of C_2F_6 into Neutral Fragments

To our knowledge there are no data on this important process, except the rough estimates discussed in Sec. 3.3.3 of Ch. 4. It might be of interest to note that measurements of neutral fragments in various low

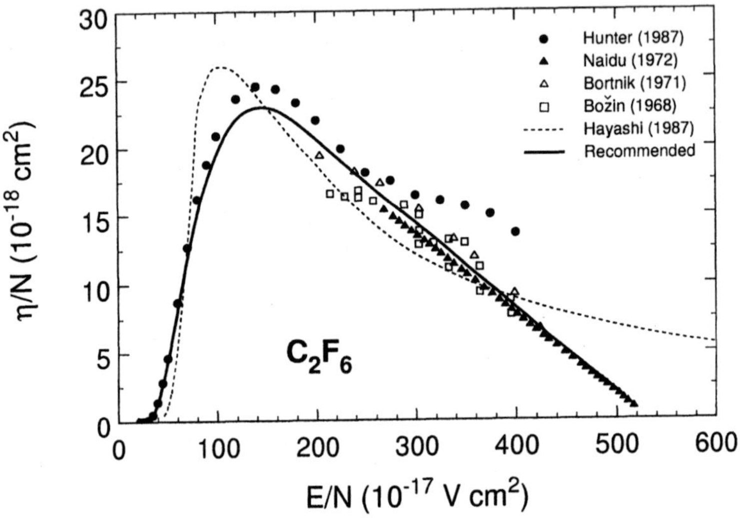

Figure 4.53. Density-normalized electron attachment coefficient, η/N, as a function of E/N (T = 293–300 K) for C_2F_6. Measurements: (•) Ref. [152]; (▲) Ref. [120]; (△) Ref. [149]; (□) Ref. [148]. Computed values: (- - -) Ref. [247]. (—) Recommended.

pressure C_2F_6 gas discharges have been made using diode laser absorption spectroscopy [275, 276].

3.5 Electron Attachment to C_2F_6

There have been a number of measurements of the electron attachment coefficient in C_2F_6. We begin this section by analyzing these measurements first because they provide some insight for understanding the electron attachment cross section data which are presented later in this section. These data are also useful in many practical applications and in computations aimed at deriving cross section sets for C_2F_6.

3.5.1 Density-Normalized Electron Attachment Coefficient, $\eta/N\,(E/N)$, of C_2F_6

The density-normalized electron attachment coefficient $\eta/N\,(E/N)$ of C_2F_6 has been measured by a number of investigators [120, 148, 149, 152], and Fig. 4.53 shows these measurements which were made at temperatures ranging from 293 to 300 K. Hunter *et al.* [152] quoted uncertainties ranging from about ±4% at low E/N to about ±7% at the highest E/N values at which they made measurements. Naidu and Prasad

[120] quoted an uncertainty of ±10% to ±20%, and the measurements of Božin and Goodyear [148] have an anticipated uncertainty of ±10%. The uncertainties of the data of Bortnik and Panov [149] are difficult to determine but they appear to be between ±5% and ±10%. Only the data of Hunter *et al.* [152] extend to low E/N. In the region where the data overlap there is considerable variation which seems to be outside of the quoted combined uncertainties. In particular, the data of Božin and Goodyear [148] are lower than the rest around $E/N = 200 \times 10^{-17}$ V cm^2 and the data of Hunter *et al.* [152] are higher than the rest above $\sim 300 \times 10^{-17}$ V cm^2. The data of Hunter *et al.* seem to indicate an enhancement in this E/N range that none of the other measurements indicate. This behavior was present in their measurements at both room temperature and at 500 K. Hunter *et al.* [152] suggested that the difference between their data using a pulsed Townsend technique and the other measurements using the steady-state Townsend method may be the neglect of secondary electron production processes in the analysis of the current growth curves in the steady-state measurements.

Figure 4.53 also shows the $\eta/N\,(E/N)$ values calculated by Hayashi and Niwa [247] from a Boltzmann code analysis. The agreement with the experimental values varies significantly with E/N.

To determine a recommended set of $\eta/N\,(E/N)$ values we performed a least-squares fit to all four sets of experimental data in Fig. 4.53 between 200×10^{-17} and 400×10^{-17} V cm^2, excluding the two lowest E/N data points of Božin and Goodyear [148] and the highest three E/N data points of Hunter *et al.* [152], since these fall outside of the combined uncertainty. We then extended the recommended curve to lower E/N by normalizing the data of Hunter *et al.* to the average value of η/N at $E/N = 200 \times 10^{-17}$ V cm^2, and to higher E/N by normalizing the data of Naidu and Prasad to the average value of η/N at 400×10^{-17} V cm^2. The solid line in Fig. 4.53 represents the curve obtained as has just been described, and recommended values from this line are given in Table 4.38.

There has been only one measurement of $\eta/N\,(E/N)$ at a temperature above ambient, namely that by Carter *et al.* [264] at 500 K. These measurements are compared in Fig. 4.54 with the room temperature data taken under similar conditions. The values of $\eta/N\,(E/N)$ at 500 K are higher than at 300 K due to the enhancement of dissociative electron attachment at the elevated temperature (Sec. 3.5.6 of Ch. 4). Values of $\eta/N_a\,(E/N)$ in mixtures of C_2F_6 with Ar or CH_4 have been reported by Hunter *et al.* [13] and by Carter *et al.* [264], and the reader is referred to these sources for relevant information.

Table 4.38. Recommended values of the density-normalized electron attachment coefficient, $\eta/N\,(E/N)$, for C_2F_6.

E/N (10^{-17} V cm^2)	η/N (10^{-18} cm^2)	E/N (10^{-17} V cm^2)	η/N (10^{-18} cm^2)
22	0.001	220	19.3
25	0.008	240	18.0
30	0.09	260	16.7
35	0.46	280	15.5
40	1.31	300	14.4
45	2.64	320	13.2
50	4.32	340	11.9
60	8.14	360	10.7
70	11.9	380	9.45
80	15.2	400	8.20
90	17.6	420	6.94
100	19.6	440	5.71
120	22.1	460	4.47
140	22.9	480	3.25
160	22.7	500	2.09
180	21.8	515	1.09
200	20.6		

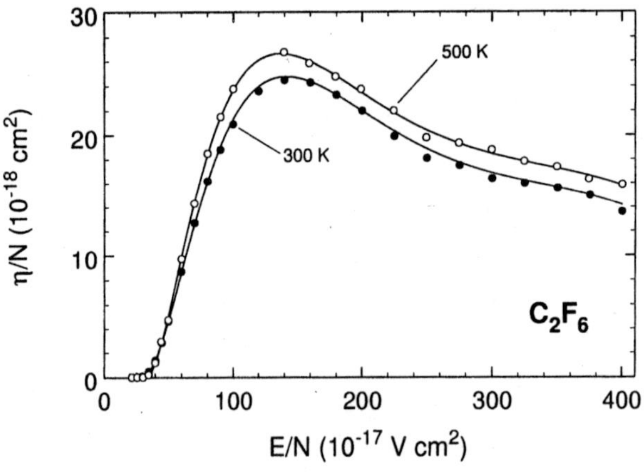

Figure 4.54. Density-normalized electron attachment coefficient, η/N, as a function of E/N and temperature for C_2F_6: (•) data of Hunter *et al.* [152] at $T = 300$ K; (○) data of Carter *et al.* [264] at $T = 500$ K.

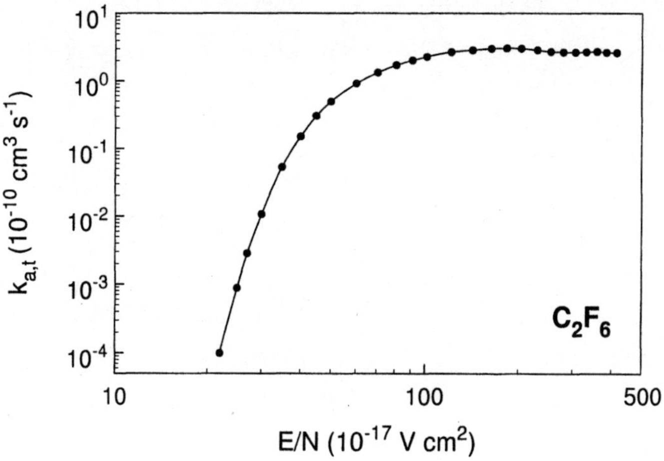

Figure 4.55. Total electron attachment rate constant as a function of E/N, $k_{a,t}(E/N)$, for C_2F_6 measured in pure C_2F_6 by Hunter *et al.* [152].

3.5.2 Total Electron Attachment Rate Constant, $k_{a,t}$, for C_2F_6

The only measurement of $k_{a,t}(E/N)$ in pure C_2F_6 is that of Hunter *et al.* [152] at $T = 300$ K. This measurement is shown in Fig. 4.55. Similar measurements have been made for mixtures of C_2F_6 with argon by Hunter and Christophorou [72] and Spyrou and Christophorou [277]. The measurements in mixtures are shown as a function of E/N in Fig. 4.56(a), and as a function of the mean electron energy $\langle \varepsilon \rangle$ in Fig. 4.56(b). The plot in Fig. 4.56(b) was possible because the measurements in the argon mixtures were made with very small concentrations of C_2F_6. This allowed determination of the mean electron energy as a function of E/N from the known electron energy distribution functions as a function of E/N in argon. Each of the two sets of data probably has a total uncertainty of $\pm 10\%$. Their average is shown by the solid line in Fig. 4.56(b), and is listed in Table 4.39 as our recommended values for the $k_{a,t}(\langle \varepsilon \rangle)$ of C_2F_6.

3.5.3 Thermal Value, $(k_{a,t})_{th}$, of the Total Electron Attachment Rate Constant, for C_2F_6

Table 4.40 lists reported values of $(k_{a,t})_{th}$ measured at $T \approx 300$ K [72, 152, 190, 191]. These values are very small ($< 1.6 \times 10^{-13}$ cm^3 s^{-1})

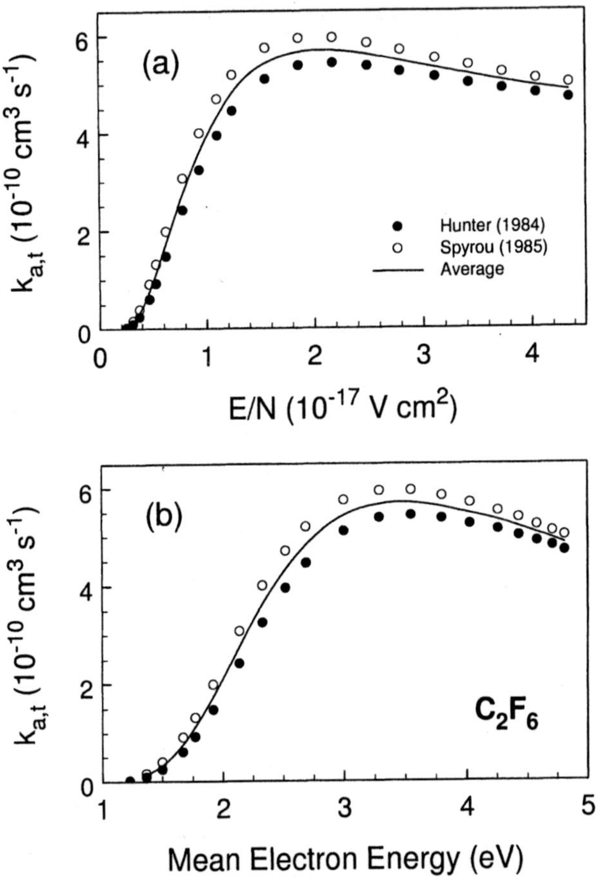

Figure 4.56. Total electron attachment rate constant, $k_{a,t}$, for C_2F_6 measured in mixtures of C_2F_6 in argon buffer gas (a) as a function of E/N and (b) as a function of the mean electron energy $\langle\varepsilon\rangle$: (•) Ref. [72]; (∘) Ref. [277]; (—) average.

and may well be due to or affected by traces of strongly electronegative impurities.

3.5.4 Total Electron Attachment Cross Section, $\sigma_{a,t}(\varepsilon)$, for C_2F_6

The main source of recent data on the total electron attachment cross section, $\sigma_{a,t}(\varepsilon)$, of C_2F_6 is the room temperature swarm-unfolded data of Christophorou and co-workers [72, 277]. These data are presented in Fig. 4.57. Figure 4.57 also shows the relative total electron attachment cross section as determined in an electron-beam experiment by Spyrou

Table 4.39. Recommended total electron attachment rate constant $k_{a,t}(\langle\varepsilon\rangle)$, as a function of the mean electron energy $\langle\varepsilon\rangle$ for C_2F_6 measured in mixtures of C_2F_6 with argon.

Mean electron energy (eV)	$k_{a,t}(\langle\varepsilon\rangle)$ (10^{-10} cm^3 s^{-1})	Mean electron energy (eV)	$k_{a,t}(\langle\varepsilon\rangle)$ (10^{-10} cm^3 s^{-1})
1.37	0.14	3.29	5.66
1.50	0.33	3.55	5.69
1.67	0.77	3.80	5.61
1.77	1.13	4.03	5.47
1.92	1.74	4.26	5.33
2.14	2.75	4.43	5.20
2.33	3.63	4.58	5.07
2.52	4.34	4.71	4.95
2.69	4.83	4.81	4.86
3.00	5.43		

Table 4.40. Thermal ($T \approx 300$ K) values of the total electron attachment rate constant, $(k_{a,t})_{th}$, for C_2F_6.

$(k_{a,t})_{th}$ (cm^3 s^{-1})	Reference
$< 1 \times 10^{-16}$	[152, 191]
$< 1 \times 10^{-13}$	[72]
$< 1.6 \times 10^{-13}$	[190]

and Christophorou [277] normalized to the peak of their swarm-unfolded total electron attachment cross section. Its shape agrees reasonably well with the swarm-unfolded data. The two major dissociative electron attachment fragment negative ions of C_2F_6 (see Sec. 3.5.5 of Ch. 4) are F^- and CF_3^-. Also available in the literature are the results of three early, low resolution, electron-beam experiments. An early beam study by Bibby and Carter [259] reported a cross section value of 1.76×10^{-20} m^2 for F^- and 0.43×10^{-20} m^2 for CF_3^- at 3.75 eV. The sum of these two peak cross section values for the two fragments is 2.19×10^{-20} m^2 which is more than ten times larger than the more recent data in Fig. 4.57 indicate. The results of a second beam study by Kurepa [260] are also more than ten times the swarm-determined total electron attachment cross sections. Neither of these measurements are considered to be reliable in light of the more recent measurements, and therefore

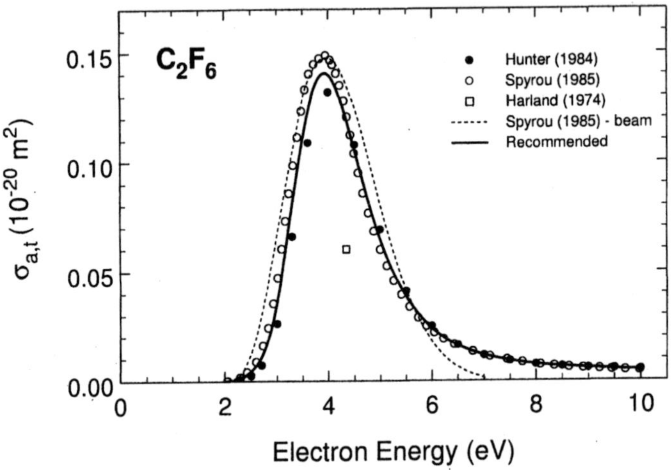

Figure 4.57. Total electron attachment cross section as a function of electron energy, $\sigma_{a,t}(\varepsilon)$, for C_2F_6. Swarm-unfolded data: (•) Ref. [72]; (○) Ref. [277]. Beam data: (□) Ref. [71]; (- - -) Ref. [277] normalized to the swarm-unfolded data at the maximum. (—) Recommended.

are not presented in Fig. 4.57. The third beam study is that of Harland and Franklin [71] who found the maximum intensity for the two fragment anions F^- and CF_3^- to be at 4.3 and 4.4 eV, respectively. At these energy values they reported the respective cross sections to be 5×10^{-22} m^2 and 1×10^{-22} m^2. Since these two ions are by far the two most intense fragment negative ions, we compare the sum (6×10^{-22} m^2) of the two with the peak values of the total electron attachment cross section (at 4.35 eV) in Fig. 4.57. It is about a factor of two lower than the other data.

The average of the two sets of swarm-unfolded measurements by Christophorou and co-workers is shown in Fig. 4.57 by the solid line. Values taken off this line are listed in Table 4.41 as our recommended values for the $\sigma_{a,t}(\varepsilon)$ of C_2F_6.

3.5.5 Dissociative Electron Attachment Fragment Negative Ions of C_2F_6

Electron beam studies of negative ion formation by electron impact on C_2F_6 have shown that below 10 eV three fragment anions (F^-, CF_3^-, and $C_2F_5^-$) are produced via a prominent broad negative ion resonance in the energy range from 2 to 7 eV. The decomposition channels were

Table 4.41. Recommended total electron attachment cross section, $\sigma_{a,t}(\varepsilon)$, for C_2F_6 ($T = 300$ K).

ε (eV)	$\sigma_{a,t}(\varepsilon)$ (10^{-20} m^2)	ε (eV)	$\sigma_{a,t}(\varepsilon)$ (10^{-20} m^2)	ε (eV)	$\sigma_{a,t}(\varepsilon)$ (10^{-20} m^2)
2.0	0.0001	4.0	0.140	7.5	0.0091
2.25	0.0007	4.5	0.105	8.0	0.0074
2.5	0.0044	5.0	0.064	8.5	0.0064
2.75	0.013	5.5	0.039	9.0	0.0057
3.0	0.036	6.0	0.024	9.5	0.0051
3.25	0.073	6.5	0.016	10.0	0.0046
3.5	0.112	7.0	0.012		

described [70, 71] as

$$C_2F_6^{-*} \rightarrow F^- + C_2F_5 \qquad (4.21a)$$

$$\rightarrow CF_3^- + CF_3 \qquad (4.21b)$$

$$\rightarrow C_2F_5^- + F. \qquad (4.21c)$$

The relative yields of the F^-, CF_3^-, and $C_2F_5^-$ fragment anions as measured by Spyrou *et al.* [70] are shown in Fig. 4.58(a). A comparison of the relative intensities of these three ions in Fig. 4.58(a) and their energetics, as have been measured by various groups is given in Table 4.42. Clearly, F^- is the most abundant species and $C_2F_5^-$ the least abundant. The relative cross sections for F^- and CF_3^- are broad, nearly identical in shape, and peak at about the same energy (4.3 eV), indicating that the two ions are generated via the same negative ion state. In contrast, the relative cross section for $C_2F_5^-$ is much narrower and peaks at a higher energy (4.8 eV) indicating that it may originate from a different negative ion state than the other two fragments anions. An energy analysis for the F^- anion by Harland and Franklin [71] revealed that the translational energy curve for F^- has a break between 4 and 5 eV. They interpreted this as resulting from the formation of F^- ions from a different negative ion state.

More recently, Weik and Illenberger [240] revisited this point and suggested that this behavior along with the observation that the $C_2F_5^-$ ions form only at energies above 4 eV indicates the existence of two energetically overlapping negative ion states in this energy range, the higher energy one being "more C–F antibonding." According to Weik and Illenberger, calculations by Weik [239] predict $3a_{2u}(\sigma^*_{CC})$ and $3a_{2u}(\sigma^*_{CF})$

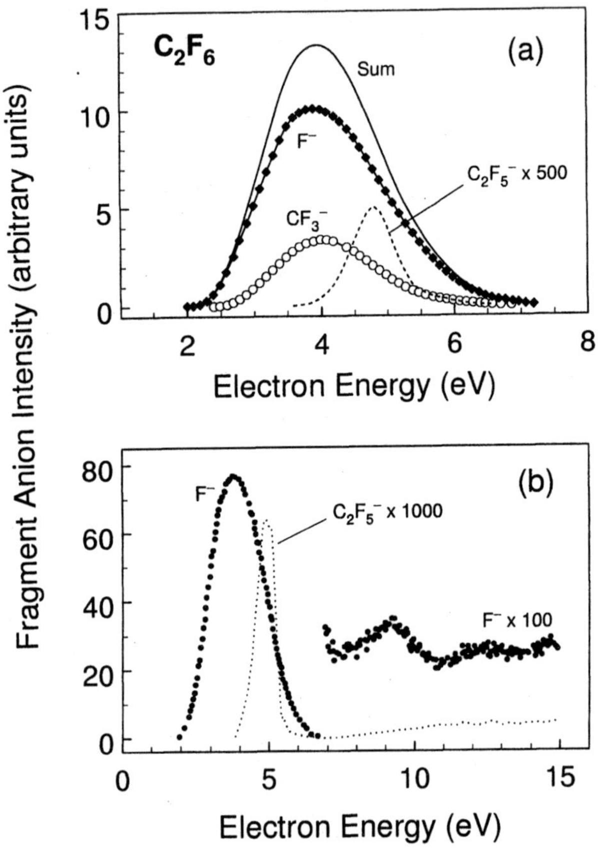

Figure 4.58. (a) Relative cross section for F^-, CF_3^-, and $C_2F_5^-$ as a function of electron energy produced by dissociative electron attachment to C_2F_6 as measured by Spyrou *et al.* [70]. (b) Relative cross section for F^- and $C_2F_5^-$ as a function of electron energy produced by dissociative electron attachment to C_2F_6 as measured by Weik and Illenberger [240].

virtual molecular orbitals in this energy range (see Table 4.26 for the assignments of Tagaki *et al.* [241] and Ishii *et al.* [238]). The data of Weik and Illenberger [240] for the production of F^- and $C_2F_5^-$ are presented in Fig. 4.58(b). They are consistent with those in Fig. 4.58(a) as to the production of F^- and $C_2F_5^-$ in the energy region between 2 and 7 eV, but they extend to higher energies. The yield of F^- in Fig. 4.58(b) indicates maxima at 9.3 and 12.5 eV. The maximum at 9.3 eV is consistent with electron scattering measurements (see Table 4.26).

No $C_2F_6^-$ parent negative ion has been observed under single collision conditions. However, a recent report [245] indicated the observation of

Table 4.42. Fragment negative ions produced by electron impact on C_2F_6, their energetics, and relative intensities.

Fragment anion	Possible reaction	Energy threshold (eV)	Energy of maximum intensity (eV)	Relative abundance (Ref.)
F^-	$C_2F_6 + e \rightarrow F^- + C_2F_5$[a]	2.1 ± 0.2	4.3 ± 0.1	100 [71]
	$C_2F_6 + e \rightarrow F^- + C_2F_5^*$	4.9 ± 0.2		
	$C_2F_6 + e \rightarrow F^- + CF_2 + CF_3$			
	$C_2F_6 + e \rightarrow F^- + F + C_2F_4$			
F^-	$C_2F_6 + e \rightarrow F^- + C_2F_5$[b]	2.0 ± 0.1	3.9 ± 0.05	100 [70]
F^-		2.2	3.75	100 [259]
CF_3^-	$C_2F_6 + e \rightarrow CF_3^- + CF_3$[b,c]	2.2 ± 0.2	4.4 ± 0.1	21 [71]
CF_3^-		2.4 ± 0.1	4.0 ± 0.05	32 [70]
CF_3^-		2.8	3.75	24 [259]
$C_2F_5^-$				1 [71]
$C_2F_5^-$		3.5 ± 0.1	4.8 ± 0.1	<0.1 [70]
F^-		2.25 ± 0.2	3.80 ± 0.2	100 [278][d]
CF_3^-		2.7 ± 0.2	3.9 ± 0.2	1.2
$C_2F_5^-$		4.0 ± 0.2	4.8 ± 0.2	0.01

[a]Harland and Franklin [71] determined a value of (5.5 ± 0.2) eV for the dissociation energy $D(F\text{–}C_2F_5)$, (4.6 ± 0.6) eV for the dissociation energy $D(CF_3\text{–}CF_3)$, (2.4 ± 0.5) eV for the electron affinity of CF_3, and ~ 2.9 eV for the excitation energy of C_2F_5. The value obtained by Harland and Franklin for the dissociation energy $D(CF_3\text{–}CF_3)$ is in reasonable agreement with the value of (4.18 ± 0.04) eV given earlier by Coomber and Whittle [279]. The Harland and Franklin values for the dissociation energies $D(F\text{–}C_2F_5)$ and $D(CF_3\text{–}CF_3)$ are also consistent with those determined from similar studies by Spyrou *et al.* [70] (see footnote *c* on this Table). The values for the electron affinity of CF_3 range from 1.36 eV to 3.25 eV and those for C_2F_5 from 2.1 eV to 3.3 eV (see Ref. [79]).

[b]Both the peak energies and the values of the energy thresholds shift toward lower energies with increasing temperature. The data given are for 300 K. The other entries given in the Table are probably also for about 300 K, although in some of the electron beam studies the gas might have been at a higher temperature than ambient via heating from the hot filament in the electron source.

[c]Spyrou *et al.* [70] estimated $D(F\text{–}C_2F_5)$ to be $\leq (5.4 \pm 0.15)$ eV from data on the reaction $C_2F_6 + e \rightarrow F^- + C_2F_5$ and $\leq (5.6 \pm 0.1)$ eV from data on the reaction $C_2F_6 + e \rightarrow F + C_2F_5^-$. From similar data on the reaction $C_2F_6 + e \rightarrow CF_3^- + CF_3$, they reported $D(CF_3\text{–}CF_3) = (3.8 \pm 0.1)$ eV.

[d]These authors also reported observation of CF^-, F_2^-, and CF_2^-.

the $C_2F_6^-$ ion in studies of electron attachment to C_2F_6 clusters. It is unlikely that this is an indication that the electron affinity of C_2F_6 is positive. Most probably the lowest negative ion state of C_2F_6 is the one with vertical energy at 4.0 eV (see Table 4.26). The $C_2F_6^-$ observed in the Lehmann *et al.* [245] study is likely to be an excited $C_2F_6^{-*}$ species formed in the lowest negative ion state that has a potential minimum and sufficiently long lifetime to be detected in the mass spectrometer. It may thus be inferred from these observations that the $C_2F_6^{-*}$ from clusters lives longer than 1 μs, while the $C_2F_6^{-*}$ formed under isolated conditions lives for just a few fs [70]. Finally, it is noted that besides the fragment anions F^-, CF_3^-, and $C_2F_5^-$, MacNeil and Thynne [278] reported observation of CF^-, F_2^-, C_2F^-, and CF_2^-.

3.5.6 Effect of Temperature on the $k_{a,t}(\langle\varepsilon\rangle)$ and $\sigma_{a,t}(\varepsilon)$ of C_2F_6

The only study on the effect of temperature on electron attachment to C_2F_6 appears to be by Spyrou and Christophorou [277]. These investigators measured the $k_{a,t}(E/N)$ of C_2F_6 in mixtures of C_2F_6 with argon for temperatures ranging from 300 to 750 K. Their results are shown in Fig. 4.59(a). The increase of the total electron attachment rate constant with increasing temperature is only modest because the dissociating negative ion state lies well above thermal (Tables 4.26 and 4.42). Spyrou and Christophorou [277] determined the total dissociative electron attachment cross section of C_2F_6 as a function of gas temperature by unfolding the corresponding rate constants in Fig. 4.59(a) and using the known electron energy distribution functions in argon. Their findings are reproduced in Fig. 4.59(b). The magnitude of $\sigma_{a,t}(\varepsilon)$ increases and its energy threshold decreases with increasing temperature. The single peak in $\sigma_{a,t}(\varepsilon)$ (due to F^- and CF_3^-) shifts from 3.9 eV at 300 K to $\sim$ 3.3 eV at 750 K and the corresponding onset shifts from 2.3 to 1.5 eV.

3.6 Electron Transport in C_2F_6

3.6.1 Electron Drift Velocity, $w(E/N)$, of C_2F_6

There have been three measurements of the electron drift velocity, $w(E/N)$, of C_2F_6. Two of these were performed at room temperature [108, 120] and the other at 500 K [264]. Figure 4.60 summarizes the room temperature data. The measurements of Hunter *et al.* [108] cover a wide range of values, while the measurements of Naidu and Prasad are

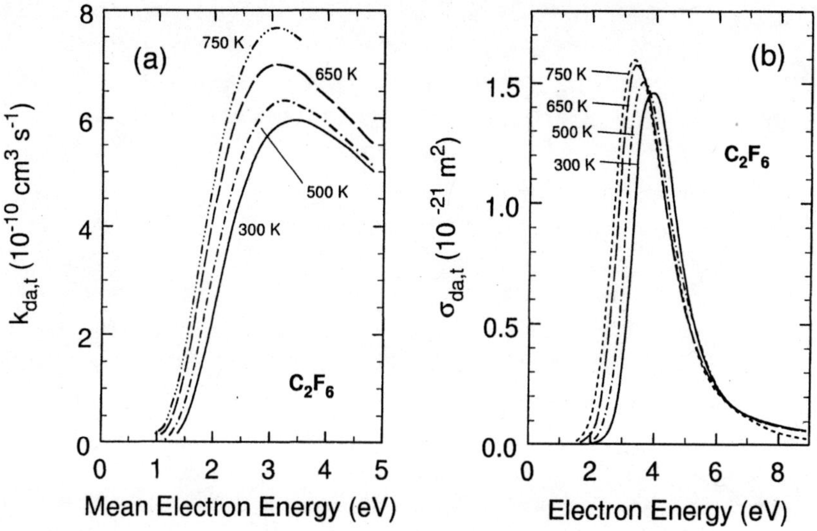

Figure 4.59. (a) Total dissociative electron attachment rate constant, $k_{da,t}(\langle\varepsilon\rangle)$, for C_2F_6 as a function of the gas temperature measured in mixtures with argon by Spyrou and Christophorou [277]. (b) Total dissociative electron attachment cross section $\sigma_{da,t}(\varepsilon)$ of C_2F_6 at various gas temperatures between 300 and 750 K, derived from the measurements in Fig. 4.59(a) [277].

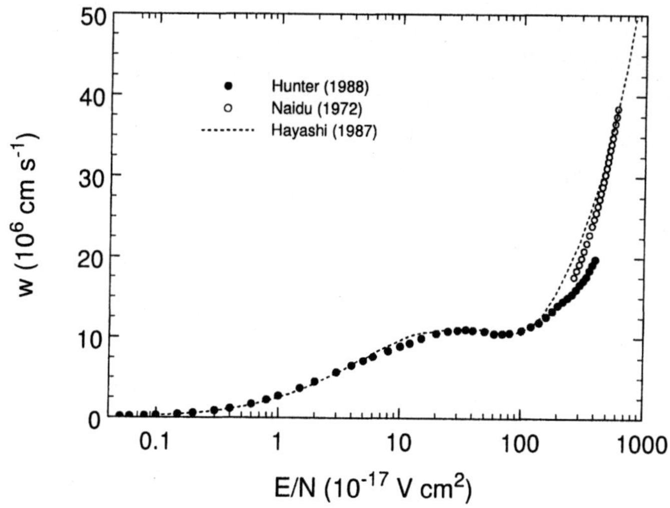

Figure 4.60. Electron drift velocity w as a function of E/N for pure C_2F_6. Measurements: (•) Ref. [108] ($T = 300$ K); (∘) Ref. [120] ($T = 293$ K). Computed values: (- - -) Ref. [247].

restricted to high E/N values (the open circles plotted in Fig. 4.60 are random points taken from the curve shown in their paper) [120]. The measurements of Hunter *et al.* [108] have stated maximum uncertainties of ±5% above $(E/N)_{\text{lim}}$, decreasing to ±2% at values of E/N below the onset of electron attachment. Hunter *et al.* corrected their data for the effects of electron attachment, ionization, and diffusion. Naidu and Prasad [120] estimated their uncertainty to be ±5%.

Considering the quoted uncertainties, there is disagreement between the two sets of data in the region of overlap. Hunter *et al.* [108] attributed the differences between their data and the data of Naidu and Prasad at high E/N to the uncertainty in the Naidu and Prasad measurement procedure. The latter measurements become uncertain at high E/N because w is large and the electron transit time is very short, making it difficult to obtain accurate transit times from oscilloscope tracings, especially when the width of the light pulse used to generate the photoelectron pulse is an appreciable fraction of the electron transit time. Another possible source of uncertainty pointed out by Hunter *et al.* [108] is the high background ion current in the Naidu and Prasad experiment.

Using the cross section set shown in Fig. 4.38, Hayashi and Niwa [247] calculated the $w(E/N)$ values represented by the broken line in Fig. 4.60, showing agreement with Hunter *et al.* at low E/N and Naidu and Prasad at high E/N. The data of Hunter *et al.* are listed in Table 4.43 as our suggested values of $w(E/N)$ for C_2F_6 in view of the effort made in this study to consider the various sources of errors and to correct for their influence on the measurements.

The electron drift velocity measurements at 300 and 500 K reported by Carter *et al.* [264] are shown in Fig. 4.61. The electron drift velocity is lower at the higher temperature. This is also evident in Table 4.44 showing the thermal values of the density-normalized electron mobility $(\mu N)_{\text{th}}$ for C_2F_6 measured as a function of temperature [122]. A rather profound decrease is observed that may be due to the effect on electron motion [122] of enhanced electron scattering from vibrationally excited $C_2F_6^*$ molecules.

Measurements have been made of the $w(E/N)$ in mixtures of C_2F_6 with various gases such as Ar and CH_4. These measurements were partially motivated by the development of fast gas mixtures (gas mixtures with very large electron drift velocities) for use in gas pulse-power switches. A sample of these data taken from Carter *et al.* [264] is shown in Fig. 4.62. It is interesting to note the negative differential conductivity (decrease in w with increasing E/N) exhibited by these mixtures for certain E/N regions which depends on mixture composition.

Table 4.43. Suggested electron drift velocities, w (T = 300 K), for C_2F_6 (data of Hunter *et al.* [108]).

(E/N) $(10^{-17}$ V cm$^2)$	$w(E/N)$ $(10^6$ cm s$^{-1})$	E/N $(10^{-17}$ V cm$^2)$	$w(E/N)$ $(10^6$ cm s$^{-1})$
0.05	0.147	30	10.9
0.06	0.176	35	11.0
0.08	0.237	40	10.9
0.10	0.295	50	10.8
0.15	0.435	60	10.5
0.2	0.580	70	10.5
0.3	0.870	80	10.6
0.4	1.15	100	10.9
0.6	1.69	120	11.4
0.8	2.23	140	11.9
1.0	2.71	160	12.6
1.5	3.69	180	13.3
2.0	4.54	200	14.0
3.0	5.68	220	14.5
4.0	6.51	240	15.0
5.0	7.13	260	15.4
6.0	7.62	280	16.0
8.0	8.36	300	16.6
10.0	8.92	320	17.1
12.0	9.30	340	17.6
15.0	9.90	360	18.4
20.0	10.5	380	19.1
25.0	10.8	400	19.8

Table 4.44. Measured density-normalized thermal-electron mobilities, $(\mu N)_{th}$, of Christophorou *et al.* [122] for C_2F_6 as a function of gas temperature.

Temperature (K)	$(\mu N)_{th}$ $(10^{23}$ V^{-1} cm^{-1} s$^{-1})$
300	2.92
400	2.20
500	1.63
600	1.23
700	0.94

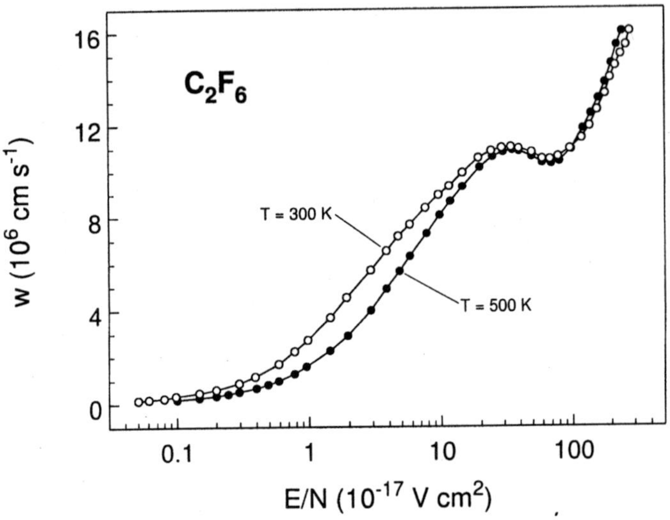

Figure 4.61. Electron drift velocity w as a function of E/N for pure C_2F_6 at 300 K (○) and 500 K (•) as measured by Carter *et al.* [264].

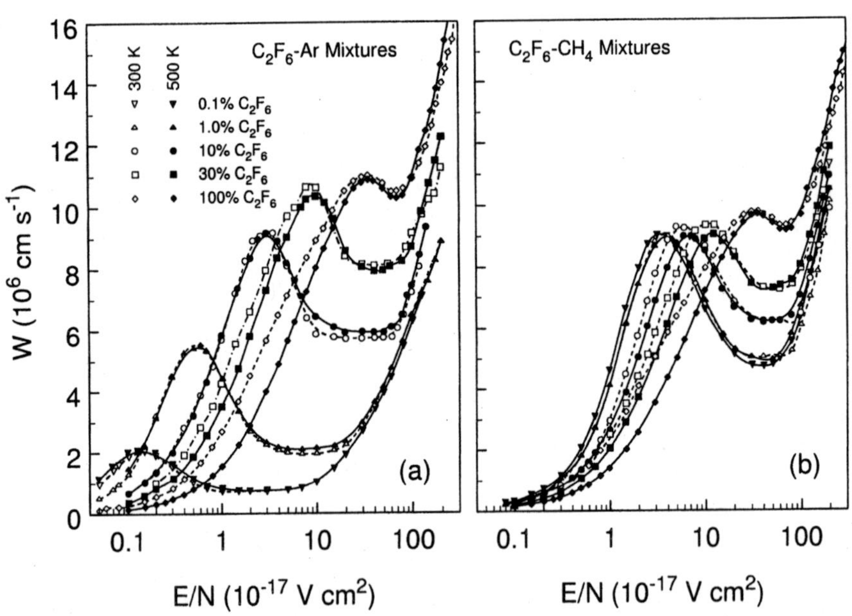

Figure 4.62. Electron drift velocity w as a function of E/N for mixtures of C_2F_6 with (a) Ar and with (b) CH_4 at two gas temperatures (300 K and 500 K). These are sample data from Carter *et al.* [264].

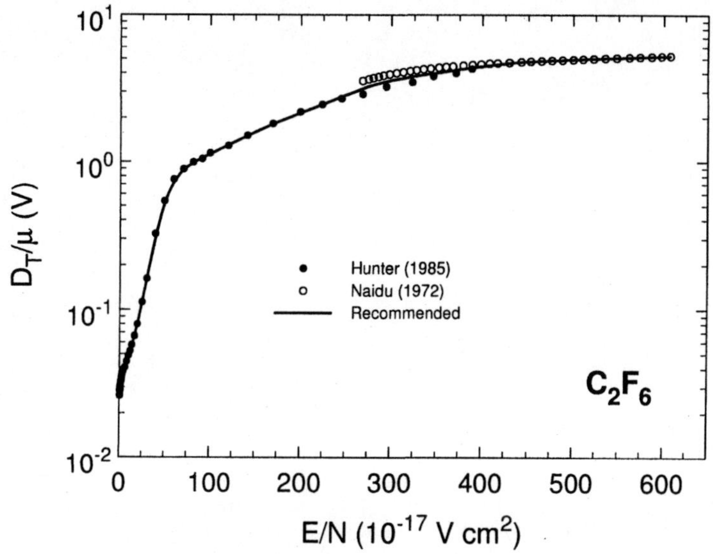

Figure 4.63. Transverse electron diffusion coefficient to electron mobility ratio as a function of E/N, $D_T/\mu\,(E/N)$, measured at 293 K for C_2F_6. (•) Ref. [13]; (○) Ref. [120]; (—) recommended.

3.6.2 Ratio of the Transverse Electron Diffusion Coefficient to Electron Mobility, $D_T/\mu\,(E/N)$, for C_2F_6

There have been two sets of measurements of $D_T/\mu\,(E/N)$ for pure C_2F_6, one by Naidu and Prasad [120] and the other by Hunter *et al.* [13]. The data of Naidu and Prasad were taken at a temperature of 293 K, pressures ≤ 0.4 kPa, and E/N values in the range 270×10^{-17} to $\sim 608 \times 10^{-17}$ V cm^2. The quoted uncertainty in their data ranges from $\pm 5\%$ at the lowest values of E/N to $\pm 3\%$ at the highest E/N values at which they made measurements. Points taken off the solid line for C_2F_6 in Fig. 2 of their paper are plotted in Fig. 4.63. Hunter *et al.* [13] measured D_T/μ for pure C_2F_6 in a lower E/N range (0.6×10^{-17} to 389×10^{-17} V cm^2). No estimate of the uncertainty of their data or the temperature at which the measurements were made was given. However, since the measurements were made using the apparatus at the Australian National University [280], the uncertainty is expected to be only a few percent and the temperature approximately 293 K. Their measurements are also plotted in Fig. 4.63. In the E/N range that the data of Hunter *et al.* overlap the values of Naidu and Prasad, the latter lie higher. We fitted the two sets of measurements as shown by the solid

Table 4.45. Recommended values of the transverse electron diffusion coefficient to electron mobility ratio, D_T/μ, as a function of E/N for C_2F_6.

E/N (10^{-17} V cm^2)	D_T/μ (V)	E/N (10^{-17} V cm^2)	D_T/μ (V)
1	0.029	60	0.712
2	0.033	70	0.868
3	0.035	80	0.967
4	0.038	90	1.05
5	0.040	100	1.13
6	0.041	150	1.59
7	0.043	200	2.15
8	0.045	250	2.83
9	0.047	300	3.56
10	0.049	350	4.00
15	0.060	400	4.43
20	0.080	450	4.77
30	0.161	500	4.93
40	0.305	550	5.07
50	0.511	600	5.20

line in Fig. 4.63. Values taken off the solid line are listed in Table 4.45 as our recommended data for the $D_T/\mu\,(E/N)$ of C_2F_6.

Measurements have also been reported [220] of the $D_T/\mu\,(E/N)$ for the mixture 10% C_2F_6/90% CH_4. These are shown in Fig. 4.64, where they are compared with the results of a Boltzmann code analysis for this mixture by Hayashi and Niwa [247]. The overall agreement is reasonable although the calculated values do not converge to the low-E/N asymptotic limit of kT/e.

3.6.3 Product, $ND_L\,(E/N)$, of the Gas Number Density N and the Longitudinal Electron Diffusion Coefficient D_L, as a Function of E/N for C_2F_6

Okumo and Nakamura [248] measured $ND_L\,(E/N)$ for mixtures containing 0.524% and 5.47% of C_2F_6 in argon. These data are shown in Fig. 4.65. Their measurements on $w(E/N)$ for these mixtures are consistent with the earlier measurements of Hunter *et al.* [13].

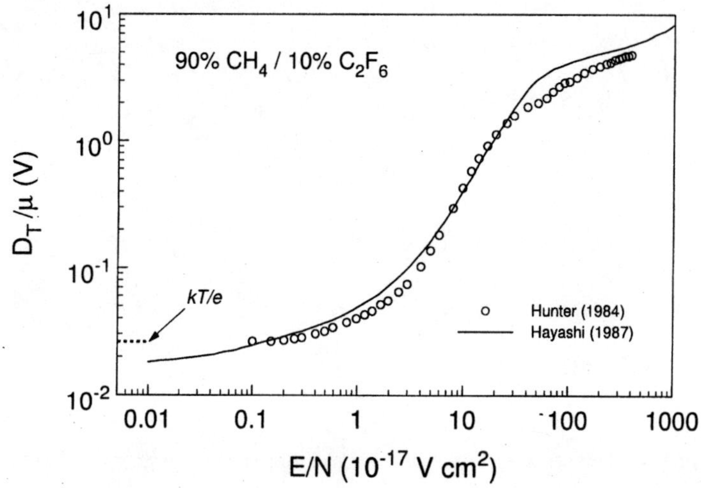

Figure 4.64. Transverse electron diffusion coefficient to electron mobility ratio as a function of E/N, $D_T/\mu\,(E/N)$, for the gas mixture 90% CH_4/10% C_2F_6. Measurements at 293 K: (○) Ref. [220]. Computed values: (—) Ref. [247].

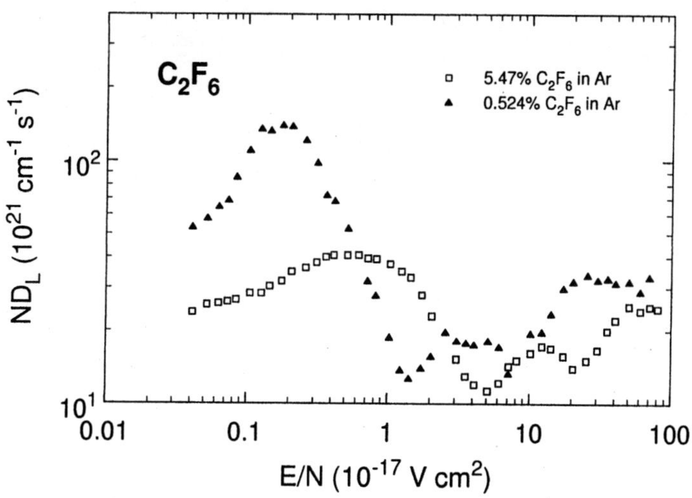

Figure 4.65. Measurements of Okumo and Nakamura [248] of $ND_L\,(E/N)$ for (□) 5.47% C_2F_6 in Ar, and (▲) 0.524% C_2F_6 in Ar.

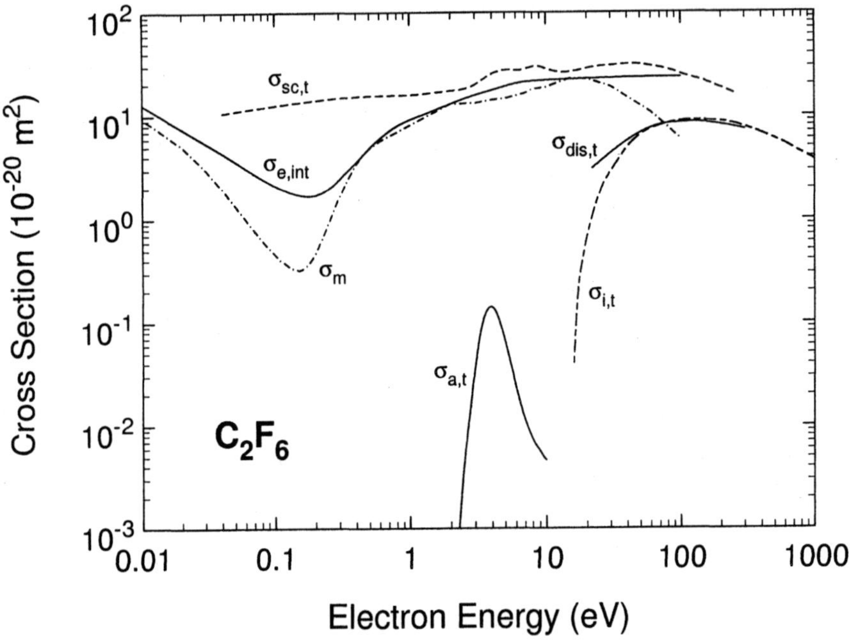

Figure 4.66. Recommended and suggested cross sections for C_2F_6.

3.7 Summary of Recommended and Suggested Electron Collision Cross Sections and Electron Transport Coefficients for C_2F_6

In Fig. 4.66 are plotted the cross sections that have been designated as recommended or suggested in Sec. 3 of this chapter. These include the:

- $\sigma_{sc,t}(\varepsilon)$ — Table 4.27, Fig. 4.39;
- $\sigma_{e,int}(\varepsilon)$ — Table 4.29, Fig. 4.41;
- $\sigma_m(\varepsilon)$ — Table 4.30, Fig. 4.42;
- $\sigma_{i,t}(\varepsilon)$ — Table 4.33, Fig. 4.47;
- $\sigma_{dis,t}(\varepsilon)$ — Table 4.34, Fig. 4.48; and
- $\sigma_{a,t}(\varepsilon)$ — Table 4.41, Fig. 4.57.

The data on the elastic differential electron scattering cross sections, $\sigma_{e,diff}(\varepsilon)$, in Table 4.28 are also suggested.

The following coefficients are recommended or suggested:

- $\alpha/N\,(E/N)$ — Table 4.35, Fig. 4.49;
- $(\alpha-\eta)/N(E/N)$ — Table 4.36, Fig. 4.50;
- $\eta/N\,(E/N)$ — Table 4.38, Fig. 4.53;
- $k_{\mathrm{a,t}}\langle\varepsilon\rangle$ — Table 4.39, Fig. 4.56 (b);
- $w(E/N)$ — Table 4.43, Fig. 4.60; and
- $D_{\mathrm{T}}/\mu\,(E/N)$ — Table 4.45, Fig. 4.63.

* * *

4. Electron Interactions with C_3F_8

Perfluoropropane (C_3F_8) is a plasma processing gas [281–286]. It is a replacement for chlorofluorocarbons because it is not harmful to stratospheric ozone. However, like other similar compounds (e.g., CF_4 and C_2F_6) it is a global warming gas. Its global warming potential over a 100-year period is 7 000 compared to that of CO_2, and its half-life in the stratosphere is 2 600 years [287]. By comparison, the respective global warming potentials of CF_4 and C_2F_6 are 6 500 and 9 200 [287], and the respective lifetimes 50 000 years and 10 000 years [26, 230, 287] (see Roehl *et al.* [288] for infrared band intensities of C_3F_8 and other perfluorinated compounds in relation to their global warming potentials). Besides plasma processing, perfluoropropane is suitable for a number of other applications. It has good thermal and chemical stability, low toxicity, relatively high vapor pressure, and is transparent to light from the infrared region down to $\sim$ 1300 Å. The magnitude and energy (or E/N) dependence of its electron attachment rate constant and electron drift velocity make it suitable for possible use in externally sustained diffuse discharge switches [12, 13, 220, 266, 289, 290], especially as the electronegative component in mixtures with buffer gases such as Ar and CH_4 [227]. Because of its high dielectric strength, it may find uses as a high voltage insulating gas [268, 271, 291].

To advance the many and diverse applications for which C_3F_8 is suited, in the following sections we review and critically evaluate existing knowledge on electron scattering and electron energy-loss processes for this molecule, and draw attention to data that are not presently available but are needed for modeling the behavior of C_3F_8 in practical uses, especially plasma processing. As for the other molecules discussed in this book, a number of collision cross sections, coefficients, and rate constants are used to quantify various processes which result from the collisions of low-energy electrons with the C_3F_8 molecule. These are defined in Table 1.3 of Ch. 1 along with the corresponding symbols and units. When possible, recommended or suggested cross sections and transport coefficients are given using the criteria and procedure discussed in Ch. 3.

4.1 Electronic and Molecular Structure of C_3F_8

McKoy *et al.* [233] calculated a small (0.07 Debye) dipole moment for the C_3F_8 molecule. Beran and Kevan [32] reported the values of 73.6×10^{-25}, 94.0×10^{-25}, and 64.7×10^{-25} cm^3 for the static polarizability of C_3F_8 depending on the method of determination they used, and McKoy *et al.* [233] calculated a value of 87.5×10^{-25} cm^3. The very small (or the absence of) electron-electric dipole scattering has a rather profound effect on the electron scattering cross section at low energies (< 1 eV) in comparison to polar gases.

A rather limited number of photoelectron and photoabsorption studies have been made for this molecule. In a mass spectrometric study of the photoionization of C_3F_8, Noutary [235] found no parent positive $C_3F_8^+$ ions. He determined photoionization thresholds for the production of $C_3F_7^+$, $C_2F_5^+$, and CF_3^+ which are listed in Table 4.46. Robin [36] reported an overall value of the photoionization onset equal to 13.70 eV based on photoelectron spectra. From studies on the photoelectron spectra of the perfluoroalkane molecules, he concluded that the uppermost molecular orbitals (MOs) in these systems are C–C σ-MOs, and that for C_3F_8 the transitions from these MOs to $3s$ orbitals (B bands) can be seen as weak excitations at 9.51 eV. The absorption spectrum of the C_3F_8 molecule has been measured by Bélanger *et al.* [296] in the gas phase for pressures varying from 13.3 to 66.7 Pa. This is reproduced in Fig. 4.67. It is structureless and peaks at 119.0 nm (10.42 eV). The value of the absorption cross section at this wavelength is $\sim 6.6 \times 10^{-22}$ m^2.

Absolute oscillator strength spectra for C_3F_8 in the C $1s$ (280–340 eV) and F $1s$ (680–740 eV) regions have been determined by Ishii *et al.* [238] from inner-shell electron energy-loss spectra using 2.5 keV energy electrons and scattering angles less than 2°. These investigators also measured the electron transmission spectrum of C_3F_8 and found negative ion resonances at 3.34 and 6.00 eV. They attributed these to σ^* molecular orbitals, since the molecule is saturated and the resonances are located well below the region in which the lowest Feshbach resonances are ex-

Table 4.46. Ionization threshold energies for C_3F_8.[a]

Ionization threshold (eV)	Ion	Method	Ref.
13.38		Photoelectron spectroscopy	[292]
13.70			[36]
15.44 ± 0.02	$C_3F_7^{+}$[b]	Photoionization	[235]
13.32 ± 0.02	$C_2F_5^{+}$		[235]
13.22 ± 0.02	CF_3^{+}		[235]
13.3 ± 0.1		Electron impact	[293]
23.5	F^{+}	Electron impact	[259]
21.0	CF^{+}		[256]
26.6	CF_2^{+}		[256]
13.4 ± 0.1[c]	CF_3^{+}	Electron impact	[294]
14.65[d]			[259]
14.4[d]			[295]
14.70[d]			[260]
15.2[e]			[256]
13.5 ± 0.1[c]	$C_2F_4^{+}$	Electron impact	[294]
14.4[e]			[256]
13.9 ± 0.1[c]	$C_2F_5^{+}$	Electron impact	[294]
15.25[d]			[259]
15.3[d]			[295]
15.3[e]			[256]
15.7 ± 0.1[c]	$C_3F_7^{+}$	Electron impact	[294]
16.5[d]			[259]
17.0[d]			[260]
17.1[d]			[295]
15.5			[256]

[a]McKoy *et al.* [233] calculated 12.55 eV and 13.68 eV for the adiabatic and the vertical ionization energies, respectively. See this reference for values of the threshold energies for other dissociative ionization channels.
[b]Reaction identified as $C_3F_8 + h\nu \rightarrow C_3F_7^{+} + F + e$.
[c]Use was made of the retarding potential difference (RPD) method to improve the electron beam energy resolution.
[d]No RPD was used; inferior electron beam energy resolution.
[e]Possibly high due to poor electron beam energy resolution.

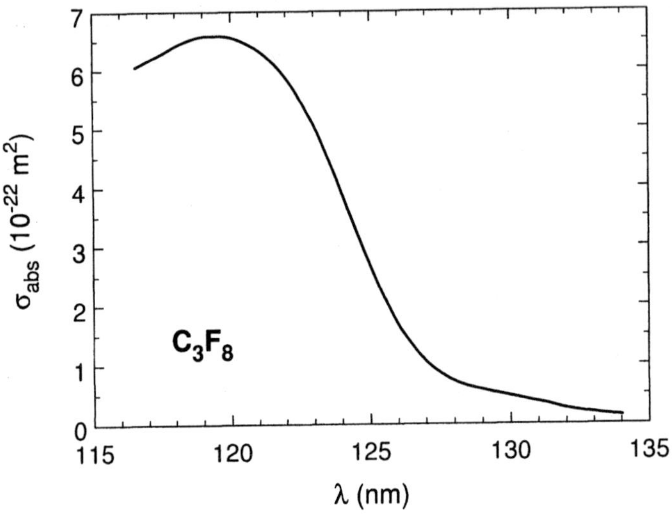

Figure 4.67. Photoabsorption spectrum of C_3F_8 from Bélanger *et al.* [296].

pected. The values they measured are in reasonable agreement with those determined from electron attachment, electron scattering, and vibrational excitation cross section experiments (Table 4.47). The energy positions of the negative ion resonances as determined from electron scattering experiments should lie at somewhat higher energy than the energy positions determined from dissociative electron attachment studies due to the effects of autodetachment on the competition between dissociation and autodetachment. From the data in Table 4.47, it can be concluded that there are at least three negative ion states of the C_3F_8 molecule at about 3.5, 6.4, and 9.0 eV (these values are the averages of the electron scattering and electron transmission data in Table 4.47). Their effects are prominently shown in the cross sections for electron scattering from the C_3F_8 molecule at energies below $\sim$ 10 eV (see Sec. 4.2 of Ch. 4).

Perfluoropropane is an electron attaching gas. It forms dissociative attachment fragment negative ions via a number of resonances lying mostly in the energy range 2–7 eV, and, in addition, it forms parent negative ions via a low lying negative ion state which is attractive and which although short-lived (lifetime $< 10^{-10}$ s) can be stabilized via collisions in high-pressure swarm experiments [300]. The most abundant dissociative attachment fragment negative ion is F^-. Interesting temperature dependencies have been observed which show that the production of parent anions decreases and the production of fragment anions increases with increasing gas temperature [293, 300].

Table 4.47. Energies of negative ion states of C_3F_8.

Energy position (eV)	Type of study	Reference
3.9 6.6 9.0	Total electron scattering	[242]
$\sim$ 3	Maximum in calculated $\sigma_m(\varepsilon)$	[252]
3.2 6.5 9.0 22	Peaks in the differential vibrational electron scattering cross section	[297, 298][a]
$\sim$ 4.5	Broad peak in the calculated vibrational inelastic scattering cross section	[252][b]
3.34 6.00	Electron transmission	[238][a]
1.4 $\sim$ 2.0 $\sim$ 4.0	SF_6 scavenger technique[c]	[299]
2.8	Maximum in total dissociative attachment cross section	[293]
3.3	Maximum in total dissociative attachment cross section	[260]
2.95	Position of dissociative attachment maximum[d]	[300]
3.15 ± 0.1 (for F^-) 3.65 ± 0.1 (for CF_3^-)	Dissociative attachment	[71]

[a] Attributed by the authors to shape resonances.
[b] The peak in the calculated vibrational inelastic electron scattering cross section at $\sim$ 0.7 eV could be attributed to direct vibrational excitation (see text).
[c] This is, in essence, a threshold electron excitation technique [301]. The peaks at 1.4 and 2.0 eV are in conflict with the rest of the data listed in the table and may reflect the fact that some of the scattered electrons which were picked up by SF_6 to form SF_6^- might have been due to direct electron scattering via vibrational excitation rather than scattering from resonances.
[d] The figure given in this table is for 300 K. The position of the resonance decreases with increasing gas temperature [300].

Table 4.47 (continued)
Energies of negative ion states of C_3F_8.

Energy position (eV)	Type of study	Reference
2.9 ± 0.1 (for F^-) 3.2 ± 0.1 (for $C_2F_5^-$) 3.3 ± 0.1 (for $C_2F_3^-$) 3.4 ± 0.1 (for CF_3^-) 3.75 ± 0.1 (for $C_3F_7^-$)	Dissociative attachment	[70]

There is evidence for direct vibrational excitation at low energies (< 1 eV) and strong indirect vibrational excitation via resonances in the energy range of about 2–10 eV. Similar to the case of CF_4 and C_2F_6, excitation of C_3F_8 to any electronic or ionic state results in fragmentation [50], and, consequently, the measured dissociation cross section for C_3F_8 is the sum of the cross sections for all these processes. The most abundant fragment positive ion is CF_3^+ (see Refs. [256, 302] and Sec. 4.3 of Ch. 4).

4.2 Electron Scattering from C_3F_8

In this section information is presented and discussed on the following cross sections for C_3F_8: total electron scattering cross section $\sigma_{sc,t}(\varepsilon)$, elastic differential electron scattering cross section $\sigma_{e,diff}(\varepsilon)$, elastic integral electron scattering cross section $\sigma_{e,int}(\varepsilon)$, momentum transfer cross section (elastic) $\sigma_m(\varepsilon)$, differential vibrational excitation cross section $\sigma_{vib,diff}(\varepsilon)$, and vibrational inelastic electron scattering cross section $\sigma_{vib,inel}(\varepsilon)$. Besides these data a self-consistent set of cross sections has been derived by Jeon and Nakamura [303] based upon multi-term Boltzmann code calculations compared with measurements they made of $w(E/N)$ and $ND_L(E/N)$ in C_3F_8-Ar mixtures. This set of cross sections is "preliminary" and is not included in this work.

4.2.1 Total Electron Scattering Cross Section, $\sigma_{sc,t}(\varepsilon)$, of C_3F_8

In Fig. 4.68 are shown the total electron scattering cross section measurements of Sanabia *et al.* [242] and Tanaka *et al.* [304]. The measurements of Tanaka *et al.* [304] have an estimated uncertainty of $< \pm 10\%$, and the uncertainties of Sanabia *et al.* [242] are expected to be about the same. The two sets of measurements agree within their combined

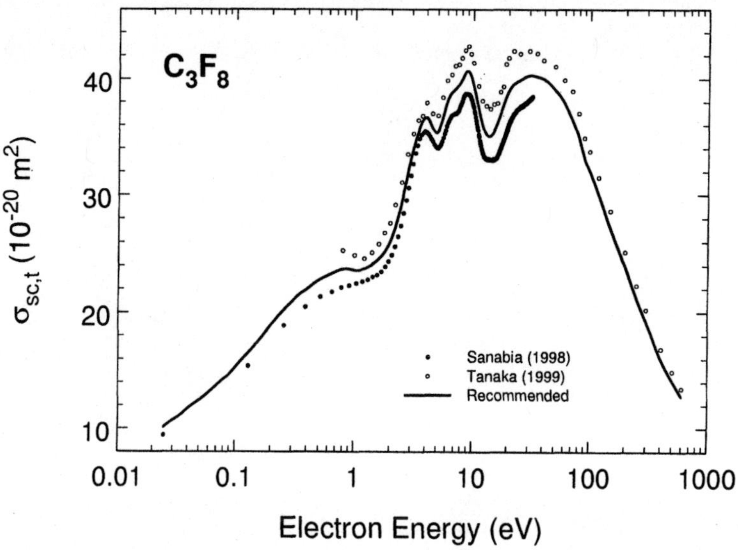

Figure 4.68. Total electron scattering cross section, $\sigma_{sc,t}(\varepsilon)$, for C_3F_8: (•) Ref. [242]; (○) Ref. [304]; (—) recommended.

uncertainties. The average of the two sets of measurements was determined over the energy range in which the two sets of measurements overlap and is shown by the solid line in the figure. This solid line was extended to lower and to higher energies by normalization, respectively, of the measurements of Sanabia *et al.* [242] to the average at 0.8 eV and the measurements of Tanaka *et al.* [304] to the average at 23 eV. In this way the range of recommended data was extended considerably. The solid line represents our recommended $\sigma_{sc,t}(\varepsilon)$ for C_3F_8, and values taken from this line are listed in Table 4.48.

These experimental cross sections show distinct structure below 30 eV. The cross section of Sanabia *et al.* [242] has a shoulder near 0.8 eV and three maxima near 3.9, 6.6, and 9.0 eV, and suggests further maxima at still higher energies. Similar structure is exhibited by the Tanaka *et al.* [304] data which show a prominent broad peak near 30 eV. The maxima below $\sim$ 10 eV are due to indirect electron scattering via the negative ion states of C_3F_8 at these energies as has also been indicated by other studies (Table 4.47). The $\sigma_{sc,t}(\varepsilon)$ for C_3F_8 declines below $\sim$ 0.8 eV as the electron energy approaches zero, due probably to the presence of a Ramsauer-Townsend minimum at these low energies. The calculations of Pirgov and Stefanov [252] indicate such a minimum in the momentum transfer cross section $\sigma_m(\varepsilon)$ near 0.07 eV (Sec. 4.2.4 of Ch. 4).

Table 4.48. Recommended total electron scattering cross section, $\sigma_{sc,t}(\varepsilon)$, for C_3F_8.

ε (eV)	$\sigma_{sc,t}(\varepsilon)$ (10^{-20} m^2)	ε (eV)	$\sigma_{sc,t}(\varepsilon)$ (10^{-20} m^2)	ε (eV)	$\sigma_{sc,t}(\varepsilon)$ (10^{-20} m^2)
0.025	10.1	0.8	23.7	30	40.3
0.030	10.7	0.9	23.7	35	40.2
0.035	11.2	1.0	23.6	40	40.0
0.040	11.7	1.5	24.3	50	39.1
0.050	12.4	2.0	26.0	60	37.9
0.060	13.1	2.5	29.3	70	36.5
0.070	13.7	3.0	33.2	80	35.0
0.080	14.3	3.5	35.7	90	33.2
0.090	14.7	4.0	36.7	100	32.1
0.10	15.2	5.0	35.4	120	30.0
0.15	17.3	6.0	38.0	150	27.2
0.20	18.9	7.0	39.1	200	23.9
0.25	20.1	8.0	39.9	250	21.2
0.30	20.9	9.0	40.7	300	19.2
0.35	21.6	10.0	39.6	400	16.0
0.40	22.0	12.5	35.6	500	14.2
0.50	22.7	15.0	35.4	600	12.8
0.60	23.2	20.0	38.9		
0.70	23.5	25.0	39.9		

The region between 0.2–2.0 eV shows an enhancement in the scattering cross section which may be due to direct vibrational excitation. This would be consistent with the peak around 0.7 eV in the vibrational inelastic electron scattering cross section calculated by Pirgov and Stefanov [252]. It also would be consistent with the broad features of the threshold-electron excitation spectrum of C_3F_8 reported by Lifshitz and Grajower [299].

4.2.2 Elastic Differential Electron Scattering Cross Sections, $\sigma_{e,diff}(\varepsilon)$, of C_3F_8

Figure 4.69 shows the elastic differential electron scattering cross sections, $\sigma_{e,diff}(\varepsilon)$, of the C_3F_8 molecule as measured by Shinohara *et al.* [297, 298] and Tanaka *et al.* [304] for incident electron energies ranging from 2 to 100 eV and for scattering angles between 20° and 130°. The uncertainty in the measurements of Tanaka *et al.* is estimated to be about ±15 to ±20%. Figure 4.69 also shows similar unpublished data by Merz and Linder [251]. These were made at lower values of the inci-

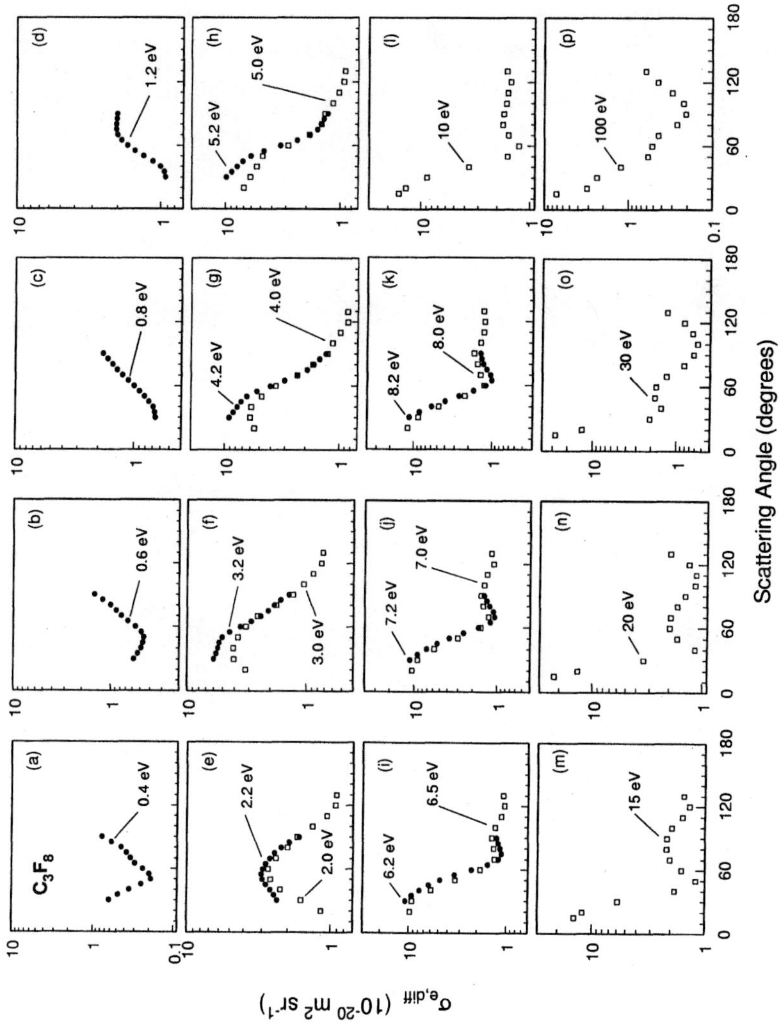

Figure 4.69. Elastic differential electron scattering cross sections, $\sigma_{e,diff}(\varepsilon)$, for C_3F_8: (□) Measurements of Tanaka and coworkers [297, 298, 304]; (•) measurements of Merz and Linder [251].

Table 4.49. Elastic differential electron scattering cross section, $\sigma_{e,diff}(\varepsilon)$, measurements of Tanaka *et al.* [304] for C_3F_8 in units of 10^{-20} m^2 sr^{-1}.

Angle	1.5 eV	2 eV	3 eV	4 eV	5 eV	6.5 eV
15°						
20°	1.22	1.19	3.27	5.76	7.10	9.77
30°	0.88	1.62	4.10	6.17	6.28	9.40
40°	1.31	2.26	4.14	6.09	5.51	5.98
50°	1.79	2.63	3.81	4.94	4.94	3.41
60°	2.03	2.72	3.25	3.74	2.95	1.90
70°	2.23	2.42	2.62	2.44	1.89	1.36
80°	2.29	2.03	1.81	1.77	1.47	1.38
90°	2.04	1.75	1.32	1.32	1.39	1.44
100°	1.64	1.37	1.08	1.18	1.18	1.32
110°	1.36	1.08	0.890	1.01	1.04	1.14
120°	1.18	0.939	0.767	0.870	0.942	1.05
130°	1.12	0.916	0.734	0.872	0.908	1.08

Angle	7 eV	8 eV	9 eV	10 eV	12 eV	15 eV
15°				16.7	15.2	14.2
20°	10.8	12.4	14.4	14.3	13.0	12.0
30°	9.31	9.10	9.23	9.05	7.67	5.93
40°	5.92	4.96	4.15	3.66	3.08	1.87
50°	3.05	2.31	1.60	1.61	1.42	1.22
60°	1.64	1.28	1.25	1.29	1.48	1.63
70°	1.30	1.43	1.61	1.55	1.77	2.05
80°	1.52	1.60	1.67	1.75	1.97	2.20
90°	1.61	1.78	1.65	1.72	1.96	2.17
100°	1.47	1.46	1.54	1.62	1.79	1.97
110°	1.34	1.33	1.42	1.59	1.61	1.59
120°	1.12	1.30	1.38	1.51	1.54	1.38
130°	1.20	1.33	1.58	1.62	1.59	1.57

dent electron energy (0.4–8.2 eV). The overall agreement between these two measurements of $\sigma_{e,diff}(\varepsilon)$ is reasonable for the overlapping energies. The data of Tanaka *et al.* [304] are presented in Table 4.49.

4.2.3 Elastic Integral Electron Scattering Cross Section, $\sigma_{e,int}(\varepsilon)$, of C_3F_8

Shinohara *et al.* [297, 298] and Tanaka *et al.* [304] extrapolated the elastic differential electron scattering cross sections they measured in

Table 4.49 (continued)
Elastic differential electron scattering cross section, $\sigma_{e,diff}(\varepsilon)$, measurements of Tanaka *et al.* [304] for C_3F_8 in units of 10^{-20} m^2 sr^{-1}.

Angle	20 eV	25 eV	30 eV	60 eV	100 eV
15°	22.0	26.1	28.1	17.7	7.25
20°	13.7	14.3	13.3	4.86	3.17
30°	3.47	2.67	2.10	2.68	2.44
40°	1.16	1.57	1.55	2.01	1.25
50°	1.69	2.01	1.84	1.28	0.596
60°	1.98	2.13	1.78	0.751	0.525
70°	1.95	1.91	1.33	0.571	0.442
80°	1.68	1.23	0.825	0.397	0.263
90°	1.42	0.813	0.640	0.334	0.207
100°	1.17	0.762	0.570	0.354	0.222
110°	1.15	0.852	0.656	0.472	0.299
120°	1.32	1.22	0.813	0.530	0.449
130°	1.93	1.65	1.283	0.942	0.632

Table 4.50. Suggested elastic integral electron scattering cross section, $\sigma_{e,int}(\varepsilon)$, for C_3F_8 from Tanaka *et al.* [304].

ε (eV)	$\sigma_{e,int}(\varepsilon)$ (10^{-20} m^2)	ε (eV)	$\sigma_{e,int}(\varepsilon)$ (10^{-20} m^2)	ε (eV)	$\sigma_{e,int}(\varepsilon)$ (10^{-20} m^2)
1.5	19.8	7	44.4	20	37.6
2.0	20.8	8	44.5	25	36.3
3.0	27.4	9	44.9	30	32.9
4.0	35.3	10	44.3	60	18.8
5.0	37.5	12	42.4	100	13.0
6.5	42.9	15	39.2		

the scattering angle range 20° and 130° to 0° and 180°. By modified phase-shift fitting and by proper integration they determined the elastic integral electron scattering cross sections plotted in Fig. 4.70. These data for $\sigma_{e,int}(\varepsilon)$ are listed in Table 4.50 as our suggested values. Similar calculations have not been made for the differential scattering data of Mann and Linder [251].

In Fig. 4.70 are also shown the high energy (1–8 keV) calculated values of $\sigma_{e,int}(\varepsilon)$. These are independent model calculation results [305] within the Born approximation. They provide high-energy asymptotic limit values for $\sigma_{e,int}(\varepsilon)$. Also shown are the static-exchange approxi-

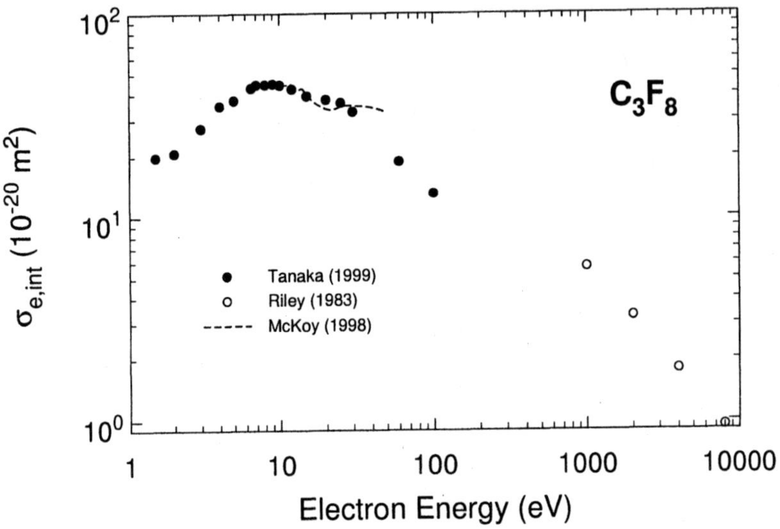

Figure 4.70. Elastic integral electron scattering cross section, $\sigma_{e,int}(\varepsilon)$, for C_3F_8: (•) measurements of Ref. [304]; (∘) calculations of Ref. [305]; (- - -) calculations of Ref. [233].

mation calculation results of McKoy *et al.* [233], that are in reasonable agreement with the measurement results.

4.2.4 Momentum Transfer Cross Section (Elastic), $\sigma_m(\varepsilon)$, of C_3F_8

There has been one determination of the momentum transfer cross section, $\sigma_m(\varepsilon)$, of C_3F_8 based on elastic differential electron scattering cross section measurements for this molecule [297, 298, 304]. Tanaka *et al.* [304] determined their momentum transfer cross section from the elastic differential electron scattering cross sections they measured (Fig. 4.69) after extrapolation to the full angle range by modified phase-shift fitting. Their data are plotted in Fig. 4.71. The measurements of Tanaka *et al.* [304] are listed in Table 4.51 as the suggested values for the $\sigma_m(\varepsilon)$ of C_3F_8.

This experimentally based cross section is compared with the $\sigma_m(\varepsilon)$ calculated by Pirgov and Stefanov [252] for the C_3F_8 molecule. These calculations are Boltzmann type and are based on the measured values of $w(E/N)$ and $D_T/\mu\,(E/N)$ in pure C_3F_8 and in mixtures of C_3F_8 with Ar. Pirgov and Stefanov used the $w(E/N)$ and $D_T/\mu\,(E/N)$ of Naidu and Prasad [120] for pure C_3F_8 in the range 270×10^{-17} to 600×10^{-17} V

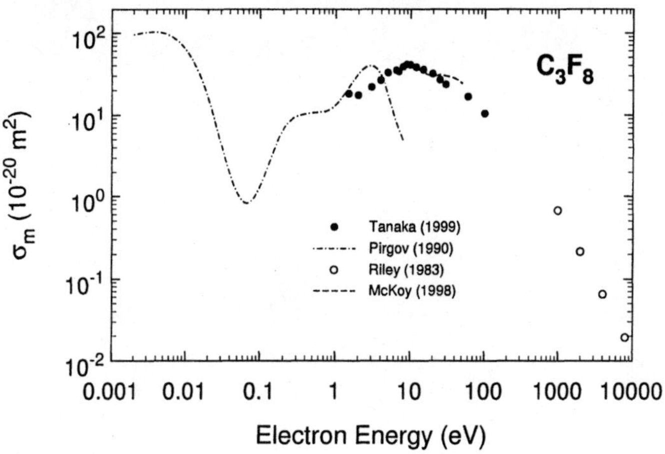

Figure 4.71. Momentum transfer cross section (elastic), $\sigma_m(\varepsilon)$, for C_3F_8: (•) measurements of Ref. [297, 304]; (○) calculations of Ref. [305]; (– · –) calculations of Ref. [252]; (– –) calculations of Ref. [233].

cm^2, and the data of Hunter *et al.* [13] for pure C_3F_8 in the range 0.1×10^{-17} to 200×10^{-17} V cm^2. They also used the $w(E/N)$ data of Hunter *et al.* [13] on mixtures of C_3F_8 with Ar or CH_4 and the data of Hunter *et al.* [13] for the dissociative electron attachment cross section of C_3F_8. For the elastic momentum transfer cross section of Ar they used the data of Miloy *et al.* [253]. Besides the distinct minimum in the momentum transfer cross section near 0.07 eV, the $\sigma_m(\varepsilon)$ of Pirgov and Stefanov shows a maximum near 3 eV due to negative ion resonances. The position of the maximum near 3 eV compares well with other data on electron scattering and electron attachment (see Table 4.47).

There has also been an independent model calculation of $\sigma_m(\varepsilon)$ within the Born approximation by Riley *et al.* [305]. The results of this calculation cover the energy range from 1–8 keV. As seen from Fig. 4.71, these calculated high-energy values of $\sigma_m(\varepsilon)$ are consistent with the measurements of Tanaka *et al.* [304] and help establish the high energy asymptotic values of $\sigma_m(\varepsilon)$. Also shown in Fig. 4.71 are the static-exchange approximation calculation results of McKoy *et al.* [233].

4.2.5 Vibrational Differential Excitation Cross Section, $\sigma_{vib,diff}(\varepsilon)$, of C_3F_8

Figure 4.72 shows the energy dependence of the vibrational differential excitation cross section, $\sigma_{vib,diff}(\varepsilon)$, for the unresolved composite modes

Table 4.51. Suggested momentum transfer (elastic) cross section, $\sigma_m(\varepsilon)$, for C_3F_8 from Tanaka *et al.* [304].

ε (eV)	$\sigma_m(\varepsilon)$ (10^{-20} m^2)	ε (eV)	$\sigma_m(\varepsilon)$ (10^{-20} m^2)	ε (eV)	$\sigma_m(\varepsilon)$ (10^{-20} m^2)
1.5	18.2	7	33.9	20	31.8
2.0	17.5	8	38.5	25	26.9
3.0	21.9	9	41.1	30	23.6
4.0	26.5	10	40.8	60	16.7
5.0	32.9	12	38.2	100	10.4
6.5	35.0	15	35.6		

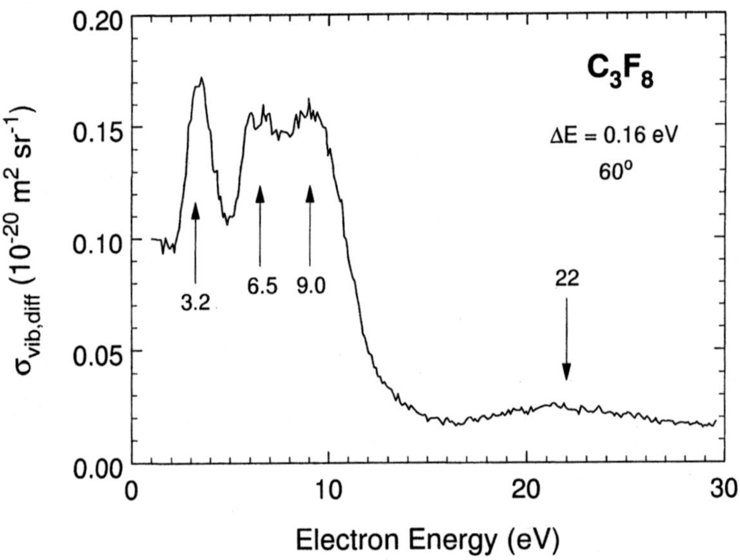

Figure 4.72. Vibrational differential excitation cross section, $\sigma_{vib,diff}(\varepsilon)$, for the unresolved composite vibrational modes of C_3F_8 at 160 meV and 60° scattering angle [297].

at 0.160 eV for a 60° scattering angle as measured by Shinohara *et al.* [297, 298]. The cross section shows peaks at 3.2, 6.5, 9.0, and 22 eV, which the authors attributed to shape resonances. The positions of the first three resonances agree well with those determined by other methods (Table 4.47).

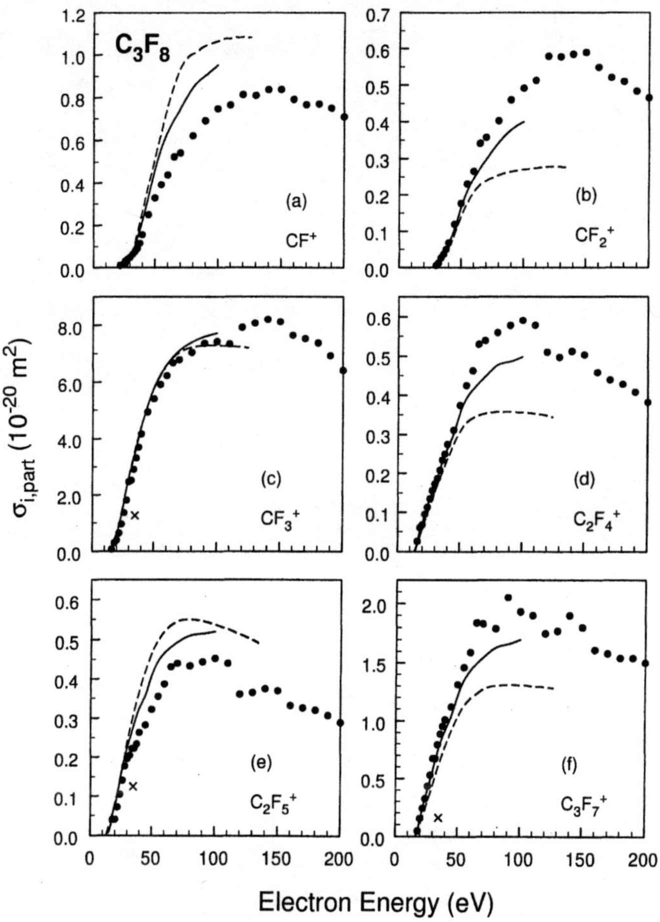

Figure 4.73. Partial ionization cross sections, $\sigma_{\text{i,part}}(\varepsilon)$, for C_3F_8: (- - -) Poll and Meichsner [256]; (•) Jiao *et al.* [302]; (×) Bibby and Carter [259]; (—) suggested.

4.3 Electron-Impact Ionization of C_3F_8

4.3.1 Partial Ionization Cross Sections, $\sigma_{\text{i,part}}(\varepsilon)$, of C_3F_8

The partial ionization cross sections, $\sigma_{\text{i,part}}(\varepsilon)$, of C_3F_8 have been measured by Poll and Meichsner [256] and by Jiao *et al.* [302]. Poll and Meichsner [256] measured the $\sigma_{\text{i,part}}(\varepsilon)$ for CF^+, CF_2^+, CF_3^+, $C_2F_4^+$, $C_2F_5^+$, and $C_3F_7^+$ produced by impact on C_3F_8 of electrons having kinetic energies in the range 12.8 to 130 eV. The CF_3^+ ion has the largest cross section of all six fragment positive ions. Their data are plotted in Fig. 4.73 as broken lines.

Table 4.52. Suggested partial ionization cross sections, $\sigma_{i,part}(\varepsilon)$, for C_3F_8 in units of 10^{-20} m^2.

ε (eV)	CF^+	CF_2^+	CF_3^+	$C_2F_4^+$	$C_2F_5^+$	$C_3F_7^+$
14					0.01	
16			0.11	0.02	0.02	
18			0.36	0.05	0.05	0.05
20			0.60	0.07	0.07	0.15
25	0.01		1.55	0.13	0.16	0.37
30	0.04	0.01	2.61	0.17	0.22	0.57
35	0.12	0.03	3.50	0.22	0.28	0.77
40	0.22	0.07	4.42	0.26	0.33	0.92
45	0.34	0.11	5.17	0.30	0.36	1.04
50	0.46	0.16	5.79	0.35	0.41	1.21
60	0.64	0.25	6.61	0.41	0.47	1.43
70	0.74	0.30	7.11	0.45	0.49	1.54
80	0.85	0.34	7.40	0.48	0.51	1.63
90	0.90	0.38	7.60	0.49	0.52	1.66
100	0.95	0.40	7.72	0.50	0.52	1.70

Similar measurements were made by Jiao *et al.* [302] for electron energies up to 200 eV with an estimated uncertainty of ±16%. Jiao *et al.* also found the CF_3^+ ion to have the largest cross section of all the six fragment positive ions they detected, and found no evidence for the production of the parent positive ion $C_3F_8^+$. The data of Jiao *et al.* are also plotted in Fig. 4.73 (solid circles). To obtain suggested values of $\sigma_{i,part}(\varepsilon)$ for each fragment ion, a fit was made to the average of the values of the two measurements for each positive-ion fragment. The sum of these fits was then normalized at each energy up to 100 eV to the recommended total ionization cross section determined in the next section. The resulting curves are shown by the solid lines in Fig. 4.73, and are listed in Table 4.52 as our suggested data for the $\sigma_{i,part}(\varepsilon)$ of the six positive-ion fragments for C_3F_8.

An early measurement of the partial ionization cross sections for the production of CF_3^+, $C_2F_5^+$, and $C_3F_7^+$ by electron impact on C_3F_8 at 35 eV by Bibby and Carter [259] gave, respectively, values equal to 1.26×10^{-20}, 0.125×10^{-20}, and 0.16×10^{-20} m^2. These values are shown by the × symbols in Fig. 4.73, and are seen to be much lower than the more recent measurements.

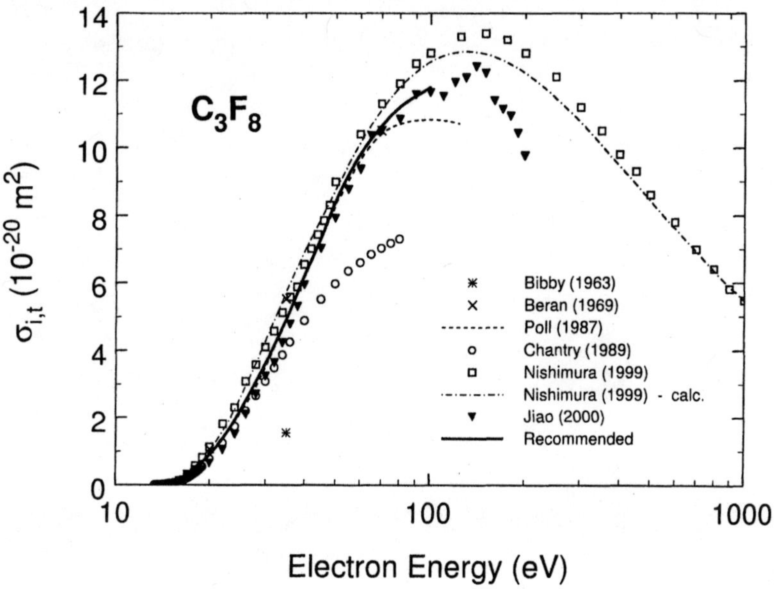

Figure 4.74. Total ionization cross section, $\sigma_{i,t}(\varepsilon)$, for C_3F_8. Measurements: (□) Ref. [137]; (▼) Ref. [302]; (- - -) Ref. [256]; (○) Ref. [293]; (×) Ref. [261]; (∗) Ref. [259]. Calculation: (— · —) Ref. [137]. (—) Recommended.

4.3.2 Total Ionization Cross Section, $\sigma_{i,t}(\varepsilon)$, of C_3F_8

There have been six measurements of the total ionization cross section, $\sigma_{i,t}(\varepsilon)$, of the C_3F_8 molecule. The first set of experimental values of $\sigma_{i,t}(\varepsilon)$ are those obtained by summation of the partial ionization cross sections measured by Poll and Meichsner [256]. Similarly, the second set of experimental values of $\sigma_{i,t}(\varepsilon)$ are those obtained by summation of the partial ionization cross sections of Jiao *et al.* [302]. The third set of measurements of $\sigma_{i,t}(\varepsilon)$ are by Nishimura *et al.* [137] who employed a method similar to that of Rapp and Englander-Golden [262]. These measurements cover the energy range from threshold to 1000 eV. All of these data are shown in Fig. 4.74.

A fourth set of measurements of $\sigma_{i,t}(\varepsilon)$ was made by Chantry and Chen [293] in the energy range 13.5–80 eV. Chantry and Chen calibrated their cross section measurements using Xe as the calibrant gas and the total ionization cross section for Xe measured by Rapp and Englander-Golden [262]. The fifth and sixth sets of measurements are the early data of Beran and Kevan [261] and Bibby and Carter [259] made only at a few values of the incident electron energy. The value of Bibby and

Table 4.53. Recommended total ionization cross section, $\sigma_{i,t}(\varepsilon)$, for C_3F_8.

ε (eV)	$\sigma_{i,t}(\varepsilon)$ (10^{-20} m^2)	ε (eV)	$\sigma_{i,t}(\varepsilon)$ (10^{-20} m^2)	ε (eV)	$\sigma_{i,t}(\varepsilon)$ (10^{-20} m^2)
13	0.001	20	0.901	60	9.80
14	0.006	25	2.22	70	10.6
15	0.054	30	3.62	80	11.2
16	0.152	35	4.92	90	11.5
17	0.323	40	6.21	100	11.8
18	0.518	45	7.31		
19	0.696	50	8.40		

Carter is the sum (1.55×10^{-20} m^2) of the ionization cross section for the production of the CF_3^+, $C_2F_5^+$, and $C_3F_7^+$ ions from C_3F_8 they measured at one value (35 eV) of the electron energy, and is clearly lower compared to the more recent measurements.

Figure 4.74 also shows the calculated values of $\sigma_{i,t}(\varepsilon)$ for the C_3F_8 molecule by Nishimura *et al.* [137]. The results of this calculation (based on the BEB method, see Ch. 2) are in good agreement with the experimental measurements.

The experimental determinations of $\sigma_{i,t}(\varepsilon)$ by Poll and Meichsner [256], Jiao *et al.* [302], and Nishimura *et al.* [137] are in reasonable agreement for electron energies up to $\sim$ 100 eV. Above this energy, the disagreement becomes significant. The data of Chantry and Chen [293] are clearly too low compared to these three measurements. For this reason, we obtained recommended values of $\sigma_{i,t}(\varepsilon)$ represented by the bold solid line in Fig. 4.74 by fitting the data of Poll and Meichsner [256], Jiao *et al.* [302], and Nishimura *et al.* [137] for energies up to 100 eV. Values taken from this fit are listed in Table 4.53 as our recommended data for the $\sigma_{i,t}(\varepsilon)$ of the C_3F_8 molecule.

4.3.3 Total Dissociation Cross Section, $\sigma_{dis,t}(\varepsilon)$, for C_3F_8

The only available measurement of the total dissociation cross section by electron impact on C_3F_8, $\sigma_{dis,t}(\varepsilon)$, is that by Winters and Inokuti [50]. This cross section represents the sum of the cross section for dissociative ionization and the cross sections for electron-impact dissociation into neutral fragments, and has a reported uncertainty of $\pm$20%. The measured values of $\sigma_{dis,t}(\varepsilon)$ are compared in Fig. 4.75 with the recommended values of $\sigma_{i,t}(\varepsilon)$ (Table 4.53). Above $\sim$ 80 eV, the recommended

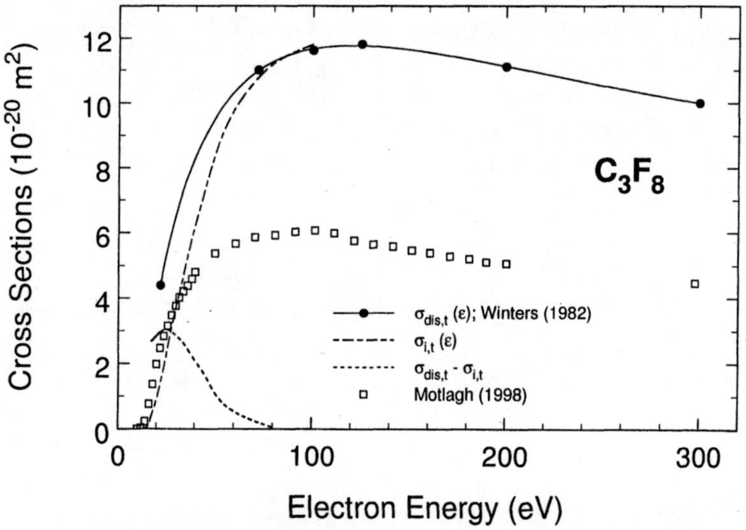

Figure 4.75. Total dissociation cross section, $\sigma_{dis,t}(\varepsilon)$, for C_3F_8 [50]. For comparison, the total ionization cross section $\sigma_{i,t}(\varepsilon)$ is shown (– - –) and the difference (- - -) $\sigma_{dis,t}(\varepsilon)-\sigma_{i,t}(\varepsilon)$, which is a gross estimate of the total cross section for dissociation into neutral fragments. Also shown are the measurements of Ref. [162] on the production of neutral CF_3 + C_2F_5 radicals by electron impact on C_3F_8 (□).

Table 4.54. Total electron-impact dissociation cross section, $\sigma_{dis,t}(\varepsilon)$, measurements of Winters and Inokuti [50] for C_3F_8.

ε (eV)	$\sigma_{dis,t}(\varepsilon)$ (10^{-20} m^2)	ε (eV)	$\sigma_{dis,t}(\varepsilon)$ (10^{-20} m^2)
22	4.39	125	11.8
72	11.0	200	11.1
100	11.6	300	10.0

values of $\sigma_{i,t}(\varepsilon)$ exceed those of $\sigma_{dis,t}(\varepsilon)$, which is physically impossible, but is the result of the relative uncertainties of the measurements. The values of $\sigma_{dis,t}(\varepsilon)$ of Winters and Inokuti are listed in Table 4.54.

There are no measurements of the cross section for dissociation of the C_3F_8 molecule into neutral fragments. An estimate of $\sigma_{dis,t}(\varepsilon)$ may be obtained by subtracting the recommended total ionization cross section $\sigma_{i,t}(\varepsilon)$, which is exclusively due to dissociative ionization, from the total dissociation cross section of Winters and Inokuti. This difference is shown in Fig. 4.75 by the short dash line. These values must be considered a gross estimate due to the uncertainties in the values of $\sigma_{i,t}(\varepsilon)$ and the relatively large stated uncertainty of $\sigma_{dis,t}(\varepsilon)$. The effect of these uncertainties becomes most apparent at energies above 80 eV, where the recommended values of $\sigma_{i,t}(\varepsilon)$ exceed the measured values of $\sigma_{dis,t}(\varepsilon)$.

Recently, Motlagh and Moore [162] measured the cross section for the production of CF_3 plus C_2F_5 radicals by electron impact on C_3F_8. Their method detects these radicals mass spectrometrically as organotellurides generated upon collision with the surface of a telluride mirror. Their results, which reflect the production of these radicals by both dissociative ionization and by dissociation into neutrals, are also shown in Fig. 4.75. The shape of the curve is similar to the $\sigma_{dis,t}(\varepsilon)$, but the experiment of Motlagh and Moore [162] does not detect all species of neutral fragments.

The state of excitation of the dissociation fragments is also of considerable interest because it affects the rates of the subsequent reactions of these products and because it produces light which may induce other processes or may be used for diagnostic purposes. Emission bands of CF_3 radicals produced by pulsed electron beam (initial energy 0.6 MeV) excitation have been studied by Hermann [306], and the formation of excited fragments in collisions of C_3F_8 with electrons having initial energies in the range 0.4–6 keV has been studied by Danilevskii *et al.* [307]. Optical emission spectra of pure C_3F_8 and of C_3F_8-O_2 plasmas have been studied by Chen and Lee [308]. The fluorine emission lines were observed and also emissions from CF_x radicals. For the C_3F_8-O_2 plasmas, the relative emission intensity of the fluorine atom and the CF_2 and CF_3 radicals depended on the percentage of O_2 in the mixture [308].

4.4 Ionization Coefficients for C_3F_8

4.4.1 Density-Normalized Ionization Coefficient, $\alpha/N\,(E/N)$, of C_3F_8

There have been three measurements [120, 152, 309] of the density-normalized ionization coefficient, $\alpha/N\,(E/N)$, for C_3F_8. These are compared in Fig. 4.76. Moruzzi and Craggs [309] made measurements at $T = 273$ K and in the E/N range of 273×10^{-17} to 910×10^{-17} V cm^2. Their data have an estimated uncertainty of $\pm 20\%$ [153]. They show some dependence on gas pressure which is not exhibited by the other two sets of measurements. Naidu and Prasad [120] made their measurements for $T = 293$ K, gas pressures in the range of 0.08–0.27 kPa, and E/N values ranging form 273×10^{-17} to 637×10^{-17} V cm^2. The

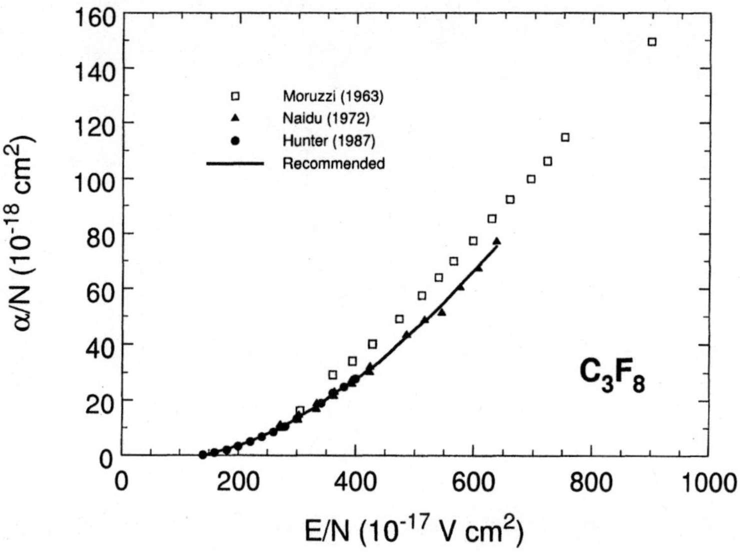

Figure 4.76. Density-normalized electron impact ionization coefficient, $\alpha/N\,(E/N)$, for C_3F_8: (•) Ref. [152]; (▲) Ref. [120]; (□) Ref. [309]; (—) recommended.

$\alpha/N\,(E/N)$ values were found to be pressure independent. They reported an overall uncertainty in their measurements of ±10% at E/N values less than $(E/N)_{\text{lim}}$ and about ±20% at the highest E/N values at which they made measurements. Hunter *et al.* [152] measured $\alpha/N\,(E/N)$ at 298 K in the E/N range 5×10^{-17} to 400×10^{-17} V cm^2 using a pulsed Townsend technique. The reported uncertainty in their measurements is less than ±10% except when one of the coefficients (electron attachment or ionization) is much smaller than the other. The values of $\alpha/N\,(E/N)$ were found to be independent of pressure in the pressure range 0.05 and 20 kPa they investigated. The overall agreement between the three sets of measurements is within the combined uncertainties, although the data of Moruzzi and Craggs are consistently higher than the other two sets of measurements. In view of the higher uncertainty of the earlier measurements, we performed a least squares fit to only the data of Naidu and Prasad [120] and Hunter *et al.* [152] which is represented in Fig. 4.76 by the solid line. Values taken off this line are listed in Table 4.55 as our recommended data for the electron impact ionization coefficient $\alpha/N\,(E/N)$ of C_3F_8.

Table 4.55. Recommended values of the density-normalized ionization coefficient, $\alpha/N\,(E/N)$, for C_3F_8.

E/N (10^{-17} V cm^2)	α/N (10^{-18} cm^2)	E/N (10^{-17} V cm^2)	α/N (10^{-18} cm^2)
140	0.12	400	27.8
160	0.99	420	30.7
180	2.02	440	33.9
200	3.34	460	37.7
220	5.02	480	41.7
240	6.88	500	45.4
260	8.84	520	49.0
280	11.1	540	52.9
300	13.4	560	57.3
320	16.2	580	62.0
340	18.9	600	66.6
360	22.1	620	71.3
380	24.9		

4.4.2 Density-Normalized Effective Ionization Coefficient, $(\alpha - \eta)/N(E/N)$, for C_3F_8

Hunter *et al.* [152] used their data on electron attachment and ionization coefficients to obtain the density-normalized effective ionization coefficient, $(\alpha - \eta)/N$, over an E/N range above and below $(E/N)_{\rm lim}$ (Fig. 4.77). Although the electron-impact ionization coefficient is independent of gas pressure (Sec. 4.4.1 of Ch. 4), the electron attachment coefficient increases with increasing pressure over the pressure range investigated by Hunter *et al.* [152], and thus the effective ionization coefficient should decrease with increasing gas pressure, as is indeed shown by the measurements of Hunter *et al.* (Fig. 4.77) and the earlier measurements of Moruzzi and Craggs [309]. One, therefore, needs to exercise caution when comparisons are made of $(\alpha - \eta)/N(E/N)$ values from various sources which might have been taken at different gas pressures. It should also be noted that since the attachment coefficient for C_3F_8 is also a function of temperature (Sec. 4.6.5 of Ch. 4) one needs to specify the gas temperature. The measurements of Hunter *et al.* were made at 298 K and those of Naidu and Prasad and Moruzzi and Craggs at 293 K. For comparison, we have plotted in Fig. 4.77 the $(\alpha - \eta)/N(E/N)$ measurements of Moruzzi and Craggs [309] and Naidu and Prasad [120] made at pressures less than 0.133 kPa. There is good agreement between the

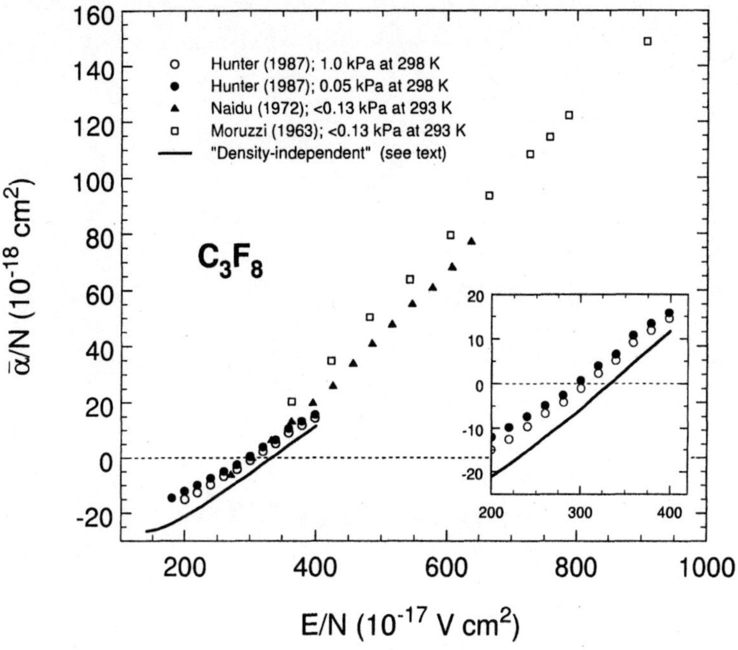

Figure 4.77. Density-normalized effective ionization coefficient, $(\alpha - \eta)/N(E/N)$, for C_3F_8: (○) Ref. [152] (1.0 kPa at 298 K); (•) Ref. [152] (0.05 kPa at 298 K); (▲) Ref. [120] ($<$ 0.13 kPa at 293 K); (□) Ref. [309] ($<$ 0.13 at 293 K); (—) density-independent values ($T \approx 298$ K) at "infinite" pressure (see text). Inset graph: Data of Ref. [152] for 1.0 and 0.05 kPa, on an expanded scale.

measurements of Hunter *et al.* [152] and Naidu and Prasad [120], but the measurements of Moruzzi and Craggs [309] are higher. The data of Hunter *et al.* and Naidu and Prasad are preferred.

Values of $(\alpha - \eta)/N(E/N)$ in Fig. 4.77 are for the temperatures 298 and 293 K and pressures < 1.0 kPa. However, since Hunter *et al.* [152] reported values of $\eta/N\,(E/N)$ at "infinite" gas pressure, we [48] used the density-independent values of Hunter *et al.* for $\eta/N\,(E/N)$ and the recommended values of $\alpha/N\,(E/N)$ to determine the room temperature ($\sim$ 298 K) density-independent values of $(\alpha-\eta)/N(E/N)$ of C_3F_8. These are shown in Fig. 4.77 as the solid line and are listed in Table 4.56.

4.4.3 $(E/N)_{\mathrm{lim}}$ for C_3F_8

In the absence of significant secondary electron loss or gain processes, $(E/N)_{\mathrm{lim}}$ can be equated with the uniform field breakdown strength of the gas. As discussed earlier in this section (see also Sec. 4.6.1 of

Table 4.56. Suggested values for the density-independent effective ionization coefficient, $(\alpha - \eta)/N(E/N)$ ($T \approx 298$ K), of C_3F_8.

E/N (10^{-17} V cm^2)	$(\alpha - \eta)/N$ (10^{-18} cm^2)	E/N (10^{-17} V cm^2)	$(\alpha - \eta)/N$ (10^{-18} cm^2)
140	−26.5	280	−8.9
160	−25.5	300	−5.8
180	−23.7	320	−2.2
200	−21.1	340	1.1
220	−18.3	360	4.8
240	−15.2	380	8.2
260	−12.0	400	11.6

Ch. 4), the electron attachment coefficient of C_3F_8 and hence $(E/N)_{\rm lim}$ are dependent upon gas pressure. The latter, therefore, cannot be directly compared with the high voltage breakdown measurements which are usually obtained at atmospheric (or higher) pressures. The data of Hunter *et al.* [152] on $(E/N)_{\rm lim}$ are in good agreement with previous measurements of $(E/N)_{\rm lim}$ based on ionization and attachment coefficients when these measurements were performed over similar pressure ranges [120, 309] These data are listed in Table 4.57. The values of $(E/N)_{\rm lim}$ obtained from breakdown strength measurements [270, 310] are also listed in Table 4.57. They overlap with the values Hunter *et al.* measured at their highest gas pressures. At higher pressures, however, the breakdown strength measurements give considerably higher $(E/N)_{\rm lim}$ values [15, 268, 270, 310].

It has been shown [311–313] that the pressure dependence that has been observed in $(E/N)_{\rm lim}$ and in breakdown strength measurements is due to the pressure dependence of the electron attachment coefficient in this gas and represents a genuine violation of Paschen's law [312]. The variation of $(E/N)_{\rm lim}$ with the C_3F_8 gas density is shown in Fig. 4.78(a). Since, moreover, the electron attachment coefficient is a function of the gas temperature, the value of $(E/N)_{\rm lim}$ will vary with gas temperature. Indeed, this has been observed to be so by Christophorou *et al.* [267, 268] [Fig. 4.78(b)]. Finally, measurements have been made of the $(E/N)_{\rm lim}$ of mixtures of C_3F_8 with CH_4 or Ar [13]. Figure 4.79 shows these measurements.

Table 4.57. Values of $(E/N)_{\rm lim}$ for C_3F_8.

$(E/N)_{\rm lim}$ (10^{-17} V cm^2)	Reference
295–330^a	[152]
302–313^b	[120]
308^b	[309]
317–358^c	[270]
331–353^d	[310]
353^e	[15]
352^f	[268]

^{a}These values are for the limited pressure range (0.05 kPa to 2 kPa) employed by Hunter *et al.* [152]. If it were possible to measure ionization and attachment coefficients at higher pressure, the values of $(E/N)_{\rm lim}$ that would be measured would most likely overlap with the experimental high voltage measurements listed in the table.
^{b}Data obtained from low pressure measurements of electron attachment and ionization coefficients.
^{c}Breakdown measurements at pressures ranging from 10 kPa to 210 kPa.
^{d}Breakdown measurements at pressures ranging from 27 kPa to 285 kPa.
^{e}Breakdown measurements at a pressure of 150 kPa.
^{f}Breakdown measurements at a pressure of 69.3 kPa.

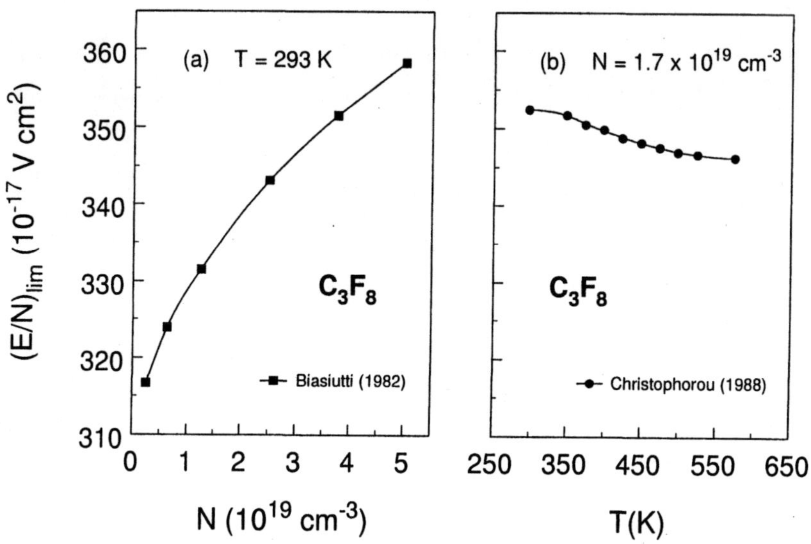

Figure 4.78. Observed variation of the $(E/N)_{\rm lim}$ of C_3F_8 with (a) gas number density [270] and (b) temperature [268].

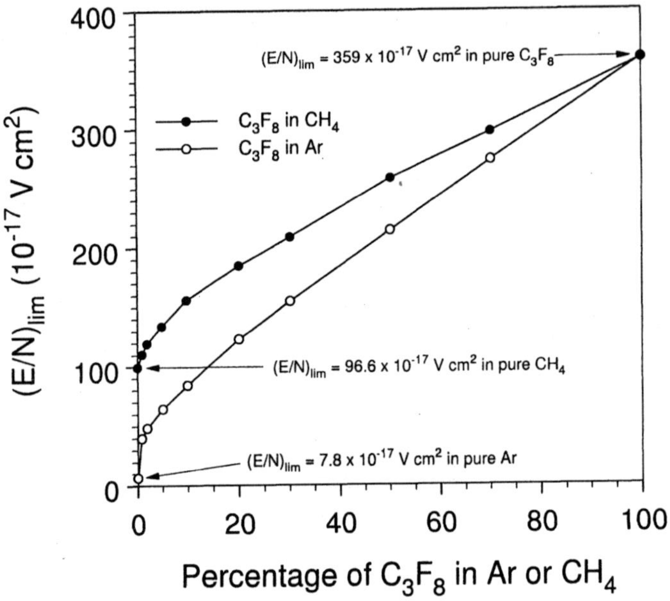

Figure 4.79. Values of $(E/N)_{\text{lim}}$ as a function of the percentage of C_3F_8 in (○) Ar or (•) CH_4 [13]. The total pressure was 109 kPa and the temperature 298 K. The $(E/N)_{\text{lim}}$ values for pure Ar, CH_4, and C_3F_8 are shown in the figure.

4.4.4 Average Energy to Produce an Electron-Ion Pair, W, for C_3F_8

The average energy to produce an electron-ion pair, W, for α particles (initial energy $\sim$ 5.1 MeV) has been measured by Reinking *et al.* [156] for pure C_3F_8 and found to be 34.4 eV per ion pair. This value is almost identical with those measured by Reinking *et al.* for CF_4 and C_2F_6. They are large compared to the W values of other polyatomic molecules [273] reflecting the high ionization threshold energies for these perfluorocarbon molecules and the considerable amount of energy going into transnational and/or internal energy of the fragments that accompany the processes of dissociative ionization in these molecules.

The W values of binary mixtures of C_3F_8 with Ar and C_2H_2 and the ternary mixture $C_3F_8/Ar/C_2H_2$ have also been measured by Reinking *et al.* [156]. Figure 4.80 shows the W values of the binary mixtures C_3F_8/Ar and C_3F_8/C_2H_2. Interestingly, as noted by Nakanishi *et al.* [272] and Reinking *et al.* [156], the perfluorocarbon-containing gas mixtures show no "Jesse effect" (i.e., an abrupt decrease in the W of the gas mixture as small amounts of C_2H_2 are added to C_3F_8), although a

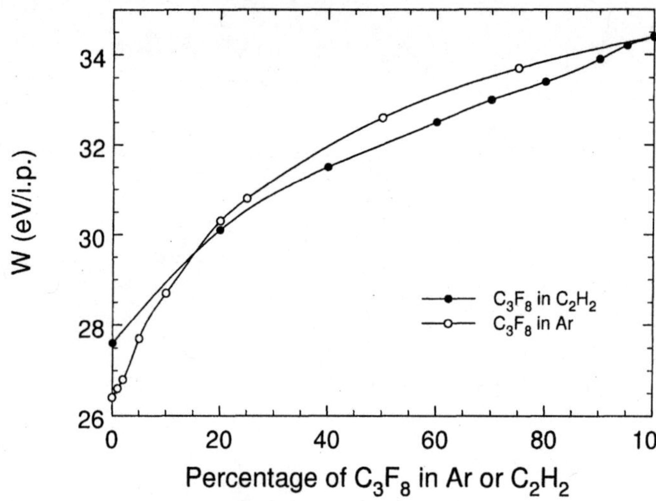

Figure 4.80. Average energy needed to produce an electron-ion pair in mixtures of C_3F_8 in Ar or C_2H_2 for α particles measured by Reinking *et al.* [156].

number of excited electronic states of the C_3F_8 molecule exist above the ionization threshold of C_2H_2. This has been attributed [156, 272] to the fast dissociation of the electronically excited $C_3F_8^*$ molecules.

4.5 Electron-Impact Dissociation of C_3F_8 Producing Neutrals

To our knowledge there are no data on this important process, except the rough estimates discussed in Sec. 4.3.3 of this chapter.

4.6 Electron Attachment to C_3F_8

4.6.1 Density-Normalized Electron Attachment Coefficient, $\eta/N\,(E/N)$, of C_3F_8

In contrast to CF_4 and C_2F_6, the electron attachment coefficient (and hence the electron attachment rate constant and cross section) of C_3F_8 depends on gas pressure. Electron attachment to CF_4 and C_2F_6 is entirely due to dissociative electron attachment, and thus the $\eta/N\,(E/N)$ for these gases is independent of gas pressure [70, 300]. In contrast, at room temperature electron attachment to C_3F_8 is partly due to nondissociative electron attachment producing parent negative ions and partly due to dissociative electron attachment producing fragment negative ions

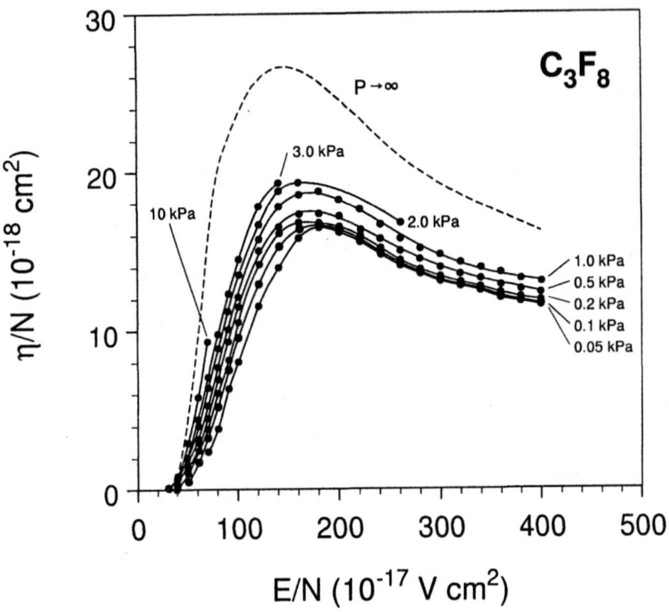

Figure 4.81. Density-normalized electron attachment coefficient, $\eta/N\,(E/N)$, for C_3F_8 at various C_3F_8 pressures (T = 298 K) measured by Hunter *et al.* [152]. The broken line refers to the extrapolated values of $\eta/N\,(E/N)$ at "infinite" gas pressure.

[72, 300]. The former are normally collision-stabilized species since the autodetachment lifetimes of the transient parent anion $C_3F_8^{-*}$ are believed to be in the range 10^{-11} to 10^{-8} s [72, 300].

The pressure dependence of $\eta/N\,(E/N)$ for C_3F_8 is evident in early $\eta/N\,(E/N)$ measurements [120, 309], and in the more recent and detailed results of Hunter *et al.* [152] reproduced in Fig. 4.81. The broken curve designated by $P \rightarrow \infty$ refers to the values of $\eta/N\,(E/N)$ at "infinite" pressure, that is, when all parent anions are stabilized. The data cover the range of pressure from 0.05 to 10.0 kPa and were taken at 298 K. They have a quoted uncertainty of about ±10% except when one of the coefficients (for electron attachment or ionization) is considerably larger than the other.

Earlier data of Naidu and Prasad [120] have an uncertainty between ±10% at $E/N < (E/N)_{\rm lim}$ and about ±20% at the highest E/N values at which they made measurements. The uncertainty in the measurements of Moruzzi and Craggs [309] is probably ±20% [153]. The fact that the $\eta/N\,(E/N)$ for C_3F_8 varies with pressure, makes it difficult to compare its values as measured by various groups, and thus the data of Moruzzi and Craggs and Naidu and Prasad are not shown in Fig. 4.81.

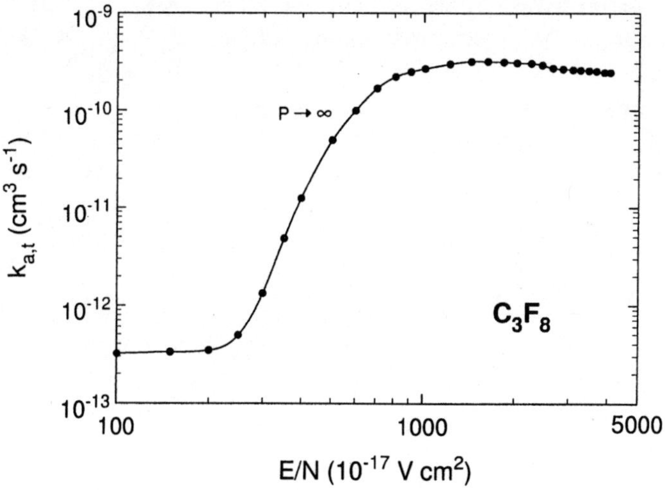

Figure 4.82. $k_{a,t}(E/N)$ measured at room temperature in pure C_3F_8 extrapolated to infinite total gas pressure by Hunter *et al.* [152].

However, a comparison of the measurements of Naidu and Prasad [120] at 0.22 kPa (at 293 K) with those of Hunter *et al.* [152] at 0.20 kPa (298 K) indicates agreement within the stated uncertainties.

4.6.2 Total Electron Attachment Rate Constant, $k_{a,t}$, for C_3F_8

Since the $k_{a,t}(E/N)$ of C_3F_8 depends on gas pressure, care must be taken to specify the number density and the nature of the medium in which the measurement is made, unless the data are extrapolated in some fashion to infinite number density. Measurements have been made of $k_{a,t}(E/N)$ in both pure C_3F_8 and in binary mixtures of C_3F_8 with the buffer gases N_2, Ar, and CH_4. These measurements are presented and discussed below.

$k_{a,t}(E/N)$ Measured in Pure C_3F_8. The only measurements of $k_{a,t}(E/N)$ in pure C_3F_8 are those of Hunter *et al.* [152] which are reproduced in Fig. 4.82. The data shown are the values of the electron attachment rate constant extrapolated to "infinite" gas pressure [152]. They represent the sum of the rate constants for both fragment and parent negative ions. The data cannot be plotted as a function of the mean electron energy because the electron energy distribution functions at the E/N values employed are not known for pure C_3F_8.

$k_{a,t}(E/N)$ Measured in Binary Mixtures of C_3F_8 with Buffer Gases. There have been three [72, 300, 314] room temperature measurements of $k_{a,t}(E/N)$ for C_3F_8 in argon buffer gas [Fig. 4.83(a)]. The measurements of Hunter and Christophorou [72] and Spyrou and Christophorou [300] were made over a large E/N range and at a number of buffer gas pressures, in contrast with those of Wang and Lee [314] which were made over a limited E/N range and for only one total gas pressure (50.66 kPa). The data of Christophorou and coworkers [72, 300] are the values of $k_{a,t}(E/N)$ which were extrapolated to infinite gas pressure. Christophorou and coworkers [72, 300] measured $k_{a,t}(E/N)$ as a function of the ratio, R, of the attaching to buffer gas pressure and used the value of $k_{a,t}(E/N)$ extrapolated to $R = 0$ so as to correct for the effect of the attaching gas on the electron energy distribution function of pure argon.

Although Wang and Lee [314] mention in their paper that they followed a similar procedure, the large discrepancy between their data and the rest of the measurements in Fig. 4.83(a) would indicate that the procedure they followed did not entirely compensate for the effect of the attaching gas pressure on the electron energy distribution function in pure argon. While their lower values for the rate constants could be partly due to the fact that their measurements were made at only one buffer gas pressure (and not extrapolated to infinite buffer gas pressure), the difference in the E/N dependence between the Wang and Lee data and those of Christophorou and coworkers indicates that the data of Wang and Lee are affected by both the disturbance of the electron energy distribution of pure argon by the attaching gas and the incomplete stabilization of the anions for the total pressure they used.

Both of these effects will be less pronounced as the buffer gas becomes more complex, as is indeed seen from the comparison of the two sets of measurements in N_2 buffer gas shown in Fig. 4.83(b). The data of Christophorou and coworkers are preferred over those of Wang and Lee for both of these buffer gases, because they have lower uncertainties, were corrected for the effect of total pressure, and cover a much larger range of E/N. For the more complex buffer gas CH_4, these effects are expected to be significantly reduced, and the Wang and Lee data in Fig. 4.83(c) are expected to be least affected by the factors just discussed. Overall, the uncertainty in the measurements of Wang and Lee is about $\pm 20\%$ and that of Christophorou and coworkers is less than $\pm 10\%$.

$k_{a,t}(\langle\varepsilon\rangle)$ of C_3F_8. The measurements of $k_{a,t}$ made as a function of E/N using Ar and N_2 as buffer gases can be plotted as a function of the mean electron energy $\langle\varepsilon\rangle$. This is possible because the electron energy

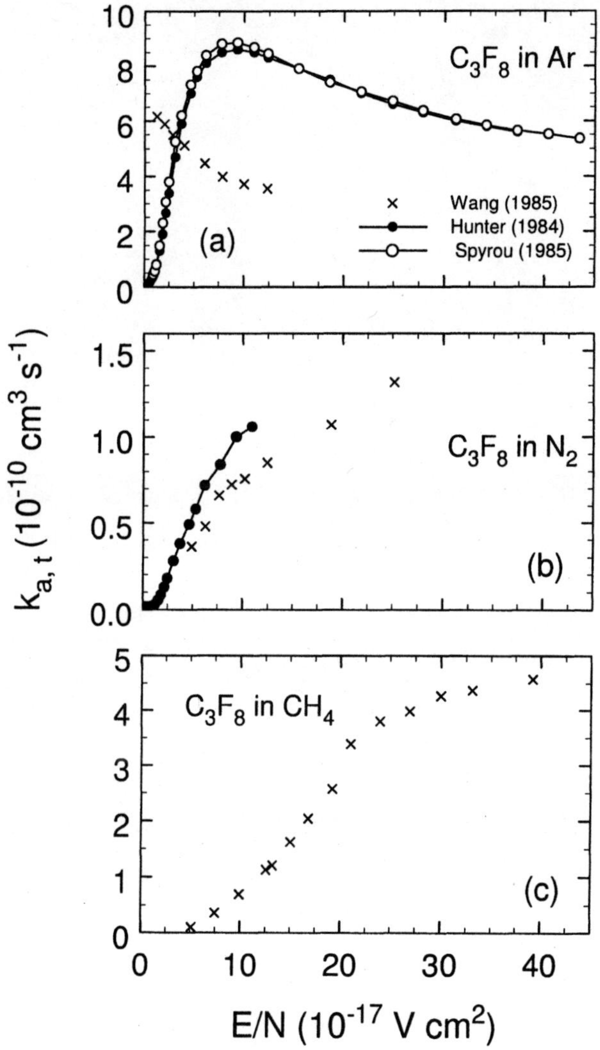

Figure 4.83. (a) $k_{a,t}(E/N)$ of C_3F_8 measured at room temperature in mixtures of C_3F_8 with Ar buffer: (•) Ref. [72]; (○) Ref. [300]; (×) Ref. [314]. The data of Refs. [72, 300] were corrected for both the effect of the C_3F_8 partial and the total gas pressure on the measured rate constants. The data of Ref. [314] were corrected only for the former. (b) $k_{a,t}(E/N)$ measured at room temperature in mixtures of C_3F_8 with N_2 buffer: (•) Ref. [72]; (×) Ref. [314]. The data of Hunter and Christophorou [72] were corrected for both the effect of the partial and total pressure on the measured rate constants. The data of Wang and Lee [314] were corrected only for the former. (c) $k_{a,t}(E/N)$ measured at room temperature in mixtures of C_3F_8 with CH_4 buffer: (×) Ref. [314].

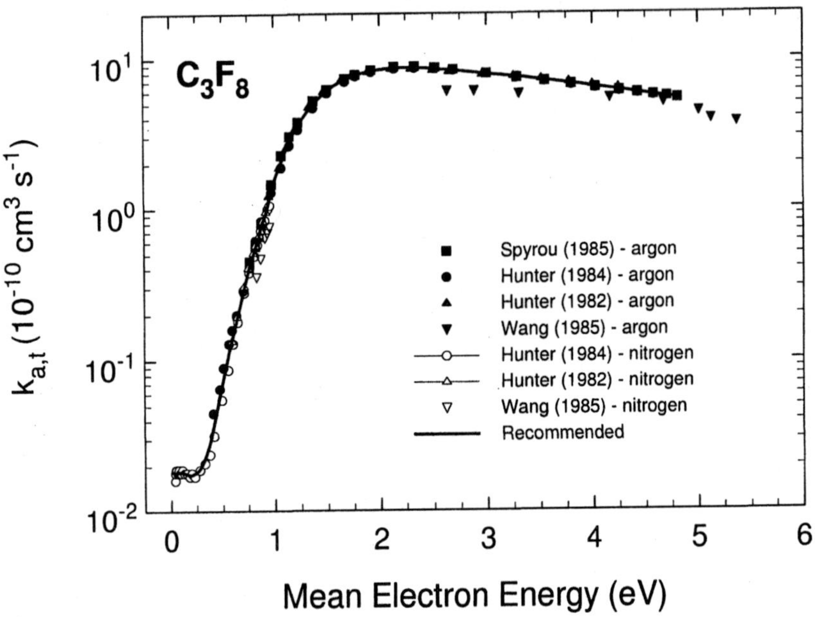

Figure 4.84. Total electron attachment rate constant as a function of mean electron energy, $k_{a,t}(\langle\varepsilon\rangle)$, for C_3F_8 ($T = 298$ K). Data from mixtures in Ar: (■) Ref. [300]; (•) Ref. [72]; (▲) Ref. [311]; (▼) Ref. [314]. Data from mixtures in N_2: (○) Ref. [72]; (△) Ref. [311]; (▽) Ref. [314]. (—) Recommended.

distribution functions for these buffer gases are known at each value of E/N at which measurements of $k_{a,t}$ were made and because the experimental conditions were such that the electron energy distributions were characteristic of the buffer gas alone. The latter condition was certainly met in the studies by Hunter and Christophorou [72] and Spyrou and Christophorou [300] as was indicated earlier in this section. Figure 4.84 shows the results for $k_{a,t}(\langle\varepsilon\rangle)$ plotted this way. The data for the buffer gases N_2 and Ar agree well. The data of Wang and Lee are lower, probably because they were not extrapolated to infinite total gas pressure. The data of Christophorou and coworkers plotted in Fig. 4.84 are the values of the total electron attachment rate constant (for "infinitely" dilute mixtures of C_3F_8 in Ar or N_2 buffer gases) extrapolated to "infinitely" large buffer gas density. Figure 4.85 shows an example of the dependence of $k_{a,t}(\langle\varepsilon\rangle)$ on the argon gas number density N_{Ar}. The values of $k_{a,t}(\langle\varepsilon\rangle)$ for $N_{Ar} \to \infty$ are shown in the figure by the curve designated k_1.

The solid line in Fig. 4.84 represents a fit to all the shown data, excluding the measurements of Wang and Lee [314]. This represents

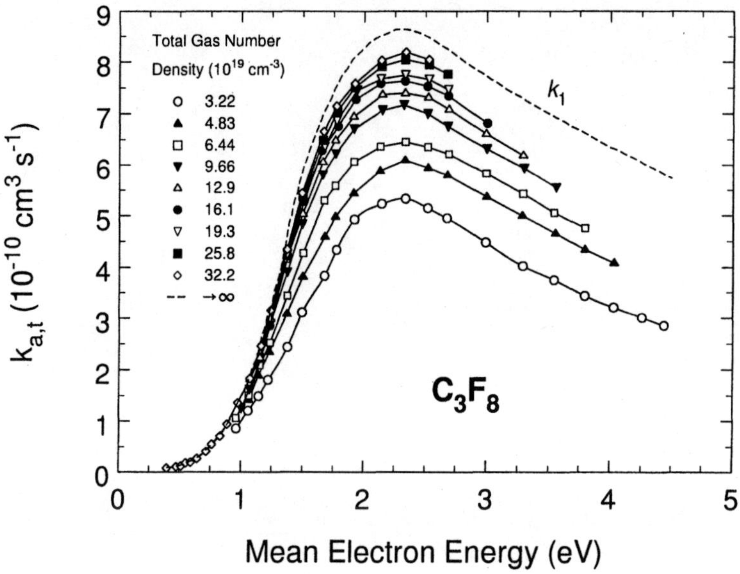

Figure 4.85. Total electron attachment rate constant $k_{a,t}(\langle\varepsilon\rangle)$ for C_3F_8 measured as a function of mean electron energy $\langle\varepsilon\rangle$ and total gas number density in a buffer gas of argon by Hunter and Christophorou [72]. The values of $k_{a,t}(\langle\varepsilon\rangle)$ are those for infinitely dilute mixtures. The broken curve designated k_1 are the extrapolated values of $k_{a,t}(\langle\varepsilon\rangle)$ at "infinite" argon pressure, i.e., under conditions for which all transient $C_3F_8^{-*}$ ions can be stabilized by collision.

our recommended values for the $k_{a,t}(\langle\varepsilon\rangle)$ (T = 300 K) of C_3F_8, and these values are listed in Table 4.58.

Thermal Value, $(k_{a,t})_{th}$, of the Total Electron Attachment Rate Constant of C_3F_8. The thermal value, $(k_{a,t})_{th}$, of the total electron attachment rate constant of the C_3F_8 molecule is difficult to determine accurately due to the likely presence of traces of impurities which attach thermal electrons more efficiently than C_3F_8. The possible presence of such impurities does not, however, affect the electron attachment measurements at higher energies because their magnitude is much larger than at thermal energies. The reported [72, 152, 190, 191] measured values of $(k_{a,t})_{th}$ are listed in Table 4.59. They show that the $(k_{a,t})_{th}$ of C_3F_8 is small, $\lesssim 1.8 \times 10^{-12}$ cm^3 s^{-1}.

Table 4.58. Recommended total electron attachment rate constant, $k_{a,t}(\langle\varepsilon\rangle)$ (T = 300 K), for C_3F_8.

Mean electron energy (eV)	$k_{a,t}$ (10^{-10} cm^3 s^{-1})	Mean electron energy (eV)	$k_{a,t}$ (10^{-10} cm^3 s^{-1})
0.05	0.02	1.75	7.64
0.07	0.02	2.00	8.42
0.10	0.02	2.25	8.65
0.20	0.02	2.50	8.50
0.30	0.02	2.75	8.20
0.40	0.03	3.00	7.84
0.50	0.07	3.25	7.54
0.60	0.15	3.50	7.16
0.70	0.28	3.75	6.80
0.80	0.50	4.00	6.44
0.90	0.85	4.25	6.10
1.00	1.54	4.50	5.74
1.25	3.80	4.75	5.45
1.50	6.10		

Table 4.59. Measured thermal (T = 300 K) values, $(k_{a,t})_{th}$, of the total electron attachment rate constant for C_3F_8.

$(k_{a,t})_{th}$ (cm^3 s^{-1})	Reference
$< 10^{-15}$	[191]
$\leq 3 \times 10^{-13}$	[152]
$< 1.2 \times 10^{-12}$	[190]
1.8×10^{-12}	[72]

4.6.3 Total Electron Attachment Cross Section, $\sigma_{a,t}(\varepsilon)$, and Total Dissociative Electron Attachment Cross Section, $\sigma_{da,t}(\varepsilon)$, of C_3F_8

Since electron attachment to the C_3F_8 molecule is total pressure dependent, care must be exercised to distinguish between the total electron attachment cross section $\sigma_{a,t}(\varepsilon)$ which is pressure dependent and the total dissociative electron attachment cross section $\sigma_{da,t}(\varepsilon)$ which is pressure independent. The former can be deduced from the total electron attachment rate constant measurements made in swarm exper-

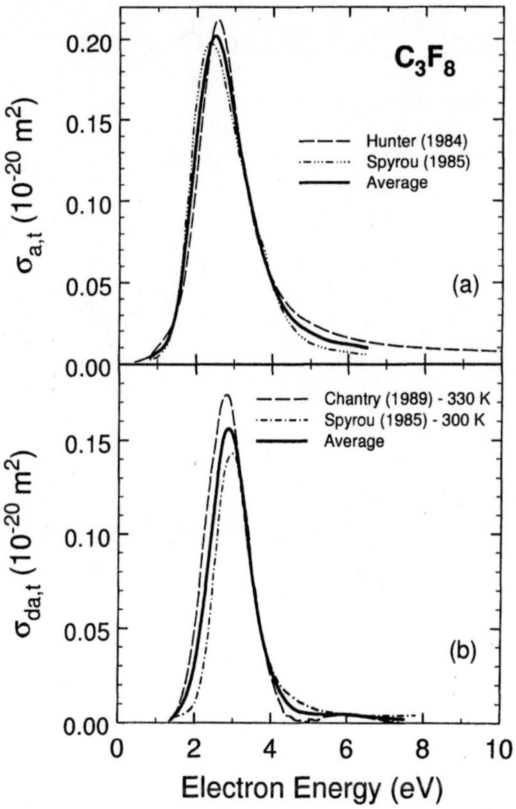

Figure 4.86. (a) Total electron attachment cross section, $\sigma_{a,t}(\varepsilon)$, for C_3F_8 (T = 300 K) unfolded from swarm data: (– –) Ref. [72]; (– · · · –) Ref. [300]; (—) average. (b) Total dissociative electron attachment cross section, $\sigma_{da,t}(\varepsilon)$, for C_3F_8: (– · –) swarm data of Ref. [300] at T = 300 K; (– –) beam data of Ref. [293] at T = 330 K; (—) average of the two sets of measurements.

iments (Sec. 4.6.2 of Ch. 4), and the later normally from electron beam studies.

Figure 4.86(a) shows two sets of values of the total electron attachment cross section $\sigma_{a,t}(\varepsilon)$ unfolded by Christophorou and collaborators [72, 300] from their room temperature $k_{a,t}(\langle\varepsilon\rangle)$ data. The uncertainty of these cross sections is $> \pm 10\%$. The average of the two independent determinations of $\sigma_{a,t}(\varepsilon)$ is shown by the solid line. Values taken off this line are listed in Table 4.60 as our suggested room temperature values for the $\sigma_{a,t}(\varepsilon)$ of the C_3F_8 molecule.

Figure 4.86(b) shows the total dissociative electron attachment cross section $\sigma_{da,t}(\varepsilon)$, deduced from electron swarm experiments by Spyrou

Table 4.60. Suggested total electron attachment cross section, $\sigma_{a,t}(\varepsilon)$ ($T = 300$ K), and total dissociative electron attachment cross section, $\sigma_{da,t}(\varepsilon)$ ($T = 300$ K), for C_3F_8.

ε (eV)	$\sigma_{a,t}(\varepsilon)$ (10^{-20} m^2)	$\sigma_{da,t}(\varepsilon)$ (10^{-20} m^2)
0.80	0.004	
1.00	0.007	
1.40	0.025	0.002
1.50	0.035	0.004
1.70	0.070	0.010
1.80	0.091	0.015
2.00	0.139	0.031
2.30	0.192	0.078
2.50	0.202	0.116
2.70	0.190	0.149
2.85	0.171	0.156
3.00	0.151	0.151
3.50	0.090	0.077
4.00	0.049	0.029
4.50	0.028	0.009
5.00	0.019	0.005
5.50	0.014	0.005
6.00	0.012	0.005
6.50	0.010	0.004
7.00		0.003
7.50		0.002

and Christophorou [300] at $T = 300$ K and the electron beam measurements of Chantry and Chen [293] at a somewhat higher temperature (330 K). Chantry and Chen [293] obtained their $\sigma_{da,t}(\varepsilon)$ cross section by normalization to the cross section data for the production of O^- from N_2O of Rapp and Briglia [315]. The $\sigma_{da,t}(\varepsilon)$ of Chantry and Chen has a value of 0.175×10^{-20} m^2 at 2.8 eV. The agreement between the electron swarm and the electron beam data is reasonable considering the difference in temperature and technique. The average of the two sets of values for $\sigma_{da,t}(\varepsilon)$ is shown by the solid line. Data taken off this line are listed in Table 4.60 as our suggested values for the $\sigma_{da,t}(\varepsilon)$ of the C_3F_8 molecule near 300 K.

In addition to the data just discussed, there are three earlier measurements of $\sigma_{da,t}(\varepsilon)$ which vary by large factors from the more recent measurements and are not considered reliable. The older of these measurements is by Bibby and Carter [259]. These workers reported observation of F^-, CF_3^- and $C_2F_5^-$ ions with cross section maxima at, respectively, 3.0, 3.4, and 3.2 eV and peak cross section values, respectively, equal to 3.65×10^{-20}, 0.23×10^{-20}, and 0.27×10^{-20} m^2. The sum of these values is 4.15×10^{-20} m^2. This value is about 26 times larger than the peak cross section value of $\sigma_{\text{da,t}}(\varepsilon)$ in Fig. 4.86(b). The second earlier measurement is that of Kurepa [260] who observed two maxima in $\sigma_{\text{da,t}}(\varepsilon)$ near 3.3 and 6.5 eV, with respective cross section values equal to 2.38×10^{-20} and 0.11×10^{-20} m^2. The peak cross section value at 3.3 eV is more than a factor of ten higher than the maximum value of $\sigma_{\text{da,t}}(\varepsilon)$ in Fig. 4.86(b). The third measurement of $\sigma_{\text{da,t}}(\varepsilon)$ is that of Harland and Franklin [71] who found the cross section maximum for F^- and CF_3^- at, respectively, (3.15 ± 0.1) eV and (3.65 ± 0.1) eV with cross section values at these energies equal to 0.5×10^{-21} m^2 and 0.05×10^{-21} m^2, respectively. The sum of these values (0.55×10^{-21} m^2) is about a factor of three lower than the value of $\sigma_{\text{da,t}}(\varepsilon)$ at 3.2 eV in Fig. 4.86(b).

4.6.4 Dissociative Electron Attachment Fragment Negative Ions of C_3F_8

A number of studies have dealt with the identification, energetics, relative abundance, and energy dependence of the fragment negative ions formed in collisions of low energy electrons with the C_3F_8 molecule. The pertinent findings of these investigations [70, 71, 259, 294] are summarized in Table 4.61. Bibby and Carter [259] observed F^-, CF_3^-, and $C_2F_5^-$ fragment anions using the electron impact method with a broad electron energy beam. The appearance onsets of these fragment anions and the respective positions of their maximum intensity are in reasonable agreement with the values of other researchers. Lifshitz and Grajower [294] used the retarding potential difference method (RPD) to improve the energy resolution of the electron beam and reported observation of F^-, F_2^-, CF_3^-, $C_2F_3^-$, $C_2F_5^-$ and $C_3F_7^-$. The peak positions and the energy thresholds for these ions are listed in Table 4.61. They are generally lower than the rest of the measurements. Harland and Franklin [71] reported thresholds and energies of maximum intensity for only F^- and CF_3^-. Their values are within the combined experimental uncertainties of the other data.

The most recent and most complete study of negative ion formation by electron impact on C_3F_8 is that of Spyrou *et al.* [70] who also employed the RPD method to improve the electron beam energy resolution. They employed an unfolding technique to correct the relative cross sections

Table 4.61. Fragment negative ions produced by electron impact on C_3F_8, their energetics, and relative intensities.

Fragment anion	Possible reaction	Energy threshold (eV)	Energy of maximum intensity (eV)	Relative abundance (Refs.)
F^-	$C_3F_8 + e \rightarrow F^- + C_3F_7$ [a]	1.7 ± 0.2	2.9 ± 0.1	100 [70]
	$C_3F_8 + e \rightarrow F^- + n - C_3F_7$	2.0 ± 0.1	3.15 ± 0.1	— [71]
	$C_3F_8 + e \rightarrow F^- + C_3F_7^*$	$\sim 4.0 \pm 0.1$		— [71]
	$\rightarrow F^- + CF_3 + C_2F_4$			
	$\rightarrow F^- + CF_2 + C_2F_5$			
		1.8 ± 0.1	3.1 ± 0.1	100 [316]
		4.1 ± 0.1		
		1.35 ± 0.1	~ 2.4	— [294]
		1.8	3.0	— [259]
CF_3^-	$C_3F_8 + e \rightarrow CF_3^- + C_2F_5$ [b,c]	2.4 ± 0.2	3.4 ± 0.1	5.4 [70]
		> 5.0	> 5.5 [d]	
		2.55 ± 0.2	3.65 ± 0.1	— [71]
		2.5 ± 0.1	3.6 ± 0.1	2.2 [316]
		$> 5.2 \pm 0.1$	5.7 ± 0.1	— [316]
		2.0 ± 0.1	~ 2.9	— [294]
		2.2	3.4	— [259]
$C_2F_3^-$		1.1 ± 0.1	3.3 ± 0.1	~ 0.2 [70]
				< 0.01 [316]
$C_2F_5^-$	$C_3F_8 + e \rightarrow C_2F_5^- + CF_3$ [b]	2.1 ± 0.2	3.2 ± 0.1	6.6 [70]
		2.4 ± 0.1	3.4 ± 0.1	1.9 [316]
		1.7 ± 0.1	~ 2.9	— [294]
		2.1	3.2	— [259]
$C_3F_7^-$		2.5 ± 0.2	3.75 ± 0.1	~ 0.2 [70]
		2.9 ± 0.1	3.9 ± 0.1	0.03 [316]
		2.4 ± 0.1	~ 3.2	— [294]

[a] From their measurements on the energetics of this reaction, Spyrou *et al.* [70] estimated the dissociation energy $D(F\text{–}C_3F_7)$ to be ≤ 5.15 eV $\pm$ 0.2 eV. This value is in very good agreement with the value $D(F\text{–}C_3F_7) \leq 5.2$ eV $\pm$ 0.1 eV obtained earlier by Harland and Thynne [316].
[b] From their measurements on the energetics of this reaction, Spyrou *et al.* [70] estimated the dissociation energy $D(CF_3\text{–}C_2F_5)$ to be equal to 3.7 eV $\pm$ 0.2 eV.
[c] From their measurements on the energetics of this reaction Harland and Franklin [71] estimated the dissociation energy $D(CF_3\text{–}C_2F_5)$ to be 4.6 eV $\pm$ 0.3 eV, and the electron affinity of the CF_3 radical to be 2.05 eV $\pm$ 0.2 eV.
[d] Unresolved structure observed.

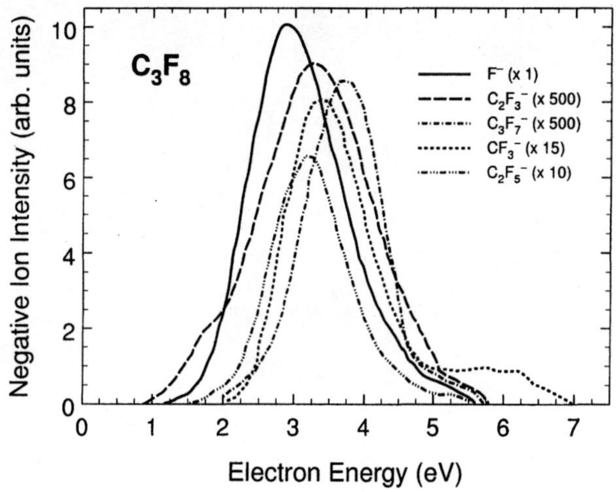

Figure 4.87. Relative intensity of fragment negative ions produced by electron impact on C_3F_8 from Spyrou *et al.* [70]. The data shown have been corrected for the finite width of the electron pulse using an unfolding technique.

for the width of the electron pulse. Spyrou *et al.* observed five fragment negative ions from low energy electron impact on C_3F_8: F^-, CF_3^-, $C_2F_3^-$, $C_2F_5^-$, and $C_3F_7^-$. Their energy onsets, energies of peak intensity, and relative abundances are also given in Table 4.61 and they are in general agreement with the results of the other studies. The results of this study are shown in Fig. 4.87. The most significant anion is F^- produced by the breaking of C–F bonds (production of F^- or $C_3F_7^-$) and the complementary anions CF_3^- and $C_2F_5^-$ produced by the breaking of C–C bonds. The complementary ions CF_3^- and $C_2F_5^-$ and the ion $C_2F_3^-$ have their resonance maxima at (3.3 ± 0.2) eV. The peak position of the predominant ion, F^-, and its weak complementary ion $C_3F_7^-$ are shifted to lower and higher energies, respectively, relative to the position of a common resonance at (3.3 ± 0.2) eV. There is evidence for a second resonance at energies > 5.5 eV.

Finally, Harland and Thynne [316] identified the fragment anions produced when 70 eV electrons interacted with C_3F_8. As expected, they observed more fragments than those produced at low energy by resonance electron attachment processes. Besides the ions listed in Table 4.61, they reported observation of a number of weaker (by factors of $\leq 10^{-4}$ compared to the intensity of F^-) anions: C^-, C_2^-, CF^-, C_3^-, F_2^-, C_2F^-, CF_2^-, $C_2F_2^-$, $C_3F_3^-$, and $C_3F_6^-$.

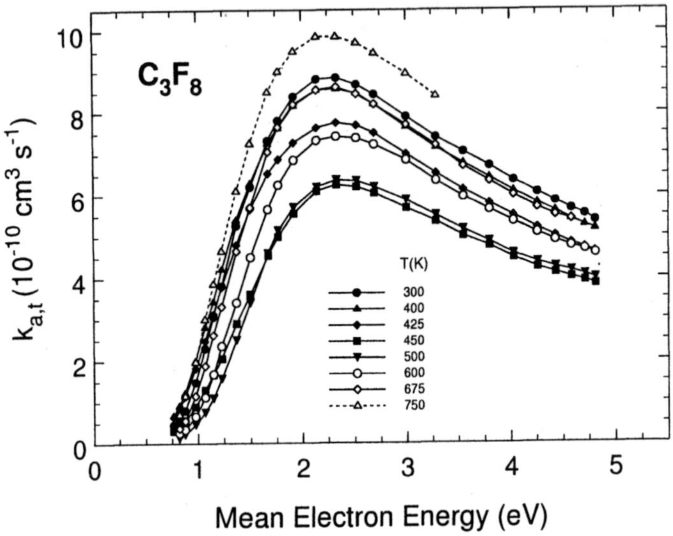

Figure 4.88. Total electron attachment rate constant as a function of the mean electron energy, $k_{a,t}(\langle\varepsilon\rangle)$, for C_3F_8 at temperatures ranging from 300 to 750 K measured by Spyrou *et al.* [300]. The data plotted were taken in mixtures of C_3F_8 with Ar and correspond to a very small pressure of C_3F_8 in a very large pressure of Ar, i.e., to the k_1 values shown in Fig. 4.85.

4.6.5 Effect of Temperature on the $k_{a,t}(\langle\varepsilon\rangle)$ and $\sigma_{a,t}(\varepsilon)$ of C_3F_8

The temperature dependence of low-energy electron attachment processes in C_3F_8 is rather complicated, but understood. As can be seen from Fig. 4.88, $k_{a,t}(\langle\varepsilon\rangle)$ first decreases and then increases with increasing temperature above ambient. This is because low-energy electron attachment to C_3F_8 under swarm conditions leads to the formation of both parent and fragment negative ions. The rate constant for the former processes normally decreases with increasing gas temperature due to increased autodetachment from the transient anion, and the rate constant for the latter processes normally increases with increasing temperature due to increase autodissociation of the transient anion [317–319]. In light of the data in Fig. 4.88 the total electron attachment cross section of the C_3F_8 molecule is expected to first decrease and then increase with increasing temperature above 300 K. This is indeed the case as can be seen from the data in Fig. 4.88.

In Fig. 4.89(a) are plotted the data on the total electron attachment cross section $\sigma_{a,t}(\varepsilon)$ for temperatures (300–450 K) for which the cross

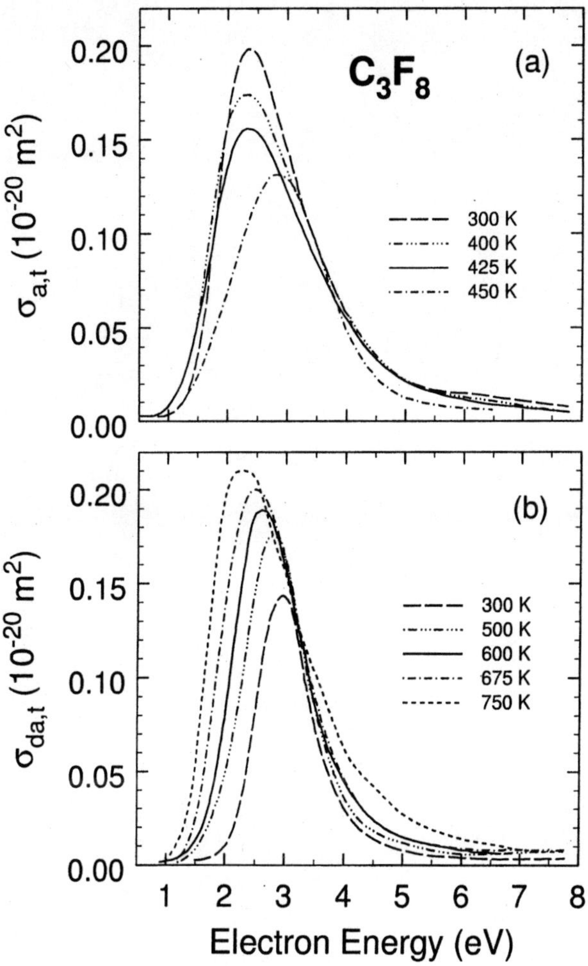

Figure 4.89. Total electron attachment cross section, $\sigma_{a,t}(\varepsilon)$, for C_3F_8 unfolded by Spyrou and Christophorou [300] from their $k_{a,t}(\langle\varepsilon\rangle)$ data shown in Fig. 4.88 at (a) 300, 400, 425, and 450 K, and (b) 500, 600, 675, and 750 K. The 300 K data shown in the figure are the dissociative electron attachment part of the total electron attachment cross section at this temperature.

section has a contribution from both parent and fragment anions. In Fig. 4.89(b) the cross sections plotted are only for dissociative electron attachment. The data shown for 300 K are the dissociative electron attachment part of the total electron attachment cross section at this temperature. The data for temperatures between 500 and 750 K are for the total dissociative electron attachment cross section $\sigma_{da,t}(\varepsilon)$ since at

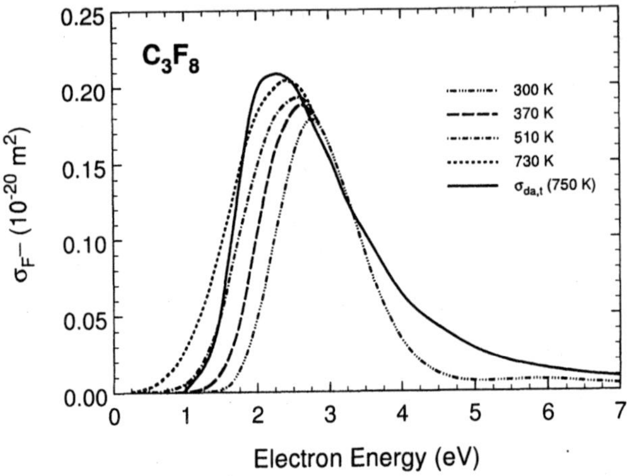

Figure 4.90. Cross section for the production of F^- by electron impact on C_3F_8 at gas temperatures of 300, 370, 510, and 730 K as measured in an electron beam experiment by Chantry and Chen [293]. For comparison the $\sigma_{da,t}(\varepsilon)$ for $T = 750$ K from Fig. 4.89(b) is also shown in the figure.

these temperatures there is no contribution to the cross section from the production of parent negative ions [300]. Electron beam measurements [293] of the formation of F^- by electron impact on C_3F_8 as a function of temperature shown in Fig. 4.90 are consistent with the swarm results in Fig. 4.89.

4.6.6 Negative Ions in C_3F_8 Plasmas

Measurement of negative ion densities in radio-frequency plasmas of C_3F_8 have been made by Haverlag and coworkers [194, 284] using laser photodetachment and subsequent detection of the photodetached electrons. Under their experimental conditions [13.56 MHz rf plasmas generated in a quasiparallel electrode system at pressures between 4 and 16 Pa (30–120 mTorr) and power densities up to 0.25 W cm^{-2}], they found the negative ion density to be more than a factor of 20 larger than the electron density. Such copious quantities of negative ions in the plasma may have origins other than the parent unexcited molecule. They most likely include fragment anions from electron attachment to radicals or from larger molecules formed by polymerization, or electron attachment to "hot" or electronically excited molecules or radicals for which dissociative electron attachment is normally significantly enhanced compared to the unexcited species [319]. Indeed, evidence for negative ion formation

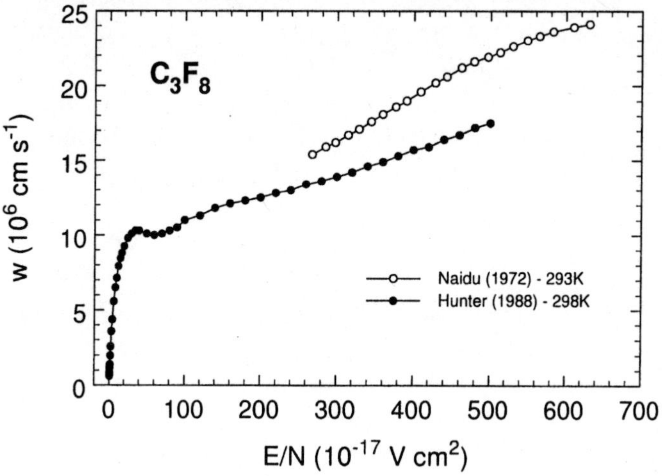

Figure 4.91. Electron drift velocity, $w(E/N)$, in pure C_3F_8: (•) Ref. [108] for T = 298 K; (○) Ref. [120] for T = 293 K.

enhancement via these processes in plasmas containing CF_4 and CHF_3 has been observed [320, 321].

In view of the increasing use of laser photodetachment to probe the negative ion concentrations in plasma reactors, measurements are needed of the photodetachment cross sections of fragment anions for this molecule and also for other perfluorocarbon molecules of interest to plasma processing such as CF_4 and C_2F_6. Especially useful will be measurements of the photodetachment cross sections for the anions F^-, CF_2^-, CF_3^-, $C_2F_5^-$, and $C_3F_7^-$. The electron affinity EA(F) of the F atom is well known. Although the values of EA(F) range from 2.81 to 4.1 eV [79], the values of (3.398 ± 0.002) eV [322] and (3.400 ± 0.002) eV [323] are the most accurate. The electron affinities of the other fragments are not well known. The values listed by Christodoulides *et al.* [79] vary considerably: 0.20–2.65 eV for CF_2, 1.36–2.60 eV for CF_3, 2.1–3.3 eV for C_2F_5, and 2.2–2.4 eV for C_3F_7.

4.7 Electron Transport in C_3F_8

4.7.1 Electron Drift Velocity, $w(E/N)$, for C_3F_8

There have been two measurements [108, 120] of the electron drift velocity, $w(E/N)$, in pure C_3F_8. These measurements are shown in Fig. 4.91. Naidu and Prasad [120] made their $w(E/N)$ measurements in the pressure range 0.08 to 0.267 kPa, and at E/N values (270×10^{-17} to

630×10^{-17} V cm^2 at $T = 293$ K) relatively larger than those of Hunter *et al.* [108] (0.4×10^{-17} to 500×10^{-17} V cm^2). Naidu and Prasad reported no effect of gas number density on $w(E/N)$ with an overall uncertainty in their measurements of less than ±5%. Hunter *et al.* [108] employed a pulsed Townsend method and pressures in the range 0.5–3.0 kPa. The estimated total uncertainty in their w values when electron attachment and ionization are negligible is ±2%, but it rises to a maximum of ±5% when either the ionization or the attachment coefficient is large due mainly to an increased uncertainty in determining the electron transit time from the break in the voltage waveform. The temperature of their experiment was 298 K.

Contrary to the conclusion of Naidu and Prasad that $w(E/N)$ is not a function of gas density, Hunter *et al.* [108] found that w depends on gas number density at high E/N even after allowing for nonequilibrium and boundary corrections to the measured electron swarm transit time. The largest pressure dependence of w occurs at E/N values near $(E/N)_{\text{lim}}$ ($\sim 290\times10^{-17}$ V cm^2 at a gas pressure of 0.05 kPa). The pressure dependence of w decreases at lower E/N such that w becomes independent of gas pressure at $E/N < 150\times10^{-17}$ V cm^2. Hunter *et al.* attributed these changes in w with the C_3F_8 pressure to the effect of electron attachment on the electron energy distribution function resulting from increases in the electron attachment coefficient with increasing gas density. The magnitude of the change in w with pressure correlates with the magnitude of the change in the electron attachment coefficient with gas density for this molecule.

Hunter *et al.* [108] attributed the differences between their measurements and those of Naidu and Prasad [120] to the experimental uncertainties in the latter measurements. They pointed out that the determination of w made by Naidu and Prasad was limited by the finite width of the light pulse they used and by high background ion currents. Since, compared to the data of Prasad and Naidu, the measurements of Hunter *et al.* have lower uncertainties, extend over a wider range of E/N values, and are corrected for gas density and other factors, they are preferred and are listed in Table 4.62 as our suggested $w(E/N)$ values for pure C_3F_8. Interestingly, the measurements show a region of negative differential conductivity (decrease in w with increasing E/N) which is less pronounced for C_3F_8 than for CF_4 (Sec. 2.6.1 of Ch. 4) and C_2F_6 (Sec. 3.6.1 of Ch. 4).

Measurements have also been made of the $w(E/N)$ in mixtures of C_3F_8 with various gases such as Ar [12, 13] and CH_4 [13]. These measurements were partially motivated by the development of fast gas mixtures for use in gas pulse-power switches. A sample of these data taken

Table 4.62. Suggested electron drift velocities in C_3F_8 ($T = 298$ K) (data of Hunter *et al.* [108]).

E/N (10^{-17} V cm^2)	w (10^6 cm s^{-1})	E/N (10^{-17} V cm^2)	w (10^6 cm s^{-1})
0.4	0.60	80	10.3
0.5	0.75	90	10.5
0.6	0.88	100	11.0
0.8	1.14	120	11.3
1.0	1.39	140	11.8
1.5	1.98	160	12.1
2.0	2.57	180	12.3
3.0	3.57	200	12.5
4.0	4.37	220	12.8
6.0	5.57	240	13.0
8.0	6.49	260	13.4
10.0	7.14	280	13.6
12.0	7.92	300	13.9
15.0	8.45	320	14.2
17.0	8.80	340	14.6
20.0	9.25	360	14.9
25.0	9.8	380	15.3
30.0	10.1	400	15.7
35.0	10.3	420	15.9
40.0	10.3	440	16.4
50.0	10.1	460	16.7
60.0	10.0	480	17.2
70.0	10.1	500	17.5

from Hunter *et al.* [13] is shown in Fig. 4.92. It is interesting to note the negative differential conductivity exhibited by these mixtures for certain E/N regions which depend on mixture composition. Recently, measurements of $w(E/N)$ in a few C_3F_8/Ar mixtures were made [303, 324] for use in multi-term Boltzmann analysis to determine electron collision cross section sets. The transport coefficients for the mixtures serve as a sensitive probe of the consistency of the calculated cross section sets.

4.7.2 Ratio of Transverse Electron Diffusion Coefficient to Electron Mobility, $D_T/\mu\,(E/N)$, for C_3F_8

Apparently the only measurements of $D_T/\mu\,(E/N)$ for C_3F_8 are those of Naidu and Prasad [120] shown in Fig. 4.93. These measurements were

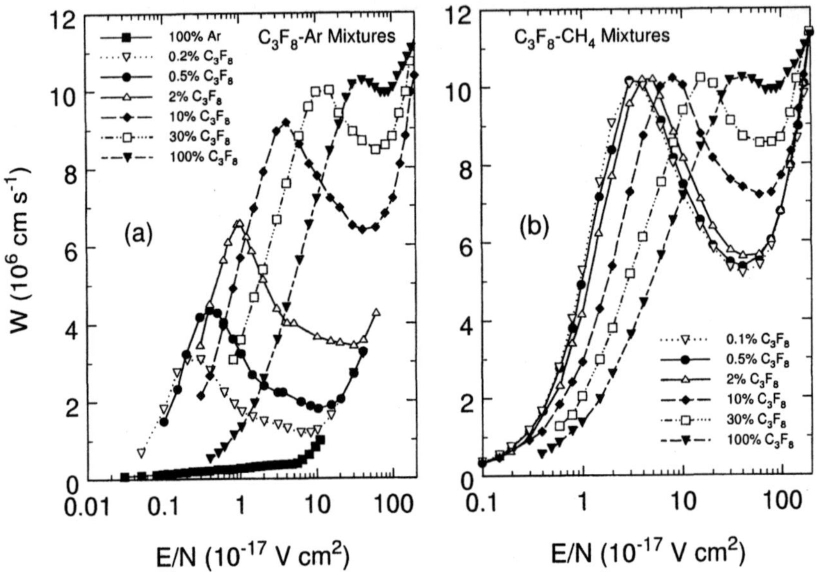

Figure 4.92. Electron drift velocity, $w(E/N)$, data of Hunter *et al.* [13] in (a) C_3F_8/Ar mixtures and (b) C_3F_8/CH_4 mixtures.

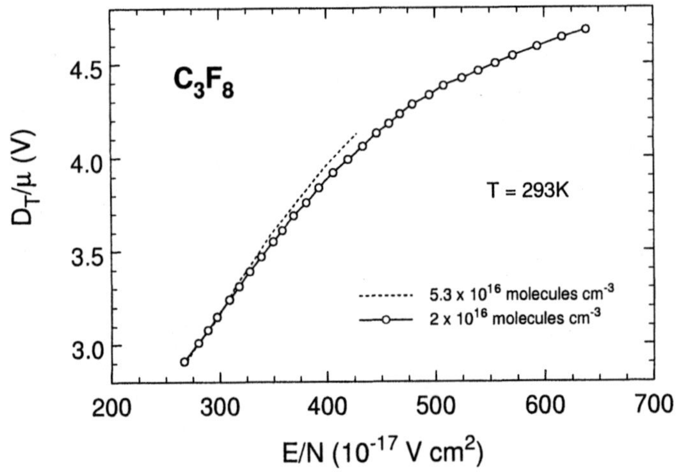

Figure 4.93. $D_T/\mu\,(E/N)$ for C_3F_8 at $T = 293$ K as obtained by Naidu and Prasad [120]: (- - -) $N = 5.3 \times 10^{16}$ molecules cm^{-3}; (○) $N = 2.0 \times 10^{16}$ molecules cm^{-3}.

made at 293 K and have an overall reported uncertainty of about ±5%. Interestingly, $D_T/\mu\,(E/N)$ was observed to exhibit a small dependence

Table 4.63. D_T/μ values for C_3F_8 ($T = 293$ K) of Naidu and Prasad [120] for a gas density of 2.0×10^{16} molecules cm^{-3}.

E/N (10^{-17} V cm^2)	D_T/μ (V)	E/N (10^{-17} V cm^2)	D_T/μ (V)
270	2.93	460	4.19
280	3.01	480	4.28
290	3.09	500	4.35
300	3.17	520	4.41
320	3.33	540	4.46
340	3.48	560	4.51
360	3.62	580	4.56
380	3.76	600	4.60
400	3.89	620	4.65
420	3.99	640	4.68
440	4.10		

on gas pressure. The measurements of Naidu and Prasad [120] for the gas density 2×10^{16} molecules cm^{-3} are listed in Table 4.63.

Recently, measurements have been made of the product $D_L N(E/N)$ for specific mixtures of C_3F_8 in Ar (0.526% [303] and 5.05% [303, 324]) for use in multi-term Boltzmann analysis to determine electron collision cross section sets. These transport coefficients along with $w(E/N)$ measurements for the same gas mixtures serve as a sensitive probe of the consistency of the calculated cross section sets.

4.8 Summary of Recommended and Suggested Electron Collision Cross Sections and Electron Transport Coefficients for C_3F_8

The cross sections that have been designated as recommended or suggested for C_3F_8 are plotted in Fig. 4.94. These include the:

- $\sigma_{sc,t}(\varepsilon)$ — Table 4.48, Fig. 4.68;
- $\sigma_{e,int}(\varepsilon)$ — Table 4.50, Fig. 4.70;
- $\sigma_m(\varepsilon)$ — Table 4.51, Fig. 4.71;
- $\sigma_{dis,t}(\varepsilon)$ — Table 4.54, Fig. 4.75;
- $\sigma_{i,t}(\varepsilon)$ — Table 4.53, Fig. 4.74;

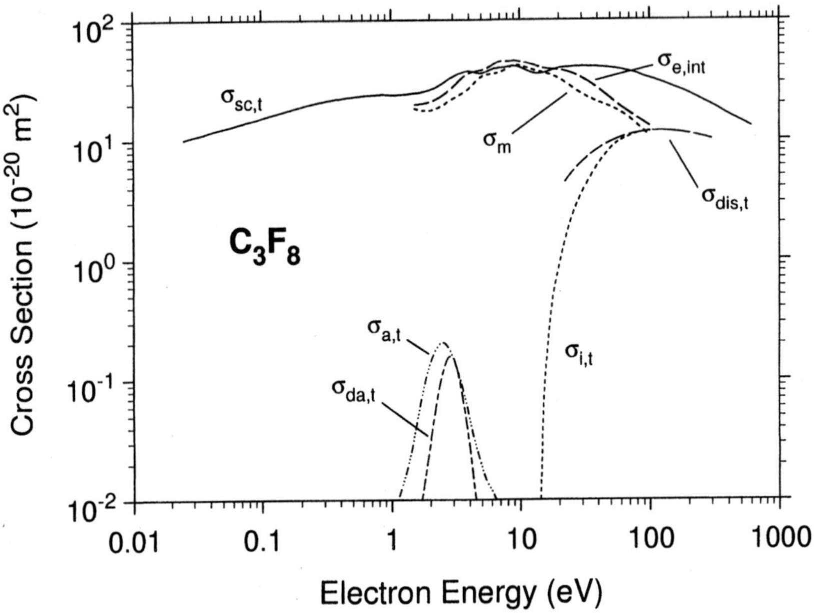

Figure 4.94. Summary of recommended and suggested cross sections for C_3F_8 (T = 298 K).

- $\sigma_{a,t}(\varepsilon)$ [T = 300 K] — Table 4.60, Fig. 4.86(a); and
- $\sigma_{da,t}(\varepsilon)$ [T = 300 K] — Table 4.60, Fig. 4.86(b).

The following coefficients are recommended or suggested:

- $\alpha/N\,(E/N)$ — Table 4.55, Fig. 4.76;
- $(\alpha - \eta)/N(E/N)$ — Table 4.56, Fig. 4.77;
- $\eta/N\,(E/N)$ — Fig. 4.81;
- $k_{a,t}(E/N)$ for pure C_3F_8 — Fig. 4.82;
- $k_{a,t}(\langle\varepsilon\rangle)$ — Table 4.58, Fig. 4.84;
- $w(E/N)$ — Table 4.62, Fig. 4.91; and
- $D_T/\mu\,(E/N)$ — Table 4.63, Fig. 4.93.

With the exception of the electron attachment cross section, there is a need for further measurements on all other cross sections, especially for the cross sections for momentum transfer, elastic integral, vibrational excitation, and dissociation into neutrals. Further, measurements of the electron transport coefficients are also indicated.

5. References

[1] L. G. Christophorou, J. K. Olthoff, and M. V. V. S. Rao, J. Phys. Chem. Ref. Data **25**, 1341 (1996).

[2] L. G. Christophorou and J. K. Olthoff, J. Phys. Chem. Ref. Data **27**, 1 (1998).

[3] L. G. Christophorou and J. K. Olthoff, J. Phys. Chem. Ref. Data **27**, 889 (1998).

[4] M. Hayashi, in *Swarm Studies and Inelastic Electron-Molecule Collisions*, L. C. Pitchford, B. V. McKoy, A. Chutjian, and S. Trajmar (Eds.) (Spinger, New York, 1987) p. 167.

[5] Y. Nakamura, in *Gaseous Electronics and Their Applications*, R. W. Crompton, M. Hayashi, D. E. Boyd, and T. Makabe (Eds.), KTK Scientific, Tokyo, Japan, 1991, p. 178.

[6] M.-C. Bordage, P. Ségur, and A. Chouki, J. Appl. Phys. **80**, 1325 (1996).

[7] M. J. Kushner, J. Appl. Phys. **53**, 2923 (1982).

[8] G. S. Oehrlein, Phys. Today **39**, 26 (1986).

[9] D. M. Manos and D. L. Flamm, *Plasma Etching*, Academic, Boston, 1989.

[10] L. E. Kline and M. J. Kushner, Crit. Rev. Solid State Mater. Sci. **16**, 1 (1989).

[11] D. Bose, D. Hash, T. R. Govindan, and M. Meyyappan, J. Phys. D **34**, 2742 (2001).

[12] L. G. Christophorou, S. R. Hunter, J. G. Carter, and R. A. Mathis, Appl. Phys. Lett. **41**, 147 (1982).

[13] S. R. Hunter, J. G. Carter, and L. G. Christophorou, J. Appl. Phys. **58**, 3001 (1985).

[14] P. Bletzinger, in *Proceedings of the 4th IEEE Pulsed Power Conference*, Albuquerque, New Mexico, June 1983, p. 37.

[15] R. E. Wootton, S. J. Dale and N. J. Zimmerman, in *Gaseous Dielectrics II*, L. G. Christophorou (Ed.), Pergamon, New York, 1980, p. 137.

[16] D. R. James, L. G. Christophorou, and R. A. Mathis, in *Gaseous Dielectrics II*, L. G. Christophorou (Ed.), Pergamom, New York, 1980, p. 115.

[17] J. Berg and E. Kuffel, in *Proceedings of the 9th International Symposium on High Voltage Engineering*, August 1995, Graz, Austria, p. 2252.

[18] L. G. Christophorou, D. L. McCorkle, D. V. Maxey, and J. G. Carter, Nucl. Instrum. Meth. **163**, 141 (1979)

[19] L. G. Christophorou, D. V. Maxey, D. L. McCorkle, and J. G. Carter, Nucl. Instrum. Meth. **171**, 491 (1980)

[20] M. K. Kopp, K. H. Valentine, L. G. Christophorou, and J. G. Carter, Nucl. Instrum. Meth. **201**, 395 (1982).

[21] Y. Yamashita, H. Kurashige, M. M. Morii, T. T. Nakamura, N. Nomura, N. Sasao, K. Shibata, Y. Fukushima, Y. Ikegami, H. Kobayashi, and T. Taniguchi, Nucl. Istrum. Meth. in Phys. Res. A **317**, 213 (1992).

[22] J. B. Gerardo and J. T. Verdeyen, *Low-Temperature Plasma Physics: Its Importance and Potential in Technology and Commerce*, Sandia National Laboratory Report No. SAND 87-1875 (1987).

[23] G. Ecker and K. U. Riemann, Exp. Tech. Phys. **35**, 119 (1987).

[24] K. H. Becker, in *Electron Collisions with Molecules, Clusters, and Surfaces*, H. Ehrhardt and L. A. Morgan (Eds.), Plenum, New York, 1994, p.127.

[25] H. Tanaka and M. Inokuti, Adv. At. Mol. Opt. Phys. **43**, 1 (2000).

[26] Intergovernmental Panel on Climate Change (IPCC), *The 1994 Report of the Scientific Assessment Working Group of IPCC.*

[27] V. Ramanathan, L. Callis, R. Cess, J. Hansen, I. Isaksen, W. Kuhn, A. Lacis, F. Luther, J. Mahlman, R. Reck, and M. Schlesinger, Rev. Geophys. **25**, 1441 (1987).

[28] R. S. Stolarski and R. D. Rundel, Geophys. Res. Lett. **2**, 443 (1975).

[29] R. A. Morris, T. M. Miller, A. A. Viggiano, J. F. Paulson, S. Solomon, and G. Reid, J. Geophys. Res. **100**, 1287 (1995).

[30] J. G. Owens, in *Gaseous Dielectrics IX*, L. G. Christophorou and J. K. Olthoff (Eds.), Kluwer Academic/Plenum Publishers, New York, 2001, p. 91.

[31] C. R. Brundle, M. B. Robin, and H. Basch, J. Chem. Phys. **53**, 2196 (1970).

[32] J. A. Beran and L. Kevan, J. Phys. Chem. **73**, 3860 (1969).

[33] J. W. Au, G. R. Burton, and C. E. Brion, Chem. Phys. **221**, 151 (1997).

[34] K. Kuroki, D. Spence, and M. A. Dillon, J. Chem. Phys. **96**, 6318 (1992).

[35] J. A. Stephens, D. Dill, and J. L. Dehmer, J. Chem. Phys. **84**, 3638 (1986).

[36] M. B. Robin, *Higher Excited States of Polyatomic Molecules*, Academic, New York, 1974, Vol. I, pp. 178-191.

[37] L. Boesten, H. Tanaka, A. Kobayashi, M. A. Dillon, and M. Kimura, J. Phys. B **25**, 1607 (1992).

[38] G. C. King and J. W. McConkey, J. Phys. B **11**, 1861 (1978).

[39] W. R. Harshbarger, M. B. Robin, and E. N. Lassettre, J. Electron Spectroscopy and Related Phenomena **1**, 319 (1972/73).

[40] T. A. Carlson, A. Fahlman, W. A. Svensson, M. O. Krause, T. A. Whitley, F. A. Grimm, M. N. Piancastelli, and J. W. Taylor, J. Chem. Phys. **81**, 3828 (1984).

[41] G. R. Cook and B. K. Ching, J. Chem. Phys. **43**, 1794 (1965).

[42] L. C. Lee, X. Wang, and M. Suto, J. Chem. Phys. **85**, 6294 (1986).

[43] G. J. Verhaart, W. J. van der Hart, and H. H. Brongersma, Chem. Phys. **34**, 161 (1978).

[44] I. Iga, M. C. A. Lopes, and J. U. Galdino, in *19th International Conference on the Physics of Electronic and Atomic Collisions Abstracts of Scientific Papers*, J. B. A. Mitchell, J. W. McCorkey, and C. E. Brion (Eds.), Whistler, British Columbia, Canada, July 1995, p. 435.

[45] L. C. Lee, E. Phillips, and D. L. Judge, J. Chem. Phys. **67**, 1237 (1977).

[46] R. E. Rebbert and P. Ausloos, J. Res. Natl. Bur. Stand. Sect. A **75**, 481 (1971).

[47] W. Zhang, G. Cooper, T. Ibuki, and C. E. Brion, Chem. Phys. **137**, 391 (1989).

[48] L. G. Christophorou, and J. K. Olthoff, J. Phys. Chem. Ref. Data **28**, 967 (1999).

[49] H. F. Winters, in *Swarm Studies and Inelastic Electron-Molecule Collisions*, L. C. Pitchford, B. V. McKoy, A. Chutjian, and S. Trajmar (Eds.), Springer, New York, 1987, p. 347.

[50] H. F. Winters and M. Inokuti, Phys. Rev. A **25**, 1420 (1982).

[51] K. Stephan, H. Deutsch, and T. D. Märk, J. Chem. Phys. **83**, 5712 (1985).

[52] B. Brehm, R. Frey, A. Küstler, and J. H. D. Eland, Int. J. Mass Spectrom. Ion Phys. **13**, 251 (1974).

[53] T. Fiegele, G. Hanel, I. Torres, M. Lezius, and T. D. Märk, J. Phys. B **33**, 4263 (2000).

[54] Y. J. Kime, D. C. Driscoll, and P. A. Dowben, J. Chem. Soc., Faraday Trans. II **83**, 403(1987).

[55] M. Schmidt, R. Seefeldt, and H. Deutsch, Int. J. Mass Spectrom. Ion Processes **93**, 141 (1989).

[56] Y. J. Kime and P. A. Dowben, J. Phys. Chem. **93**, 6881 (1989).

[57] G. Hagenow, W. Denzer, B. Brutschy, and H. Baumgärtel, J. Phys. Chem. **92**, 6487 (1988).

[58] L. M. Sverdlov, M. A. Kovner, and E. P. Krainov, *Vibrational Spectra of Polyatomic Molecules*, Wiley, New York, 1974.

[59] N. V. Mantzaris, A. Boudouris, and E. Gogolides, J. Appl. Phys. **77**, 6169 (1995).

[60] D. Field, J. P. Ziesel, P. M. Guyon, and T. R. Govers, J. Phys. B **17**, 4565 (1984).

[61] M. G. Curtis, and I. C. Walker, J. Chem. Soc. Faraday Trans. II **85**, 659 (1989).

[62] E. Illenberger, Chem. Phys. Lett. **80**, 153 (1981).

[63] T. Oster, A. Kühn, and E. Illenberger, Int. J. Mass Spectrom. Ion Processes **89**, 1 (1989).

[64] E. Illenberger, in *Linking the Gaseous and Condensed Phases of Matter*, L. G. Christophorou, E. Illenberger, and W. F. Schmidt (Eds.), Plenum, New York, 1994, p. 49.

[65] Y. Le Coat, J.-P. Ziesel, and J.-P. Guillotin, J. Phys. B **27**, 965 (1994).

[66] A. Mann and F. Linder, J. Phys. B **25**, 545 (1992).

[67] A. Mann and F. Linder, J. Phys. B **25**, 533 (1992).

[68] R. K. Jones, J. Chem. Phys. **84**, 813 (1986).

[69] C. Szmytkowski, A. M. Krzysztofowicz, P. Janicki, and L. Rosenthal, Chem. Phys. Lett. **199**, 191 (1992).

[70] S. M. Spyrou, I. Sauers, and L. G. Christophorou, J. Chem. Phys. **78**, 7200 (1983).

[71] H. P. W. Harland and J. L. Franklin, J. Chem. Phys. **61**, 1621 (1974).

[72] S. R. Hunter and L. G. Christophorou, J. Chem. Phys. **80**, 6150 (1984).

[73] R. Hashemi, A. Kühn, and E. Illenberger, Int. J. Mass Spectrom. Ion Processes **100**, 753 (1990).

[74] W. M. Huo, Phys. Rev. A **38**, 3303 (1988).

[75] J. A. Tossell and J. W. Davenport, J. Chem. Phys. **80**, 813 (1984); Erratum, J. Chem. Phys. **83**, 4824 (1985).

[76] M. T. N. Varella, C. Winstead, V. McKoy, M. Kitajima, and H. Tanaka, Phys. Rev. A **65**, 022702 (2002).

[77] H. J. T. Preston and J. J. Kaufman, Chem. Phys. Lett. **50**, 157 (1977).

[78] M. J. S. Dewar and H. S. Pzepa, J. Am. Chem. Soc. **100**, 784 (1978).

[79] A. A. Christodoulides, D. L. McCorkle, and L. G. Christophorou, in *Electron-Molecule Interactions and Their Applications*, L. G. Christophorou (Ed.), Academic, New York, 1984, Vol. 2, Chap. 6.

[80] J. Lotter, A. Kühn, and E. Illenberger, Chem. Phys. Lett. **157**, 171 (1989).

[81] J. Lotter and E. Illenberger, J. Phys. Chem. **94**, 8951 (1990).

[82] R. A. Bonham and M. R. Bruce, in *Proceedings of the Joint Symposium on Electron and Ion Swarms and Low-Energy Electron Scattering*, Gold Coast, Australia, July 1991, p. 5.

[83] T. Nakano and H. Sugai, Jpn. J. Appl. Phys. **31**, 2919 (1992).

[84] R. L. Asher and B. Ruscic, J. Chem. Phys. **106**, 210 (1997).

[85] A. Hansel, Ch. Scheiring, M. Glantschning, W. Lindinger, and E. E. Ferguson, J. Chem. Phys. **109**, 1748 (1998).

[86] M. Tichy, G. Javahery, N. D. Twiddy, and E. E. Ferguson, Int. J. Mass Spectrom. Ion Processes **79**, 231 (1987).

[87] E. R. Fisher and P. B. Armentrout, Int. J. Mass Spectrom. and Ion Processes **101**, R1 (1990).

[88] I. Powis, Mol. Phys. **39**, 311 (1980).

[89] G. K. Jarvis and R. P. Tuckett, Chem. Phys. Lett. **295**, 145 (1998).

[90] W. L. Morgan, JILA Data Center Report No. 34, June 1, 1991.

[91] W. L. Morgan, Plasma Chemistry and Plasma Processing **12**, 477 (1992).

[92] R. A. Bonham, Jpn. J. Appl. Phys. **33**, 4157 (1994).

[93] G. P. Karwasz, R. S. Brusa, and A. Zecca, Rivista Del Nuovo Cimento **24** (No. 4), 1 (2001).

[94] A. Zecca, G. P. Karwasz, and R. S. Brusa, Phys. Rev. A **46**, 3877 (1992).

[95] O. Sueoka, in *Atomic Physics with Positrons*, J. W. Humberston and E. A. G. Armour (Eds.), Plenum, New York, 1987, pp. 41-54.

[96] H.-X. Wan, Ph.D. dissertation, University of Maryland, 1990.

[97] O. Sueoka, S. Mori, and A. Hamada, J. Phys. B **27**, 1453 (1994).

[98] H. Nishimura, K. Okuda, and Y. Nakamura, in *Proceedings of the International Symposium on Electron-Molecule Collisions and Swarms*, I. Fabrikant, G. Gallup, and P. Burrow (Eds.), July 2001, Lincoln, NE, p. 105.

[99] K. L. Baluja, A. Jain, V. Di Martino, and F. A. Gianturco, Europhys. Lett. **17**, 139(1992).

[100] Y. Jiang, J. Sun, and L. Wan, Phys. Rev. A **52**, 398 (1995).

[101] S. Mori, Y. Katayama, and O. Sueoka, At. Coll. Res. Japan Progr. Rept. **11**, 19 (1985).

[102] O. Sueoka and S. Mori, J. Phys. Soc. Jpn. **53**, 2491 (1984).

[103] O. Sueoka and S. Mori, J. Phys. B **19**, 4035 (1986).

[104] O. Sueoka, H. Takaki, A. Hamada, H. Sato, and M. Kimura, Chem. Phys. Lett. **288**, 124 (1998).

[105] S. L. Lunt, J. Randell, J.-P. Ziesel, G. Mrotzek, and D. Field, J. Phys. B **31**, 4225 (1998).

[106] M. G. Curtis, I. C. Walker, and K. J. Mathieson, J. Phys. D **21**, 1271 (1988).

[107] B. Stefanov, N. Popkirova, and L. Zarkova, J. Phys. B **21**, 3989 (1988).

[108] S. R. Hunter, J. G. Carter, and L. G. Christophorou, Phys. Rev. A **38**, 58 (1988).

[109] K. Masek, L. Laska, R. D'Agostino, and F. Cramarossa, Contrib. Plasma Phys. **27**, 15 (1987).

[110] M.-C. Bordage, P. Ségur, L. G. Christophorou, and J. K. Olthoff, J. Appl. Phys. **86**, 3558 (1999).

[111] R. Tice and D. Kivelson, J. Chem. Phys. **46**, 4743 (1967).

[112] T. Sakae, S. Sumiyoshi, E. Murakami, Y. Matsumoto, K. Ishibashi, and A. Katase, J. Phys. B **22**, 1385 (1989).

[113] W. A. Isaacs, C. W. McCurdy, and T. N. Rescigno, Phys. Rev. A **58**, 309 (1998).

[114] A. P. P. Natalense, M. H. F. Bettega, L. G. Ferreira, and M. A. P. Lima, Phys. Rev. A **52**, R1(1995).

[115] C. Winstead, Q. Sun, and V. McKoy, J. Chem. Phys. **98**, 1105 (1993).

[116] F. A. Gianturco, R. R. Lucchese, and N. Sanna, J. Chem. Phys. **104**, 6482 (1996).

[117] F. A. Gianturco, R. R. Lucchese, and N. Sanna, J. Chem. Phys. **100**, 6464 (1994).

[118] D. Raj, J. Phys. B **24**, L431 (1991).

[119] R. Curik, F. A. Gianturco, and N. Sanna, J. Phys. B **33**, 615 (2000).

[120] M. S. Naidu and A. N. Prasad, J. Phys. D **5**, 983 (1972).

[121] C. S. Lakshminarasimha, J. Lucas, and D. A. Price, Proc. IEE **120**, 1044 (1973).

[122] L. G. Christophorou, P. G. Datskos, and J. G. Carter, Chem. Phys. Lett. **186**, 11 (1991).

[123] W. G. Golden, C. Marcott, and J. Overend , J. Chem. Phys. **68**, 2081 (1978).

[124] Y. Itikawa, J. Phys. Soc. Japan **36**, 1127 (1974).

[125] C. Ma, M. R. Bruce, and R. A. Bonham, Phys. Rev. A **44**, 2921 (1991); Erratum:Phys. Rev. A **45**, 6932 (1992).

[126] T. Masuoka and A. Kobayashi, J. Chem. Phys. **113**, 1559 (2000).

[127] M. R. Bruce, C. Ma, and R. A. Bonham, Chem. Phys. Lett. **190**, 285 (1992).

[128] M. R. Bruce, L. Mi, C. R. Sporleder, and R. A. Bonham, J. Phys. B **27**, 5773 (1994).

[129] K. Codling, L. J. Frasinski, P. A. Heatherly, M. Stankiewicz, and F. P. Larkins, J. Phys. B **24**, 951 (1991).

[130] H. U. Poll, C. Winkler, D. Margreiter, V. Grill, and T. D. Märk, Int. J. Mass Spectrom. Ion Processes, **112**, 1 (1992).

[131] T. D. Märk, in *Electron-Molecule Interactions and Their Applications*, L. G. Christophorou (Ed.), Academic, Orlando, Florida, 1984, Vol. 1, Ch. 3.

[132] M. R. Bruce and R. A. Bonham, Int. J. Mass Spectrom. Ion Processes **123**, 97 (1993).

[133] M. R. Bruce and R. A. Bonham, J. Molecular Structure **352/353**, 235 (1995).

[134] H. Nishimura, in *Proceedings of the 8th Symposium on Plasma Processing*, K. Tachibana (Ed.), Nagoya, Japan, 1991, pp. 333-336.

[135] M. V. V. S. Rao and S. K. Srivastava, Private communication, 1996.

[136] M. V. V. S. Rao and S. K. Srivastava, in *Proceedings of the Twentieth International Conference on the Physics of Electronic and Atomic Collisions*, Vienna, Austria, July 1997, Vol. II, p. MO150.

[137] H. Nishimura, W. M. Huo, M. A. Ali, and Y.-K. Kim, J. Chem Phys. **110**, 3811 (1999).

[138] D. R. Sieglaff, R. Rejoub, B. G. Lindsay, and R. F. Stebbings, J. Phys. B **34**, 799 (2001).

[139] I. Torres, R. Martinez, and F. Castaño, J. Phys. B **35**, 2423 (2002).

[140] D. Margreiter, H. Deutsch, M. Schmidt, and T. D. Märk, Int. J. Mass Spectrom. Ion Processes **100**, 157 (1990).

[141] M. Gryzinski, Phys. Rev A **138**, 305 (1965).

[142] H. Deutsch, K. Becker, and T. D. Märk, Int. J. Mass Spectrom. Ion Processes **167/168**, 503 (1997).

[143] H. Deutsch, K. Becker, S. Matt, and T. D. Märk, Int. J. Mass Spectrom. Ion Processes **197**, 37 (2000).

[144] Y.-K. Kim, W. Hwang, M. A. Ali, and M. E. Rudd, in *Proceedings Twentieth International Conference on the Physics of Electronic and Atomic Collisions*, Vienna, Austria, July 1997, Vol. II, P. WE103.

[145] W. M. Huo, Phys. Rev. A **64**, 042719 (2001).

[146] K. Leiter, K. Stephan, H. Deutsch, and T. D. Märk, in *Proceedings of the Symposium on Atomic and Surface Physics* (1984).

[147] W. M. Huo, C. E. Dateo, and G. D. Fletcher, in *22nd International Conference on Photonic, Electronic, and Atomic Collisions, Abstracts of Contributed Papers*, S. Datz, M. E. Bannister, H. F. Krause, L. H. Saddiq, D. R. Schultz, and C. R. Vane (Eds.), Santa Fe, NM, July 2001.

[148] S. E. Božin and C. C. Goodyear, J. Phys. D **1**, 327 (1968).

[149] I. M. Bortnik and A. A. Panov, Sov. Phys. Tech. Phys. **16**, 571 (1971).

[150] C. S. Lakshminarasimha, J. Lucas, and R. A. Snelson, Proc. Inst. Electr. Eng. **122**, 1162 (1975).

[151] M. Shimozuma, H. Tagashira, and H. Hasegawa, J. Phys. D **16**, 971 (1983).

[152] S. R. Hunter, J. G. Carter, and L. G. Christophorou, J. Chem. Phys. **86**, 693 (1987).

[153] J. W. Gallagher, E. C. Beaty, J. Dutton, and L. C. Pitchford, J. Phys. Chem. Ref. Data **12**, 109 (1983).

[154] P. G. Datskos, J. G. Carter, and L. G. Christophorou, J. Appl. Phys. **71**, 15 (1992).

[155] L. G. Christophorou, P. G. Datskos, and J. G. Carter, Nucl. Instrum. Methods in Phys. Research A **309**, 160 (1991).

[156] G. F. Reinking, L. G. Christophorou, and S. R. Hunter, J. Appl. Phys. **60**, 499 (1986).

[157] H. Sugai, H. Toyoda, T. Nakano, and M. Goto, Contrib. Plasma Phys. **35**, 415 (1995).

[158] V. Tarnovsky, P. Kurunczi, D. Rogozhnikov, and K. Becker, Int. J. Mass Spectrom. Ion Processes **128**, 181 (1993).

[159] V. Tarnovsky and K. Becker, J. Chem. Phys. **98**, 7868 (1993).

[160] W. M. Huo, V. Tarnovsky, K. H. Becker, Chem. Phys. Lett. **358**, 328 (2002).

[161] L. Mi and R. A. Bonham, J. Chem. Phys. **108**, 1910 (1998).

[162] S. Motlagh and J. H. Moore, J. Chem. Phys. **109**, 432 (1998).

[163] P. J. Van der Burgt and J. W. McConkey, J. Phys. B **24**, 4821 (1991).

[164] H. A. Van Sprang, H. H. Brongersma, and F. J. de Heer, Chem. Phys. **35**, 51 (1978).

[165] U. Müller, T. Bubel, G. Schulz, A. Sevilla, J. Dike, and K. Becker, Z. Phys. D **24**, 131 (1992).

[166] K. A. Blanks and K. Becker, J. Phys. B **20**, 6157 (1987).

[167] K. A. Blanks, A. E. Tabor, and K. Becker, J. Chem. Phys. **86**, 4871 (1987).

[168] I. Torres, R. Martínez, M. N. Sánchez Rayo, J. A. Fernández, and F. Castaño, J. Phys. B **32**, 5437 (1999).

[169] M. B. Roque, R. B. Siegel, K. E. Martus, V. Tarnovsky, and K. Becker, J. Chem. Phys. **94**, 341 (1991).

[170] S. Wang, J. L. Forand, and J. W. McConkey, Can. J. Phys. **67**, 699 (1989).

[171] I. R. Lambert, S. M. Mason, and R. P. Tuckett, J. Chem. Phys. **89** 2683 (1988).

[172] M. Suto, N. Washida, H. Akimoto, and M. Nakamura, J. Chem. Phys. **78**, 1019 (1983).

[173] L. C. Lee, J. L. Han, C. Ye, and M. Suto, J. Chem. Phys. **92**, 133 (1990).

[174] K. H. Becker, Comments At. Mol. Phys. **30**, 261 (1994).

[175] C. Suzuki and K. Kadota, Appl. Phys. Lett. **67**, 2569 (1995).

[176] C. Suzuki, K. Sasaki, and K. Kadota, J. Appl. Phys. **82**, 5321 (1997).

[177] J. P. Booth, G. Hancock, N. D. Perry, and M. J. Toogood, J. Appl. Phys. **66**, 5251 (1989).

[178] J. P. Booth, G. Cunge, P. Chabert, and N. Sadeghi, J. Appl. Phys. **85**, 3097 (1999).

[179] L. D. B. Kiss, J.-P. Nicolai, W. T. Conner, and H. H. Sawin, J. Appl. Phys. **71**, 3186 (1992).

[180] I. Iga, M. V. V. S. Rao, S. K. Srivastava, and J. C. Nogueira, Z. Phys. D **24**, 111 (1992).

[181] C. Lifshitz and R. Grajower, Int. J. Mass Spectrom. Ion Processes **10**, 25 (1972/1973).

[182] J. Dutton, A. Goodings, A. K. Lucas, and A. W. Williams, J. Phys. D **20**, 1322 (1987).

[183] K. A. G. MacNeil and J. C. J. Thynne, Int. J. Mass Spectrom. Ion Phys. **3**, 455 (1975).

[184] J. W. Hastie and J. L. Margrave, J. Phys. Chem. **73**, 1105 (1969).

[185] Y. Le Coat, N. M. Hedhili, R. Azria, M. Tronc, F. Weik, and E. Illenberger, Int. J. Mass Spectrom. and Ion Processes **164**, 231(1997).

[186] K. Nagesha and L. Sanche, J. Appl. Phys. **88**, 5211 (2000).

[187] A. D. Bass, J. Gamache, L. Parenteau, and L. Sanche, J. Phys. Chem. **99**, 6123 (1995).

[188] L. Mi, C. R. Sporleder, and R. A. Bonham, Chem. Phys. Lett. **251**, 252 (1996).

[189] K. Mitsuke, S. Suzuki, T. Imamura, and I. Koyano, J. Chem. Phys. **95**, 2398 (1991).

[190] F. J. Davis, R. N. Compton, and D. R. Nelson, J. Chem. Phys. **59**, 2324 (1973).

[191] R. W. Fesssenden and K.M. Bansal, J. Chem. Phys. **53**, 3468 (1970).

[192] R. Schumacher, H. -R. Sprünken, A. A. Christodoulides, and R. N. Schindler, J. Phys. Chem. **82**, 2248 (1978).

[193] J. L. Jauberteau, G. J. Meeusen, M. Haverlag, G. M. W. Kroesen, and F. J. de Hoog, J. Phys. D **24**, 261 (1991).

[194] A. Kono, M. Haverlag, G. M. W. Kroesen, and F. J. de Hoog, J. Appl. Phys. **70**, 2939 (1991).

[195] E. Gogolides, M. Stathakopoulos, and A. Boudouris, J. Phys. D **27**, 1878 (1994).

[196] N. Gee and G. R. Freeman, J. Chem. Phys. **95**, 102 (1991).

[197] J. Va'vra, P. Coyle, J. Kadyk, and J. Wise, Nucl. Instrum. Methods in Phys. Research A **324**, 113 (1993).

[198] O. Kiselev, O. Prokofiev, and A. Vorobyov, GEM Report No. TN-93-417, 1993

[199] B. Schmidt and Polenz, Nucl. Instr. Methods in Phys. Research A **273**, 488 (1988).

[200] R. A. Snelson, Ph. D. thesis, University of Liverpool, 1974.

[201] V. A. Lisovskiy and V. D. Yegorenkov, in *Proceedings of the International Symposium on Electron-Molecule Collisions and Swarms*, Y. Hatano, H. Tanaka, and N. Kouchi (Eds.), Tokyo, Japan, July 1999, p.156.

[202] S. R. Hunter and L. G. Christophorou, in *Electron-Molecule Interactions and Their Applications*, L. G. Christophorou (Ed.), Academic, Orlando, Florida,1984, Vol. 2, Ch. 3.

[203] A. V. Vasenkov, J. Appl. Phys. **88**, 626 (2000).

[204] Y. Hayashi and Y. Nakamura, in *Proceedings of the International Conference on Atomic and Molecular Data and Their Applications*, NIST Special Publication 926, W. L. Wiese and P. J. Mohr (Eds.), U. S. Department of Commerce, Gaithersburg, MD, 1998, p. 248.

[205] L. G. Christophorou, *Atomic and Molecular Radiation Physics*, Wiley-Interscience, New York, 1971, Ch. 4.

[206] L. G. Christophorou and J. K. Olthoff, Adv. Atom. Molec. Opt. Phys. **44**, 59 (2000).

[207] H. Deutsch, T. D. Märk, V. Tarnovsky, K. Becker, C. Cornelissen, L. Cespiva, and V. Bonacic-Koutecky, Int. J. Mass Spectrom. Ion Processes **137**, 77 (1994).

[208] V. McKoy, W. L. Morgan, P. D. Haaland, C. Winstead, W. K. Trail, and C. Jiao, Sematech Technology Transfer Report # 00023909A-TR, March 31, 2000.

[209] V. McKoy, W. L. Morgan, P. D. Haaland, C. Winstead, W. K. Trail, and C. Jiao, Sematech Technology Transfer Report # 00124057A-TR, January 15, 2001.

[210] Y.-K. Kim and K. K. Irikura, *Atomic and Molecular Data and Their Applications*, K. A. Berrington and K. L. Bell (Eds.), American Institute of Physics 1-56396-971-8, Conference Proceedings CP543, 2000, p. 220.

[211] W. M. Huo, V. Tarnovsky, and K. Becker, in *XXII International Conference on Photonic, Electronic, and Atomic Collisions, Abstracts of Contributed Papers*, S. Datz, M. E. Bannister, H. F. Krause, L. H. Saddiq, D. R. Schultz, and C. R. Vane (Eds.), Santa Fe, NM, July 2001, p. 263.

[212] R. C. Wetzel, F. A. Biaocchi, and R. S. Freund, Bull. Amer. Phys. Soc. **30**, 147 (1985).

[213] A. V. Vasenkov, in *Proceedings of the International Conference on Phenomena in Ionized Gases*, P. Pisarczyk, T. Pisarczyk, and J. Wolowski (Eds.), Contributed papers, Vol. IV, Warsaw, Poland, July 1999, p. 159.

[214] J. W. Coburn and E. Kay, J. Vac. Sci. Technol. **16**, 407 (1979).

[215] D. L. Flamm, in *Gaseous Dielectrics V*, L. G. Christophorou and D. W. Bouldin (Eds.), Pergamon, New York, 1987, p. 317.

[216] M. O. de Beeck, M. Goethals, and L. Van den hove, J. Electrochem. Soc. **139**, 2644 (1992).

[217] M. Haverlag, A. Kono, G. M. W. Kroesen, and F. J. de Hoog, Mater. Sci. Forum **140–142**, 235 (1993).

[218] H.-H. Park, K.-H. Kwon, J.-L. Lee, K.-S. Suh, O.-J. Kwon, K.-I. Cho, and S.-C. Park, J. Appl. Phys. **76**, 4596 (1994).

[219] P. Ho, J. E. Johannes, R. J. Buss, and E. Meeks, J. Vac. Sci. Technol. A **19**, 2344 (2001).

[220] S. R. Hunter, J. G. Carter, L. G. Christophorou, and V. K. Lakdawala, in *Gaseous Dielectrics IV*, L. G. Christophorou and M. O. Pace (Eds.), Pergamon, New York, 1984, p. 224.

[221] V. E. Scherrer, R. J. Commisso, R. F. Fernsler, and I. M. Vitkovitsky, in *Gaseous Dielectrics IV*, L. G. Christophorou and M. O. Pace (Eds.), Pergamon, New York, 1984, p. 238.

[222] R. J. Commisso, R. F. Fernsler, V. E. Scherrer, and I. M. Vitkovitsky, Appl. Phys. Lett. **47**, 1056 (1985).

[223] G. Schaefer, K. H. Schoenbach, M. Kristiansen, B. E. Strickland, R. A. Korzekwa, and G. Z. Hutcheson, Appl. Phys. Lett. **48**, 1776 (1986).

[224] G. Schaefer and K. H. Schoenbach, IEEE Trans. Plasma Science **PS-14**, 561 (1986).

[225] G. Schaefer, K. H. Schoenbach, R. A. Korzekwa, and M. Kristiansen, in *Gaseous Dielectrics V*, L. G. Christophorou and D. W. Bouldin (Eds.), Pergamon, New York, 1987, p. 374.

[226] S. R. Hunter, in *Gaseous Dielectrics V*, L. G. Christophorou and D. W. Bouldin (Eds.), Pergamon, New York, 1987, p. 363.

[227] L. G. Christophorou, D. L. McCorkle, and S. R. Hunter, in *Gaseous Dielectrics V*, L. G. Christophorou and D. W. Bouldin (Eds.), Pergamon, New York, 1987, p. 381.

[228] S. Arai, T. Watanabe, Y. Ishikawa, T. Oyama, O. Hayashi, and T. Ishii, Chem. Phys. Lett. **112**, 224 (1984).

[229] E. Cook, World Resources Institute, Washington, D.C., February (1995).

[230] A. R. Ravishankara, S. Solomon, A. A. Turnipseed, and R. F. Warren, Science **259**, 194 (1993).

[231] P. Maroulis, J. Langan, A. Johnson, R. Ridgeway, and H. Withers, Semiconductor International, November 1994, Pub. No. 325-9502.

[232] A. L. McClellan, *Tables of Experimental Dipole Moments*, W. H. Freeman and Company, San Francisco, CA, 1963, p. 51.

[233] V. McKoy, C. Winstead, W. L. Morgan, and P. D. Haaland, Sematech Technology Transfer Report # 98063515A-TR, June 30, 1998.

[234] P. Sauvageau, J. Doucet, R. Gilbert, and C. Sandorfy, J. Chem. Phys. **61**, 391 (1974).

[235] C. J. Noutary, J. Res. Nat. Bur. Stands., **72A**, 479 (1968).

[236] I. G. Simm, C. J. Danby, and J. H. D. Eland, J. Chem. Soc. Chem. Commun., 832 (1973).

[237] I. G. Simm, C. J. Danby, and J. H. D. Eland, Int. J. Mass Spectrom. Ion Phys. **14**, 285 (1974).

[238] I. Ishii, R. McLaren, A. P. Hitchcock, K. D. Jordan, Y. Choi, and M. B. Robin, Can. J. Chem. **66**, 2104 (1988).

[239] F. Weik, quoted in [240].

[240] F. Weik and E. Illenberger, J. Chem. Phys. **103**, 1406 (1995).

[241] T. Takagi, L. Boesten, H. Tanaka, and M. A. Dillon, J. Phys. B. **27**, 5389 (1994).

[242] J. Sanabia, G. D. Cooper, J. A. Tossell, and J. H. Moore, J. Chem. Phys. **108**, 389 (1998).

[243] C. Szmytkowski, P. Mozejko, G. Kasperski, and E. Ptasińska-Denga, J. Phys. B **33**, 15 (2000).

[244] E. Lindholm and J. Li, J. Phys. Chem. **92**, 1731 (1988).

[245] L. Lehmann, S. Matejcik, and E. Illenberger, Ber. Bunsenges. Phys. Chem. **101**, 287 (1997).

[246] T. Shimanouchi, *Tables of Molecular Vibrational Frequencies Consolidated Volume I*, NSRDS-NBS 39, June 1972, p. 95.

[247] M. Hayashi and A. Niwa, in *Gaseous Dielectrics V*, L. G. Christophorou and D. W. Bouldin (Eds.), Pergamon, New York, 1987, p. 27.

[248] H. Okumo and Y. Nakamura, in *Proceedings of the International Conference on Atomic and Molecular Data and Their Applications*, NIST Special Publication 926, W. L. Wiese and P. J. Mohr (Eds.), U. S. Department of Commerce, Gaithersburg, MD, 1998, p. 265.

[249] O. Sueoka, H. Takaki, A. Hamada, and M. Kimura, in *Twentieth International Conference on the Physics of Electronic and Atomic Collisions*, Vienna, Austria, July 1997; Abstracts of Contributed Papers, Vol. 1, F. Aumayr, G. Betz, and H. P. Winter (Eds.), Paper WEO57.

[250] V. McKoy, C. Winstead, and W. L. Morgan, Sematech Technology Transfer Report # 97043274A-TR, August 22, 1997.

[251] R. Merz and F. Linder, quoted in Ref. [2].

[252] P. Pirgov and B. Stefanov, J. Phys. B **23**, 2879 (1990).

[253] H. B. Milloy, R. W. Crompton, J. A. Rees, and A. J. Robertson, Aust. J. Phys. **30**, 61 (1977).

[254] R. A. Morris, C. J. Patrissi, D. J. Sardella, P. Davidovits, and D. L. McFadden, Chem. Phys. Lett. **102**, 41 (1983).

[255] H. T. Davis and L. D. Schmidt, Chem. Phys. Lett. **16**, 260 (1972).

[256] H. U. Poll and J. Meichsner, Contrib. Plasma Phys. **27**, 359 (1987).

[257] C. Q. Jiao, A. Garscadden, and P. D. Haaland, Chem. Phys. Lett. **310**, 52 (1999).

[258] R. Basner, M. Schmidt, E. Denisov, H. Deutsch, P. Lopata, V. Tarnovsky, and K. Becker, in *Proceedings of the XXIVth International Conference on Phenomena in Ionized Gases*, Warsaw, Poland, July 1999; Contributed Papers, P. Pisarczyk, T. Pisarczyk, and J. Wolowski (Eds.), Vol. IV, p. 17.

[259] M. M. Bibby and G. Carter, Trans. Faraday Soc. **59**, 2455 (1963).

[260] M. V. Kurepa, in *3rd Czechoslovakian Conference on Electronics and Vacuum Physics Transactions*, Prague, September 1965, p. 107.

[261] J. A. Beran and L. Kevan, J. Phys. Chem. **73**, 3866 (1969).

[262] D. Rapp and P. Englander-Golden, J. Chem. Phys. **43**, 1464 (1965).

[263] H. Deutsch, K. Becker, R. Basner, M. Schmidt, and T. D. Märk, J. Phys. Chem. A **102**, 8819 (1998).

[264] J. G. Carter, S. R. Hunter, and L. G. Christophorou, in *Gaseous Dielectrics V*, L. G. Christophorou and D. W. Bouldin (Eds.), Pergamon, New York, 1987, p. 47.

[265] W. W. Byszewski, M. J. Enright, and J. M. Proud, in *Gaseous Dielectrics IV*, L. G. Christophorou and M. O. Pace (Eds.), Pergamon, New York, 1984, p. 53.

[266] W. W. Byszewski, M. J. Enright, and J. M. Proud, in *Gaseous Dielectrics IV*, L. G. Christophorou and M. O. Pace (Eds.), Pergamon, New York, 1984, p. 255.

[267] L. G. Christophorou, R. A. Mathis, S. R. Hunter, and J. G. Carter, in *Gaseous Dielectrics V*, L. G. Christophorou and D. W. Bouldin (Eds.) Pergamon, New York, 1987, p. 88.

[268] L.G. Christophorou, R. A. Mathis, S. R. Hunter, and J. G. Carter, J. Appl. Phys. **63**, 52 (1988).

[269] W. A. Wilson, J. H. Simons, and T. J. Brice, J. Appl. Phys. **21**, 203 (1950).

[270] G. Biasiutti, in *Gaseous Dielectrics III*, L. G. Christophorou (Ed.), Pergamon, New York, 1982, p. 174.

[271] G. Camilli, T. W. Liao, and R. E. Plump, Trans. AIEE **74** (Pt I), 637 (1955).

[272] K. Nakanishi, L. G. Christophorou, J. G. Carter, and S. R. Hunter, J. Appl. Phys. **58**, 633 (1985).

[273] L. G. Christophorou, *Atomic and Molecular Radiation Physics*, Wiley-Interscience, New York, 1971, Ch. 2.

[274] S. S. Nagra and D. A. Armstrong, Can. J. Chem. **53**, 3305 (1975).

[275] S. C. Sun, V. Patel, E. A. Whittaker, B. Singh, and J. H. Thomas III, J. Vac. Sci. Technol. A **11**, 1193 (1993).

[276] K. Miyata, M. Hori, and T. Goto, J. Vac. Sci. Technol. A **14**, 2343 (1996).

[277] S. M. Spyrou and L. G. Christophorou, J. Chem. Phys. **82**, 2620 (1985).

[278] K. A. G. MacNeil and J. C. J. Thynne, Int. J. Mass Spectrom. and Ion Phys. **2**, 1 (1969).

[279] J. W. Coomber and E. Whittle, Trans. Faraday Soc. **63**, 1394 (1967).

[280] R. W. Crompton and R. L. Jory, Aust. J. Phys. **15**, 451 (1962).

[281] E. Kay, A. Dilks, and D. Seybold, J. Appl. Phys. **51**, 5678 (1980).

[282] C. Bubeck and E. Kay, J. Vac. Sci. Technol. **18**, 360 (1981).

[283] R. d'Agostino, F. Cramarossa, V. Colaprico, and R. d'Ettole, J. Appl. Phys. **54**, 1284 (1983).

[284] M. Haverlag, A. Kono, D. Passchier, G. M. W. Kroesen, W. J. Goedheer, and F. J. de Hoog, J. Appl. Phys. **70**, 3472 (1991).

[285] Y. Haque and B. D. Ratner, J. Polym. Sci.: Part B: Polym. Phys. **26**, 1237 (1988).

[286] L. Zazzera, W. Reagen, and A. Cheng, J. Electrochem. Soc. **144**, 3597 (1997).

[287] J. T. Houghton, L. G. Meira Filho, B. A. Callander, N. Harris, A. Kattenberg, and K. Maskell, *Climate Change 1995, The Science of Climate Change*, Cambridge University Press, New York, 1996, p. 22.

[288] C. M. Roehl, D. Boglu, C. Brühl, and G. K. Moortgat, Geophys. Res. Lett. **22**, 815 (1995).

[289] P. Bletzinger, in *XVI International Conference on Phenomena in Ionized Gases*, Contributed Papers, Vol. 2, W. Bötticher, H. Wenk, and E. Schulz-Gulde (Eds.), Duesseldorf, Germany, August 29-September 2, 1983, p. 218.

[290] L. G. Christophorou and S. R. Hunter, in *Electron-Molecule Interactions and Their Applications*, L. G. Christophorou (Ed.) Academic, New York, 1984, Vol. 2, Ch. 5.

[291] L. G. Christophorou, Nucl. Instr. Meth. Phys. Res. A **268**, 424 (1988).

[292] M. J. S. Dewar and S. D. Worley, J. Chem. Phys. **50**, 654 (1969).

[293] P. J. Chantry and C. L. Chen, J. Chem. Phys. **90**, 2585 (1989).

[294] C. Lifshitz and R. Grajower, Int. J. Mass Spectrom. Ion Phys. **3**, 211 (1969).

[295] C. Lifshitz and F. A. Long, J. Phys. Chem. **69**, 3746 (1965).

[296] G. Bélanger, P. Sauvageau, and C. Sandorfy, Chem. Phys. Lett. **3**, 649 (1969).

[297] T. Shinohara, Y. Tachibana, M. Yuri, H. Tanaka, and L. Boesten, in *Proceedings of the International Conference on the Physics of Electronic and Atomic Collisions*, Whistler, British Columbia, Canada, 26 July – 1 August 1995; Abstracts of Contributed Papers, J. B. A. Mitchell, J. W. McConkey, and C. E. Brion (Eds.), p. 13.

[298] H. Tanaka, Private communication, 1997.

[299] C. Lifshitz and R. Grajower, Int. J. Mass Spectrom. Ion Phys. **4**, 92 (1970).

[300] S. M. Spyrou and L. G. Christophorou, J. Chem. Phys. **83**, 2829 (1985).

[301] L. G. Christophorou, *Atomic and Molecular Radiation Physics*, Wiley-Interscience, New York, 1971, Ch. 5.

[302] C. D. Jiao, A. Garscadden, and P. D. Haaland, Chem. Phys. Lett. **325**, 203 (2000).

[303] B.-H. Jeon and Y. Nakamura, in *International Conference on Atomic and Molecular Data and Their Applications*, 29 September – 2 October 1997, National Institute of Standards and Technology, Gaithersburg, MD, USA. Paper MO35; Book of Abstracts, p. 45; Private communication, October 1997.

[304] H. Tanaka, Y. Tachibana, M. Kitajima, O. Sueoka, H. Takaki, A. Hamada, and M. Kimura, Phys. Rev. A **59**, 2006 (1999).

[305] M. E. Riley, C. J. MacCallum, and F. Biggs, At. Data Nucl. Data Tables **15**, 443 (1975); Erratum 28, 379 (1983).

[306] R. Hermann, Radiat. Phys. Chem. **34**, 369 (1989).

[307] N. P. Danilevskii, I. Yu. Rapp, V. T. Koppe, and A. G. Koval, Opt. Spectrosc. (USSR) **60**, 441 (1986).

[308] M.-M. Chen and Y. H. Lee, Proc. Electrochem. Soc. **83**, 3 (1983).

[309] J. L. Moruzzi and J. D. Craggs, Proc. Phys. Soc. **82**, 979 (1963).

[310] K. Nakanishi, D. R. James, H. Rodrigo, and L. G. Christophorou, J. Phys. D **17**, L73 (1984).

[311] S. R. Hunter, L. G. Christophorou, D. R. James, and R. A. Mathis, in *Gaseous Dielectrics III*, L. G. Christophorou (Ed.), Pergamon, New York, 1982, p. 7.

[312] S. R. Hunter and L. G. Christophorou, J. Appl. Phys. **57**, 4377 (1985).

[313] S. R. Hunter and L. G. Christophorou, in *Gaseous Dielectrics IV*, Pergamon, New York, L. G. Christophorou and M. O. Pace (Eds.), 1984, p. 115.

[314] W. C. Wang and L. C. Lee, J. Appl. Phys. **58**, 184 (1985).

[315] D. Rapp and D. D. Briglia, J. Chem. Phys. **43**, 1480 (1965).

[316] P. W. Harland and J. C. J. Thynne, Int. J. Mass Spectrom. Ion Phys. **9**, 253 (1972).

[317] L. G. Christophorou, L. A. Pinnaduwage, and P. G. Datskos, in *Linking the Gaseous and Condensed Phases of Matter*, L. G. Christophorou, E. Illenberger and W. Schmidt (Eds.), Plenum, New York, 1994, p. 415.

[318] L. G. Christophorou and P. G. Datskos, Int. J. Mass Spectrom. and Ion Processes **149/150**, 59 (1995).

[319] L. G. Christophorou and J. K. Olthoff, Adv. At. Mol. Opt. Phys. **44**, 155 (2000).

[320] W. W. Stoffels, E. Stoffels, and K. Tachibana, Jpn. J. Appl. Phys., Part 1 **36**, 4638 (1997).

[321] E. Stoffels, W. W. Stoffels, D. Vender, M. Haverlag, G. M. W. Kroesen, and F. J. de Hoog, Contrib. Plasma Phys. **35**, 331 (1995).

[322] R. Milstein and R. S. Berry, J. Chem. Phys. **55**, 4146 (1971).

[323] H.-P. Popp, Z. Naturforsch. A **20**, 642 (1965).

[324] B.-H. Jeon, S.-C. Ha, and Y. Nakamura, in *Proceedings of the XXIII International Conference on Phenomena in Ionized Gases*, Contributed Papers, M. C. Bordage and A. Gleizes (Eds.), Toulouse, France, July 1977, Vol. 5, p. V-4.

Chapter 5

ELECTRON INTERACTIONS WITH CHF_3, CF_3I, AND c-C_4F_8

1. Introduction

In this chapter we assess and synthesize the available information on the cross sections and the rate coefficients for collisional interactions of trifluoromethane (CHF_3) [1], trifluoroiodomethane (CF_3I) [2], and perfluorocyclobutane (c-C_4F_8) [3] with low-energy electrons. Definitions of the symbols we use to describe the various collision processes for these three gases are as discussed in Ch. 1 (Table 1.3). All three gases are used in manufacturing semiconductor devices and have a number of other important applications.

2. Electron Interactions with CHF_3

Trifluoromethane or fluoroform (CHF_3) is a plasma processing gas. As a fluorohydrocarbon, it does not contain atomic chlorine and is, thus, not destructive of stratospheric ozone. Trifluoromethane is, however, a greenhouse gas. It has a global warming potential over a 100-year time period of 12 100 relative to the global warming potential of CO_2 over the same time period [4]. The global warming potential of CHF_3 is about twice that of CF_4, which is also used in plasma processing (Ch. 4). However, the lifetime of CHF_3 in the atmosphere is much shorter (250 years compared to the 50 000-year lifetime of CF_4 [4]). This makes CHF_3 a desirable substitute for CF_4 in some industrial applications.

In this section, we assess and synthesize the available information on the cross sections and the rate coefficients for collisional interactions of electrons with CHF_3. This information is based on our two earlier papers on this subject [1, 5] and on work that has appeared since.

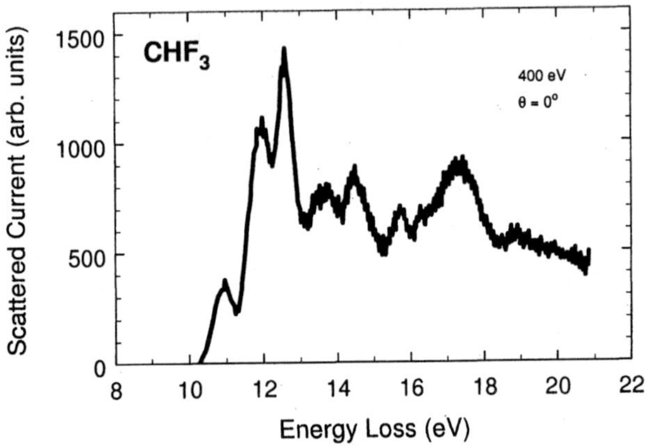

Figure 5.1. Electron-impact energy loss spectrum of CHF_3 obtained by Harshbarger *et al.* [19] with 400 eV incident energy electrons scattered at an angle $\theta = 0°$.

The cross sections discussed in this section for CHF_3 have been used by various authors as inputs in Boltzmann code analyses in efforts to obtain cross section sets consistent with the recent electron transport data which are also discussed in this section. The reader is referred to Refs. [6–8] for details.

2.1 Electronic and Molecular Structure of CHF_3

The structure of the CHF_3 molecule is a symmetric top of the C_{3v} point group. The outer electronic orbital structure [9–12] of CHF_3 is $\ldots(4a_1)^2\,(5a_1)^2\,(3e)^4\,(4e)^4\,(5e)^4\,(1a_2)^2\,(6a_1)^2$.

In contrast to CF_4 which is nonpolar, the CHF_3 molecule is polar. Its electric dipole moment has been reported to be 5.504×10^{-30} C m (= 1.65 D) [13–15]. As a consequence of its dipole moment, CHF_3 has a strong rotational spectrum (e.g., see Cazzoli *et al.* [16]). The CHF_3 molecule also has a sizeable static polarizability. Beran and Kevan [17] calculated the static polarizability of CHF_3 using three different methods and gave the values 26.9×10^{-25}, 27.8×10^{-25}, and 35.5×10^{-25} cm^3. Sutter and Cole [15] obtained the value 35.7×10^{-25} cm^3 from dielectric constant measurements. For information on frequency-dependent polarizabilities of CHF_3, see Kobayashi *et al.* [18].

There have been a number of electron-impact energy loss, photoelectron, and photoabsorption studies of CHF_3, and we only refer to representative data here. Figure 5.1 shows the electron-impact energy loss spectrum [19] of CHF_3 up to 21 eV. The spectrum was obtained us-

Table 5.1. Vertical excitation and ionization energies of CHF_3 with assignments. The data are from Harshbarger *et al.* [19], unless otherwise noted.

Excitation energy (eV)	Ionization energy (eV)	Assignment
10.92; 11.1[a]		$6a_1 \rightarrow 3s$
11.95; 11.9[a]		$6a_1 \rightarrow 3p$
12.58; 12.7[a]		$5e \rightarrow 3s$
13.65; 13.7[a]		$5e \rightarrow 3p$
		$4e \rightarrow 3s$
14.49		$4e \rightarrow 3p$
	14.8	$6a_1$
	15.5	$1a_2$
15.76		$4e \rightarrow 3d$
	16.2	$5e$
16.42		$3e \rightarrow 3s$
		$5a_1 \rightarrow 3s$
	17.24	$4e$
17.4		$3e \rightarrow 3p$
		$5a_1 \rightarrow 3p$
19.14		$3e \rightarrow 3d$
		$5a_1 \rightarrow 3d$
	20.84	$3e$
		$5a_1$

[a]Data of Ying and Leung [12].

ing electrons with 400 eV incident energy scattered at an angle $\theta = 0°$. Table 5.1 lists the measured vertical excitation and ionization energies of CHF_3 with assignments as given by Harshbarger *et al.* [19]. Table 5.1 also lists similar data derived from measurements of the generalized oscillator strength of valence electronic transitions of CHF_3 as a function of energy loss (0 to 15 eV) using angle-resolved electron energy-loss spectroscopy with 2.5 keV incident energy electrons [12]. A more recent *ab initio* configuration interaction calculation [20] confirmed the Rydberg nature of the electronic transitions that constitute the absorption spectrum of CHF_3, but maintained that "most of the transitions correspond to relatively unlocalized charge transfers and that the distinction between types *s*, *p*, or *d* Rydberg states is difficult to establish."

The vacuum ultraviolet (VUV) absorption spectrum of CHF_3 has been reported by Sauvageau *et al.* [21] in the wavelength range 60–120 nm, and by Wu *et al.* [10] in the wavelength range 18–72 nm using synchrotron radiation. As an example of this type of measurements, we reproduce in Fig. 5.2 the data of Wu *et al.* [10]. Adiabatic and vertical

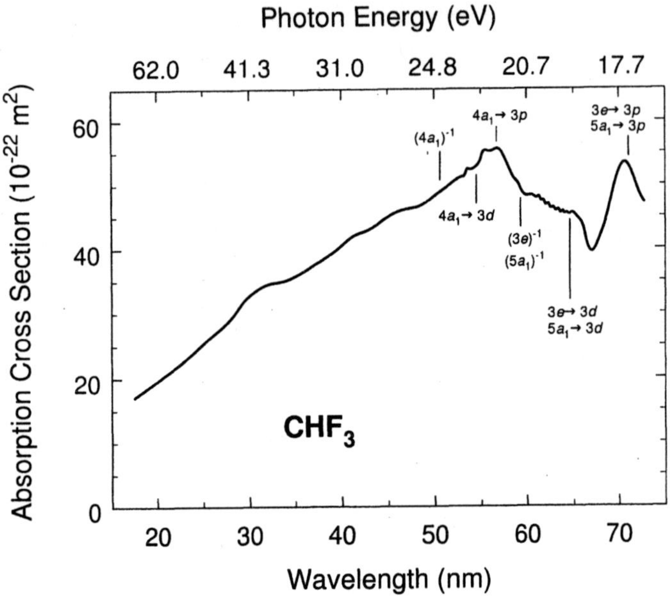

Figure 5.2. Absorption cross section of CHF_3 in the wavelength range 17.5–76 nm of Wu *et al.* [10]. Indicated in the figure are the positions of the probable Rydberg states and their ion states.

ionization energies [9, 22–24] of CHF_3 determined from photoelectron spectra are given in Table 5.2 (see also a relevant review by Bieri *et al.* [25]). Interestingly, in optical studies, vibrational fine structure has been observed for some excited electronic states [10, 21, 22], showing that "stable" excited electronic states exist for this molecule.

The energies of the six fundamental vibrational modes ν_1, ν_2, ν_3, ν_4, ν_5, and ν_6 of CHF_3 are 0.3763, 0.1415, 0.0864, 0.1708, 0.1435, and 0.063 eV, respectively [26–28].

2.2 Electron Scattering from CHF_3

2.2.1 Total Electron Scattering Cross Section, $\sigma_{sc,t}(\varepsilon)$, of CHF_3

There have been four recent measurements of the total electron scattering cross section, $\sigma_{sc,t}(\varepsilon)$, of CHF_3 which are compared in Fig. 5.3. These are the measurements of Sanabia *et al.* [29] between $\sim$ 0.1 and 20 eV, the measurements of Sueoka *et al.* [30] which extend from $\sim$ 0.8 to 500 eV, the measurements of Iga *et al.* [31] in the range 300–1000 eV,

Table 5.2. Adiabatic and vertical ionization energies of CHF_3 determined from the photoelectron spectra of Brundle *et al.* [9] unless otherwise indicated.

Orbital	Adiabatic ionization energy (eV)	Vertical ionization energy (eV)
$6a_1$	≥ 13.8; 13.84;[a] 14.19;[b] 13.86[c]	14.8; 14.77[c]
$1a_2$		15.5; 15.46[c]
$5e$		16.2; 16.16[c]
$4e$	17.11; 17.13[c]	17.24; 17.26[c]
$3e$		
	20.6;[d] 20.05[c]	19.84;[d] ~ 21[c]
$5a_1$		
$4a_1$	≥ 24.34	24.44

[a]Ref. [22].
[b]Ref. [23].
[c]Ref. [24].
[d]Unresolved blend of two or more transitions.

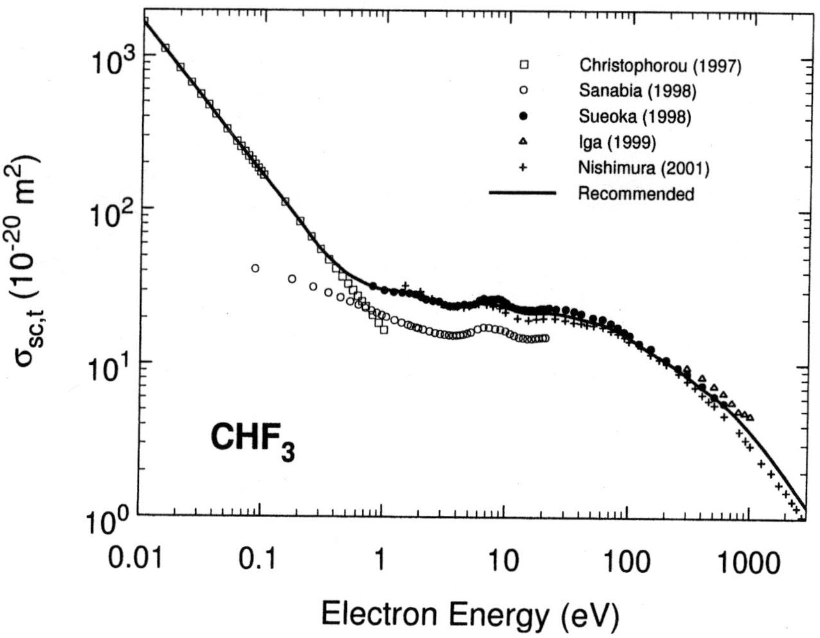

Figure 5.3. Total electron scattering cross section, $\sigma_{sc,t}(\varepsilon)$, for CHF_3: (○) Ref. [29]; (●) Ref. [30]; (△) Ref. [31]; (+) Ref. [32]; (□) calculations from Ref [1]; (—) recommended.

and the measurements of Nishimura *et al.* [32] which extend the data from about 1 to 3000 eV. The uncertainty of these data is difficult to establish, but it is likely to exceed ±10%. The use of a linear transmission apparatus in these measurements may affect the data at low energies due to forward scattering. The data of Nishimura *et al.* [32] have been corrected for forward scattering. The data of the three measurements by Sueoka *et al.* [30], Iga *et al.* [31], and Nishimura *et al.* [32] are all in good agreement, while the values of Sanabia *et al.* [29] fall below.

Because the CHF_3 molecule has a sizeable electric dipole moment, estimates can be made of the momentum transfer and the total electron scattering cross sections at low electron energies (thermal and near thermal) based upon the Born approximation theory of Altshuler [33]. In this approximation the momentum transfer cross section $\sigma_{\mathrm{m}}(v)$ varies with the incident electron velocity as

$$\sigma_{\mathrm{m}}(v) = 1.72D^2/v^2 \tag{5.1}$$

where the dipole moment, D, is in Debye units (1 D $= 3.3356 \times 10^{-30}$ C m) and the electron velocity, v, is in cm s^{-1}. In addition, Christophorou and Christodoulides [34] have found that for a large number of polar molecules they studied with electric dipole moments in the range $2 \times 10^{-30}\,\mathrm{C\,m} < D < 13.68 \times 10^{-30}$ C m, the experimental cross sections for electrons having a Maxwellian energy distribution lie within 50% of those calculated via Eq. (5.1).

Christophorou *et al.* [1] employed Eq. (5.1) and obtained the momentum transfer cross section (see Sec. 2.2.4 of Ch. 5) using 5.504×10^{-30} C m (= 1.65 D) for the D of CHF_3 [13–15]. They also followed Christophorou and Christodoulides [34, 35] and approximated the total electron scattering cross section for thermal and near-thermal energy electrons using

$$\sigma_{\mathrm{sc,t}}(v) \approx A_{\mathrm{I}}/v^2, \tag{5.2}$$

where A_{I} is a constant equal to $7.7574 \times 10^5 S^{-1}$ and $S = w/(E/P)$. The slope, S, is determined in the low E/P region where the electron drift velocity, w, varies linearly with E/P (E is the applied uniform electric field and P is the total gas pressure at the temperature of the experiment). Using for S the value 1.773×10^7 cm^2 Pa V^{-1} s^{-1} ($= 13.3 \times 10^4$ cm^2 Torr V^{-1} s^{-1}) reported by Christophorou *et al.* [36] for $E/N < 3.69 \times 10^{-17}$ V cm^2 ($E/P < 1.2$ V cm^{-1} Torr^{-1}), they estimated the electron scattering cross section, $\sigma_{\mathrm{sc,t}}(\varepsilon)$, at low electron energies shown by the open squares in Fig. 5.3. In this low-energy region (< 1 eV), the calculated values of $\sigma_{\mathrm{sc,t}}(\varepsilon)$ can be useful as an estimate of this quantity in the absence of accurate direct measurements. They are, however, approximate and are expected to fall below the true values of $\sigma_{\mathrm{sc,t}}(\varepsilon)$,

Table 5.3. Recommended total electron scattering cross section, $\sigma_{sc,t}(\varepsilon)$, for CHF_3.

ε (eV)	$\sigma_{sc,t}(\varepsilon)$ (10^{-20} m^2)	ε (eV)	$\sigma_{sc,t}(\varepsilon)$ (10^{-20} m^2)	ε (eV)	$\sigma_{sc,t}(\varepsilon)$ (10^{-20} m^2)
0.005	3321.0	1.5	29.3	40	19.8
0.01	1661.0	2.0	27.9	50	19.0
0.02	830.0	2.5	26.1	60	18.2
0.05	332.0	3.0	24.9	70	17.3
0.10	166.0	4.0	24.1	80	16.5
0.15	111.0	5.0	24.5	90	15.6
0.2	83.0	6.0	24.9	100	14.8
0.3	55.5	7.0	25.0	150	12.0
0.4	44.4	8.0	24.7	200	10.4
0.5	38.2	9.0	24.1	300	8.38
0.6	35.3	10.0	23.5	400	7.03
0.7	33.3	15.0	21.6	500	6.18
0.8	31.9	20.0	21.7	1000	3.65
0.9	31.0	25.0	21.5	2000	1.84
1.0	30.6	30.0	21.0	3000	1.19

progressively by larger amounts as the electron energy approaches 1 eV, as is seen in Fig. 5.3. Similarly, in the case of H_2O ($D = 6.17 \times 10^{-30}$ C m) and NH_3 ($D = 4.87 \times 10^{-30}$ C m), the $\sigma_{sc,t}(\varepsilon)$ determined via Eq. (5.2) is comparable with the experimental measurements below approximately 2 eV, but begins to fall below the experimental values progressively by larger amounts as the electron energy is increased above $\sim$ 2 eV [36, 37].

We have determined recommended values for the $\sigma_{sc,t}(\varepsilon)$ of CHF_3 which are represented in Fig. 5.3 by the solid line, and are listed in Table 5.3. These values were derived by fitting the three sets of experimental values [30–32] that are in agreement between 1 and 600 eV. This curve then is merged with the estimated values below 1 eV. Above 600 eV, the data of Nishimura *et al.* [32] are used after being normalized to the average value at 600 eV.

2.2.2 Elastic Differential Electron Scattering Cross Section, $\sigma_{e,diff}(\varepsilon)$, of CHF_3

The elastic differential electron scattering cross section, $\sigma_{e,diff}(\varepsilon)$, of CHF_3 has been measured by Tanaka *et al.* [38, 39] for incident electron energies 1.5, 2, 3, 5, 6.5, 7, 8, 9, 10, 15, 20, 30, 60, and 100 eV and scattering angles between 15° and 135°. These data are reproduced

Table 5.4. Elastic differential electron scattering cross section, $\sigma_{\mathrm{e,diff}}(\varepsilon)$, measurements of Tanaka and coworkers [38, 39] as presented in Ref. [40] for CHF_3 in units of 10^{-20} m^2 sr^{-1}.

Angle	1.5 eV	2 eV	3 eV	5 eV	6.5 eV	7 eV
20°	9.69	7.50	5.43	6.72	6.44	6.48
30°	6.09	4.81	3.84	4.08	4.66	4.50
40°	4.03	3.27	3.15	3.16	3.31	3.16
50°	3.24	2.41	2.29	2.58	2.44	2.17
60°	2.43	2.12	2.05	2.06	1.78	1.55
70°	1.94	1.83	1.56	1.45	1.24	1.00
80°	1.68	1.58	1.35	1.14	0.884	0.681
90°	1.66	1.42	1.18	0.828	0.577	0.567
100°	1.28	1.21	1.08	0.673	0.488	0.471
110°	1.19	1.10	0.924	0.556	0.527	0.531
120°	1.09	1.01	0.860	0.541	0.544	0.605
130°	1.05	0.964	0.744	0.630	0.772	0.737

Angle	8 eV	9 eV	10 eV	15 eV	20 eV	30 eV
15°				8.46	9.50	12.1
20°	6.51	6.59	7.31	6.29	7.21	8.68
30°	4.13	4.77	4.72	4.28	3.80	3.47
40°	3.20	3.20	3.23	2.68	2.09	1.15
50°	2.28	2.25	2.13	1.62	1.12	0.795
60°	1.34	1.43	1.37	1.03	0.782	0.749
70°	0.901	0.899	0.856	0.753	0.855	0.768
80°	0.603	0.630	0.659	0.800	0.836	0.578
90°	0.536	0.633	0.638	0.754	0.732	0.321
100°	0.530	0.581	0.681	0.763	0.567	0.222
110°	0.524	0.615	0.701	0.791	0.530	0.287
120°	0.624	0.662	0.751	0.751	0.585	0.453
130°	0.756	0.821	0.850	0.863	0.762	0.631

in Fig. 5.4. The experimental errors in these measurements have been reported to be between ±15% and ±30% [38, 39]. The numerical values of these measurements for electron energies from 1.5 to 30 eV have been tabulated in Ref. [40], and are reproduced in Table 5.4.

There have also been a number of recent calculations [8, 40–42] of the $\sigma_{\mathrm{e,diff}}(\varepsilon)$ of CHF_3 using the Schwinger multichannel method for various values of the incident electron energy. The results of these calculations are compared with the experimental data in Fig. 5.4. The computed values by Diniz *et al.* [41] using the Schwinger multichannel method at the static-exchange level (broken lines in Fig. 5.4) are in general agree-

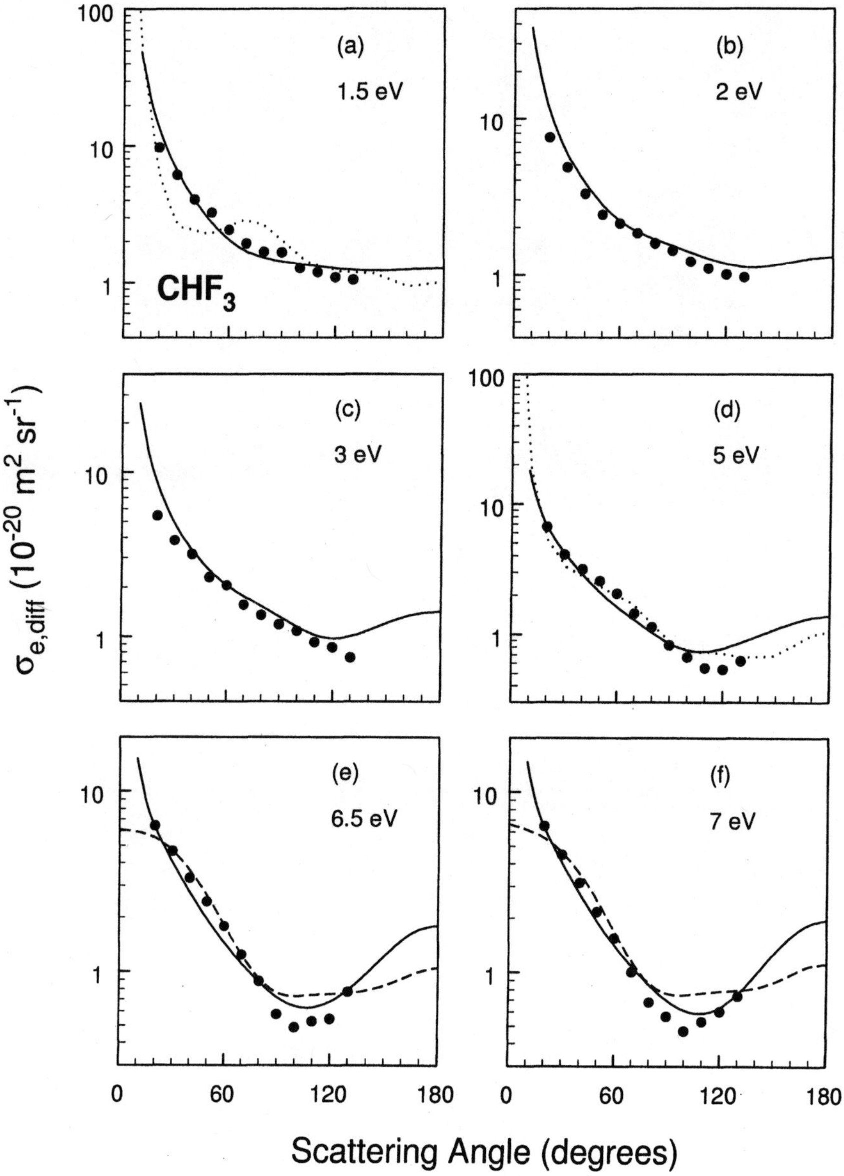

Figure 5.4. Elastic differential electron scattering cross sections, $\sigma_{e,\text{diff}}(\varepsilon)$, for CHF_3. Experimental data: (•) data of Tanaka and coworkers [38, 39] as presented in Ref. [40]. Computed values: (– –) results of Diniz *et al.* [41]; (—) results of Varella *et al.* [40]; (· · ·) results of Morgan *et al.* [8].

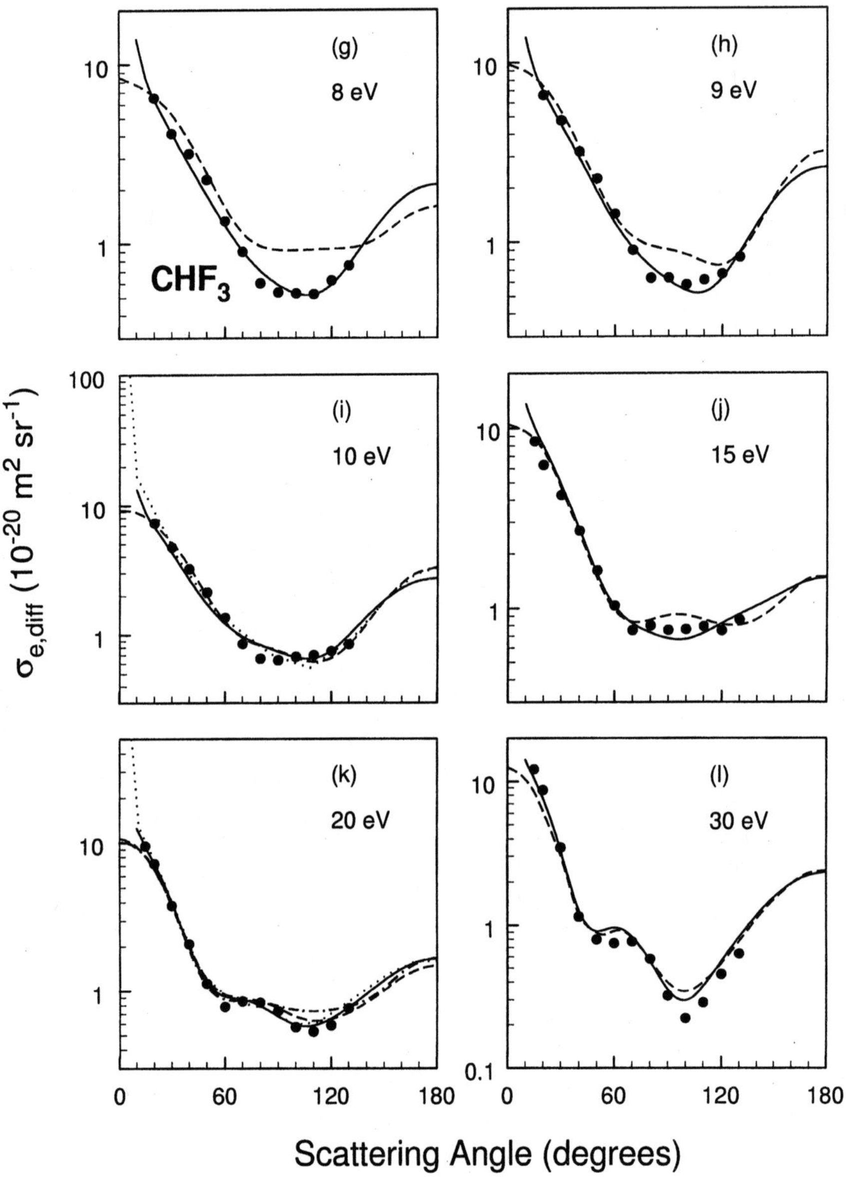

Figure 5.4 (continued)
Elastic differential electron scattering cross sections, $\sigma_{e,diff}(\varepsilon)$, for CHF_3. Experimental data: (•) data of Tanaka and coworkers [38, 39] as presented in Ref. [40]. Computed values: (– · –) results of Natalense *et al.* [42]; (– –) results of Diniz *et al.* [41]; (—) results of Varella *et al.* [40]; (···) results of Morgan *et al.* [8].

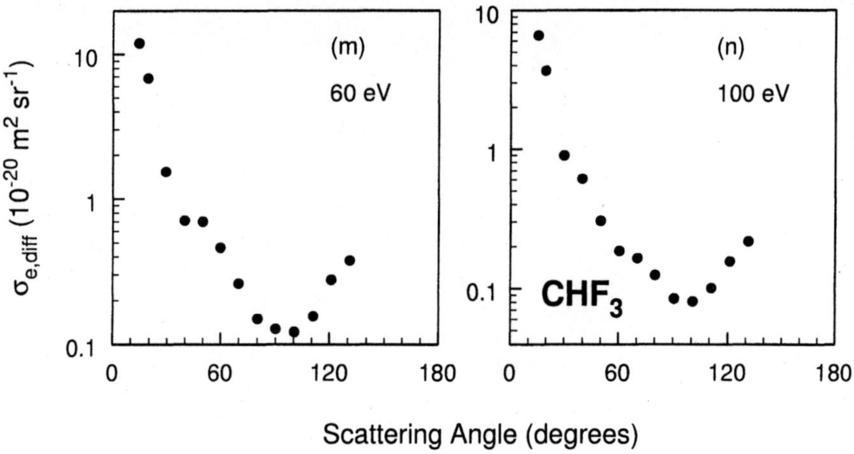

Figure 5.4 (continued)
Experimental elastic differential electron scattering cross section data, $\sigma_{\text{e,diff}}(\varepsilon)$, for CHF_3 at 60 and 100 eV as measured by Tanaka and coworkers [38, 39].

ment with the experimental data, especially for scattering angles smaller than 60°. For larger scattering angles, the agreement between theory and experiment improves as the incident electron energy increases. Diniz *et al.* [41] interpreted this to mean that because CHF_3 is a polar molecule, the electron-dipole interaction is dominant at small scattering angles and masks other important effects at small angles. As a result, long and short range correlation effects will play a more important role at larger scattering angles.

Similar agreement is seen for the computed values of Varella *et al.* [40], Morgan *et al.* [8], and Natalense *et al.* [42] at energies $\geq$ 5 eV. The values of Morgan *et al.* [8] exhibit large discrepancies between both the other calculated values and the experimental data at 1.5 eV and at small scattering angles due to the omission of polarization effects in the calculation of Morgan *et al.* [8].

2.2.3 Elastic Integral Electron Scattering Cross Section, $\sigma_{\text{e,int}}(\varepsilon)$, of CHF_3

Varella *et al.* [40] and Morgan *et al.* [8] computed the $\sigma_{\text{e,int}}(\varepsilon)$ of CHF_3 using the Schwinger multichannel method and their results are shown in Fig. 5.5. According to Morgan *et al.* [8] the sharp increase of $\sigma_{\text{e,int}}(\varepsilon)$ at the lowest electron energies shown in their computed values is a common artifact of the static exchange approximation. However, as they have

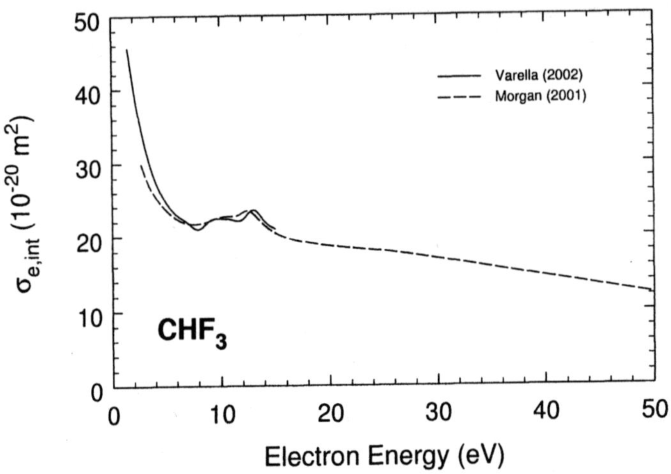

Figure 5.5. Computed elastic integral electron scattering cross section, $\sigma_{\mathrm{e,int}}(\varepsilon)$, for CHF_3: (—) Ref. [40]; (- -) Ref. [8].

noted, CHF_3 is a polar molecule, and long-range electron scattering by the dipole potential which is fully included in their calculation will lead to such an increase.

Unfortunately, there are no experimental data with which to compare these results. No calculation of $\sigma_{\mathrm{e,int}}(\varepsilon)$ has been performed for CHF_3 based upon the measured values of $\sigma_{\mathrm{e,diff}}(\varepsilon)$.

2.2.4 Momentum Transfer Cross Section, $\sigma_{\mathrm{m}}(\varepsilon)$, of CHF_3

As discussed in Sec. 2.2.1 of this chapter, Eq. (5.1) can be employed to obtain approximate values of $\sigma_{\mathrm{m}}(\varepsilon)$ at low electron energies (below $\sim$ 0.5 eV) using the known dipole moment of CHF_3. Values obtained in this manner by Christophorou *et al.* [1] are plotted in Fig. 5.6. Also plotted in this figure are values of $\sigma_{\mathrm{m}}(\varepsilon)$ computed by Natalense *et al.* [42], Varella *et al.* [40], and Morgan *et al.* [8] using the Schwinger multichannel method. There are no experimental data with which to compare these calculated results.

2.2.5 Electron-Impact Excitation Cross Section, $\sigma_{\mathrm{exc}}(\varepsilon)$, of CHF_3

Morgan *et al.* [8] computed integral cross sections for electron-impact excitation of the $\tilde{a}^3A_1$, $\tilde{A}\,^1A_1$, $\tilde{b}\,^3E$, and $\tilde{B}\,^1E$ states of CHF_3 using

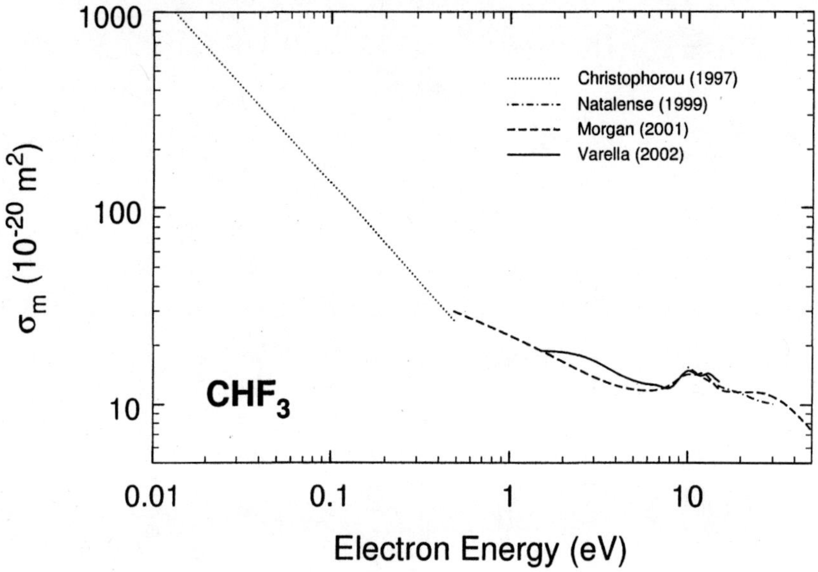

Figure 5.6. Calculated values of the momentum transfer cross section, $\sigma_m(\varepsilon)$, for CHF_3: ($\cdots$) Ref. [1]; (– · –) Ref. [42]; (– –) Ref. [8]; (—) Ref. [40].

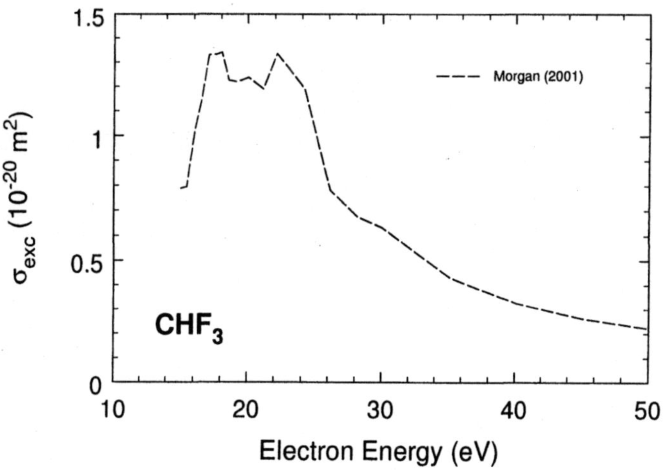

Figure 5.7. Sum of the computed integral cross sections for electron-impact excitation of the $\tilde{a}^3A_1$, $\tilde{A}^1A_1$, $\tilde{b}^3E$, and $\tilde{B}^1E$ states of CHF_3 [8].

the Schwinger multichannel method and their sum is shown in Fig. 5.7. These results were used by Morgan *et al.* [8] in a Boltzmann code as a rough estimate of the total electronic excitation cross section of CHF_3.

2.3 Electron-Impact Ionization of CHF_3

2.3.1 Partial Electron-Impact Ionization Cross Sections, $\sigma_{i,part}(\varepsilon)$, of CHF_3

There have been five measurements [43–48] of the partial ionization cross sections, $\sigma_{i,part}(\varepsilon)$, of CHF_3, all of which are shown in Fig. 5.8. Some of the studies [43, 47] were unable to discriminate between ions that differed by one H atom. Only measurements where the identity of the ion was uniquely identified are shown in Fig. 5.8

The first set of measurements of $\sigma_{i,part}(\varepsilon)$ of CHF_3 is that of Poll and Meichsner [43] for the ions CF_3^+, CHF_2^+ + CF_2^+, and CF^+. Their data presented here were obtained from Fig. 2(j) of their paper [43]. The second set of measurements is that of Goto *et al.* [44] for the ions CF_3^+, CHF_2^+, CF_2^+, CHF^+, CF^+, CH^+, and F^+. The third set of measurements is that of Jiao *et al.* [45, 46] for the production of the ions CF_3^+, CHF_2^+, CF_2^+, and CF^+. These $\sigma_{i,part}(\varepsilon)$ measurements were made using Fourier-transform mass spectrometry and have a quoted estimated uncertainty of ±10%. The fourth set of measurements of the $\sigma_{i,part}(\varepsilon)$ of CHF_3 is that by Iga *et al.* [47] who observed the ions CF_3^+, CHF_2^+, CF_2^+, CHF^+, CF^+, F^+, CH^+, and C^+, and measured the $\sigma_{i,part}(\varepsilon)$ for the ions CF_3^+, CHF_2^+ + CF_2^+, CHF^+ + CF^+, F^+, and CH^+ + C^+. For mass/charge selection these workers used both a quadrupole and a time-of-flight mass spectrometer and for absolute cross section determination they employed a relative flow technique [49] with Ar as the reference gas. The relative error in these $\sigma_{i,part}(\varepsilon)$ measurements is estimated [47] to be ±13% for (CHF_2^+ + CF_2^+), ±21% for F^+, and ±16% for the rest of the ions. The fifth set of measurements of the $\sigma_{i,part}(\varepsilon)$ of CHF_3 is from Torres *et al.* [48] for the ions CF_3^+, CHF_2^+, CF_2^+, CHF^+, CF^+, HF^+, F^+, CH^+, C^+, and H^+. These measurements were made with reference to Ar by probing the supersonic expansion of CHF_3 (+ Ar) with a monoenergetic electron beam (0 to 100 eV incident energy) within a time-of-flight mass spectrometer. The overall uncertainty of these data is estimated to be $< \pm10\%$ for ions with a mass similar to that of the reference gas (Ar) and between ±15% and ±20% for the light fragments.

A significant amount of scatter exists in the data presented in Fig. 5.8. This lack of agreement between measurements and the lack of complete data for all ions makes it difficult to determine which data to use in the determination of suggested values of $\sigma_{i,part}(\varepsilon)$ for CHF_3. As mentioned previously, Poll and Meichsner [43] and Iga *et al.* [47] measured the sum of cross sections for the production of ions that differed by a single H

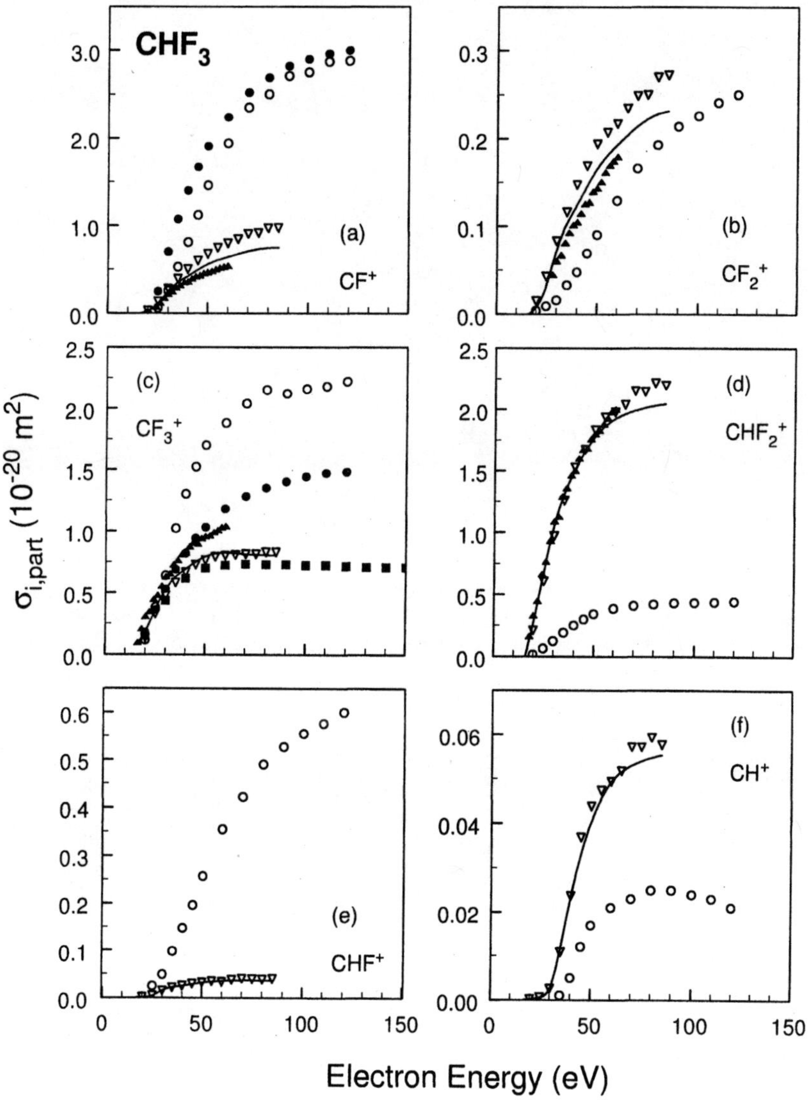

Figure 5.8. Partial electron-impact ionization cross sections, $\sigma_{i,part}(\varepsilon)$, of CHF_3 for the production of the indicated ions: (•) data of Poll and Meichsner [43]; (○) data of Goto *et al.* [44]; (▲) data of Jiao *et al.* [45, 46]; (■) data of Iga *et al.* [47]; (▽) data of Torres *et al.* [48]. Not all authors measured all ions, and some studies were not able to resolve ions that differed by a single H atom (see text and Fig. 5.9). The suggested cross section for each product ion is shown by a solid line.

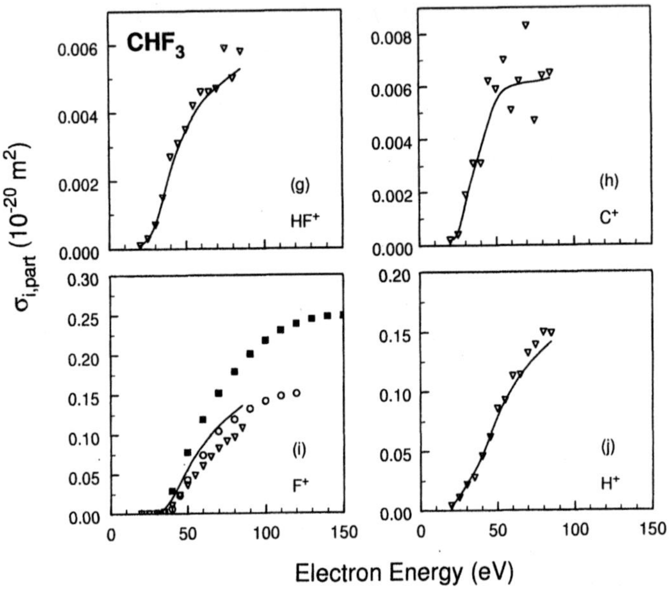

Figure 5.8 (continued)
Partial electron-impact ionization cross sections, $\sigma_{i,part}(\varepsilon)$, of CHF_3 for the production of the indicated ions: (•) data of Poll and Meichsner [43]; (○) data of Goto *et al.* [44]; (▲) data of Jiao *et al.* [45, 46]; (■) data of Iga *et al.* [47]; (▽) data of Torres *et al.* [48]. Not all authors measured all ions, and some studies were not able to resolve ions that differed by a single H atom (see text and Fig. 5.9). The suggested cross section for each product ion is shown by a solid line.

atom, because their instrumentation did not have sufficient mass resolution to differentiate between the ions of interest (i.e., $CHF_2^+ + CF_2^+$, $CHF^+ + CF^+$, and CH^+ and C^+). A comparison of these measured cross sections with the sum of the individual cross sections measured in the other studies is shown in Fig. 5.9. This comparison shows excellent agreement between the measurements of Jiao *et al.* [45, 46], Iga *et al.* [47], and Torres *et al.* [48] for all but the lightest ions, which is generally consistent with the partial ionization data shown in Fig. 5.8, and the total ionization data discussed later.

Therefore, the partial ionization data of Jiao *et al.* [45, 46], Iga *et al.* [47], and Torres *et al.* [48] are accepted for determining our suggested partial ionization cross sections. The available cross sections for each ion from these three studies were fit, summed, and then normalized at each energy to the recommended total ionization cross section derived in the next section. The resulting suggested cross sections are shown in Fig. 5.8 by the solid lines and are listed in Table 5.5.

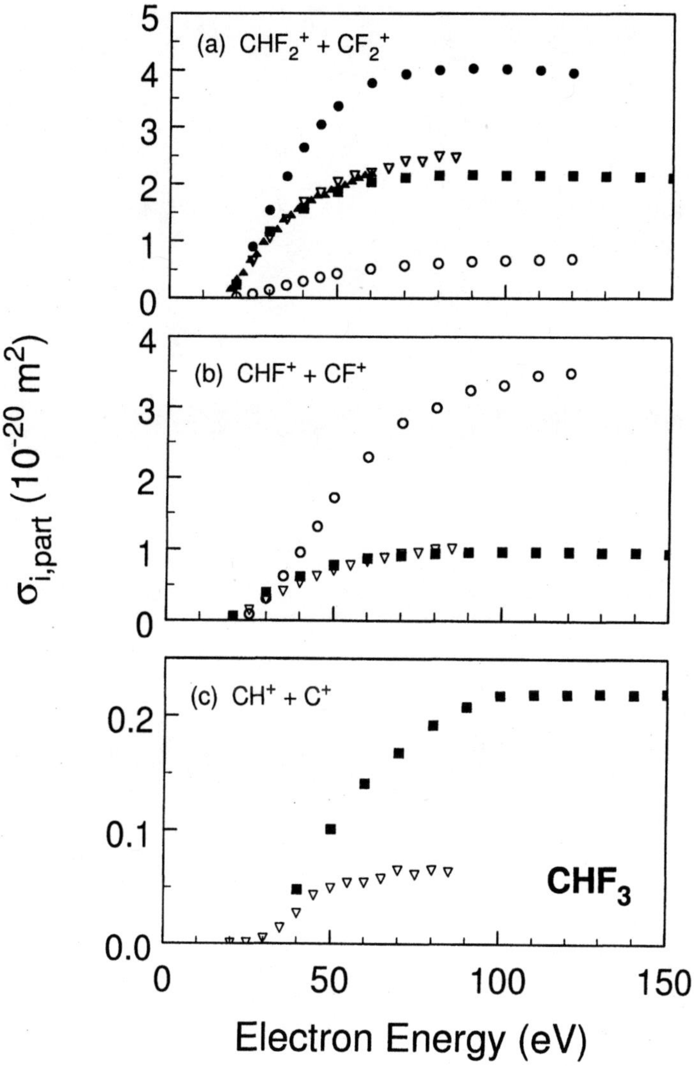

Figure 5.9. Partial electron-impact ionization cross sections, $\sigma_{\mathrm{i,part}}(\varepsilon)$, of CHF_3 for the sum of the production of the two indicated ions: (•) data of Poll and Meichsner [43]; (○) data of Goto *et al.* [44]; (▲) data of Jiao *et al.* [45, 46]; (■) data of Iga *et al.* [47]; (▽) data of Torres *et al.* [48]. The data from Refs. [43, 47] were measured as the sum of two ions, while the data from the other sources are calculated from the sum of the two respective cross sections presented in Fig. 5.8.

Table 5.5. Suggested values of the partial electron-impact ionization cross sections, $\sigma_{i,part}(\varepsilon)$, of CHF_3 in units of 10^{-20} m^2 (see the text).

ε (eV)	CF^+	CF_2^+	CF_3^+	CHF_2^+	CHF^+
16			0.080		
20	0.027	0.008	0.185	0.262	0.001
25	0.112	0.032	0.362	0.675	0.004
30	0.250	0.073	0.524	1.05	0.015
40	0.423	0.124	0.694	1.50	0.025
50	0.558	0.166	0.794	1.78	0.032
60	0.640	0.193	0.824	1.92	0.036
70	0.708	0.214	0.813	1.99	0.037
80	0.744	0.228	0.807	2.03	0.037
85	0.747	0.231	0.804	2.05	0.037

ε (eV)	CH^+	HF^+	C^+	F^+	H^+
20	0.0002	0.0001	0.0002		0.004
25	0.0006	0.0003	0.0005	0.0002	0.012
30	0.0027	0.0007	0.0016	0.0004	0.023
40	0.024	0.0023	0.0037	0.020	0.047
50	0.040	0.0035	0.0054	0.057	0.079
60	0.049	0.0043	0.0060	0.087	0.10
70	0.053	0.0047	0.0061	0.11	0.12
80	0.055	0.0051	0.0062	0.13	0.14
85	0.055	0.0053	0.0063	0.14	0.14

2.3.2 Total Electron-Impact Ionization Cross Section, $\sigma_{i,t}(\varepsilon)$, of CHF_3

The sum, $\sigma_{i,part,t}(\varepsilon)$, of the $\sigma_{i,part}(\varepsilon)$ of each of the five independent measurements [43–48] discussed in the previous section can be taken to represent an estimate of the total electron-impact ionization cross section, $\sigma_{i,t}(\varepsilon)$, of the CHF_3 molecule as measured by the five groups of investigators. The five sets of these values of $\sigma_{i,t}(\varepsilon)$ are plotted in Fig. 5.10. While the uncertainty of these data is difficult to assess, it is likely to exceed ±10%, although Iga *et al.* [47] quoted a total uncertainty for their $\sigma_{i,part}(\varepsilon)$ values of ±9%. As with the data for $\sigma_{i,part}(\varepsilon)$ discussed in the previous section, the values of $\sigma_{i,t}(\varepsilon)$ derived from the data of Jiao *et al.* [45, 46], Iga *et al.* [47], and Torres *et al.* [48] are in good agreement,

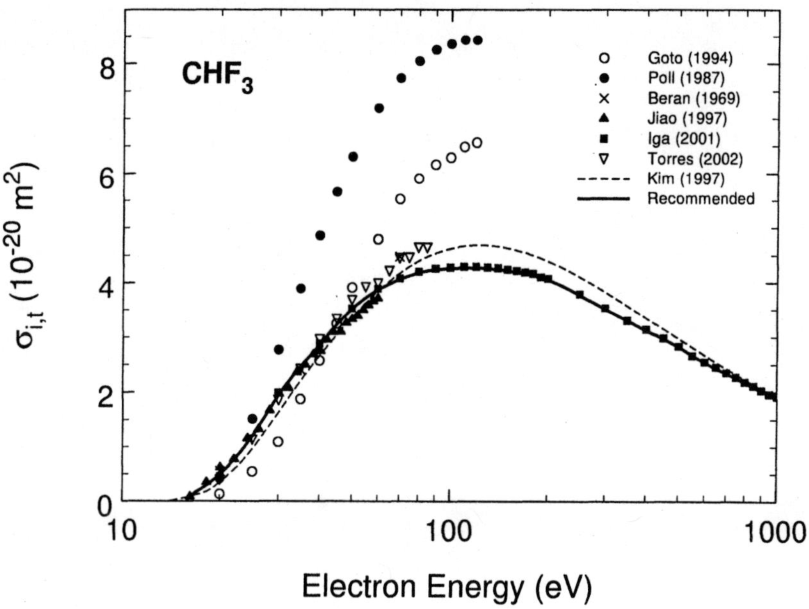

Figure 5.10. Total ionization cross section, $\sigma_{i,t}(\varepsilon)$, for CHF_3 as obtained by summing up the partial ionization cross sections, $\sigma_{i,part}(\varepsilon)$: (•) Ref. [43]; (○) Ref. [44]; (▲) Ref. [45]; (■) Ref. [47]); (▽) Ref. [48]; (×) Ref. [50]. Also shown (– –) are the calculated results of Kim *et al.* [51], and the recommended values (—).

while the values derived from Poll and Meichsner [43] and Goto *et al.* [44] lie significantly higher.

Also shown in Fig. 5.10 are the earlier measurements of $\sigma_{i,t}(\varepsilon)$ by Beran and Kevan [50] which were made at only three values of the incident electron energy. Additionally, the result of a calculation by Kim *et al.* [51, 52] based on a model [53, 54] which combines binary encounter theory and the Bethe theory of electron-impact ionization (see Ch. 2), is shown in the figure. Both the early measurement and the calculated values of $\sigma_{i,t}(\varepsilon)$ are in good agreement with the measurements of Jiao *et al.* [45, 46], Iga *et al.* [47], and Torres *et al.* [48].

Recommended values for the $\sigma_{i,t}(\varepsilon)$ of the CHF_3 molecule have been determined by accepting the values of $\sigma_{i,t}(\varepsilon)$ from Jiao *et al.* [45, 46], Iga *et al.* [47], and Torres *et al.* [48], and fitting their data up to 70 eV. Above 70 eV, the recommended data are extended to 1000 eV by normalizing the data of Iga *et al.* [47] to the average value at 70 eV. These values are shown in Fig. 5.10 by the solid line, and are listed in Table 5.6.

Table 5.7 lists the energy thresholds of a number of reactions leading to positive ions by electron impact on CHF_3 [44, 48, 55–57]. The larger

Table 5.6. Recommended values of the total electron-impact ionization cross section, $\sigma_{i,t}(\varepsilon)$, for CHF_3.

ε (eV)	$\sigma_{i,t}(\varepsilon)$ (10^{-20} m^2)	ε (eV)	$\sigma_{i,t}(\varepsilon)$ (10^{-20} m^2)	ε (eV)	$\sigma_{i,t}(\varepsilon)$ (10^{-20} m^2)
16	0.080	80	4.18	400	3.12
20	0.486	90	4.24	500	2.82
25	1.20	100	4.26	600	2.54
30	1.94	120	4.27	700	2.34
40	2.86	150	4.22	800	2.17
50	3.51	200	4.05	900	2.02
60	3.87	250	3.77	1000	1.91
70	4.06	300	3.51		

positive ions have lower energy thresholds. By selectively choosing the electron energy, one may eliminate the production of ions with higher energy onsets using the information in this table.

Measurements of absolute fluxes and energy distributions of ions in inductively coupled plasmas have been made for a number of mixtures containing CHF_3 (e.g., see Kim *et al.* [58] for CHF_3/Ar and $CHF_3/Ar/O_2$ mixtures). Measurements have also been made of the cross sections for ion-molecule reactions involving ions generated in CHF_3 discharges [59].

2.3.3 Density-Normalized Effective Ionization Coefficient, $(\alpha - \eta)/N\,(E/N)$, of CHF_3

De Urquijo *et al.* [60, 61] reported measurements of the density-normalized effective ionization coefficient, $(\alpha - \eta)/N\,(E/N)$, for CHF_3. These data are shown in Fig. 5.11 and are listed in Table 5.8 as our suggested values for the $(\alpha-\eta)/N\,(E/N)$ of this gas. Interestingly, the data in Fig. 5.11 indicate an $(E/N)_{lim}$ value for this gas of $< 80\times10^{-17}$ V cm^2, which would be somewhat lower than the uniform field breakdown voltage ($\sim 97.5 \times 10^{-17}$ V cm^2) of CHF_3 reported by Christophorou *et al.* [36]. It might be noted that CHF_3 has been used as a buffer gas in mixtures with strongly electron-attaching gases for gaseous dielectrics purposes [36, 62–65]. In this regard, also, Hernández-Ávila *et al.* [66] reported measurements of $(\alpha - \eta)/N\,(E/N)$ and $(E/N)_{lim}$ for mixtures of CHF_3 with SF_6 with compositions ranging from 1% to 50% SF_6.

Table 5.7. Threshold energies for "appearance" of positive ions by electron impact on CHF_3.

Reaction	Threshold energy (eV)	Reference
$CHF_3 + e \rightarrow CF_3^+ + H$	15.2	[44]
	14.67	[55]
	14.53	[56]
	14.42	[57]
	13.9 ± 0.6	[48]
$CHF_3 + e \rightarrow CHF_2^+ + F$	16.8	[44]
	16.4	[55]
	15.75	[57]
	15.7 ± 0.5	[48]
$CHF_3 + e \rightarrow CF_2^+ + HF$	17.6	[44]
$\rightarrow CF_2^+ + H + F$	17.5	[55]
	19.5 ± 0.5	[48]
	20.2	[57]
$CHF_3 + e \rightarrow CHF^+ + F_2$	19.8	[44]
$\rightarrow CHF^+ + 2F$	20.7 ± 0.5	[48]
$CHF_3 + e \rightarrow CF^+ + HF + F$	20.9	[44]
	20.75	[57]
	20.2	[55]
$\rightarrow CF^+ + H + F_2$	20.9 ± 0.5	[48]
$CHF_3 + e \rightarrow HF^+ + C + F_2$	28.8 ± 0.5	[48]
$CHF_3 + e \rightarrow CH^+ + F_2 + F$	33.5	[44]
$\rightarrow CH^+ + 3F$	29.9 ± 0.5	[48]
$CHF_3 + e \rightarrow C^+ + H + 3F$	33.8 ± 0.5	[48]
$CHF_3 + e \rightarrow F^+ + \ldots$	37.0 ± 0.5	[48]
	37.0	[44]

2.4 Electron-Impact Dissociation of CHF_3 into Neutral Fragments

2.4.1 Partial Cross Sections for Electron-Impact Dissociation of CHF_3 into Neutral Fragments, $\sigma_{\mathrm{dis,neut,part}}(\varepsilon)$

Goto *et al.* [44] reported measurement of the partial cross sections, $\sigma_{\mathrm{dis,neut,part}}(\varepsilon)$, for dissociation of CHF_3 by electron impact into the

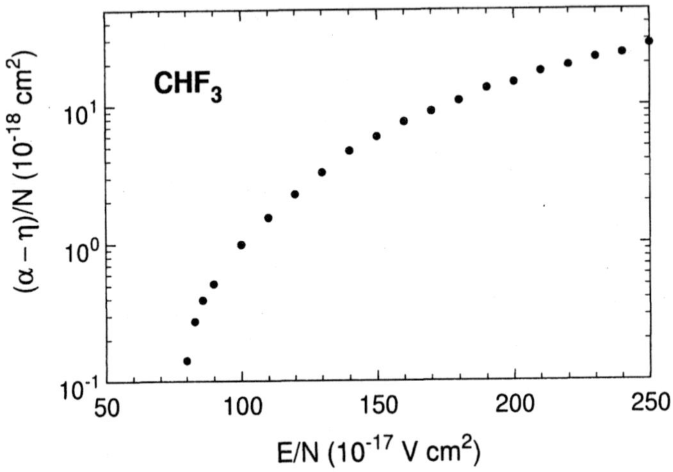

Figure 5.11. Values of the density-normalized effective ionization coefficient, $(\alpha - \eta)/N\,(E/N)$, for CHF_3 measured by de Urquijo *et al.* [60, 61].

Table 5.8. Values of the density-normalized effective ionization coefficient, $\bar{\alpha}/N(E/N) \equiv (\alpha - \eta)/N\,(E/N)$, for CHF_3 measured by de Urquijo *et al.* [60, 61].

E/N (10^{-17} V cm^2)	$\bar{\alpha}/N$ (10^{-18} cm^2)	E/N (10^{-17} V cm^2)	$\bar{\alpha}/N$ (10^{-18} cm^2)
80	0.142	160	7.61
83	0.273	170	9.04
86	0.391	180	10.8
90	0.512	190	13.3
100	0.985	200	14.6
110	1.53	210	17.5
120	2.27	220	19.3
130	3.27	230	22.1
140	4.68	240	23.7
150	5.93	250	27.7

neutral fragments CF_3, CHF_2, CF_2, CHF, and CF. These cross section data have subsequently been revised by Sugai *et al.* [67] using the cross sections for electron-impact dissociation of the CF_x (x = 1–3) radicals reported by Tarnovsky *et al.* [68, 69]. The revised data of Sugai *et al.* [67] for the CF_3, CF_2, and CF radicals are plotted in Fig. 5.12(a) and indicate that the most abundant neutral fragments generated by electron

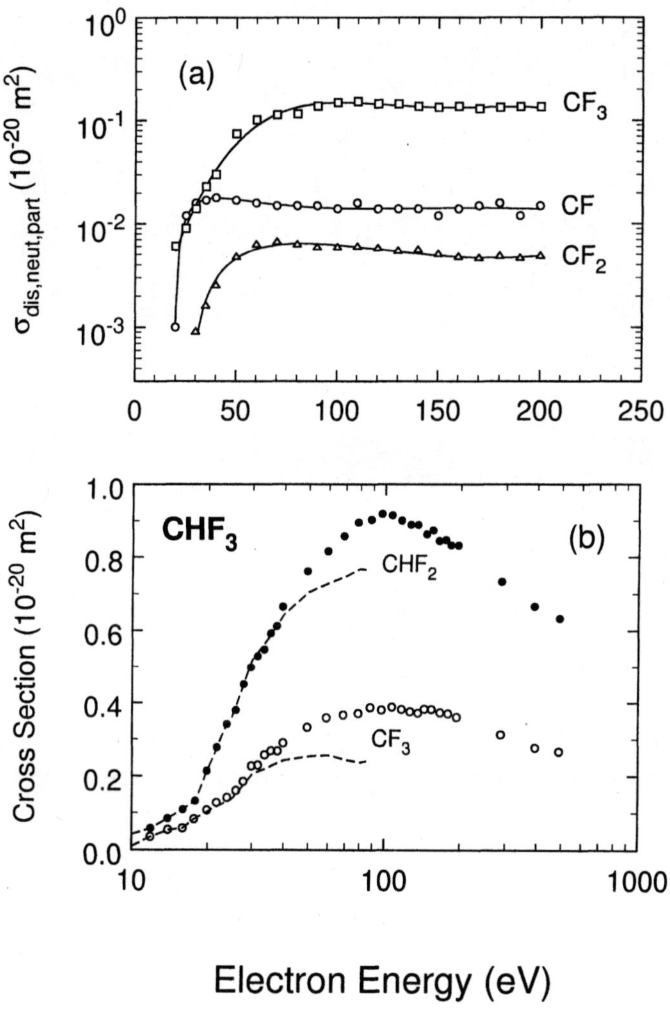

Figure 5.12. (a) Partial cross sections, $\sigma_{\text{dis,neut,part}}(\varepsilon)$, for electron-impact dissociation of CHF_3 into the neutral fragments CF_3, CF_2, and CF as measured by Goto *et al.* [44] and revised by Sugai *et al.* [67]. (b) Cross sections for the production of CHF_2 (•) and CF_3 (○) by neutral dissociation and dissociative ionization from electron impact on CHF_3 as measured by Motlagh and Moore [70]. The dashed lines show the measurements of Motlagh and Moore [70] for the production of CHF_2 and CF_3 minus the suggested cross sections for production of F^+ and H^+, respectively. These represent the cross sections for the production of CHF_2 and CF_3 by dissociation into neutral species only.

Table 5.9. Threshold energies for electron-impact dissociation of CHF_3 into neutral species[a] determined by Goto *et al.* [44].

Reaction	Threshold energy (eV)
$CHF_3 + e \to CF_3 + H + e$	$11.0^{b,c}$
$CHF_3 + e \to CHF_2 + F + e$	$13.0^{b,c}$
$CHF_3 + e \to CF_2 + H + F + e$	15.0^{b}
$CHF_3 + e \to CF_2 + HF + e$	9.1^{b}
$CHF_3 + e \to CHF + 2F + e$	16.9^{b}
$CHF_3 + e \to CHF + F_2 + e$	16.1^{b}
$CHF_3 + e \to CF + H + 2F + e$	20.5^{b}
$CHF_3 + e \to CF + F_2 + H + e$	19.7^{b}
$CHF_3 + e \to CF + HF + F + e$	14.6^{b}

[a]The CF_3–H bond dissociation energy is (4.46 ± 0.17) eV [56].
[b]Expected value.
[c]Measured value; experimental uncertainty ± 0.5 eV.

impact on CHF_3 are the radicals CF_3 and CF. The revised cross sections are very small and differ from the original data by factors of up to 16.

More recently, Motlagh and Moore [70] reported relative measurements of the cross section for electron-impact dissociation of CHF_3 into CHF_2 and CF_3 neutral fragments (produced by neutral dissociation and by dissociative ionization). They detected these radicals mass spectrometrically as organotellurides produced by their reaction at the surface of a tellurium mirror. Via a number of assumptions regarding calibration factors, Motlagh and Moore [70] put their measurements on an absolute scale as shown in Fig. 5.12(b). Their estimated uncertainty is $\pm 30\%$. Their data can be further analyzed to determine the production of CHF_2 and CF_3 solely by neutral dissociation by subtracting the $\sigma_{\mathrm{i,part}}(\varepsilon)$ for production of F^+ and H^+, respectively, from their measurements. We have done this, and the results are shown by the broken lines in Fig. 5.12(b).

A comparison of the two sets of data for the common neutral species CF_3 shows that the values of Sugai *et al.* [67] are low (see also Sec. 2.4.2 of this chapter). This was noted previously for both CF_4 [5] and CHF_3 [1, 5]. It should also be noted that there is another significant discrepancy between the data of Motlagh and Moore [70] and Goto *et al.* [44] in that Motlagh and Moore observe the CHF_2 signal to significantly exceed that of the CF_3 radical, while Goto *et al.* observed almost no CHF_2 signal.

Table 5.9 lists the threshold energies for electron-impact dissociation of CHF_3 into neutral species as given by various authors. They help in

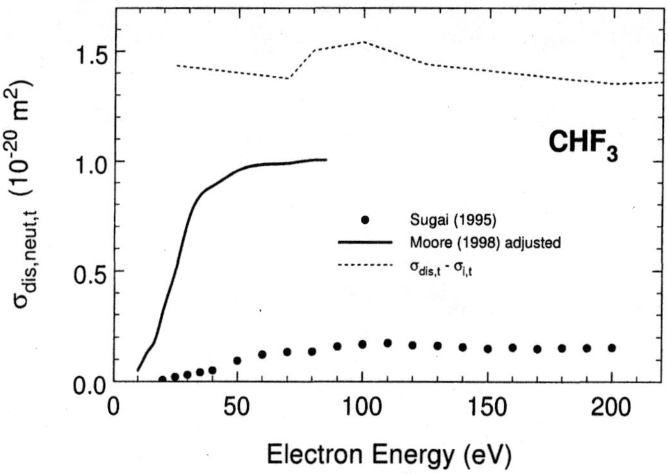

Figure 5.13. Estimates of the total cross section, $\sigma_{\rm dis,neut,t}(\varepsilon)$, for electron-impact dissociation of CHF_3 into neutral fragments (see text): (•) derived by Sugai *et al.* [67] from the sum of the revised data of Goto *et al.* [44] presented in Fig. 5.12(a); (—) derived from the sum of the data of Motlagh and Moore [70] for production of CHF_2 and CF_3 shown in Fig. 5.12(b). Also shown (- - -) are values derived from the difference $\sigma_{\rm dis,t}(\varepsilon) - \sigma_{\rm i,t}(\varepsilon)$ (see text).

understanding the energetics of the various dissociation processes leading to neutral fragments.

2.4.2 Total Cross Section for Electron-Impact Dissociation of CHF_3 into Neutral Fragments, $\sigma_{\rm dis,neut,t}(\varepsilon)$

Fig. 5.13 shows the sums of the partial cross sections, $\sigma_{\rm dis,neut,t}(\varepsilon)$, for dissociation of CHF_3 by electron impact into neutral fragments as determined from the data in Figs. 5.12(a) and 5.12(b). These sums of the partial dissociation cross sections into neutrals represent lower-limit estimates of the total cross section for electron-impact dissociation of the CHF_3 molecule into neutral fragments, $\sigma_{\rm dis,neut,t}(\varepsilon)$, because they do not include contributions from the production of all possible neutral species.

Also plotted in Fig. 5.13 is the difference $\sigma_{\rm dis,t}(\varepsilon) - \sigma_{\rm i,t}(\varepsilon)$ which is approximately equal to $\sigma_{\rm dis,neut,t}(\varepsilon)$. We estimated this difference using the limited measurements of $\sigma_{\rm dis,t}(\varepsilon)$ shown in Fig. 5.14 in the next section and the presently suggested values of $\sigma_{\rm i,t}(\varepsilon)$ from Sec. 2.3.2 of this chapter. While the uncertainty of $\sigma_{\rm dis,neut,t}(\varepsilon)$ determined this way (dashed line in Fig. 5.13) is large, these values confirm that the data

Table 5.10. Values of the total electron-impact dissociation cross section, $\sigma_{dis,t}(\varepsilon)$, for CHF_3 measured by Winters and Inokuti [71].

ε (eV)	$\sigma_{dis,t}(\varepsilon)$ (10^{-20} m^2)	ε (eV)	$\sigma_{dis,t}(\varepsilon)$ (10^{-20} m^2)
22	2.4	125	5.7
72	5.5	200	5.4
100	5.8	300	4.9

of Sugai *et al.* [67] grossly underestimate the magnitude of $\sigma_{dis,neut,t}(\varepsilon)$. The measurements of Motlagh and Moore [70] are consistent with the calculation of $\sigma_{dis,neut,t}(\varepsilon)$ as their measurement accounts for the contribution of only two fragments. Thus at this time the data of Motlagh and Moore must be treated as the most reliable lower limit of $\sigma_{dis,neut,t}(\varepsilon)$ for CHF_3. Clearly, the need for more work is indicated.

2.4.3 Total Cross Section for Electron-Impact Dissociation, $\sigma_{dis,t}(\varepsilon)$, of CHF_3

The CHF_3 molecules predominantly dissociate or predissociate upon electronic excitation. Although the observation of vibrational structure for some electronic transitions of the CHF_3 molecule would indicate stable excited electronic states, dissociation is nonetheless the dominant process that follows electronic excitation of this molecule. Thus, Winters and Inokuti [71] argued that the total dissociation cross section of CHF_3 is equal to the total cross section for electronic excitation (the sum of the cross sections for excitation to all electronic and ionic states).

The limited data of Winters and Inokuti [71] on the total dissociation cross section of CHF_3 up to 300 eV are listed in Table 5.10 and are plotted in Fig. 5.14.

2.5 Electron Attachment to CHF_3

2.5.1 Formation of Fragment Negative ions by Low-Energy Electron Impact on CHF_3

There have been a number of investigations [72–82] of electron attachment to CHF_3, but the data are still fragmentary and no absolute cross sections as a function of the electron energy have been reported for any of the dissociative electron attachment negative ion fragments. The

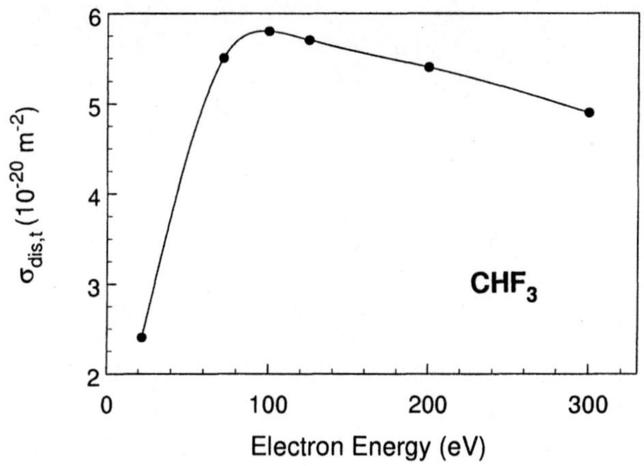

Figure 5.14. Total electron-impact dissociation cross section, $\sigma_{\mathrm{dis,t}}(\varepsilon)$, of CHF_3 from Winters and Inokuti [71].

three mass spectrometric studies [72, 75, 81] are in agreement in that the predominant dissociative electron attachment channels are those leading to the production of F^-. Actually, Reese *et al.* [72] and Scheunemann *et al.* [81] observed only F^- while MacNeil and Thynne [75] reported the observation of F^-, H^-, C^-, CH^-, CF^-, F_2^-, CF_2^-, and CF_3^-, with F^- being by far the most abundant negative ion.

Appearance onsets and peak energies for F^- production from CHF_3 are given in Table 5.11. Table 5.11 also gives the vertical attachment energies measured in an electron transmission experiment [82] along with the vertical attachment energies obtained via bound and continuum multiple scattering Xα calculations [82]. Figure 5.15 shows the yield for the formation of F^- from CHF_3 as a function of electron energy [81]. Clearly, at least three negative ion resonances lead to the production of F^- below 14 eV. Depending on the energetics, a number of dissociative electron attachment processes [75] can lead to the production of F^-:

$$CHF_3 + e \rightarrow F^- + CHF_2 \tag{5.3a}$$
$$\rightarrow F^- + CHF + F \tag{5.3b}$$
$$\rightarrow F^- + CF_2 + H \tag{5.3c}$$
$$\rightarrow F^- + 2F + CH. \tag{5.3d}$$

According to Scheunemann *et al.* [81], the first resonance (near 5 eV) is due to Reaction (5.3a). The higher energy resonances near 10 and 12 eV can be due to a multiplicity of fragmentation channels. It is interesting to note that the cross section for the production of F^- via

Table 5.11. Negative ion states of CHF_3.

Energy (eV)	Method of observation
2.9^a	Dissociative electron attachment[b]
3.7^a	
9.3^a	
0.04	Electron swarm[c]
$\sim 4.5\ (2.2 \pm 0.3)^d$	Dissociative electron attachment[e]
$10.1 \pm 0.1\ (8.2 \pm 0.2)^d$	
$12.3 \pm 0.2\ (\leq 11)^d$	
$6.3(?)^f$	Electron transmission[g]
9.3^f	
9.10^f	Multiple scattering $X\alpha$ (MS-$X\alpha$) bound-state calculations[g]
10.01^f	
9.44^f	Continuum MS-$X\alpha$ (CMS-$X\alpha$) method[g]
9.64^f	

[a]Thresholds of dissociative electron attachment processes.
[b]Ref. [75].
[c]Data of Lee [73] as analyzed by Blaunstein and Christophorou [74].
[d]The first value is the energy position of the cross section maximum and the value in parenthesis is the respective energy threshold.
[e]Ref. [81].
[f]Vertical attachment energy.
[g]Ref. [82].

the 10 eV resonance far exceeds that from the lower lying resonance. By comparison to the electron attachment data on CF_4 (see Sec. 2.5 of Ch. 4) and methane [79, 80], dissociative electron attachment to CHF_3 is expected to have a peak cross section value $> 10^{-23}$ m^2. The rise in the F^- yield beyond 14 eV has been attributed [81] to the positive ion-negative ion pair process

$$CHF_3 + e \rightarrow F^- + CHF_2^+ + e \tag{5.4}$$

for which Scheunemann *et al.* [81] have calculated a threshold of (11.5 ± 0.3) eV. The mass spectrometric data shown in Fig. 5.15, clearly show the existence of at least three negative-ion resonances leading to F^- formation near 4.5, 10.1, and 12.3 eV. On the basis of these findings, it would be expected that the electron scattering cross sections (elastic and inelastic) of CHF_3 in this energy range will exhibit structure associated

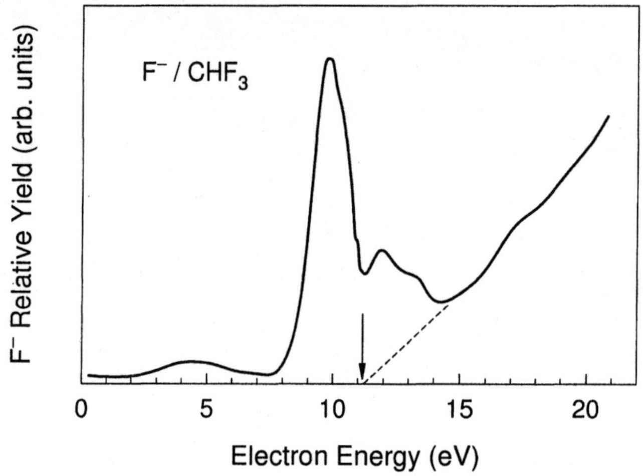

Figure 5.15. Relative cross section for F^- produced by dissociative electron attachment to CHF_3 from Scheunemann *et al.* [81]. The rising part with an apparent onset near 11.5 eV (indicated by the arrow) is attributed to the nonresonant production of F^- via the negative ion-positive ion pair process.

with these negative ion resonances. However the measured $\sigma_{sc,t}(\varepsilon)$ for CHF_3 only exhibits a peak in this energy range near 9 eV (see Fig. 5.3).

Negative ions have been detected in rf plasmas of CHF_3 using a laser photodetachment method [83]. It is important to note that under high gas pressure conditions, a multitude of secondary negative ions and negative ion clusters can form [59, 84]. For instance, the primary negative ion F^- which is formed via the dissociative attachment of electrons to CHF_3 undergoes rapid ion-molecule reactions yielding the cluster anions $(HF)_nF^-$ ($n = 1$–3) [84].

2.5.2 Total Electron Attachment Rate Constant, $k_{a,t}(E/N)$, of CHF_3

Wang *et al.* [85, 86] measured the total electron attachment rate constant $k_{a,t}(E/N)$ in CHF_3 to be approximately 13×10^{-14} cm^3 s^{-1}, which was found to be almost independent of E/N for $E/N < 50 \times 10^{-17}$ V cm^2. This low value is consistent with a recent measurement [87] of $(k_{a,t})_{th}$ and also with earlier values of $(k_{a,t})_{th}$ obtained using the electron swarm method (see Table 5.12). It is also consistent with the findings of a recent electron-swarm mass spectrometric measurement [88]. A much higher value for $(k_{a,t})_{th}$ determined [74] from measure-

Table 5.12. Values of the thermal ($T \approx 295$ K) electron attachment rate constant, $(k_{a,t})_{th}$, for CHF_3.

$(k_{a,t})_{th}$ (10^{-14} cm^3 s^{-1})	Reference
4.6	[76]
< 6.2	[77]
3.6	[78]
1.1×10^5	[73, 74]

ments made by Lee [73] is considered to be in error. Also, Wang *et al.* [85] conjectured that a value of $(k_{a,t})_{th}$ indicated by the work of Clark *et al.* [89], which is 10 times higher than the value Wang *et al.* [85], may have been influenced by traces of electronegative impurities.

2.6 Electron Transport Coefficients for CHF_3

2.6.1 Electron Drift Velocity, $w\,(E/N)$, in CHF_3

There have been three recent sets of measurements of the electron drift velocity, $w\,(E/N)$, in 100% CHF_3 [60, 85, 86, 90]. Two of these measurements [60, 85, 86] were made using a pulsed Townsend technique, and the third [90] was made using a double shutter drift tube with variable drift distance. The data of the three investigations are compared in Fig. 5.16(a) and are seen to be in good agreement. Below 75×10^{-17} V cm^2 the solid line is fit to the data of Wang *et al.* [85, 86] and de Urquijo *et al.* [60]. An extrapolation of the fit based upon the shape of the data of de Urquijo *et al.* [60] and Mitani and Nakamura [90] is used to extend the line for E/N values greater than 75×10^{-17} V cm^2. In the low E/N region where $w\,(E/N)$ varies linearly with E/N, the data agree with the early measurements of Christophorou *et al.* [36]. Recommended values of $w\,(E/N)$ taken off the solid line in Fig. 5.16(a) are listed in Table 5.13. It is interesting to observe the differences between the $w\,(E/N)$ in CF_4 (Ch. 4) and in CHF_3 due to the large permanent electric dipole moment of the latter molecule.

Two sets of measurements of $w\,(E/N)$ have been made in mixtures of CHF_3 with argon [85, 86, 89]. The data of Wang *et al.* [85, 86] are more extensive and are reproduced in Fig. 5.16(b). Comparison of the data for the mixtures from the two sources [85, 86, 89] can be made for only two mixture compositions (0.1% and 1%) for which they are in

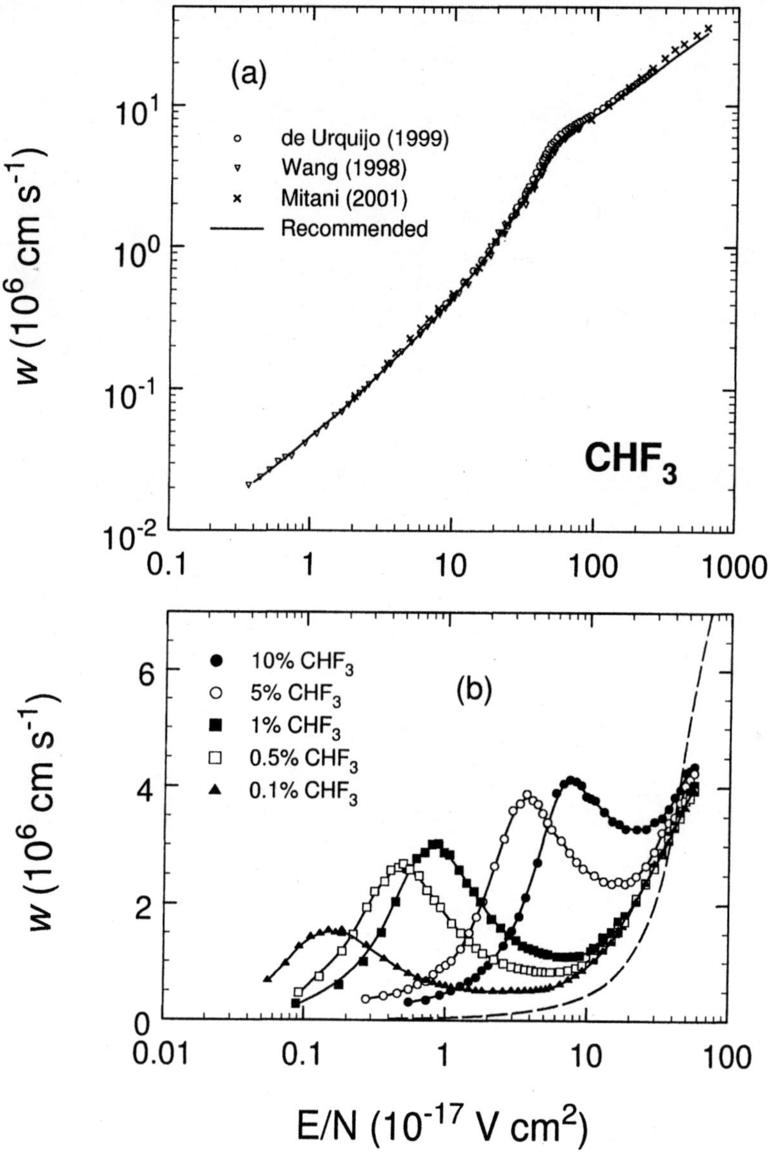

Figure 5.16. (a) Electron drift velocity, $w(E/N)$, in 100% CHF_3 ($T \approx 298$ K): (○) Ref. [60]; (▽) Ref. [85, 86]; (×) Ref. [90]; (—) recommended values. (b) Electron drift velocity, $w(E/N)$, in mixtures of CHF_3 with argon at the indicated compositions ($T \approx 298$ K, total pressure of 1.13 kPa). The solid lines are the weighted-least-squares fits to the data of Wang *et al.* [85, 86]. The dashed line is the recommended value of $w(E/N)$ in 100% CHF_3 from Fig. 5.16(a).

Table 5.13. Recommended electron drift velocities, $w\,(E/N)$, for CHF_3 ($T \approx 298$ K).

E/N (10^{-17} V cm^2)	w (10^6 cm s^{-1})	E/N (10^{-17} V cm^2)	w (10^6 cm s^{-1})
0.40	0.022	15	0.720
0.45	0.024	20	1.09
0.50	0.026	25	1.52
0.60	0.030	30	2.02
0.70	0.034	40	3.38
0.80	0.038	50	4.98
0.90	0.042	60	6.10
1.0	0.046	70	6.92
1.5	0.065	80	7.49
2.0	0.085	90	8.12
2.5	0.105	100	8.66
3.0	0.125	150	11.6
4.0	0.166	200	14.3
5.0	0.208	250	16.9
6.0	0.252	300	19.6
7.0	0.296	400	24.5
8.0	0.342	500	28.9
9.0	0.390	600	33.0
10.0	0.439		

reasonable agreement. For $w\,(E/N)$ in mixtures of CHF_3 with SF_6, see Hernández-Ávila *et al.* [66].

2.6.2 Product of Gas Number Density and Longitudinal Electron Diffusion Coefficient, $ND_L\,(E/N)$, for CHF_3

The only measurement of $ND_L\,(E/N)$ is by Mitani and Nakamura [90] using a double shutter drift tube with variable drift distance technique over the E/N range of 3.5×10^{-17} to 600×10^{-17} V cm^2 at room temperature ($T \approx 298$ K). These data are shown in Fig. 5.17.

2.7 Electron-Impact Induced Light Emission from CHF_3

A number of investigations have been made of light emission from CHF_3 under electron impact [91–93]. The first such study was by Van Sprang *et al.* [91] who used electrons with 100 eV incident energy. The

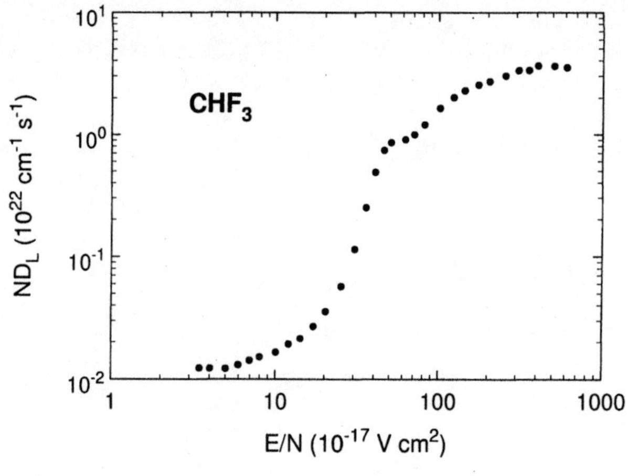

Figure 5.17. Product of gas number density and longitudinal electron diffusion coefficient, $ND_L(E/N)$, for CHF_3 ($T \approx 298$ K) measured by Mitani and Nakamura [90].

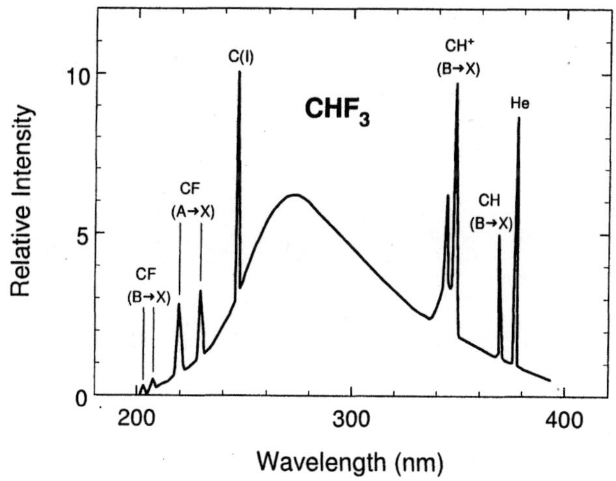

Figure 5.18. Electron-impact induced emission spectrum of CHF_3 in the wavelength range 200–400 nm measured by Van Sprang *et al.* [91].

emission spectrum they obtained in the wavelength range 200–400 nm is reproduced in Fig. 5.18. Table 5.14 lists the emission cross sections measured by Van Sprang *et al.* [91] for the various atomic F and H lines for 100 eV incident energy electrons. Van Sprang *et al.* [91] also reported relative cross sections for the continuous emission from CHF_3 excited by electrons having kinetic energies from 20 to 2 000 eV. They ascribed the

Table 5.14. Emission cross sections, σ_{em} (100 eV), for various atomic F and H lines induced by impact of 100 eV electrons on CHF_3 from Van Srang *et al.* [91].

Emission line	Wavelength (nm)	σ_{em} (10^{-19} cm^2)^a
F(I) $^2P_0 - {}^2P$	703.7	1.5
	712.7	1.1
F(I) $^2D_0 - {}^2P$	760.7	0.7
	775.4	4.7
	780	1.9
F(I) $^2S_0 - {}^2P$	731.1	2.0
F(I) $^4P_0 - {}^4P$	733.1	1.8
	739.8	3.9
	742.5	1.1
F(I) $^4D_0 - {}^4P$	677.3	2.0
	685.6	5.9
	690.2	4.8
H_α	656.2	2.7
H_β	486.1	5.7

[a]Quoted uncertainty ±10%.

observed continua to the $C\,^2E - X\,^2A_1$ transition in CHF_3^+. (Continuous emission from F_2 also occurs [91] in the wavelength range 200–400 nm, because F_2 is a dissociation product of CHF_3). In contrast with the Van Sprang *et al.* assignment of the continuous emission, Aarts [92] assigned the 230–350 nm emission band (the maximum intensity of this broad emission is around 270 nm) to the CF_2 ($\tilde{A} \rightarrow X$) transition and this assignment was later confirmed by Creasy *et al.* [93]. The threshold for the continuum emission was found to be (14.1 ± 0.5) eV [92]. Aarts used electrons with incident energies of up to 1 keV, and Creasy *et al.* used electron, photon, and metastable-atom impact methods. Creasy *et al.* [93] observed fluorescence which they associated with the $D\,^2A_1$ state of CHF_3^+ (adiabatic ionization energy 20.08 eV) (see also Lee *et al.* [94] concerning observation of fluorescence from CHF_3^{+*} under photoexcitation). Aarts and Creasey reported emission spectra similar to those in Fig. 5.18.

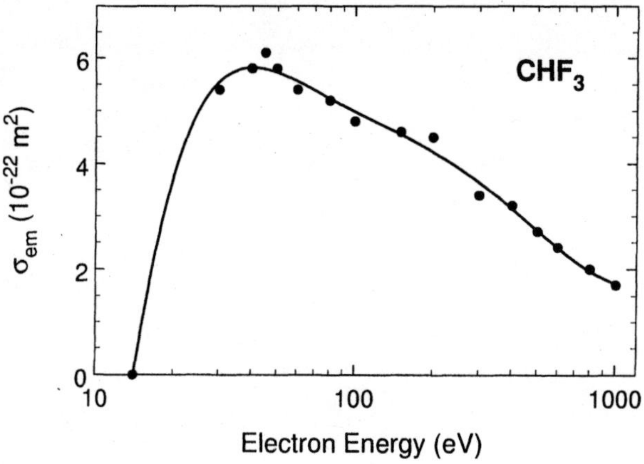

Figure 5.19. Emission cross section for the 270 nm band in electron impact on CHF_3 as a function of incident electron energy from Aarts [92].

The emission cross section for the 270 nm band produced by electron impact on CHF_3 has been measured as a function of electron energy [92] and is plotted in Fig. 5.19. Relative cross sections as a function of electron energy have been also reported by Van Sprang *et al.* [91].

Another important study dealing with light emission induced by electron impact on CHF_3 has been made by Wang and McConkey [95]. These investigators measured the emission spectrum of CHF_3 in the vacuum ultraviolet range (50–130 nm) under single collision conditions for incident electron energies up to 600 eV. The spectrum showed contributions from neutral and singly ionized fluorine and carbon fragments and also from neutral hydrogen atoms. Wang and McConkey [95] made absolute measurements of the cross sections for the F I and F II emissions in the range 75.1–97.77 nm, the C I and C II emissions in the range 68.7–127.7 nm, and the hydrogen atom emissions in the range 91.46–121.6 nm following impact of 200 eV incident electrons on CHF_3. The carbon emission cross sections are generally lower than the fluorine emission cross sections. The emission spectrum is dominated by the F I and H I emissions with the strongest lines at 78, 80.8, 95.5, 97.4 and 97.6 nm for F I, and the Ly α and Ly β for H I. The cross sections at these wavelengths are 2.2×10^{-19}, 3.3×10^{-19}, 9.3×10^{-19}, 5.9×10^{-19}, 2.9×10^{-19}, 27.1×10^{-19}, and 6.5×10^{-19} cm^2, respectively. All the major F I lines involve transitions of a $3s$ or $3d$ electron decaying to the ground state $1s^2\,2s^2\,2p^5\,{}^2P_{1/2,3/2}$ (for further details see Ref. [95]).

Danilevskii *et al.* [96] also studied light emission from excited CHF_3 fragments formed by electron impact on CHF_3. They have reported mea-

Table 5.15. Cross sections of emissions from excited F atoms, Balmer series lines of the hydrogen atom, and the CH molecular fragment ($A^2\Delta \rightarrow X\,^2\Pi$) formed by impact of 4 keV energy electrons on CHF_3 measured by Danilevskii *et al.* [96].

Wavelength (nm)	Transition	Cross Section (10^{-21} cm^2)
F atom		
623.96	$3s^4P \rightarrow 3p^4S^0$	14.2
634.85	$3s^4P \rightarrow 3p^4S^0$	10.9
641.36	$3s^4P \rightarrow 3p^4S^0$	10.0
669.05	$3s^4P \rightarrow 3p^2D^0$	5.4
677.4	$3s^4P \rightarrow 3p^4D^0$	7.0
683.4	$3s^4P \rightarrow 3p^4D^0$	6.0
685.6	$3s^4P \rightarrow 3p^4D^0$	17.3
687	$3s^4P \rightarrow 3p^4D^0$	7.0
690.98	$3s^4P \rightarrow 3p^4D^0$	10.2
733.2	$3s^4P \rightarrow 3p^4P^0$	3.9
739.86	$3s^4P \rightarrow 3p^4P^0$	8.5
703.75	$3s^2P \rightarrow 3p^2P^0$	3.1
712.79	$3s^2P \rightarrow 3p^2P^0$	2.6
720.26	$3s^2P \rightarrow 3p^2P^0$	1.7
731.1	$3s^2P \rightarrow 3p^2P^0$	4.9
775.47	$3s^2P \rightarrow 3p^2D^0$	36.2
CH ($A^2\Delta \rightarrow X^2\Pi$)		20.4
H_α		184.6
H_β		19.4
H_γ		17.0
H_δ		9.7

surements of the effective excitation cross sections of excited F atoms, the Balmer series of the hydrogen atom, and the CH molecular fragment ($A\,^2\Delta \rightarrow X\,^2\Pi$) in the wavelength range 300–850 nm using 0.4 to 6 keV incident energy electrons. This incident electron energy is much higher than the incident electron energies used by Van Sprang *et al.* [91] and Wang and McConkey [95]. Thus, one would expect the cross sections of Danilevskii *et al.* [96] to be smaller. Their measured emission cross sections for 4 keV incident electrons are listed in Table 5.15. Interestingly, Danilevskii *et al.* [96] observed emission from only excited fragments (see also Keller *et al.* [97]).

The emission spectra of CF_3 radicals produced in the photolysis of CHF_3 have also been studied (e.g., see Suto and coworkers [98–101]).

2.8 Electron Interactions with CHF_3 Neutral Fragments

Electron-impact dissociation of the CHF_3 molecule produces CF_3, CF_2, and CF radicals which are common radicals produced by electron impact on CF_4. For these three radicals cross sections for electron impact-ionization producing the respective parent positive ions have been reported [68, 69, 102] and have been summarized and discussed in Ch. 4. Similarly, cross sections for electron-impact dissociative ionization of these three radicals have been measured by Tarnovsky *et al.* [69] and have also been summarized in Ch. 4. No information exists on the other two possible fragments, namely CHF_2 and CHF, produced by electron impact on CHF_3.

Measurements of the concentrations of the CF_3, CF_2, and CF radicals in rf CHF_3 etching plasma have recently been made as a function of the microwave power, the gas pressure, and the time after discharge initiation using infrared diode laser absorption spectroscopy [103–106]. The reader is referred to these sources for details.

2.9 Summary of Recommended and Suggested Cross Sections and Coefficients for CHF_3

The cross sections that have been designated as recommended or suggested for CHF_3 are plotted in Fig. 5.20. These include the:

- $\sigma_{\mathrm{sc,t}}(\varepsilon)$ — Table 5.3, Fig. 5.3;
- $\sigma_{\mathrm{i,t}}(\varepsilon)$ — Table 5.6, Fig. 5.10; and
- $\sigma_{\mathrm{dis,t}}(\varepsilon)$ — Table 5.10, Fig. 5.14.

The total cross section for dissociation into neutral fragments, $\sigma_{\mathrm{dis,neut,t}}(\varepsilon)$, is best estimated at this time by the data of Motlagh and Moore [70] shown in Fig. 5.13. Additionally, the partial ionization cross section data of Table 5.5 (Fig. 5.8) are suggested. Fig. 5.20 clearly shows the limitations of the existing data and the large gaps in the present database.

Data also have been designated as suggested or recommended for the following electron transport parameters:

- $(\alpha - \eta)/N\,(E/N)$ — Table 5.8, Fig. 5.11; and
- $w(E/N)$ — Table 5.13, Fig. 5.16(a).

Basic measurements and calculations are still needed for virtually all elastic and inelastic electron scattering processes, namely, cross sections

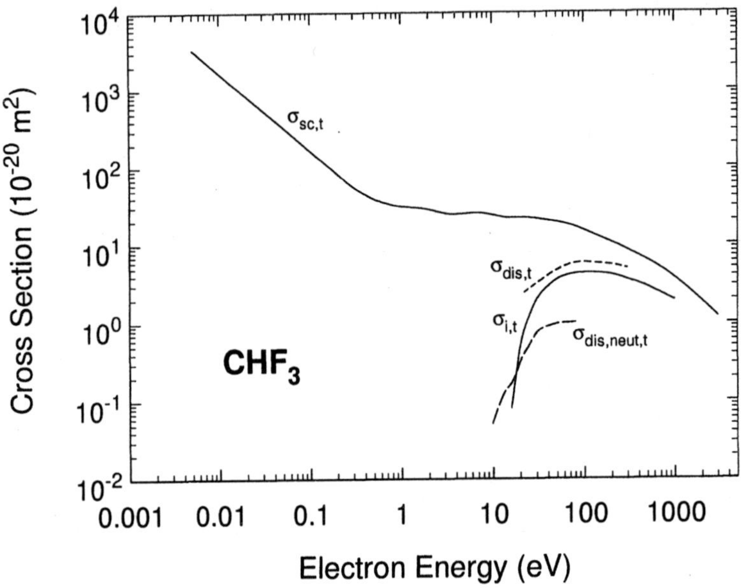

Figure 5.20. Recommended and suggested cross sections for CHF_3.

for total, differential, and elastic electron scattering, cross section for momentum transfer, and cross sections for direct and indirect electronic, vibrational, and rotational excitation. Also, additional data are needed for electron-impact dissociation and ionization to resolve the apparent discrepancies between the direct measurement of $\sigma_{\text{dis,neut,t}}(\varepsilon)$ and $\sigma_{\text{i,t}}(\varepsilon)$ and $\sigma_{\text{dis,t}}(\varepsilon)$. Measurements are necessary for all electron transport, attachment, and ionization coefficients over wide ranges of E/N, with the exception of the electron drift velocity.

* * *

3. Electron Interactions with CF_3I

Trifluoroiodomethane (CF_3I) has been considered as an alternative plasma etching gas (e.g., see Refs. [107–112]). It produces copious quantities of the radical CF_3 via a strong dissociative electron attachment process at energies near 0.0 eV. It is also an "environmentally friendly" gas. Its overall atmospheric half-life is very short ($\sim$ 1 day [113]; $<$ 2 days [114]; photolysis lifetime from several hours at the equator to a few days at high latitudes [115]). Consequently, its global warming potential is extremely low ($<$ 1 for a 100-year and 500-year time period relative to CO_2 [116]; < 3 [116], < 5 [113] relative to CO_2 for a 20-year time period). This is in sharp contrast with the much higher global warming potentials of other plasma etching gases, such as the fully fluorinated hydrocarbons and SF_6 [116]. Because of its short atmospheric lifetime, the CF_3I molecule does not contribute significantly to ozone depletion [113].

In this section, we assess, synthesize, and discuss the available information on the cross sections and rate coefficients for collisional interactions of electrons with the CF_3I molecule. Low-energy electron collision data for this plasma processing gas are sparse. Limited cross section data are available for only total and elastic differential electron scattering, electron-impact ionization, and electron attachment. There is a need for further scrutiny of these data and for measurements of the cross sections of other electron-collision processes. Also, there is a need to measure the coefficients for electron transport, electron-impact ionization, and electron attachment for this molecule.

3.1 Electronic Structure and Basic Properties of CF_3I

The trifluoroiodomethane (CF_3I) is a polar molecule belonging to the C_{3v} point group. McClellan [117] lists two values for its electric dipole moment: 0.92 D and (1.0 ± 0.1) D ($1\,D = 3.3356 \times 10^{-30}$ C m). Of its six fundamental vibrational frequencies (ν_1, …, ν_6), ν_4 and ν_5 are degenerate and ν_1, ν_2, and ν_4 are infrared (IR) active [118]. The energies of the six vibrational modes of CF_3I as given by Shimanouchi [118] are, respectively, 1080 cm^{-1} (0.133 eV), 742 cm^{-1} (0.092 eV), 286 cm^{-1} (0.036 eV), 1187 cm^{-1} (0.147 eV), 537 cm^{-1} (0.067 eV), and 260 cm^{-1} (0.032 eV) (see also Bürger *et al.* [119] for vibrational frequencies of $^{12}CF_3I$ and $^{13}CF_3I$).

The CF_3I molecule has a positive (adiabatic) electron affinity. Atom-impact crossed-molecular beam studies have provided values for the electron affinity (EA) of the CF_3I molecule. These are listed in Table 5.16 along with other pertinent fundamental data for CF_3I and CF_3I^-. The lowest unoccupied molecular orbital (LUMO) of CF_3I is an antibonding σ-type orbital of a_1 symmetry that is almost entirely localized on the C–I bond. The small dissociation energy and relatively long length of the $F_3C–I^-$ bond indicate that this system can be regarded as a weakly bound radical anion [120].

In Table 5.17 are listed the positions of the negative ion states of the CF_3I molecule, as they have been determined from electron scattering and electron attachment studies (see Secs. 3.2 and 3.4 of Ch. 5).

Table 5.16. Basic data on CF_3I and CF_3I^-.

Physical quantity	Value	Method and reference
CF_3I		
Electron affinity (adiabatic)	1.278 eV	Calculation [120]
	1.29 ± 0.1 eV	Charge exchange in Cs-CF_3I collisions [121]
	1.4 ± 0.2 eV	Charge exchange in Cs-CF_3I collisions [122]
	1.54 ± 0.2 eV	Charge exchange in Na-CF_3I collisions [123]
	1.6 ± 0.2 eV	Charge exchange in Cs-CF_3I collisions [123]
	2.2 ± 0.2 eV	Charge exchange in K-CF_3I collisions [124]
D (CF_3–I)	1.9 eV	Electron attachment mass spectrometry [125]
	2.0 ± 0.2 eV	Charge exchange in Na-CF_3I collisions [123]
	2.05 ± 0.2 eV	Charge exchange in Cs-CF_3I collisions [122]
	2.2 ± 0.2 eV	Charge exchange in K-CF_3I collisions [124]
	2.33 ± 0.17 eV	Reaction kinetics [126]
	2.39 ± 0.04 eV	Photodissociation [127]
	2.681 eV	Calculation [120]
	(2.4 eV)[a]	
R (C–I)	2.122 ± 0.037 Å	Electron diffraction [128]
	2.134 ± 0.02 Å	Microwave spectroscopy [129]
	2.135 ± 0.033 Å	Electron diffraction [130]
	2.137 ± 0.032 Å	Electron diffraction [130]
	2.140 Å	Calculation [120]
	2.1438 Å	Microwave spectroscopy [131]
R (C–F)	1.328 ± 0.026 Å	Electron diffraction [128]
	1.3285 Å	Microwave spectroscopy [131]
	1.334 ± 0.015 Å	Electron diffraction [130]
	1.340 ± 0.021 Å	Electron diffraction [130]
	1.343 Å	Calculation [120]
∠F–C–F	108.1°	Calculation [120]
	$108.2 \pm 1.6°$	Electron diffraction [130]
	$108.3 \pm 2°$	Electron diffraction [128]
	$108.4 \pm 1.9°$	Electron diffraction [130]
	108.42°	Microwave spectroscopy [131]

[a]Roszak *et al.* in Ref. [120] point out that the spin-orbit effect could be important in computing the bond dissociation energy and the electron affinity of the CF_3I molecule, since the $^2P_{3/2} - ^2P_{1/2}$ splitting of the I atom is 0.94 eV. The value in parentheses includes spin-orbit correction.

Table 5.16 (continued)
Basic data on CF_3I and CF_3I^-.

Physical quantity	Value	Method and reference
CF_3I^-		
D (CF_3–I^-)	0.29 ± 0.2 eV	Charge exchange in Cs-CF_3I collisions [123]
	0.35 ± 0.2 eV	Charge exchange in Na-CF_3I collisions [123]
	0.38 ± 0.1 eV^b	Charge exchange in Cs-CF_3I collisions [122]
	0.610 eV	Calculation [120]
R (C–I)	2.881 Å	Calculation [120]
R (C–F)	1.367 Å	Calculation [120]
∠F–C–F	106.0°	Calculation [120]

^{b}Obtained using EA(I) = 3.063 eV.

The energy positions of these resonances explain the structure observed in the various electron scattering cross sections. The presence of the zero-energy dissociative electron attachment resonance makes the contribution of this process to the total electron scattering cross section substantial.

In Fig. 5.21 is shown the room-temperature total photoabsorption cross section, $\sigma_{\text{pa,t}}(\lambda)$, of CF_3I as a function of the photon wavelength, λ, in the wavelength range 160–390 nm [113, 115, 144, 145]. The spectrum consists of a continuous absorption band with a maximum cross section value of $\sim 6.0 \times 10^{-19}$ cm^2 at 267 nm (4.64 eV). The absorption band is due to $n \rightarrow \sigma^*$ transitions involving excitation of a nonbonding p electron on the I atom to an antibonding σ^* orbital involving the C and I atoms [115] (see also Robin [146] and Herzberg [147]). Photoabsorption measurements were also reported as a function of temperature [113, 115, 144, 145]. The measurements of Brouwer and Troe [144] were made at 300, 625, and 1050 K and show considerable high-temperature broadening. They also indicate that the CF_3I molecule thermally decomposes above $\sim$ 1000 K (see also Ref. [148]).

Recent electron energy-loss spectra of CF_3I at small scattering angles (3°, 5°, and 10°) and 30, 60, and 100 eV incident energy electrons [141–143] are consistent with the photoabsorption results. Figure 5.22 shows these electron energy-loss spectra, which cover the energy-loss range of 4–14 eV. Taken at the forward scattering angles of 3°, 5°, and 10°, these spectra exhibit little variation with incident electron energy. They

Table 5.17. Energy positions of the negative ion states of the CF_3I molecule.

Energy position (eV)	Reaction mechanism/ Possible assignment	Comments and references
0.0	$e + CF_3I \rightarrow I^- + CF_3$ [a]	Electron attachment using mass spectrometry [125]
0.0	$e + CF_3I \rightarrow I^- + CF_3$	Dissociative attachment producing I^- [133–135]
0.0	$e + CF_3I \rightarrow I^- + CF_3$ $a_1(\text{C–I}\,\sigma^*)$	Total dissociative attachment cross section measurement[b] [136]
~0.0	$e + CF_3I \rightarrow I^- + CF_3$	Total dissociative attachment cross section measurement[b] [137]
0.4	$a_1(\text{C–I}\,\sigma^*)$	Calculation [138]
~1.0	$e + CF_3I \rightarrow F^- + CF_2I$[c]	Total dissociative attachment cross section measurement [137]
~1.5	$e + CF_3I \rightarrow F^- + CF_2I$	Electron attachment using mass spectrometry [125]
1–2	$e + CF_3I^* \rightarrow F^- + CF_2I$	Formation of F^- from CF_3I at 500 K and 700 K[d] [139]
3.8	$e + CF_3I \rightarrow F^- + CF_2I$[e]	Dissociative attachment producing F^- [133, 134]
3.8	$e + CF_3I \rightarrow CF_3^- + I$	Dissociative attachment producing CF_3^- [133–135]
~4.5	$a_1(\text{C–F}\,\sigma^*)$	Total dissociative attachment cross section measurement [136]
~4.9	$a_1(\text{C–F}\,\sigma^*)$ and $e(\text{C–I}\,\pi^*)$	Total electron scattering cross section measurement / Calculation [138]
~5.5	$a_1(\text{C–F}\,\sigma^*)$	Vibrational excitation cross section function measurements [140–143]
~8	$e(\text{C–I}\,\pi^*)$	Vibrational excitation cross section function measurements [140–143]

[a]Dibeler *et al.* [132] give the onset of this process as ~ 0.0 eV, but the maximum intensity at the unacceptably high value of 1.8 eV.
[b]Due almost entirely to the respective reaction shown in Column 2 of this Table.
[c]Suggested possible reaction in Ref. [137].
[d]Found to be strongly temperature dependent.
[e]Dibeler *et al.* [132] give a threshold value of 3.6 ± 0.3 eV for this process, and a maximum-abundance energy of 5.6 eV.

show distinct structure below the first ionization threshold energy (near 10.4 eV, see Table 5.18) with peaks at 4.7, 7.4, 8.1, 9.0, and 9.8 eV. The

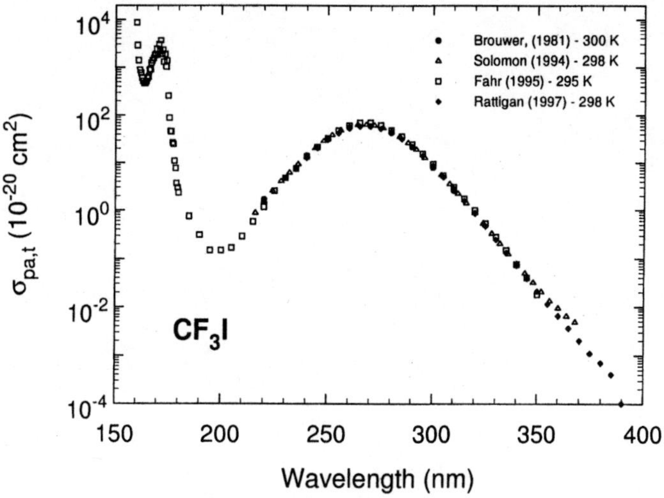

Figure 5.21. Total photoabsorption cross section, $\sigma_{\mathrm{pa,t}}(\lambda)$, as a function of the photon wavelength, λ, for CF_3I: (•) Ref. [144]; (△) Ref. [113]; (□) Ref. [145]; (♦) Ref. [115].

structure at 7.4 eV in the electron-energy-loss spectrum is consistent with that in the $\sigma_{\mathrm{pa,t}}(\lambda)$ around 170 nm.

Photoionization, photodissociation, and electron-impact studies have provided data for the ionization threshold energies for the formation of the parent positive ion and the various positive-ion fragments. These are listed in Table 5.18. It should be noted that some of the values listed for the case of CF_3I^+ do not distinguish between the two ground states, $X\,^2E_{3/2}$ and $X\,^2E_{1/2}$, of CF_3I^+ which are separated by ~ 0.6 eV.

3.2 Electron Scattering Cross Sections for CF_3I

3.2.1 Total Electron Scattering Cross Section, $\sigma_{\mathrm{sc,t}}(\varepsilon)$, of CF_3I

Underwood-Lemons *et al.* [138] measured the total electron scattering cross section of CF_3I from ~ 0.5 to 12 eV. Figure 5.23 shows their data. The rise in the cross section below ~ 1 eV was attributed to the existence of a resonance of $[a_1(\text{C–I}\,\sigma^*)]$ symmetry near zero energy, and the peak near 4.9 eV was tentatively ascribed to two negative-ion resonances of $[a_1(\text{C–F}\,\sigma^*)]$ and $[e(\text{C–I}\,\pi^*)]$ symmetries. Electron attachment studies [136] are consistent with the positions and assignments of these resonances, although Oster *et al.* [134] attributed the broad dissociative

Table 5.18. Threshold energies for the production of the parent and fragment positive ions by photon and electron impact on CF_3I.

Positive ion	Threshold energy (eV)	Reference	Method
CF_3I^+ $(X\,^2E_{3/2})^{a,b}$	10.23	[149, 150]	MSPI[c]
	10.29	[151, 152]	
	10.32 ± 0.03	[153]	ICR/PI[d]
	10.45	[154]	PES[e]
	10.61	[155]	
	10.81	[156]	PES
	$\leq 10.9^f$	[157]	EI[g]
CF_3I^+ $(X\,^2E_{1/2})$	10.91	[151, 152]	
	11.18	[154]	PES
	11.37	[156]	PES
CF_3^+	10.89	[149]	MSPI
	10.91^h	[158]	DPI[i]
	11.36 ± 0.03	[153]	
	11.85^j	[158]	DPI
	≤ 11.7	[157]	EI
$I^+(^3P_2)$	12.70	[158]	DPI
$I^+(^3P_0)$	13.50	[158]	DPI
$I^+(^3P_1)$	13.58	[158]	DPI
	≤ 14.5	[157]	EI
CF_2I^+	13.40	[158]	DPI
	≤ 14.3	[157]	EI
CF_2^+	17.62	[158]	DPI

[a]The CF_3I^+ ground state is a doublet, $X\,^2E_{3/2}$ and $X^2E_{1/2}$. The two states are separated by ~ 0.6 eV.
[b]Associated with the removal of an I-atom lone-pair electron.
[c]MSPI = Mass Spectrometric Photoionization Study.
[d]ICR/PI = Ion Cyclotron Resonance/Photoionization Technique.
[e]PES = Photoelectron Spectra.
[f]The electron energy resolution was not adequate to allow separation of the $X\,^2E_{3/2}$ and $X^2E_{1/2}$ states of CF_3I^+.
[g]EI = Electron Impact.
[h]For the products CF_3^+ + I $(^2P_{3/2})$.
[i]DPI = Dissociative Photoionization.
[j]For the products CF_3^+ + I $(^2P_{1/2})$.

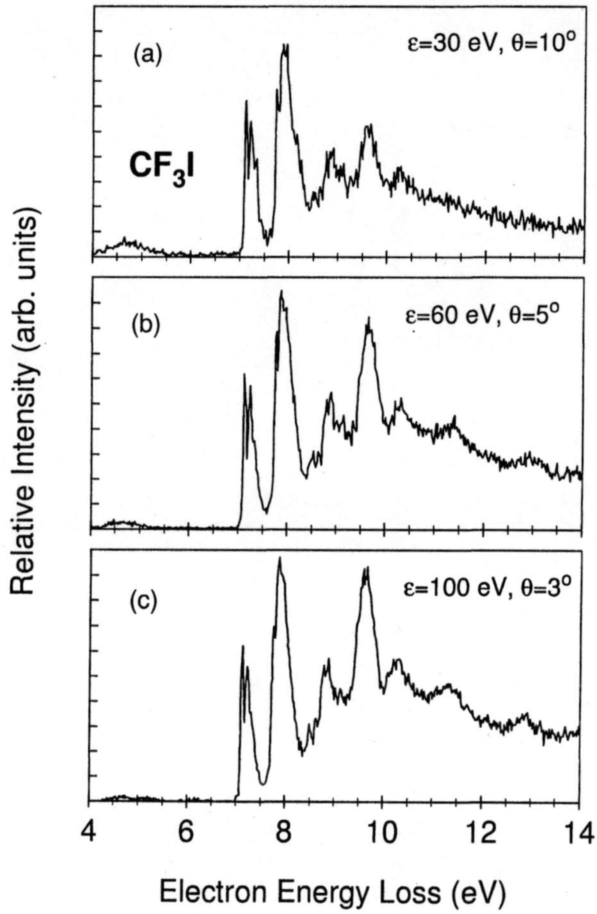

Figure 5.22. Electron energy-loss spectrum of CF_3I at 10°, 5°, and 3° scattering angles, respectively, for 30, 60, and 100 eV incident-electron energies measured by Kitajima *et al.* [141–143].

electron attachment resonance they observed near 4 eV to a temporary negative-ion state resulting from a two-particle (core excited) resonance associated with occupation of the lowest unoccupied molecular orbital (LUMO) of the neutral molecule.

Kawada *et al.* [159, 160] have also measured the total electron scattering cross section for CF_3I in the energy range 0.8–600 eV. These data are also presented in Fig. 5.23, and although they fall systematically higher than the data of Underwood-Lemons *et al.*, the energy dependence of the two sets of measurements is in overall agreement.

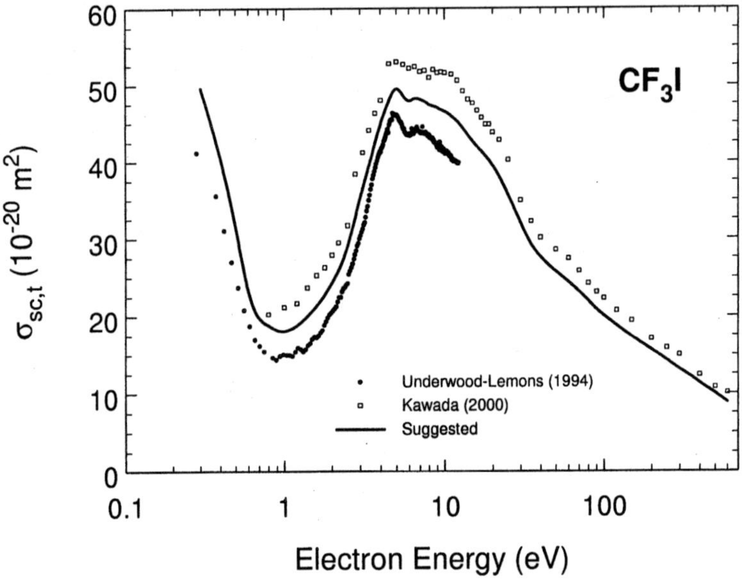

Figure 5.23. Total electron scattering cross section as a function on electron energy, $\sigma_{sc,t}(\varepsilon)$, for CF_3I: (•) Ref. [138]; (□) Ref. [159]; (—) suggested.

We have derived suggested values for the $\sigma_{sc,t}(\varepsilon)$ of CF_3I by fitting a curve to the average of the two available sets of measurements between 1 and 11 eV. The suggested data are then extended to higher and lower energies by normalizing the low-energy data of Underwood-Lemons *et al.* to the average at 1 eV, and by normalizing the high-energy data of Kawada *et al.* to the average at 11 eV. These values are shown by the solid line in Fig. 5.23. Values derived from this solid line are listed in Table 5.19 as our suggested data for the $\sigma_{sc,t}(\varepsilon)$ of the CF_3I molecule.

3.2.2 Elastic Differential Electron Scattering Cross Section, $\sigma_{e,diff}(\varepsilon)$, of CF_3I

Elastic differential electron scattering cross sections, $\sigma_{e,diff}(\varepsilon)$, have been measured by Kitajima *et al.* [141–143] for a number of electron energies between 1.5 and 60 eV over a scattering-angle range of 20° to 130° using a crossed-beam experiment. These measurements are shown in Fig. 5.24 and are listed in Table 5.20. Kitajima *et al.* indicated that the increases in the $\sigma_{e,diff}(\varepsilon)$ at low electron energies and small scattering angles result from electron-dipole scattering due to the large permanent electric dipole moment of the CF_3I molecule. It was also noted by

Table 5.19. Suggested values for the total electron scattering cross section, $\sigma_{sc,t}(\varepsilon)$, of the CF_3I molecule.

ε (eV)	$\sigma_{sc,t}(\varepsilon)$ (10^{-20} m^2)	ε (eV)	$\sigma_{sc,t}(\varepsilon)$ (10^{-20} m^2)	ε (eV)	$\sigma_{sc,t}(\varepsilon)$ (10^{-20} m^2)
0.30	49.6	5.5	48.6	45	26.5
0.40	40.5	6.0	47.9	50	25.7
0.50	31.5	6.5	48.2	60	24.4
0.60	23.3	7.0	48.1	70	23.1
0.70	19.8	7.5	47.8	80	21.9
0.80	18.8	8.0	47.4	90	20.8
0.90	18.2	8.5	47.2	100	20.0
1.00	18.1	9.0	47.0	150	17.3
1.25	19.1	9.5	46.7	200	15.6
1.50	20.7	10.0	46.5	250	14.3
1.75	22.4	11.0	45.9	300	13.1
2.0	24.3	12.0	45.2	350	12.2
2.5	29.0	15.0	42.6	400	11.3
3.0	35.3	20.0	39.4	450	10.6
3.5	40.5	25.0	35.7	500	9.97
4.0	44.9	30.0	32.0	600	8.82
4.5	47.9	35.0	29.2		
5.0	49.5	40.0	27.6		

Okamoto *et al.* [140] that the largest contribution to the $\sigma_{e,diff}(\varepsilon)$ comes from the I atom.

3.2.3 Momentum Transfer Cross Section, $\sigma_m(\varepsilon)$, of CF_3I

To our knowledge no measurements have been reported of the momentum transfer cross section, $\sigma_m(\varepsilon)$, of the CF_3I molecule. However, since the CF_3I molecule is polar, an estimate of its momentum transfer cross section at low electron energies (say, below ~ 0.5 eV) can be made using the Altshuler [33] formula (Eq. 5.1) and the experimental value of the dipole moment of the CF_3I molecule (Sec. 3.1 of Ch. 5). If we take for the dipole moment of the CF_3I molecule the average (0.96 D) of the two values given by McClellan [117] and use Eq. (5.1), we obtain the $\sigma_m(\varepsilon)$ values listed in Table 5.21.

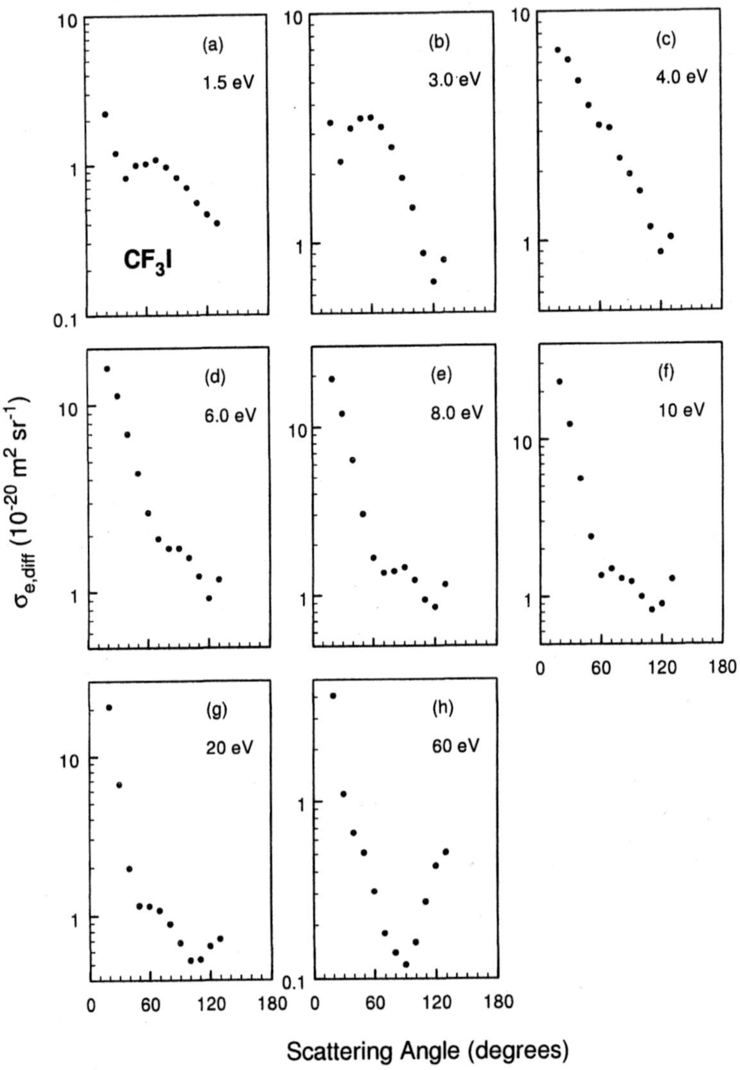

Figure 5.24. Elastic differential electron scattering cross sections, $\sigma_{e,\text{diff}}(\varepsilon)$, for CF_3I at a number of incident electron energies as measured by Kitajima *et al.* [141–143].

3.2.4 Differential Vibrational Excitation Cross Section, $\sigma_{\text{vib,diff}}(\varepsilon)$, of CF_3I

Okamoto *et al.* [140] and Kitajima *et al.* [141–143] used a crossed-beam experiment and measured the differential vibrational excitation cross section, $\sigma_{\text{vib,diff}}(\varepsilon)$, of the CF_3I molecule for an electron-energy

Table 5.20. Elastic Differential electron scattering cross sections, $\sigma_{e,diff}(\varepsilon)$, of CF_3I (in units of 10^{-20} m^2 sr^{-1}) for a number of incident electron energies as measured by Kitajima *et al.* [141–143].

	Electron energy							
Angle	1.5 eV	3 eV	4 eV	6 eV	8 eV	10 eV	20 eV	60 eV
20°	2.22	3.35	6.79	15.7	19.3	23.0	20.8	4.06
30°	1.21	2.27	6.16	11.2	12.0	12.4	6.62	1.10
40°	0.82	3.16	4.96	7.01	6.40	5.62	1.98	0.66
50°	1.00	3.49	3.88	4.33	3.03	2.39	1.16	0.51
60°	1.02	3.52	3.18	2.65	1.67	1.36	1.15	0.31
70°	1.09	3.20	3.10	1.92	1.36	1.50	1.08	0.18
80°	0.97	2.61	2.28	1.70	1.39	1.30	0.89	0.14
90°	0.82	1.91	1.94	1.70	1.46	1.24	0.68	0.12
100°	0.70	1.41	1.63	1.52	1.23	1.00	0.53	0.16
110°	0.55	0.90	1.14	1.21	0.94	0.82	0.54	0.27
120°	0.46	0.68	0.89	0.92	0.85	0.90	0.65	0.43
130°	0.40	0.84	1.03	1.16	1.15	1.29	0.72	0.51

Table 5.21. Calculated values of $\sigma_m(\varepsilon)$ for CF_3I determined via Eq. (5.1) and using the value of 0.96 D for the dipole moment of CF_3I.

ε (eV)	$\sigma_m(\varepsilon)$ (10^{-20} m^2)
0.005	902.0
0.01	451.0
0.05	90.2
0.1	45.1
0.5	9.0

loss of 0.14 eV and a scattering angle of 60° as a function of the incident electron energy from 1.5 to 17 eV (Fig. 5.25). This energy loss corresponds to the energies of the ν_1 and ν_2 stretching modes of CF_3I. The differential vibrational excitation cross section for the ν_1 and ν_2 stretching modes shows enhancements near 5.5 and 8 eV which were attributed to negative ion resonances [140, 141, 143]. The sharp increase in the cross section with decreasing electron energy below $\sim$ 2 eV sig-

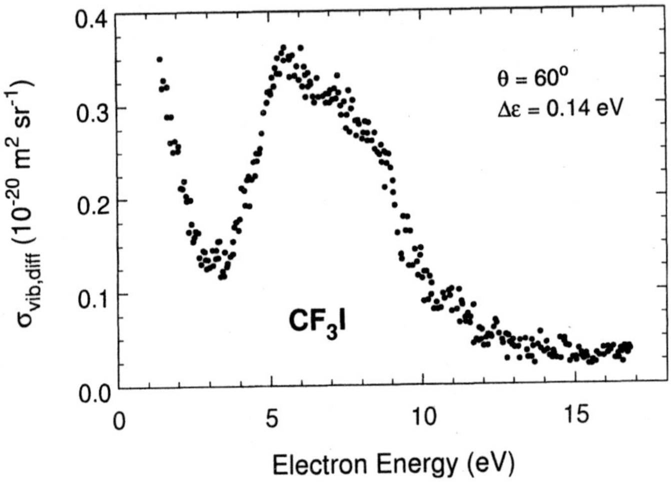

Figure 5.25. Differential vibrational excitation cross section as a function of the incident electron energy, $\sigma_{\text{vib,diff}}(\varepsilon)$, for CF_3I measured by Kitajima *et al.* [141–143]. The measurements are for a scattering angle of 60° and an electron-energy loss of 0.14 eV.

nifies the existence of another resonance at these energies, in agreement with electron attachment data (Table 5.17).

3.3 Electron-Impact Ionization of CF_3I

3.3.1 Partial Ionization Cross Sections, $\sigma_{\text{i,part}}(\varepsilon)$, of CF_3I

There have been two measurements of the partial ionization cross sections, $\sigma_{\text{i,part}}(\varepsilon)$, of the CF_3I molecule, which were both made using Fourier transform mass spectrometry. The first of these measurements was made by Riehl [157], and the other by Jiao *et al.* [161, 162]. The measurements of Riehl were calibrated using the data of Wetzel *et al.* [163] for Xe at an energy of 25 eV and have a quoted uncertainty in excess of ±30%. The measurements of Jiao *et al.* were normalized to the data of Wetzel *et al.* for Ar at 50 eV and have an uncertainty of about ±22% (the uncertainty in the energy scale in both studies is about ±0.5 eV). Since both of these sets of measurements were made at the same laboratory, and since the more recent data have a smaller uncertainty, we show in Fig. 5.26 only the more recent measurements for the production of CF_3I^+, CF_2I^+, CF_3^+, CF_2^+, CF^+, and I^+. For the electron energy range covered by these measurements ($\leq$ 70 eV), CF_3I^+ is the most abundant positive ion followed by I^+, CF_3^+, and CF_2I^+. The cross sections for

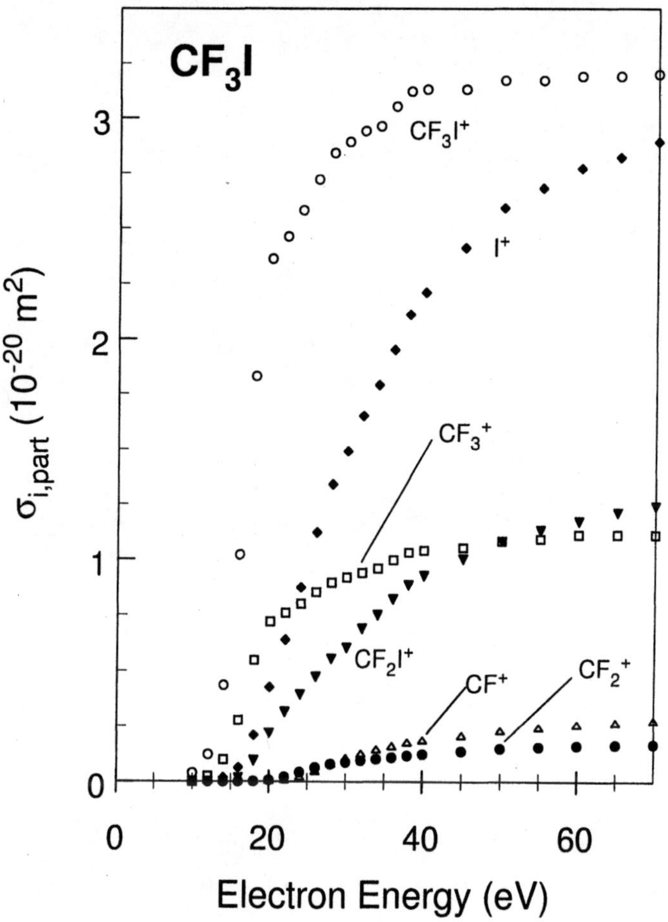

Figure 5.26. Partial ionization cross sections, $\sigma_{\mathrm{i,part}}(\varepsilon)$, for CF_3I measured by Jiao *et al.* [161, 162].

CF_2^+ and CF^+ are about a factor of 10 smaller. Riehl [157] estimated the cross section for production of F^+ to be less than 2×10^{-22} m^2 for electron energies below 50 eV.

In Table 5.22 are listed the $\sigma_{\mathrm{i,part}}(\varepsilon)$ data of Jiao *et al.* [161, 162] as our presently suggested values for the partial ionization cross sections for this molecule.

The relative abundances of the various positive ions formed by electron impact on CF_3I as measured by Jiao *et al.* at 70 eV, differ from the 70 eV "cracking pattern" of the CF_3I molecule reported in Heller and Milne [164]. The ratio of the Jiao *et al.* partial ionization cross sections at

Table 5.22. Suggested values for the partial ionization cross sections, $\sigma_{i,part}(\varepsilon)$, of CF_3I from Jiao *et al.* [161, 162].

	$\sigma_{i,part}(\varepsilon)$ (10^{-20} m^2)					
ε (eV)	CF^+	CF_2^+	CF_3^+	I^+	CF_2I^+	CF_3I^+
10						0.039
12			0.025			0.121
14			0.098	0.018		0.435
16			0.277	0.063	0.016	1.02
18			0.547	0.208	0.093	1.83
20		0.006	0.720	0.425	0.219	2.36
22	0.005	0.019	0.759	0.640	0.313	2.46
24	0.018	0.042	0.800	0.874	0.392	2.58
26	0.042	0.063	0.852	1.12	0.473	2.72
28	0.076	0.078	0.896	1.34	0.555	2.84
30	0.101	0.086	0.919	1.49	0.604	2.89
32	0.121	0.095	0.940	1.65	0.690	2.94
34	0.140	0.100	0.958	1.79	0.750	2.96
36	0.155	0.107	0.995	1.95	0.820	3.05
38	0.171	0.115	1.03	2.11	0.885	3.12
40	0.181	0.122	1.04	2.21	0.925	3.13
45	0.202	0.135	1.05	2.41	1.00	3.13
50	0.225	0.147	1.08	2.59	1.08	3.17
55	0.240	0.154	1.09	2.68	1.13	3.17
60	0.253	0.160	1.11	2.77	1.17	3.19
65	0.261	0.164	1.11	2.82	1.21	3.19
70	0.271	0.167	1.11	2.89	1.24	3.20

70 eV for the ions $CF_3I^+ : I^+ : CF_3^+ : CF_2^+ : CF^+$ is 100 : 90.3 : 34.7 : 5.2 : 8.5, while Heller and Milne list, respectively, 100 : 95.6 : 77.4 : 7.3 : 11.6. The higher relative intensities of the lower-mass ions measured by Heller and Milne, as compared to the measurements of Jiao *et al.*, indicate possible discrimination against high-mass ions common to quadrupole mass spectrometers.

For ion-molecule reactions involving I^+ and CF_3^+ with CF_3I see Berman *et al.* [153], and for formation of CF_3I^+ in photoinduced charge transfer at surfaces see Sun *et al.* [165].

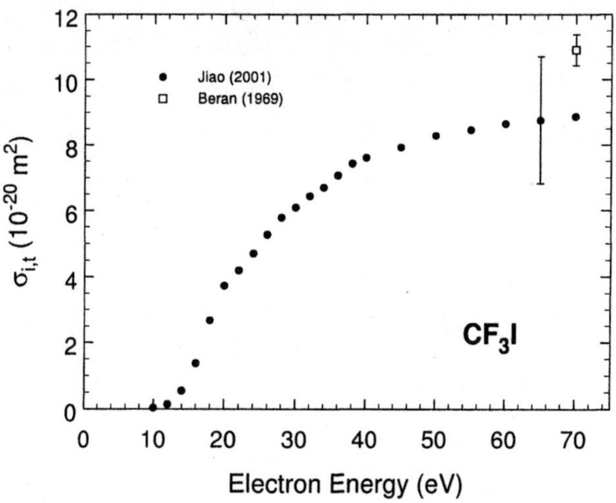

Figure 5.27. Total ionization cross section, $\sigma_{i,t}(\varepsilon)$, for CF_3I: (•) Ref. [161, 162]; (□) Ref. [50].

3.3.2 Total Ionization Cross Section, $\sigma_{i,t}(\varepsilon)$, of CF_3I

The total ionization cross section, $\sigma_{i,t}(\varepsilon)$, of CF_3I obtained by Jiao *et al.* [162] by summing their partial ionization cross sections is shown in Fig. 5.27. The error bar shown in the figure indicates the ±22% uncertainty quoted by the authors for the partial ionization cross sections. Also plotted in Fig. 5.27 is the value $[(10.9 \pm 0.48) \times 10^{-20}\ m^2]$ of $\sigma_{i,t}(\varepsilon)$ of CF_3I measured by Beran and Kevan [50] using 70 eV incident energy electrons. The data are seen to be just within the combined uncertainties at this energy. The data of Jiao *et al.* are listed in Table 5.23 as our presently suggested values for the $\sigma_{i,t}(\varepsilon)$ of CF_3I.

To our knowledge no measurements of the electron-impact ionization coefficient have been made for CF_3I.

3.4 Electron Attachment to CF_3I

3.4.1 Relative Cross Sections for the Production of the Fragment Negative Ions I^-, F^-, CF_3^-, and FI^- by Electron Attachment to CF_3I

A number of studies have investigated the process of electron attachment to CF_3I. These include electron-beam measurements of the relative cross sections for the formation of the various fragment negative ions

Table 5.23. Suggested values for the total ionization cross section, $\sigma_{i,t}(\varepsilon)$, of CF_3I from Jiao *et al.* [161, 162].

ε (eV)	$\sigma_{i,t}(\varepsilon)$ (10^{-20} m^2)	ε (eV)	$\sigma_{i,t}(\varepsilon)$ (10^{-20} m^2)	ε (eV)	$\sigma_{i,t}(\varepsilon)$ (10^{-20} m^2)
10	0.039	26	5.27	45	7.93
12	0.146	28	5.78	50	8.29
14	0.551	30	6.09	55	8.46
16	1.38	32	6.44	60	8.65
18	2.68	34	6.70	65	8.76
20	3.73	36	7.08	70	8.88
22	4.20	38	7.43		
24	4.71	40	7.61		

[125, 132–135, 139, 166], electron-beam measurements of the total electron attachment cross section [136, 137], electron-swarm measurements of the thermal-electron attachment rate constant [167–170], very-low-energy electron-beam total electron attachment cross-section measurements [171], and measurements of the rate constant of bound-electron attachment at thermal and sub-thermal electron energies [172, 173]. Based on the results of these studies, electron attachment to the CF_3I molecule produces I^-, F^-, CF_3^-, and FI^-, principally via two dissociating negative-ion states, one near 0.0 eV and another near 3.8 eV. This is seen from the measurements of Oster *et al.* [134] shown in Fig. 5.28. The zero-energy process generates I^- via the reaction

$$e + CF_3I \rightarrow I^- + CF_3, \tag{5.5}$$

and is exothermic by $-(0.67\pm0.1)$ eV [134]. The 3.8 eV process generates F^-, CF_3^-, and FI^-. The F^- and CF_3^- fragment anions are generated via the reactions

$$e + CF_3I \rightarrow F^- + CF_2I \tag{5.6a}$$

$$\rightarrow CF_3^- + I. \tag{5.6b}$$

Reactions (5.6a) and (5.6b) are endothermic, respectively, by about (1.5±0.3) eV and (0.7±0.3) eV [134]. According to Oster *et al.*, the weak F^- signal near 0.0 eV does not arise from the direct dissociative electron attachment process (5.6a), but rather from an ion-molecule reaction of the form $I^- + CF_3I \rightarrow F^- + CF_2I_2$.

The formation of the FI^- ion has an "appearance energy" of (2.2 ± 0.2) eV and requires a multiple dissociation process [133]. Interestingly,

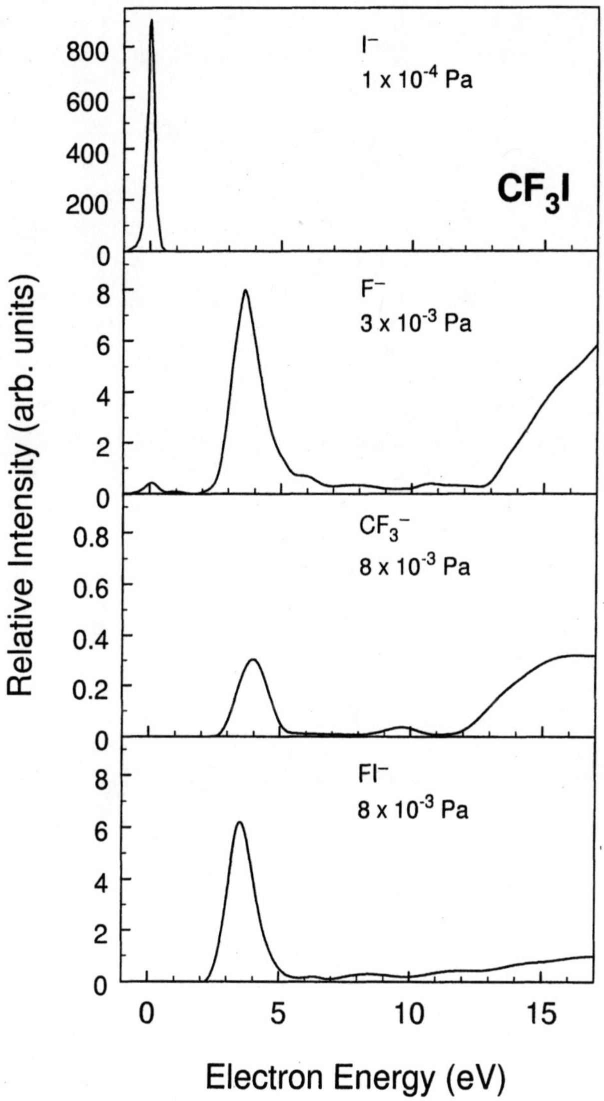

Figure 5.28. Relative cross sections for the formation of I^-, F^-, CF_3^-, and FI^- by electron impact on CF_3I for $T = 300$ K (data of Oster *et al.* [134]). Note the pressures in comparing the magnitudes of the relative cross sections of the various negative ions. The work of Oster *et al.* also indicates possible production of I^- at higher energies besides the zero-energy peak.

the complementary reaction to Reaction (5.6b), that is, the production of $I^- + CF_3$, was not observed from the 3.8 eV negative ion state [133, 134].

Reactions (5.5) and (5.6b) result from breaking the C–I bond as a result of localization of the extra electron into a molecular orbital with antibonding $\sigma^*_{\mathrm{C-I}}$ character. This is supported by theoretical calculations [138] and also by the fact that in the formation of CF_3^- at threshold, the total excess energy of the transient anion appears as kinetic energy of the dissociation products [133]. This is also consistent with photoabsorption and photodissociation studies. For instance, Suh *et al.* [174] found that during the photodissociation of CF_3I at 266 nm (4.66 eV), 85% of the available energy is partitioned into translational energy.

In addition to the dissociative electron attachment processes (5.5) and (5.6), F^- and CF_3^- have been observed, respectively, at energies above 12.3 and 11.8 eV [133, 134], and their production at these energies has been attributed to the nonresonant ion-pair formation processes

$$e + CF_3I \rightarrow CF_3I^* + e \tag{5.7a}$$

$$\hookrightarrow F^- + CF_2I^+ \tag{5.7b}$$

$$\hookrightarrow CF_3^- + I^+. \tag{5.7c}$$

The CF_3I^- parent negative ion has not been observed in any of the gas-phase electron attachment studies. However, it is known to be formed in charge-transfer reactions involving K, Na, and Cs atoms [122–124] and in electron attachment to CF_3I clusters [134]. Interestingly, in a study of the desorption of CF_3^- following the interaction of low-energy (0–10 eV) electrons with a CF_3I film [135], the desorption cross section was found to be more than two orders of magnitude greater than the corresponding gas-phase dissociative electron attachment cross section.

3.4.2 Effect of Temperature on the Production of F^- by Dissociative Electron Attachment to CF_3I

Figure 5.29 shows the relative cross section for the formation of F^- by dissociative electron attachment to CF_3I at three temperatures (350, 500 and 700 K) as reported by Hahndorf and Illenberger [139]. The weak F^- signal shown near 0.0 eV in the room temperature measurements (Fig. 5.28) increases substantially with increasing temperature above ambient. However, even at 700 K the relative cross section for the formation of F^- near 0.0 eV is small compared with that for the formation of I^- at this energy.

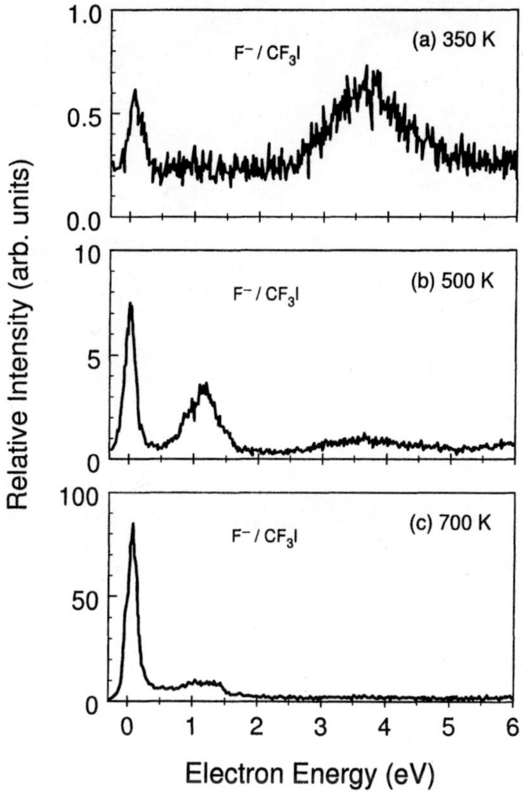

Figure 5.29. Variation with gas temperature of the relative cross section for the formation of F^- by dissociative electron attachment to CF_3I measured by Hahndorf and Illenberger [139].

In view of the magnitude and energy dependence of the cross section for I^- production, it is not expected that the I^- signal will exhibit a significant temperature dependence [175]. Hence, for temperatures in this range, the total electron attachment cross section as measured by Underwood-Lemons *et al.* [136], and the total thermal-electron attachment-rate constant as measured by various groups should not be considerably affected by the gas temperature, as is indeed the case (see Sec. 3.4.5 of Ch. 5). Moreover, the data in Fig. 5.29 indicate that the production of F^- from CF_3I has another maximum between 1 and 2 eV. This is consistent with earlier observations of a negative-ion peak around 1 eV by Marriott and Craggs [125] and by Buchel'nikova [137] (see Table 5.17).

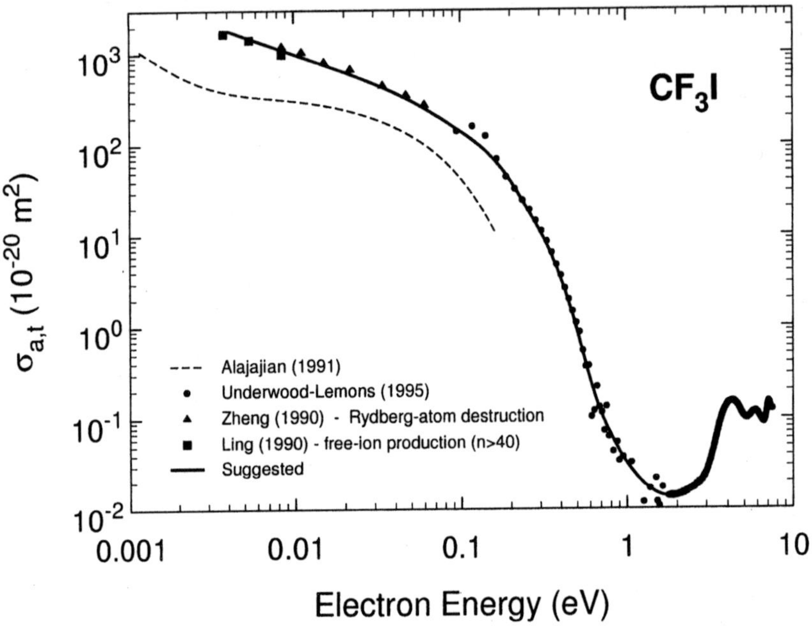

Figure 5.30. Total electron attachment cross section as a function of electron energy, $\sigma_{a,t}(\varepsilon)$, for CF_3I. Free-electron attachment: (- - -) Ref. [171]; (•) Ref. [136]. Bound-electron attachment: (▲) Rydberg-atom destruction data from Ref. [173]; (■) free-ion production data ($n > 40$) from Ref. [172]; (—) suggested.

3.4.3 Total Electron Attachment Cross Section as a Function of Electron Energy, $\sigma_{a,t}(\varepsilon)$, of CF_3I

The total electron attachment cross section, $\sigma_{a,t}(\varepsilon)$, of CF_3I has been measured by Underwood-Lemons [136, 176] for electron energies ≤ 7.5 eV under single-collision conditions and temperatures between 393 and 563 K. Figure 5.30 shows the measurements at 393 K. No significant temperature effects were observed. The peak energy and symmetry assignment of the negative-ion resonances observed in a total electron-scattering cross section study [138] are given in Table 5.17. The steep rise of the cross section toward zero energy is attributed to electron capture into the LUMO of CF_3I with a_1 symmetry and σ^* (C–I) bonding character. Thus the fragment ion is I^- [Reaction (5.5)]. In contrast, the fragment negative-ion peak near 4.5 eV is associated with the next higher unoccupied orbital a_1 (C–F σ^*). Thus the primary ion associated with this resonance is F^- [Reaction (5.6a)].

Also plotted in Fig. 5.30 is the total electron attachment cross section of Alajajian *et al.* [171] determined using the Kr photoionization method

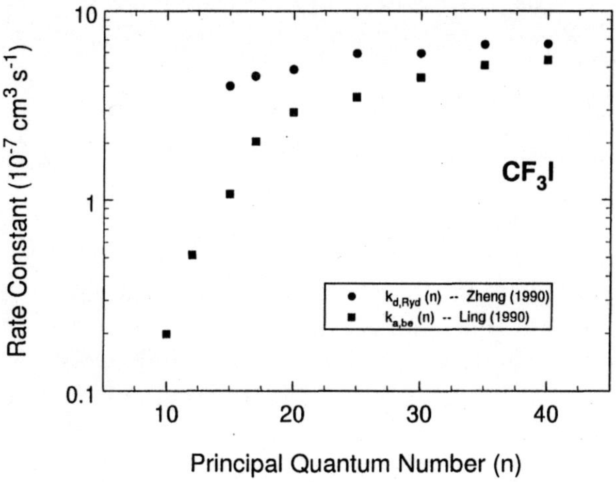

Figure 5.31. Rate constants, $k_{\mathrm{d,Ryd}}(n)$, for Rydberg-atom destruction (•, Ref. [173]), and, $k_{\mathrm{a,be}}(n)$, for negative-ion production (■, Ref. [172]) in collisions of $K^*(nd)$ with CF_3I as a function of n (see text).

for the production of monoenergetic low-energy electrons. The relative cross section values obtained by this method were put on an absolute scale by normalization to the thermal ($T = 300$ K) electron attachment-rate-constant value of Shimamori and Nakatani [167] (see Table 5.26).

Besides the data of Alajajian *et al.* [171], two other measurements of $\sigma_{\mathrm{a,t}}(\varepsilon)$ for CF_3I in the low-energy range have been plotted in Fig. 5.30. The first is derived from the rate constant, $k_{\mathrm{d,Ryd}}(n)$, for the destruction of the Rydberg atoms $K^*(nd)$ in collisions with CF_3I [173]. The second is derived from the rate constant for negative-ion formation, $k_{\mathrm{a,be}}(n)$, in collisions of $K^*(nd)$ with CF_3I [172]. In both cases n is the principal quantum number of the Rydberg electron. The rate constant $k_{\mathrm{a,be}}(n)$ refers to the reaction

$$K^*(nd) + CF_3I \rightarrow K^+ + CF_3I^{-*} \rightarrow K^+ + CF_3 + I^-, \qquad (5.8)$$

and is less than or equal to $k_{\mathrm{d,Ryd}}(n)$ since the probability of escape of I^- can be less than one. As can be seen from Fig. 5.31 the difference between $k_{\mathrm{d,Ryd}}(n)$ and $k_{\mathrm{a,be}}(n)$ depends on the value of n. As n becomes small, postattachment electrostatic attraction between the product ions (I^- and K^+) becomes important causing larger differences between the two rate constants.

The rate constants $k_{\mathrm{d,Ryd}}(n)$ [173] and $k_{\mathrm{a,be}}(n)$ [172] were converted to functions of the electron energy, using the relationship $\varepsilon = R/v^2$, where R is the Rydberg constant (13.6 eV). These values were then used to determine the bound-electron attachment cross section, $\sigma_{a,be}(\varepsilon)$ (and similarly $\sigma_{d,Ryd}(\varepsilon)$), from

$$\sigma_{a,be}(v) = k_{a,be}(n)/v_{rms}, \tag{5.9}$$

where v_{rms} is the root-mean-square velocity of the Rydberg electron. All values of $\sigma_{d,Ryd}(\varepsilon)$ determined this way are shown in Fig. 5.30. However, for $\sigma_{a,be}(\varepsilon)$ we have plotted only the data for $n > 40$ since a number of studies (e.g., see Refs. [177–180]) have indicated that Eq. (5.9) is expected to be valid only for large values of n ($\gtrsim 30$). Figure 5.30 shows that the low-energy bound-electron attachment data agree well with the higher-energy electron-beam measurements of Underwood-Lemons *et al.* [136]. However, both sets of data disagree with the data of Alajajian *et al.* [171] which were obtained by normalization to the electron swarm attachment rate constant. This may indicate a difference in the magnitude of the electron attachment rate constant near thermal energies as determined by the electron swarm method and by techniques employing bound electron capture processes (see discussion in the next section).

The data in Fig. 5.30 show that at thermal energy (~ 0.038 eV, $T = 300$ K), the total electron attachment cross section exceeds 10^{-18} m^2. This value is more than two orders of magnitude higher than the value of 7.8×10^{-21} m^2 measured near 0 eV by Buchel'nikova [137], which is not shown in Fig. 5.30.

The data of Underwood-Lemons *et al.* [136], Zheng *et al.* [173], and Ling *et al.* [172] are in agreement, and the solid line in Fig. 5.30 is a fit to these three sets of data. Values taken from this curve are listed in Table 5.24 as our suggested data for the $\sigma_{a,t}(\varepsilon)$ of CF_3I.

3.4.4 Total Electron Attachment Rate Constant as a Function of the Mean Electron Energy, $k_{a,t}(\langle\varepsilon\rangle)$, of CF_3I

The measurements of $k_{a,t}(\langle\varepsilon\rangle)$ by Shimamori *et al.* [168] and Sunagawa and Shimamori [170] are shown in Fig. 5.32. A pulse radiolysis microwave-cavity method was employed with provisions for electron heating above thermal and an indirect method of determining the mean electron energy (see Ref. [168] for details). Also shown in the figure are the $k_{a,t}(\langle\varepsilon\rangle)$ measurements of Alajajian *et al.* [171] who used the Kr photoionization technique for the production of monoenergetic low-energy electrons and the room temperature ($T = 300$ K) thermal-electron attachment-rate constant, $(k_{a,t})_{th}$, of Shimamori and Nakatani [167] (1.7×10^{-7} cm^3 s^{-1}, see Table 5.26 in the following section) for normalization of their relative data. The average of the four "room tem-

Table 5.24. Suggested values for the total electron attachment cross section, $\sigma_{a,t}(\varepsilon)$, of CF_3I.

ε (eV)	$\sigma_{a,t}(\varepsilon)$ (10^{-20} m^2)	ε (eV)	$\sigma_{a,t}(\varepsilon)$ (10^{-20} m^2)	ε (eV)	$\sigma_{a,t}(\varepsilon)$ (10^{-20} m^2)
0.004	1820	0.10	139.0	4.00	0.138
0.005	1575	0.15	76.1	4.25	0.148
0.006	1398	0.20	39.4	4.50	0.140
0.007	1265	0.25	19.9	4.75	0.122
0.008	1157	0.30	11.1	5.00	0.104
0.009	1070	0.40	3.38	5.25	0.100
0.010	997	0.50	0.930	5.50	0.107
0.015	757	0.60	0.266	5.75	0.115
0.020	623	0.70	0.114	6.00	0.112
0.025	532	0.80	0.067	6.25	0.102
0.030	464	0.90	0.044	6.50	0.092
0.040	369	1.0	0.032	6.75	0.098
0.050	302	1.5	0.015	7.00	0.135
0.060	251	2.0	0.014	7.25	0.134
0.070	212	2.5	0.017	7.50	0.118
0.080	183	3.0	0.026		
0.090	158	3.5	0.070		

perature" values of $(k_{a,t})_{th}$ is 1.9×10^{-7} cm^3 s^{-1} (Table 5.26, following section), and this value is also plotted in Fig. 5.32. Additionally, we have plotted the values of $k_{d,Ryd}(n)$ measured by Zheng *et al.* [173], and the values of $k_{a,be}(n)$ for $n > 40$ measured by Ling *et al.* [172].

The bound-electron attachment rate-constant measurements indicate higher values than the free-electron attachment rate-constant measurements made in electron swarm experiments. At thermal (T $\sim$ 300 K) energies the electron swarm data are lower by about a factor of 2. This difference is not understood and complicates the determination of suggested values of the $k_{a,t}(\langle\varepsilon\rangle)$ for CF_3I. Because all four independent measurements of the thermal value of $k_{a,t}(\langle\varepsilon\rangle)$ are in essential agreement (they are within the quoted or expected uncertainties), we choose to fit only the electron swarm data in Fig. 5.32 in an effort to determine suggested values for $k_{a,t}(\langle\varepsilon\rangle)$. Our fit to the electron swarm data is shown by the solid line, and values from this line are listed in Table 5.25 as our suggested data for the of the $k_{a,t}(\langle\varepsilon\rangle)$ of the CF_3I molecule.

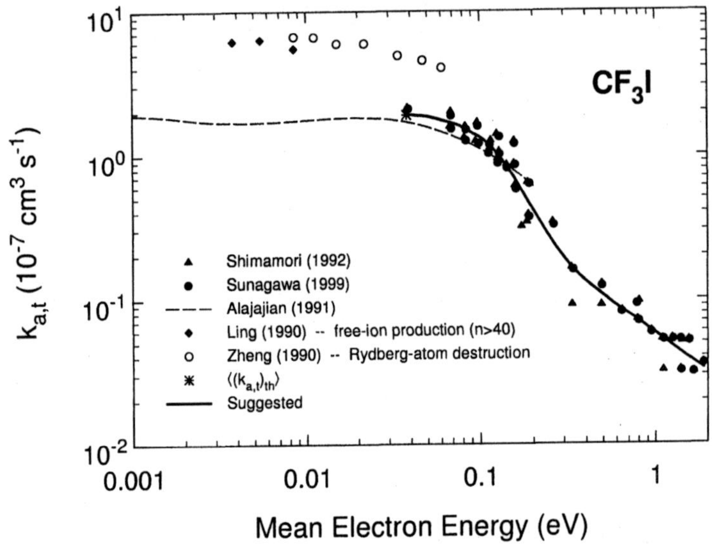

Figure 5.32. Total electron attachment rate constant as a function of the mean electron energy, $k_{a,t}(\langle\varepsilon\rangle)$, for CF_3I. Swarm data: (▲) Ref. [168]; (•) Ref. [170]. Low-energy electron beam data: (- - -) Ref. [171]. Bound-electron attachment data: (○) Rydberg-atom destruction data from Ref. [173]; (♦) free-ion production data ($n > 40$) from Ref. [172]. (∗) Average thermal value (1.9×10^{-7} cm^3 s^{-1}) of the room temperature electron swarm measurements (see text). (—) Suggested values for $k_{a,t}(\langle\varepsilon\rangle)$.

Table 5.25. Suggested values for the total electron attachment rate constant, $k_{a,t}(\langle\varepsilon\rangle)$, of CF_3I.

$\langle\varepsilon\rangle$ (eV)	$k_{a,t}$ (10^{-7} cm^3 s^{-1})	$\langle\varepsilon\rangle$ (eV)	$k_{a,t}$ (10^{-7} cm^3 s^{-1})
0.038	1.90	0.4	0.132
0.05	1.85	0.5	0.106
0.06	1.75	0.6	0.089
0.07	1.66	0.7	0.079
0.08	1.56	0.8	0.070
0.09	1.43	0.9	0.063
0.10	1.33	1.0	0.058
0.15	0.754	1.5	0.040
0.20	0.410	1.9	0.033
0.30	0.193		

Table 5.26. Thermal values, $(k_{a,t})_{th}$, of the total electron attachment rate constant of CF_3I.

$(k_{a,t})_{th}$ (10^{-7} cm^3 s^{-1})	T (K)	Reference
1.5	250	[167]
1.7 ± 0.2	~ 300	[167]
2.0	350	[167]
1.9 ± 0.1	300	[181]
2.0 ± 0.2	~ 300	[168]
2.2	293	[169]
2.4	615	[169]
2.2	777	[169]
1.5	1022[a]	[169]

[a]Data suspected for possible error due to possible thermal decomposition of the gas.

3.4.5 Thermal Electron Attachment Rate Constant, $(k_{a,t})_{th}$, of CF_3I

The available data on the thermal value of the total electron attachment rate constant, $(k_{a,t})_{th}$, are listed in Table 5.26. The room-temperature values of $(k_{a,t})_{th}$ are within the combined quoted uncertainties. Their average is 1.9×10^{-7} cm^3 s^{-1}.

Interestingly, the measurements of $(k_{a,t})_{th}$ in Fig. 5.33 show a small increase with temperature from 250 to 500 K and a decline beyond 600 K. This is consistent with other observations [175] involving strong dissociative electron attachment processes peaking near 0 eV when the potential energy curve of the dissociating negative ion state crosses that for the ground state of the neutral molecule in such a way that population of vibrational levels higher than the $\nu = 0$ of the neutral molecule results in the initial state of the neutral molecule lying above the dissociating negative ion state.

3.5 Optical Emission from Electron Impact on CF_3I

Martínez *et al.* [182] performed an experimental study of the visible and ultraviolet emissions following pulsed electron impact on CF_3I. For electron-impact energies of 100 eV, the spectra in the 200–600 nm region show a number of narrow lines superimposed on two broad bands centered at 300 and 470 nm. The emission bands have been attributed to

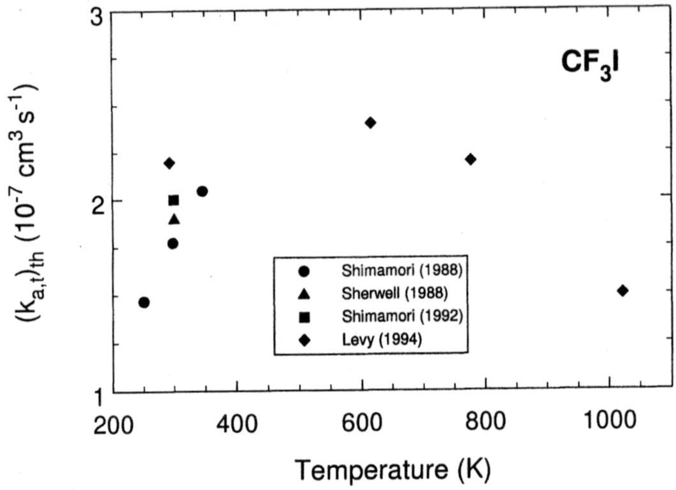

Figure 5.33. Temperature dependence of the thermal electron attachment rate constant, $(k_{a,t})_{th}$, of CF_3I: (•) Ref. [167]; (▲) Ref. [181]; (■) Ref. [168]; (♦) Ref. [169].

the CF_3^* and IF^* species and the atomic lines to the I_I^* and I_{II}^* fragments [182].

3.6 Summary of Suggested Cross Sections and Coefficients for CF_3I

The meager data on the electron scattering cross sections for the CF_3I molecule are summarized in Fig. 5.34. The following are suggested values for

- $\sigma_{sc,t}(\varepsilon)$ — Table 5.19, Fig. 5.23;
- $\sigma_{i,t}(\varepsilon)$ — Table 5.23, Fig. 5.27; and
- $\sigma_{a,t}(\varepsilon)$ — Table 5.24, Fig. 5.30.

The $\sigma_{e,diff}$ (Table 5.20) and $\sigma_{i,part}(\varepsilon)$ (Table 5.22) are also suggested but are not plotted in the figure. The suggested values of $k_{a,t}(\langle\varepsilon\rangle)$ are listed in Table 5.25.

With the exception of the total electron scattering cross section, total electron-impact ionization cross section, and total electron attachment cross section, there is need for measurement of the cross sections for all other electron collision processes for CF_3I. No experimental data exist for the electron transport, ionization, and attachment coefficients of this molecule.

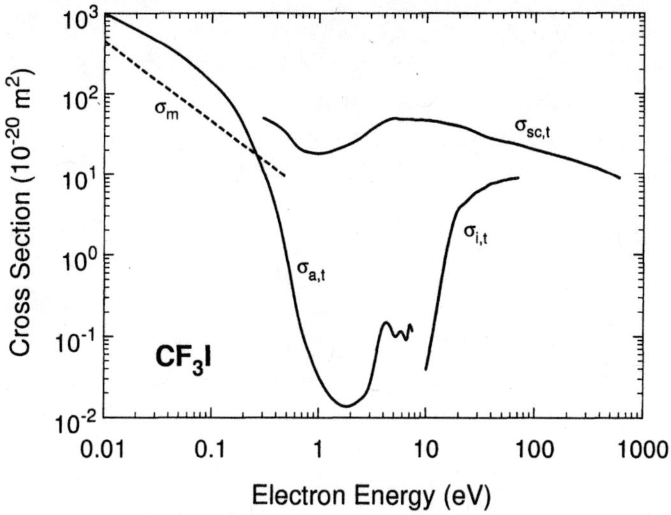

Figure 5.34. Summary of suggested electron collision cross sections for CF_3I.

* * *

4. Electron Interactions with c-C_4F_8

Perfluorocyclobutane (c-C_4F_8) is a plasma processing gas employed in silicon etching (e.g., see Refs. [112, 183–188]). Electron impact on c-C_4F_8 generates large quantities of CF_2 radicals, which in c-C_4F_8 plasmas form a polymer coating on silicon and account for the high etching selectivity of SiO_2 over Si (e.g., see Refs. [112, 185, 187, 188]). The c-C_4F_8 molecule also decomposes thermally above about 633 K [189], principally via the unimolecular decomposition reaction c-$C_4F_8 \rightarrow 2\,C_2F_4$ (see Refs. [189, 190] and references cited therein), and thus CF_2 radicals may be formed in a c-C_4F_8 plasma by electron-impact dissociation of the C_2F_4 byproduct. Consistent with this decomposition mechanism are infrared multiphoton dissociation studies [191–194] of c-C_4F_8 which show that photodissociation occurs via the process c-$C_4F_8 + nh\nu \rightarrow 2\,C_2F_4$.

Besides its use in plasma etching, c-C_4F_8 has many other applications ranging from its use in retinal detachment surgery [195, 196] to its possible utilization as a gaseous dielectric, especially in gas mixtures (e.g., see Refs. [36, 197–199]). Perfluorocyclobutane is also of environmental interest because it is a global warming gas [200]. Its half-life in the atmosphere, based on the assumption that photolysis is the dominant atmospheric loss process for c-C_4F_8, is very long (3 200 years) [200].

However, Morris *et al.* [201] argue that the atmospheric lifetime of c-C_4F_8 is significantly reduced (to 1 400 years) if consideration is given to its atmospheric removal by electron interactions. The long atmospheric lifetime of c-C_4F_8 partly accounts for its high global warming potential, which is 8 700 times that of CO_2 over a 100-yr time period [200].

In this section a number of electron collision cross section, coefficient, and rate-constant data for c-C_4F_8 are synthesized, assessed, and discussed. These include cross sections for total electron scattering, elastic differential electron scattering, partial and total ionization, dissociation into neutral fragments, and electron attachment. Also discussed are data on electron transport, ionization, and attachment coefficients.

The available data on both the electron collision cross sections and the electron transport coefficients need further scrutiny. Measurements are needed of the momentum transfer and elastic integral cross sections, and of the cross sections for other significant low-energy electron collision processes such as vibrational and electronic excitation. In addition, electron transport data over wider E/N ranges are needed.

Assessment of the data on the electron affinity of c-C_4F_8, and the electron attachment to and electron scattering by c-C_4F_8, suggests the existence of negative ion states at about -0.6, 4.9, 6.9, 9.0, and 10.5 eV. Besides the experimental data summarized, assessed, and discussed in this section, there have been two Boltzmann-code calculations [202, 203] of various coefficients, but these results are not discussed here.

4.1 Structural and Electronic Properties of c-C_4F_8

Electron diffraction studies [204–209] have shown that the carbon atoms in the c-C_4F_8 molecule are not planar. Similarly, infrared studies [210–213] are consistent with a nonplanar molecular structure. The c-C_4F_8 molecule has a puckered structure and belongs to the D_{2d} symmetry point group. Table 5.27 lists information on its structural properties and also values of the electron affinity, ionization threshold energy, dissociation energy, and polarizability of the c-C_4F_8 molecule.

The c-C_4F_8 molecule forms parent negative ions below $\sim$ 1 eV and this property is consistent with the molecule having a positive electron affinity (EA). Besides the "zero-energy" electron attachment process which is associated with the negative ion state located at –EA, fragment negative ions are formed via a number of negative ion states which lie between 0 and $\sim$ 12 eV. The energy positions of these negative ion states, as indicated by the experimental data on negative ion formation, are listed in Table 5.28 and are plotted in Fig. 5.35. Similar results

Table 5.27. Physical and structural data on the c-C_4F_8 molecule.

Physical quantity	Value	Ref.	Method
Electron affinity	$\sim$0.63 eV	[214]	Electron attachment/ detachment studies
	0.61 eV	[215]	Electron attachment
	$\geq$0.4 eV	[216]	Endothermic negative-ion charge-transfer reactions
Ionization threshold energy	11.6 ± 0.2 eV	[217]	TPEPICO spectroscopy[a]
	12.1 ± 0.1 eV[b]	[218]	Electron impact
	12.23 eV[c]	[11]	
	12.25 eV[d]	[220]	Electron impact
Dissociation energy, D(C–F)	$\sim 4.4 \pm 0.2$ eV	[218]	Electron impact
Polarizability	73.7×10^{-25} cm^3	[221]	From liquid density and refractive index using the Lorenz-Lorentz equation
	92.4×10^{-25} cm^3	[221]	Ion Cyclotron Resonance
	74.9×10^{-25} cm^3	[221]	Semi-empirical calculation
	104.3×10^{-25} cm^3	[221]	Semi-empirical calculation
	124.7×10^{-25} cm^3	[221]	Semi-empirical calculation
Molecular structure[e] parameters			
C–C internuclear distance	1.63 ± 0.02 Å	[206]	Electron diffraction
	1.60 ± 0.04 Å	[205]	Electron diffraction
	1.58 Å	[209]	Electron diffraction
	1.566 ± 0.008 Å	[207]	Electron diffraction
	1.560 ± 0.009 Å	[208]	Electron diffraction
	1.54 Å	[204]	Electron diffraction
C–F internuclear distance	1.38 Å	[204]	Electron diffraction
	1.333 ± 0.002 Å	[207]	Electron diffraction
	1.333 Å	[209]	Electron diffraction
	1.332 ± 0.005 Å	[206]	Electron diffraction
	1.33 ± 0.02 Å	[205]	Electron diffraction
	1.324 ± 0.005 Å	[208]	Electron diffraction

[a]TPEPICO = Threshold photoelectron-photoion coincidence technique.
[b]This is the value of the threshold energy for the formation of $C_3F_5^+$, the lowest threshold value for the positive ions listed in Ref. [218].
[c]Vertical value.
[d]The same (lowest) value is listed for the "appearance" of $C_3F_5^+$ and $C_2F_4^+$.
[e]Chang *et al.* in Ref. [207] give the tilt angle for CF_2 to be $-5.4°$.

Table 5.27 (continued)
Physical and structural data on the c-C_4F_8 molecule.

Physical quantity	Value	Ref.	Method
F–C–F angle	110°	[209]	Electron diffraction
	109.9° ± 0.3°	[207]	Electron diffraction
	109.5° ± 3°	[205]	Electron diffraction
	109° ± 1°	[208]	Electron diffraction
	108° ± 2°	[206]	Electron diffraction
C–C–C angle	89.3° ± 0.3°	[207]	Electron diffraction
	∼89°	[205]	Electron diffraction
	88.8°	[209]	Electron diffraction
Dihedral angle	10° ± 3°	[206]	Electron diffraction
	17.4° ± 0.3°	[207]	Electron diffraction

obtained from electron scattering experiments are also listed in Table 5.28 and plotted in Fig. 5.35.

In the last column of Fig. 5.35 are shown the suggested energy positions of the lowest negative ion states of c-C_4F_8: −0.62, 4.9, 6.9, 9.0 and 10.5 eV. These suggested values were determined as follows: The value −0.62 eV is based on the average value (0.62 eV) of the highest two electron affinity values listed in Table 5.27. The third (lower) value was not considered since it is only a lower limit. The 4.9 and 9.0 eV values are those from electron scattering experiments, which are expected to lie somewhat higher than the corresponding values determined from dissociative electron attachment experiments. The 6.9 and 10.5 eV values are averages of values measured by various groups (see Table 5.28) for the energies at which the cross section for the production of F^- by dissociative electron attachment to c-C_4F_8 exhibits a maximum.

4.2 Electron Scattering Cross Sections for c-C_4F_8

4.2.1 Total Electron Scattering Cross Section, $\sigma_{sc,t}(\varepsilon)$, of c-C_4F_8

There have been two recent measurements [29, 228, 229] of the total electron scattering cross section, $\sigma_{sc,t}(\varepsilon)$, of c-C_4F_8 which are compared in Fig. 5.36. The measurements of Sanabia *et al.* [29] cover the low-energy range of 1–20 eV, and those of Nishimura [228, 229] cover a wider energy range of 2–3000 eV. Sanabia *et al.* did not quote the

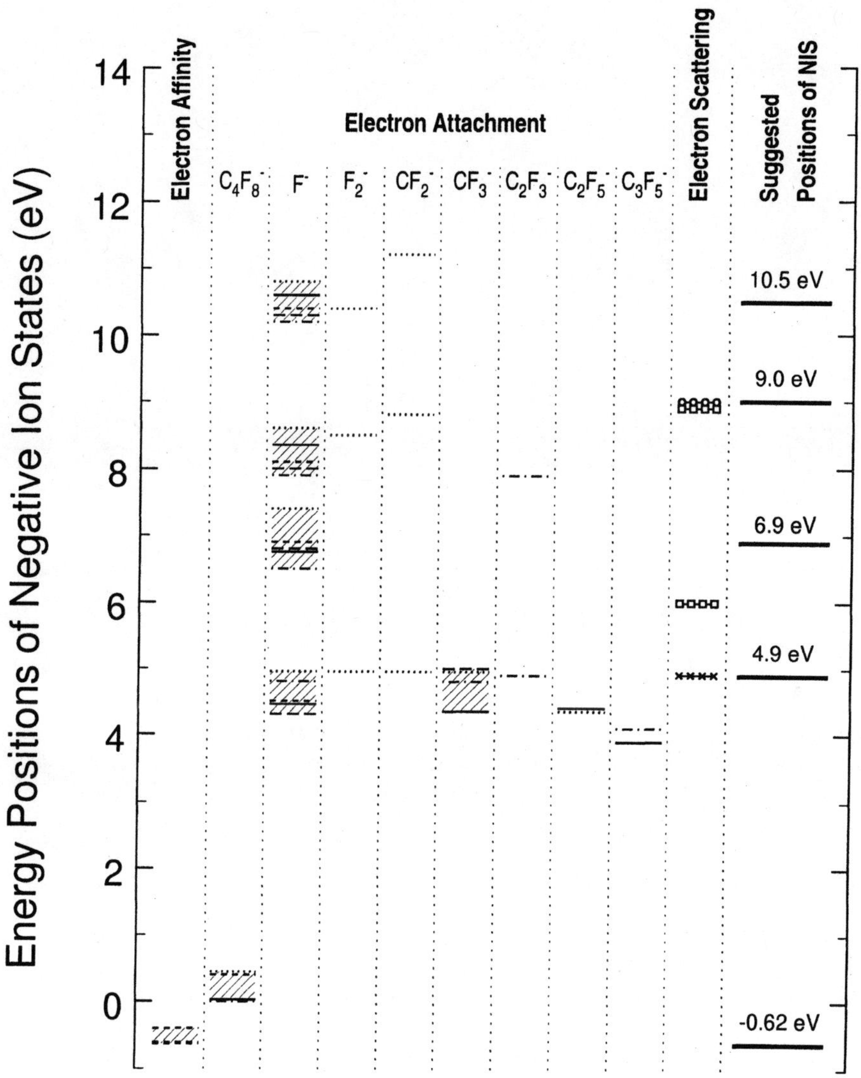

Figure 5.35. Energy positions, E_{NIS}, of the negative ion states of c-C_4F_8 below 12 eV as obtained from electron affinity, electron attachment, and electron scattering studies. Electron affinity values: see Table 5.27 and the text. Electron attachment values: (···) Ref. [225]; (—) Ref. [218]; (– · –) Ref [222]; (- - -) Ref. [224]; (– –) Ref. [220]. Electron scattering values: (– × –) Ref. [227]; (–□–) Ref. [29]; (– ○ –) Refs. [228, 229].

Table 5.28. Negative ion states of c-C_4F_8.

Energy position (eV)	Type of measurement	Reference
~ 0.0	Nondissociative electron attachment	[222]
0.03^a $(0.23)^b$	Nondissociative electron attachment	[218, 223]
0.4 ± 0.08	Total electron attachment	[224]
0.45 ± 0.1	Nondissociative electron attachment	[225]
1.75^a $(1.9)^b$	Dissociative electron attachment producing F^-	[218]
1.75^a $(1.9)^b$	Dissociative electron attachment producing $C_3F_5^-$	[218]
3.75^a $(4.05)^b$	Dissociative electron attachment producing $C_3F_5^-$	[218]
4.1	Dissociative electron attachment producing $C_3F_5^-$	[222]
4.3	Dissociative electron attachment producing F^-	[220]
4.3^a $(4.6)^b$	Dissociative electron attachment producing F^-	[218]
4.35^a	Dissociative electron attachment producing CF_3^-	[218]
4.35 ± 0.1	Dissociative electron attachment producing $C_2F_5^-$	[225]
4.4^a	Dissociative electron attachment producing $C_2F_3^-$	[218]
4.5 ± 0.08	Dissociative electron attachment producing F^-	[224]
4.8	Dissociative electron attachment producing F^-	[222]
4.8	Dissociative electron attachment producing CF_3^-	[222]
4.9	Dissociative electron attachment producing $C_2F_3^-$	[222]
4.95 ± 0.1	Dissociative electron attachment producing F^-	[225, 226]
4.95 ± 0.1	Dissociative electron attachment producing F_2^-	[225]
4.95 ± 0.1	Dissociative electron attachment producing CF_2^-	[225]
4.95 ± 0.1	Dissociative electron attachment producing CF_3^-	[225, 226]
5.0	Dissociative electron attachment producing CF_3^-	[220]
4.91	Derivative electron transmission spectrum	[227]
~ 6	Total electron scattering cross section	[29]
6.5	Dissociative electron attachment producing F^-	[222]
6.75^a $(7.1)^b$	Dissociative electron attachment producing F^-	[218]
6.8	Dissociative electron attachment producing F^-	[220]
6.9 ± 0.08	Dissociative electron attachment producing F^-	[224]
7.4 ± 0.2	Dissociative electron attachment producing F^-	[225, 226]

[a]Calibration made using the production of SF_6^- from SF_6 (peak at 0.0 eV).
[b]Calibration made using the production of O^- from CO (peak at 9.8 eV).

uncertainty of their data, but Nishimura gives the combined systematic and statistical uncertainties of his measurements to be between ±2.8% and ±5.9% depending on the electron energy. Representative error bars for the data of Nishimura are shown in Fig. 5.36. Generally, the data of Nishimura lie higher than those of Sanabia *et al.* [29].

Table 5.28 (continued)
Negative ion states of c-C_4F_8.

Energy position (eV)	Type of measurement	Reference
7.9	Dissociative electron attachment producing F^-	[222]
7.9	Dissociative electron attachment producing $C_2F_3^-$	[222]
8.0	Dissociative electron attachment producing F^-	[220]
8.1 ± 0.08	Dissociative electron attachment producing F^-	[224]
8.2^a $(8.45)^b$	Dissociative electron attachment producing F^-	[218]
8.5 ± 0.3	Dissociative electron attachment producing F_2^-	[225]
8.6 ± 0.2	Dissociative electron attachment producing F^-	[225, 226]
8.8 ± 0.1	Dissociative electron attachment producing CF_2^-	[225]
~ 9	Total electron scattering cross section	[228, 229]
~ 9	Total electron scattering cross section	[29]
10.2	Dissociative electron attachment producing F^-	[222]
10.3	Dissociative electron attachment producing F^-	[220]
10.4 ± 0.08	Dissociative electron attachment producing F^-	[224]
10.4 ± 0.3	Dissociative electron attachment producing F_2^-	[225]
10.5^a $(10.8)^b$	Dissociative electron attachment producing F^-	[218]
10.8 ± 0.2	Dissociative electron attachment producing F^-	[225, 226]
11.2 ± 0.2	Dissociative electron attachment producing CF_2^-	[225]

[a] Calibration made using the production of SF_6^- from SF_6 (peak at 0.0 eV).
[b] Calibration made using the production of O^- from CO (peak at 9.8 eV).

The cross section of Sanabia *et al.* [29] exhibits a pronounced minimum near 4 eV, and rises steeply as the energy is decreased below $\sim$ 2 eV. This latter feature is consistent with the large electron attachment cross section observed near zero energy (see Sec. 4.5.4 of Ch. 5). The cross section of Sanabia *et al.* also indicates an enhancement near 6.0 eV, which is consistent with the existence of a negative ion resonance at 5.9 eV, as detected in the derivative electron transmission spectrum [227] and in dissociative electron attachment studies (Table 5.28). Both the cross section data of Sanabia *et al.* [29] and those of Nishimura [228, 229] show another cross section enhancement near 9 eV, which is in agreement with dissociative electron attachment studies indicating a negative ion state at this energy (Table 5.28). In addition, the data of Nishimura show "humps" at about 20 and 40 eV. These energies are outside the energy region where cross-section enhancements due to negative ion states are normally expected.

The disparity between the two sets of cross section measurements makes it difficult to recommend cross section values for the $\sigma_{sc,t}(\varepsilon)$ of

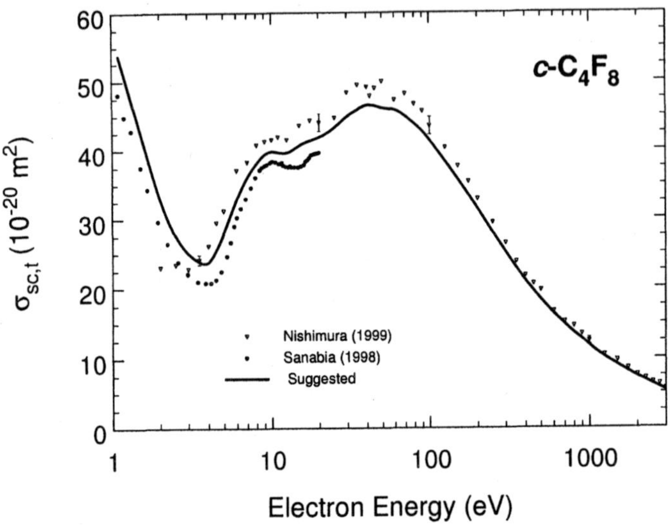

Figure 5.36. Total electron scattering cross section, $\sigma_{sc,t}(\varepsilon)$, of c-C_4F_8: (•) Ref. [29]; (▽) Refs. [228, 229]; (—) suggested.

c-C_4F_8. However, we obtain suggested values for $\sigma_{sc,t}(\varepsilon)$ between 1.1 and 3000 eV by determining the least squares average of the two sets of measurements in the overlapping energy range of 4–20 eV, and by extending the suggested values to higher and lower energies by normalizing the high energy ($\geq$ 20 eV) cross section data of Nishimura to the average at 20 eV and the low energy ($\leq$ 4 eV) cross section data of Sanabia *et al.* to the average at 4 eV. The resultant suggested cross section $\sigma_{sc,t}(\varepsilon)$ is shown in Fig 5.36 by the solid line, and values determined from this fit are listed in Table 5.29.

4.2.2 Elastic Differential Electron Scattering Cross Sections, $\sigma_{e,diff}(\varepsilon)$, of c-C_4F_8

Figure 5.37 compares the recent measurements of $\sigma_{e,diff}(\varepsilon)$ of c-C_4F_8 by Okamoto and co-workers [140, 142] and the calculated values of this quantity by McKoy and co-workers [230, 231]. The agreement between the calculated and the measured values of $\sigma_{e,diff}(\varepsilon)$ improves as the electron energy increases. The $\sigma_{e,diff}(\varepsilon)$ functions show pronounced variations with scattering angle at all electron energies (1.5–100 eV) employed in these investigations. The experimental data for $\sigma_{e,diff}(\varepsilon)$ are listed in Table 5.30.

Table 5.29. Suggested values for the total electron scattering cross section, $\sigma_{sc,t}(\varepsilon)$, of c-$C_4F_8$.

ε (eV)	$\sigma_{sc,t}(\varepsilon)$ (10^{-20} m^2)	ε (eV)	$\sigma_{sc,t}(\varepsilon)$ (10^{-20} m^2)	ε (eV)	$\sigma_{sc,t}(\varepsilon)$ (10^{-20} m^2)
1.1	53.8	13	39.8	150	35.7
1.5	42.6	15	40.7	200	31.4
2.0	32.9	17	41.3	300	25.1
2.5	27.7	20	41.9	400	21.0
3.0	25.1	25	43.3	500	18.3
3.5	23.9	30	44.9	600	16.3
4.0	23.8	35	45.9	700	14.8
4.5	25.6	40	46.5	800	13.6
5.0	28.0	45	46.4	900	12.7
6.0	32.8	50	46.1	1000	11.8
7.0	36.1	60	45.8	1500	9.00
8.0	38.1	70	44.9	2000	7.35
9.0	39.4	80	43.8	2500	6.20
10.0	39.8	90	42.7	3000	5.29
12.0	39.6	100	41.4		

4.2.3 Vibrational Differential Excitation Cross Section, $\sigma_{vib,diff}(\varepsilon)$, of c-$C_4F_8$

Figure 5.38 shows the recent measurements of Tanaka and co-workers [142] of the vibrational differential excitation cross section for an energy loss $\Delta\varepsilon = 0.15$ eV and a scattering angle of $\theta = 50°$. The major peak near 7.5 eV, the rise of the cross section below 3 eV, and the cross section enhancements near 4.9 and 10.5 eV are attributed to the existence of negative ion states at these energies in agreement with the electron attachment and other electron scattering data listed in Table 5.28. The data in Fig. 5.38 show the significance of indirect vibrational excitation of the c-C_4F_8 molecule by electron impact.

4.3 Electron-Impact Ionization of c-C_4F_8

4.3.1 Partial Ionization Cross Sections, $\sigma_{i,part}(\varepsilon)$, of c-$C_4F_8$

There have been two measurements of the partial ionization cross sections, $\sigma_{i,part}(\varepsilon)$, of c-$C_4F_8$ — the first by Toyoda *et al.* [232] and the second by Jiao *et al.* [233, 234]. Toyoda *et al.* [232] used quadrupole mass spectrometry, and measured the cross sections for dissociative ionization

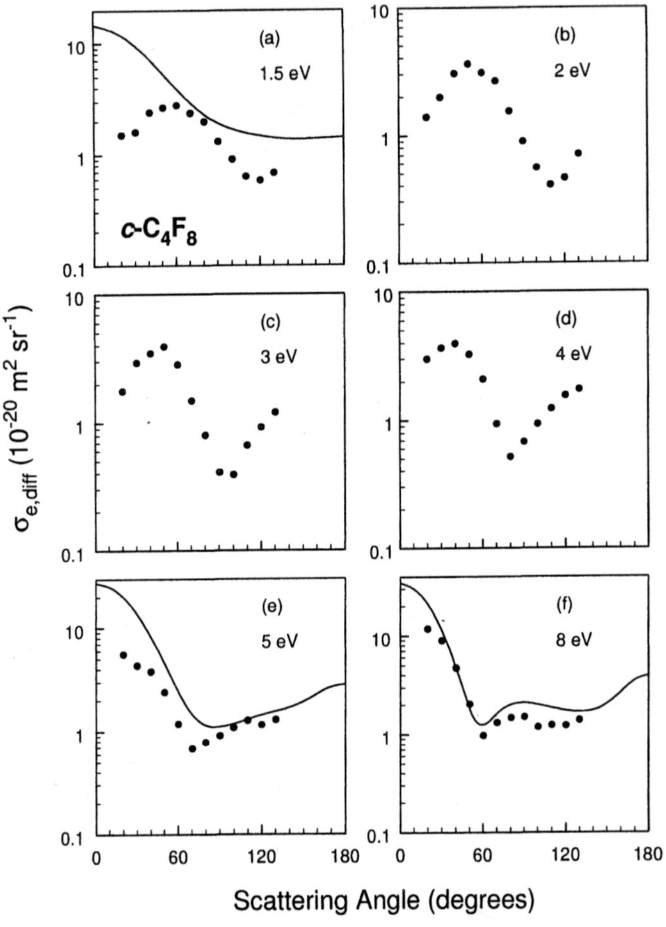

Figure 5.37. Elastic differential electron scattering cross sections, $\sigma_{\mathrm{e,diff}}(\varepsilon)$, for c-$C_4F_8$: (•) Okamoto *et al.* [140, 142]; (—) Winstead and McKoy [230, 231].

of the c-C_4F_8 molecule by electron impact leading to the formation of CF^+, CF_2^+, CF_3^+, $C_2F_3^+$, $C_2F_4^+$, and $C_3F_5^+$ from threshold to 117.8 eV. Their cross section measurements have an uncertainty of about ±10% and plotted in Fig. 5.39. Jiao *et al.* [233, 234] measured the $\sigma_{\mathrm{i,part}}(\varepsilon)$ of c-C_4F_8 from 16 to 200 eV using Fourier transform mass spectrometry. They detected 13 fragment positive ions as opposed to the six reported by Toyoda *et al.* The absolute values of their partial ionization cross sections were obtained by normalization to those of Wetzel *et al.* [163] for argon and probably have an uncertainty of about ±20%. They are compared in Fig. 5.39 with those of Toyoda *et al.* and with some early

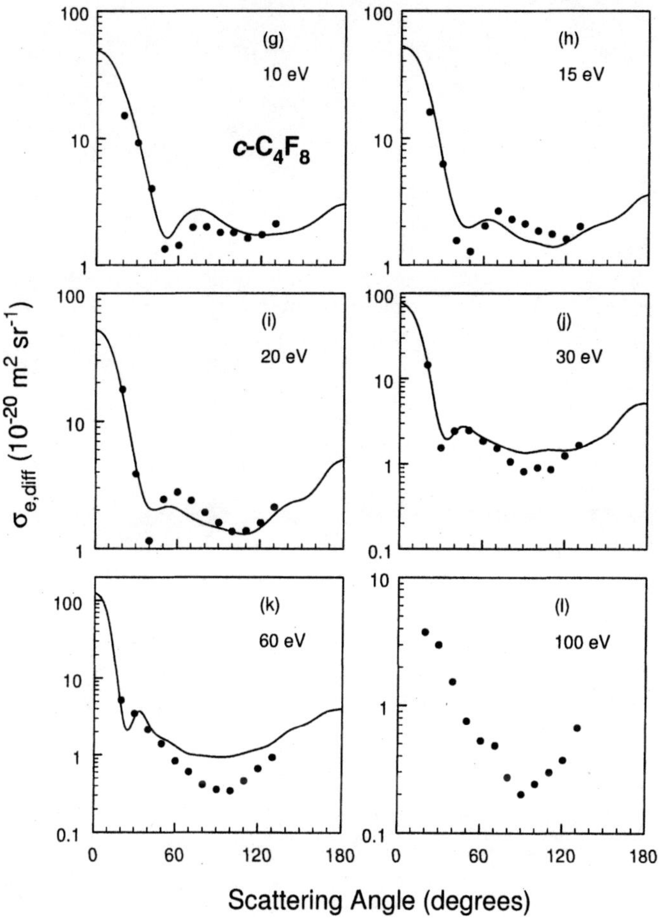

Figure 5.37 (continued)
Elastic differential electron scattering cross sections, $\sigma_{e,\text{diff}}(\varepsilon)$, for c-$C_4F_8$: (•) Okamoto *et al.* [140, 142]; (—) Winstead and McKoy [230, 231].

measurements for the production of CF_3^+, $C_2F_4^+$, and $C_3F_5^+$ at 35 eV by Bibby and Carter [220].

There are substantial differences between the two sets of measurements which depend on the particular positive ion fragment involved. However, suggested partial ionization cross sections can be determined by fitting the averages of the cross sections for the ions measured by both Toyoda *et al.* [232] and Jiao *et al.* [234], and by fitting the data for the ions that were measured only by Jiao *et al.* The sum of the fitted data was then normalized at each energy to the suggested value of $\sigma_{i,t}(\varepsilon)$

Table 5.30. Elastic differential electron scattering cross sections, $\sigma_{e,diff}(\varepsilon)$, for c-C_4F_8 in units of 10^{-20} m^2 sr^{-1}. The data are from Okamoto and co-workers [140, 142], and the numbers presented here were obtained by digitizing data appearing in figures.

Angle	1.5 eV	2 eV	3 eV	4 eV	5 eV	8 eV
20°	1.52	1.40	1.77	3.00	5.59	11.8
30°	1.62	1.99	2.95	3.66	4.37	8.98
40°	2.43	3.04	3.49	3.96	3.83	4.73
50°	2.68	3.59	3.93	3.25	2.42	2.04
60°	2.81	3.09	2.84	2.09	1.19	0.978
70°	2.38	2.65	1.49	0.941	0.683	1.33
80°	1.99	1.55	0.803	0.519	0.784	1.49
90°	1.37	0.897	0.407	0.683	0.913	1.52
100°	0.910	0.555	0.388	0.936	1.10	1.20
110°	0.633	0.406	0.666	1.23	1.29	1.25
120°	0.583	0.460	0.923	1.56	1.17	1.24
130°	0.683	0.707	1.20	1.74	1.30	1.41

Angle	10 eV	15 eV	20 eV	30 eV	60 eV	100 eV
20°	15.0	16.0	17.7	14.5	5.14	3.75
30°	9.20	6.22	3.85	1.54	3.46	2.99
40°	3.99	1.55	1.15	2.45	2.14	1.53
50°	1.33	1.27	2.44	2.50	1.39	0.755
60°	1.41	2.02	2.76	1.87	0.837	0.526
70°	1.97	2.65	2.38	1.53	0.608	0.484
80°	1.99	2.28	1.92	1.06	0.416	0.273
90°	1.79	2.10	1.59	0.813	0.358	0.201
100°	1.79	1.85	1.36	0.907	0.345	0.242
110°	1.62	1.74	1.37	0.862	0.464	0.298
120°	1.73	1.59	1.59	1.25	0.667	0.372
130°	2.10	2.01	2.11	1.67	0.933	0.667

derived in the next section. The resulting suggested values for $\sigma_{i,part}(\varepsilon)$ are shown by the solid lines in Fig. 5.39 and are listed in Table 5.31.

The parent positive ion c-$C_4F_8^+$ was not detected in either of these electron-impact studies [232, 234] or in the photon-impact study of Jarvis *et al.* [217], indicating that the ground state of the c-$C_4F_8^+$ ion is not bound in the Franck-Condon region. Nevertheless, Smith and Kevan [235] reported observation of a weak c-$C_4F_8^+$ signal in studies of the total and dissociative charge-transfer cross sections of Xe^+ with c-C_4F_8. These latter observations may indicate a minimum in the potential energy surface of c-$C_4F_8^+$ at large internuclear separation distances.

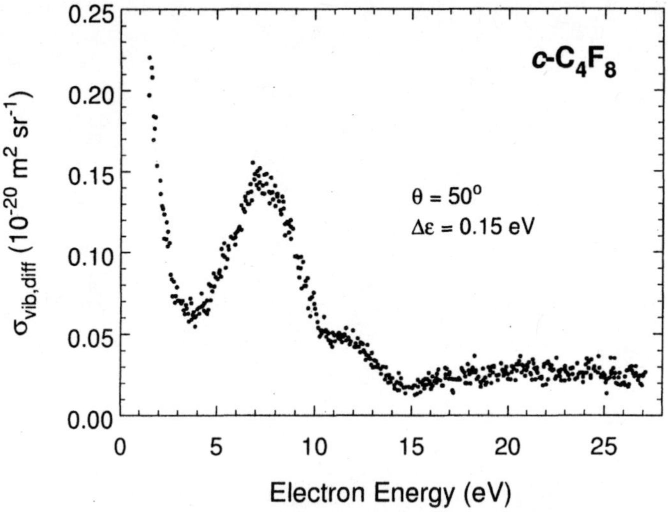

Figure 5.38. Vibrational differential electron scattering cross section, $\sigma_{\text{vib,diff}}(\varepsilon)$, for c-$C_4F_8$ measured by Tanaka *et al.* [142].

In Table 5.32 are listed the energy thresholds for the appearance of a number of positive ions by electron (and photon) impact on c-C_4F_8.

4.3.2 Total Ionization Cross Section, $\sigma_{i,t}(\varepsilon)$, of c-$C_4F_8$

In Fig. 5.40 is plotted the sum of the partial ionization cross sections of Toyoda *et al.* [232] and, similarly, the sum of the partial ionization cross sections of Jiao *et al.* [234] (the "structure" in the data of Jiao *et al.* may not be real). In addition, Bibby and Carter [220] made an early measurement of the partial ionization cross section for the production of CF_3^+, $C_2F_4^+$, and $C_3F_5^+$ from c-C_4F_8 at only one value (35 eV) of the incident electron energy. The sum of the partial ionization cross sections of these three positive-ion fragments is 5.35×10^{-20} m^2 and is shown in Fig. 5.40. Also shown are two earlier measurements of the total ionization cross section, one by Kurepa [224] and the other by Beran and Kevan [50]. The former measurements were made at incident electron energies ranging from threshold to 100 eV and the latter at an incident electron energy of 70 eV.

The differences between the various measurements is probably within the combined uncertainty considering the fact that the values of Toyoda *et al.* do not include the contributions of some of the weaker ions measured by Jiao *et al.* (The sum of the cross sections, at 70 eV, of the

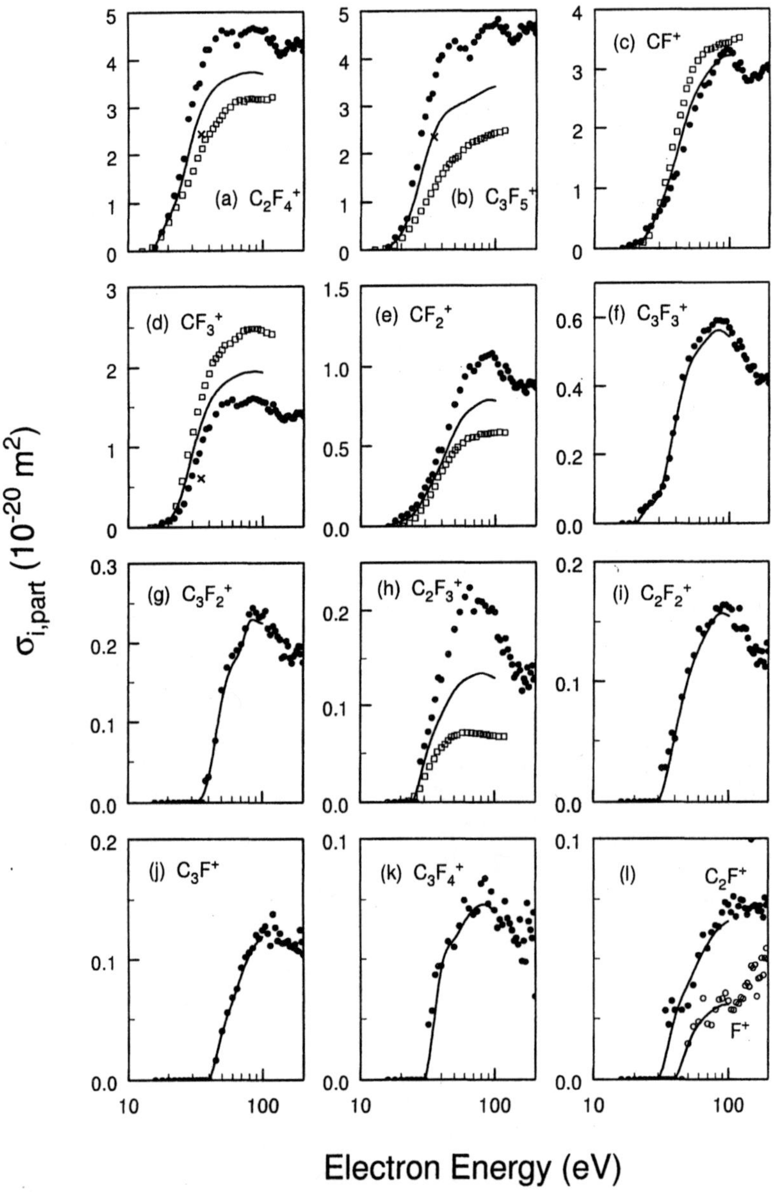

Figure 5.39. Partial ionization cross sections, $\sigma_{\mathrm{i,part}}(\varepsilon)$, of c-C_4F_8: (□) Toyoda *et al.* [232]; (•, ○) Jiao *et al.* [234]; (×) Bibby and Carter [220]; (—) suggested.

Table 5.31. Suggested partial ionization cross sections, $\sigma_{i,part}(\varepsilon)$, of c-$C_4F_8$.

ε (eV)	$C_2F_4^+$	$C_3F_5^+$	CF^+	CF_3^+	CF_2^+	$C_3F_3^+$	$C_3F_2^+$
16	0.114	0.059	0.005	0.007			
20	0.680	0.351	0.069	0.086	0.034		
25	1.44	1.07	0.283	0.407	0.087	0.048	
30	2.35	1.87	0.689	0.908	0.191	0.084	
40	3.25	2.74	1.54	1.54	0.397	0.298	0.031
50	3.54	2.98	2.33	1.77	0.587	0.453	0.124
60	3.65	3.09	2.73	1.87	0.698	0.509	0.171
70	3.71	3.20	2.96	1.92	0.744	0.541	0.195
80	3.74	3.29	3.11	1.94	0.776	0.561	0.226
90	3.74	3.37	3.20	1.95	0.787	0.557	0.228
100	3.71	3.41	3.23	1.94	0.782	0.543	0.224

ε (eV)	$C_2F_3^+$	$C_2F_2^+$	C_3F^+	$C_3F_4^+$	C_2F^+	F^+
16						
20						
25	0.003					
30	0.042					
40	0.089	0.054		0.046	0.025	
50	0.115	0.101	0.038	0.058	0.039	0.014
60	0.127	0.128	0.065	0.066	0.049	0.023
70	0.132	0.143	0.087	0.071	0.056	0.027
80	0.134	0.154	0.103	0.073	0.061	0.029
90	0.132	0.157	0.114	0.072	0.064	0.031
100	0.129	0.155	0.119	0.070	0.066	0.031

seven fragment ions reported by Jiao *et al.* but not by Toyoda *et al.* is 1.1×10^{-20} m^2). A suggested cross section $\sigma_{i,t}(\varepsilon)$ of c-C_4F_8 has been derived by a fit to the two most recent measurements. The suggested cross section is shown in Fig. 5.40 by the solid line and values derived from this fit are listed in Table 5.33.

It should be noted that since CF_2 is the most abundant radical in the electron-impact dissociation of the c-C_4F_8 molecule into neutral fragments, electron-impact ionization of this radical may also be important in c-C_4F_8 plasmas. The electron-impact ionization cross section of the CF_2 radical has been measured by Tarnovsky and Becker [68] (see also Ch. 4).

Table 5.32. Energy thresholds for the appearance of positive-ion fragments by electron impact on c-C_4F_8.

Positive ion fragment	Energy threshold (eV)	Reference
$C_3F_5^+$	11.6 ± 0.2^a	[217]
	12.1 ± 0.1	[218]
$C_2F_4^+$	11.8^a	[217]
	12.35 ± 0.1	[218]
CF_3^+	14.4 ± 0.2	[218]
	18.1	[232]
CF_2^+	20.3	[232]
CF^+	18.4 ± 0.2	[218]
	19.2	[232]

[a]Photon-impact value, the rest of the data listed are electron-impact results.

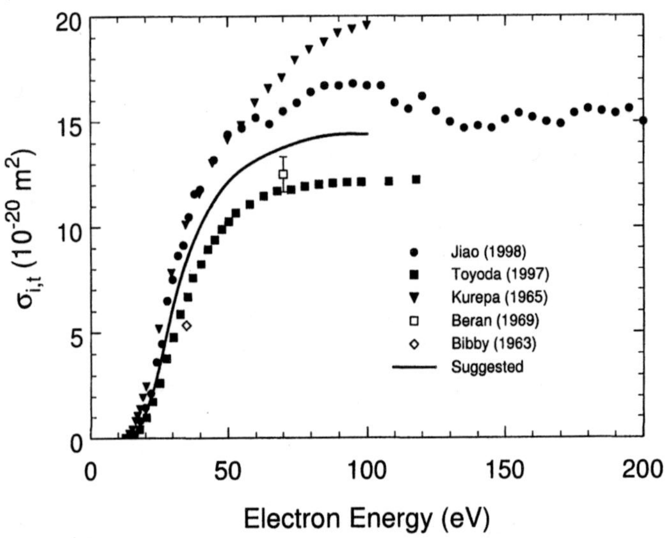

Figure 5.40. Total ionization cross section, $\sigma_{i,t}(\varepsilon)$, of c-C_4F_8: (●) Ref. [234]; (■) Ref. [232]; (▼) Ref. [224]; (□) Ref. [50]; (◊) Ref. [220]; (—) suggested.

Table 5.33. Suggested values for the total ionization cross section, $\sigma_{i,t}(\varepsilon)$, of c-$C_4F_8$.

ε (eV)	$\sigma_{i,t}(\varepsilon)$ (10^{-20} m^2)	ε (eV)	$\sigma_{i,t}(\varepsilon)$ (10^{-20} m^2)	ε (eV)	$\sigma_{i,t}(\varepsilon)$ (10^{-20} m^2)
16	0.185	40	10.0	80	14.2
20	1.22	50	12.1	90	14.4
25	3.34	60	13.2	100	14.4
30	6.14	70	13.8		

Table 5.34. Values of the density-normalized ionization coefficient, $\alpha/N\,(E/N)$, for c-C_4F_8 derived from a fit to the data of Naidu *et al.* [236].

E/N (10^{-17} V cm^2)	α/N (10^{-18} cm^2)	E/N (10^{-17} V cm^2)	α/N (10^{-18} cm^2)
300	1.60	500	35.5
350	8.27	550	45.5
400	16.7	600	56.2
450	26.1	650	66.6

4.3.3 Density-Normalized Ionization Coefficient, $\alpha/N\,(E/N)$, of c-C_4F_8

There have been two experimental determinations of the density-normalized ionization coefficient $\alpha/N\,(E/N)$ of this molecule, one by Naidu *et al.* [236] over the approximate E/N range from 300×10^{-17} to 650×10^{-17} V cm^2, and the other by Tagashira *et al.* [237] for three values of E/N ($394 \times 10^{-17}, 425 \times 10^{-17}$, and 455×10^{-17} V cm^2). The measurements of Naidu *et al.* [236] were conducted at a gas temperature of 293 K and at two values of the gas pressure, namely, 0.084 and 0.133 kPa. No pressure dependence was observed in their measurements of this quantity. According to Naidu *et al.*, the uncertainty of their data is about ±10% at values of E/N below $(E/N)_{\text{lim}}$ and about ±20% at higher E/N values. These data are plotted in Fig. 5.41. The solid line is a least squares fit to the data, and values obtained from the fit are listed in Table 5.34. These values should be treated with caution until they are validated by additional measurements. If these data are used in conjunction with the density-normalized electron attachment coefficient, $\eta/N\,(E/N)$, of Naidu *et al.* [236] to determine the density-normalized

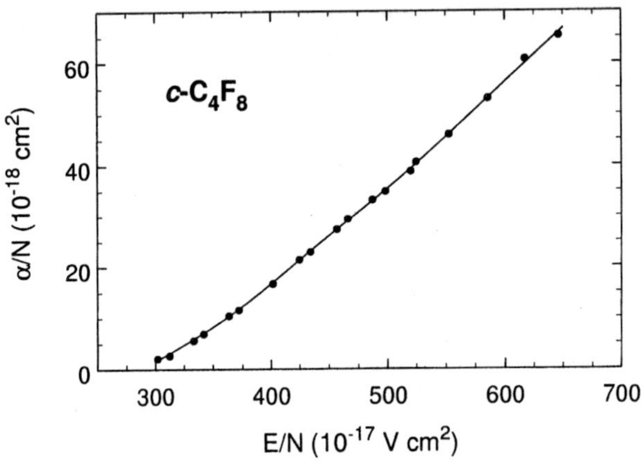

Figure 5.41. Density-normalized ionization coefficient, $\alpha/N\,(E/N)$, of $c\text{-}C_4F_8$ measured by Naidu *et al.* [236].

effective ionization coefficient, $(\alpha - \eta)/N\,(E/N)$, the resultant values differ from the two available direct measurements of $(\alpha - \eta)/N\,(E/N)$ (see Fig. 5.42).

The limited measurements of Tagashira *et al.* [237] exhibited a pronounced decrease of α/N with increasing $c\text{-}C_4F_8$ gas pressure (from 0.33 to 5.3 kPa). No explanation was provided for this observation, which, if valid, would imply that the data of Naidu *et al.* [236] are appropriate only for the gas pressures used in their experiments. The measurements of Tagashira *et al.* need to be repeated, expanded, and further scrutinized to confirm this reported pressure dependence.

4.3.4 Density-Normalized Effective Ionization Coefficient, $(\alpha - \eta)/N\,(E/N)$, of $c\text{-}C_4F_8$

There have been two direct measurements of the density-normalized effective ionization coefficient, $(\alpha - \eta)/N\,(E/N)$, of $c\text{-}C_4F_8$ using the pulsed Townsend method — one by Wen and Wetzer [238] and the other by de Urquijo [239, 240]. These are compared in Fig. 5.42. The measurements of Wen and Wetzer [238] were conducted at a gas pressure of 0.133 kPa and those by de Urquijo [239, 240] at gas pressures ranging from 0.08 to 1 kPa. The latter data indicate that $(\alpha - \eta)/N\,(E/N)$ is pressure dependent in the pressure range covered by this study. The decrease in $(\alpha - \eta)/N\,(E/N)$ with N seems to be consistent with the decrease of α/N observed by Tagashira *et al.* [237]. Although this pressure dependence may originate from "secondary" electron attachment

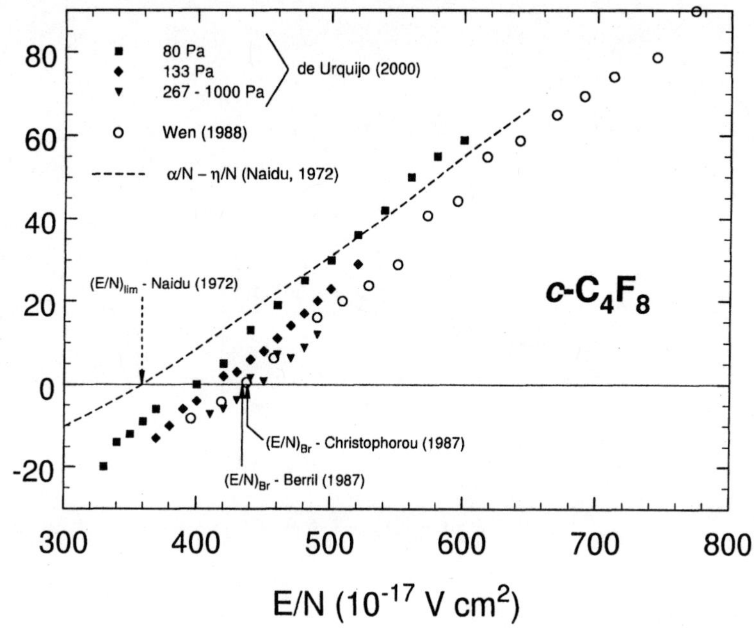

Figure 5.42. Density-normalized effective ionization coefficient, $(\alpha-\eta)/N\,(E/N)$, of c-C_4F_8: (○) Ref. [238]; (■, ♦, ▼) Refs. [239, 240]; (- - -) Ref. [236]. The broken arrow shows the $(E/N)_{\rm lim}$ value of Ref. [236] and the solid arrows represent the values of $(E/N)_{\rm Br}$ as measured in Refs. [241, 242].

processes (electron scattering via high-lying negative ion states and attachment of the slowed-down electrons by c-C_4F_8 via the near-zero electron attachment process forming c-$C_4F_8^-$) and from the effect of such indirect electron scattering processes on the electron energy distribution function, presently there are no quantitative measurements to support such an interpretation.

Also shown in Fig. 5.42 are $(\alpha-\eta)/N\,(E/N)$ values obtained using the data of Naidu *et al.* [236] for $\alpha/N\,(E/N)$ and $\eta/N\,(E/N)$ (see Sec. 4.5.8 of Ch. 5) which were measured using the steady-state Townsend method. These data differ from the direct measurements of $(\alpha-\eta)/N\,(E/N)$. The data of Naidu *et al.* give for $(E/N)_{\rm lim}$ a value of 359×10^{-17} V cm^2 which is much lower than the value of 434×10^{-17} V cm^2 determined by Wen and Wetzer [238] and those inferred from the measurements of de Urquijo and Basurto [240] (see Fig. 5.42). It is also much lower than the room-temperature values of the uniform-field breakdown strength, $(E/N)_{\rm Br}$, measured by Berril *et al.* [241] and Christophorou *et al.* [242] for this gas. The measurements of Berril *et al.* [241] of $(E/N)_{\rm Br}$ ranged from

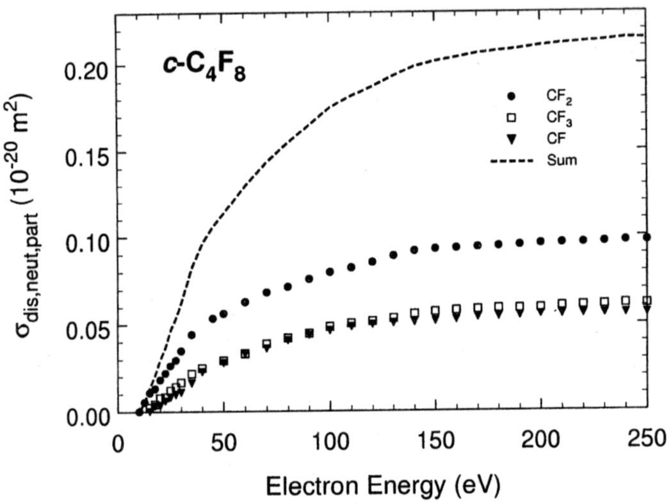

Figure 5.43. Partial cross sections, $\sigma_{\text{dis,neut,part}}(\varepsilon)$, for electron-impact dissociation of c-C_4F_8 into the neutral fragments CF (▼), CF_2 (•), and CF_3 (□). The data are from Toyoda *et al.* [232] and the broken curve (- - -) represents the sum of the partial cross sections for the three fragments.

428×10^{-17} to 432×10^{-17} V cm^2 and were made at pressures between 10.1 and 60.6 kPa. The value of $(E/N)_{\text{Br}}$ measured by Christophorou *et al.* [242] is 438×10^{-17} V cm^2 and was made at 69.3 kPa. These comparisons would indicate that the measurements of Naidu *et al.* [236] for both $\alpha/N\,(E/N)$ and $\eta/N\,(E/N)$ require validation by independent measurements.

4.4 Electron-Impact Dissociation of c-C_4F_8 into Neutral Fragments

Toyoda *et al.* [232] used appearance mass spectrometry in a dual electron beam system to measure the absolute cross sections, $\sigma_{\text{dis,neut,part}}(\varepsilon)$, for electron-impact dissociation of c-C_4F_8 into the neutral radicals CF, CF_2, and CF_3 from threshold to 250 eV. These cross sections, along with their sum, are shown in Fig. 5.43 and are listed in Table 5.35. The sum of these partial dissociation cross sections may be taken as a lower limit to the total cross section, $\sigma_{\text{dis,neut,t}}(\varepsilon)$, for dissociation of c-C_4F_8 into neutrals by electron impact, because other neutral fragments for which the cross sections have not been measured will contribute to $\sigma_{\text{dis,neut,t}}(\varepsilon)$. For instance, Toyoda *et al.* [232] observed the C_3F_5 radical, but reported only the relative cross section for its formation. The stated

Table 5.35. Cross sections, $\sigma_{\text{dis,neut,part}}(\varepsilon)$, for electron-impact dissociation of c-C_4F_8 into the neutral fragments CF, CF_2, and CF_3 measured by Toyoda *et al.* [232]. The last column gives the sum of the cross sections in the preceding three columns.

ε (eV)	CF	CF_2	CF_3	Sum
10		0.0002		0.0002
12.5		0.0055		0.0055
15	0.0002	0.0108	0.0024	0.0134
17.5	0.0034	0.0131	0.0042	0.0206
20	0.0035	0.0183	0.0077	0.0295
22.5	0.0062	0.0221	0.0082	0.0365
25	0.0085	0.0264	0.0120	0.0469
27.5	0.0098	0.0295	0.0141	0.0534
30	0.0110	0.0349	0.0165	0.0624
35	0.0163	0.0444	0.0216	0.0823
40	0.0227		0.0247	0.0964
45		0.0536		0.106
50	0.0279	0.0563	0.0292	0.113
60	0.0332	0.0630	0.0332	0.129
70	0.0362	0.0685	0.0388	0.144
80	0.0409	0.0716	0.0422	0.155
90	0.0439	0.0759	0.0448	0.165
100	0.0464	0.0798	0.0485	0.175
110	0.0486	0.0823	0.0505	0.181
120	0.0498	0.0854	0.0516	0.187
130	0.0506	0.0890	0.0533	0.193
140	0.0511	0.0917	0.0556	0.198
150	0.0517	0.0928	0.0570	0.202
160	0.0526	0.0933	0.0577	0.204
170	0.0538	0.0939	0.0583	0.206
180	0.0541	0.0944	0.0588	0.207
190	0.0547	0.0950	0.0588	0.209
200	0.0554	0.0958	0.0591	0.210
210	0.0554	0.0961	0.0601	0.212
220	0.0557	0.0963	0.0605	0.213
230	0.0558	0.0967	0.0608	0.213
240	0.0559	0.0973	0.0614	0.215
250	0.0560	0.0973	0.0612	0.215

absolute and relative uncertainties of the cross sections $\sigma_{\text{dis,neut,part}}(\varepsilon)$ are, respectively, ±100% and ±20% [232]. However, similar cross sections obtained by the same group for other molecules, such as CF_4 and CHF_3, have been shown to be significantly smaller than their true values (see Ch. 4 and Sec. 2.4 of this chapter). The threshold energies for

Table 5.36. Autodetachment lifetime of c-$C_4F_8^{-*}$.

Lifetime(s)	Method of measurement	Reference
10×10^{-6}	Time-of-flight mass spectrometry	[222]
12×10^{-6}	Time-of-flight mass spectrometry	[244]
14.8×10^{-6}	Time-of-flight mass spectrometry	[225]
$\sim 200 \times 10^{-6}$	Ion cyclotron resonance	[245]
500×10^{-6}	Ion cyclotron resonance	[246]

the production of the neutral fragments CF, CF_2, and CF_3 by electron-impact on c-C_4F_8 as determined by Toyoda *et al.* [232] are, respectively, 14.5, 10.5, and 12.8 eV.

A model calculation predicting radical composition in c-C_4F_8 plasmas was conducted by Kazumi *et al.* [243]. This calculation predicts the production of a number of other radicals besides those detected in the experimental study of Toyoda *et al.* [232].

4.5 Electron Attachment to c-C_4F_8

4.5.1 Electron Beam Determined Total Electron Attachment Cross Section, $\sigma_{a,t}(\varepsilon)$, of c-C_4F_8

A number of electron beam studies [218, 222–225] have shown that at electron energies below $\sim$ 1 eV, the parent negative ion c-$C_4F_8^-$ is produced. The c-$C_4F_8^{-*}$ initially formed by electron capture, is unstable with respect to autodetachment [79]. Under single-collision conditions (low gas pressure) the lifetime of the isolated c-$C_4F_8^{-*}$ ion toward autodetachment was found to be between 10 and 15 μs when measured using time-of-flight mass spectrometry [222, 225, 244], and longer than $\sim$ 200 μs when measured using ion-cyclotron resonance (ICR) techniques [245, 246] (Table 5.36). In the latter case the electrons normally have lower kinetic energies than in the former, and this may explain the longer lifetime of c-$C_4F_8^{-*}$ in the ICR measurements. For a number of other long-lived metastable molecular negative ions it has been found that their autodetachment lifetime decreases as the energy of the attached electron increases (e.g., see Refs. [79, 247]).

Kurepa [224] reported a total electron attachment cross section, $\sigma_{a,t}(\varepsilon)$, for c-C_4F_8 which below $\sim$ 1 eV should represent the cross section for the formation of the parent negative ion c-$C_4F_8^-$ (see discussion later in this section). This cross section is shown in Fig. 5.44. In plotting these data

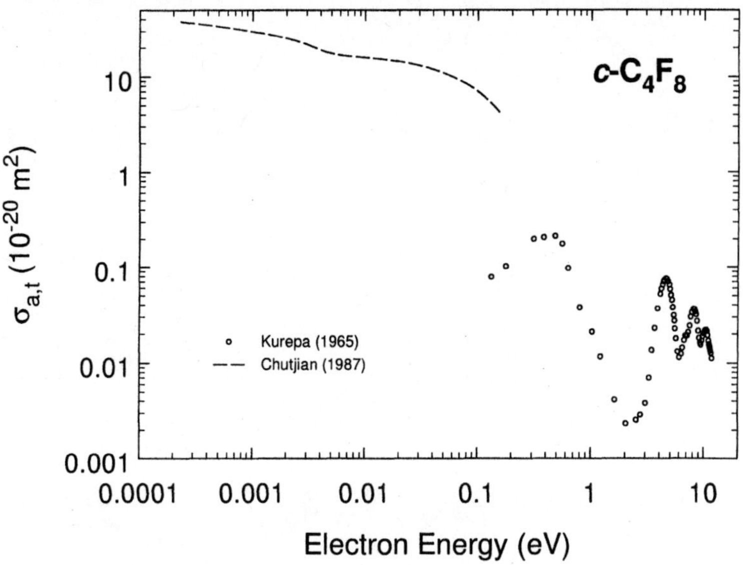

Figure 5.44. Total electron attachment cross section, $\sigma_{a,t}(\varepsilon)$, as a function of electron energy for c-C_4F_8 measured using electron beam techniques: (o) Ref. [224]; (- - -) Ref. [248]. The cross section below ~ 1 eV is due to the formation of c-$C_4F_8^-$.

we assigned the maximum value of the cross section for the formation of c-$C_4F_8^-$ at 0.4 eV the value of 0.21×10^{-20} m^2 reported by Kurepa [224] in Table 5 of his paper. Also plotted in Fig. 5.44 is the low energy, electron-beam-determined cross section for the formation of c-$C_4F_8^-$ by Chutjian and Alajajian [248]. These investigators used their krypton photoionization technique and put their relative measurements on an absolute scale by normalization to the swarm data of Christodoulides *et al.* [249]. These cross sections are discussed further in Sec. 4.5.5 of this chapter where they are compared with the total electron attachment cross sections obtained from electron swarm data.

Prior to presenting and discussing the electron swarm data, we present and discuss here the relative cross section data obtained by electron beam investigations of the production of the various fragment negative ions formed by electron impact on c-C_4F_8. These studies [72, 218, 220, 222, 224–226] show the formation of many fragment negative ions via a number of negative ion states in the electron energy range between 2 and 15 eV. The most systematic measurements of the relative cross sections for the production of the various fragment negative ions as a function of the electron energy in absolute relation to each other were made by Sauers *et al.* [222] using time-of-flight mass spectrometry. These

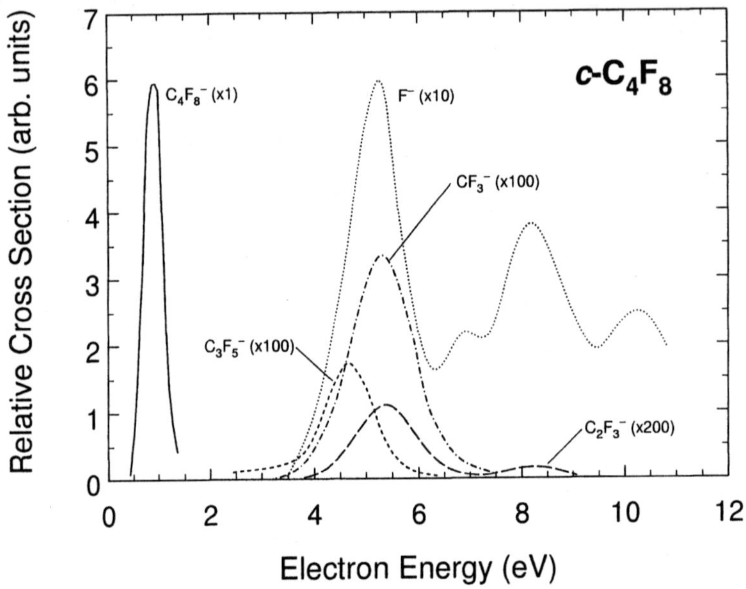

Figure 5.45. Relative cross sections for the formation of the parent c-$C_4F_8^-$ and the fragment F^-, CF_3^-, $C_2F_3^-$, and $C_3F_5^-$ negative ions by electron attachment to c-C_4F_8 (note the multiplication factors) (data of Sauers *et al.* [222]).

relative cross sections are shown in Fig. 5.45 along with the relative cross section for the production of the parent negative ion c-$C_4F_8^{-*}$ at electron energies below 1.5 eV. The most abundant fragment negative ion is seen to be F^-. This ion is formed via a number of negative ion states. Recent measurements of the negative-ion density in a high density c-C_4F_8 plasma using a laser photodetachment technique indicate the major negative ion species to be F^- [250], consistent with these findings.

The other observed negative ion fragments (CF_3^-, $C_2F_3^-$, and $C_3F_5^-$) have much lower cross sections than F^- as they require multiple bond breaking and molecular rearrangement. All four fragment negative ions (F^-, CF_3^-, $C_2F_3^-$, and $C_3F_5^-$) are produced at electron energies between 4 and 5 eV, indicating the existence of a common negative ion state in this energy range. There is general agreement among the electron beam studies [72, 218, 220, 222, 224–226] as to the energy positions of the various peaks in the relative yields of the various negative ions (see Table 5.28) except for the peak at 1.75 eV reported by Lifshitz and Grajower [218], which has not been observed in any of the other electron beam studies.

Table 5.37. Peak cross section values for negative ion fragments formed by electron impact on c-C_4F_8 (see text).

Peak cross section value (10^{-22} m^2)	Peak energy (eV)	Fragment ion	Reference
236	4.3	F^-	[220]
0.9	4.95	F^-	[226]
7.56	4.5	total[a]	[224]
0.2	7.4	F^-	[226]
1.90	6.9	total[a]	[224]
0.4	8.6	F^-	[226]
3.60	8.1	total[a]	[224]
0.5	10.8	F^-	[226]
2.25	10.4	total[a]	[224]
11.1	5.0	CF_3^-	[220]
0.001	4.95	CF_3^-	[226]

[a] About equal to that for the production of F^- at this energy.

Three of the electron beam studies [220, 224, 226] reported absolute cross sections for some ions at their peak energies. These are summarized in Table 5.37. They disagree considerably with each other and with the electron swarm data discussed later in the section.

4.5.2 Total Electron Attachment Rate Constant as a Function of E/N, $k_{a,t}(E/N)$, of c-C_4F_8

There have been two measurements of the total electron attachment rate constant, $k_{a,t}(E/N)$, as a function of the density-reduced electric field E/N employing electron-swarm techniques and mixtures of c-C_4F_8 with the buffer gases of N_2 and Ar [249, 251]. These room temperature ($T = 298$ K) measurements are shown in Fig. 5.46.

4.5.3 Total Electron Attachment Rate Constant as a Function of the Mean Electron Energy, $k_{a,t}(\langle\varepsilon\rangle)$, of c-$C_4F_8$

The total electron attachment rate constant, $k_{a,t}(\langle\varepsilon\rangle)$, as a function of the mean electron energy, $\langle\varepsilon\rangle$, can be determined from the electron

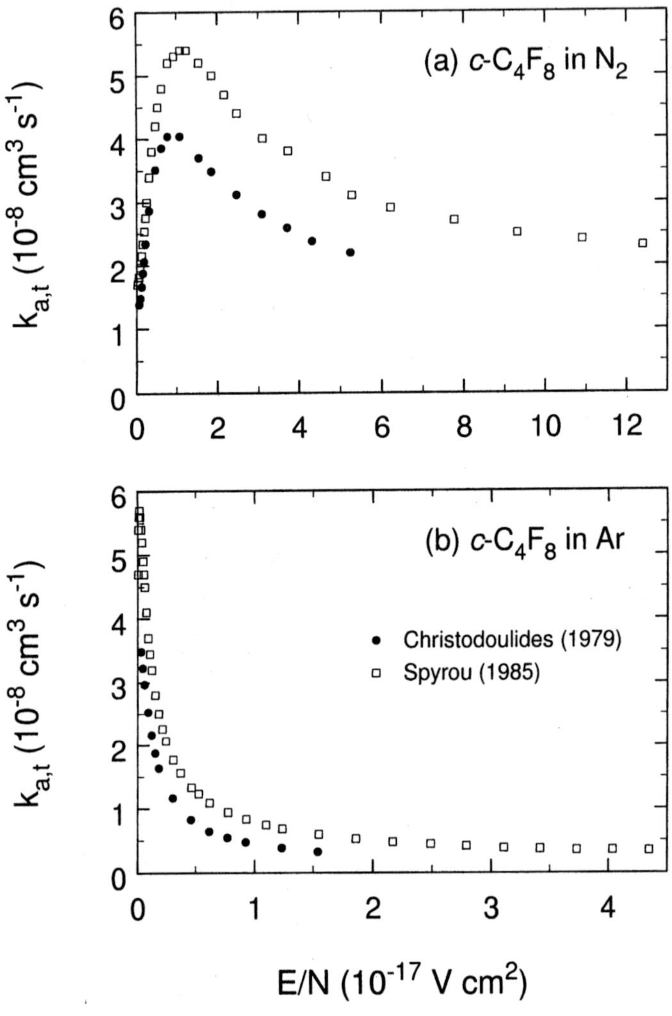

Figure 5.46. Total electron attachment rate constant, $k_{a,t}(E/N)$, as a function of E/N for c-C_4F_8 measured in the buffer gas (a) N_2 and (b) Ar: (•) Ref. [249]; (□) Ref. [251].

swarm measurements of $k_{a,t}(E/N)$ in the buffer gases N_2 and Ar shown in Fig. 5.46 since the electron energy distribution function is known for each value of E/N at which the $k_{a,t}$ was measured in the buffer gases N_2 and Ar. These room-temperature (T = 298 K) measurements of $k_{a,t}(\langle\varepsilon\rangle)$ are shown in Fig. 5.47 along with two other measurements of $k_{a,t}(\langle\varepsilon\rangle)$ [252, 253]. Although the differences between these four independent measurements of $k_{a,t}(\langle\varepsilon\rangle)$ from Christophorou's group are gen-

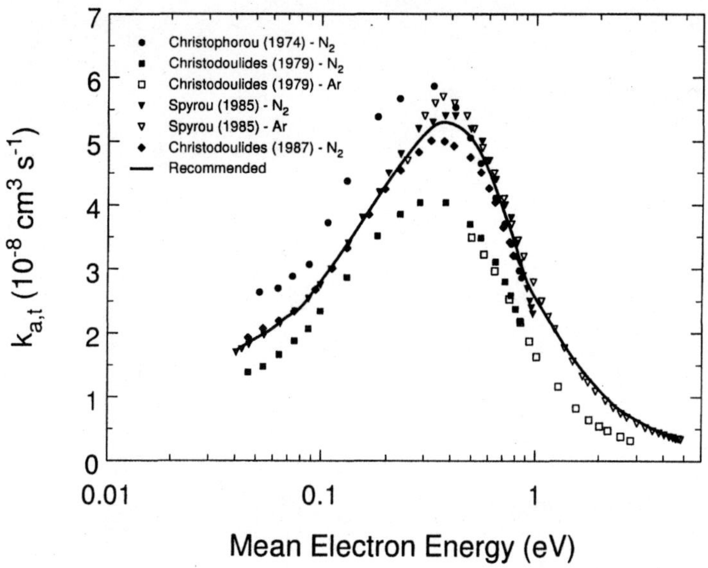

Figure 5.47. Total electron attachment rate constant, $k_{a,t}(\langle\varepsilon\rangle)$, as a function of $\langle\varepsilon\rangle$ for c-C_4F_8: (•) Ref. [252] in N_2; (■) Ref. [249] in N_2; (□) Ref. [249] in Ar; (▼) Ref. [251] in N_2; (▽) Ref. [251] in Ar; (♦) Ref. [253] in N_2; (—) recommended.

erally within the combined uncertainties (the individual uncertainties are about ±10%), the more recent values [251, 253] are preferred. The solid line in Fig. 5.47 is a least squares fit to these data (taken in both N_2 and Ar buffer gases) and represents our recommended values for the $k_{a,t}(\langle\varepsilon\rangle)$ of c-C_4F_8. Values obtained from this line are listed in Table 5.38.

4.5.4 Swarm-Determined Total Electron Attachment Cross Section, $\sigma_{a,t}(\varepsilon)$, of c-$C_4F_8$

In Fig. 5.48 are shown the values of the total electron attachment cross section, $\sigma_{a,t}(\varepsilon)$, for c-$C_4F_8$ at energies below 1.5 eV. These have been obtained by Christophorou and coworkers [249, 251, 252] from an unfolding of their data on $k_{a,t}(E/N)$ measured in mixtures of c-C_4F_8 with the buffer gases N_2 [249, 252] and Ar [251]. The respective electron energy distribution functions were used for each buffer gas. While the cross section from the more recent measurement exhibits a peak near 0.4 eV, the earlier data of Christophorou *et al.* [252] indicate a "double peak" with a maximum near 0.2 eV. It may be possible that the peak near 0.2 eV in the earliest data resulted from unfolding errors or from

Table 5.38. Recommended values of the total electron attachment rate constant, $k_{a,t}(\langle\varepsilon\rangle)$, as a function of $\langle\varepsilon\rangle$ for c-C_4F_8.

$\langle\varepsilon\rangle$ (eV)	$k_{a,t}$ (10^{-8} cm^3 s^{-1})	$\langle\varepsilon\rangle$ (eV)	$k_{a,t}$ (10^{-8} cm^3 s^{-1})
0.04	1.72	0.50	5.03
0.05	1.92	0.60	4.53
0.06	2.08	0.70	3.92
0.07	2.26	0.80	3.37
0.08	2.40	0.90	2.84
0.09	2.61	1.00	2.55
0.10	2.82	1.50	1.62
0.15	3.69	2.00	1.11
0.20	4.34	2.50	0.80
0.25	4.80	3.00	0.64
0.30	5.12	3.50	0.52
0.35	5.29	4.00	0.43
0.40	5.28	4.50	0.35

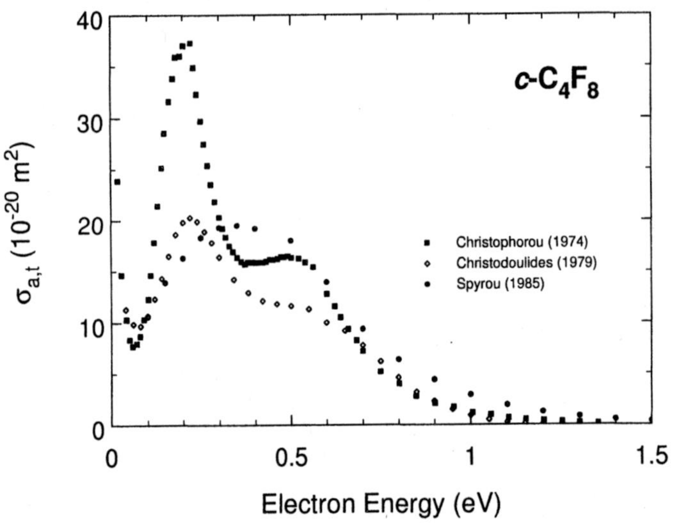

Figure 5.48. Swarm-determined total electron attachment cross section, $\sigma_{a,t}(\varepsilon)$, as a function of electron energy ($\leq$ 1.5 eV) for c-C_4F_8: (■) Ref. [252]; (◊) Ref. [249]; (•) Ref. [251].

Table 5.39. Suggested values of the total electron attachment cross section, $\sigma_{a,t}(\varepsilon)$, of c-$C_4F_8$.

ε (eV)	$\sigma_{a,t}(\varepsilon)$ (10^{-20} m^2)	ε (eV)	$\sigma_{a,t}(\varepsilon)$ (10^{-20} m^2)	ε (eV)	$\sigma_{a,t}(\varepsilon)$ (10^{-20} m^2)
0.010	15.9	0.40	15.8	5.50	0.022
0.015	15.2	0.45	15.1	6.00	0.012
0.020	14.5	0.50	14.2	6.50	0.015
0.030	13.2	0.60	11.5	6.75	0.018
0.040	12.0	0.70	8.22	7.00	0.019
0.050	10.9	0.80	5.35	7.25	0.021
0.060	10.0	0.90	3.24	7.5	0.026
0.070	9.67	1.0	1.97	8.0	0.035
0.080	9.83	1.5	0.235	8.5	0.032
0.090	10.2	2.0	0.049	9.0	0.020
0.10	10.8	2.5	0.022	9.5	0.016
0.15	14.6	3.0	0.014	10.0	0.020
0.20	17.5	3.5	0.016	10.5	0.022
0.25	18.4	4.0	0.047	11.0	0.018
0.30	17.8	4.5	0.075	11.5	0.013
0.35	16.7	5.0	0.055		

an uncertainty in the electron energy distribution functions of N_2 that were used at the time. For this reason we shall not consider further this cross section.

4.5.5 Comparison of the Values of the $\sigma_{a,t}(\varepsilon)$ of c-C_4F_8 Derived from Electron Swarm and Electron Beam Experiments

In Fig. 5.49(a) are compared the values of $\sigma_{a,t}(\varepsilon)$ derived from electron swarm experiments [249, 251, 252] with the values of $\sigma_{a,t}(\varepsilon)$ derived from electron-beam experiments by Kurepa [224] and Chutjian and Alajajian [248]. The data from the various sources differ considerably. Nonetheless, we have determined suggested values for the $\sigma_{a,t}(\varepsilon)$ of c-C_4F_8 which are shown by the solid line in Fig. 5.49(b) and are listed in Table 5.39.

These suggested values were obtained from a least-squares fit to the data in Fig. 5.49(b). These data are as follows: Below 0.06 eV the electron beam data of Chutjian and Alajajian [248]; between 0.06 and 1.0 eV the electron swarm data of Christodoulides *et al.* [249] and Spyrou *et al.* [251]; between 1 and 3 eV the data of Spyrou *et al.* [251] derived from measurements in Ar buffer gas and normalized to the average of

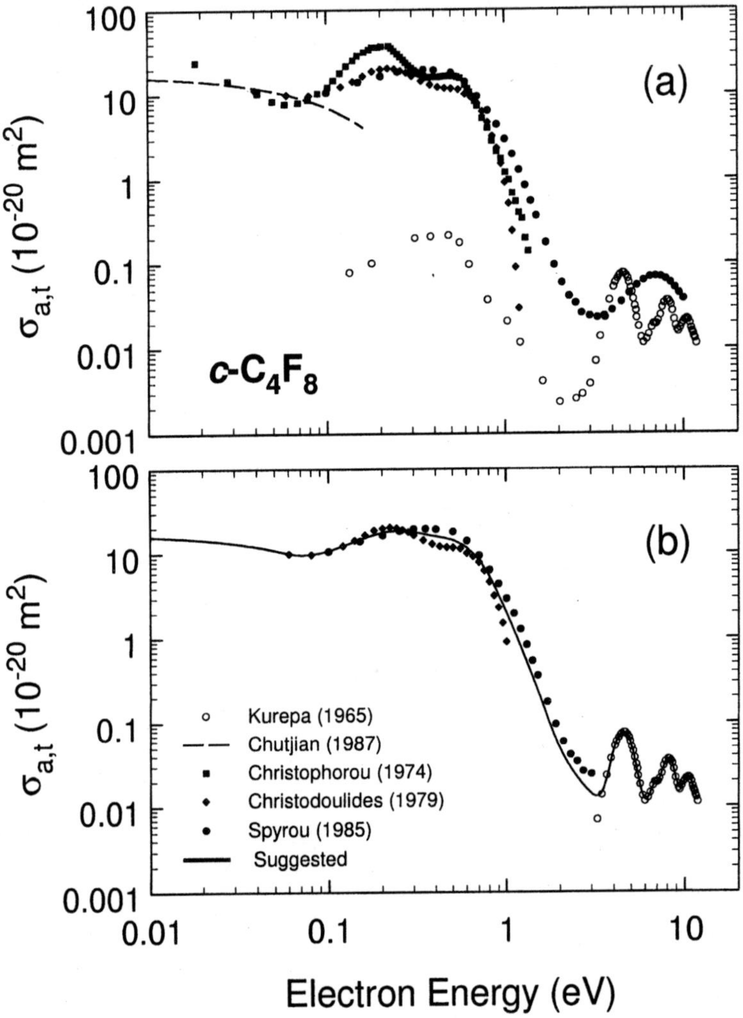

Figure 5.49. (a) Comparison of electron swarm and electron beam data for the total electron attachment cross section, $\sigma_{a,t}(\varepsilon)$, of c-C_4F_8. Electron-swarm data: (■) Ref. [252]; (♦) Ref. [249]; (•) Ref. [251]. Electron-beam data: (○) Ref. [224]; (– –) Ref. [248]. (b) (– –) Data of Chutjian and Alajajian [248] below 0.06 eV (hidden behind solid line); (♦) data of Christodoulides *et al.* Ref. [249] between 0.06 and 1.0 eV; (•) data of Spyrou *et al.* [251] between 0.06 and 3.0 eV; (○) data of Kurepa [224] above 3 eV; (—) suggested (see text).

the data of Christodoulides *et al.* [249] and Spyrou *et al.* [251] at 1 eV; above 3 eV the electron beam data of Kurepa [224].

Table 5.40. Thermal values, $(k_{a,t})_{th}$, of the total electron attachment rate constant for c-C_4F_8.

$(k_{a,t})_{th}$ (10^{-9} cm^3 s^{-1})	T (K)	Reference
0.4 ± 0.1[a]	298	[246]
8–9[a,b]	300	[254, 255]
11.1	300	[77]
12	298	[249]
12	298	[256]
13	298	[197]
15	300	[201]
16	298	[251]
16 ± 5	375	[214]
18.1	301	[253]
21	298	[252]

[a]Measurements made using the ion cyclotron resonance technique. The rest of the data were obtained using electron swarm methods.
[b]This value supersedes an earlier much higher value of 110×10^{-9} cm^3 s^{-1} published by the same group using the same method [215].

The low energy data of Kurepa [224] were not considered in view of the difficulty of the early electron beam methods in obtaining accurate cross sections at low electron energies. The electron swarm data of Christodoulides *et al.* [249] were not considered at energies in excess of 1 eV since these were determined from measurements made only in mixtures with N_2. Similarly, the $\sigma_{a,t}(\varepsilon)$ data of Christophorou *et al.* [252] were not included for the reasons discussed earlier in Section 4.5.4.

4.5.6 Thermal Value, $(k_{a,t})_{th}$, of the Total Electron Attachment Rate Constant of c-C_4F_8

The published experimental data on $(k_{a,t})_{th}$ are listed in Table 5.40. If the lowest two measurements are excluded, the average of the remaining eight room-temperature measurements is 1.5×10^{-8} cm^3 s^{-1}. The lowest two values in Table 5.40 were measured using the ICR technique in which the electron energies are generally lower than the "thermal" electron energies in the electron swarm experiments. Thus, the ICR-measured magnitudes of $(k_{a,t})_{th}$ may be lower than the electron swarm values because the electron attachment rate constant of c-C_4F_8 decreases as the electron energy approaches 0.0 eV (Fig. 5.47).

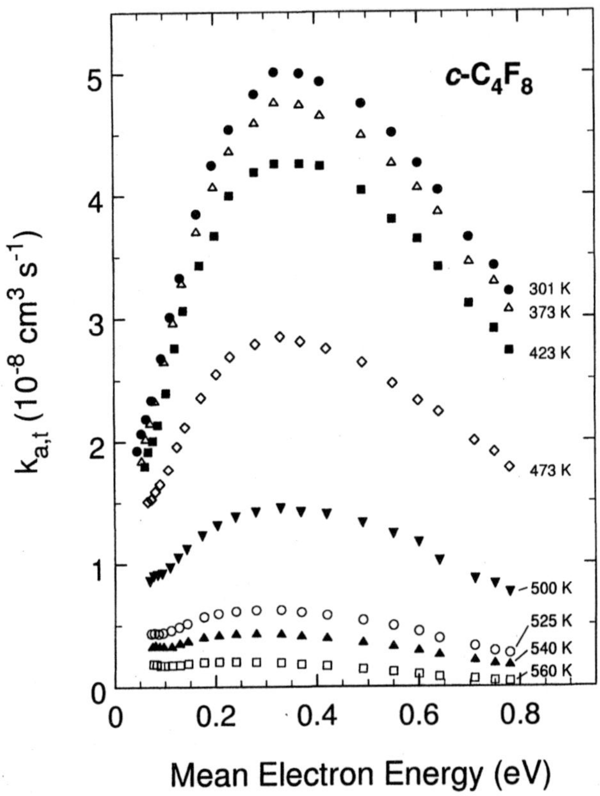

Figure 5.50. Total electron attachment rate constant as a function of $\langle\varepsilon\rangle$ and T, $k_{a,t}(\langle\varepsilon\rangle, T)$ for c-C_4F_8 measured by Christodoulides *et al.* [253].

4.5.7 Total Electron Attachment Rate Constant of c-C_4F_8 as a Function of the Mean Electron Energy and Gas Temperature, $k_{a,t}(\langle\varepsilon\rangle, T)$

Christodoulides *et al.* [253] measured the total electron attachment rate constant as a function of E/N in N_2 buffer gas at a number of gas temperatures from 301 to 560 K. The $k_{a,t}(E/N, T)$ data were reported as $k_{a,t}(\langle\varepsilon\rangle, T)$ and are reproduced in Fig. 5.50. It is seen that the total electron attachment rate constant decreases with increasing gas temperature. Subsequent similar studies for other molecules which form parent negative ions and have electron affinities $\lesssim 0.5$ eV have shown that this "negative" temperature dependence is largely due to thermally induced electron detachment from the parent negative ion [257].

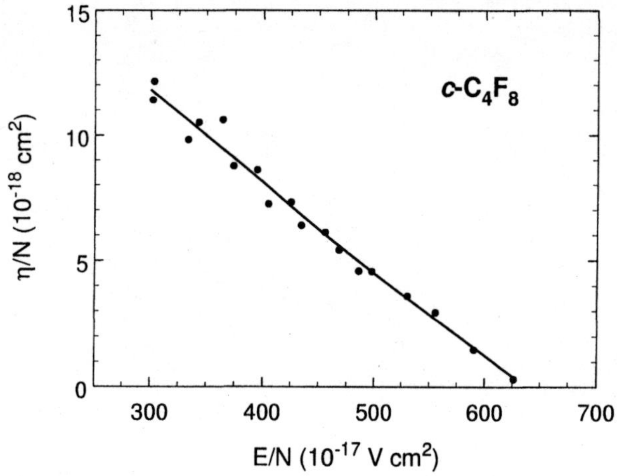

Figure 5.51. Density-normalized electron attachment coefficient, $\eta/N\,(E/N)$, for c-C_4F_8 measured by Naidu *et al.* [236].

Table 5.41. Values of $\eta/N\,(E/N)$ for c-C_4F_8 ($T = 293$ K) derived from a fit to the data of Naidu *et al.* [236].

E/N (10^{-17} V cm^2)	$\eta/N\,(E/N)$ (10^{-18} cm^2)	E/N (10^{-17} V cm^2)	$\eta/N\,(E/N)$ (10^{-18} cm^2)
300	11.8	500	4.49
350	9.98	550	2.87
400	8.10	600	1.25
450	6.22	625	0.43

4.5.8 Density-Normalized Electron Attachment Coefficient, $\eta/N\,(E/N)$, of c-C_4F_8

There has been only one experimental determination of the density-normalized electron attachment coefficient, $\eta/N\,(E/N)$, of c-C_4F_8 over an extended range of E/N, namely, that by Naidu *et al.* [236] shown in Fig. 5.51. These measurements were made at a temperature of 293 K and at two values of the gas pressure (0.084 and 0.133 kPa). They have a quoted uncertainty of about ±10% for E/N values below $(E/N)_{\rm lim}$ and about ±20% for E/N values above $(E/N)_{\rm lim}$. The solid line shown in Fig. 5.51 is a least squares fit to the data. Values taken from this fit are listed in Table 5.41. For the reasons discussed in Sec. 4.3.4 of this

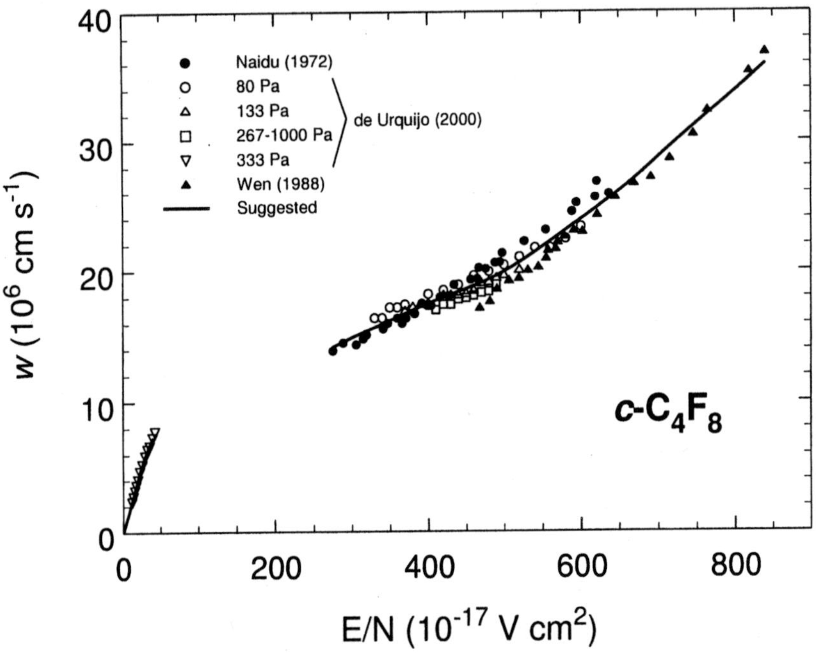

Figure 5.52. Electron drift velocity, $w\,(E/N)$, in c-C_4F_8 at room temperature: (•) Ref. [236]; (○, △, □, ▽) Ref. [240]; (▲) Ref. [238]; (—) suggested.

chapter, and the fact that the limited measurements of Takashira *et al.* [237] indicate a decrease in η/N with gas pressure, these values require further validation.

4.6 Electron Transport Coefficients for c-C_4F_8

4.6.1 Electron Drift Velocity, $w\,(E/N)$, in c-C_4F_8

In Fig. 5.52 are shown the measurements of Naidu *et al.* [236] for the electron drift velocity, $w\,(E/N)$, in c-C_4F_8 made at 293 K. The overall uncertainty in these data was quoted by Naidu *et al.* to be less than ±5%. Also shown in Fig. 5.52 are the room temperature measurements of $w\,(E/N)$ by Wen and Wetzer [238] and de Urquijo and Basurto [240]. The measurements by de Urquijo and Basurto [240] were obtained for a number of gas pressures (see figure legend), and seem to indicate a small decrease of the electron drift velocity with increasing gas pressure. The solid line shown in Fig. 5.52 is a least squares fit to the data of Naidu *et al.*, Wen and Wetzer, and de Urquijo and Basurto. Values obtained from

Table 5.42. Suggested room-temperature values of the electron drift velocity, $w\,(E/N)$, for c-C_4F_8.

E/N (10^{-17} V cm^2)	w (10^6 cm s^{-1})	E/N (10^{-17} V cm^2)	w (10^6 cm s^{-1})
0	0.00	450	18.6
10	1.91	500	19.9
20	4.02	550	21.8
30	5.83	600	23.9
40	7.28	650	26.1
–	–	700	28.7
275	14.2	750	31.3
300	14.9	800	33.8
350	16.2	840	35.9
400	17.5		

this fit are listed in Table 5.42 as the suggested values for the $w\,(E/N)$ of c-C_4F_8.

Measurements of $w\,(E/N)$ in two mixtures of c-C_4F_8 with Ar at relative concentrations of 0.468% and 4.91% have been reported by Yamaji *et al.* [258]. They may be of interest to those using Boltzmann codes for the calculation of electron transport parameters for c-C_4F_8.

4.6.2 Ratio of the Lateral Electron Diffusion Coefficient to Electron Mobility, $D_T/\mu\,(E/N)$, for c-C_4F_8

The $D_T/\mu\,(E/N)$ measurements by Naidu *et al.* [236] seem to be the only available data for this gas. They were obtained at 293 K and for two gas pressures (0.084 and 0.133 kPa). Their overall uncertainty is quoted by Naidu *et al.* to vary "between about ±5% at the lowest E/N and about ±3% at the high E/N values." These data are shown in Fig. 5.53. Although the data indicate a slight pressure dependence of $D_T/\mu\,(E/N)$, this variation is within the combined uncertainty of the measurements made at the two pressures. Thus, we performed a least squares fit to all the data in Fig. 5.53 which is shown by the solid line in the figure. Values from this fit are listed in Table 5.43 as the suggested data for the $D_T/\mu\,(E/N)$ of c-C_4F_8.

For measurements of $D_T/\mu\,(E/N)$ in two mixtures of c-C_4F_8 with Ar at relative concentrations of 0.468% and 4.91% see Yamaji *et al.* [258].

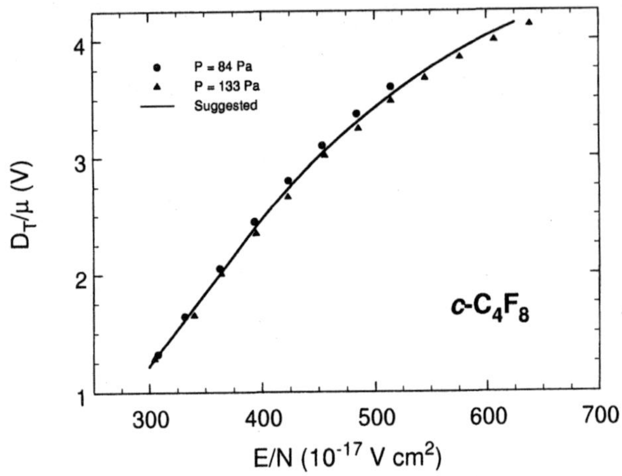

Figure 5.53. Ratio of the lateral electron diffusion coefficient to electron mobility, $D_T/\mu\,(E/N)$, for c-C_4F_8 ($T = 293$ K) from Naidu *et al.* [236].

Table 5.43. Values of $D_T/\mu\,(E/N)$ for c-C_4F_8 ($T = 293$ K) derived from the data of Naidu *et al.* [236].

E/N (10^{-17} V cm^2)	D_T/μ (V)	E/N (10^{-17} V cm^2)	D_T/μ (V)
300	1.22	500	3.42
350	1.84	550	3.76
400	2.47	600	4.03
450	3.00	625	4.14

4.7 Ion-Molecule Reactions in c-C_4F_8

There are three studies of negative ion-molecule reactions involving c-C_4F_8 which might be of interest to its use as a plasma processing gas. The first study is by Smith and Kevan [235] who measured the cross sections for total and dissociative charge transfer in collisions of rare-gas positive ions with c-C_4F_8. They found that in such collisions the predominant ion is $C_2F_4^+$ at recombination energies of up to 16 eV. Only for Xe^+ bombardment did they detect a small parent positive ion c-$C_4F_8^+$. The second study is by Su and Kevan [259] on ion-molecule reactions in c-C_4F_8 using ICR mass spectrometry. These studies have shown that fluoride transfer and collision-induced dissociation reactions

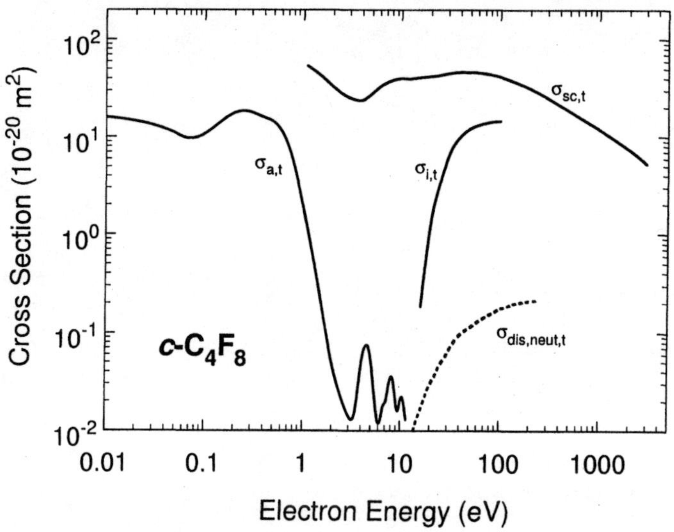

Figure 5.54. Summary of suggested cross sections for c-C_4F_8 (see text).

predominate. The third study is by Morris *et al.* [260] who measured the rate constants and product branching ratios for reactions of the atmospheric ions O^+, O_2^+, O^-, O_2^-, NO^+, H_3O^+, CO_3^-, and NO_3^- with c-C_4F_8. They found that the last four ions are unreactive.

4.8 Summary of Suggested and Recommended Cross Sections and Coefficients for c-C_4F_8

The suggested values for the electron scattering cross sections of the c-C_4F_8 molecule are summarized in Fig. 5.54. The suggested values for

- $\sigma_{sc,t}(\varepsilon)$ — Table 5.29, Fig. 5.36;
- $\sigma_{i,t}(\varepsilon)$ — Table 5.33, Fig. 5.40; and
- $\sigma_{a,t}(\varepsilon)$ — Table 5.39, Fig. 5.49(b)

are shown by the solid lines, and the lower-limit values of $\sigma_{dis,neut,t}(\varepsilon)$ (Fig. 5.43; Table 5.35) are shown by the broken line.

Also suggested or recommended are the measurements of

- $\sigma_{e,diff}(\varepsilon)$ — Table 5.30, Fig. 5.37;
- $\sigma_{i,part}(\varepsilon)$ — Table 5.31, Fig. 5.39;

- $k_{a,t}(\langle\varepsilon\rangle)$ — Table 5.38, Fig. 5.47;
- $\eta/N\,(E/N)$ — Table 5.41, Fig. 5.51;
- $\alpha/N\,(E/N)$ — Table 5.34, Fig. 4.41;
- $w\,(E/N)$ — Table 5.42, Fig. 5.52 ; and
- $D_T/\mu\,(E/N)$ — Table 5.43, Fig. 5.53.

The available data on both the electron collision cross sections and the electron transport coefficients need further scrutiny. In addition, measurements are needed of the momentum transfer and elastic integral cross sections, as well as of the cross sections of other significant low-energy electron collision processes such vibrational and electronic excitation. Measurements of electron transport coefficients over a wider range of E/N values are also indicated.

* * *

5. References

[1] L. G. Christophorou, J. K. Olthoff, and M. V. V. S. Rao, J. Phys. Chem. Ref. Data **26**, 1 (1997).

[2] L. G. Christophorou and J. K. Olthoff, J. Phys. Chem. Ref. Data **29**, 553 (2000).

[3] L. G. Christophorou and J. K. Olthoff, J. Phys. Chem. Ref. Data **30**, 449 (2001).

[4] Intergovernmental Panel on Climate Change (IPCC), *The 1994 Report of the Scientific Assessment Working Group of IPCC*, p. 28.

[5] L. G. Christophorou and J. K. Olthoff, J. Phys. Chem. Ref. Data **28**, 967 (1999).

[6] M. J. Kushner and D. Zhang, J. Appl. Phys. **88**, 3231 (2000).

[7] M. C. Bordage and P. Ségur, in *Proceedings of the XXV International Conference on Phenomena in Ionized Gases*, T. Goto (Ed.), Nagoya, Japan, July 2001, Vol. 3, p. 253.

[8] W. L. Morgan, C. Winstead, and V. McKoy, J. Appl. Phys. **90**, 2009 (2001).

[9] C. R. Brundle, M. B. Robin, and H. Basch, J. Chem. Phys. **53**, 2196 (1970).

[10] C. Y. R. Wu, L. C. Lee, and D. L. Judge, J. Chem. Phys. **71**, 5221 (1979).

[11] M. B. Robin, *Higher Excited States of Polyatomic Molecules*, Academic, New York, 1974, Vol. I, pp. 178–191.

[12] J. F. Ying and K. T. Leung, Phys. Rev. A **53**, 1476 (1996).

[13] R. D. Nelson, Jr., D. R. Lide, Jr., and A. A. Maryott, *Selected Values of Electric Dipole Moments for Molecules in the Gas Phase*, NSRDS-NBS 10, U.S. GPO, Washington, DC, 1967.

[14] W. L. Meerts and I. Ozier, J. Chem. Phys. **75**, 596 (1981).

[15] H. Sutter and R. H. Cole, J. Chem. Phys. **52**, 132 (1970).

[16] G. Cazzoli, L. Cludi, G. Cotti, L. Dore, C. D. Esposti, M. Bellini, and P. de Natale, J. Mol. Spectrosc. **163**, 521 (1994).

[17] J. A. Beran and L. Kevan, J. Phys. Chem. **73**, 3860 (1969).

[18] R. Kobayashi, R. D. Amos, H. Koch, and P. Jørgensen, Chem. Phys. Lett. **253**, 373 (1996).

[19] W. R. Harshbarger, M. B. Robin, and E. N. Lassettre, J. Electron Spectrosc. Relat. Phenom. **1**, 319 (1972/1973).

[20] C. Larrieu, M. Chaillet, and A. Dargelos, J. Chem. Phys. **94**, 1327 (1991).

[21] P. Sauvageau, R. Gilbert, P. P. Berlow, and C. Sandorfy, J. Chem. Phys. **59**, 762 (1973).

[22] S. Stokes and A. B. F. Duncan, J. Am. Chem. Soc. **80**, 6177 (1958).

[23] F. C.-Y. Wang and G. E. Leroi, Ann. Isr. Phys. Soc. **6**, 210 (1983).

[24] B. P. Pullen, T. A. Carlson, W. E. Moddeman, G. K. Schweitzer, W. E. Bull, and F. A. Grimm, J. Chem. Phys. **53**, 768 (1970).

[25] G. Bieri, L. Åsbrink, and W. von Niessen, J. Electron Spectrosc. Relat. Phenom. **23**, 281 (1981).

[26] A. Campargue and F. Stoeckel, J. Chem. Phys. **85**, 1220 (1986).

[27] K. Kim and C. W. Park, J. Mol. Struct. **161**, 297 (1987.

[28] B. Galabov, T. Dudev, and W. J. Orville-Thomas, J. Mol. Struct. **145**, 1(1986).

[29] J. E. Sanabia, G. D. Cooper, J. A. Tossell, and J. H. Moore, J. Chem. Phys. **108**, 389 (1998).

[30] O. Sueoka, H. Takaki, A. Hamada, H. Sato, and M. Kimura, Chem. Phys. Lett. **288**, 124 (1998).

[31] I. Iga, P. R. Pinto, and M. G. P. Homem, in *Proceedings of the International Symposium on Electron-Molecule Collisions and Swarms*, Y. Hatano, H. Tanaka, and N. Kouchi (Eds.), Tokyo, Japan, July 1999, p. 123.

[32] H. Nishimura, K. Okuda, and Y. Nakamura, in *Proceedings of the International Symposium on Electron-Molecule Collisions and Swarms*, I. Fabrikant, G. Gallup, and P. Burrow (Eds.), Lincoln, NE, July 2001, p. 105.

[33] S. Altshuler, Phys. Rev. **107**, 114 (1957).

[34] L. G. Christophorou and A. A. Christodoulides, J. Phys. B **2**, 71 (1969).

[35] L. G. Christophorou, *Atomic and Molecular Radiation Physics*, Wiley-Interscience, New York, 1971, Ch. 4.

[36] L. G. Christophorou, D. R. James, and R. A. Mathis, J. Phys. D **14**, 675 (1981).

[37] J. A. LaVerne and A. Mozumder, Radiat. Res. **96**, 219 (1983).

[38] H. Tanaka, T. Masai, M. Kimura, T. Nishimura, and Y. Itikawa, Phys. Rev. A **56**, R3338 (1997).

[39] H. Tanaka, private communication, 1998.

[40] M. T. do N. Varella, C. Winstead, V. McKoy, M. Kitajima, and H. Tanaka, Phys. Rev. A **65**, 022702 (2002).

[41] R. B. Diniz, M. A. P. Lima, and F. J. da Paixão, J. Phys. B **32**, L539 (1999).

[42] A. P. P. Natalense, M. H. F. Bettega, L. G. Ferreira, and M. A. P. Lima, Phys. Rev. A **59**, 879 (1999).

[43] H. U. Poll and J. Meichsner, Contrib. Plasma Phys. **27**, 359 (1987).

[44] M. Goto, K. Nakamura, H. Toyoda, and H. Sugai, Jpn. J. Appl. Phys. **33**, 3602 (1994).

[45] C. Q. Jiao, R. Nagpal, and P. D. Haaland, Chem. Phys. Lett. **269**, 117 (1997).

[46] C. Q. Jiao, private communication, January 1999.

[47] I. Iga, I. P. Sanches, S. K. Srivastava, and M. Mangan, Int. J. Mass Spectrom. **208**, 159 (2001).

[48] I. Torres, R. Martinez, and F. Castaíño, J. Phys. B **35**, 2423 (2002).

[49] E. Krishnakumar and S. K. Srivastava, J. Phys. B **21**, 1055 (1988).

[50] J. A. Beran and L. Kevan, J. Phys. Chem. **73**, 3866 (1969).

[51] Y.-K. Kim, W. Hwang, M. A. Ali, and M. E. Rudd, in *Proceedings of the Twentieth International Conference on the Physics of Electronic and Atomic Collisions*, Vienna, Austria, 1997, Vol. II, p. WE103.

[52] Y.-K. Kim, private communication, 1996.

[53] Y.-K. Kim and M. E. Rudd, Phys. Rev. A **50**, 3954 (1994).

[54] W. Hwang, Y.-K. Kim, and M. E. Rudd, J. Chem. Phys. **104**, 2956 (1996).

[55] D. L. Hobrock and R. W. Kiser, J. Phys. Chem. **68**, 575 (1964)

[56] J. B. Farmer, I. H. S. Henderson, F. P. Lossing, and D. G. H. Marsden, J. Chem. Phys. **24**, 348 (1956).

[57] C. Lifshitz and F. A. Long, J. Phys. Chem. **69**, 3731 (1965).

[58] J. S. Kim, M. V. V. S. Rao, M. A. Cappelli, S. P. Sharma, and M. Meyyappan, Plasma Sources Sci. Technol. **10**, 191 (2001).

[59] B. L. Peko, R. L. Champion, M. V. V. S. Rao, and J. K. Olthoff, J. Appl. Phys. **92**, 1657 (2002).

[60] J. de Urquijo, E. Basurto, C. Cisneros, and I. Alvarez, in *Proceedings of the XXIVth International Conference on Phenomena in Ionized Gases, Contributed Papers*, P. Pisarczyk, T. Pisarczyk, and J. Wołowski (Eds.), Warsaw, Poland, July 1999, Vol. I, p. 73.

[61] J. de Urquijo, I. Alvarez, and C. Cisneros, Phys. Rev. E **60**, 4990 (1999).

[62] L. G. Christophorou, D. R. James, and R. Y. Pai, in *Applied Atomic Collision Physics*, Academic, New York, Vol. 5, 1982, p. 87.

[63] L. G. Christophorou, D. R. James, I. Sauers, M. O. Pace, R. Y. Pai, and A. Fatheddin, in *Gaseous Dielectrics III*, Pergamon, New York, 1982, p. 151.

[64] EPRI Report EPRI El-2620, September 1982.

[65] R. S. Nema, S. V. Kulkarni, and E. Husain, IEEE Trans. Electr. Insul. **EI-17**, 434 (1982).

[66] J. L. Hernández-Ávila, E. Basurto, and J. de Urquijo, in *Proceedings of the XXVth International Conference on Phenomena in Ionized Gases*, T. Goto (Ed.), Nagoya, Japan, July 2001, Vol. 3, p. 215.

[67] H. Sugai, H. Toyoda, T. Nakano, and M. Goto, Contrib. Plasma Phys. **35**, 415 (1995).

[68] V. Tarnovsky and K. Becker, J. Chem. Phys. **98**, 7868 (1993).

[69] V. Tarnovsky, P. Kurunczi, D. Rogozhnikov, and K. Becker, Int. J. Mass Spectrom. Ion Processes **128**, 181 (1993).

[70] S. Motlagh and J. H. Moore, J. Chem. Phys. **109**, 432 (1998).

[71] H. F. Winters and M. Inokuti, Phys. Rev. A **25**, 1420 (1982).

[72] R. M. Reese, V. H. Dibeler, and F. L. Mohler, J. Res. Natl. Bur. Stand. **57**, 367 (1956).

[73] T. G. Lee, J. Phys. Chem. **67**, 360 (1963).

[74] R. P. Blaunstein and L. G. Christophorou, J. Chem. Phys. **49**, 1526 (1968).

[75] K. A. G. MacNeil and J. C. J. Thynne, Int. J. Mass Spectrom. Ion Phys. **2**, 1 (1969).

[76] R. W. Fessenden and K. M. Bansal, J. Chem. Phys. **53**, 3468 (1970).

[77] F. J. Davis, R. N. Compton, and D. R. Nelson, J. Chem. Phys. **59**, 2324 (1973).

[78] A. A. Christodoulides, R. Schumacher, and R. N. Schindler, Int. J. Chem. Kinet. **10**, 1215 (1978).

[79] L. G. Christophorou, D. L. McCorkle, and A. A. Christodoulides, in *Electron-Molecule Interactions and Their Applications*, L. G. Christophorou (Ed.), Academic, New York, 1984, Vol. 1, Ch. 6.

[80] L. G. Christophorou, Environ. Health Perspectives **36**, 3 (1980).

[81] H.-U. Scheunemann, M. Heni, E. Illenberger, and H. Baumgärtel, Ber. Bunsenges. Phys. Chem. **86**, 321 (1982).

[82] A. Modelli, F. Scagnolari, G. Distefano, D. Jones, and M. Guerra, J. Chem. Phys. **96**, 2061 (1992).

[83] M. Haverlag, A. Kono, D. Passchier, G. M. W. Kroesen, W. J. Goedheer, and F. J. de Hoog, J. Appl. Phys. **70**, 3472 (1991).

[84] T. O. Tiernan, C. Chang, and C. C. Cheng, Environ. Health Perspectives **36**, 47 (1980).

[85] Y. Wang, L. G. Christophorou, J. K. Olthoff, and J. K. Verbrugge, in *Gaseous Dielectrics VIII*, L. G. Christophorou and J. K. Olthoff (Eds.), Plenum, New York, 1998, p. 39.

[86] Y. Wang, L. G. Christophorou, J. K. Olthoff, and J. K. Verbrugge, Chem. Phys. Lett. **304**, 303 (1999).

[87] A. Rosa and I. Szamrej, J. Phys. Chem. A **104**, 67 (2000).

[88] G. K. Jarvis, C. A. Mayhew, L. Singleton, and S. M. Spyrou, Int. J. Mass Spectrom. Ion Processes **164**, 207 (1997).

[89] J. D. Clark, B. W. Wright, J. D. Wrbanek, and A. Garscadden, in *Gaseous Dielectrics VIII*, L. G. Christophorou and J. K. Olthoff (Eds.), Plenum, New York, 1998, p. 23.

[90] M. Mitani and Y. Nakamura, in *Proceedings of the XXVth International Conference on Phenomena in Ionized Gases*, T. Goto (Ed.), July 2001, Nagoya, Jpn., Vol. 3., p. 281.

[91] H. A. Van Sprang, H. H. Brongersma, and F. J. De Heer, Chem. Phys. **35**, 51 (1978).

[92] J. F. M. Aarts, Chem. Phys. **95**, 443 (1985).

[93] J. C. Creasey, I. R. Lambert, R. P. Tuckett, and A. Hopkirk, Molec. Phys. **71**, 1355 (1990).

[94] L. C. Lee, J. C. Han, C. Ye, and M. Suto, J. Chem. Phys. **92**, 133 (1990).

[95] S. Wang and J. W. McConkey, Can. J. Phys. **67**, 694 (1989).

[96] N. P. Danilevskii, I. Y. Rapp, V. T. Koppe, and A. G. Koval, Opt. Spectr. **60**, 441 (1986).

[97] O. Keller, C. Mang, M. Nikola, and G. Schulz, Eur. Phys. J. D **4**, 309 (1998).

[98] M. Suto and N. Washida, J. Chem. Phys. **78**, 1007 (1983).

[99] M. Suto and N. Washida, J. Chem. Phys. **78**, 1012 (1983).

[100] M. Suto, N. Washida, H. Akimoto, and M. Nakamura, J. Chem Phys. **78**, 1019 (1983).

[101] M. Suto and L. C. Lee, J. Chem. Phys. **79**, 1127 (1983).

[102] H. Deutsch, T. D. Märk, V. Tarnovsky, K. Becker, C. Cornelissen, L. Cespiva, and V. Bonacic-Koutecky, Int. J. Mass Spectrom. Ion Processes **137**, 77 (1994).

[103] K. Takahashi, M. Hori, K. Maruyama, S. Kishimoto, and T. Goto, Jpn. J. Appl. Phys. **32**, L694 (1993).

[104] K. Takahashi, M. Hori, and T. Goto, Jpn. J. Appl. Phys. **33**, 4745 (1994).

[105] K. Maruyama, K. Ohkouchi, Y. Ohtsu, and T. Goto, Jpn. J. Appl. Phys. **33**, 4298 (1994).

[106] K. Takahashi, M. Hori, S. Kishimoto, and T. Goto, Jpn. J. Appl. Phys. **33**, 4181(1994).

[107] S. M. Karecki, L. C. Pruette, and L. R. Reif, Mat. Res. Soc. Symp. Proc. **447**, 67 (1997).

[108] S. M. Karecki, L. C. Pruette, and L. R. Reif, J. Vac. Sci. Technol. A **16**, 755 (1998).

[109] R. A. Levy, V. B. Zaitsev, K. Aryusook, C. Ravindranath, V. Sigal, A. Misra, S. Kesari, D. Rufin, J. Sees, and L. Hall, J. Mater. Res. **13**, 2643 (1998).

[110] F. Fracassi and R. d'Agostino, J. Vac. Sci. Technol. B **16**, 1867 (1998).

[111] S. Samukawa and K. Tsuda, Jpn. J. Appl. Phys. **37**, L1095 (1998).

[112] S. Samukawa and T. Mukai, in *Proceedings of the International Symposium on Electron-Molecule Collisions and Swarms*, Y. Hatano, H. Tanaka, and N. Kouchi (Eds.), Tokyo, Japan, July 1999, p. 76.

[113] S. Solomon, J. B. Burkholder, A. R. Ravishankara, and R. R. Garcia, J. Geophys. Res. D: Atmos. **99**, 20929 (1994).

[114] *Climate Change 1995; The Science of Climate Change* (Published for the Intergovernmental Panel on Climate Change), J. T. Houghton, L. G. M. Filho, B. A. Callander, N. Harris, A. Kattenberg, and K. Maskell (Eds.), Cambridge University Press, Cambridge, U. K., 1996, p. 93.

[115] O. V. Rattigan, D. E. Shallcross, and R. A. Cox, J. Chem. Soc. Faraday Trans. **93**, 2839 (1997).

[116] *Climate Change 1995; The Science of Climate Change* (Published for the Intergovernmental Panel on Climate Change), J. T. Houghton, L. G. M. Filho, B. A. Callander, N. Harris, A. Kattenberg, and K. Maskell (Eds.), Cambridge University Press, Cambridge, U. K., 1996, p. 121.

[117] A. L. McClellan, *Tables of Experimental Dipole Moments*, W. H. Freeman and Company, San Francisco, 1963, p. 36.

[118] T. Shimanouchi, J. Phys. Chem. Ref. Data **3**, 269 (1974).

[119] H. Bürger, K. Burczyk, H. Hollenstein, and M. Quack, Mol. Phys. **55**, 255 (1985).

[120] S. Roszak, W. S. Koski, J. J. Kaufman, and K. Balasubramanian, J. Chem. Phys. **106**, 7709 (1997).

[121] C. B. Leffert, Ph.D. dissertation, Wayne State University, 1974; as quoted in Ref. [122].

[122] S. Y. Tang, B. P. Mathur, E. W. Rothe, and G. P. Reck, J. Chem. Phys. **64**, 1270 (1976).

[123] R. N. Compton, P. W. Reinhardt, and C. D. Cooper, J. Chem. Phys. **68**, 4360 (1978).

[124] P. E. McNamee, K. Lacman, and D. R. Herschbach, Faraday Discuss. Chem. Soc. **55**, 318 (1973).

[125] J. Marriott and J. D. Craggs, J. Electronics **1**, 405 (1956).

[126] C. A. Goy, A. Lord, and H. O. Pritchard, J. Phys. Chem. **71**, 1086 (1967).

[127] P. Felder, Chem. Phys. **143**, 141 (1990).

[128] H. J. M. Bowen, Trans. Faraday Soc. **50**, 444 (1954).

[129] J. Sheridan and W. Gordy, J. Chem. Phys. **20**, 591 (1952).

[130] C.-H. Wong and V. Schomaker, J. Chem. Phys. **28**, 1010 (1958).

[131] A. P. Cox, G. Duxbury, J. A. Hardy, and Y. Kawashima, J. Chem. Soc. Faraday 2 **76**, 339 (1980).

[132] V. H. Dibeler, R. M. Reese, and F. L. Mohler, J. Res. Natl. Bur. Stand. **57**, 113 (1956).

[133] M. Heni and E. Illenberger, Chem. Phys. Lett. **131**, 314 (1986).

[134] T. Oster, O. Ingólfsson, M. Meinke, T. Jaffke, and E. Illenberger, J. Chem. Phys. **99**, 5141 (1993).

[135] Y. Le Coat, R. Azria, M. Tronc, O. Ingólfsson, and E. Illenberger, Chem. Phys. Lett. **296**, 208 (1998).

[136] T. Underwood-Lemons, T. J. Gergel, and J. H. Moore, J. Chem. Phys. **102**, 119 (1995).

[137] I. S. Buchel'nikova, J. Exptl. Theoret. Phys. (U.S.S.R.) **35**, 1119 (1958) [Soviet Physics JETP **8**, 783 (1959)].

[138] T. Underwood-Lemons, D. C. Winkler, J. A. Tossell, and J. H. Moore, J. Chem. Phys. **100**, 9117 (1994).

[139] I. Hahndorf and E. Illenberger, Int. J. Mass Spectrom. Ion Processes **167/168**, 87 (1997).

[140] M. Okamoto, M. Hoshino, Y. Sakamoto, S. Watanabe, M. Kitajima, H. Tanaka, and M. Kimura, in *Proceedings of the International Symposium on Electron-Molecule Collisions and Swarms*, Y. Hatano, H. Tanaka, and N. Kouchi (Eds.), Tokyo, Japan, July 1999, p. 191.

[141] M. Kitajima, M. Okamoto, H. Tanaka, and S. Samukawa, (manuscript communicated to the authors by Professor H. Tanaka [142]).

[142] H. Tanaka, private communication, January 2000.

[143] M. Kitajima, M. Okamoto, K. Sunohara, H. Tanaka, H. Cho, S. Samukawa, S. Eden, and N. J. Mason, J. Phys. B **35**, 3257 (2002).

[144] L. Brouwer and J. Troe, Chem. Phys. Lett. **82**, 1 (1981).

[145] A. Fahr, A. K. Nayak, and R. E. Huie, Chem. Phys. **199**, 275 (1994).

[146] M. B. Robin, *Higher Excited States of Polyatomic Molecules*, Academic, New York, Vol. I, 1974.

[147] G. Herzberg, *Molecular Spectra and Molecular Structure III: Electronic Spectra and Electronic Structure of Polyatomic Molecules*, D. Van Nostrand, Princeton, NJ, 1966, p. 532.

[148] S. S. Kumaran, M.-C. Su, K. P. Lim, and J. V. Michael, Chem. Phys. Lett. **243**, 59 (1995).

[149] C. J. Noutary, J. Res. Natl. Bur. Stand. **72** A, 479 (1969).

[150] S. G. Lias, J. E. Bartmess, J. F. Liebman, J. L. Holmes, R. D. Levin, and W. G. Mallard, J. Phys. Chem. Ref. Data **17**, 60 (1988).

[151] I. Novak, as quoted in Ref. [152].

[152] L. R. Thorne and J. L. Beauchamp, J. Chem. Phys. **74**, 5100 (1981).

[153] D. W. Berman, J. L. Beauchamp, and L. R. Thorne, Int. J. Mass Spectrom. Ion Phys. **39**, 47 (1981).

[154] T. Cvitaš, H. Güsten, L. Klasinc, I. Novadj, and H. Vančik, Z. Naturforsch. **33a**, 1528 (1978).

[155] M. B. Robin, *Higher Excited States of Polyatomic Molecules*, Academic, Orlando, Florida, Vol. III, 1985, p. 27.

[156] R. A. A. Boschi and D. R. Salahub, Can. J. Chem. **52**, 1217 (1974).

[157] K. B. Riehl, Ph.D. dissertation, Air Force University, 1992; Air Force Technical Report No. AFIT/DS/ENP/92-3, December 1992.

[158] I. Powis, O. Dutuit, M. Richard-Viard, and P. M. Guyon, J. Chem. Phys. **92**, 1643 (1990).

[159] M. K. Kawada, O. Sueoka, and M. Kimura, Chem. Phys. Lett. **330**, 34 (2000).

[160] M. Kimura, private communication, June 2000.

[161] H. Jiao, private communication, December 1999.

[162] C. Q. Jiao, B. Ganguly, C. A. DeJoseph, Jr., A. Garscadden, Int. J. Mass Spectrom. **208**, 127 (2001).

[163] R. C. Wetzel, F. A. Baiocchi, T. R. Hayes, and R. S. Freund, Phys. Rev. A **35**, 559 (1987).

[164] S. R. Heller and G. W. A. Milne (Eds.), *EPA/NIH Mass Spectra Data Base*, U.S.A., GPO, Washington, 1978.

[165] Z.-J. Sun, A. L. Schwaner, and J. M. White, Chem. Phys. Lett. **219**, 118 (1994).

[166] O. Ingólfsson and E. Illenberger, Int. J. Mass Spectrom. Ion Processes **155**, 1 (1996).

[167] H. Shimamori and Y. Nakatani, Chem. Phys. Lett. **150**, 109 (1988).

[168] H. Shimamori, Y. Tatsumi, Y. Ogawa, and T. Sunagawa, J. Chem. Phys. **97**, 6335 (1992).

[169] R. G. Levy, S. J. Burns, and D. L. McFadden, Chem. Phys. Lett. **231**, 132 (1994).

[170] T. Sunagawa and H. Shimamori, in *Proceedings of the International Symposium on Electron-Molecule Collisions and Swarms*, Y. Hatano, H. Tanaka, and N. Kouchi (Eds.), Tokyo, Japan, July 1999, p. 181.

[171] S. H. Alajajian, K.-F. Man, and A. Chutjian, J. Chem. Phys. **94**, 3629 (1991).

[172] X. Ling, M. A. Durham, A. Kalamarides, R. W. Marawar, B. G. Lindsay, K. A. Smith, and F. B. Dunning, J. Chem. Phys. **93**, 8669 (1990).

[173] Z. Zheng, X. Ling, K. A. Smith, and F. B. Dunning, J. Chem. Phys. **92**, 285 (1990).

[174] M. Suh, W. Sung, S.-U. Heo, and H. J. Hwang, J. Phys. Chem. A **103**, 8365 (1999).

[175] L. G. Christophorou and J. K. Olthoff, Adv. At. Mol. Opt. Phys. **44**, 155 (2000).

[176] J. H. Moore, private communication, June 2000.

[177] L. G. Christophorou and J. K. Olthoff, J. Phys. Chem. Ref. Data **29**, 267 (2000).

[178] B. G. Zollars, C. Higgs, F. Lu, C. W. Walter, L. G. Gray, K. A. Smith, F. B. Dunning, and R. F. Stebbings, Phys. Rev. A **32**, 3330 (1985).

[179] X. Ling, B. G. Lindsay, K. A. Smith, and F. B. Dunning, Phys. Rev. A **45**, 242 (1992).

[180] F. B. Dunning, J. Phys. B **28**, 1645 (1995).

[181] J. Sherwell, R. Cooper, D. C. Nguyen, and S. P. Mezyk, Aust. J. Chem. **41**, 1491 (1988).

[182] R. Martínez, J. Terrón, I. Merelas, M. N. S. Rayo, and F. Castaño, J. Phys. B **31**, 1793 (1998).

[183] J. W. Coburn, Plasma Chem. and Plasma Process. **2**, 1 (1982).

[184] M. Miyamura, O. Tsukakoshi, and S. Komiya, J. Vac. Sci. Technol. **20**, 986 (1982).

[185] K. Nojiri and E. Iguchi, J. Vac. Sci. Technol. B **13**, 1451 (1995).

[186] T. Maruyama, N. Fujiwara, K. Shiozawa, and M. Yoneda, J. Vac. Sci. Technol. A **13**, 810 (1995).

[187] H. Kimura, K. Shiozawa, K. Kawai, H. Miyatake, and M. Yoneda, Jpn. J. Appl. Phys. **34**, 2114 (1995).

[188] K. Miyata, M. Hori, and T. Goto, Jpn. J. Appl. Phys., Part 1 **36**, 5340 (1997).

[189] J. N. Butler, J. Amer. Chem. Soc. **84**, 1393 (1962).

[190] J. M. Simmie, W. J. Quiring, and E. Tschuikow-Roux, J. Phys. Chem. **73**, 3830 (1969).

[191] J. M. Preses, R. E. Weston, Jr., and G. W. Flynn, Chem. Phys. Lett. **46**, 69 (1977).

[192] M. Quack and G. Seyfang, Ber. Bunsenges. Phys. Chem. **86**, 504 (1982).

[193] E. Pochon, R. E. Weston, Jr., and G. W. Flynn, J. Phys. Chem. **89**, 86 (1985).

[194] A. Yokoyama, K. Yokoyama, and G. Fujisawa, Chem Phys. Lett. **237**, 106 (1995).

[195] E. W. D. Norton, Trans. Amer. Acad. Ophalmol. Otolaryngol. **77**, 85 (1973).

[196] S. Brubaker, G. A. Peyman, and C. Vygantas, Arch. Ophthalmol. **92**, 324 (1974).

[197] L. G. Christophorou, R. A. Mathis, D. R. James, and D. L. McCorkle, J. Phys. D **14**, 1889 (1981).

[198] L. G. Christophorou, R. A. Mathis, S. R. Hunter, and J. G. Carter, J. Appl. Phys. **63**, 52 (1988).

[199] L. G. Christophorou, Nucl. Instrum. Methods Phys. Res. A **268**, 424 (1988).

[200] J. T. Houghton, L. G. M. Filho, B. A. Callander, N. Harris, A. Kattenberg, and K. Maskell, *Climate Change 1995*, Cambridge University Press, New York, 1996, p. 22.

[201] R. A. Morris, T. M. Miller, A. A. Viggiano, J. F. Paulson, S. Solomon, and G. Reid, J. Geophys. Res. **100**, 1287 (1995).

[202] J. P. Novak and M. F. Fréchette, J. Appl. Phys. **63**, 2570 (1988).

[203] H. Itoh, T. Miyachi, M. Kawaguchi, Y. Nakao, and H. Tagashira, J. Phys. D **24**, 277 (1991).

[204] O. Bastiansen, O. Hassel, and L. K. Lund, Acta Chem. Scand. **3**, 297 (1949).

[205] H. P. Lemaire and R. L. Livingston, J. Am. Chem. Soc. **74**, 5732 (1952).

[206] N. V. Alekseev, I. A. Ronova, and P. P. Barzdain, Zh. Strukt. Khim. **9**, 1073 (1968).

[207] C. H. Chang, R. F. Porter, and S. H. Bauer, J. Mol. Struct. **7**, 89 (1971).

[208] N. V. Alekseev and P. P. Barzdain, Zh. Strukt. Khim. **15**, 181 (1972).

[209] B. Beagley, R. Calladine, R. G. Pritchard, and S. F. Taylor, J. Mol. Struct. **158**, 309 (1987).

[210] R. P. Bauman and B. J. Bulkin, J. Chem. Phys. **45**, 496 (1966).

[211] F. A. Miller and R. J. Capwell, Spectrochim. Acta **27A**, 1113 (1971).

[212] C. Mao, C.-S. Nie, and Z.-Y. Zhu, Spectrochim. Acta **44A**, 1093 (1988).

[213] G. Fischer, R. L. Purchase, and D. M. Smith, J. Mol. Struct. **405**, 159 (1997).

[214] T. M. Miller, R. A. Morris, A. E. S. Miller, A. A. Viggiano, and J. F. Paulson, Int. J. Mass Spectrom. Ion Processes **135**, 195 (1994).

[215] K. G. Mothes and R. N. Schindler, Ber. Bunsenges. Phys. Chem. **75**, 938 (1971).

[216] C. Lifshitz, T. O. Tiernan, and B. M. Hughes, J. Chem. Phys. **59**, 3182 (1973).

[217] G. K. Jarvis, K. J. Boyle, C. A. Mayhew, and R. P. Tuckett, J. Phys. Chem. A **102**, 3230 (1998).

[218] C. Lifshitz and R. Grajower, Int. J. Mass Spectrom. Ion Phys. **10**, 25 (1972/1973).

[219] M. B. Robin, *Higher Excited States of Polyatomic Molecules*, Academic, New York, 1974, Vol. I, p.180.

[220] M. M. Bibby and G. Carter, Trans. Faraday Soc. **59**, 2455 (1963).

[221] L. Kevan and J. H. Futrell, J. Chem. Soc., Faraday Trans. 2 **68**, 1742 (1972).

[222] I. Sauers, L. G. Christophorou, and J. G. Carter, J. Chem. Phys. **71**, 3016 (1979).

[223] R. Grajower and C. Lifshitz, Israel J. Chem. **6**, 847 (1968).

[224] M. V. Kurepa, *3rd Cz. Conference on Electronics and Vacuum Physics Transactions*, 1965, p. 107.

[225] P. W. Harland and J. C. J. Thynne, Int. J. Mass Spectrom. Ion Phys. **10**, 11 (1972/1973).

[226] P. W. Harland and J. L. Franklin, J. Chem. Phys. **61**, 1621 (1974).

[227] I. Ishii, R. McLaren, A. P. Hitchcock, K. D. Jordan, Y. Choi, and M. B. Robin, Can. J. Chem. **66**, 2104 (1988).

[228] H. Nishimura, in *Proceedings of the International Symposium on Electron-Molecule Collisions and Swarms*, Y. Hatano, H. Tanaka, and N. Kouchi, (Eds.), Tokyo, Japan, July 1999, p. 103.

[229] H. Nishimura, private communication, November 1999.

[230] V. McKoy, as quoted in [142].

[231] C. Winstead and V. McKoy, J. Chem. Phys. **114**, 7407 (2001).

[232] H. Toyota, M. Ito, and H. Sugai, Jpn. J. Appl. Phys. **36**, 3730 (1997).

[233] C. Q. Jiao, A. Garscadden, and P. D. Haaland, in *Gaseous Dielectrics VIII*, L. G. Christophorou and J. K. Olthoff (Eds.), Kluwer Academic/Plenum Publishers, New York, 1998, p. 57.

[234] C. Q. Jiao, A. Garscadden, and P. D. Haaland, Chem. Phys. Lett. **297**, 121 (1998).

[235] D. L. Smith and L. Kevan, J. Chem. Phys. **55**, 2290 (1971).

[236] M. S. Naidu, A. N. Prasad, and J. D. Craggs, J. Phys. D **5**, 741 (1972).

[237] T. Tagashira, Y. Miyamoto, and M. Shimozuma, in *Proceedings of the XVIII International Conference on Phenomena in Ionized Gases*, Swansea, U.K., July 1987, p. 680.

[238] C. Wen and J. M. Wetzer, in *IX International Conference on Gas Discharges and Their Applications*, Venezia, Italy, September 1988, p. 367.

[239] J. de Urquijo, private communication, July 2000.

[240] J. de Urquijo and E. Basurto, J. Phys. D **34**, 1352 (2001).

[241] J. Berril, M. J. Christensen, and I. W. McAllister, in *Gaseous Dielectrics V*, L. G. Christophorou and D. W. Bouldin (Eds.), Pergamon, New York, 1987, p. 304.

[242] L. G. Christophorou, R. A. Mathis, S. R. Hunter, and J. G. Carter, in *Gaseous Dielectrics V*, L. G. Christophorou and D. W. Bouldin (Eds.), Pergamon, New York, 1987, p. 88.

[243] H. Kazumi, R. Hamasaki, and K. Tago, Plasma Sources Sci. Technol. **5**, 200 (1996).

[244] W. T. Naff, C. D. Cooper, and R. N. Compton, J. Chem. Phys. **49**, 2784 (1968).

[245] J. M. S. Henis and C. A. Mabie, J. Chem. Phys. **53**, 2999 (1970).

[246] R. L. Woodin, M. S. Foster, and J. L. Beauchamp, J. Chem. Phys. **72**, 4223 (1980).

[247] L. G. Christophorou, Adv. Electron. Electron Phys. **46**, 55 (1978).

[248] A. Chutjian and S. H. Alajajian, J. Phys. B **20**, 839 (1987).

[249] A. A. Christodoulides, L. G. Christophorou, R. Y. Pai, and C. M. Tung, J. Chem. Phys. **70**, 1156 (1979).

[250] A. Kono and K. Kato, Appl. Phys. Lett. **77**, 495 (2000).

[251] S. M. Spyrou, S. R. Hunter, and L. G. Christophorou, J. Chem. Phys. **83**, 641 (1985).

[252] L. G. Christophorou, D. L. McCorkle, and D. Pittman, J. Chem. Phys. **60**, 1183 (1974).

[253] A. A. Christodoulides, L. G. Christophorou, and D. L. McCorkle, Chem. Phys. Lett. **139**, 350 (1987).

[254] E. Schultes, Ph. D. dissertation, University of Bonn, 1973; as quoted in Ref. [255].

[255] A. A. Christodoulides, E. Schultes, R. Schumacher, and R. N. Schindler, Z. Naturforschg. **29a**, 389 (1974).

[256] K. M. Bansal and R. W. Fessenden, J. Chem. Phys. **59**, 1760 (1973).

[257] L. G. Christophorou and P. G. Datskos, Int. J. Mass Spectrom. Ion Processes **149/150**, 59 (1995).

[258] M. Yamaji, Y. Okada, and Y. Nakamura, in *Proceedings of the International Symposium on Electron-Molecule Collisions and Swarms*, Y. Hatano, H. Tanaka, and N. Kouchi (Eds.), Tokyo, Japan, July 1999, p. 151.

[259] T. Su and L. Kevan, J. Phys. Chem. **77**, 148 (1973).

[260] R. A. Morris, A. A. Viggiano, S. T. Arnold, and J. F. Paulson, Int. J. Mass Spectrom. Ion Processes **149/150**, 287 (1995).

Chapter 6

ELECTRON INTERACTIONS WITH Cl_2, CCl_2F_2, BCl_3, AND SF_6

1. Introduction

In this chapter we assess and synthesize the available information on the cross sections and the rate coefficients for collisional interactions of chlorine (Cl_2) [1], dichlorodifluoromethane (CCl_2F_2) [2], boron trichloride (BCl_3) [3], and sulfur hexafluoride (SF_6) [4]. Definitions of the symbols we use to describe the various collision processes for these gases are as discussed in Ch. 1 (Table 1.3). All four gases are used in the fabrication of semiconductor devices and have a number of other important applications (for instance, SF_6 is the most commonly used gaseous dielectric [5]).

2. Electron Interactions with Cl_2

Molecular chlorine (Cl_2) is a plasma processing gas (e.g., Refs. [6–29]). It is used in plasma etching of semiconductors where the Cl atoms produced in a gas discharge efficiently etch silicon surfaces. The dominant primary electron interaction processes are taken to be single-step electron-impact ionization of Cl_2 and Cl, dissociation of Cl_2 into neutrals, and dissociative electron attachment to Cl_2 [6, 15, 18, 22]. The basic species involved in Cl_2 plasmas, then, are the three molecular species: Cl_2, Cl_2^-, and Cl_2^+, and the three atomic species: Cl, Cl^-, and Cl^+. Although recent work on the interactions of Cl_2 with slow electrons is largely motivated by plasma etching technology, considerable work on electron interactions with the Cl_2 molecule was done in the 1970s and 1980s motivated by gas ultraviolet (UV) laser applications [30–36]. In this latter application the fundamental process of interest is dissociative electron attachment producing halogen atomic negative ions (Cl^-) which efficiently recombine with the rare-gas positive ions [34, 35] to form the lasing species (e.g., ArF*, KrF*, XeCl* excimers) in rare-gas-halide lasers [32, 33].

Molecular chlorine is also of atmospheric and environmental interest [37]. It is a potential atmospheric reservoir of chlorine atoms [38] which are released photolytically via the reaction $Cl_2 + h\nu$ ($\lambda \leq 500$ nm) $\rightarrow$ 2Cl.

In this section low-energy electron interactions with the Cl_2 molecule are reviewed. Information is synthesized and assessed on the cross sections for total electron scattering, total rotational excitation, total elastic electron scattering, momentum transfer, total vibrational excitation, electronic excitation, total dissociation into neutrals, total ionization, total electron attachment, and ion-pair formation. Similar data on the density-normalized ionization coefficient, density-normalized electron attachment coefficient, density-normalized effective ionization coefficient, electron transport coefficients, and electron attachment rate constant are also synthesized and critically evaluated. Cross sections are suggested for total electron scattering, total elastic electron scattering, total ionization, dissociation into neutrals, electron attachment, and ion-pair formation. A cross section is derived for the total vibrational excitation cross section via low-lying negative ion resonances. Data are suggested for the coefficients of electron attachment, ionization, and effective ionization, and for the rate constant of electron attachment. While progress has been made regarding our knowledge on electron-chlorine interactions at low energies (< 100 eV), there is still a need for (i) improvement in the uncertainties of all suggested cross sections, (ii) measurement of the cross sections for momentum transfer, vibrational excitation, electronic excitation, and dissociative ionization, and (iii) accurate measurement of the electron transport coefficients in pure Cl_2 and in mixtures with rare gases.

Also, since the Cl_2 molecule is one of the simplest reactive gases used in plasma processing (often in mixtures with Ar), we consider it desirable from the point-of-view of this application to also provide relevant information on the most likely discharge byproducts, namely, Cl_2^-, Cl_2^+, Cl, Cl^-, and Cl^+. In this way, one may have more complete information about the key species and processes. Therefore, while the emphasis in this section is on low-energy electron interactions with the neutral Cl_2 molecule, pertinent information is also provided for the Cl_2^+ and Cl_2^- molecular ions, and for the atomic species Cl, Cl^-, and Cl^+.

An early attempt to critically evaluate low-energy electron-impact cross section data for Cl_2 was made by Morgan [36]. In addition, a number of investigators have used Boltzmann codes and electron transport data to calculate cross sections and rate coefficients for some electron collision processes in Cl_2 discharges [6, 15, 39–41]. However, the values resulting from these calculations are questionable because of the limited measurements of electron transport parameters upon which they are based, and because of the model dependent nature of the calculated cross sections. The results of two such calculations by Rogoff *et al.* [6] and Pinhão and Chouki [40] are compared with other data later in this section.

2.1 Electronic and Molecular Structure of Cl_2

2.1.1 Cl_2

The electronic structure of the outermost (valence) shell of the Cl_2 molecule in its ground electronic state [42–44] is $\ldots(\sigma_g\,3p)^2$, $(\pi_u\,3p)^4$, $(\pi_g\,3p)^4$, $(\sigma_u\,3p)^0$ and it has $^1\Sigma_g^+$ symmetry. The first four excited electronic states of Cl_2 listed by Huber and Herzberg [45] are $A'\,^3\Pi_{2u}$, $A\,^3\Pi_{1u}$, $B\,^3\Pi_{0u}^+$, and $C\,^1\Pi_u$. A number of theoretical and experimental studies have located many other excited electronic states (see below).

There are three types of sources of information of interest to the present study regarding the electronic structure of the chlorine molecule: calculations, photoabsorption and photoelectron studies, and electron energy-loss investigations. Although our interest is focused on the third type of information, basic information provided by the other two types of investigations is included in this section as complementary.

The most useful theoretical work concerning the electronic states of chlorine is the *ab initio* calculations of Peyerimhoff and Buenker [46]. These workers calculated potential energy curves for the ground and excited electronic states of the chlorine molecule and for its positive and negative ions using the multireference single and double excitation with configuration interaction (MRD-CI) method. They considered all states which correlate with the lowest atomic limit $[Cl(^2P_u) + Cl(^2P_u)]$ and many others which go into ionic $Cl^+ + Cl^-$ or Rydberg $Cl^* + Cl$ asymptotes. All singlet states which correlate with the ground atomic products were found to be repulsive. Among the triplet states of Cl_2 which dissociate into the ground state atoms, only the $^3\Pi_u$ state is not repulsive. The potential energy curves calculated by Peyerimhoff and Buenker are reproduced in Fig. 6.1. The potential-energy curves shown in the figure are for the electronic states of Cl_2 which dissociate into the lowest atomic limit $[Cl(^2P_u) + Cl(^2P_u)]$. In Fig. 6.1 are also shown the potential-energy curves for the lowest electronic states of Cl_2^+ with

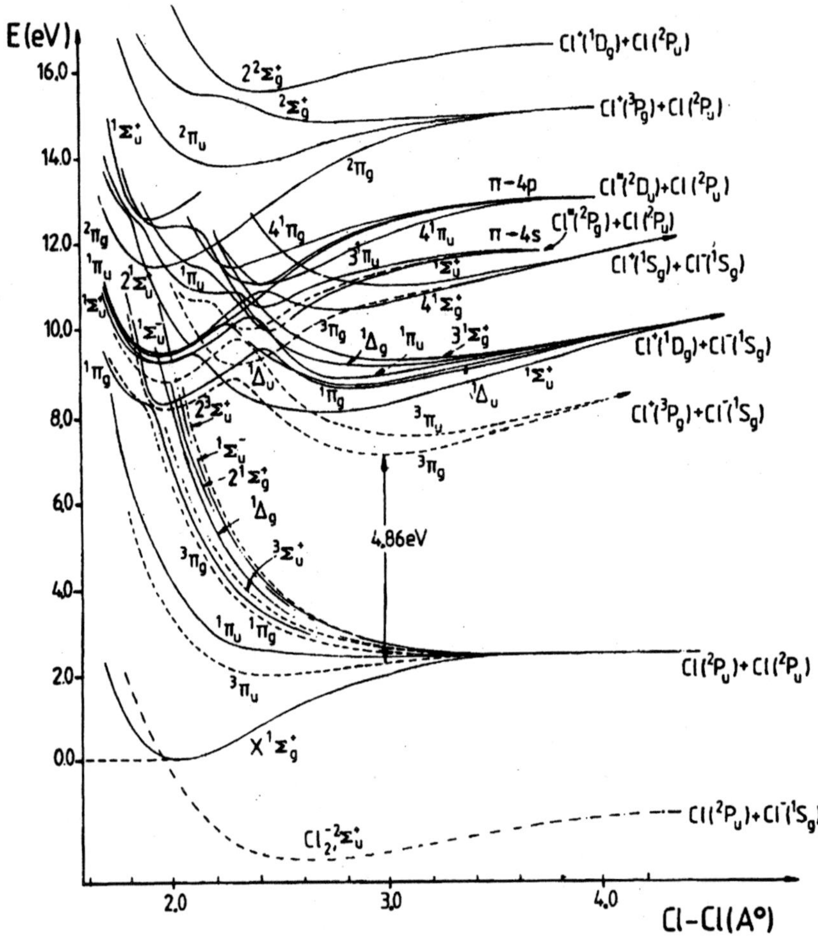

Figure 6.1. Composite potential-energy diagrams for most of the electronic states of the Cl_2 molecule as calculated by Peyerimhoff and Buenker [46].

various asymptotic limits and a potential-energy curve for the ground state of Cl_2^- for the asymptotic limit $Cl\,(^2P_u) + Cl^-\,(^1S_g)$. As will be seen from subsequent discussion in this section, the potential-energy curves for Cl_2, Cl_2^-, and Cl_2^+ in Fig. 6.1 are most helpful in understanding the low-energy electron interaction processes with the Cl_2 molecule. Peyerimhoff and Buenker calculated for the dissociation energy D_0, the vertical ionization energy, and the electron affinity of Cl_2 , respectively, the values 2.455, 11.48, and 2.38 eV. These values are in good agreement

Table 6.1. Vertical electronic energies (MRD-CI values) from the ground state of Cl_2 to various excited states as calculated by Peyerimhoff and Buenker [46].

State	Excitation	Vertical energy (eV)
$X\,^1\Sigma_g^+$		0.00
$1\,^3\Pi_u$	$\pi_g \rightarrow \sigma_u$	3.24
$1\,^1\Pi_u$	$\pi_g \rightarrow \sigma_u$	4.04
$1\,^3\Pi_g$	$\pi_u \rightarrow \sigma_u$	6.23
$1\,^1\Pi_g$	$\pi_u \rightarrow \sigma_u$	6.86
$1\,^3\Sigma_u^+$	$\sigma_g \rightarrow \sigma_u$	6.80
$^1\Delta_g$	$\pi_g^2 \rightarrow \sigma_u^2$	8.25
$2\,^3\Pi_g$	$\pi_g \rightarrow 4s$	8.34
$2\,^1\Pi_g$	$\pi_g \rightarrow 4s$	8.38
$^1\Delta_g$	$\pi_g^2 \rightarrow \sigma_u^2$	8.25
$2\,^1\Sigma_g^+$	$\pi_g^2 \rightarrow \sigma_u^2$	8.35
$2\,^3\Pi_u$	$\pi_g \rightarrow 4p\sigma$	8.80
$2\,^1\Pi_u$	$\pi_g \rightarrow 4p\sigma$	9.22
$1\,^1\Sigma_u^+$	$\pi_g \rightarrow 4p\pi$	9.32
$1\,^1\Sigma_u^-$	$\pi_g \rightarrow 4p\pi$	9.58
$1\,^1\Delta_u$	$\pi_g \rightarrow 4p\pi$	9.62
$2\,^1\Sigma_u^-$	$\pi_u\pi_g \rightarrow \sigma_u^2$	9.67
$2\,^3\Sigma_u^+$	$\pi_u\pi_g \rightarrow \sigma_u^2$	9.75
$^1\Delta_g$	$\pi_g \rightarrow 4d\pi$	9.92
$^1\Pi_g$	$\pi_g \rightarrow 4d\sigma$	10.01
$^1\Pi_g$	$\pi_g \rightarrow 4d\delta$	10.10
$2\,^1\Sigma_u^+$	$\sigma_g \rightarrow \sigma_u;\, \pi_g \rightarrow 4p\pi$	10.34
$3\,^3\Pi_u$	$\pi_u \rightarrow 4s$	11.33
$3\,^1\Pi_u$	$\pi_u \rightarrow 4s$	11.51
$^1\Pi_u$	$\pi_u \rightarrow 4d\delta$	12.74

with experimental values discussed later in this section. The estimated electronic-state energies (MRD-CI values) are listed in Table 6.1.

There have been many photoabsorption and photoionization studies [38, 45, 47–66], as well as a number of photoelectron studies [67–73] of Cl_2. The data on the total photoabsorption cross section, $\sigma_{pa,t}(\lambda)$, of Cl_2 have been discussed and summarized by a number of groups (e.g., see Refs. [38, 64, 66]). In Fig. 6.2 are plotted the measurements of $\sigma_{pa,t}(\lambda)$ of a number of investigators [38, 49, 51, 57, 59, 62, 63, 65, 66] in the wavelength range 15.5–600 nm. Between 250 and $\sim$ 500 nm the agreement among the various measurements is good. The extensive measurements of Samson and Angel [63] cover the wavelength range 15.5–103.8 nm and are recommended for this energy range. The measurements of Samson and Angel of the total photoionization cross section $\sigma_{pi,t}(\lambda)$ show that

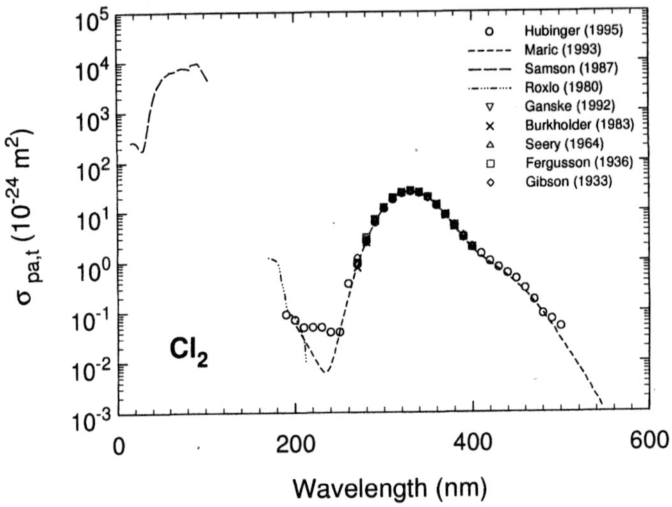

Figure 6.2. Total photoabsorption cross section as a function of photon wavelength, $\sigma_{\mathrm{pa,t}}(\lambda)$, for Cl_2: (— —) Ref. [63]; (– · · · –) Ref. [59]; (○) Ref. [66]; (- - -) Ref. [38]; (▽) Ref. [65]; (×) Ref. [62]; (△) Ref. [57]; (□) Ref. [51]; (◇) Ref. [49].

$\sigma_{\mathrm{pi,t}}(\lambda)$ is equal to the total photoabsorption cross section $\sigma_{\mathrm{pa,t}}(\lambda)$ except in the wavelength range 82.5–77.0 nm where it is up to 10% lower, depending on the value of the wavelength. The decrease of the photoionization efficiency below 1.0 in this wavelength range has been attributed to photoabsorption processes leading to the production of neutrals [63].

Measurements have also been made by Samson and Angel [63] of the partial photoionization cross section, $\sigma_{\mathrm{pi,part}}(\lambda)$, for the production of Cl_2^+ and Cl^+ by photon impact on Cl_2. These are shown in Fig. 6.3. Also plotted are the results of Samson and Angel for the production of Cl_2^{++} by photon impact on Cl_2. The data in Fig. 6.3 show that for photon wavelengths down to $\sim$ 80.0 nm, the cross section for the production of the Cl_2^+ ion is about equal to the total, that is, it far exceeds the cross section for the production of the Cl^+ ion (dissociative photoionization has a much lower probability than nondissociative photoionization). At progressively shorter wavelengths, dissociative photoionization becomes more probable. The cross section for double electron ejection is negligible down to about 40 nm. Carlson *et al.* [71] and Gallagher *et al.* [64] have published measurements of the production of Cl_2^+ in the ionic states $(2\pi_{\mathrm{g}}^{-1})X\,^2\Pi_{\mathrm{g}}$, $(2\pi_{\mathrm{u}}^{-1})A\,^2\Pi_{\mathrm{u}}$, and $(5\sigma_{\mathrm{g}}^{-1})B\,^2\Sigma_{\mathrm{g}}$ by photon impact on Cl_2.

Data on photoionization energetics are given in Table 6.2 where they are compared with data obtained using other methods. For further

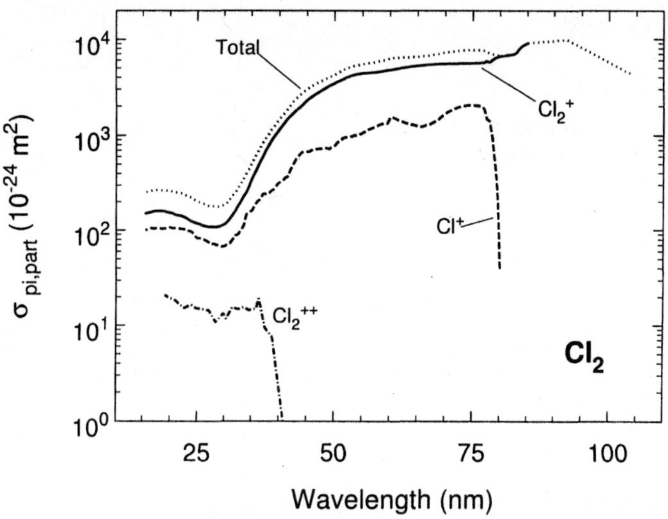

Figure 6.3. Partial photoionization cross sections as a function of photon wavelength, $\sigma_{pi,part}(\lambda)$, for the production of (—) Cl_2^+, (– –) Cl^+, and (– · –) Cl_2^{++} from Cl_2 (data of Samson and Angel [63]). The total cross section is shown by the dotted line.

spectroscopic investigations of the electronic structure of the chlorine molecule see Lee and Walsh [54], Iczkowski *et al.* [55], Douglas *et al.* [56], Frost *et al.* [67], Bondybey and Fletcher [81], Huber and Herzberg [45], Douglas [61], Moeller *et al.* [82], Burkholder and Bair [62], McLoughlin *et al.* [83], Lonkhuyzen and Lange [72], and Frost *et al.* [77].

There have been three major electron-impact studies of the electronic structure of Cl_2: the threshold-electron excitation study of Jureta *et al.* [43] which covered the excitation energy range up to 11.5 eV, the electron energy-loss study of Spence *et al.* [44] which covered the energy-loss range 5.5–14.5 eV, and the electron energy-loss study of Stubbs *et al.* [84] which covered the electron energy-loss range up to 14.252 eV. Threshold-electron excitation methods are best suited for locating optically-forbidden states, while electron energy-loss spectra using sufficiently energetic electrons give spectra similar to photoabsorption.

Figure 6.4(a) shows the threshold-electron excitation spectrum of Cl_2 obtained by Jureta *et al.* [43] in the region of Rydberg excitation between about 7.5 and 11.5 eV, and Fig. 6.4(b) shows an electron energy-loss spectrum of Cl_2 obtained by Spence *et al.* [44] at a scattering angle of 3° using 200 eV incident energy electrons. The threshold-electron excitation technique of Jureta *et al.* had an energy resolution [full-width-at-half-maximum (FWHM)] of about 35 meV and the energy-loss experi-

Table 6.2. Dissociation energy, vibrational energy, equilibrium internuclear separation, spin-orbit splitting, electron affinity, energy position of negative ion states, ionization threshold energy, dissociative ionization threshold energy, energy threshold for double ionization, and energy threshold for ion-pair formation of Cl_2.

Physical Quantity	Value/Method/Reference
Dissociation energy Cl_2 ($X^1\Sigma_g^+$) (eV)	$D_0 = 2.4793$ [45] $D_0 = 2.4794$ [56] $D_e = 2.5139$ [56]
Vibrational energy (eV)	0.0694 [45, 56]
Equilibrium internuclear separation (Å)	1.9879 [45] 1.9878 [56]
Spin-orbit splitting (eV)	0.080 ± 0.002 (for the $(1\pi_g)^{-1}$ state) [68]
Electron affinity (eV)	2.45[a] [74]
Energy position of negative ion states (eV)	See Table 6.11
Ionization threshold energy (eV)	
Cl_2^+ ($X^2\Pi_{g,3/2}$)	*Adiabatic* 11.50 (photoelectron) [45] 11.48 ± 0.01 (photoionization) [53] 11.49[b] (photoelectron) ($^2\Pi_g$) [69] 11.48 (photoelectron) [72] 11.480 ± 0.005 eV [75] 11.50 (photoelectron) ($^2\Pi_g$) [67] 11.5 (photoelectron) [70] 11.51 ± 0.01 (photoelectron) [68] *Vertical* 11.48 ± 0.01 (electron impact) [76] 11.559 (photoelectron) [72] 11.59 ± 0.01 (photoelectron) [68]
Cl_2^+ ($^2\Pi_{g,1/2}$)	11.56 [75] 11.63 (vertical, electron momentum spectroscopy, $2\pi_g$) [77] ~11.6 (electron impact) [76]

[a]Thirteen values are listed by Christodoulides *et al.* [74]. If we exclude the lowest three as too low and the highest one as too high, the average of the other nine values is 2.45 eV which is within the combined quoted uncertainty of the averaged values.
[b]0-0 band.

Table 6.2 (continued)
Dissociation energy, vibrational energy, equilibrium internuclear separation, spin-orbit splitting, electron affinity, energy position of negative ion states, ionization threshold energy, dissociative ionization threshold energy, energy threshold for double ionization, and energy threshold for ion-pair formation of Cl_2.

Physical Quantity	Value/Method/Reference
Cl_2^+ ($^2\Pi_{u,3/2}$)	~ 11.8 (electron impact) [76]
	11.80 (electron impact) [78]
	11.80 ± 0.1 (electron impact) [79]
Cl_2^+ ($^2\Pi_{u,1/2}$)	~ 11.9 (electron impact) [76]
Cl_2^+ ($^2\Pi_u$)	*Adiabatic*
	13.96 ± 0.02 (photoelectron) [68]
	14.0 (photoelectron) [70]
	14.0[c] (photoelectron) [69]
	14.04 (photoelectron, $^2\Pi_{u,3/2}$) [72]
	14.11 (photoelectron) [67]
	Vertical
	14.33 (photoelectron) [70]
	14.39 (photoelectron, $^2\Pi_{u,3/2}$) [72]
	14.40 ± 0.02 (photoelectron) [68]
	14.43[d] (photoelectron) [69]
	14.41 (electron momentum spectroscopy, $2\pi_u$) [77]
Cl_2^+ ($^2\Sigma_g^+$)	*Adiabatic*
	15.72 ± 0.02 (photoelectron) [68]
	15.70[c] (photoelectron) [72]
	15.8 (photoelectron) [70]
	15.8[c] (photoelectron) [69]
	Vertical
	14.09 ± 0.03 (electron impact) [76]
	15.94 (photoelectron) [67]
	16.082 (photoelectron) [72]
	16.08 ± 0.02 (photoelectron) [68]
	16 (photoelectron) [70]
	16.10[d] (photoelectron) [69]
	16.18 (electron momentum spectroscopy, $5\sigma_g$) [77]

[c]Onset.
[d]Band maximum.

Table 6.2 (continued)
Dissociation energy, vibrational energy, equilibrium internuclear separation, spin-orbit splitting, electron affinity, energy position of negative ion states, ionization threshold energy, dissociative ionization threshold energy, energy threshold for double ionization, and energy threshold for ion-pair formation of Cl_2.

Physical Quantity	Value/Method/Reference
Cl_2^+ $(^2\Sigma_u^+)$	*Vertical* 20.61 ± 0.06 (electron impact) [76] 21.8, 24.0 (electron momentum spectroscopy, $4\sigma_u$) [77] 27.3 (electron momentum spectroscopy, $4\sigma_g$) [77]
Dissociative ionization threshold energy (eV) ($Cl_2 + e \rightarrow Cl^+ + Cl + 2e$)	15.45 (adiabatic) [70] 15.7 ± 0.3 (electron impact) [79]
Energy threshold for double ionization (eV)	30.5 (photoionization) [63] 31.13 [Cl_2^{++} $(X^3\Sigma_g^-, \nu = 0)$, threshold photoelectron spectroscopy] [73]
Energy threshold for ion-pair formation (eV) ($Cl_2 + e \rightarrow Cl^+ + Cl^- + e$)	11.9 ± 0.2 [79, 80]

ment of Spence *et al.* had an energy resolution of 50–60 meV. As expected, the spectra they obtained at small scattering angles [Fig. 6.4(b)] differ from the threshold-electron excitation spectra, and correspond closely to the photoabsorption spectra. The most prominent features of the energy-loss spectra arise from excitation of optically allowed Rydberg states. Larger-angle scattering data showed additional structures due to excitation of optically forbidden states. The spectra also showed the presence of hot bands. Such observation of hot bands in electron scattering spectra is unusual, but because the ground-state vibrational spacing of Cl_2 is small (0.0694 eV) and the Franck-Condon overlappings particularly favorable, such structures become relatively strong for some electronic transitions. Stubbs *et al.* [84] had a better electron beam energy resolution (FWHM $\approx$ 18 meV) than the other two studies. This allowed them to obtain highly resolved electron energy-loss spectra up to 14.252 eV, using electrons with incident energy between 10 and 120 eV. These incident electron energies are lower than used by Spence *et al.* (200 eV).

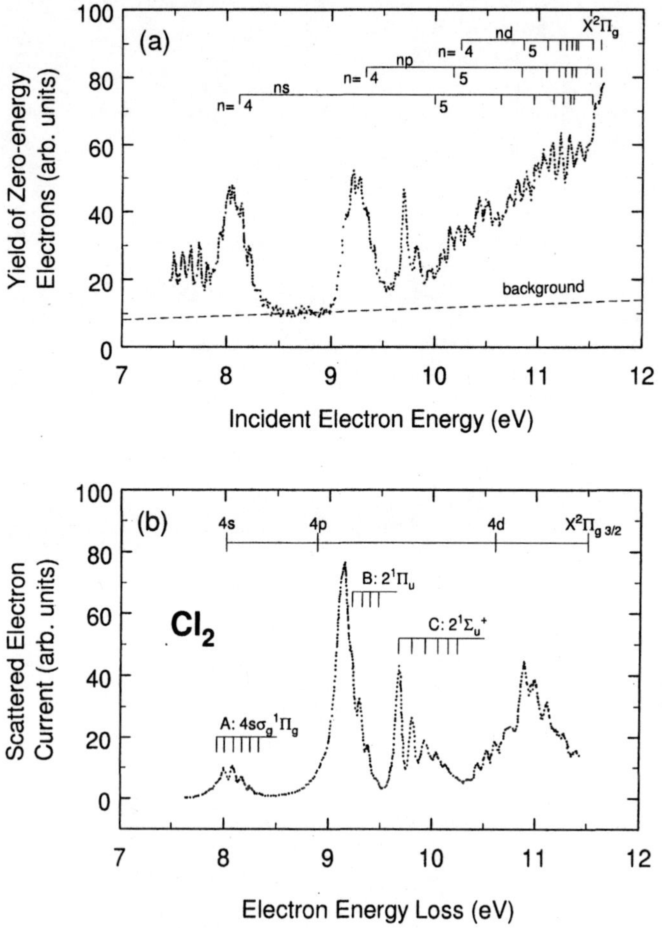

Figure 6.4. (a) Threshold-electron excitation spectrum of Cl_2 from Jureta *et al.* [43]. Here the electron energy is varied and the excitation spectrum reflects the relative probability of electrons having energy from about 7.5 eV to about 11.5 eV to lose all their energy in a collision with a Cl_2 molecule. The experiment detects only the "zero-energy electrons." (b) Electron energy-loss spectrum of Cl_2 between 7.5 and 11.5 eV using 200 eV incident energy electrons and a scattering angle of 3°. The scale above the spectrum shows the expected locations of Rydberg states due to excitations of a π_g $3p$ electron to $4s$, $4p$, and $4d$ orbitals (from Spence *et al.* [44]).

The strongest structures in the energy-loss spectrum of chlorine lie between 9 and 10 eV under all scattering conditions. They primarily consist of two vibrational series comprising transitions that are allowed by electric-dipole selection rules, and have been previously reported in both photoabsorption and electron impact studies. In Table 6.3 are listed the

Table 6.3. Comparison of the energies of the $4s\sigma_g\,{}^3\Pi_g$, $4s\sigma_g\,{}^1\Pi_g$, $2\,{}^3\Pi(1u)$, $2\,{}^1\Pi_u$, $2\,{}^1\Sigma_u^+$, and ${}^1\Pi_g(?)$ states of Cl_2. The electronic configuration and term symbol are as given by Spence *et al.* [44].

Name/ Assignment	Vibrational level	Energy (eV) Energy-loss experiment Ref. [44]	Energy (eV) Threshold-electron excitation experiment Ref. [43]	Energy (eV) Photoabsorption experiments[a] Ref. [82]
$4s\sigma_g\,{}^3\Pi_g$	hot band		(7.83)[b]	
	$\nu = 0$		7.91	
	$\nu = 1$		7.99	
	$\nu = 2$		(8.07)	
	$\nu = 3$		(8.15)	
$4s\sigma_g\,{}^1\Pi_g$	hot band		(7.87)	
	$\nu = 0$	7.939	(7.95)	
	$\nu = 1$	8.019	8.03	
	$\nu = 2$	8.101	8.11	
	$\nu = 3$	8.186	8.19	
	$\nu = 4$	8.270	8.27	
	$\nu = 5$	8.354	(8.35)	
$2\,{}^3\Pi(1u)$	$\nu = 0$		9.130	9.116
	$\nu = 1$		9.190	9.193
$2\,{}^1\Pi_u$	$\nu = 0$	9.225	9.250	9.230
	$\nu = 1$	9.305	9.320	9.307
	$\nu = 2$	9.381	9.395	9.384
	$\nu = 3$	9.455	9.465	9.459
	$\nu = 4$		9.530	9.534
$2\,{}^1\Sigma_u^+$	hot band		(9.620)	
	$\nu = 0$	9.682	9.695[c]	9.688
	$\nu = 1$	9.815	9.815[c]	9.807
	$\nu = 2$	9.938	9.930[c]	9.928
	$\nu = 3$	10.046		10.028
	$\nu = 4$	10.141		
${}^1\Pi_g(?)$		9.900		
		9.966		
		10.025		

[a]Other photoabsorption data can be found in Refs. [54, 55, 61].
[b]Numbers in parentheses represent unresolved components.
[c]May be due to the presence of a nearby triplet state.

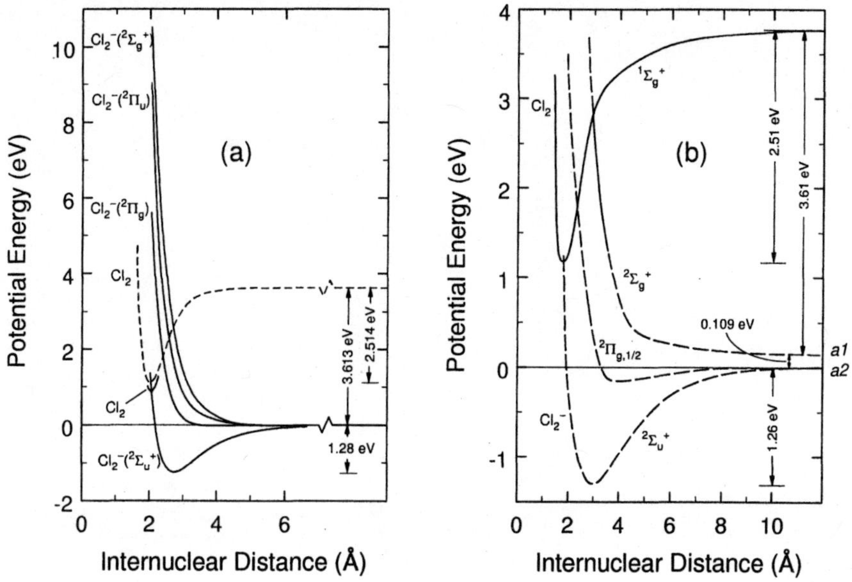

Figure 6.5. (a) Potential-energy curves for the lowest four negative ion states ($^2\Sigma_u^+$, $^2\Pi_g$, $^2\Pi_u$, $^2\Sigma_g^+$) of Cl_2^- as calculated by Gilbert and Wahl [85] in the molecular orbital self-consistent-field approximation. (The broken and solid curves for Cl_2 are two different determinations by Gilbert and Wahl.) (b) Potential-energy curves for the states $^2\Sigma_u^+$, $^2\Pi_{g,1/2}$, and $^2\Sigma_g^+$ of Cl_2^- determined by Lee *et al.* [89] using their photodissociation cross section measurements for Cl_2^- and the potential-energy curves of Gilbert and Wahl in Fig. 6.5(a). The symbols *a*1 and *a*2 represent, respectively, the asymptotic limits Cl ($^2P_{3/2}$) + Cl^- and Cl ($^2P_{1/2}$) + Cl^-.

energy positions of the energy-loss peaks observed in the electron-impact studies of Spence *et al.* [44] and Jureta *et al.* [43]. For comparison, photoabsorption data from Moeller *et al.* [82] are also shown along with possible identification of the states responsible for the observed energy losses. A comparison of the values of the energies of the various states as determined from the energy-loss spectra and from the threshold-electron excitation spectra shows excellent agreement (Table 6.3).

2.1.2 Cl_2^-

The Cl_2^- negative ion consisting of Cl^- ($^1S_0$) and Cl ($^2P_{3/2,1/2}$) has four electronic states. These states can be arranged [80, 85–91] in order of increasing energy as: $^2\Sigma_u^+$, $^2\Pi_g$, $^2\Pi_u$, $^2\Sigma_g^+$. In Fig. 6.5(a) are shown the potential-energy curves for these states as calculated by Gilbert and Wahl [85] in the molecular-orbital self-consistent-field (SCF) approxima-

Table 6.4. Some physical parameters for Cl_2^-.

Quantity	Value	Method/Ref.
Dissociation energy, D_e	1.28 eV 1.24 eV	Calculation [85] Calculation [86]
Dissociation energy, D_0	1.26 eV	Ref. [45]
Equilibrium internuclear distance, R_e	2.65 Å 2.71 Å	Calculation [85] Calculation [86]
Fundamental vibrational frequency	0.0322 eV 0.0320 eV	Calculation [85] Calculation [86]
Transition energies[a]	3.46 eV ($^2\Sigma_g^+ \rightarrow {}^2\Sigma_u^+$) 2.89 eV ($^2\Pi_u \rightarrow {}^2\Sigma_u^+$) 1.78 eV ($^2\Pi_g \rightarrow {}^2\Sigma_u^+$)	Calculation [86]
Ionization energy of Cl_2^- [b]	2.39 eV	Ref. [45]
Electron affinity of Cl_2 [b]	2.45 eV[c]	Ref. [74]
Negative ion states	~0.0 eV ($^2\Sigma_u^+$) 2.5 eV ($^2\Pi_g$) 5.5 eV ($^2\Pi_u$) 7.5 eV ($X\,^2\Pi_g)(4s\sigma)^2\,[^2\Pi_{1/2,3/2}]$	From Table 6.11

[a] At the ground state equilibrium bond length.
[b] These two quantities should be the same and have both adiabatic and vertical values. The vertical values normally exceed the adiabatic.
[c] Thirteen values are listed by Christodoulides *et al.* [74]. If we exclude the lowest three values listed in this reference as too low and the highest one as too high, the average of the remaining nine values is 2.45 eV. This value is within the combined quoted uncertainty of the values used in the averaging.

tion. Similar curves are shown in Fig. 6.5(b) for the negative ion states $^2\Sigma_u^+$, $^2\Pi_{g,1/2}$, and $^2\Sigma_g^+$ as they have been determined by Lee *et al.* [89] using their photodissociation cross section measurements for Cl_2^- and adjusted potential energy curves for Cl_2^- based on the calculations of Gilbert and Wahl [85]. The numerical values shown in the Fig. 6.5(b) for the various quantities are those used by Lee *et al.*, and the designations $a1$ and $a2$ refer, respectively, to the Cl ($^2P_{3/2}$) and Cl ($^2P_{1/2}$) asymptotic limits. Data for a number of physical parameters for the Cl_2^- ion are given in Table 6.4.

Table 6.5. Some physical parameters for Cl_2^+.

Parameter	Value	Ref.
Equilibrium separation (Å)	1.890 ($^2\Pi_{g,3/2}$)	[72]
Dissociation energy (eV)	3.99 ($^2\Pi_g$)	[67]
	3.966 ($^2\Pi_{g,3/2}$)	[72]
	3.876 ($^2\Pi_{g,1/2}$)	[72]
	1.38 ($^2\Pi_u$)	[67]
	1.41 ($^2\Pi_{u,3/2}$)	[72]
	1.32 ($^2\Pi_{u,1/2}$)	[72]
D_0 (eV)	3.95	[45]
Energy of fundamental vibration (eV)	0.08004 ($^2\Pi_{g,3/2}$)	[45]
	~0.080 ($^2\Pi_{g,3/2}$)	[72]
	0.07994 ($^2\Pi_{g,1/2}$)	[45]
	~0.0459 ($^2\Pi_{u,3/2}$)	[72]
	~0.0347 ($^2\Sigma_g^+$)	[72]
Energies to various ionic states (adiabatic/vertical) (eV)	See Table 6.2	

2.1.3 Cl_2^+

Photoelectron studies have shown [70] that the known states of the Cl_2^+ ion correspond to the ejection of one electron from one of the occupied orbitals of the outer orbital structure of the Cl_2 molecule [$(\sigma_g)^2$, $(\pi_u)^4$, $(\pi_g)^4$]. The $^2\Sigma_g^+$ state of Cl_2^+ lies above the first dissociation limit [70] (Table 6.2). Optical emission from the excited $A\,^2\Pi_u$ state of Cl_2^+ to the ground state $X\,^2\Pi_g$ of Cl_2^+ is known [70], but emission from the $^2\Sigma_g^+$ state to the $A\,^2\Pi_u$ state, although allowed by the selection rules, has not been observed, possibly because the $^2\Sigma_g^+$ state is entirely predissociated [70]. The photodissociation spectra [83] of Cl_2^+ obtained in the range 1.80–2.55 eV showed vibrational structure indicating that the dissociation of these ions involves a predissociation mechanism. Data on low-lying ionic states of Cl_2^+ derived from optical emission and photoelectron spectroscopy investigations are listed in Table 6.5.

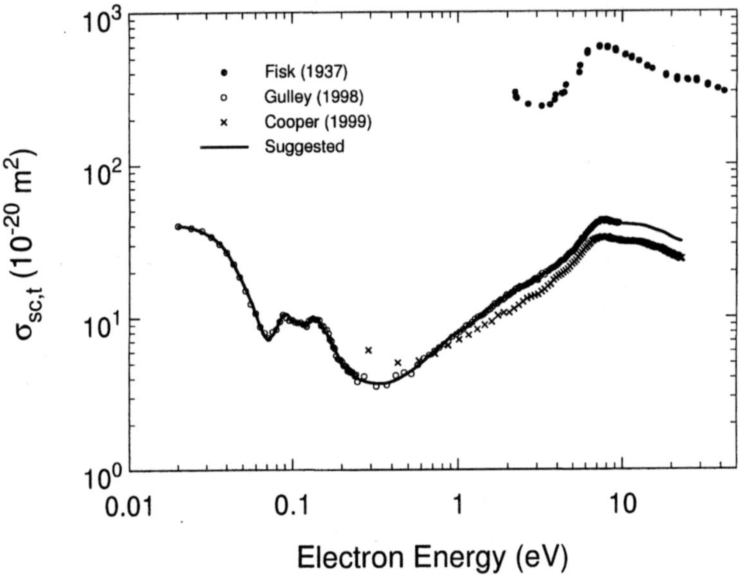

Figure 6.6. Total electron scattering cross section, $\sigma_{sc,t}(\varepsilon)$, of Cl_2: (•) Ref. [92]; (○) Ref. [93]; (×) Ref. [94]; (—) suggested values.

2.2 Electron Scattering from Cl_2

2.2.1 Total Electron Scattering Cross Section, $\sigma_{sc,t}(\varepsilon)$, of Cl_2

Up until recently, the only data on the total electron scattering cross section, $\sigma_{sc,t}(\varepsilon)$, of Cl_2 were the 1937 measurements of Fisk [92] which are very large (see Fig. 6.6). The absence of reliable experimental data on $\sigma_{sc,t}(\varepsilon)$, coupled with the lack of calculations of this quantity, led to two recent measurements [93, 94] of the $\sigma_{sc,t}(\varepsilon)$ of Cl_2. The measurements of Gulley *et al.* [93] covered the energy range 0.02–9.5 eV and those of Cooper *et al.* [94] covered the range 0.3–23 eV. They are both plotted in Fig. 6.6 and are seen to be very much smaller than the old measurements of Fisk. The uncertainties are estimated to be ±20% in the measurements of Cooper *et al.* and ±8% in the measurements of Gulley *et al.* While the shapes of the cross sections determined by the two recent measurements are similar, the magnitude of $\sigma_{sc,t}(\varepsilon)$ as measured by Cooper *et al.* is systematically lower than that measured by Gulley *et al.* at all but the lowest energies (below ~ 0.7 eV). The magnitude of the data of Gulley *et al.* are consistent with the total rotational excitation cross sections (see Sec. 2.2.2 of Ch. 6). Cooper *et al.* [94] pointed out that the lower values of their $\sigma_{sc,t}(\varepsilon)$ measurements may in part result from the fact that electrons scattered into small angles ($\lesssim 2°$) with little energy loss are detected as "unscattered" in their apparatus, and since the measurements of Gote and Ehrhardt [95] on rotational scattering from Cl_2 showed (see Sec. 2.2.5 of Ch. 6) that forward scattering is appreciable, this may be a significant cause of error in determining the value of $\sigma_{sc,t}(\varepsilon)$ of Cl_2.

In the energy range covered by the two recent experimental studies, the $\sigma_{sc,t}(\varepsilon)$ for Cl_2 has two distinct features: It shows a minimum around 0.4 eV and structure that can be attributed to resonance-enhanced electron scattering. In connection with the latter, the peaks at low energies in the data of Gulley *et al.* [93] and the bump (or unresolved peak) in the data of Cooper *et al.* [94] at 2.5 eV correspond to the energy positions of the negative ion states identified in electron attachment studies near 0 and 2.5 eV (see Sec. 2.5.1 of Ch. 6 and Refs. [87, 96, 97]). Similarly, the strong peak near 7 eV (Fig. 6.6) corresponds to the negative ion state (see Sec. 2.5.1 of Ch. 6 and Refs. [87, 96, 97]) at 5.5 eV overlapping with the lowest electron excited Feshbach resonance of Cl_2 at 7.5 eV identified by Spence [98] in an electron transmission experiment. The spacing of the peaks and inflections in the data of Gulley *et al.* at 0.09, 0.14, and 0.2 eV may be associated with indirect (resonance enhanced) vibrational excitation of Cl_2 via the near 0 eV negative ion state of Cl_2^- (see discussion in subsequent sections).

In view of the fact that the data of Gulley *et al.* [93] exhibit lower uncertainty, superior electron energy resolution, a more extensive energy range, and consistency with rotational excitation cross section data [95], they are deemed to be the most reliable. We performed a least squares fit to the measurements of Gulley *et al.* which we extended to 23 eV using the shape of the Cooper *et al.* cross section between 9.5 and 23 eV. The solid line in Fig. 6.6 is a plot of this least-squares fit, and represents our suggested $\sigma_{sc,t}(\varepsilon)$ for Cl_2. Data taken off this line are listed in Table 6.6.

2.2.2 Total Rotational Electron Scattering Cross Section, $\sigma_{rot,t}(\varepsilon)$, of Cl_2

Gote and Ehrhardt [95] measured the absolute differential cross sections for electron-impact rotational excitation of Cl_2. These data [1, 95] allowed determination of the total cross section for rotational scattering (rotational elastic plus rotational inelastic) as a function of electron-impact energy, $\sigma_{rot,t}(\varepsilon)$. In Fig. 6.7 is plotted (open circles) the cross section $\sigma_{rot,t}(\varepsilon)$ as determined (summed over all rotational states j and

Table 6.6. Suggested total electron scattering cross section, $\sigma_{sc,t}(\varepsilon)$, for Cl_2.

ε (eV)	$\sigma_{sc,t}(\varepsilon)$ (10^{-20} m^2)	ε (eV)	$\sigma_{sc,t}(\varepsilon)$ (10^{-20} m^2)	ε (eV)	$\sigma_{sc,t}(\varepsilon)$ (10^{-20} m^2)
0.02	40.0	0.22	4.44	3.5	19.9
0.03	35.2	0.25	4.00	4.0	21.9
0.04	26.8	0.30	3.75	4.5	24.2
0.05	17.0	0.35	3.70	5.0	26.8
0.06	10.7	0.40	3.80	6.0	34.5
0.07	7.36	0.50	4.32	7.0	41.2
0.08	8.50	0.60	5.00	8.0	42.8
0.09	10.6	0.70	5.83	9.0	41.0
0.10	9.68	0.80	6.55	10.0	40.3
0.11	9.26	0.90	7.36	12.0	39.7
0.12	9.06	1.00	7.97	14.0	38.6
0.13	9.76	1.20	9.06	16.0	36.7
0.14	9.89	1.50	11.1	18.0	35.1
0.15	8.90	2.00	13.9	20.0	33.0
0.17	7.19	2.50	16.0	22.0	31.5
0.20	5.09	3.00	17.9	23.0	31.0

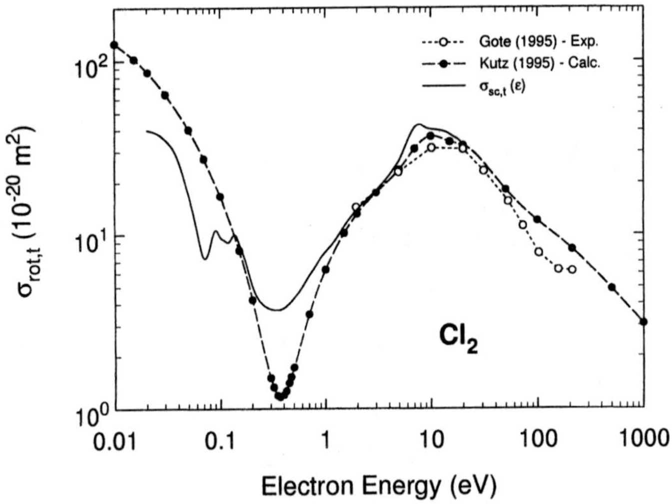

Figure 6.7. Total cross section for rotational scattering, $\sigma_{rot,t}(\varepsilon)$, for Cl_2: (∘) Values calculated by Kutz and Meyer [99] from the measurements of Gote and Ehrhardt [95]; (•) *ab initio* calculations by Kutz and Meyer [99]. For comparison, the suggested $\sigma_{sc,t}(\varepsilon)$ (solid line in Fig. 6.6) is also plotted.

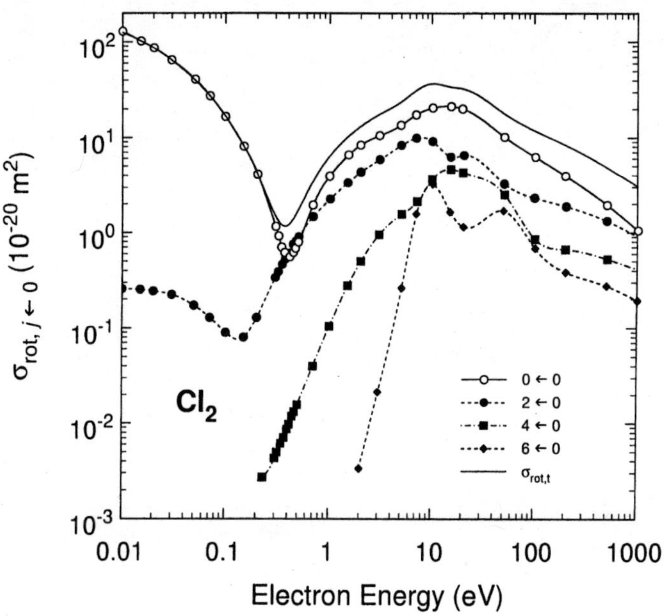

Figure 6.8. Integrated (over angle) excitation cross sections, $\sigma_{rot,j \leftarrow 0}(\varepsilon)$, for Cl_2 for the rotational excitation channels (○) $0 \leftarrow 0$, (•) $2 \leftarrow 0$, (■) $4 \leftarrow 0$, and (♦) $6 \leftarrow 0$ of Cl_2 calculated by Kutz and Meyer [99]. Also shown for comparison are the calculated values of $\sigma_{rot,t}(\varepsilon)$.

scattering angles) by Kutz and Meyer [99] from the data of Gote and Ehrhardt [95]. Also plotted in Fig. 6.7 are the full-potential calculation results of Kutz and Meyer (solid circles) which extend over a much larger energy range. There is good agreement between theory and experiment in the overlapping energy range.

The experimental and calculated values of $\sigma_{rot,t}(\varepsilon)$ are compared in Fig. 6.7 with the suggested value of $\sigma_{sc,t}(\varepsilon)$ (solid line in Fig. 6.6). From the figure it can be seen that the measured value of $\sigma_{rot,t}(\varepsilon)$ slightly exceeds the total scattering cross section near 2 eV. This is physically impossible, but the discrepancy is within the combined uncertainties of the two measurements. It is interesting to note the deep minimum shown by the calculated $\sigma_{rot,t}(\varepsilon)$ is also present in the measured $\sigma_{sc,t}(\varepsilon)$. Below this minimum the calculated values for $\sigma_{rot,t}(\varepsilon)$ exceed the measured $\sigma_{sc,t}(\varepsilon)$, which brings into question the calculated values of $\sigma_{rot,t}(\varepsilon)$ for $\varepsilon < 0.2$ eV.

In Fig. 6.8 are shown the various contributions to $\sigma_{rot,t}(\varepsilon)$, that is, the integrated (over angle) excitation cross sections, $\sigma_{rot,j \leftarrow 0}(\varepsilon)$, for $j =$

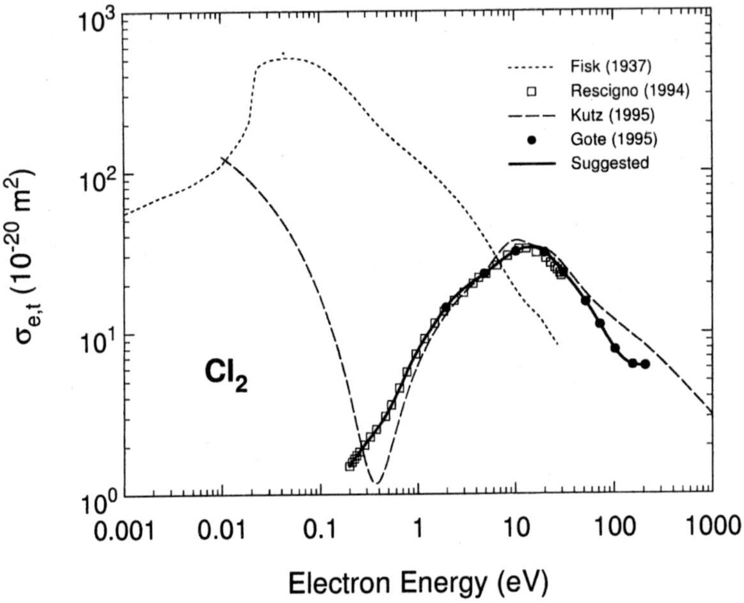

Figure 6.9. Total elastic electron scattering cross section, $\sigma_{e,t}(\varepsilon)$, for Cl_2: (- - -) calculated total elastic electron scattering cross section, $\sigma_{e,t}(\varepsilon)$ [92]; (□) calculated total elastic electron scattering cross section, $\sigma_{e,t}(\varepsilon)$ [100]; (•) measured $\sigma_{rot,t}(\varepsilon)$ [data of Kutz and Meyer [99] based on measurements by Gote and Ehrhardt [95]]; (– –) calculated total rotational electron scattering cross section, $\sigma_{rot,t}(\varepsilon)$ [99]; (—) suggested values.

$0, 2, 4, 6$. Clearly, the rotationally elastic electron scattering channel ($j = 0$) dominates over all energies, especially below the minimum.

2.2.3 Total Elastic Electron Scattering Cross Section, $\sigma_{e,t}(\varepsilon)$, of Cl_2

There are no measurements of the total elastic electron scattering cross section, $\sigma_{e,t}(\varepsilon)$, of Cl_2. There have been, however, two calculations of this cross section, the old phase-shift calculation of Fisk [92], and the more recent close-coupling calculation of Rescigno [100]. These results are shown in Fig. 6.9. The Fisk result is clearly unacceptable. We have also plotted in Fig. 6.9 the total rotational scattering cross section $\sigma_{rot,t}(\varepsilon)$ as calculated by Kutz and Meyer [99]. Similarly, we have plotted the $\sigma_{rot,t}(\varepsilon)$ determined by Kutz and Meyer from the measurements of Gote and Ehrhardt [95]. From an experimental perspective, $\sigma_{rot,t}(\varepsilon)$, may be considered equivalent to $\sigma_{e,t}(\varepsilon)$ since $\sigma_{rot,t}(\varepsilon)$ contains a large

Table 6.7. Suggested total elastic electron scattering cross section, $\sigma_{e,t}(\varepsilon)$, for Cl_2.

ε (eV)	$\sigma_{e,t}(\varepsilon)$ (10^{-20} m^2)	ε (eV)	$\sigma_{e,t}(\varepsilon)$ (10^{-20} m^2)	ε (eV)	$\sigma_{e,t}(\varepsilon)$ (10^{-20} m^2)
0.20	1.50	2.5	16.9	20	30.9
0.22	1.64	3.0	18.6	22	29.5
0.25	1.82	3.5	19.9	23	28.8
0.30	2.11	4.0	21.1	25	27.3
0.35	2.38	4.5	22.1	30	24.0
0.40	2.66	5.0	23.2	40	19.4
0.50	3.30	6.0	25.2	50	16.1
0.60	4.10	7.0	27.1	60	13.6
0.70	4.98	8.0	28.8	70	11.6
0.80	5.99	9.0	30.2	80	10.1
0.90	6.89	10.0	31.3	90	8.87
1.0	7.77	12.0	32.7	100	7.99
1.2	9.34	14.0	33.1	150	6.31
1.5	11.4	16.0	32.9	200	6.16
2.0	14.6	18.0	32.1		

elastic component and the energy loss of rotational excitations is small ($< 10^{-4}$ eV [99]). The $\sigma_{rot,t}(\varepsilon)$ based on the experimental data of Gote and Ehrhardt [95] and the $\sigma_{e,t}(\varepsilon)$ calculated by Rescigno are similar in shape and comparable in magnitude over a large energy range. The solid line in Fig. 6.9 represents a fit to these two data sets, and values obtained from this fit are given in Table 6.7 as our suggested data for the $\sigma_{e,t}(\varepsilon)$ of molecular chlorine.

2.2.4 Momentum Transfer Cross Section, $\sigma_m(\varepsilon)$, of Cl_2

There are no measurements of the momentum transfer cross section, $\sigma_m(\varepsilon)$, for Cl_2. The results of two Boltzmann-code analyses [6, 40] are questionable, in part because they were hindered by the lack of accurate and sufficient electron transport coefficient measurements. The two Boltzmann analyses used the early data of Bailey and Healey [101] for the electron drift velocity and the characteristic energy for a 20%Cl_2/80%He mixture by volume, and the ionization and attachment coefficients of Božin and Goodyear [102]. In Fig. 6.10 the Boltzmann-calculation results are compared with the close-coupling calculation result of Rescigno [100]. These cross sections differ substantially, especially at low energies, stressing the need for a direct measurement of $\sigma_m(\varepsilon)$. They also

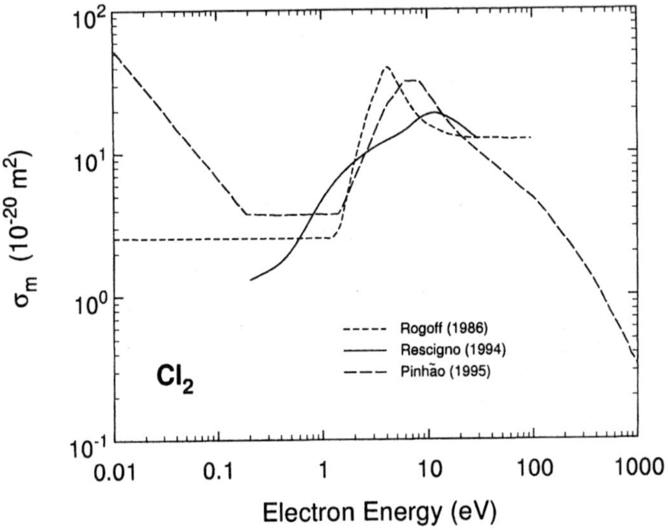

Figure 6.10. Calculated momentum transfer cross section, $\sigma_m(\varepsilon)$, for Cl_2: (- - -) Ref. [6]; (—) Ref. [100]; (– –) Ref. [40].

indicate that there is a need for measurements of electron transport coefficients that would allow a more reliable Boltzmann-code analysis. The need for such measurements is made more apparent because the cross sections of Rogoff *et al.* [6] have been used in various discharge models. Of the available values for $\sigma_m(\varepsilon)$, the *ab initio* results of Rescigno [100] are preferred because they are not model dependent and because of the agreement between Rescigno's calculations and measured values of $\sigma_{e,t}(\varepsilon)$ and $\sigma_{dis,neut,t}(\varepsilon)$ (see Secs. 2.2.3 and 2.4 of Ch. 6, respectively).

2.2.5 Inelastic Electron Scattering Cross Section, $\sigma_{inel}(\varepsilon)$, for Cl_2

Rotational Excitation Cross Section, $\sigma_{rot}(\varepsilon)$, for Cl_2. Rotational excitation of Cl_2 by electron impact can be either direct or indirect via the formation of short-lived negative ion states. The experimental measurements by Gote and Ehrhardt [95] of the differential cross sections for rotational excitation of Cl_2 by electron impact at energies between 2 and 200 eV and in the angular range 10°–160° have clearly shown that rotational excitation of the Cl_2 molecule in its vibrational and electronic ground states by slow electrons is an efficient electron scattering process.

Cross sections exceeding 10^{-20} m^2 have been measured. As discussed earlier in this section, Gote and Ehrhardt reported rotationally summed cross sections and partial rotational excitation cross sections (i.e., cross sections for excitation to various rotational levels) as the percentage of their relative contribution to the rotationally summed cross sections. The cross sections in the forward direction belong mostly to rotationally elastic electron scattering. Above a scattering angle of about 30°, the scattering is dominated by rotationally inelastic processes. Kutz and Meyer's [99] close-coupling calculation of the rotational excitation of Cl_2 by electron impact over the energy range of 0.01–1000 eV, neglecting vibrational, resonant, and electronic excitation, shows two different excitation mechanisms, the importance of which depends on the electron energy. At low electron energies only a few rotational quanta are exchanged and the differential cross section decreases exponentially with Δj. At high electron energies the excitation spectrum shows a rotational rainbow, i.e., the differential cross section has a maximum at a relatively high Δj. The location of the maximum depends upon the electron energy and the scattering angle. Large scattering angles, in addition to high electron energies, are needed for the observation of a rotational rainbow. The scattering angle can only be large when, classically speaking, the impact parameter is small, i.e., when the impacting electron penetrates the electron cloud and comes near the core of the molecule. At low incident energy, the electron essentially interacts with the long-range parts of the potential of the target. For homonuclear molecules these are the quadrupole and polarization potentials [99]. In their calculation Kutz and Meyer [99] used the polarizabilities $\alpha_0 = 24.42$ a.u. and $\alpha_2 = 16.293$ a.u. (1 a.u. $= 0.1482 \times 10^{-24}$ cm^3).

In Fig. 6.11 are compared the close-coupling rotationally summed differential electron scattering cross sections calculated by Kutz and Meyer [99] (solid line) with the experimental values of Gote and Ehrhardt [95] for various incident electron energies. The agreement is good, adding credence to the calculation and the underlying assumptions.

The full-model potential calculation results of Kutz and Meyer for the integrated excitation cross section and for the first four rotational excitation channels were shown previously in Fig. 6.8. The total electron scattering cross section (for all scattering channels) has a minimum at about 0.5 eV which was found to be very sensitive to small changes of the potential. The integrated cross section decreases with the final rotational state j. At low scattering angles and electron energies only a few rotational quanta are transferred ("normal" excitation mechanism), whereas at high scattering angles and electron energies many rotational quanta can be exchanged (rotational rainbow mechanism).

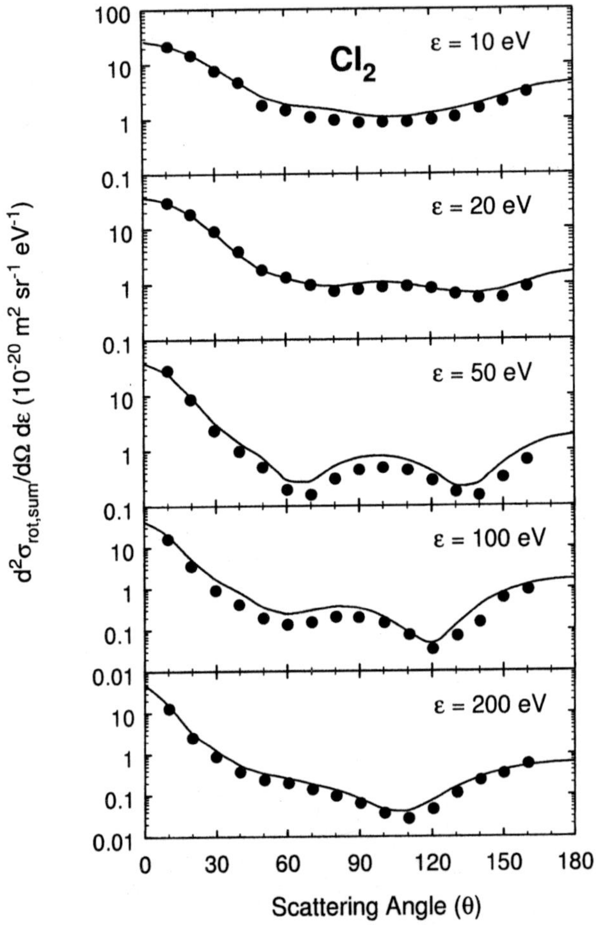

Figure 6.11. Comparison of experimental and calculated rotationally summed differential electron scattering cross sections $d^2\sigma_{rot,sum}/d\Omega d\varepsilon$, for Cl_2 at incident electron energies of 10, 20, 50, 100, and 200 eV: (•) experimental data of Gote and Ehrhardt [95]; (—) close-coupling calculation results of Kutz and Meyer [99].

Another calculation of rotational excitation of Cl_2 was performed by Ernesti *et al.* [103] within the two-center Coulomb-scattering approximation. This study predicted a rainbow scattering pattern which is consistent both with the close-coupling result and with the experimental data.

Total Vibrational Excitation Cross Section, $\sigma_{vib,t}(\varepsilon)$, of Cl_2. There are no experimentally determined total vibrational excitation cross

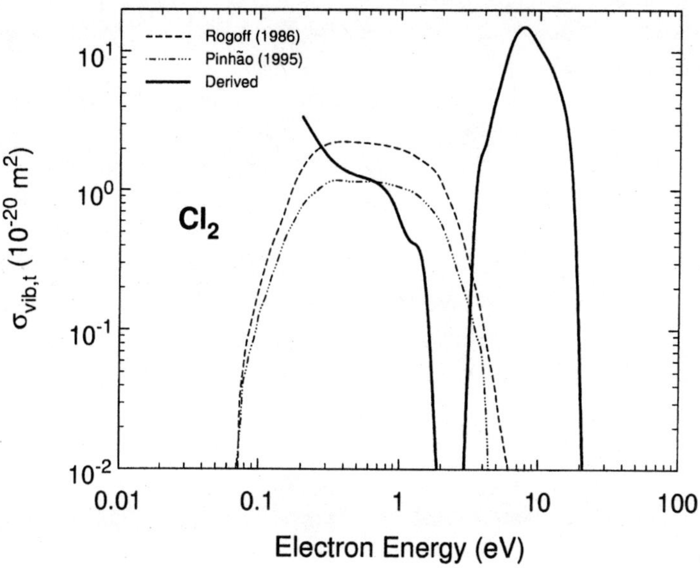

Figure 6.12. Total vibrational excitation cross section, $\sigma_{\mathrm{vib,t}}(\varepsilon)$, for Cl_2. Results of Boltzmann-code analyses: (- - -) [6]; (— · · · —) Ref. [40]. (—) Estimate of $\sigma_{\mathrm{vib,t}}(\varepsilon)$, derived from present analysis described in text.

sections, $\sigma_{\mathrm{vib,t}}(\varepsilon)$, for Cl_2. There are only the results of two Boltzmann-code calculations [6, 40] based on limited experimental data. These results are compared in Fig. 6.12. Their assumed energy dependence is similar (although there is no experimental evidence to support such a shape), and their magnitudes differ. Thus, there is a need for a direct measurement of the vibrational excitation cross section for this molecule, and there is also a need for more and better electron transport data to enhance the usefulness of the $\sigma_{\mathrm{vib,t}}(\varepsilon)$ calculated from Boltzmann codes.

Vibrational excitation cross sections are important in efforts to model plasma reactors due to their large effect on the electron energy distribution function (see, for example, Refs. [104, 105]). For this reason, we have attempted to deduce a rough estimate of $\sigma_{\mathrm{vib,t}}(\varepsilon)$ from the available cross sections for other processes. We assumed the suggested values for $\sigma_{\mathrm{sc,t}}(\varepsilon)$ (Table 6.6, Fig. 6.6), $\sigma_{\mathrm{e,t}}(\varepsilon)$ (Table 6.7, Fig. 6.9), $\sigma_{\mathrm{i,t}}(\varepsilon)$ (Table 6.8, Fig. 6.14), $\sigma_{\mathrm{dis,neut,t}}(\varepsilon)$ (Table 6.10, Fig. 4.16), and $\sigma_{\mathrm{da,t}}(\varepsilon)$ (Table 6.12, Fig. 6.17), and took the difference

$$\sigma_{\mathrm{sc,t}}(\varepsilon) - [\sigma_{\mathrm{e,t}}(\varepsilon) + \sigma_{\mathrm{i,t}}(\varepsilon) + \sigma_{\mathrm{dis,neut,t}}(\varepsilon) + \sigma_{\mathrm{da,t}}(\varepsilon)] \approx \sigma_{\mathrm{vib,t}}(\varepsilon) \approx \sigma_{\mathrm{vib,indir}}(\varepsilon) \quad (6.1)$$

to be a measure of $\sigma_{\text{vib,t}}(\varepsilon)$. Since direct vibrational excitation for a homopolar molecule such as Cl_2 is expected to be small [106, 107], $\sigma_{\text{vib,t}}(\varepsilon)$ may be taken, in this case, to be the cross section for indirect (resonance enhanced) vibrational excitation, $\sigma_{\text{vib,indir}}(\varepsilon)$, of the Cl_2 molecule via its negative ion states. Values of $\sigma_{\text{vib,indir}}(\varepsilon)$ derived in this way are shown in Fig. 6.12 (solid line), where the two Boltzmann computed values of $\sigma_{\text{vib,t}}(\varepsilon)$ are also shown. The $\sigma_{\text{vib,indir}}(\varepsilon)$ deduced in this study bares no similarity to the computed values of $\sigma_{\text{vib,t}}(\varepsilon)$. In spite of the uncertainty involved in the derivation of $\sigma_{\text{vib,indir}}(\varepsilon)$, this deduced cross section shows that the indirect vibrational excitation cross section of Cl_2 is potentially very large. In the absence of any direct measurements of $\sigma_{\text{vib,t}}(\varepsilon)$, the present derived cross section $\sigma_{\text{vib,indir}}(\varepsilon)$ is preferred to those provided by the Boltzmann codes.

Electronic Excitation Cross Sections, $\sigma_{\text{elec}}(\varepsilon)$, for Cl_2. There have been no measurements of the cross sections for electron-impact excitation of any of the electronic states of Cl_2. However, there have been three calculations of cross sections for some of the lowest excited electronic states of Cl_2. Rogoff *et al.* [6] reported cross sections for electron impact excitation of the electronic states $^3\Pi_u$, $^1\Pi$, and the sum $2^1\Pi_u + 2^1\Sigma_u^+$ that are derived from a Boltzmann-code analysis. Another Boltzmann-code calculation by Pinhão and Chouki [40] reported cross sections for electronic excitation of $^3\Pi_u$ + $^1\Pi$, $^3\Sigma_u$ + $^3\Pi_g$ + $^1\Pi_g$, and $2^1\Pi_u$ + $2^1\Sigma_u^+$. Also, Rescigno [100] performed close-coupling calculations using the complex Kohn variational method and reported excitation cross sections for $^3\Pi_u$, $^1\Pi_u$, $^3\Pi_g$, $^1\Pi_g$, and $^3\Sigma_u^+$. Rescigno refers indiscriminately to the cross sections he calculated for these five states as cross sections for excitation or as cross sections for dissociation, the implication being that all excitations to these states lead to dissociation. This would be consistent with the potential energy curves for the excited states calculated by Peyerimhoff and Buenker [46] (Fig. 6.1). He also calculated the total cross sections for electron-impact excitation of the $^1\Pi_u$ and $^1\Sigma_u$ Rydberg states of Cl_2 using the Born-dipole approximation, and found that the Born-dipole cross sections far exceeded those he calculated using the close-coupling method.

Since the excitation cross sections of Rogoff *et al.* [6] have been used in various plasma models, we compared them with the results of the other two calculations in the few cases where this is possible. In Fig. 6.13(a) the cross sections estimated by the three studies for $^3\Pi_u + {}^1\Pi_u$ are compared. In Fig. 6.13(b) a similar comparison is made for $2^1\Pi_u + 2^1\Sigma_u^+$. In Fig. 6.13(c) the cross sections of Rogoff *et al.* [6] and of Rescigno [100] for electron-impact excitation of the electronic states $^3\Pi_u$ and $C\,^1\Pi_u$ are

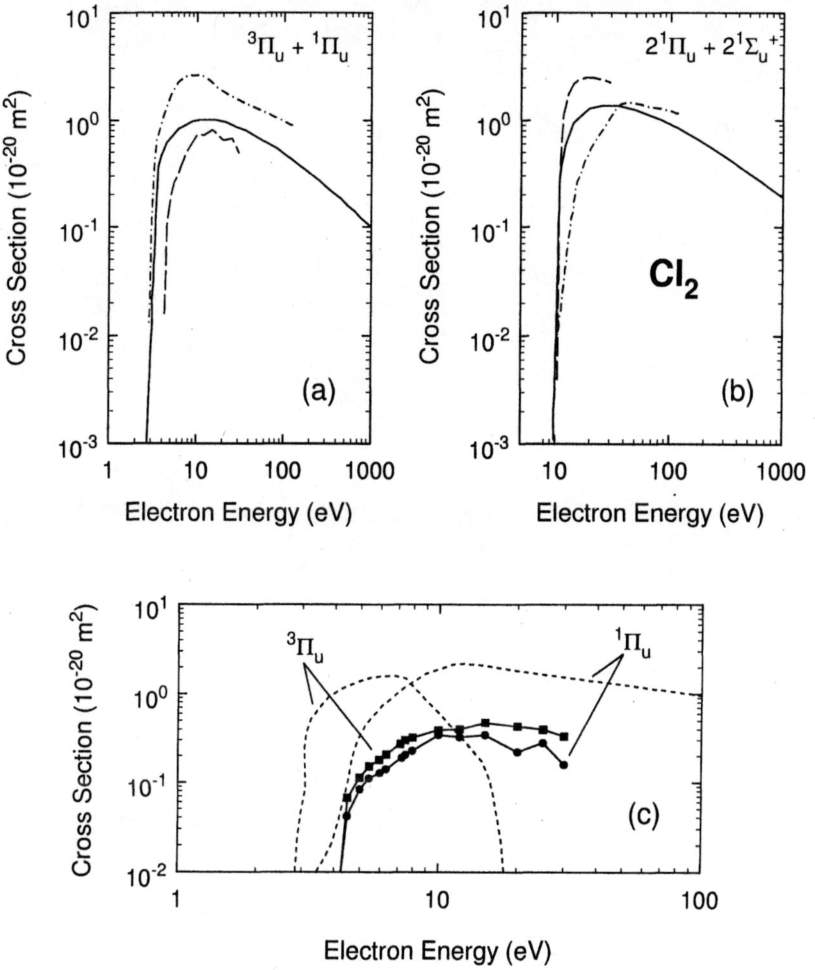

Figure 6.13. Comparison of calculated cross sections for electronic excitation, $\sigma_{elec}(\varepsilon)$, of Cl_2. (a) Excitation of $^3\Pi_u + {}^1\Pi_u$: (– · –) Rogoff *et al.* [6]; (—) Pinhão and Chouki [40]; (– –) Rescigno [100]. (b) Excitation of $2^1\Pi_u + 2^1\Sigma_u^+$: (– · –) Rogoff *et al.* [6]; (—) Pinhão and Chouki [40]; (– –) Rescigno [100]. (c) Excitation of $^3\Pi_u$ and $^1\Pi_u$: (- - -) Rogoff *et al.* [6]; (■, •) Rescigno [100].

compared. The vertical excitation energies of $^3\Pi_u$ and $^1\Pi_u$ are, respectively, 3.31 and ~ 4.05 eV (see Table 6.1). The agreement between the Boltzmann-code-deduced electronic excitation cross sections and those of Rescigno is poor. Clearly, more experimental and computational work is indicated.

2.3 Electron-Impact Ionization of Cl_2

2.3.1 Total Ionization Cross Section, $\sigma_{i,t}(\varepsilon)$, of Cl_2

In Fig. 6.14 are compared the available data on the electron-impact total ionization cross section, $\sigma_{i,t}(\varepsilon)$, of Cl_2. These include the measurements by Center and Mandl [108], Kurepa and Belić [96], Stevie and Vasile [109], and Srivastava and Boivin [110, 111]. The cross section measurements of Center and Mandl were made using argon as the calibrant gas, and normalizing to the ionization cross section for Ar of Rapp and Englander-Golden [116]. The stated uncertainty of these measurements is ±15%. The cross section measurements of Kurepa and Belić were made in the energy range of 10–100 eV and have a reported relative error of ±20%. Below ~ 50 eV they are higher than the values obtained by Center and Mandl. The third set of measurements were made by Stevie and Vasile [109] in the energy range 12–100 eV using a mass spectrometer and a modulated molecular beam. These determinations of $\sigma_{i,t}(\varepsilon)$ were made relative to those of the three calibrant gases Ar, O_2, and Kr for which they used the respective data of Rapp and Englander-Golden [116]. The values plotted in the figure are the averages of the data using each of the three calibrant gases. The authors indicated an uncertainty in their data for 70 eV as shown in Fig. 6.14 of approximately ±20%. Their measurements agree with those of Kurepa and Belić [96] near the threshold, but they are considerably higher for energies greater than ~ 15 eV. Clearly, these three sets of data differ not only in magnitude, but also in the measured energy dependence of $\sigma_{i,t}(\varepsilon)$. The more recent relative measurements of Srivastava and Boivin [110, 111] are also shown in Fig. 6.14. These cover a broader energy range, from threshold to 700 eV, and were arbitrarily normalized to the 70 eV point of the "suggested" curve discussed later and shown by the solid line in Fig. 6.14. Interestingly, the cross section of Srivastava and Boivin shows structure near 25 eV which, although not as evident, is nonetheless indicated by some of the other measurements, and may be due to autoionization.

In Fig. 6.14 also are shown the results of three recent calculations by Kim [112], Deutsch *et al.* [113], and Joshipura and Limbachiya [114]. The calculation by Kim includes multiple ionization but not autoionization. The results of all three calculations are in reasonable agreement with the measurements of Kurepa and Belić [96] and Stevie and Vasile [109]. We have averaged the measured values of Kurepa and Belić [96] and Stevie and Vasile [109], even though the differences in their magnitudes exceed their combined uncertainties, and take this to be our suggested

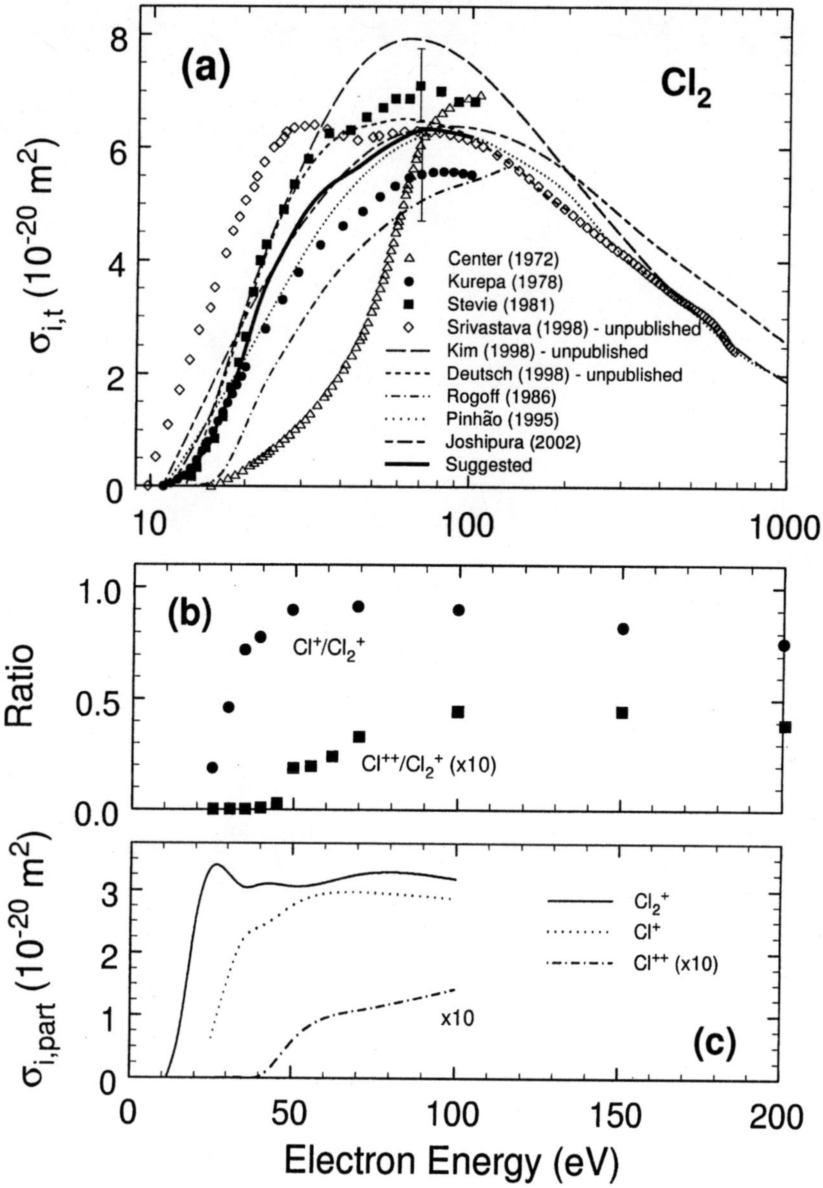

Figure 6.14. (a) Electron-impact total ionization cross section, $\sigma_{i,t}(\varepsilon)$, of Cl_2: (△) Ref. [108]; (•) Ref. [96]; (■) Ref. [109]; (◇) Ref. [110, 111]; (— —) Ref. [112]; (- - -) Ref. [113]; (– – –) Ref. [114]; (–·–) Ref. [6]; (···) Ref. [40]; (—) suggested. (b) Ratios of Cl^+/Cl_2^+ and Cl^{++}/Cl_2^+ as a function of electron energy measured by Calandra *et al.* [115]. (c) Partial ionization cross sections calculated by normalizing the ratio data of Calandra *et al.* [115] with the suggested total ionization cross section derived in Fig. 6.14(a).

Table 6.8. Suggested total ionization cross section, $\sigma_{i,t}(\varepsilon)$, for Cl_2.

ε (eV)	$\sigma_{i,t}(\varepsilon)$ (10^{-20} m^2)	ε (eV)	$\sigma_{i,t}(\varepsilon)$ (10^{-20} m^2)	ε (eV)	$\sigma_{i,t}(\varepsilon)$ (10^{-20} m^2)
11.5	0.03	22	3.25	60	6.15
12.0	0.11	24	3.79	65	6.25
13.0	0.25	26	4.17	70	6.32
14.0	0.43	28	4.51	75	6.33
15.0	0.69	30	4.80	80	6.31
16.0	0.99	35	5.26	85	6.28
17.0	1.32	40	5.49	90	6.25
18.0	1.67	45	5.68	95	6.22
19.0	2.06	50	5.87	100	6.19
20.0	2.47	55	6.03		

values for the $\sigma_{i,t}(\varepsilon)$ of Cl_2. We have not included the values of Center and Mandl [108] due to the obviously inconsistent shape of their cross section when compared to the other [96, 109] measured values. These average values are shown by the bold line in Fig. 6.14 and are listed in Table 6.8.

The model-dependent total ionization cross section of Rogoff *et al.* [6] and Pinhão and Chouki [40] deduced from modeling of chlorine discharges are also plotted in Fig. 6.14. While the Pinhão and Chouki cross section is in general agreement with the most reliable measurements, that of Rogoff *et al.* is not. However, such a comparison is biased by the input cross section assumed by each calculation.

Threshold ionization energies leaving the Cl_2^+ ion in various states of excitation have been given in Table 6.2. Also listed in Table 6.2 are the values for the threshold energy for dissociative ionization ($Cl_2 + e \rightarrow Cl^+ + Cl + 2e$) and for double ionization.

Recently, Calandra *et al.* [115] have measured relative partial ionization cross sections for the formation of Cl_2^+, Cl^+, and Cl^{++} from Cl_2 from 25 to 200 eV. The ratios of Cl^+ and Cl^{++} production to Cl_2^+ production are shown in Fig. 14(b) over this energy range. These measurements show that Cl_2^+ is the most abundant ion formed at all energies, but Cl^+ production becomes comparable in magnitude at electron energies $\gtrsim 50$ eV. This finding is consistent with the photoabsorption measurements, which show that the production of Cl_2^+ far exceeds the production of Cl^+ for dissociative photoionization (see Fig. 6.3).

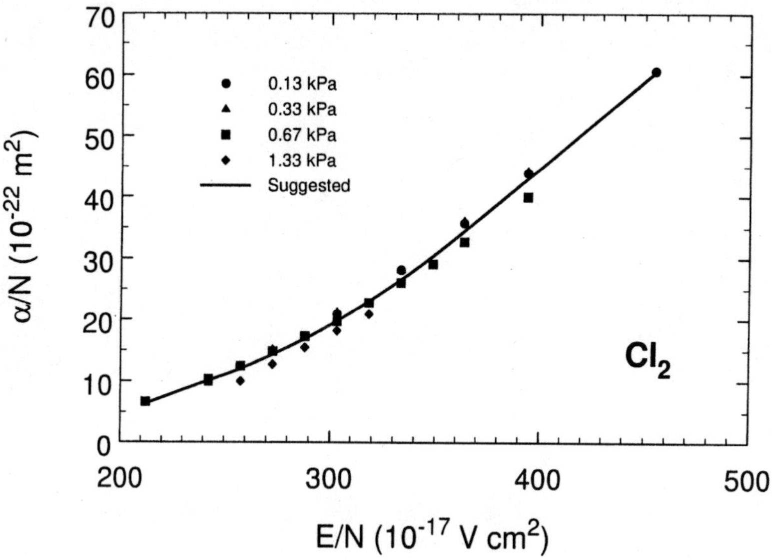

Figure 6.15. Density-normalized electron-impact ionization coefficient, $\alpha/N\,(E/N)$, for Cl_2 at various gas pressures measured by Božin and Goodyear [102]. The solid line is a least-squares fit to all the data points and represents the suggested values for $\alpha/N\,(E/N)$.

In Fig. 6.14(c) we have put the relative partial ionization cross section measurements of Calandra *et al.* [115] on an absolute scale by normalizing them to the suggested values of $\sigma_{\mathrm{i,t}}(\varepsilon)$ shown in Fig. 6.14(a) and listed in Table 6.8. Below 25 eV it was assumed that all ion production was attributable to Cl_2^+, which could account for some of the structure observed in the cross section for Cl_2^+ production near 25 eV. These partial ionization cross sections represent our best estimates of $\sigma_{\mathrm{i,part}}(\varepsilon)$ for Cl_2 until direct measurements are available.

2.3.2 Density-Normalized Electron-Impact Ionization Coefficient, $\alpha/N\,(E/N)$, of Cl_2

The only measurement of the density-normalized electron-impact ionization coefficient, $\alpha/N\,(E/N)$, of Cl_2 is that of Božin and Goodyear [102] shown in Fig. 6.15. These measurements were made at $T = 293$ K for Cl_2 pressures of 0.13, 0.33, 0.67, and 1.33 kPa. From a least-squares fit to the data in Fig. 6.15, we obtained the values listed in Table 6.9, which represent our suggested values for the $\alpha/N\,(E/N)$ of Cl_2.

Table 6.9. Suggested density-normalized electron-impact ionization coefficient, $\alpha/N\,(E/N)$, for Cl_2 from Božin and Goodyear [102].

E/N (10^{-17} V cm^2)	α/N (10^{-22} m^2)	E/N (10^{-17} V cm^2)	α/N (10^{-22} m^2)
213	6.45	340	28.2
220	7.34	360	33.4
240	9.82	380	39.0
260	12.4	400	44.7
280	15.5	420	50.4
300	19.2	440	56.2
320	23.4	450	59.1

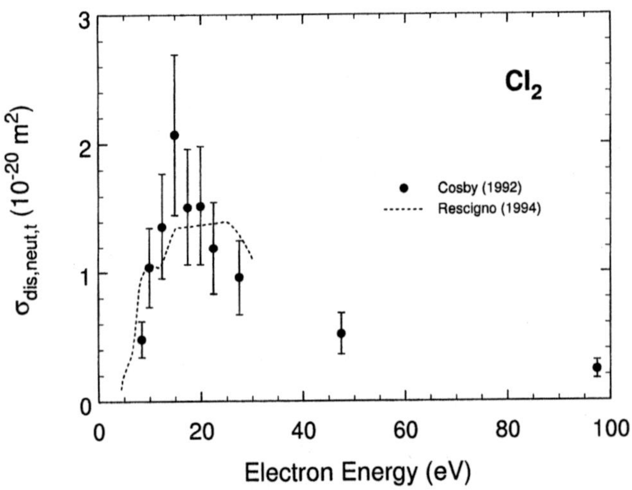

Figure 6.16. Total cross section for electron-impact dissociation into neutral fragments, $\sigma_{\mathrm{dis,neut,t}}(\varepsilon)$, for Cl_2: (•) measurements by Cosby and Helm [118, 119]; (- - -) calculations by Rescigno [100] of the sum of the cross sections for electronic excitation of the lowest five electronic states ($^3\Pi_u$, $^1\Pi_u$, $^3\Pi_g$, $^1\Pi_g$, $^3\Sigma_u^+$) of Cl_2.

2.4 Electron-Impact Dissociation of Cl_2 into Neutral Fragments

There has been only one measurement [117–119] of the total cross section for electron-impact dissociation of Cl_2 into neutrals, $\sigma_{\mathrm{dis,neut,t}}(\varepsilon)$. The data of Cosby and Helm [117–119] are shown in Fig. 6.16. Also shown is the sum of the cross sections calculated by Rescigno [100] for

Table 6.10. Total cross section for electron-impact dissociation into neutral fragments, $\sigma_{\text{dis,neut,t}}(\varepsilon)$, for Cl_2 from Cosby and Helm [118].

ε (eV)	$\sigma_{\text{dis,neut,t}}(\varepsilon)$ (10^{-20} m^2)	ε (eV)	$\sigma_{\text{dis,neut,t}}(\varepsilon)$ (10^{-20} m^2)
8.4	0.48 ± 0.14	19.9	1.52 ± 0.46
9.9	1.04 ± 0.31	22.4	1.19 ± 0.36
12.4	1.36 ± 0.41	27.4	0.96 ± 0.29
14.9	2.07 ± 0.62	47.4	0.52 ± 0.16
17.4	1.51 ± 0.45	97.4	0.24 ± 0.07

the lowest five excited electronic states ($^3\Pi_u$, $^1\Pi_u$, $^3\Pi_g$, $^1\Pi_g$, $^3\Sigma_u^+$) of Cl_2 which are reached by promoting an occupied valence electron into the antibonding ($5\sigma_u$) orbital. The calculation by Rescigno showed that the total dissociation cross section is the largest for the $^3\Pi_u$ state up to the highest energy (30 eV) he investigated. The agreement between Rescigno's calculations and the experimental data is good, supporting the premise that all electronic excitations of Cl_2 result in dissociation. The experimental data of Cosby and Helm [117–119] are listed in Table 6.10 as the presently suggested values for $\sigma_{\text{dis,neut,t}}(\varepsilon)$ for the Cl_2 molecule.

In an earlier study, Wells and Zipf [120] observed dissociative excitation of Cl_2 and identified the fragments as, in part, atomic chlorine in long-lived high-Rydberg excited states [Cl* (HR)] produced through

$$e + Cl_2 \rightarrow e + Cl + Cl^* \text{ (HR)} \tag{6.2}$$

and

$$e + Cl_2 \rightarrow 2e + Cl^+ + Cl^* \text{ (HR)}. \tag{6.3}$$

They associated an energy threshold for reactions (6.2) and (6.3), respectively, equal to (14.8 ± 1) eV and (29.2 ± 5) eV.

Another process for neutral fragment production is dissociative recombination ($e + Cl_2^+ \rightarrow Cl + Cl$). No data exit on this process (see Mitchel [121] for data on this process for other species).

2.5 Electron Attachment to Cl_2

As we discussed in Sec. 2.1.2 of this chapter, the Cl_2^- negative ion consisting of Cl^- ($^1S_0$) and Cl ($^2P_{3/2,1/2}$) has four electronic states whose order of increasing energy is: $^2\Sigma_u^+$, $^2\Pi_g$, $^2\Pi_u$, $^2\Sigma_g^+$ (Fig. 6.5). The participation of these states in dissociative electron attachment of Cl_2 depends on the way their potential-energy curves cross the ground-state potential energy curve $X\,^1\Sigma_g^+$ of the neutral Cl_2 molecule. On the basis of the Cl_2^- potential-energy curves in Fig. 6.5, one would expect formation of the parent anion Cl_2^- at zero energy, and the formation of Cl^- at near-zero energy and in three higher-energy ranges below 10 eV. For the fragment negative ion Cl^-, the dissociative electron attachment reactions

$$Cl_2(X\,^1\Sigma_g^+) + e \rightarrow Cl_2^{-*} \rightarrow Cl^-(^1S_0) + Cl(^2P_{3/2,1/2}) \qquad (6.4)$$

involve the ground state $X\,^1\Sigma_g^+$ of Cl_2 in the $\nu = 0$ and perhaps $\nu = 1$ state, and the four $^2\Sigma_u^+$, $^2\Pi_g$, $^2\Pi_u$, $^2\Sigma_g^+$ negative ion states of Cl_2 which are correlated with the dissociation limit $Cl^-\,(^1S_0) + Cl\,(^2P_{3/2,1/2})$. This limit lies 1.10 eV [Cl $(^2P_{1/2})$] and 1.21 eV [Cl $(^2P_{3/2})$] below the minimum of the potential energy curve for the ground state of Cl_2 [see Fig. 6.5(b)].

Three electron-beam experimental studies [87, 96, 97] have shown that the yield of Cl^- from Cl_2 exhibits three peaks, near 0 eV, 2.5 eV, and 5.5 eV (Table 6.11). These were ascribed [96, 97] to the $^2\Sigma_u^+$, $^2\Pi_g$, and $^2\Pi_u$ resonant states of Cl_2^-, respectively. The ground state, $^2\Sigma_u^+$, of Cl_2^- is formed by addition of an extra electron to the lowest unfilled $(\sigma_u 3p)$ Cl_2 orbital of the ground-state electron configuration of Cl_2: $[(\ldots)(\sigma_g 3p)^2(\pi_u 3p)^4(\pi_g 3p)^4]$. The core-excited $^2\Pi_g$ and $^2\Pi_u$ states of Cl_2^- are formed by exciting one electron of the $^2\Sigma_u^+$ shape resonance from the $\pi_g 3p$ and $\pi_u 3p$ to the $\sigma_u 3p$ orbital, respectively. An electron-transmission study by Spence [98] located the lowest-lying electron-excited Feshbach resonance in Cl_2 at 7.46 eV. He associated this resonance with Rydberg states having symmetry $(X\,^2\Pi_g)(4s\,\sigma)^2[^2\Pi_{1/2,3/2}]$. The derivative of the transmitted current in Cl_2 between 7.0 and 9.0 eV showed a progression of six resonances starting at 7.46 eV, with an average spacing between adjacent resonances of 80 meV. A recent high resolution ($\sim$ 60 meV FWHM) electron beam study [125, 126] of dissociative electron attachment to Cl_2 between $\sim$ 0.0 and 0.7 eV showed two resonances at 0.03 and 0.250 eV. The former peak has been attributed [125] to dissociative electron attachment via the $^2\Sigma_u^+$ of Cl_2^-. The latter may be due to dissociative electron attachment via one of the excited $^2\Pi$ states of Cl_2^- [125].

The parent negative ion Cl_2^- is not normally formed in the gas phase. The transient anion in the lowest negative ion state, $Cl_2^{-*}\,(^2\Sigma_u^+)$, must be collisionally stabilized before it breaks up by dissociative electron attachment. Since, moreover, dissociative electron attachment occurs at subpicosecond times, collisional stabilization of Cl_2^{-*} can only take place at high gas densities when the collisional stabilization time becomes comparable to, or shorter than, the dissociative electron attachment

Table 6.11. Negative ion states of Cl_2.

Energy (eV)	Assigned symmetry of corresponding negative ion state	Method/Reference
0.03 ± 0.03 2.5 ± 0.15 5.5 ± 0.15	$^2\Pi_g$ [a] $^2\Pi_u$ [a] $^2\Sigma_g^+$ [a]	Maxima in the dissociative electron attachment cross section measured in an electron-impact mass-spectrometric study [87]
0.0 2.5 ± 0.05 5.75 ± 0.05 9.7[b]	$^2\Sigma_u^+$ $^2\Pi_g$ $^2\Pi_u$ $^2\Sigma_g^+$	Maxima in the dissociative electron attachment cross section measured in an electron-impact mass-spectrometric study [80, 96]
0 2.5 5.5	$^2\Sigma_u^+$ $^2\Pi_g$ $^2\Pi_u$	Dissociative electron attachment using a crossed-beam electron impact spectrometer [97][c]
0.07	$^2\Sigma_u^+$	Electron swarm [122]
2.4 ± 0.1		Electron-impact mass spectrometry [76]
7.46 ($\nu = 0$) and subsequent peaks separated by 0.08 eV corresponding to $\nu = 1$–5	Electron-excited Feshbach resonances formed by addition of two $ns\sigma$ electrons to the $X\,^2\Pi_g$ positive-ion core	Electron transmission [98]
~ 2[d] ~ 5[d]	$^2\Pi_g$ $^2\Pi_u$	Studies of Cl^- ions produced by dissociative electron attachment from *condensed* Cl_2 [123]

[a]These assignments are incorrect, see text.
[b]Azria *et al.* [97]) did not observe the 9.7 eV resonance indicated by the data of Kurepa and Belić [96].
[c]Assignments based on angular distribution analysis of the Cl^- ions. According to Azria *et al.* [97], there may be a small contribution of the $^2\Sigma_g^+$ state of Cl_2^- to the Cl^- formation at the low-energy side of the resonance at 5.5 eV.
[d]These values are about 0.5 eV lower than the corresponding gaseous data. This may be due to the effect of the polarization energy of condensed Cl_2 on the negative-ion states of the isolated Cl_2 molecule [123, 124].

time, or in the condensed phase. No parent negative ions have been observed in electron attachment studies in the gas phase. They, however, have been observed in gas-phase negative-ion charge transfer reactions

[127, 128] and in the condensed phase [123]. With regard to the latter-type investigation, Azria *et al.* [123] studied the production of Cl^- by dissociative electron attachment in electron-stimulated desorption from Cl_2 condensed on a platinum substrate. They found that the energy dependence of the Cl^- signal exhibits two peaks at about 2 and 5 eV which they attributed to the $^2\Pi_g$ and $^2\Pi_u$ Cl_2^- resonant states. Thus, in the condensed phase (in the chlorine lattice on the surface of the substrate) the dissociation dynamics of Cl_2^- are similar to those in the gas phase except possibly with a 0.5 eV downward shift in the resonance energy positions. (See Christophorou [124, 129] for a discussion of the effect of the medium and state of matter on the energetics of negative ion states).

2.5.1 Total Dissociative Electron Attachment Cross Section, $\sigma_{da,t}(\varepsilon)$, of Cl_2

Dissociative electron attachment to Cl_2 is rather simple in its products: only Cl^- is produced directly. Thus, electron beam experiments with mass analysis and total electron attachment experiments without mass analysis should yield the same results. In spite of this, it seems that the only absolute measurement of the total dissociative electron attachment cross section, $\sigma_{da,t}(\varepsilon)$, of Cl_2 is that of Kurepa and Belić [96]. Their data are shown in Fig. 6.17 for the energy range from 0 to 13 eV with an uncertainty of $\pm 20\%$. The data indicate that dissociative electron attachment to Cl_2 proceeds mainly via three negative-ion states located near 0 eV, (2.5 ± 0.05) eV, and (5.75 ± 0.05) eV. A weak process they observed between 9 and 11.5 eV was not observed by others [97] (see Table 6.11).

In Fig. 6.17 is also plotted, for comparison, the relative cross section for the production of Cl^- from Cl_2 as determined in a higher-energy resolution study by Tam and Wong [87]. Here the data of Tam and Wong have been normalized to the Kurepa and Belić cross section at 2.5 eV. Other than the small differences in the shape and energy position of the resonance near 5 eV, the overall shapes of the two cross sections are in reasonable agreement. The sharp peak at zero energy is worth noting as it is consistent with the electron swarm data (Sec. 2.5.2 of Ch. 6). The energy positions and assignments of the negative ion resonances as determined in the study of Tam and Wong [87] are compared with other data in Table 6.11. Comparison with other studies indicates that the assignments of Tam and Wong are apparently incorrect. The sequence of their assignments is in error because their calculations show the potential

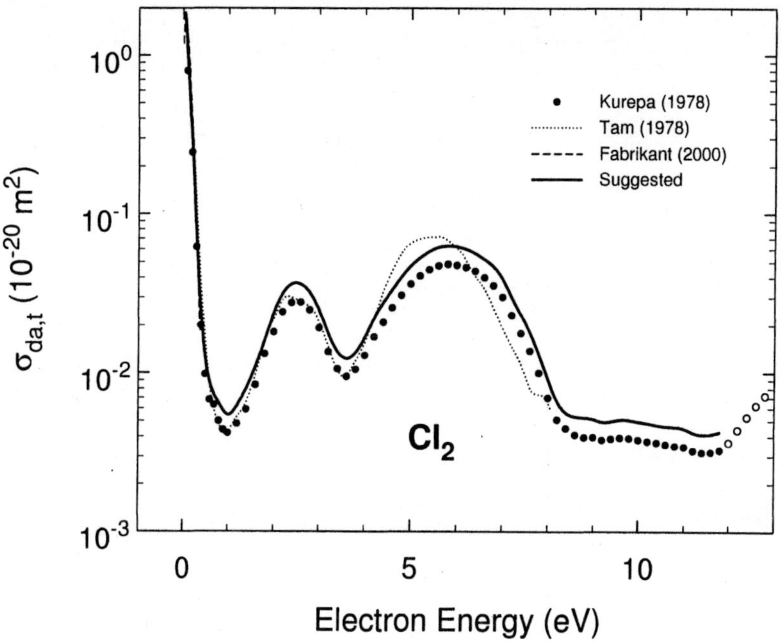

Figure 6.17. Total dissociative electron attachment cross section, $\sigma_{da,t}(\varepsilon)$, for Cl_2: (•) measurements of Kurepa and Belić [96]; (···) relative cross section for the production of Cl^- from Cl_2 measured by Tam and Wong [87] normalized to the Kurepa and Belić cross section at 2.5 eV; (—) cross section of Kurepa and Belić [96] adjusted upwards by 30%; (– –) calculation by Fabrikant *et al.* [130] (nearly hidden behind the solid line for $\varepsilon \leq 0.45$ eV). The open symbols represent the contribution to the measured cross section attributed to ion-pair production.

energy curve for the $^2\Sigma_u^+$ anionic state not crossing the potential energy curve for the $X\,^1\Sigma_g^+$ ground state of the Cl_2 molecule.

Based on the analysis of the total electron attachment rate constant in Sec. 2.5.2 below, the values of $\sigma_{da,t}(\varepsilon)$ given by Kurepa and Belić appear to be approximately 30% lower than indicated by the electron swarm measurements. Thus we have adjusted their cross section upwards by this percentage for our suggested values of $\sigma_{da,t}(\varepsilon)$. This adjusted cross section is shown in the figure by the solid line, and values taken from this curve are listed in Table 6.12 as our suggested data for the $\sigma_{da,t}(\varepsilon)$ of the Cl_2 molecule. The results of a recent calculation by Fabrikant *et al.* [130] of the dissociative electron attachment cross section of Cl_2 for electron energies ≤ 0.45 eV are in agreement with this adjustment of the data of Kurepa and Belić.

Table 6.12. Suggested total dissociative electron attachment cross section, $\sigma_{da,t}(\varepsilon)$, for Cl_2.

ε (eV)	$\sigma_{da,t}(\varepsilon)$ (10^{-20} m^2)	ε (eV)	$\sigma_{da,t}(\varepsilon)$ (10^{-20} m^2)	ε (eV)	$\sigma_{da,t}(\varepsilon)$ (10^{-20} m^2)
0.05	1.83	3.0	0.025	7.6	0.018
0.10	1.04	3.2	0.018	8.0	0.0091
0.20	0.319	3.6	0.012	8.2	0.0066
0.30	0.081	4.0	0.017	8.6	0.0053
0.40	0.026	4.2	0.022	9.0	0.0051
0.50	0.013	4.6	0.033	9.2	0.0049
0.60	0.0088	5.0	0.047	9.6	0.0051
0.80	0.0065	5.2	0.053	10.0	0.0049
1.0	0.0055	5.6	0.062	10.2	0.0048
1.2	0.0062	6.0	0.062	10.6	0.0046
1.6	0.011	6.2	0.060	11.0	0.0045
2.0	0.024	6.6	0.052	11.2	0.0042
2.2	0.032	7.0	0.039	11.6	0.0041
2.6	0.036	7.2	0.030	11.8	0.0043

Furthermore, a recent study [131] of dissociative electron attachment to Cl_2 at very low energies (0–0.195 eV) using the laser photoelectron attachment method (see Sec. 2.5.1 of Ch. 2) has shown that the dissociative electron attachment cross section near-zero energy exhibits a *p*-wave behavior.

2.5.2 Total Electron Attachment Rate Constant as a Function of the Density-Reduced Electric Field E/N, $k_{a,t}(E/N)$, and the Mean Electron Energy $\langle\varepsilon\rangle$, $k_{a,t}(\langle\varepsilon\rangle)$, for Cl_2

$k_{a,t}(E/N)$ for Cl_2. McCorkle *et al.* [122] measured the total electron attachment rate constant, $k_{a,t}(E/N)$, of Cl_2 using mixtures of Cl_2 with N_2. Their measurements covered the E/N range of 0.06×10^{-17} to 4×10^{-17} V cm^2 with a probable uncertainty of ±10%. The measurements were made at room temperature (298 K) and also at other temperatures above and below ambient (see later in this section). The total gas number density in their experiments was 6.48×10^{19} molecules cm^{-3} and the Cl_2 gas number density was in the range (0.2–2.3) $\times 10^{15}$ molecules cm^{-3}. The rate constant was found to be independent of both

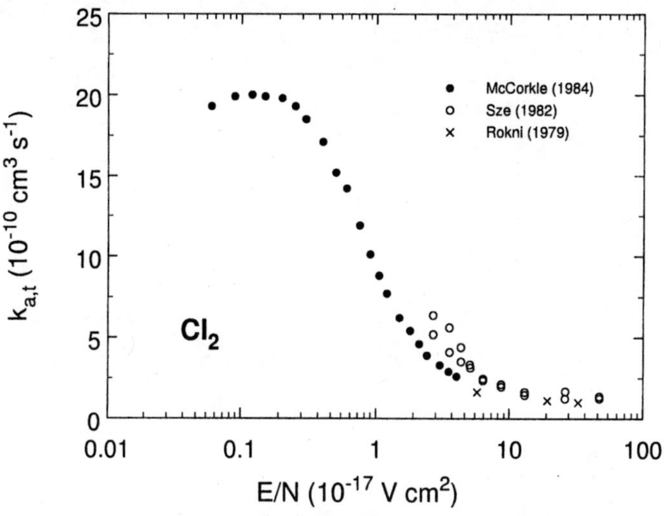

Figure 6.18. Total electron attachment rate constant as a function of E/N, $k_{a,t}(E/N)$, for Cl_2 ($T \sim$ 298–300 K):(•) Ref. [122]; (○) Ref. [132]; (×) Ref. [88].

the total and the electron attaching gas pressures. The measurements at room temperature are plotted in Fig. 6.18.

Another measurement of $k_{a,t}(E/N)$ was made by Sze *et al.* [132] using mixtures of Cl_2 with N_2. These measurements were made at 300 K and for only one mixture composition [the Cl_2 gas number density in the mixture was 260 parts per million (ppm)]. These measurements are also plotted in Fig. 6.18, along with the limited measurements made by Rokni *et al.* [88] at 300 K. With the exception of the measurements of Sze *et al.* below $\sim 5 \times 10^{-17}$ V cm^2, these data are not incompatible with those of McCorkle *et al.* Besides their measurements in mixtures with N_2, Sze *et al.* [132] also reported $k_{a,t}(E/N)$ for one mixture of Cl_2 in Ar. These data are not considered here since the measurements were made for only one mixture concentration (260 ppm) and the effect of Cl_2 on the electron energy distribution function in pure Ar could not be assessed. For the same reason, early measurements by Bradbury [133] on the probability of electron attachment per collision in a Cl_2/Ar mixture are not included.

$k_{a,t}(\langle\varepsilon\rangle)$ for Cl_2. McCorkle *et al.* [122] determined the $k_{a,t}(\langle\varepsilon\rangle)$ for Cl_2 using their measurements of $k_{a,t}(E/N)$ and the electron energy distribution functions for N_2 they calculated using a Boltzmann code at each E/N for which they measured $k_{a,t}$. These derived data are shown

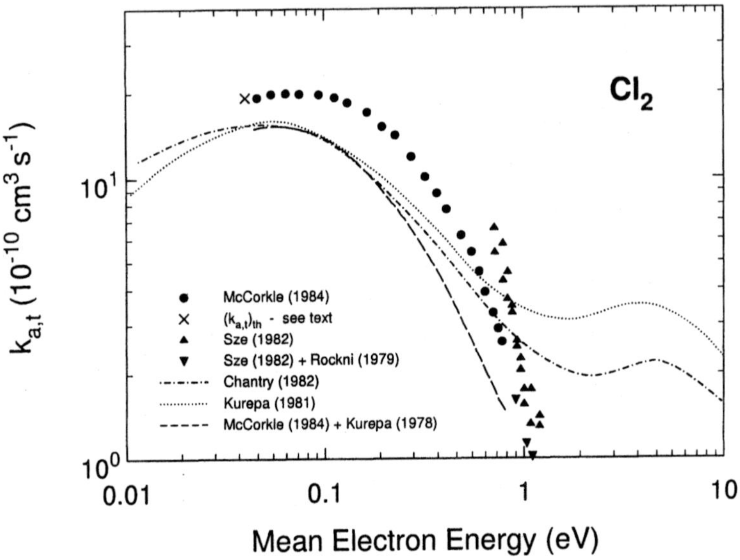

Figure 6.19. Total electron attachment rate constant as a function of the mean electron energy $\langle\varepsilon\rangle$, $k_{a,t}(\langle\varepsilon\rangle)$, for Cl_2 ($T \sim 298$ K): (•) Ref. [122]; (×) $(k_{a,t})_{th}$ determined from the average of the two most recent values of the thermal ($T \sim 300$ K) electron attachment rate (Table 6.14); (▲) Ref. [132]; (▼) Ref. [132] using the rate constants measured by Rockni *et al.* [88]; (—·—) Ref. [135] using the $\sigma_{a,t}(\varepsilon)$ of Kurepa and Belić [96] and a Maxwellian distribution function for the electron energies; (···) Ref. [136] using the $\sigma_{a,t}(\varepsilon)$ of Kurepa and Belić [96] and a Maxwellian distribution function for the electron energies; (- - -) Ref. [122] using the $\sigma_{a,t}(\varepsilon)$ of Kurepa and Belić [96] and the electron energy distribution functions they calculated for N_2.

in Fig. 6.19 for $T = 298$ K. In this figure is plotted also the thermal value, $(k_{a,t})_{th}$, of $k_{a,t}(\langle\varepsilon\rangle)$ as given by the average of the two most recent measurements [122, 134] of this quantity (see later in this section). In addition, values of $k_{a,t}(\langle\varepsilon\rangle)$ reported by the following four groups of investigators are plotted in the figure: (1) Values reported by Sze *et al.* [132] determined from their $k_{a,t}(E/N)$ measurements and also from the measurements of Rokni *et al.* [88]. These are in fair agreement with the McCorkle *et al.* [122] data. (2) Values estimated by McCorkle *et al.* using the electron energy distributions in N_2 and the total electron attachment cross section of Kurepa and Belić [96] (Fig. 6.17). While the Kurepa and Belić-based $k_{a,t}(\langle\varepsilon\rangle)$ have a similar energy dependence to the directly measured rate constants, they are lower in magnitude (by $\sim 30\%$ for a mean electron energy of 0.08 eV) indicating that the cross sections of Kurepa and Belić are lower than their true values. (3) Values of $k_{a,t}(\langle\varepsilon\rangle)$ determined by Chantry [135] and by Kurepa *et al.* [136]

Table 6.13. Suggested total electron attachment rate constant, $k_{a,t}(\langle\varepsilon\rangle)$ (T = 298 K), for Cl_2 measured by McCorkle *et al.* [122].

$\langle\varepsilon\rangle$ (eV)	$k_{a,t}$ (10^{-10} cm^3 s^{-1})	$\langle\varepsilon\rangle$ (eV)	$k_{a,t}$ (10^{-10} cm^3 s^{-1})
0.054	19.9 (0.9)[a]	0.322	10.1 (1.2)
0.064	20.0 (0.9)	0.368	8.8 (1.2)
0.075	19.9 (1.1)	0.411	7.7 (1.4)
0.094	19.8 (1.2)	0.487	6.2 (1.1)
0.113	19.3 (1.3)	0.550	5.4 (0.9)
0.131	18.5 (1.3)	0.599	4.6 (0.7)
0.165	17.1 (1.5)	0.640	3.9 (0.5)
0.196	15.2 (1.4)	0.704	3.3 (0.5)
0.228	14.2 (1.5)	0.745	2.9 (0.5)
0.275	11.9 (1.5)	0.779	2.6 (0.5)

[a]Values in parentheses are standard deviations as given by McCorkle *et al.* [122].

using the total electron attachment cross section of Kurepa and Belić [96] (Fig. 6.17) and a Maxwellian electron energy distribution function. The assumption of a Maxwellian distribution function for the electron energies is unrealistic at high E/N, as is shown by the large difference between the calculated $k_{a,t}(\langle\varepsilon\rangle)$ and the experimental measurements of $k_{a,t}(\langle\varepsilon\rangle)$. The data of McCorkle *et al.* [122] are listed in Table 6.13 as our suggested values for the $k_{a,t}(\langle\varepsilon\rangle)$ of Cl_2 at 298 K.

Values of $k_{a,t}(\langle\varepsilon\rangle)$ derived from limited measurements in mixtures of Cl_2 with argon [132, 88] are uncertain and are not included in this work.

Thermal Value, $(k_{a,t})_{th}$, of the Total Electron Attachment Rate Constant for Cl_2. In Table 6.14 are listed the values of the electron attachment rate constant at thermal energies, $(k_{a,t})_{th}$ ($T \approx 300$ K). These are independent measurements by various groups [122, 134, 137–140] and they vary significantly. The two most recent measurements [122, 134] are consistent with each other and we take their average, 19.3×10^{-10} cm^3 s^{-1}, as the best present estimate of $(k_{a,t})_{th}$ ($T \approx 300$ K) (plotted in Fig. 6.19 as a $\times$ symbol).

Effect of Temperature on the Electron Attachment Rate Constant, $k_{a,t}(\langle\varepsilon\rangle, T)$, for Cl_2. There have been two measurements [122, 134] of the dependence of the total electron attachment rate constant $k_{a,t}$ of Cl_2 on gas temperature. The measurements of McCorkle *et al.* [122] were made at various mean electron energies from thermal to

Table 6.14. Thermal values, $(k_{a,t})_{th}$, of the total electron attachment rate constant for Cl_2 near room temperature.

$(k_{a,t})_{th}$ (10^{-10} cm^3 s^{-1})	Temperature (K)	Reference
2.8 ± 0.4	300	[137]
3.1	293	[138]
11.0	300	[139]
18.6 ± 1.2	298	[122]
20.0 ± 3.0	300	[134]
37 ± 17	350	[140]

Table 6.15. Variation of $(k_{a,t})_{th}$ of Cl_2 with temperature.

Temperature (K)	$(k_{a,t})_{th}$ (10^{-10} cm^3 s^{-1})	Reference
213	12.2	[122]
233	13.5	
253	15.1	
273	16.7	
298	18.6	
323	21.4	
203	< 10	[134]
300	20	
455	33	
590	48	

0.78 eV, and the measurements of Smith *et al.* [134] were made at only thermal energies. The former results are reproduced in Fig. 6.20(a), and the latter are compared with the former in Fig. 6.20(b). All of the data for $(k_{a,t})_{th}(T)$ are tabulated in Table 6.15.

2.5.3 Density-Normalized Electron Attachment Coefficient, $\eta/N\,(E/N)$, for Cl_2

The early measurements of $\eta/N\,(E/N)$ by Bailey and Healey [101] in Cl_2 at 288 K are not in agreement with the more recent measurements of Božin and Goodyear [102] made at 293 K (Fig. 6.21). Božin and

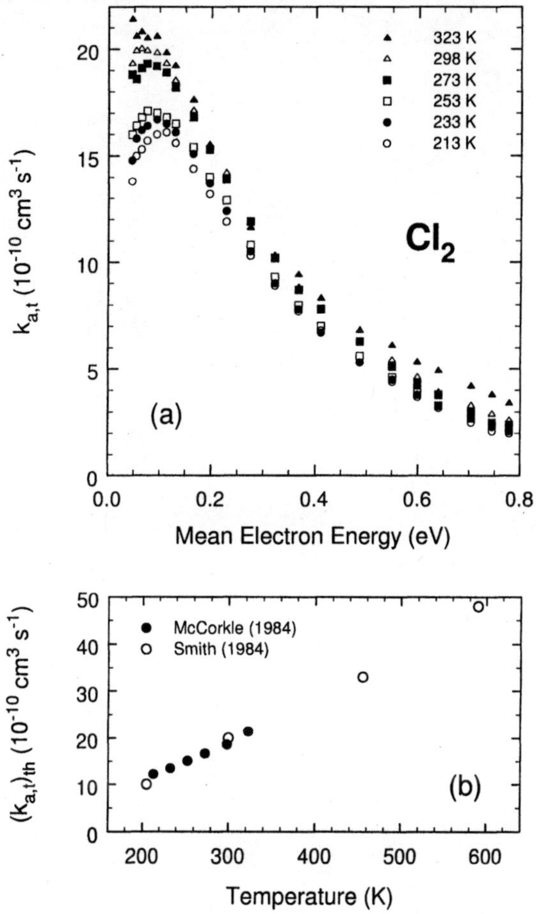

Figure 6.20. (a) Variation of $k_{a,t}(\langle\varepsilon\rangle)$ of Cl_2 with temperature measured by McCorkle *et al.* [122]. (b) Variation of $(k_{a,t})_{th}$ of Cl_2 with temperature: (•) Ref. [122]; (○) Ref. [134].

Goodyear indicated an uncertainty of $\pm 10\%$, but the average uncertainty of their data is more likely twice this value. They also indicated that their measurements for 1.33 kPa may be more uncertain than those at the other three pressures (see Fig. 6.21). Therefore, the solid line in Fig. 6.21 is a least squares fit to the data of Božin and Goodyear at pressures of 0.13, 0.33, and 0.67 kPa, and data taken off this curve are listed in Table 6.16. In the absence of other measurements, these values are presently suggested for the η/N (E/N) of Cl_2, but clearly there is a need for further measurements.

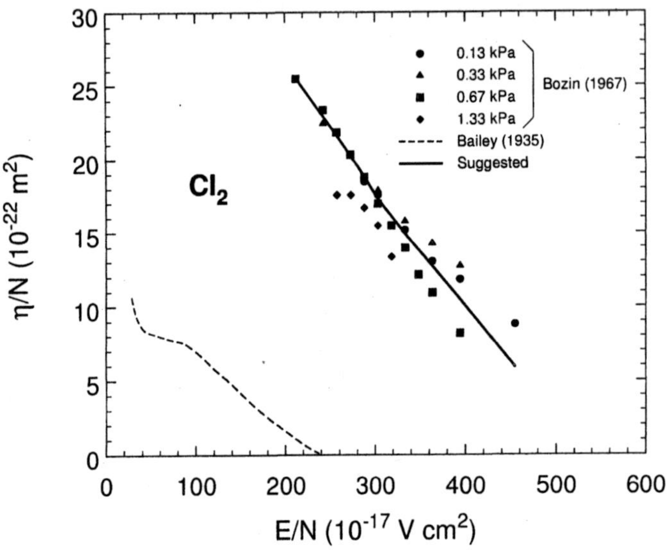

Figure 6.21. Density-normalized electron attachment coefficient, $\eta/N\,(E/N)$, for Cl_2: (●, ▲, ■, ♦) Ref. [102]; (- - -) Ref. [101]; (—) suggested.

Table 6.16. Suggested values for the density-normalized electron attachment coefficient, $\eta/N\ (E/N)$, for Cl_2 derived from the data of Božin and Goodyear [102].

E/N (10^{-17} V cm^2)	η/N (10^{-22} m^2)	E/N (10^{-17} V cm^2)	η/N (10^{-22} m^2)
215	25.3	350	13.7
225	24.4	375	11.9
250	22.3	400	10.0
275	20.0	425	8.14
300	17.6	450	6.26
325	15.6		

2.5.4 Density-Normalized Effective Ionization Coefficient, $(\alpha - \eta)/N\ (E/N)$, for Cl_2

Božin and Goodyear [102] reported measurements of the density-normalized effective ionization coefficient $(\alpha-\eta)/N\ (E/N)$ for Cl_2 made at room temperature ($T = 293$ K) for gas pressures of 0.13, 0.33, 0.67, and 1.33 kPa. Figure 6.22 shows their data which have a stated uncertainty of ±10%. The solid line is a least-squares fit to the data at all

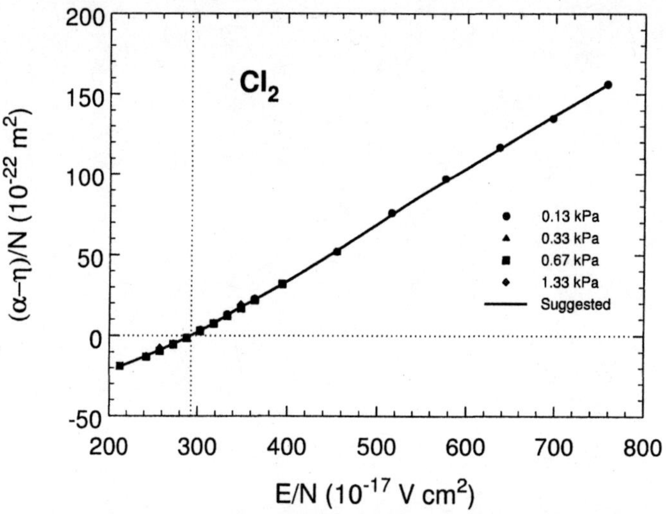

Figure 6.22. Density-normalized effective ionization coefficient, $(\alpha - \eta)/N\,(E/N)$, for Cl_2 measured by Božin and Goodyear [102]. The solid line represents the suggested values.

Table 6.17. Suggested values of the density-normalized effective ionization coefficient, $(\alpha - \eta)/N\,(E/N)$, for Cl_2 derived from Božin and Goodyear [102].

E/N (10^{-17} V cm^2)	$(\alpha - \eta)/N$ (10^{-22} m^2)	E/N (10^{-17} V cm^2)	$(\alpha - \eta)/N$ (10^{-22} m^2)
215	−18.5	500	69.1
250	−10.7	550	86.8
300	2.13	600	103.7
350	18.0	650	120.3
400	33.7	700	136.9
450	50.9	750	153.4

pressures, and values taken off this curve are listed in Table 6.17 as the presently suggested values of $(\alpha - \eta)/N\,(E/N)$ for Cl_2.

2.5.5 Cross Section for Ion-Pair Formation, $\sigma_{ip}(\varepsilon)$, for Cl_2

Besides the formation of negative ions via the resonant electron attachment processes discussed in the preceding sections, negative ions have been observed to form by electron impact on Cl_2 via the direct

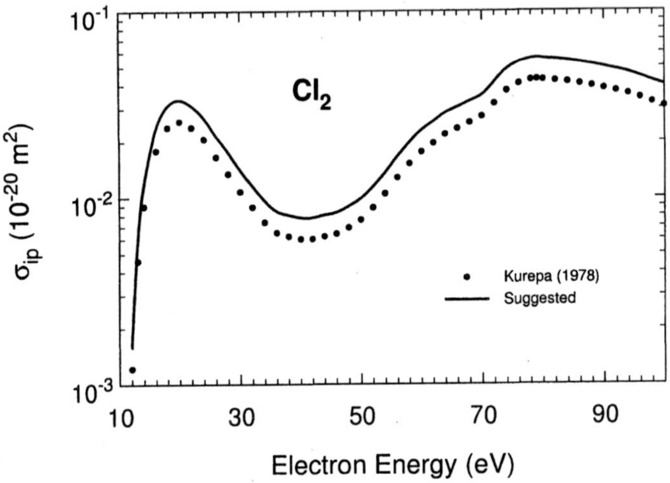

Figure 6.23. Cross section for ion-pair formation, $\sigma_{\text{ip}}(\varepsilon)$, for Cl_2 of Kurepa and Belić [96]. The solid line is the same data adjusted upward by 30% as discussed in Secs. 2.5.1 and 2.5.2 of Ch. 6.

process of polar dissociation, that is, via the ion-pair process

$$Cl_2 + e \rightarrow Cl^+ + Cl^- + e. \tag{6.5}$$

The energy onset for Process (6.5) is 11.9 eV (Table 6.2). The cross section values of Kurepa and Belić [96] for negative-ion formation above 11.9 eV (see Fig. 6.17) are due largely to the ion-pair process, Reaction (6.5), although in some energy regions contributions from indirect electron attachment processes are possible. The data of Kurepa and Belić have a quoted relative uncertainty of ±20% and are plotted in Fig. 6.23. As discussed in Secs. 2.5.1 and 2.5.2 of Ch. 6, these data need to be adjusted upward by 30%, and the so-adjusted data are shown in Fig. 6.23 by the solid line. Data from this line are listed in Table 6.18 as our suggested values for the $\sigma_{\text{ip}}(\varepsilon)$ of Cl_2.

2.5.6 Negative Ions in Cl_2 Discharges

There have been a number of studies dealing with negative ions in Cl_2 gas discharges. By way of example we refer in this section to only two such studies [141, 142] dealing with laser-induced photodetachment of negative ions and its use to infer the density of negative ions in the plasma. Han *et al.* [141] described a technique for sampling negative ions in the hollow cathode and hollow anode of Cl_2/N_2 discharges. The

Table 6.18. Suggested cross section for negative ion-positive ion pair production, $\sigma_{ip}(\varepsilon)$, in Cl_2 between 12 eV and 100 eV (adjusted data of Kurepa and Belić [96].)

ε (eV)	$\sigma_{ip}(\varepsilon)$ (10^{-20} m^2)	ε (eV)	$\sigma_{ip}(\varepsilon)$ (10^{-20} m^2)
12	0.0016	52	0.0114
13	0.0060	56	0.0166
14	0.0117	60	0.0229
16	0.0234	62	0.0255
18	0.0312	66	0.0304
20	0.0335	70	0.0354
22	0.0312	72	0.0413
26	0.0216	76	0.0528
30	0.0140	80	0.0553
32	0.0116	82	0.0546
36	0.0085	86	0.0525
40	0.0078	90	0.0497
42	0.0078	92	0.0481
46	0.0083	96	0.0442
50	0.0099	100	0.0399

photoelectron transient signals which were induced by laser photodetachment of the negative ions present in the discharge were employed to probe the ion concentration. The observed negative-ion transient signal allowed a study of the kinetics of the three negative ions (Cl^-, Cl_2^-, and Cl_3^-) they observed in the discharge. Interestingly, the authors concluded from their measurements that the Cl_3^- ion is likely to be due to the recombination of Cl^- and Cl_2, and the Cl_2^- ion is likely to be the result of three-body electron attachment to Cl_2. Hebner [142] also employed laser photodetachment spectroscopy to infer the density of chlorine negative ions in inductively coupled chlorine plasmas.

2.6 Electron Transport in Cl_2

2.6.1 Electron Drift Velocity, $w(E/N)$, in Cl_2

The only known measurements of the electron drift velocity, $w(E/N)$, in Cl_2 are those made in 1935 by Bailey and Healey [101] using the volume mixtures 20%Cl_2/80%He, 20%Cl_2/80%CO_2, and 40%Cl_2/60%CO_2. Bailey and Healey also showed a curve for w vs E/N which they identified as the w for Cl_2. This curve is reproduced in Fig. 6.24(a), but it is considered uncertain because of its indirect determination from the drift

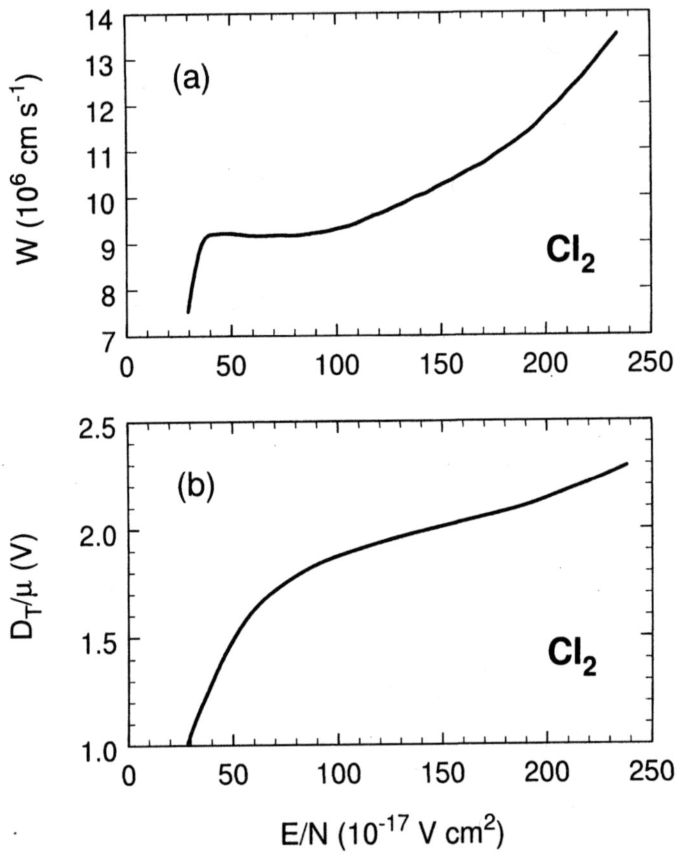

Figure 6.24. (a) Electron drift velocity, $w\,(E/N)$, for Cl_2 ($T = 288$ K) measured by Bailey and Healey [101]. (b) Ratio of the lateral electron diffusion coefficient to the electron mobility, D_T/μ, for Cl_2 ($T = 288$ K) derived from data given by Bailey and Healey [101]. Both sets of data are uncertain.

velocities they measured in the mixtures just mentioned. Clearly, measurements of $w\,(E/N)$ for Cl_2 are needed, but are difficult as is generally the case for strongly electron attaching gases.

2.6.2 Lateral Electron Diffusion Coefficient to Electron Mobility Ratio, $D_T/\mu\,(E/N)$, for Cl_2

There are no measurements of this quantity for Cl_2. Bailey and Healey [101] reported measurements of the quantity k_M (mean energy of agitation of electrons in terms of the mean energy of agitation of the gas molecules at 288 K) which we used to determine the ratio of the lateral electron diffusion coefficient D_T to the electron mobility μ, via the relationship $D_T/\mu = k_M(kT/e)$. The values of D_T/μ determined this way are shown in Fig. 6.24(b). They should be considered uncertain. Measurements of D_T/μ over a wide range of E/N are needed.

2.7 Optical Emission from Cl_2 Gas Discharges

There have been a number of studies of light emission from chlorine excited in an electrical discharge (see, for instance, Refs. [143–149]). It is clear from these investigations that the emission spectrum from a chlorine vapor discharge consists, in addition to the atomic line spectrum, of a large number of red-degraded bands extending from about 640 to 340 nm which were generally assigned to Cl_2^+.

2.8 Suggested Cross Sections and Coefficients for Cl_2

Due to the paucity of confirmed data, no cross sections are considered "recommended" at this time for Cl_2. However, a significant amount of data exist which are "suggested" as the best data presently available. These include:

- $\sigma_{sc,t}(\varepsilon)$ — Table 6.6, Fig. 6.6;
- $\sigma_{e,t}(\varepsilon)$ — Table 6.7, Fig. 6.9;
- $\sigma_{i,t}(\varepsilon)$ — Table 6.8, Fig. 6.14(a);
- $\sigma_{dis,neut,t}(\varepsilon)$ — Table 6.10, Fig. 4.16;
- $\sigma_{da,t}(\varepsilon)$ — Table 6.12, Fig. 6.17; and
- $\sigma_{ip}(\varepsilon)$ — Table 6.18, Fig. 6.23.

These cross sections are plotted in Fig. 6.25. Also shown in Fig. 6.25 is the derived $\sigma_{vib,t}(\varepsilon)$ (Fig. 6.12) for which we do not provide tabulated values due to the potential for large uncertainties inherent in the derivation method used. It should be observed that the values of $\sigma_{e,t}(\varepsilon)$ exceed those of $\sigma_{sc,t}(\varepsilon)$ near 2 eV. While this is physically impossible, the amount that $\sigma_{e,t}(\varepsilon)$ exceeds $\sigma_{sc,t}(\varepsilon)$ is less than the quoted uncertainties of the two measurements. It should be noted also that although cross sections have been suggested for total electron scattering, total elastic, vibrational excitation, ionization, dissociation into neutrals, dissociative electron attachment, and ion-pair formation, there is a need to improve

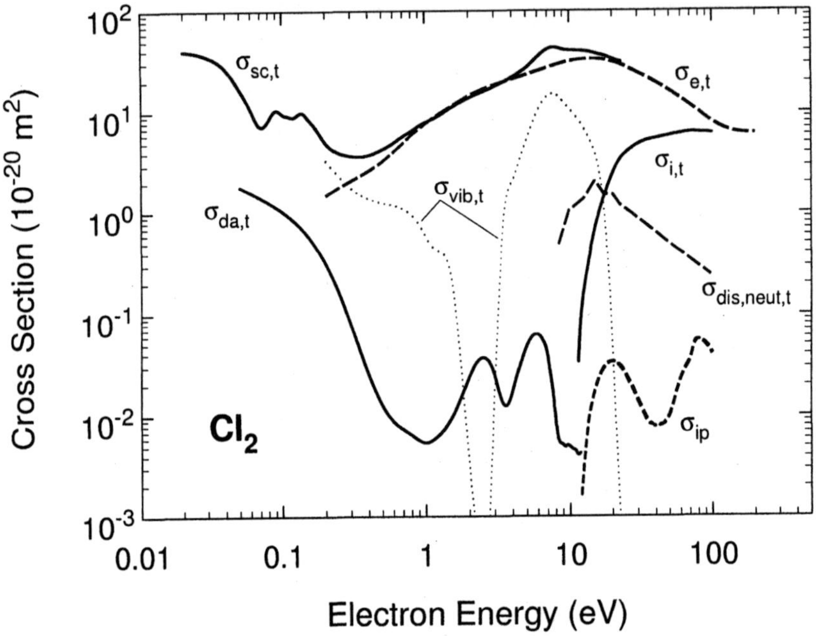

Figure 6.25. Suggested cross sections for Cl_2.

the accuracy and reliability of all these cross sections. Furthermore, there is a need for measurements of the cross sections for momentum transfer, dissociative ionization, vibrational excitation, and electronic excitation, for which no data exist at this time.

Also suggested are

- $k_{a,t}(\langle\varepsilon\rangle)$ — Table 6.13, Fig. 6.19;
- $\alpha/N\,(E/N)$ — Table 6.9, Fig. 6.15;
- $\eta/N\,(E/N)$ — Table 6.16, Fig. 6.21; and
- $(\alpha-\eta)/N\,(E/N)$ — Table 6.17, Fig. 6.22.

There is a need for measurement of all other coefficients.

2.9 Electron Collision Data for Cl and Cl$^+$

2.9.1 Cl

Atomic chlorine is an open-shell atom with a ground-state configuration of $1s^2\,2s^2\,2p^6\,3s^2\,3p^5\,(^2P_{3/2})$. Its electron affinity is well-established.

Table 6.19. Ionization energy of Cl ($^2P_{3/2}$) for the production of Cl^+ ($^3P_{2,1,0}$), Cl^+ ($^1D_2$), and Cl^+ ($^1S_0$).

Ionic state	Ionization energy (eV)	References
$^3P_2$	12.97	[154][a]
	12.97 ± 0.02	[153][a]
	12.967 ± 0.001	[155, 156][b]
$^3P_1$	13.06	[154]
	13.06 ± 0.02	[153]
	13.053	[155, 156]
$^3P_0$	13.1	[154]
	13.090	[155, 156]
$^1D_2$	14.42	[154]
	14.41 ± 0.02	[153]
	14.412	[155, 156]
$^1S_0$	16.42	[154]
	16.42 ± 0.02	[153]
	16.423	[155]

[a]He I (584 Å) photoelectron spectra data.
[b]Spectroscopic data.

Of the 38 values listed by Christodoulides *et al.* [74], those obtained using the photodetachment method are the most accurate. These are: (3.613 ± 0.003) eV [150], (3.610 ± 0.002) eV [151], and (3.616 ± 0.003) eV [152]. A value of 3.613 eV is recommended. Studies of He I photoelectron spectra [153, 154] of Cl ($^2P_{3/2}$) gave the ionization threshold energies listed in Table 6.19 for the production of Cl^+ in the ionic states $^3P_{2,1,0}$, $^1D_2$, and $^1S_0$.

De Lange *et al.* [157] used electron modulation spectroscopy and measured the photoionization cross section of Cl at the He I wavelength (584 Å) normalized to that for HCl and HBr at this wavelength. The cross section for ionization of the Cl atom into the ionic states Cl^+ ($^3P_{2,1,0}$), Cl^+ ($^1D_2$), and Cl^+ ($^1S_0$) were measured to be $(19.7 \pm 2.5) \times 10^{-18}$ cm^2, $(11.4 \pm 1.5) \times 10^{-18}$ cm^2, and $(2.16 \pm 0.28) \times 10^{-18}$ cm^2, respectively. The photoionization cross section as a function of photon wavelength, $\sigma_{pi,Cl}(\lambda)$, of the Cl atom was measured by Samson *et al.* [158] from 755 to 158 Å (16.4–75 eV) with an overall estimated uncertainty of ±8%. Their data are plotted in Fig. 6.26.

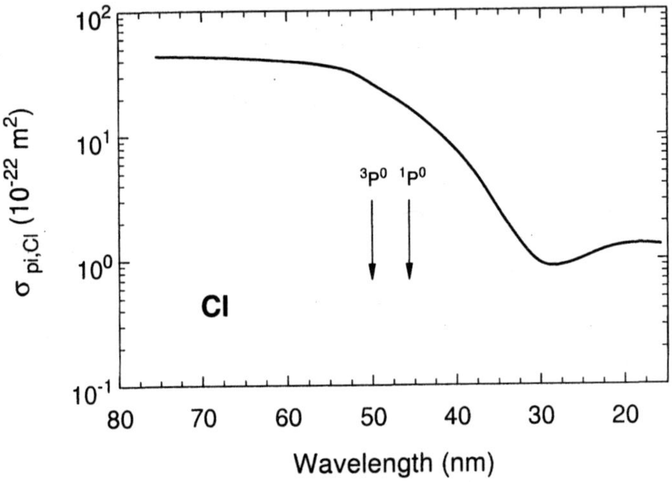

Figure 6.26. Photoionization cross section as a function of photon wavelength, $\sigma_{\text{pi,Cl}}(\lambda)$, for atomic chlorine from the measurements of Samson *et al.* [158]. The arrows show the $^3P^0$ and $^1P^0$ limits.

Total Electron Scattering Cross Section, $\sigma_{\text{sc,t,Cl}}(\varepsilon)$, of Cl. There are no measurements or calculations of the total electron scattering cross section, $\sigma_{\text{sc,t,Cl}}(\varepsilon)$, of atomic chlorine. However, since we are dealing with an atomic species, below the threshold for electronic excitation of the chlorine atom at 8.90 eV, the total electron scattering cross section $\sigma_{\text{sc,t,Cl}}(\varepsilon)$ is equal to the total elastic electron scattering cross section $\sigma_{\text{e,t,Cl}}(\varepsilon)$. Above the ionization onset of the Cl atom at 12.97 eV, $\sigma_{\text{sc,t,Cl}}(\varepsilon) = \sigma_{\text{e,t,Cl}}(\varepsilon) + \sigma_{\text{exc,t,Cl}}(\varepsilon) + \sigma_{\text{i,t,Cl}}(\varepsilon)$, where $\sigma_{\text{exc,t,Cl}}(\varepsilon)$ and $\sigma_{\text{i,t,Cl}}(\varepsilon)$ are, respectively, the total cross sections for electronic excitation and electron-impact ionization of the Cl atom. Between 8.90 and 12.97 eV, $\sigma_{\text{sc,t,Cl}}(\varepsilon) = \sigma_{\text{e,t,Cl}}(\varepsilon) + \sigma_{\text{exc,t,Cl}}(\varepsilon)$. Thus, in principle, the cross section $\sigma_{\text{sc,t,Cl}}(\varepsilon)$ for the chlorine atom for the three energy regions mentioned above can be constructed using the expressions indicated for each energy region and data on $\sigma_{\text{e,t,Cl}}(\varepsilon)$, $\sigma_{\text{i,t,Cl}}(\varepsilon)$, and $\sigma_{\text{exc,t,Cl}}(\varepsilon)$. Unfortunately, this exercise is not feasible at the present time since, as will be seen later in this section, only the cross section for single ionization $\sigma_{\text{i,Cl}}(\varepsilon)$ is known with reasonable accuracy.

Momentum Transfer Cross Section, $\sigma_{\text{m,Cl}}(\varepsilon)$, of Cl. There have been two calculations of the momentum transfer cross section, $\sigma_{\text{m,Cl}}(\varepsilon)$, of the Cl atom, the R-matrix calculation of Griffin *et al.* [159] and the multiconfiguration Hartree-Fock calculation of Saha [160]. Figure 6.27

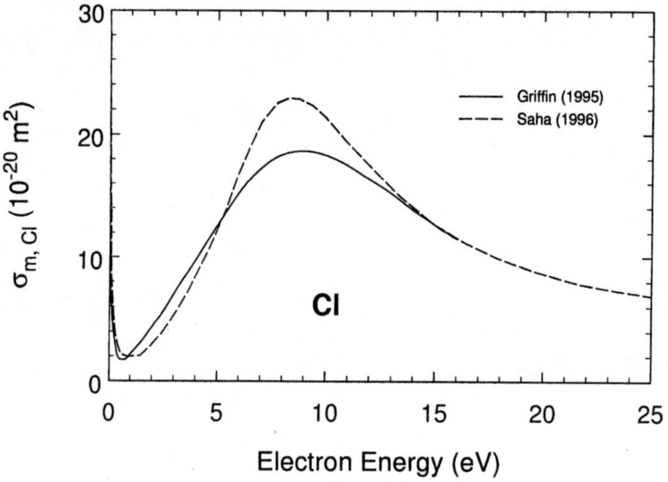

Figure 6.27. Calculated momentum transfer cross section, $\sigma_{m,Cl}(\varepsilon)$, for atomic chlorine: (—) Ref. [159]; (- - -) Ref. [160].

compares the results of these two calculations. Both calculations show the presence of a minimum in $\sigma_{m,Cl}(\varepsilon)$ (at 0.95 eV [160], at 0.75 eV [159]). This minimum is similar to the well-known Ramsauer-Townsend minimum in the scattering cross section of the neighboring rare-gas Ar atom.

Total Elastic Electron Scattering Cross Section, $\sigma_{e,t,Cl}(\varepsilon)$, of Cl. There are four calculations of the total elastic electron scattering cross section, $\sigma_{e,t,Cl}(\varepsilon)$, of the Cl atom [159–162], but no measurements. These are compared in Fig. 6.28. They all show the existence of a Ramsauer-Townsend minimum at low energies ($\sim$ 0.7 eV [161], $\sim$ 0.4 eV [162], 0.75 eV [159], and 0.95 eV [160]). Robinson and Geltman [161] performed a plane-wave calculation, Fabrikant [162] used the method of extrapolation of potential parameters along the isoelectronic sequence of positive ions to obtain scattering lengths for *e*-Cl scattering, Griffin *et al.* [159] used the R-matrix method, and Saha [160] performed a multiconfiguration Hartree-Fock calculation. The agreement between these calculated cross sections is reasonable.

Electron-Impact Excitation Cross Section, $\sigma_{exc,Cl}(\varepsilon)$, of Cl. Ganas [163] calculated cross sections, $\sigma_{exc,Cl}(\varepsilon)$, for electron-impact excitation of the 4*s*, 5*s*, 6*s*, 4*p*, 5*p*, 3*d*, 4*d*, and 5*d* states of the chlorine atom from its ground state $3p(^2P)$. These are shown in Fig. 6.29. Similarly,

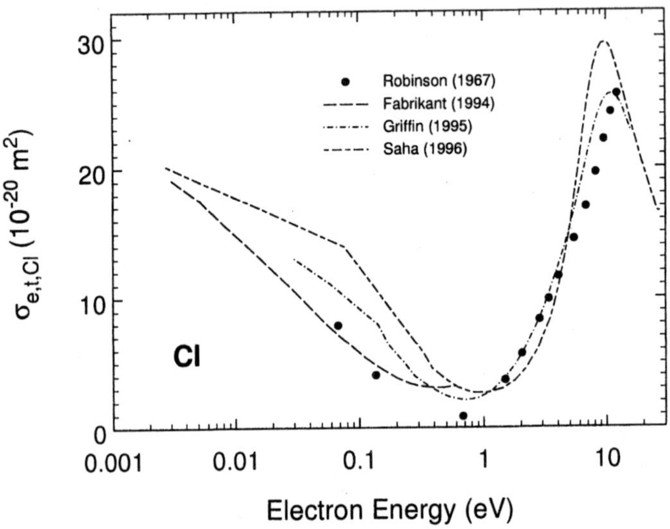

Figure 6.28. Calculated total elastic electron scattering cross section, $\sigma_{e,t,Cl}(\varepsilon)$, for atomic chlorine: (•) Ref. [161]; (– –) Ref. [162]; (– · –) Ref. [159]; (- - –) Ref. [160].

Griffin *et al.* [159] calculated electron-impact excitation cross sections of Cl to the $3p^4\,4s\,{}^4P_{5/2}$ level using the R-matrix method, but the result of their calculation was found to depend on the number of states they considered. For this reason it is not considered here.

Electron-Impact Single-Ionization Cross Section, $\sigma_{i,Cl}(\varepsilon)$, of Cl. Hayes *et al.* [164] measured the electron-impact single-ionization cross section, $\sigma_{i,Cl}(\varepsilon)$, of the Cl atom from the ionization threshold to 200 eV with an uncertainty of ±14%. Their data are plotted in Fig. 6.30 and are listed in Table 6.20 as our suggested data since these are the only experimental measurements with a specified uncertainty. Also shown in Fig. 6.30 are the $\sigma_{i,Cl}(\varepsilon)$ calculated by Lotz [165], Gopaljee *et al.* [166], and Joshipura and Limbachiya [114]. Lotz calculated $\sigma_{i,Cl}(\varepsilon)$ using an empirical formula and estimated an error of +40% and −30%, Gopaljee *et al.* [166] used the binary encounter approximation, and Joshipura and Limbachiya [114] employed what they called a "complex scattering potential-ionization contribution" method. Not included in Fig. 6.30 are the distorted-wave calculation results of Griffin *et al.* [159] because they were found to vary considerably with the details of the calculation. Lennon *et al.* [167] also reviewed and recommended data for $\sigma_{i,Cl}(\varepsilon)$ and other positive ions of the Cl atom up to 16^+.

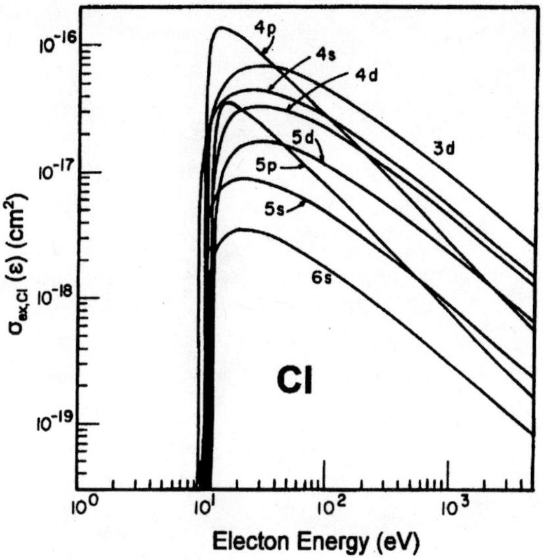

Figure 6.29. Calculated cross sections for electron-impact excitation of the $4s$, $5s$, $6s$, $4p$, $5p$, $3d$, $4d$, and $5d$ states of the chlorine atom from the ground state $3p(^2P)$ by Ganas [163].

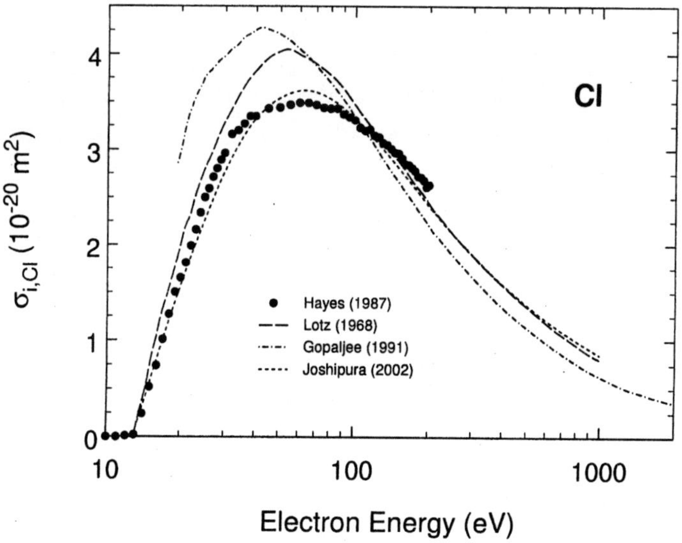

Figure 6.30. Electron-impact single-ionization cross section, $\sigma_{\mathrm{i,Cl}}(\varepsilon)$, for the Cl atom: (•) measurements from Ref. [164]; (– –) calculations from Ref. [165]; (– · –) calculations from Ref. [166]; (- - -) calculations from [114].

Table 6.20. Cross section, $\sigma_{i,Cl}(\varepsilon)$, for single ionization of Cl by electron impact from Hayes *et al.* [164].

ε (eV)	$\sigma_{i,Cl}(\varepsilon)$ (10^{-20} m^2)	ε (eV)	$\sigma_{i,Cl}(\varepsilon)$ (10^{-20} m^2)	ε (eV)	$\sigma_{i,Cl}(\varepsilon)$ (10^{-20} m^2)
11	0.00	34	3.20	105	3.23
12	0.01	36	3.27	110	3.20
13	0.02	38	3.35	115	3.21
14	0.24	40	3.35	120	3.15
15	0.52	45	3.43	125	3.13
16	0.74	50	3.44	130	3.07
17	1.01	55	3.47	135	3.05
18	1.27	60	3.49	140	3.01
19	1.50	65	3.49	145	2.97
20	1.65	70	3.47	150	2.96
22	1.99	75	3.44	155	2.91
24	2.34	80	3.43	160	2.85
26	2.59	85	3.43	170	2.81
28	2.80	90	3.37	180	2.72
30	2.96	95	3.34	190	2.68
32	3.16	100	3.31	200	2.63

Radiative Attachment to Cl. Radiative attachment to the Cl atom, viz.,

$$\mathrm{Cl}(^3P_{3/2,1/2}) + e \rightarrow \mathrm{Cl}^-(^1S_0) + h\nu \tag{6.6}$$

and the resulting radiative attachment continuum $h\nu$ (also known as the affinity spectrum) has long been investigated (e.g., see Refs. [151, 152, 168–170]). The cross section for Process (6.6) is expected to be very small [170, 171]. In Reaction (6.6) the photon energy consists of the electron affinity of the Cl atom and the kinetic energy of the attached electron. Because the kinetic energy of a free electron in, say, a plasma has a continuous range of values, the emission spectrum resulting from Process (6.6) is continuous. From its long-wavelength limit (i.e., for the case where the kinetic energy of the captured electron is zero) the electron affinity (EA) of the Cl atom has been accurately determined. Thus, Pietsch and Rehder [151] obtained $\lambda(P_{3/2}) = (343.4 \pm 0.2)$ nm, corresponding to an EA for Cl ($P_{3/2}$) of (3.610 ± 0.002) eV, and $\lambda(P_{1/2}) = (333.1 \pm 0.4)$ nm, corresponding to an EA for Cl ($P_{1/2}$) of (3.722 ± 0.005) eV. Similarly, the radiative attachment continuum was found by Mück and Popp [152] to begin at 342.8 nm yielding an EA for Cl of 3.616 eV.

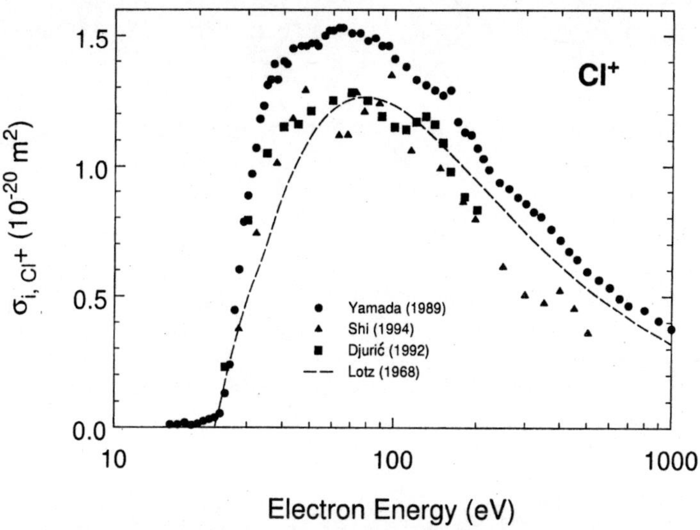

Figure 6.31. Cross section, $\sigma_{i,Cl^+}(\varepsilon)$, for single ionization of Cl^+ by electron impact ($Cl^+ + e \rightarrow Cl^{++} + 2e$): (•) measurements from Ref. [172]; (▲) measurements from Ref. [173]; (■) measurements from Ref. [174]; (– –) semiempirical calculation from Ref. [165, 172].

2.9.2 Cl^+

In Fig. 6.31 is shown the electron-impact ionization cross section as a function of electron energy, $\sigma_{i,Cl^+}(\varepsilon)$, for Cl^+ as measured in three crossed-beam experiments [172–174]. The results of Yamada *et al.* [172] extend from threshold to 1000 eV and have estimated total systematic uncertainties of −8% to +10%. The measurements of Shi *et al.* [173] cover the electron-impact energy range from 30 to 500 eV and have a reported uncertainty of ±13%. Similarly, the data of Djurić *et al.* [174] stretch from threshold to 200 eV and have a systematic uncertainty of ±10%. The measurements of Yamada *et al.* are consistently ~ 25% higher than the other two sets of measurements, possibly because of detector efficiency problems [173, 174]. Consistent with the measurements of Djurić *et al.* and Shi *et al.* is the prediction of the semiempirical calculation of Lotz [165] (see Fig. 6.31).

For electron- impact ionization cross section data on Cl^{++} see Mueller *et al.* [175]. See also, the review by Lennon *et al.* [167] for ionization cross sections and ionization rate coefficients for multiply charged positive ions of Cl.

2.10 Electron Detachment, Electron Transfer, and Recombination and Diffusion Processes Involving Cl_2

2.10.1 Electron Detachment Involving Cl_2^- and Cl^-

The large cross section for dissociative electron attachment to the Cl_2 molecule makes the dissociative electron attachment process for chlorine an efficient mechanism to remove slow electrons in chlorine-containing plasma gases. Due to the depletion of free electrons, a higher electric field strength is required to increase the source of ionization [39] and sustain the ionization balance. In the active discharge, although electron detachment processes involve both Cl_2^- and Cl^-, those involving Cl^- are by far more significant in view of the larger abundance of Cl^- (see Sec. 2.5 of Ch. 6).

Photodestruction (Photodetachment and Photodissociation) of Cl_2^-. The interaction of light with Cl_2^- may result in either photodetachment

$$Cl_2^- + h\nu \rightarrow Cl_2 + e, \tag{6.7}$$

or photodissociation

$$Cl_2^- + h\nu \rightarrow Cl + Cl^- \tag{6.8}$$

of the Cl_2^- ion. These processes can be discussed and understood with reference to the potential energy curves shown in Fig. 6.5 for the ground state of Cl_2 ($^1\Sigma_g^+$) and Cl_2^- ($^2\Sigma_u^+$) and the excited states of Cl_2^- ($^2\Sigma_g^+$ and $^2\Pi_g$). Photodetachment from Cl_2^- ($^2\Sigma_u^+$, $\nu = 0$) should be observed at a minimum energy corresponding to the EA of Cl_2 (Table 6.2). The cross section for photodetachment depends on the threshold law for photodetachment and the Franck-Condon factors which describe the overlap of the $\nu = 0$ level of Cl_2^- ($^2\Sigma_u^+$) with the vibrational levels of the Cl_2 ($^1\Sigma_g^+$) ground state. Vibrational excitation in the molecular ion will also have an effect on the probability of photodetachment. Because of the large difference in the bond length of Cl_2^- and Cl_2 (Fig. 6.5, Tables 6.2 and 6.4), photodetachment will occur to high-lying vibrational levels of Cl_2 with low probability. Photodissociation is expected to result from excitation of Cl_2^- ($^2\Sigma_u^+$) into the repulsive excited states of Cl_2^-. The total photodestruction cross section is a combination of the two processes.

About 20 years ago, Processes (6.7) and (6.8) were the subject of a few investigations [89, 176–178]. In Fig. 6.32 are shown the absolute measurements of Lee *et al.* [89] of the cross section, $\sigma_{\mathrm{pdest},Cl_2^-}(\lambda)$, for the photodestruction of the Cl_2^- ion. These were made over the wavelength

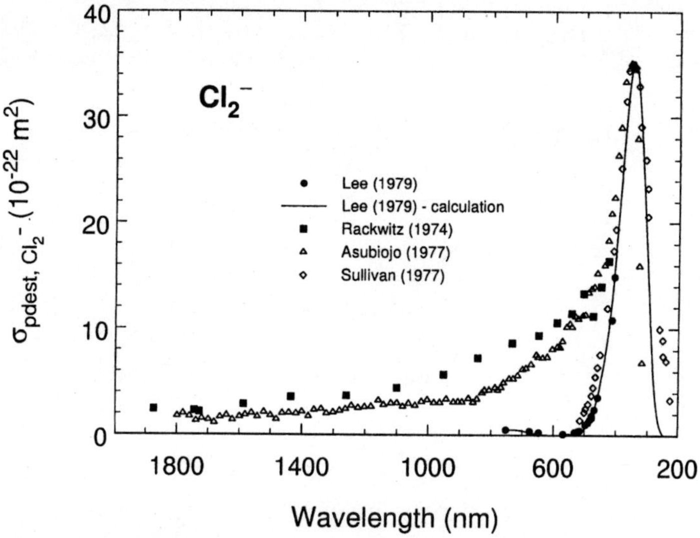

Figure 6.32. Cross section for photodestruction of Cl_2^- as a function of photon wavelength, $\sigma_{pdest,Cl_2^-}(\lambda)$: (•) measurements of Lee *et al.* [89]; (—) calculation by Lee *et al.* [89]; (■) measurements of Rackwitz *et al.* [176]; (△) relative measurements of Asubiojo *et al.* [177] normalized to the data of Lee *et al.* at 354 nm; (◊) relative measurements of Sullivan *et al.* [178] normalized to the data of Lee *et al.* at 354 nm.

range 350–760 nm using a drift-tube mass spectrometer-laser apparatus. The solid circles are their experimental measurements and the solid curve is their calculated fit to their data. The strong peak in the photodestruction cross section was attributed [89] to the electronic transition $^2\Sigma_u^+ \rightarrow {}^2\Sigma_g^+$. In the experiments of Lee *et al.*, the Cl_2^- ion was probably produced via a three-body electron attachment process to Cl_2 and was converted to Cl_3^- in collisions with Cl_2. Also plotted in Fig. 6.32 are the absolute photodestruction cross section measurements of Rackwitz *et al.* [176] made in the photon energy range from 0.5 to 3.0 eV, and the relative photodestruction cross section of Asubiojo *et al.* [177] made between about 400 and 1800 nm and normalized to the data of Lee *et al.* [89] at 354 nm. The results of Rackwitz *et al.* suggest that the Cl_2^- ion formed by electron impact is vibrationally excited and this has a rather significant influence on the photodestruction cross section in the threshold region. This is supported by the work of Sullivan *et al.* [178] who examined photoinduced reactions of Cl_2^- in the gas phase using ion cyclotron resonance techniques. They found that the Cl_2^- ion undergoes photodissociation in preference to photodetachment, and that the photodissociation spectrum of Cl_2^- exhibits one broad peak in the wavelength region from 220 to 700 nm with a maximum at (350 ± 10) nm which they attributed to the $^2\Sigma_u^+ \rightarrow {}^2\Sigma_g$ transition. This cross section has also been plotted in Fig. 6.32 after it has been normalized to the Lee *et al.* data at 354 nm.

It can thus be concluded [89] from the results of these four investigations that the Cl_2^- ion photodissociates rather than photodetaches, that the cross section for photodestruction depends on the electronic excitation of Cl_2^- upon photon impact, that the cross section threshold shifts to energies lower than the dissociation energy limit (1.26 eV, Table 6.4) of Cl_2^- into Cl^- + Cl when the anion is vibrationally excited, and that the differences in the band widths between the four studies probably reflect differences in the vibrational temperature in the four experimental methods employed.

Electron-Induced and Collisional Detachment of Cl_2^-. We know of no data on electron-induced detachment or collisional detachment involving the Cl_2^- ion.

Photodetachment of Cl^-. For Cl^- the most significant reactions and parameters are those involving the removal of the attached electron. These processes have been discussed by many authors (see for instance, Refs. [179, 180]). In this subsection we briefly discuss data on photodetachment of the Cl^- ion, and in the next subsection we present data on collisional detachment of the Cl^- ion.

The Cl^- has a complete $3p^6$ subshell. Thus, the photodetachment process involves the removal of an electron from the p orbital and can be represented by

$$Cl^-(^1S_0) + h\nu \rightarrow Cl(^2P_{3/2,1/2}) + e. \tag{6.9}$$

An early review of the experimental and theoretical data on the cross section, $\sigma_{pd,Cl^-}(\lambda)$, for Reaction (6.9) was given by Popp [170]. In Fig. 6.33 are compared the experimental [152, 168, 169, 181–184] and the calculated [161, 185–187] data on $\sigma_{pd,Cl^-}(\lambda)$ for Reaction (6.9). Most of these results were obtained over 20 years ago. The uncertainties in the experimental measurements are as follows: the single measurement of Berry *et al.* [181] at 336 nm (15×10^{-22} m^2) has a quoted uncertainty of $+12 \times 10^{-22}$ m^2 and -5×10^{-22} m^2; Mück and Popp's [152] and Mandl's [182] uncertainties were quoted as ±25%; Roth's [168] and Pietsch's [169] uncertainties are as shown by the typical error bars in Fig. 6.33; Wang and Lee [183, 184] reported photodetachment cross section values for Cl^- equal to 25×10^{-22} m^2 and 10×10^{-22} m^2 at 193 and 248 nm, respectively, but gave no uncertainties. On the calculation side, Robinson and

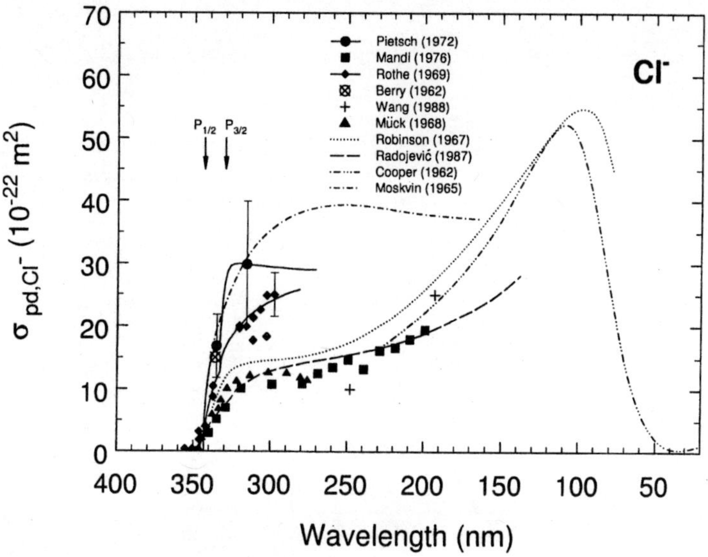

Figure 6.33. Photodetachment cross section for Cl^-, $\sigma_{pd,Cl^-}(\lambda)$, as a function of photon wavelength, λ. Measurements: (•) Ref. [169]; (■) Ref. [182]; (♦) Ref. [168]; (⊗) Ref. [181]; (+) Ref. [183]; (▲) Ref. [152]. Calculations: (···) Ref. [161]; (– –) Ref. [187]; (–···–) Ref. [185]; (–·–) Ref. [186]. Typical error bars are shown in the figure for only the measurements of Refs. [168, 169]. See the text for the reported uncertainties of the other measurements. The energy positions of the $P_{1/2}$ and $P_{3/2}$ photodetachment thresholds are also shown in the figure.

Geltman [161] quoted an uncertainty of ±20%. It should be noted that the relativistic random-phase approximation result of Radojević *et al.* [187] extends to 100 eV and that Radojević *et al.* shifted their calculated curve from the theoretical threshold to the experimental value. It is seen from Fig. 6.33 that the spread in the experimental data is outside of the quoted uncertainties. The limited recent measurements of Wang and Lee [183] are consistent with the earlier measurements of Mandl [182] and Mück and Popp [152], but all three measurements are lower (often by a factor of 2 or more) than the data of Rothe [168], Pietsch [169], and Berry *et al.* [181]. On the theoretical side, the calculated values of $\sigma_{pd,Cl^-}(\lambda)$ by Moskvin [186] differ substantially from the results of the other three calculations [161, 185, 187].

Collisional Detachment of Cl^-. Collisional detachment reactions fall into three groups [179, 180]: direct detachment, detachment with excitation (of autodetaching levels, or of a neutral product, or via charge transfer to a negative ion state of the target), and detachment with bond-

Table 6.21. Associative detachment thermal rate constants involving Cl^-.

Reactants	Temperature (K)	Associative detachment thermal rate constant (cm^3 $molecule^{-1}$ s^{-1})	Ref.
$Cl^- + H \rightarrow HCl + e$	296	9.6×10^{-10}	[188]
	296	10.0×10^{-10}	[189]
	296	9.0×10^{-10}	[190]
$Cl^- + O \rightarrow ClO + e$	300	$< 1 \times 10^{-11}$	[191]
$Cl^- + N \rightarrow ClN + e$	300	$< 1 \times 10^{-11}$	[191]
$Cl^- + Cl_2\ (+He) \rightarrow Cl_3^-$	ambient	0.9×10^{-29} (cm^6 $molecule^{-2}$ s^{-1})	[192]

ing (reactive collision with detachment, or associative detachment). The magnitude and the dependence of the collisional detachment cross section, $\sigma_{cd}(\mathcal{E})$, on the energy, $\mathcal{E}$, of the reactants varies with the type of the detachment process. (Note that $\mathcal{E}$ refers to the energy of the reactants, i.e., the projectile and the neutral target.) Thus, the rising parts of $\sigma_{cd}(\mathcal{E})$ as the kinetic energy of the reactants increases are principally due to direct collisional detachment, while the rising parts of $\sigma_{cd}(\mathcal{E})$ as the kinetic energy of the reactants decreases toward thermal energy are due to associative detachment. Generally, there is a threshold for the direct collisional detachment process which occurs (when the reactants are in their ground states) when their kinetic energy is equal to the EA of the species carrying the extra electron, although in certain cases such as for the reactions $Cl^- + M$ (where M is a molecule), the $\sigma_{cd}(\mathcal{E})$ increases rapidly from the threshold which itself is considerably greater than the EA of the Cl atom. The associative detachment process, besides being responsible for the large cross sections at thermal and near-thermal energies, also accounts for maxima often seen in the $\sigma_{cd}(\mathcal{E})$ functions at higher energies due to negative ion resonances. In Table 6.21 are listed values of the thermal ($T \sim 300$ K) rate constants for the associative detachment reactions

$$Cl^- + X \rightarrow ClX + e \tag{6.10}$$

where X = H, O, N, or Cl_2.

The threshold for collisional detachment can be low, and the cross section for collisional detachment can be very large [179, 180] indeed, in many cases, much larger than the cross section for photodetachment.

Table 6.22. Energy threshold for the detachment of Cl^- in collisions with various target gases as reported by Doverspike *et al.* [193].

Reactants	Threshold energy (eV)
$Cl^- + H_2$	5.5 ± 0.1
$Cl^- + D_2$	5.5 ± 0.1
$Cl^- + N_2$	7.6 ± 0.1
$Cl^- + O_2$ [a]	4.4 ± 0.2
$Cl^- + CO$	7.1 ± 0.2
$Cl^- + CO_2$	7.3 ± 0.2
$Cl^- + CH_4$	6.2 ± 0.2

[a] In addition to direct detachment there are several other processes which may contribute to the products of this reaction at energies below 4.4 eV, such as the charge-transfer reaction $Cl^- + O_2 \rightarrow Cl + O_2^-$ which is endothermic by $\sim$ 3.1 eV and the associative detachment reaction $Cl^- + O_2 \rightarrow ClO_2 + e$ which is endothermic by 3.4 eV (see Ref. [193]).

When the associative detachment reactions (6.10) are exothermic, that is, when the so-called energy defect (the energy difference between the dissociation energy of ClX and the EA of Cl) is positive, and the reactions are not hindered by geometric or other factors, the thermal values of the rate constants are large ($\sim 10^{-9}$ cm^3 s^{-1}) and close to the values of the orbiting Langevin collision rate constants. Collisional detachment, then, especially when it is field assisted, can be a dominant electron release mechanism in electrically stressed gases.

Doverspike *et al.* [193] measured the total electron detachment cross sections for collisions of Cl^- with a number of molecular targets X (X = H_2, D_2, O_2, N_2, CO, CO_2, and CH_4) for collision energies below the threshold for detachment to several hundred eV. The reaction studied is

$$Cl^- + X \rightarrow Cl + X + e. \tag{6.11}$$

In all such collisions the detachment thresholds were found to exceed the electron affinity of the Cl atom. Table 6.22 lists the threshold values for collisional detachment as reported by Doverspike *et al.* [193]. The results of Doverspike *et al.* are shown in Fig. 6.34(a) for energies near threshold and in Fig. 6.34(b) for higher energies.

Huq *et al.* [195] measured the total cross sections for charge transfer and electron detachment of Cl^- on Cl_2. In Fig. 6.35 are shown their measurements of the total cross sections for electron detachment and for "slow" ion production (via charge transfer). The quoted uncertainty is about ±10%. In Fig. 6.35 are also shown the earlier measurements by Hasted and Smith [196] who reported cross sections for electron de-

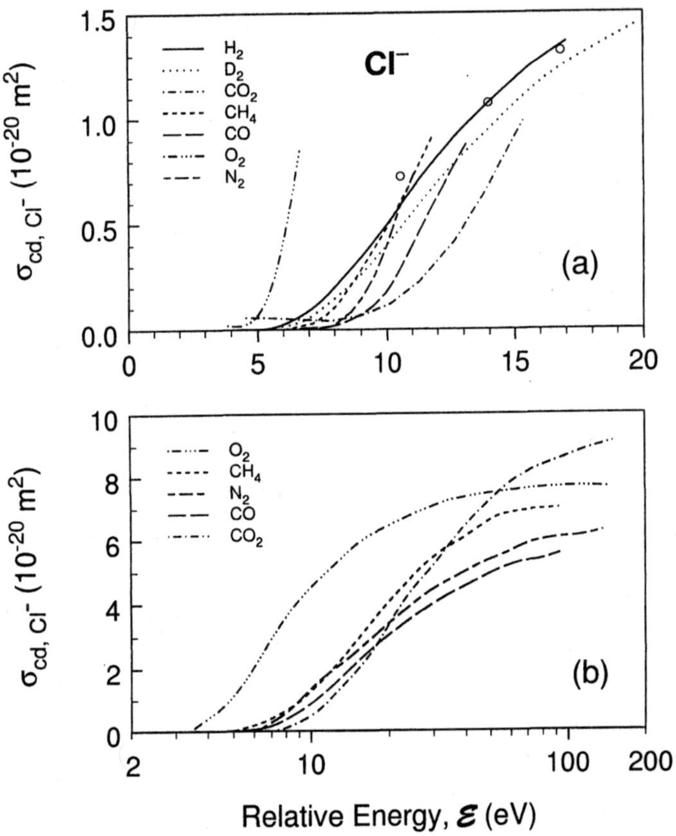

Figure 6.34. Collisional detachment cross sections, $\sigma_{cd,Cl^-}(\mathcal{E})$, as a function of the relative energy of the reactants, $\mathcal{E}$, involving Cl^- and various molecular targets. (a) near threshold energies and (b) over a wider energy range. All data are from Doverspike *et al.* [193] except for the three data points (o) for the $Cl^- + H_2$ reaction which are those of Bydin and Dukel'skii [194].

tachment in collisions of Cl^- with Cl_2 in the energy range from 10 to 2500 eV. According to Huq *et al.* [195], it appears that, at the lowest energies, the Hasted and Smith study did not fully resolve ions from electrons.

Measurements of the translational energy thresholds for electron transfer reactions for various atomic negative ions to Cl_2 at room temperature [127, 128] allowed determination of the electron affinity of the Cl_2 molecule. Thus, from measurements of the energy thresholds for the endothermic electron transfer reactions $I^- + Cl_2$ and $Cl^- + Cl_2$, Hughes *et al.* [128] obtained a value of (2.62 ± 0.2) eV for the EA of Cl_2. Similarly,

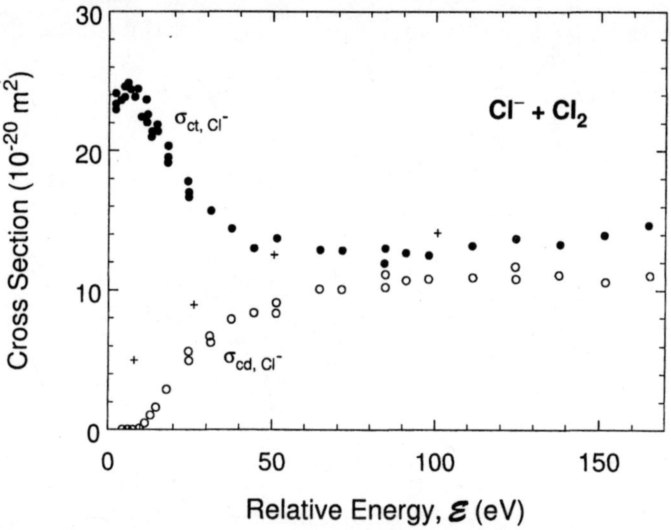

Figure 6.35. Cross section, $\sigma_{ct,Cl^-}(\mathcal{E})$, for charge transfer as a function of the relative energy of the reactants, $\mathcal{E}$, in collisions of Cl^- with Cl_2: (•) data of Huq *et al.* [195]. For comparison the cross section $\sigma_{cd,Cl^-}(\mathcal{E})$ is also shown: (○) data of Huq *et al.* [195]; (+) data of Hasted and Smith [196].

from the room temperature relative cross sections for the reactions of I^-, Br^-, and Cl^- with Cl_2, Chupka *et al.* [127] obtained for the EA of Cl_2 the value of (2.38 ± 0.10) eV.

2.10.2 Electron Transfer in Collisions of Cl_2^- and Cl^- with Cl_2

While the reaction

$$Cl_2^- + Cl_2 \rightarrow Cl_3^- + Cl \tag{6.12}$$

is endoergic when the reactants are thermalized [89], Hughes *et al.* [128] found that it becomes exoergic at energies in excess of 0.3 eV and has a rate constant at this energy equal to 0.0084×10^{-10} cm^3 molecule^{-1} s^{-1}. Similarly, the addition reaction

$$Cl^- + Cl_2 \rightarrow Cl_3^- \tag{6.13}$$

was found by Babcock and Streit [192] to have a three-body rate constant (with He as the third body) of 0.9×10^{-29} cm^6 molecule^{-2} s^{-1}.

The electron affinity of the chlorine molecule can be determined from measurements of the translational energy thresholds for electron-transfer reactions for various atomic negative ions (e.g., I^- and Cl^-) to Cl_2. Thus, Hughes *et al.* [128] and Chupka *et al.* [127] determined via such reactions the electron affinity of the Cl_2 molecule to be, respectively, (2.32 ± 0.1) eV and (2.38 ± 0.1) eV.

2.10.3 Recombination and Diffusion Processes Involving Cl_2^+, Cl^- and Cl

Recombination of Cl_2^+ and Cl^-. Positive ion-negative ion recombination measurements in flowing afterglow plasmas by Church and Smith [197] gave the value of 5.0×10^{-8} cm^3 molecule^{-1} s^{-1} for the rate constant of the reaction $Cl_2^+ + Cl^- \rightarrow$ products.

Recombination of Cl. Boyd and Burns [198] compared recombination and dissociation rate constants for halogens obtained by a variety of experimental techniques. The Cl-Cl recombination is exothermic ($\Delta H \sim -1.1$ eV) and requires a third body, M, i.e.,

$$\mathrm{Cl} + \mathrm{Cl} + \mathrm{M} \rightarrow \mathrm{Cl_2} + \mathrm{M}. \tag{6.14}$$

Boyd and Burns observed that the three-body recombination rate constant for Reaction (6.14) decreases with increasing temperature and that the Cl_2 molecules are not efficient third bodies at any temperature. Measurements of atomic chlorine concentration in Cl_2 plasmas using infrared absorption spectroscopy by Richards and Sawin [7] showed that gas-phase recombination is an insignificant Cl loss mechanism. For the temperature of their experiment (770 K), the rate constant for Reaction (6.14) ($M = Cl_2$) is $\sim 2.8 \times 10^{-32}$ cm^6 molecule^{-2} s^{-1} [198]. Richards and Sawin thus concluded that the major mechanism for Cl loss is likely to be recombination on the electrode surfaces.

Diffusion of Cl and Cl^- in Gases. Chang *et al.* [199] measured the diffusion coefficient of atomic chlorine in molecular chlorine. They reported a value for the diffusion coefficient of chlorine atoms in chlorine molecules of (0.149 ± 0.025) cm^2 s^{-1} at 298 K and 1 atm. Similarly, Hwang *et al.* [200] measured the diffusion coefficients of atomic chlorine in rare gases via radiative recombination reactions. At 296 K and 101.33 kPa (1 atm) of rare-gas pressure, the values of the diffusion constant for Cl in He, Ne, Ar, Kr, and Xe were measured to be, respectively, (0.75 ± 0.12) cm^2 s^{-1}, (0.32 ± 0.05) cm^2 s^{-1}, (0.19 ± 0.03) cm^2 s^{-1}, (0.14 ± 0.02) cm^2 s^{-1}, and (0.12 ± 0.02) cm^2 s^{-1}.

Eisele *et al.* [201] measured the longitudinal diffusion coefficients for Cl^- ions in Ne, Ar, Kr, and Xe as a function of E/N. Measurements

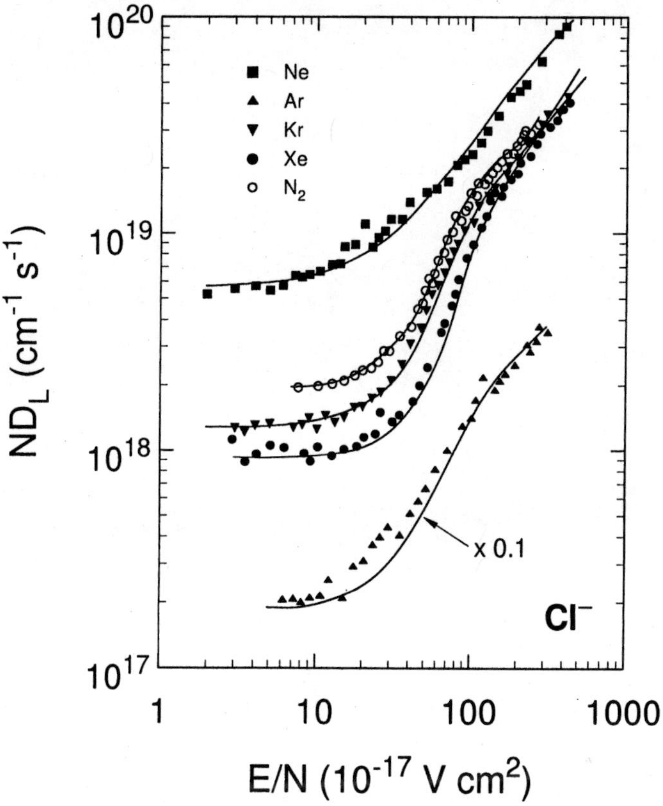

Figure 6.36. The product, $D_L N\,(E/N)$, of the longitudinal diffusion coefficient D_L and the neutral gas number density N as a function of E/N for Cl^- in Ne, Ar, Kr, and Xe (data from Eisele *et al.* [201]). Also shown are measurements made in N_2 (data of Thackston *et al.* [205]). The points are experimental data and the curves are calculated results using the generalized Einstein relation between D_L and K. Note that the Ar data have been multiplied by 0.1 for the convenience of display.

were made at about 300 K and at gas pressures below 0.067 kPa. They are shown in Fig. 6.36. As $E/N \to 0$, the ions are in thermal equilibrium with the gas molecules and the diffusion coefficient is isotropic, related to the ionic mobility K by the relation $K = eD/kT$, where e is the ionic charge, k is Boltzmann's constant, and T is the gas temperature. For larger values of E/N, this relation is not valid because the diffusion coefficient has components that refer to the directions parallel and perpendicular to the electric field (for computational techniques allowing the calculation of the diffusion coefficient at any value of E/N from knowledge of the ionic mobility at that E/N see Refs. [202–204]).

Finally, Thackston *et al.* [205] reported measurements of the longitudinal diffusion coefficients for Cl^- in N_2. These measurements are also shown in Fig. 6.36 and have a reported uncertainty of $\pm 7\%$ at all E/N.

2.11 Summary of Data for Other Species and Processes Involving Cl_2 Plasmas

With the exception of the limited measurements on electron-impact ionization of Cl and Cl^+, no measurements are known to have been made for other electron collision processes for the species Cl, Cl^-, Cl^+, and Cl_2^-. With regard to data on other important processes in Cl_2 plasmas, data have been summarized in the preceding section on the photodetachment of Cl^-, charge transfer reactions involving Cl^- and various molecular partners, and diffusion coefficients for Cl^- in rare gases and N_2. Much work in needed on electron collision and other processes involving the main species in Cl_2 plasmas.

* * *

3. Electron Interactions with CCl_2F_2

Dichlorodifluoromethane (CCl_2F_2) is an industrial gas that was widely used as a refrigerant, a foam blowing agent, and an aerosol propellant. It is also a plasma processing gas [206] and has been used as an additive in gas dielectric mixtures. Dichlorodifluoromethane is of atmospheric and environmental interest because it is an ozone depleting gas due to solar photolysis releasing chlorine atoms which enter into catalytic stratospheric-ozone-depleting reactions [207, 208]. Its residence time in the environment is about 102 years [209] and its global warming potential over a 100-year period has been reported to be 8 500 compared to the global warming potential of one for CO_2 [209]. Although CCl_2F_2 decomposes under low energy electron impact by dissociative ionization (Sec. 3.3 of Ch. 6) and by dissociative electron attachment (Sec. 3.5 of Ch. 6), to our knowledge these modes of decomposition of CCl_2F_2 have not been considered in studies aimed at the determination of its environmental impact. In this connection, it is worth noting that the electron-impact decomposition of CCl_2F_2 and other chlorofluorocarbons via dissociative electron attachment processes increases considerably with increasing rovibrational excitation, that is, with increasing temperature. This may be of interest in its use in plasma reactors and in its environmental impact and removal from the environment.

The available information on the cross sections and rate coefficients for collisional interactions of CCl_2F_2 with electrons is critically evaluated, synthesized, and discussed in this section.

3.1 Electronic and Molecular Structure of CCl_2F_2

The dichlorodifluoromethane molecule has C_{2v} symmetry. The valence shell independent-particle electronic configuration of the ground state can be written [210–212] as: $(1a_1)^2\,(1b_1)^2\,(2a_1)^2\,(1b_2)^2\,(3a_1)^2\,(2b_2)^2\,(4a_1)^2\,(2b_1)^2\,(5a_1)^2\,(1a_2)^2\,(3b_1)^2\,(3b_2)^2\,(6a_1)^2\,(2a_2)^2\,(4b_1)^2\,(4b_2)^2 : {}^1A_1$.

High resolution photoelectron spectroscopy [210, 211] gives 12.26, 12.53, 13.11, 13.45, 14.36, 15.9, 16.30, 16.9, 19.3, 20.4, and 22.4 eV, for the vertical ionization energies of the outer valence orbitals $4b_2$, $4b_1$, $2a_2$, $6a_1$, $3b_2$, $3b_1$, $1a_2$, $5a_1$, $(2b_1 + 4a_1)$, $2b_2$, and $3a_1$, respectively. The CCl_2F_2 molecule has a dipole moment of 1.835×10^{-30} C m (0.55 D) [213]. Beran and Kevan [214] estimated three values (59.2×10^{-25}, 67.7×10^{-25}, and 64.3×10^{-25} cm^3) for the static polarizability of CCl_2F_2 using three different methods of calculation.

A number of researchers investigated the electronic structure of the CCl_2F_2 molecule (see, for example, Refs. [210–212, 215–220]). Especially well investigated are the photoabsorption, photoionization, and photofragmentation properties of CCl_2F_2 [207, 211, 215–218]. In Fig. 6.37 the results of four photoabsorption cross section measurements [207, 215–217] are compared with cross sections obtained from differential oscillator strength measurements using electrons with 100 eV [221] and 8 keV [210] energy. With the exception of the photoabsorption cross sections of Jochims *et al.* [215], the rest of the data are in reasonable agreement. The data of Jochims *et al.* have a quoted uncertainty of up to ±20%. The cross section determined from the differential oscillator strength data of Huebner *et al.* [221] was determined by normalization to the absorption cross section of Person *et al.* [222] at 12.22 eV and has a stated uncertainty of ±30%. Zhang *et al.* [210] pointed out that their technique has constant energy resolution (1 eV FWHM) which is independent of the energy loss, whereas photoabsorption techniques have an energy resolution which becomes lower with increasing photon energy. This large difference in energy resolution complicates the comparison between the results obtained by their technique and the photoabsorption measurements. The absolute oscillator strength measurements of Zhang *et al.* [210] are quoted with an uncertainty of ±5%. Above ∼ 24 eV the data of Zhang *et al.* [210] are in agreement with the photoabsorption cross section values of Wu *et al.* [217]. See further discussion and comparison with other data in Zhang *et al.* [210]. Also see Refs. [210–212,

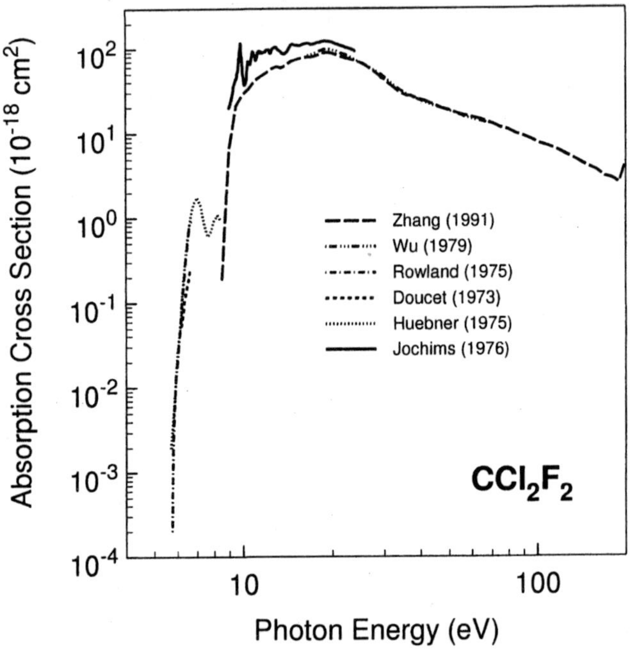

Figure 6.37. Photoabsorption cross sections as a function of photon energy for CCl_2F_2. Photoabsorption studies: (– · –) Ref. [207]; (—) Ref. [215]; (- - -) Ref. [216]; (– · · · –) Ref. [217]. Differential oscillator strength measurements: (· · ·) Ref. [221]; (– –) Ref. [210].

215–220] for information on reactions, fragmentation patterns, energy positions, and cross sections for specific ions.

The electron energy-loss spectrum of CCl_2F_2 has been measured by King and McConkey [219] using electrons with 500 eV initial kinetic energy. Figure 6.38 shows the energy-loss spectrum of CCl_2F_2 they obtained between 6 and 16 eV at 0° scattering angle. Table 6.23 lists the energy positions of the main features of the energy-loss spectrum as given by King and McConkey [219]. The energy positions of the main peaks are in good agreement with photoabsorption and other energy-loss studies [215, 218, 222].

Table 6.24 lists the vertical ionization energies for CCl_2F_2. This molecule has four close-lying ionization thresholds due to ionization from the molecular orbitals formed by the chlorine lone pairs [218]. Their values are 12.3, 12.6, 13.2 and 13.5 eV [216]. The vacuum ultraviolet (VUV) spectra up to 9.9 eV are also due to transitions from such orbitals [216]. The lowest CCl ionization onset is 14.4 eV [216] although

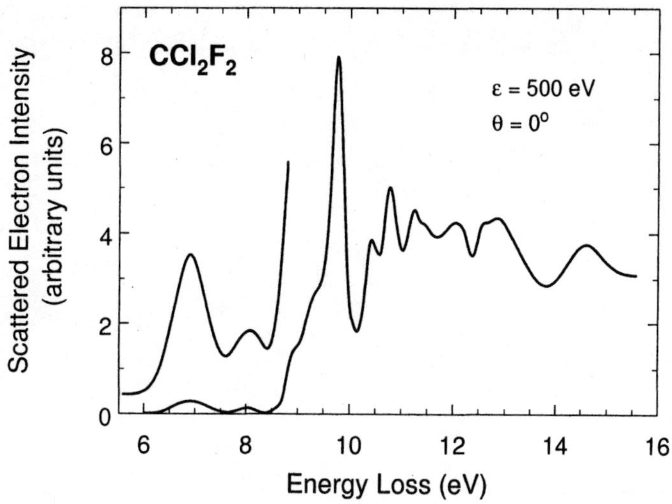

Figure 6.38. Electron energy-loss spectrum of CCl_2F_2 at an incident electron energy of 500 eV and 0° scattering angle from King and McConkey [219].

King and McConkey [219] argued that above 12 eV essentially all processes involve direct or dissociative ionization. Zhang *et al.* [210] showed that the ionization efficiency is equal to one for electron energies above $\sim$ 17.5 eV.

Absolute dipole differential oscillator strengths for inner shell spectra have been determined by Zhang *et al.* [220] from high resolution energy-loss studies using electrons with 3 keV incident energy and zero degree mean scattering angle. They also measured absolute photoabsorption oscillator strengths in the equivalent photon energy range 8.5–200 eV, and ionic photofragmentation branching ratios and the photoionization efficiency at equivalent photon energies from the first ionization threshold to 70 eV. Absolute partial photoionization oscillator strengths for dissociative photoionization have been obtained by them.

The CCl_2F_2 molecule has nine nondegenerate fundamental vibrational modes, $\nu_1, \ldots, \nu_9$. Their energies, nuclear motion, symmetry, and infrared activity as summarized by Mann and Linder [223] are given in Table 6.25 (see, also, Shimanouchi [224]).

Dichlorodifluoromethane is an electronegative gas. Its electron attachment properties are discussed in Sec. 3.5 of this chapter. However, to aid the understanding of the data on electron scattering from CCl_2F_2, we present here a summary of the work on the energies of the negative ion states of CCl_2F_2. The results obtained by various methods [184,

Table 6.23. Energies of peaks or shoulders, in eV, in the photoabsorption or energy-loss spectrum of CCl_2F_2.

Ref. [219]	Ref. [218]	Ref. [215]	Ref. [222][a]
6.95 (bp)[b]			
8.13 (bp)			
8.9 (s)	9.17 (s)		
9.35 (s)			
9.59 (s)			
9.77 (p)	9.81 (p)	9.8 (p)	
9.86 (s)			
10.45 (p)	10.46 (p)	10.5 (p)	
10.79 (p)	10.78 (p)	10.80 (p)	
11.29 (p)	11.29 (p)	11.24 (p)	
11.49 (p)		11.50 (p)	
11.75 (bp)		11.80 (p)	
12.06 (p)	12.10 (p)		12.10 (p)
12.18 (s)			12.25 (p)
12.64 (p)		12.6 (p)	
12.76 (p)	12.88 (p)	12.75 (p)	12.73 (p)
12.93 (p)		12.90 (p)	12.95 (p)
13.29 (s)			13.3 (s)
14.69 (bp)		14.70 (bp)	14.7 (bp)
15.4 (bp)			15.4 (bp)
16.34 (bp)		16.50 (bp)	16.0 (s)
			16.4 (bp)
18.0 (s)			18.0 (s)
19.2 (bp)		19.20 (bp)	19.2 (bp)
24.4 (bp)			
26.9 (bp)			

[a] As reported in Ref. [219].
[b] The symbols p, bp, and s, indicate peak, broad peak, and shoulder in the spectra, respectively.

225–244] are summarized in Table 6.26. The adiabatic electron affinity (EA) of CCl_2F_2 has been determined by Dispert and Lacmann [225] to be (0.4 ± 0.3) eV using a potassium-atom beam to create the parent negative ion $CCl_2F_2^-$ via electron transfer in binary potassium-CCl_2F_2 collisions. A multiple scattering $X\alpha$ calculation [226, 227] has also given a positive value for the EA of the CCl_2F_2 molecule equal to 0.4 eV. A more recent quantum mechanical calculation [228] gave a value of 0.67 eV for the adiabatic electron affinity of CCl_2F_2.

Besides the potassium-atom charge exchange collision technique of Dispert and Lacmann [225], five other types of experimental methods

Table 6.24. Vertical ionization energies of CCl_2F_2, in eV, obtained from photoelectron data.

Ref. [215]	Ref. [216]	Ref. [211][a]	Assignment [216, 218]
12.25	12.3	12.26	Cl lone pair
12.50	12.6	12.53	Cl lone pair
13.20	13.2	13.11	Cl lone pair
13.50	13.5	13.45	Cl lone pair
14.35	14.4	14.36	Lowest C–Cl molecular orbital
16.25		15.9	
		16.30	
		16.9	
19.2		19.3	
20.0		20.4	
		22.4	

[a]Vertical ionization energies of the outer valence orbitals.

Table 6.25. Vibrational modes of CCl_2F_2 from Mann and Linder [223].

Mode	Energy (meV)	Nuclear motion	Symmetry	Activity
ν_1	136.1	CF_2 symmetric stretch	A_1	IR
ν_2	82.7	CF_2 bending	A_1	IR
ν_3	56.6	CCl_2 symmetric stretch	A_1	
ν_4	32.5	CCl_2 bending	A_1	IR
ν_5	39.9	torsion	A_2	
ν_6	144.7	CF_2 asymmetric stretch	B_1	IR
ν_7	55.3	CF_2 plane rocking	B_1	
ν_8	114.4	CCl_2 asymmetric stretch	B_2	IR
ν_9	53.9	CCl_2 plane rocking	B_2	

(threshold electron attachment technique [229], swarm electron attachment technique [184, 230–233], electron beam/mass spectrometric techniques for electron attachment [234–238], electron scattering [239, 240], electron transmission [241, 244]), and one calculation [242] provided information on the negative ion states of CCl_2F_2. We have also determined the energies of the negative ion states of CCl_2F_2 from the positions of the maxima in the total indirect inelastic electron scattering cross section using data reported by Mann and Linder [223] (see Sec. 3.2 of Ch. 6).

Table 6.26. Energies, E_{NIS}, of the negative ion states of CCl_2F_2.

Energy (eV)[a]	Method	Ref.
~ 0.0	Threshold attachment technique	[229]
< 0.1 ~ 0.18 (shoulder) 1.05	Swarm-unfolded total attachment cross section using N_2 as the buffer gas	[230]
0.07 0.30 0.93	Swarm-unfolded total attachment cross section using N_2 as the buffer gas	[231]
~ 0 0.8 3.8	Swarm-unfolded total attachment cross section using N_2 and Ar as buffer gases	[232]
~ 0.0 ~ 0.7 ~ 3.5	Energies where the total attachment rate constant measured in N_2 and Ar buffer gases shows maxima	[184, 233]
~ 0.0 ~ 0.6 3.5	Electron beam measurement of the total attachment cross section	[234]
0.55 (Cl^-) 0.65 (Cl_2^-) 2.85 (FCl^-) 3.1 (F^-) 3.55 ($CFCl_2^-$) The energy dependence of the sum of all negative ions gives peaks at 0.6 eV and at 3.2 eV	Mass spectrometric study of dissociative attachment using a trochoidal monochromator	[235–237]
~ 0.0 0.3 0.95 3.6	Mass spectrometric study of dissociative attachment using a trochoidal monochromator	[238]

[a]The following values have been reported for the EA of the CCl_2F_2 molecule: (0.4 ± 0.3) eV [225], 0.4 eV [226, 227], and 0.67 eV [228].

The energies of the negative ion states as determined by the various methods are shown in Fig. 6.39. It should be noted that the energies of the negative ion states as determined from the resonance energies

Table 6.26 (continued)
Energies, E_{NIS}, of the negative ion states of CCl_2F_2.

Energy (eV)	Method	Ref.
0.71 3.55	Mass spectrometric study of dissociative attachment using a trochoidal monochromator	[244]
0.7 (Cl^-) 3.2 (F^-) 3.7 ($CFCl_2^-$)	ICR study of dissociative attachment	[239]
1.0 2.6 4.0 5.9	Total electron scattering cross section measurements	[240]
1.2 3.4 4.6 6.4	Estimates[b] of resonance energies determined from electron scattering experiments	[242]
0.8 3.1 5.1 6.7	Theoretical estimates[c]	[242]
0.98 (0.97)[d] 2.35 (2.36)[d] 3.88 (3.86)[d]	Vertical electron affinity values determined by electron transmission spectroscopy	[241, 244]
1.0 2.5 4.0 6.0 9.0	Maxima in the total indirect inelastic electron scattering cross section	Present analysis based on the work of Ref. [223]; See Fig. 6.48 in Sec. 3.2 of Ch. 6

[b]Taken from Fig. 5 of Ref. [242].
[c]Calculated term values taken from Fig. 5 of Ref. [242].
[d]Values in parenthesis are from Ref. [244].

in the electron attachment cross sections are normally lower than the corresponding energy positions determined from electron scattering due to the effect of autodetachment on the former. In the last column of Fig. 6.39 are given the "average" energies of the negative ion states of CCl_2F_2 based on the data given in the figure. On the basis of this

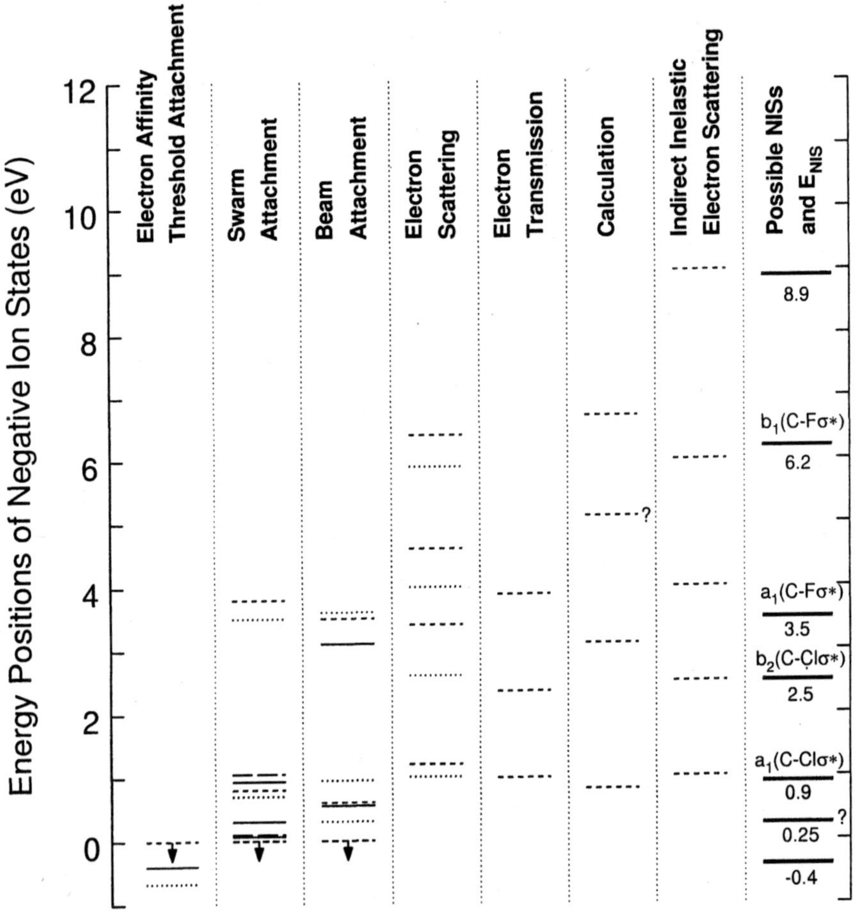

Figure 6.39. Energy positions of the negative ion states of CCl_2F_2 below 10 eV as obtained by various methods. Column 1: (- - -) threshold attachment technique [229]; (—) potassium-atom charge-exchange collision technique [225] and calculation [226, 227]; (···) calculation [228]. Column 2: Electron swarm attachment techniques: (- - -) Ref. [232]; (···) Ref. [184]; (– –) Ref. [230]; (—) Ref. [231]. Column 3: Electron beam attachment techniques: (- - -) Ref. [234]; (—) Ref. [235]; (···) Ref. [238]. Column 4: Electron scattering: (···) Ref. [240]; (- - -) Ref. [242]. Column 5: Electron transmission: (- - -) Ref. [241]. Column 6: Calculation: (- - -) Ref. [242]. Column 7: Indirect inelastic electron scattering: (- - -) present (see text). Column 8: Possible negative ion states and their energies and assignments (see text).

figure, there exist at least five negative states above the zero level whose average energy positions are: 0.9, 2.5, 3.5, 6.2, and 8.9 eV. The 8.9 eV resonance lies in the region of electronic excitation (see Table 6.23) and most likely is associated with excited electronic states. The additional peak indicated by some experiments [230, 231, 238] near 0.25 eV is questionable; its existence is indicated by two electron swarm [230, 231] and one total electron attachment beam study [238], but it is absent in the cross section function obtained in a similar electron beam study [234] and in the cross section functions obtained in electron beam scattering experiments (see Table 6.26). It may be associated with vibrationally excited molecules, but further work is needed to clarify the situation. The calculation of Underwood-Lemons *et al.* [242] located a negative ion state at 5.1 eV, but most likely this state is associated with that at 3.5 eV since no experimental evidence exists for a resonance at this energy from any other source, and since the calculation predicts four negative ion states which can be rationalized with the findings of other studies and molecular orbital assignments (see further discussion in Sec. 3.2 of Ch. 6).

3.2 Electron Scattering from CCl_2F_2

In this section the existing information is presented and discussed for the following electron collision cross sections: total electron scattering cross section, momentum transfer cross section, elastic differential electron scattering cross section, elastic integral electron scattering cross section, and inelastic electron scattering cross section for rotational and vibrational (direct and indirect) excitation. When possible, the results of a model-based cross section set for CCl_2F_2 reported by Hayashi [245] are compared with experimental data.

3.2.1 Total Electron Scattering Cross Section, $\sigma_{sc,t}(\varepsilon)$, of CCl_2F_2

There are three sets of measurements [240, 242, 243] of the $\sigma_{sc,t}(\varepsilon)$ of CCl_2F_2 below 50 eV, one set of measurements [246] between 75 and 4000 eV, and one calculation [247] between 10 and 1000 eV. Figure 6.40 compares these data. Jones [240] quoted a most probable error of ±4.6% for energies < 4.0 eV, about +4.5% and −4.2% for energies between 4.2 and 15.0 eV, +4.9% and −4.2% for energies between 16 and 25 eV, and +7.9% and −4.6% for energies between 26 and 500 eV. Underwood-Lemons *et al.* [242] do not explicitly quote the total uncertainty of their data, but indicate that two principal sources of error in determining the magnitude of the cross section in their experiments are the length of the electron trajectory through the target and the target number density. According to Underwood-Lemons *et al.* the former "introduces an uncertainty in the cross section of as large as ±43% at 0.2 eV, declining

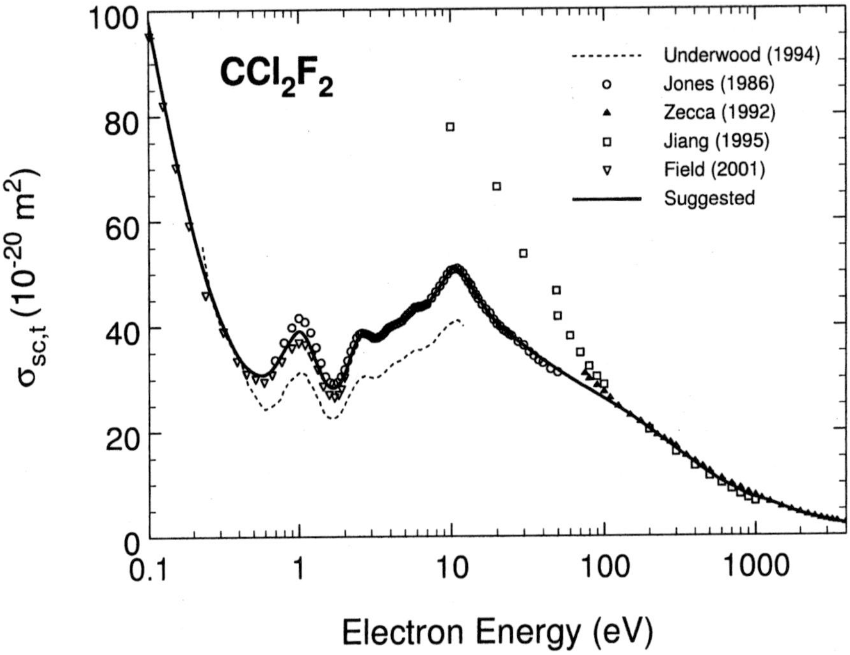

Figure 6.40. Total electron scattering cross section, $\sigma_{sc,t}(\varepsilon)$, for CCl_2F_2. Experimental: (- - -) Ref. [242]; (○) Ref. [240]; (▽) Ref. [243]; (▲) Ref. [246]. Calculated: (□) Ref. [247]. (—) Suggested.

to ±17% at 0.5 eV and to less than ±10% above 1 eV," and the latter "implies an uncertainty of about ±11%." Field *et al.* [243] estimate their uncertainty to be ±5% and indicate that their measurements (Fig. 6.40) lie systematically ≤ 10% lower than those of Jones [240] in the resonance region near 1 eV "probably due to their poorer ability to detect forward scattering associated with vibrationally inelastic collisions in this energy region." The shapes of the three low-energy experimental sets of data [240, 242, 243] are nearly identical. The magnitudes of the data of Jones [240] and Underwood-Lemons *et al.* [242] differ by about 25%. The higher energy data of Zecca *et al.* [246] have estimated systematic errors of less than ±3%. The calculation by Jiang *et al.* [247] employed the additivity rule and the complex optical potential, and is not expected to give accurate results at low energies. Clearly, the calculated cross sections are much larger than the experimental data below about 80 eV, but are in reasonable agreement with the measurements of Zecca *et al.* [246] above this energy.

Table 6.27. Suggested $\sigma_{sc,t}(\varepsilon)$ for CCl_2F_2.

ε (eV)	$\sigma_{sc,t}(\varepsilon)$ (10^{-20} m^2)	ε (eV)	$\sigma_{sc,t}(\varepsilon)$ (10^{-20} m^2)	ε (eV)	$\sigma_{sc,t}(\varepsilon)$ (10^{-20} m^2)
0.10	98.1	5.0	41.6	100	26.2
0.20	57.6	6.0	43.4	150	23.0
0.30	41.5	7.0	44.1	200	20.4
0.35	37.2	8.0	46.3	250	18.3
0.40	34.1	9.0	48.5	300	16.6
0.45	32.3	10.0	50.5	350	15.1
0.50	31.3	12.5	49.0	400	13.9
0.60	30.9	15.0	45.3	450	12.8
0.70	32.6	20.0	40.5	500	11.9
0.80	35.4	25.0	38.0	600	10.4
0.90	37.7	30.0	36.2	700	9.4
1.0	39.0	35.0	34.7	800	8.5
1.5	29.4	40.0	33.5	900	7.7
2.0	31.0	45.0	32.5	1000	7.3
2.5	38.1	50.0	31.6	2000	4.3
3.0	38.0	60.0	30.1	3000	3.0
3.5	38.2	70.0	28.9	4000	2.2
4.0	39.9	80.0	27.9		
4.5	40.6	90.0	27.0		

The experimental data of Jones [240], Zecca *et al.* [246], and Field *et al.* [243] were fitted using a least squares calculation, and the result is shown in Fig. 6.40 by the solid line. The values of Underwood-Lemons *et al.* [242] have not been used because of their higher stated uncertainties, and because their magnitudes are generally lower than the other two measurements over the same energy range. Values for the fitted cross section curve are listed in Table 6.27 as our presently suggested values of the $\sigma_{sc,t}(\varepsilon)$ of CCl_2F_2.

3.2.2 Momentum Transfer Cross Section, $\sigma_m(\varepsilon)$, of CCl_2F_2

There are no measurements of the momentum transfer cross section, $\sigma_m(\varepsilon)$, of CCl_2F_2. There have only been two calculations of $\sigma_m(\varepsilon)$ using the two-term approximation to the Boltzmann equation and various electron swarm and electron beam data [245, 248]. The results of these calculations are presented in Fig. 6.41 and differ significantly. Clearly an experimental measurement of $\sigma_m(\varepsilon)$ is needed.

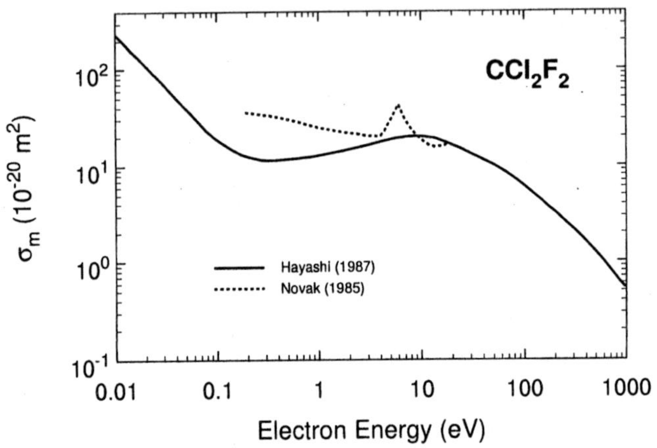

Figure 6.41. Calculated values of the momentum transfer cross section, $\sigma_m(\varepsilon)$, for CCl_2F_2: (—) Ref. [245]; (- - -) Ref. [248].

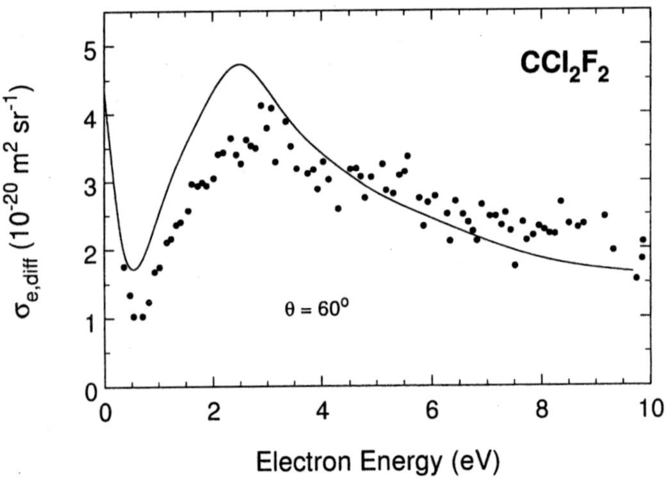

Figure 6.42. Elastic differential electron scattering cross section, $\sigma_{e,diff}(\varepsilon)$, of CCl_2F_2: (—) data of Rohr [249] as quoted in Mann and Linder [223]; (•) Mann and Linder [223].

3.2.3 Elastic Differential Electron Scattering Cross Section, $\sigma_{e,diff}(\varepsilon)$, of CCl_2F_2

Mann and Linder [223] and Rohr [249] measured cross sections for vibrationally elastic electron-CCl_2F_2 scattering. Figure 6.42 compares their results for a 60° scattering angle. The pronounced minimum at

Table 6.28. Differential cross sections, $\sigma_{e,diff}(\varepsilon)$, for elastic electron scattering from CCl_2F_2 in units of 10^{-20} m^2 sr^{-1} from Mann and Linder [223].

	Scattering angle								
ε (eV)	20°	30°	40°	50°	60°	70°	80°	90°	100°
0.5	3.98	3.07	2.20	1.67	1.21	0.80	0.79	0.80	0.73
1.0	3.38	2.27	1.80	1.67	1.69	1.79	1.83	1.76	1.65
1.5	2.62	2.34	2.39	2.74	2.58	2.40	2.30	1.71	1.61
2.0	3.48	2.92	3.29	3.67	3.18	2.69	2.48	1.73	1.60
2.5	5.30	3.61	4.21	4.33	3.55	2.73	2.44	1.73	1.67
3.0	7.39	4.25	4.83	4.71	3.68	2.59	2.32	1.69	1.74
3.5	9.57	4.73	5.14	4.97	3.55	2.42	2.19	1.66	1.80
4.0	11.38	5.11	5.34	5.20	3.30	2.29	2.12	1.66	1.87
4.5	12.23	5.52	5.57	5.35	3.11	2.22	2.11	1.66	1.89
5.0	13.13	5.99	5.78	5.28	2.99	2.18	2.12	1.69	1.89
5.5	13.52	6.57	5.83	5.03	2.86	2.12	2.16	1.73	1.91
6.0	13.32	7.37	5.74	4.76	2.66	2.03	2.24	1.76	1.96
6.5	13.93	8.25	5.79	4.53	2.49	1.93	2.35	1.78	2.04
7.0	15.46	8.57	5.86	4.35	2.38	1.86	2.47	1.80	2.10
7.5	16.83	8.43	5.74	4.14	2.30	1.81	2.52	1.81	2.13
8.0	18.26	8.14	5.55	3.81	2.23	1.81	2.48	1.77	2.13
8.5	20.23	8.02	5.44	3.48	2.14	1.82	2.46	1.71	2.15
9.0	22.31	8.10	5.32	3.17	1.98	1.81	2.41	1.65	2.24
9.5	24.56	8.07	5.23	3.01	1.83	1.75	2.35	1.59	2.49

$\sim$ 0.5 eV has been interpreted [223] as a Ramsauer-Townsend minimum in the elastic channel. Randell *et al.* [250], however, give the position of the Ramsauer-Townsend minimum to be in the range of 0.04–0.06 eV (see later in this section).

The differential cross sections of Mann and Linder [223] for elastic electron-CCl_2F_2 scattering are listed in Table 6.28. The measurements show that the cross section for elastic electron scattering in the forward direction increases with increasing electron energy.

3.2.4 Elastic Integral Electron Scattering Cross Section, $\sigma_{e,int}(\varepsilon)$, of CCl_2F_2

Mann and Linder [223] obtained the elastic integral electron scattering cross section, $\sigma_{e,int}(\varepsilon)$, by extrapolating their elastic differential cross sections (Table 6.28), weighted by $\sin\theta$, towards 0° and 180°. Their $\sigma_{e,int}(\varepsilon)$, taken from Fig. 3 of their article [223], is shown in Fig. 6.43.

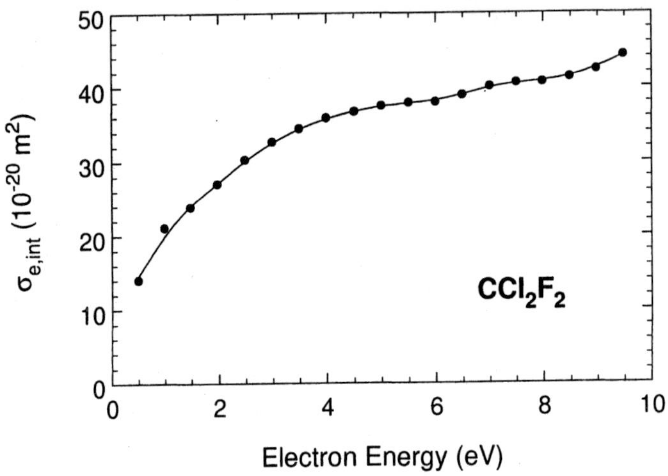

Figure 6.43. Elastic integral electron scattering cross section, $\sigma_{e,int}(\varepsilon)$, for CCl_2F_2 from Mann and Linder [223].

Table 6.29. Elastic integral electron scattering cross section, $\sigma_{e,int}(\varepsilon)$, for CCl_2F_2 derived from a fit to the data of Mann and Linder [223].

ε (eV)	$\sigma_{e,int}(\varepsilon)$ (10^{-20} m^2)	ε (eV)	$\sigma_{e,int}(\varepsilon)$ (10^{-20} m^2)	ε (eV)	$\sigma_{e,int}(\varepsilon)$ (10^{-20} m^2)
0.5	14.7	2.0	27.2	6.0	38.2
0.6	15.8	2.5	30.2	7.0	39.9
0.7	16.9	3.0	32.6	8.0	40.8
0.8	18.0	3.5	34.4	9.0	42.6
0.9	19.0	4.0	35.7	9.5	44.2
1.0	20.0	4.5	36.7		
1.5	24.1	5.0	37.4		

Numerical values from a fit to their data are listed in Table 6.29. Interestingly, $\sigma_{e,int}(\varepsilon)$ is structureless in this energy range although the total electron scattering cross section $\sigma_{sc,t}(\varepsilon)$ shows distinct structure due to the many negative ion resonances in this energy region (see Fig. 6.40). As noted by Mann and Linder, the structure in the total electron scattering cross section is due to inelastic electron scattering (see Sec. 3.2.5 of Ch. 6).

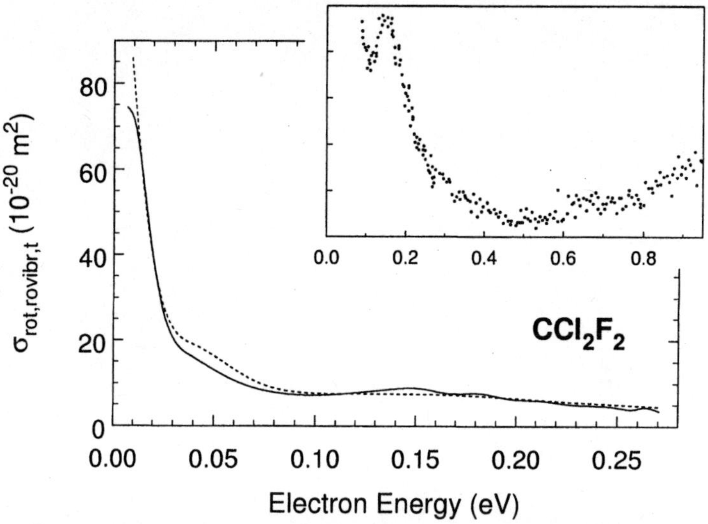

Figure 6.44. Cross section for pure rotational and rovibrational electron scattering, $\sigma_{rot,rovibr}(\varepsilon)$, of CCl_2F_2. The experimental data (—) were arbitrarily scaled to fit the theoretical value (- - -) at 10 meV (Randell *et al.* [250]). The insert shows the scattering cross section (in arbitrary units) up to 0.95 eV (from Randell *et al.* [250], see this reference for details on the determination of these cross sections).

3.2.5 Inelastic Electron Scattering Cross Section, $\sigma_{inel}(\varepsilon)$, of CCl_2F_2

The CCl_2F_2 molecule shows significant vibrational excitation by electron impact at low energies. Furthermore, due to its permanent electric dipole moment, it shows considerable rotational excitation. Both, rotational and vibrational excitation are enhanced by indirect electron scattering through negative ion resonances [250].

Rotational Excitation of CCl_2F_2. The work of Randell *et al.* [250] on very-low-energy electron scattering by CCl_2F_2 has shown pure rotational excitation and rovibrational excitation of this molecule. Figure 6.44 shows the cross section for pure rotational and rovibrational electron scattering they deduced from their "backward scattering cross section" measurements and the adiabatic-rotation approximation under a number of assumptions [250]. The rapid rise below 0.1 eV has been attributed by Randell *et al.* [250] to pure rotational scattering, and the cross section between 0.1 and 0.2 eV to threshold rovibrational excitation. The rovibrational scattering is weaker than the pure rotational scattering. Their analysis indicated the existence of a weak Ramsauer-Townsend (R-T)

Table 6.30. Total direct vibrational excitation cross section, $\sigma_{\mathrm{vib,dir,t}}(\varepsilon)$, for CCl_2F_2 from Mann and Linder [223].

ε (eV)	$\sigma_{\mathrm{vib,dir,t}}(\varepsilon)$ (10^{-20} m^2)	ε (eV)	$\sigma_{\mathrm{vib,dir,t}}(\varepsilon)$ (10^{-20} m^2)	ε (eV)	$\sigma_{\mathrm{vib,dir,t}}(\varepsilon)$ (10^{-20} m^2)
0.030	0.20	0.25	11.1	2.5	2.73
0.035	0.25	0.30	10.6	3.0	2.38
0.040	0.29	0.35	10.0	3.5	2.08
0.045	0.34	0.40	9.48	4.0	1.84
0.050	0.38	0.45	8.91	4.5	1.65
0.060	0.43	0.5	8.39	5.0	1.48
0.070	0.48	0.6	7.53	6.0	1.27
0.080	0.52	0.7	6.83	7.0	1.09
0.090	0.61	0.8	6.29	8.0	0.94
0.100	0.88	0.9	5.78	9.0	0.85
0.125	3.92	1.0	5.39	10.0	0.80
0.150	7.59	1.5	4.05		
0.20	10.9	2.0	3.23		

minimum at about 40–60 meV, that is, at a much lower energy than the 500 meV minimum (see Fig. 6.42) indicated by the data of Mann and Linder [223].

Vibrational Excitation of CCl_2F_2. The CCl_2F_2 molecule has nine nondegenerate vibrational modes of which five (ν_1, ν_2, ν_4, ν_6, and ν_8) are infrared (IR) active (see Table 6.25). The Born dipole approximation for the total vibrational excitation cross section (the sum of the cross sections for ν_1, ν_2, ν_4, ν_6, and ν_8) has been determined by Mann and Linder [223] and is shown in Fig. 6.45(a). This sum (Table 6.30) represents the total direct vibrational excitation cross section, $\sigma_{\mathrm{vib,dir,t}}(\varepsilon)$, for this molecule.

Hayashi [245] obtained vibrational cross sections designated by him as $\sigma_{\nu 3}$ and $\sigma_{\nu 81}$ from a Boltzmann code analysis. The sum of these cross sections is compared in Fig. 6.45(b) with the $\sigma_{\mathrm{vib,dir,t}}(\varepsilon)$ [Fig. 6.45(a)] and $\sigma_{\mathrm{sc,t}}(\varepsilon) - \sigma_{\mathrm{e,int}}(\varepsilon)$. The sum $\sigma_{\nu 3} + \sigma_{\nu 81}$ of the Hayashi cross sections should contain both direct and indirect vibrational excitation and, thus, it should be larger than $\sigma_{\mathrm{vib,dir,t}}(\varepsilon)$, but it should not exceed $\sigma_{\mathrm{sc,t}}(\varepsilon) - \sigma_{\mathrm{e,int}}(\varepsilon)$. There is a large difference between $\sigma_{\nu 3} + \sigma_{\nu 81}$ and $\sigma_{\mathrm{sc,t}}(\varepsilon) - \sigma_{\mathrm{e,int}}(\varepsilon)$ indicating the large uncertainty in these data.

The excitation functions for some of the vibrational modes of CCl_2F_2 have been measured by Mann and Linder [223] and are shown in Fig. 6.46.

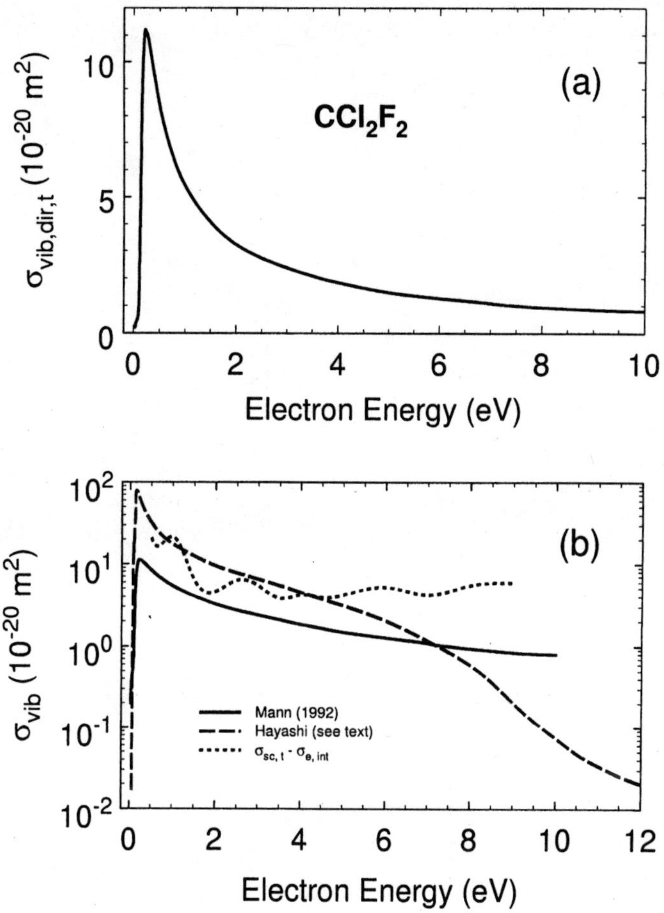

Figure 6.45. (a) Born dipole approximation for the total direct vibrational excitation cross section, $\sigma_{\mathrm{vib,dir,t}}(\varepsilon)$, for CCl_2F_2: (—) sum of ν_1, ν_2, ν_4, ν_6, and ν_8 (derived from Fig. 3 of Mann and Linder [223]). (b) Comparison of vibrational excitation cross sections: (—) $\sigma_{\mathrm{vib,dir,t}}(\varepsilon)$ from Fig. 6.45(a); (- - -) $\sigma_{\mathrm{sc,t}}(\varepsilon) - \sigma_{\mathrm{e,int}}(\varepsilon)$ [determined using the values of $\sigma_{\mathrm{sc,t}}(\varepsilon)$ and $\sigma_{\mathrm{e,int}}(\varepsilon)$ in Tables 6.27 and 6.29, respectively]; (– –) from Hayashi [245].

Below 1 eV direct electron scattering leads to strong excitation of the infrared active stretching modes. In the energy range between about 0.5 and 10 eV the predominant mode of vibrational excitation is indirect electron scattering through the decay of the negative ion resonances of CCl_2F_2 in this energy range (Table 6.26, Fig. 6.39). This has been demonstrated by Mann and Linder who measured differential cross sec-

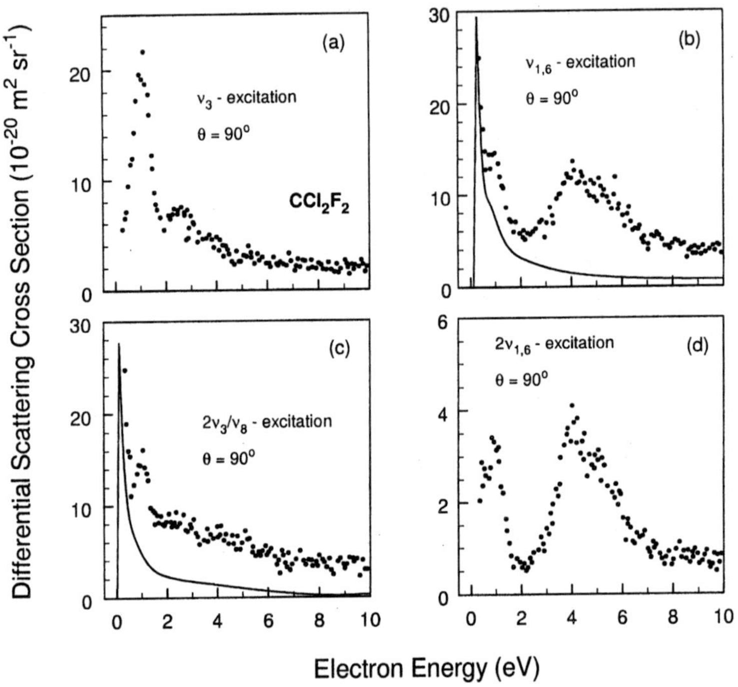

Figure 6.46. Differential electron scattering cross sections as a function of electron energy for the most important energy-loss processes in electron-CCl_2F_2 collisions below 10 eV, at a scattering angle of 90°. The solid lines are the Born dipole cross sections for excitation of $\nu_{1,6}$ and ν_8 (from Mann and Linder [223]).

tions for vibrational excitation of CCl_2F_2 in the energy range 0.5–10 eV (see below).

Indirect Excitation of CCl_2F_2 Via Negative Ion Resonances. Based on information provided by various techniques (see also discussion in Sec. 3.1 of Ch. 6 and Fig. 6.39), we can conclude that there exist at least four negative ion states of the CCl_2F_2 molecule below the electronic excitation threshold (near 7 eV) of the molecule: the lowest-lying one with a minimum located 0.4 eV (= EA) below that of the neutral and a vertical attachment energy of 0.9 eV, and three higher-lying ones at 2.5, 3.5, and 6.2 eV. The assumption that the potential energy surface of the lowest negative ion state of CCl_2F_2 has a minimum at the adiabatic value of the electron affinity (0.4 eV, Table 6.26) and rises steeply at small internuclear separations, such that it is responsible for the resonance observed in electron attachment and electron scattering experiments at 0.9 eV, is based on the calculations of Tossell [228] (see

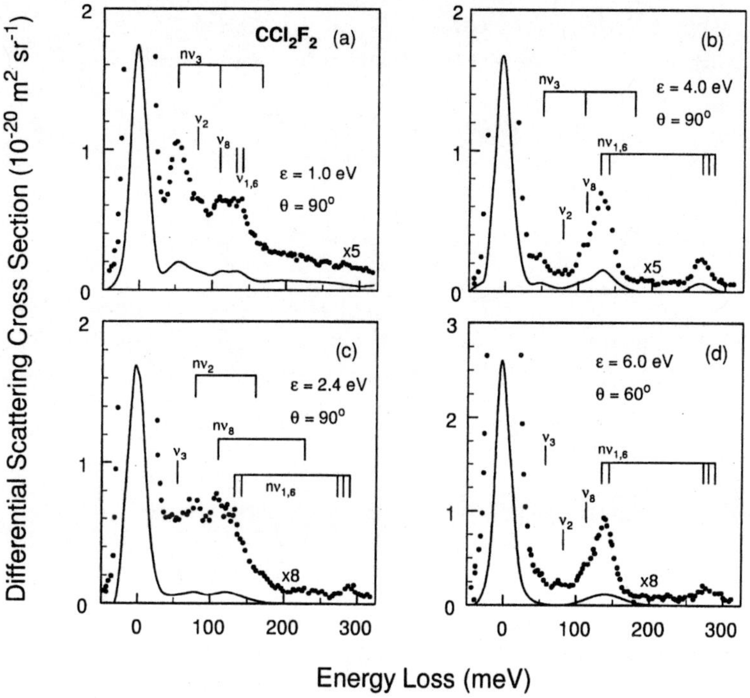

Figure 6.47. Energy-loss spectra for electron-CCl_2F_2 scattering at scattering angles of 90° and 60° and incident electron energies of 1.0, 2.4, 4.0, and 6.0 eV (from Mann and Linder [223]).

also Ref. [242]) which show profound changes between the geometries of CCl_2F_2 and $CCl_2F_2^-$ (see also Sec. 3.5 of Ch. 6 regarding the large internuclear relaxation in $CCl_2F_2^-$ compared with CCl_2F_2). It is also consistent with the *ab initio* self consistent field (SCF) calculation by Burrow *et al.* [241], according to which four valence-type resonances are expected below about 5 eV which can be ascribed, in increasing energetic order, to the unoccupied orbitals a_1 (C–Cl σ^*), b_2 (C–Cl σ^*), a_1 (C–F σ^*), and b_1 (C–F σ^*). Burrow *et al.* have ascribed the resonances they detected in an electron transmission experiment at 1.0, 2.4 and 3.9 eV to the lowest three of these molecular orbitals.

With these findings in mind, Mann and Linder [223] measured the energy-loss spectra of CCl_2F_2 for electrons having initial energies equal to 1.0, 2.4, 4.0, and 6.0 eV, i.e., roughly equal to the energy positions of the observed four negative ion resonances of CCl_2F_2. In Fig. 6.47 are shown their results. For 1.0 eV, the most prominent energy loss is assigned to the excitation of the ν_3 vibration. This is consistent with the (C–Cl σ^*) character of the A_1 (C–Cl σ^*) resonance at 1.0 eV. Also consistent with this assignment of the 1.0 eV resonance are the data on dissociative electron attachment around this energy (see Table 6.26), both for the production of Cl^- involving the C–Cl bond breaking and the production of Cl_2^- involving the C–Cl_2 dissociation. In vibrational excitation, the C–Cl_2 symmetric stretch mode ν_3 is the dominant excitation process.

Additionally, the data in Fig. 6.47 show that in the A_1 (C–F σ^*) resonance at 4.0 eV, the excitation of $\nu_{1,6}$ is the dominating process. This is consistent with the (C–F σ^*) character of this resonance. The A_1 (C–F σ^*) resonance at 4.0 eV is also the appropriate precursor of the group of fragment negative ions observed at 3.5 eV (the position of the resonance is apparently shifted downward by about 0.5 eV in the dissociative electron attachment channel in comparison to the electron scattering channel). The dominant fragment anion at this energy is F^- in accord with the (C–F σ^*) character of this resonance.

With regard to the 2.5 eV resonance, it is clear that if the resonance assignments mentioned above are adopted, the B_2 (C–Cl σ^*) resonance at 2.5 eV does not seem to decay via dissociative attachment since none of the electron attachment studies have shown a peak at this energy. Peculiarly, the resonance shows up only in the vibrational excitation channel. Its energy-loss spectrum shows excitation of several modes none of which predominates (Fig. 6.47(c)).

Shown in Fig. 6.48 is the total indirect inelastic electron scattering cross section, $\sigma_{\text{inel,indir,t}}(\varepsilon)$, which we estimated by subtracting the sum $\sigma_{\text{vib,dir,t}}(\varepsilon) + \sigma_{\text{e,int}}(\varepsilon)$, as determined from Mann and Linder [223], from our suggested values of $\sigma_{\text{sc,t}}(\varepsilon)$ (solid line, Fig. 6.40). The energy positions of the maxima in $\sigma_{\text{inel,indir,t}}(\varepsilon)$ are indicated in the figure by the vertical lines and are compared in Table 6.26 with other data. The maximum at 9.0 eV indicates the location of a negative ion state associated with the excited electronic states in this energy region.

It is thus seen (Figs. 6.46 and 6.47) that the A_1 (C–Cl σ^*) and A_1(C–F σ^*) resonances are the dominant features in the vibrational excitation functions of the CCl_2F_2 molecule. As observed by Mann and Linder [223], the dominant vibrational excitation corresponds to the respective valence character of each resonance, i.e., ν_3 for the A_1 (C–Cl σ^*) resonance and ν_1 for the A_1(C–F σ^*) resonance. The observed selective excitation of the CF_2 and CCl_2 stretching modes corresponding to the respective valence character of the resonance is in agreement with the symmetry selection rules [223]. The indirect excitation of the ν_1 and ν_3 vibrational modes is to be contrasted with the direct vibrational excitation of the infrared active modes ν_1, ν_2, ν_4, ν_6, and ν_8 at lower energies.

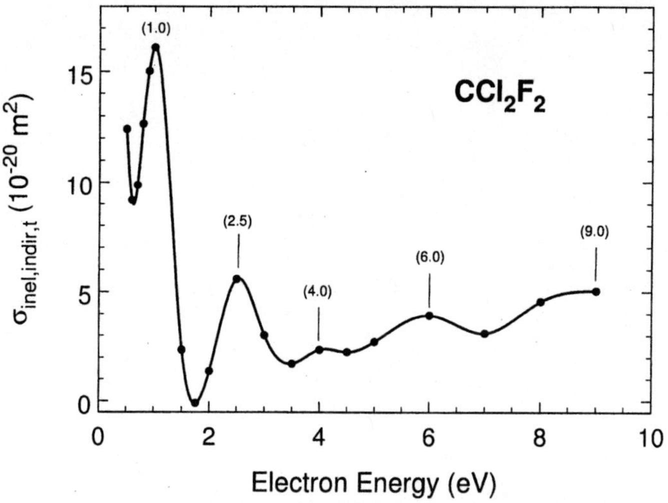

Figure 6.48. Total inelastic indirect electron scattering cross section, $\sigma_{\text{inel,indir,t}}(\varepsilon)$, for CCl_2F_2 (see text).

Electronic Excitation of CCl_2F_2. The threshold for electronic excitation of the CCl_2F_2 molecule is about 7.0 eV (Table 6.23) and the corresponding threshold for electron-impact ionization is about 12.3 eV (Table 6.24). There are no data on the cross section for electronic excitation of CCl_2F_2. An estimate of the sum $\sigma_{\text{inel,t}}(\varepsilon) + \sigma_{\text{e,int}}(\varepsilon)$ can be obtained from

$$\sigma_{\text{sc,t}}(\varepsilon) - \sigma_{\text{i,t}}(\varepsilon) - \sigma_{\text{da,t}}(\varepsilon) = \sigma_{\text{inel,t}}(\varepsilon) + \sigma_{\text{e,int}}(\varepsilon). \qquad (6.15)$$

3.3 Electron-Impact Ionization of CCl_2F_2

3.3.1 Total Ionization Cross Section, $\sigma_{\text{i,t}}(\varepsilon)$, of CCl_2F_2

There have been three experimental measurements [234, 251, 252] of the total ionization cross section, $\sigma_{\text{i,t}}(\varepsilon)$, of CCl_2F_2. The first measurements are those of Beran and Kevan [251] which were made for only three values of the incident electron energy. The second set of measurements is that of Pejčev *et al.* [234] who used a parallel plate condenser-type ionization chamber and a trochoidal electron monochromator as the electron source to cover the energy range from threshold to 250 eV. The third set of measurements is by Leiter *et al.* [252] who employed a double focusing sector field mass spectrometer. They reported partial ioniza-

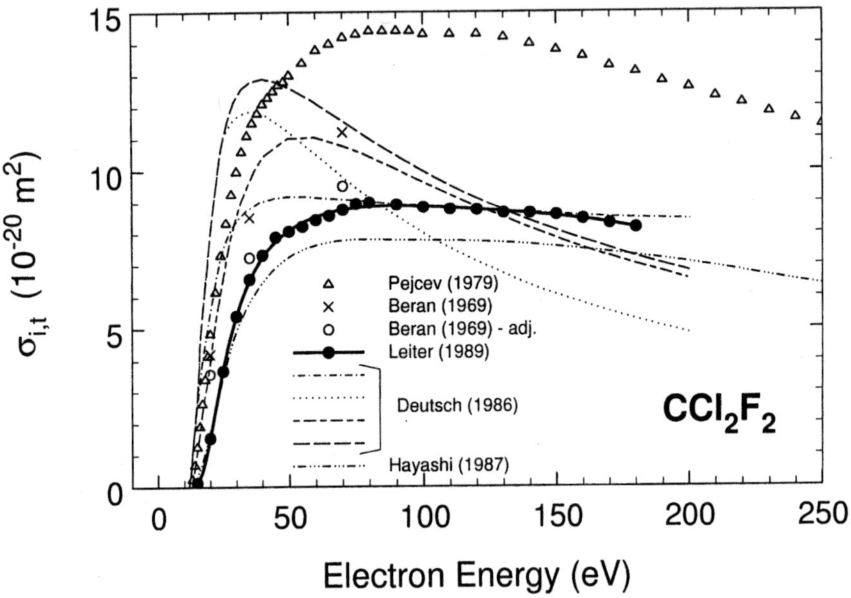

Figure 6.49. Total ionization cross section, $\sigma_{i,t}(\varepsilon)$, for CCl_2F_2. Experimental data: ($\triangle$) Ref. [234]; ($\bullet$) Ref. [252]; ($\times$) Ref. [251]; ($\circ$) Ref. [251] adjusted (see text). Calculated data as presented by Deutsch *et al.* [253]: (– –) Ref. [254]; (- – -) Ref. [255]; ($\cdots$) Ref. [256]; (– · –) Ref. [257]; (– · · · –) Ref. [245].

tion cross sections from threshold to 180 eV (see Sec. 3.3.2 of Ch. 6), and determined the absolute total ionization cross section by taking the charge-weighted sum of their partial ionization cross sections. Leiter *et al.* [252] estimated the uncertainty of their total ionization cross section measurements to be $\pm 10\%$. No uncertainty values were given by Pejčev *et al.* [234].

The results of these three groups of investigators are shown in Fig. 6.49, and there is substantial disagreement between these data. The values of Beran and Kevan [251] for a number of atomic and molecular species are consistently higher by about 15% compared to the more reliable measurements of Rapp and Englander-Golden [116]. We thus reduced the values of Beran and Kevan [251] by 15%. Their adjusted data are shown in Fig. 6.49 by the open circles and are seen to be in reasonable agreement with the measurements of Leiter *et al.* [252]. The measurements of Pejčev *et al.* [234] although similar in shape to the data of Leiter *et al.* [252] are higher by as much as a factor of 2. Although the total ionization cross section values of Leiter *et al.* may be low because their measurements have not been corrected for possible discrimination of the

Table 6.31. Total ionization cross section, $\sigma_{i,t}(\varepsilon)$, for CCl_2F_2 measured by Leiter *et al.* [252].

ε (eV)	$\sigma_{i,t}(\varepsilon)$ (10^{-20} m^2)	ε (eV)	$\sigma_{i,t}(\varepsilon)$ (10^{-20} m^2)	ε (eV)	$\sigma_{i,t}(\varepsilon)$ (10^{-20} m^2)
15	0.12	55	8.22	110	8.78
20	1.55	60	8.43	120	8.76
25	3.67	65	8.57	130	8.66
30	5.40	70	8.76	140	8.65
35	6.56	75	8.95	150	8.59
40	7.32	80	8.99	160	8.46
45	7.88	90	8.92	170	8.31
50	8.07	100	8.84	180	8.17

energetic fragment ions [258], they are in better agreement with the adjusted data of Beran and Kevan [251]. It should be noted also that since the predominant positive ion is $CClF_2^+$, the effects of ion discrimination on the value of the total ionization cross section may be small.

Also shown in Fig. 6.49 are the results of a number of calculations by Deutsch *et al.* [253] using empirically modified collision theories [254–257]. Interestingly, the calculations seem to be more consistent with the data of Pejčev *et al.* [234] below about 50 eV and with the data of Leiter *et al.* [252] at the higher energies. The total ionization cross section deduced by Hayashi [245] lies below all data, experimental and calculated. Clearly more measurements are indicated. At the present time we tentatively suggest the data of Leiter *et al.* which are listed in Table 6.31.

3.3.2 Partial Ionization Cross Sections, $\sigma_{i,part}(\varepsilon)$, of CCl_2F_2

Leiter *et al.* [252] have measured partial ionization cross sections, $\sigma_{i,part}(\varepsilon)$, for the production of the following singly ionized species by electron impact on CCl_2F_2: $CCl_2F_2^+$, CCl_2F^+, $CClF_2^+$, CCl_2^+, Cl_2^+, $CClF^+$, ClF^+, CF_2^+, CCl^+, Cl^+, CF^+, F^+, and C^+. They reported an uncertainty in these values of about ±10%.

These measurements are shown in Fig. 6.50 and are listed in Table 6.32. By far the largest cross section over the entire energy range covered is that for the $CClF_2^+$ positive ion. The cross section for this ion is nearly five times larger than the next most abundant ion, and is about 400 times larger than that for the parent positive ion $CCl_2F_2^+$. This clearly

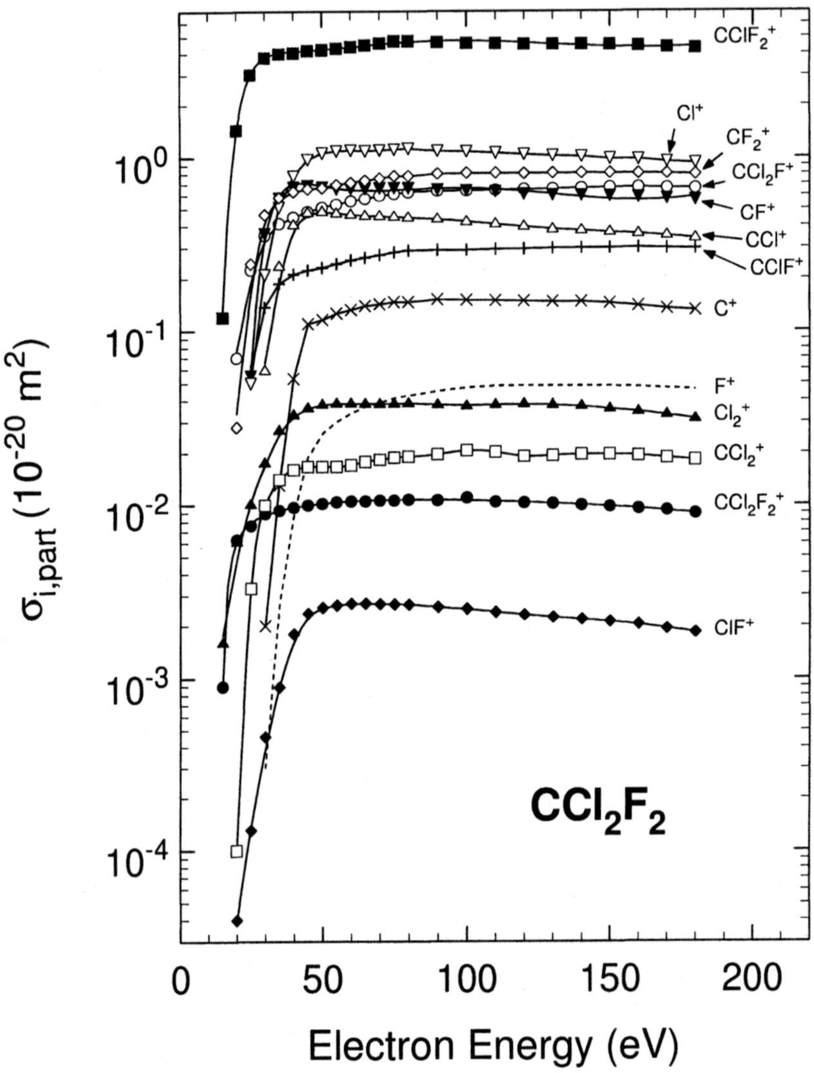

Figure 6.50. Partial ionization cross sections, $\sigma_{\mathrm{i,part}}(\varepsilon)$, for CCl_2F_2 (data of Leiter *et al.* [252]).

shows that the preponderance of ionizing collisions are dissociative for CCl_2F_2. The multiplicity of dissociation channels leads to a multiplicity of neutral and charged particles, and demonstrates the extreme fragility of this molecule toward low energy electrons. The energy dependencies of the partial ionization cross sections are rather similar, except when a number of different processes contribute to the formation of a particular

Table 6.32. Partial electron impact ionization cross sections, $\sigma_{i,part}(\varepsilon)$, in units of 10^{-20} m^2, for the production of singly ionized species from CCl_2F_2 (data of Leiter *et al.* [252]).

	$\sigma_{i,part}(\varepsilon)$					
ε (eV)	$CCl_2F_2^+$	CCl_2F^+	$CClF_2^+$	CCl_2^+	Cl_2^+	$CClF^+$
15	0.0009		0.12		0.0016	
20	0.0063	0.070	1.44	0.0001	0.0061	
25	0.0076	0.225	3.02	0.0033	0.0101	0.059
30	0.0089	0.352	3.78	0.0100	0.0175	0.137
35	0.0093	0.411	3.99	0.0140	0.0268	0.187
40	0.0097	0.447	4.05	0.0160	0.0329	0.212
45	0.0100	0.482	4.17	0.0167	0.0360	0.223
50	0.0102	0.510	4.21	0.0167	0.0378	0.231
55	0.0104	0.531	4.28	0.0167	0.0383	0.242
60	0.0105	0.556	4.39	0.0170	0.0383	0.254
65	0.0105	0.573	4.48	0.0178	0.0381	0.263
70	0.0106	0.598	4.60	0.0183	0.0380	0.272
75	0.0106	0.612	4.72	0.0188	0.0381	0.281
80	0.0107	0.626	4.72	0.0190	0.0383	0.288
90	0.0106	0.637	4.66	0.0196	0.0376	0.290
100	0.0110	0.640	4.60	0.0206	0.0370	0.290
110	0.0104	0.644	4.58	0.0202	0.0376	0.293
120	0.0103	0.651	4.54	0.0191	0.0380	0.294
130	0.0102	0.654	4.54	0.0193	0.0375	0.297
140	0.0100	0.661	4.54	0.0196	0.0370	0.298
150	0.0098	0.668	4.52	0.0196	0.0356	0.298
160	0.0095	0.668	4.44	0.0193	0.0343	0.299
170	0.0093	0.661	4.35	0.0187	0.0329	0.297
180	0.0089	0.658	4.28	0.0182	0.0311	0.295

positive ion which have different energetic onsets (e.g., Cl_2^+, CF_2^+, CCl_2^+, F^+, Cl^+). Leiter *et al.* reported no ionization threshold energies for these species. We have used their cross section data in Table 6.32 to estimate the energy thresholds for a number of singly charged positive ions which are listed in Table 6.33. Also in Table 6.33 are given the threshold energies for the production of singly positive ions via a number of photoprocesses. The electron impact values of the threshold energies for the various positive ions exceed the corresponding values obtained using photoionization methods, indicating differences in the adiabatic and vertical values of the energy thresholds.

Table 6.32 (continued)
Partial electron impact ionization cross sections, $\sigma_{i,part}(\varepsilon)$, in units of 10^{-20} m^2, for the production of singly ionized species from CCl_2F_2 (data of Leiter *et al.* [252]).

	$\sigma_{i,part}(\varepsilon)$						
ε (eV)	ClF^+	CF_2^+	CCl^+	Cl^+	CF^+	F^+	C^+
20	0.00004	0.028					
25	0.00013	0.243		0.05	0.055		
30	0.00046	0.462	0.059	0.21	0.360	0.0003	0.002
35	0.00089	0.583	0.235	0.51	0.579	0.0027	0.013
40	0.00179	0.632	0.403	0.78	0.671	0.0087	0.053
45	0.00236	0.656	0.475	0.97	0.686	0.0189	0.110
50	0.00255	0.664	0.481	1.06	0.671	0.0257	0.116
55	0.00263	0.693	0.470	1.09	0.655	0.0301	0.126
60	0.00269	0.713	0.463	1.10	0.658	0.0336	0.132
65	0.00269	0.733	0.455	1.09	0.658	0.0366	0.139
70	0.00268	0.757	0.453	1.10	0.661	0.0388	0.142
75	0.00267	0.778	0.449	1.11	0.664	0.0410	0.146
80	0.00265	0.780	0.444	1.12	0.661	0.0424	0.146
90	0.00256	0.802	0.436	1.09	0.649	0.0451	0.151
100	0.00250	0.810	0.420	1.08	0.640	0.0470	0.150
110	0.00240	0.810	0.408	1.06	0.631	0.0481	0.149
120	0.00231	0.806	0.394	1.04	0.610	0.0484	0.148
130	0.00224	0.810	0.380	1.02	0.603	0.0481	0.147
140	0.00218	0.810	0.375	1.01	0.600	0.0481	0.147
150	0.00212	0.810	0.366	0.98	0.594	0.0481	0.144
160	0.00204	0.806	0.358	0.97	0.585	0.0475	0.139
170	0.00192	0.802	0.349	0.94	0.573	0.0467	0.134
180	0.00183	0.794	0.337	0.92	0.564	0.0459	0.131

3.3.3 Multiple (Double) Ionization Cross Sections, $\sigma_{i,mult}(\varepsilon)$, of CCl_2F_2

Cross sections for doubly charged positive ions produced by electron impact on CCl_2F_2 have been reported by Leiter *et al.* [252]. These are listed in Table 6.34 and are plotted in Fig. 4.51. They have a reported uncertainty of ±20%. The cross sections for doubly ionized species are generally much smaller than the cross sections for singly ionized species. From the data of Leiter *et al.* we also estimated the energy thresholds for doubly charged ions listed in Table 6.33 by a linear extrapolation of their cross sections to zero.

Table 6.33. Threshold energies, in eV, for the production of positive ions by electron impact on CCl_2F_2.[a]

Positive ion	Threshold energy (electron impact)[b] (eV)	Threshold energy (photophysical data for the indicated reaction or positive ion) (eV)	Ref.
$CCl_2F_2^+$	13.9	11.8[d]	[210]
	~12.3[c]	11.75	[215]
		11.75	[259]
		11.87	[260]
		12.24 ± 0.01	[261]
		12.26	[262]
		12.31 ± 0.05	[263]
CCl_2F^+	17.9	14.0 ± 1	[210]
		14.15	[215]
		13.81	[259]
		13.30 ± 0.05	[264]
$CClF_2^+$	15.1	11.5 ± 1	[210]
		12.10	[215]
		11.99	[259]
		11.96 ± 0.3	[264]
CCl_2^+	21.7	46 ± 1	[210]
CF_2^+	19.1	$CF_2^+ + Cl_2$	
		14.6[d]	[210]
		14.90 ± 0.3	[264]
		$CF_2^+ + 2Cl$	
		17.1[d]	[210]
		17.5 ± 1	[210]
		17.22	[215]
		16.98	[259]
$CClF^+$	21.7	18.5 ± 1	[210]
		17.76	[259]
		18.60 ± 0.05	[264]
Cl_2^+	12.2	$Cl_2^+ + CF_2$	
		15.40 ± 0.1	[264]

[a]Photophysical data are also listed for comparison where available.
[b]Present estimates based on the data of Leiter *et al.* [252]. These estimates are very approximate and are intended to be used for guiding purposes only.
[c]From Table 6.24.
[d]Calculated by Zhang *et al.* [210] using thermochemical data under the assumption of zero kinetic energy of fragmentation.

Table 6.33 (continued)
Threshold energies, in eV, for the production of positive ions by electron impact on CCl_2F_2.[a]

Positive ion	Threshold energy (electron impact)[b] (eV)	Threshold energy (photophysical data for the indicated reaction or positive ion) (eV)	Ref.
ClF^+	18.3		
CCl^+	27.3	24 ± 1	[210]
		21.60 ± 0.1	[264]
CF^+	23.0	$CF^+ + F + Cl_2$	
		17.4^d	[210]
		17.65	[215]
		17.35 ± 0.05	[264]
		$CF^+ + F + 2Cl$	
		19.9^d	[210]
		20 ± 1	[210]
		20.20	[215]
		19.84 ± 0.05	[264]
		$CF^+ + FCl + Cl$	
		17.3^d	[210]
Cl^+	20.0	$Cl^+ + CF_2 + Cl$	
		18.5^d	[210]
		18.76 ± 0.05	[264]
		$Cl^+ + CF + FCl$	
		21.2^d	[210]
		20 ± 1	[210]
		$Cl^+ + CF + F + Cl$	
		23.8^d	[210]
		$Cl^+ + C + FCl + F$	
		26.8^d	[210]
		$Cl^+ + C + F_2 + Cl$	
		27.7^d	[210]
		$Cl^+ + C + 2F + Cl$	
		29.3^d	[210]

[a]Photophysical data are also listed for comparison where available.
[b]Present estimates based on the data of Leiter *et al.* [252]. These estimates are very approximate and are intended to be used for guiding purposes only.
[c]From Table 6.24.
[d]Calculated by Zhang *et al.* [210] using thermochemical data under the assumption of zero kinetic energy of fragmentation.

Table 6.33 (continued)
Threshold energies, in eV, for the production of positive ions by electron impact on CCl_2F_2.[a]

Positive ion	Threshold energy (electron impact)[b] (eV)	Threshold energy (photophysical data for the indicated reaction or positive ion) (eV)	Ref.
F^+	30.0	$F^+ + CF + Cl_2$	
		25.7[d]	[210]
		$F^+ + CF + 2Cl$	
		28.2[d]	[210]
		$F^+ + C + FCl + Cl$	
		31.2[d]	[210]
		$F^+ + C + F + Cl_2$	
		31.3[d]	[210]
		$F^+ + C + F + 2Cl$	
		33.8[d]	[210]
		36 ± 1	[210]
C^+	31.8	$C^+ + 2FCl$	
		20.5[d]	[210]
		$C^+ + F_2 + Cl_2$	
		23.5[d]	[210]
		$C^+ + FCl + F + Cl$	
		25.1[d]	[210]
		$C^+ + 2F + Cl_2$	
		25.1[d]	[210]
		$C^+ + F_2 + 2Cl$	
		26.0[d]	[210]
		$C^+ + 2F + 2Cl$	
		27.6[d]	[210]
		31 ± 1	[210]
Cl^{++}	52.2		
CCl^{++}	48.7		
CCl_2^{++}	40.0		
$CClF^{++}$	40.0		
$CClF_2^{++}$	37.9	38 ± 1	[210]
CCl_2F^{++}	33.9		

[a]Photophysical data are also listed for comparison where available.
[b]Present estimates based on the data of Leiter *et al.* [252]. These estimates are very approximate and are intended to be used for guiding purposes only.
[c]From Table 6.24.
[d]Calculated by Zhang *et al.* [210] using thermochemical data under the assumption of zero kinetic energy of fragmentation.

Table 6.34. Multiple (double) ionization cross sections, $\sigma_{i,mult}(\varepsilon)$, in units of 10^{-22} m^2, for the production of Cl^{++}, CCl^{++}, CCl_2^{++}, $CClF^{++}$, $CClF_2^{++}$, and CCl_2F^{++} in electron collisions with CCl_2F_2 (data of Leiter *et al.* [252]).

	$\sigma_{i,mult}(\varepsilon)$					
ε (eV)	Cl^{++}	CCl^{++}	CCl_2^{++}	$CClF^{++}$	$CClF_2^{++}$	CCl_2F^{++}
40			0.06	0.004	0.09	0.34
45			0.45	0.052	0.39	0.99
50			1.07	0.104	0.71	1.50
55	0.006	0.15	1.52	0.221	1.00	1.88
60	0.019	0.31	1.88	0.274	1.23	2.10
65	0.048	0.54	2.08	0.321	1.39	2.29
70	0.091	0.66	2.25	0.358	1.51	2.44
75	0.134	0.74	2.37	0.388	1.60	2.58
80	0.164	0.81	2.46	0.408	1.67	2.66
90	0.215	0.91	2.61	0.443	1.76	2.82
100	0.250	0.99	2.70	0.470	1.80	2.90
110	0.272	1.02	2.78	0.487	1.82	2.93
120	0.287	1.03	2.81	0.497	1.80	2.92
130	0.294	1.03	2.83	0.499	1.78	2.90
140	0.296	1.04	2.79	0.497	1.76	2.88
150	0.293	1.03	2.77	0.492	1.72	2.85
160	0.287	1.02	2.74	0.485	1.69	2.78
170	0.279	0.99	2.68	0.473	1.65	2.71
180	0.272	0.98	2.62	0.460	1.60	2.64

3.3.4 Ionization Coefficients of CCl_2F_2

Density-Normalized Ionization Coefficient, $\alpha/N\,(E/N)$, of CCl_2F_2. There have been a number of measurements of the density-normalized ionization coefficient, $\alpha/N\,(E/N)$, of CCl_2F_2 which are compared in Fig. 6.52. The agreement is generally good except at low E/N. These measurements have generally been made at temperatures between 293 and 295 K and have estimated uncertainties ranging from ±3% [267] to ±5% [266], to ±10% [268–270]. The solid line in Fig. 6.52 is a least squares fit to the six sets of experimental data. Values from this fit are listed in Table 6.35 as our recommended data for the $\alpha/N\,(E/N)$ of CCl_2F_2.

Measurements have also been reported of the density-normalized ionization coefficient $\alpha/N\,(E/N)$ of binary mixtures of CCl_2F_2 with a number of gases. Information on these has been published for the follow-

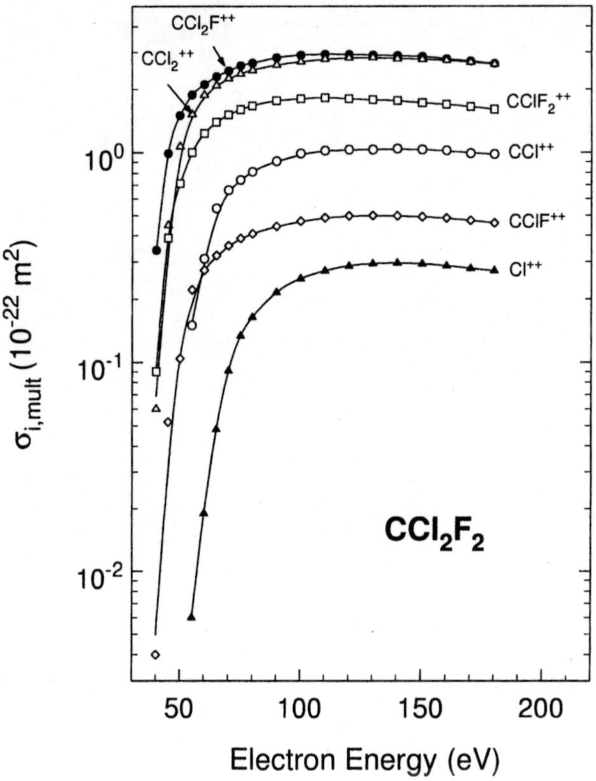

Figure 6.51. Multiple (double) ionization cross sections, $\sigma_{i,mult}(\varepsilon)$, for CCl_2F_2 measured by Leiter *et al.* [252].

ing gas mixtures: CCl_2F_2/N_2 [270–273]; CCl_2F_2/air [267]; CCl_2F_2/SF_6 [274, 275]; and CCl_2F_2/CO_2 [276, 277].

Average Energy to Produce an Electron-Ion Pair for CCl_2F_2. The average energy to produce an electron-ion pair, W, by high energy α particles (initial energy $\sim$ 5.1 MeV) slowed down in CCl_2F_2 is 29.5 eV [278].

3.4 Electron-Impact Dissociation of CCl_2F_2 into Neutral Fragments

There have been no measurements of the total dissociation cross section for formation of neutral species, $\sigma_{dis,neut,t}(\varepsilon)$, in collisions of low-energy electrons with this molecule. Based on the values of $\sigma_{dis,neut,t}(\varepsilon)$

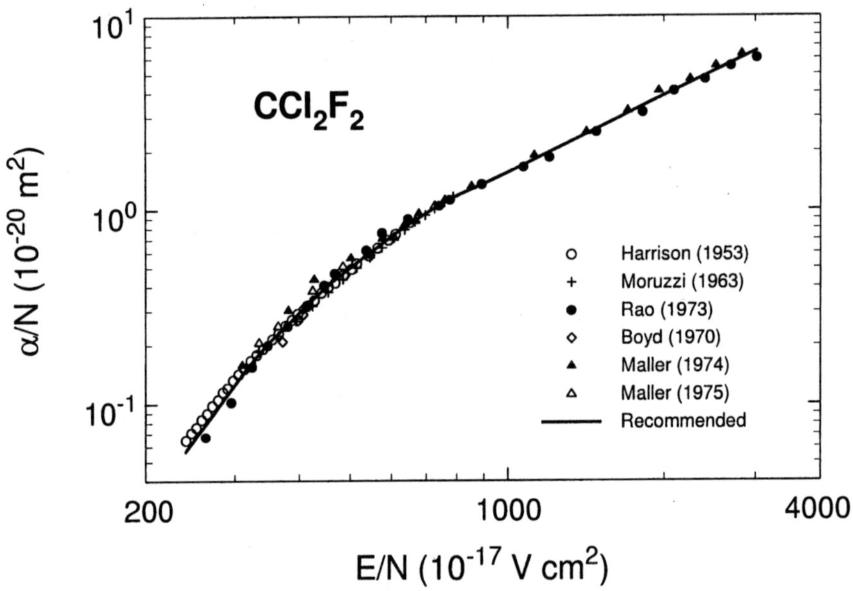

Figure 6.52. Density-normalized ionization coefficient, $\alpha/N\,(E/N)$, for CCl_2F_2: (○) Ref. [265]; (+) Ref. [266]; (•) Ref. [267]; (◇) Ref. [268]; (▲) Ref. [269]; (△) Ref. [270]; (—) recommended.

Table 6.35. Recommended values for the density-normalized ionization coefficient, $\alpha/N\,(E/N)$, of CCl_2F_2.

E/N (10^{-17} V cm^2)	α/N (10^{-20} m^2)	E/N (10^{-17} V cm^2)	α/N (10^{-20} m^2)
250	0.066	800	1.18
300	0.13	850	1.27
350	0.21	900	1.36
400	0.29	950	1.46
450	0.40	1000	1.56
500	0.50	1250	2.08
550	0.61	1500	2.65
600	0.73	2000	3.85
650	0.84	2500	5.15
700	0.96	3000	6.51
750	1.07		

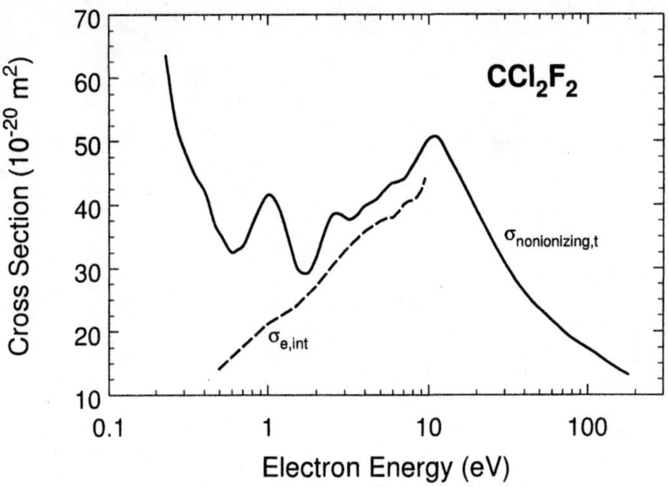

Figure 6.53. Nonionizing part, $\sigma_{\text{nonionizing,t}}(\varepsilon)$, of the total electron scattering cross section for CCl_2F_2.

for CF_4 (see Ch. 4) and CHF_3 (see Ch. 5), the $\sigma_{\text{dis,neut,t}}(\varepsilon)$ for CCl_2F_2 is expected to be much smaller than the $\sigma_{\text{sc,t}}(\varepsilon)$ of CCl_2F_2. In principle, this cross section could be determined from

$$\sigma_{\text{dis,neut,t}}(\varepsilon) = \sigma_{\text{sc,t}}(\varepsilon) - [\sigma_{\text{e,int}}(\varepsilon) + \sigma_{\text{i,t}}(\varepsilon) + \sigma_{\text{vib,dir,t}}(\varepsilon) + \sigma_{\text{vib,indir,t}}(\varepsilon) + \sigma_{\text{a,t}}(\varepsilon) + \sigma_{\text{electronic,t}}(\varepsilon)], \tag{6.16}$$

if the rest of the cross sections of Eq. (6.16) were known. In Eq. (6.16), $\sigma_{\text{electronic,t}}(\varepsilon)$ represents the total cross section for electronic excitation not leading to dissociation or ionization and is expected to be small. In the present state of the available measurements, the only quantity one can derive with confidence from the above equation is the difference $\sigma_{\text{sc,t}}(\varepsilon) - \sigma_{\text{i,t}}(\varepsilon)$ which represents the nonionizing part, $\sigma_{\text{nonionizing,t}}(\varepsilon)$, of the total electron scattering cross section $\sigma_{\text{sc,t}}(\varepsilon)$. This has been obtained for energies up to 200 eV, using the suggested values of $\sigma_{\text{sc,t}}(\varepsilon)$ in Fig. 6.40 and the data of Leiter *et al.* [252] for $\sigma_{\text{i,t}}(\varepsilon)$ and is shown in Fig. 6.53. Above 10 eV the cross sections for direct and indirect vibrational excitation and electron attachment are small, and thus in this energy range Eq. (6.16) may be written as

$$\sigma_{\text{sc,t}}(\varepsilon) - \sigma_{\text{i,t}}(\varepsilon) \approx \sigma_{\text{e,int}}(\varepsilon) + \sigma_{\text{dis,neut,t}}(\varepsilon) + \sigma_{\text{electronic,t}}(\varepsilon). \tag{6.17}$$

Since, moreover, electronic excitation predominantly leads to dissociation and the cross section for dissociation into neutral species is expected to be small compared to $\sigma_{\mathrm{sc,t}}(\varepsilon)$ and $\sigma_{\mathrm{i,t}}(\varepsilon)$, Eq. (6.17) may be further reduced to

$$\sigma_{\mathrm{sc,t}}(\varepsilon) - \sigma_{\mathrm{i,t}}(\varepsilon) \approx \sigma_{\mathrm{e,int}}(\varepsilon). \tag{6.18}$$

Unfortunately, $\sigma_{\mathrm{e,int}}(\varepsilon)$ is only known for energies less than 10 eV for this molecule. Above this energy, Eq. (6.18) can give an upper limit for $\sigma_{\mathrm{e,int}}(\varepsilon)$. This relationship seems to be consistent with the existing measurements as can be seen from Fig. 6.53.

3.5 Electron Attachment to CCl_2F_2

There have been many experimental studies of the electron attachment properties of the CCl_2F_2 molecule. We begin this section by analyzing first the available data on the electron attachment coefficient and the electron attachment rate constant for CCl_2F_2, because these data provide an insight for understanding the electron attachment cross section data for CCl_2F_2 presented later in this section.

3.5.1 Density-Normalized Electron Attachment Coefficient, $\eta/N\,(E/N)$, of CCl_2F_2

The density-normalized electron attachment coefficient, $\eta/N\,(E/N)$, of CCl_2F_2 has been measured by a number of investigators [265–272, 274]. Figure 6.54 shows these measurements which were made at temperatures ranging from 293 to 298 K. The quoted uncertainties vary from ±5% to ±15% [±5% (Ref. [266]), ±15% (Ref. [267]), ±10% (Ref. [268]), ±10% (Ref. [270])]. With the exception of the data of Siddagangappa *et al.* [271] and Harrison and Gebarlle [265], there is reasonable agreement among the measurements. The solid line in Fig. 6.54 represents a least squares fit to all of the data except those of Refs. [265, 271]. Values from this fit are given in Table 6.36 as our recommended values.

3.5.2 Total Electron Attachment Rate Constant, $k_{\mathrm{a,t}}(E/N)$, of CCl_2F_2

There have been four sets of measurements [184, 230, 231, 233] of the $k_{\mathrm{a,t}}(E/N)$ of CCl_2F_2 in N_2 buffer gas and one measurement [184] using Ar as the buffer gas. The variation of the $k_{\mathrm{a,t}}$ of CCl_2F_2 with E/N measured in the buffer gases N_2 and Ar is shown in Fig. 6.55. There is reasonable agreement amongst the measurements in N_2 whose uncertainties are all approximately ±10%.

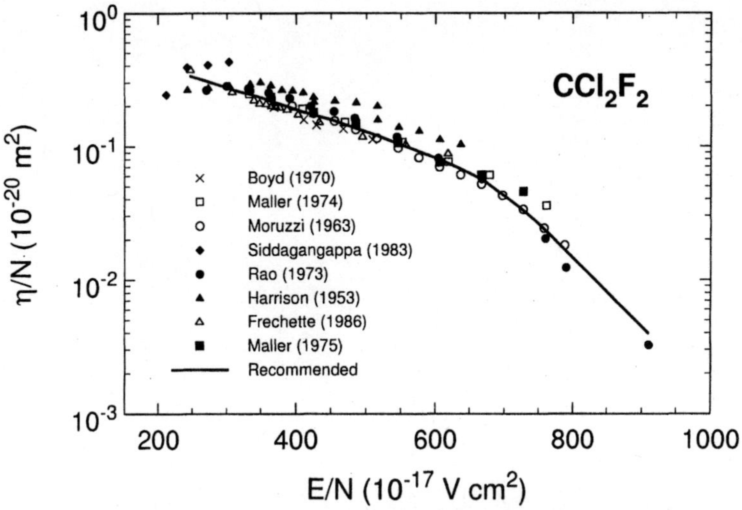

Figure 6.54. Density-normalized electron attachment coefficient, $\eta/N\,(E/N)$, for CCl_2F_2: (×) Ref. [268]; (□) Ref. [269]; (○) Ref. [266]; (♦) Ref. [271]; (•) Ref. [267]; (▲) Ref. [265]; (△) Ref. [274]; (■) Ref. [270]; (—) recommended.

Table 6.36. Recommended density-normalized electron attachment coefficient, η/N, for CCl_2F_2.

E/N (10^{-17} V cm^2)	η/N (10^{-20} m^2)	E/N (10^{-17} V cm^2)	η/N (10^{-20} m^2)
250	0.33	600	0.082
300	0.27	650	0.062
350	0.23	700	0.042
400	0.19	750	0.026
450	0.16	800	0.015
500	0.13	850	0.0081
550	0.10	900	0.0045

The data in Fig. 6.55 and those of Christophorou *et al.* [230] are plotted in Fig. 6.56 as a function of the mean electron energy, $\langle\varepsilon\rangle$, determined from the buffer gas electron energy distribution functions. The data of Christophorou *et al.* [230] lie higher than the rest of the measurements and for this reason they were excluded from the average shown in Fig. 6.57 by the broken line. In the averaging process the first two data points of McCorkle *et al.* were excluded, since none of the other three

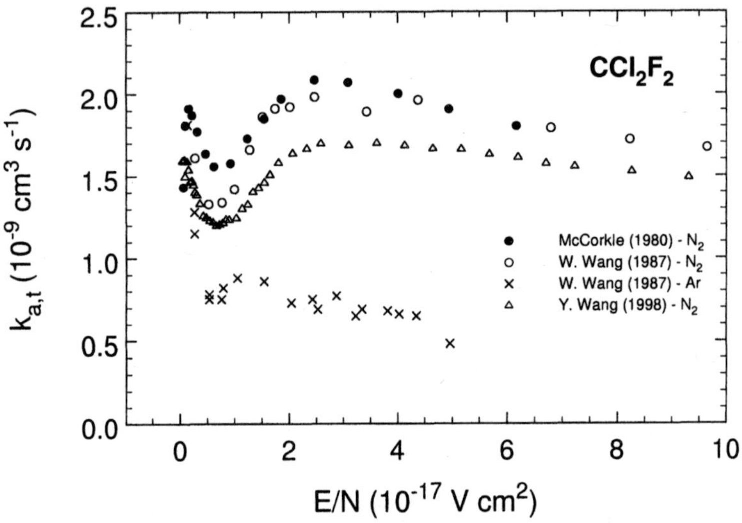

Figure 6.55. Total electron attachment rate constant, $k_{a,t}(E/N)$, for CCl_2F_2 measured in mixtures with N_2: (•) Ref. [231]; (○) Ref. [184]; (△) Ref. [233], and Ar: (×) Ref. [184].

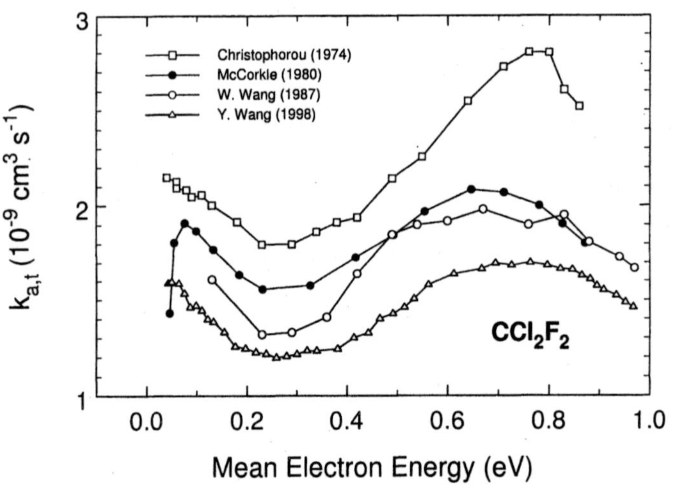

Figure 6.56. Total electron attachment rate constant, $k_{a,t}(\langle\varepsilon\rangle)$, for CCl_2F_2 measured in a buffer gas of N_2: (•) Ref. [231]; (□) Ref. [230]; (○) Ref. [184]; (△) Ref. [233].

sets of similar measurements showed a downward trend. Also plotted in Fig. 6.57 are the results of a number of studies which measured only

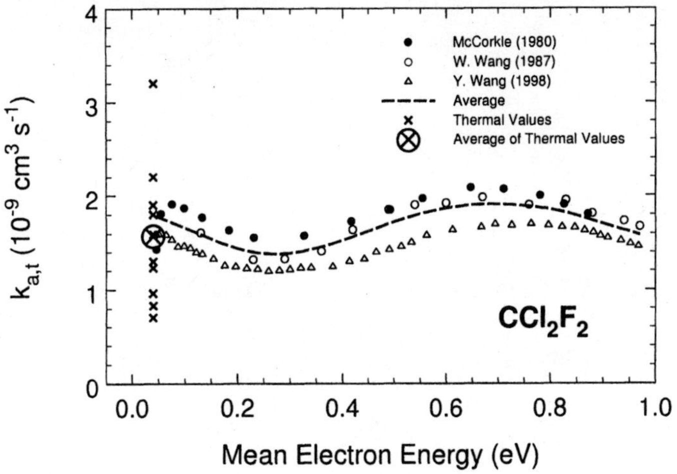

Figure 6.57. Total electron attachment rate constant, $k_{a,t}(\langle\varepsilon\rangle)$, for CCl_2F_2 measured in a buffer gas of N_2: (•) Ref. [231]; (○) Ref. [184]; (△) Ref. [233]; (- - -) least squares average of all the data; (×) thermal values of $k_{a,t}$ as measured by various groups using a number of techniques (Table 6.38); (⊗) average of the thermal values of $k_{a,t}$.

the thermal (T = 300 K) value, $(k_{a,t})_{th}$, of $k_{a,t}$. The average of all the values of $(k_{a,t})_{th}$ (see Table 6.38 in Sec. 3.5.3 of Ch. 6) is shown by ⊗.

Finally, there has been one other study [279] that reported electron attachment rate constants versus mean electron energy for CCl_2F_2 measured in mixtures with N_2. The energy dependence and the magnitude of the results of this study are at variance with the rest of the data in Fig. 6.57 and for this reason these data are not shown in the figure.

The average rate constant in Fig. 6.57 is compared in Fig. 6.58 with the data of Wang and Lee [184] in argon. The low-energy range data obtained from mixtures with N_2 and the high-energy range data obtained from measurements in mixtures with Ar merge smoothly to cover the energy range from 0.04 to 5 eV. Values obtained from a fit to these two sets of data are listed in Table 6.37, and represent our recommended values for the $k_{a,t}(\langle\varepsilon\rangle)$ of CCl_2F_2.

3.5.3 Thermal Value of the Total Electron Attachment Rate Constant, $(k_{a,t})_{th}$, of CCl_2F_2

Table 6.38 lists reported values of $(k_{a,t})_{th}$ of CCl_2F_2 measured at T = 293–300 K. The mean of these values is $(15.7 \pm 7.1) \times 10^{-10}$ cm^3 s^{-1}. These values have also been plotted in Fig. 6.57.

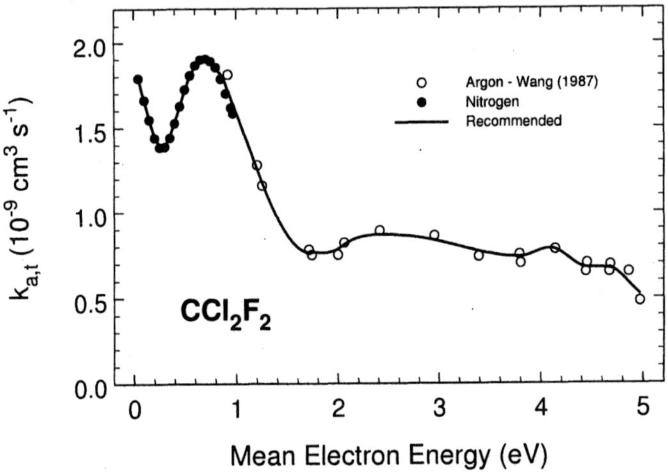

Figure 6.58. Total electron attachment rate constant, $k_{a,t}(\langle\varepsilon\rangle)$, for CCl_2F_2 measured in mixtures with N_2 and Ar: (•) average of the data in N_2 from Fig. 6.57; (∘) data of Ref. [184] for Ar mixtures. The solid line through the data points is a least squares fit to the data.

3.5.4 Effect of Temperature on the $k_{a,t}(E/N)$ of CCl_2F_2

The data presented in Figs. 6.54–6.58 show that the CCl_2F_2 molecule attaches electrons with energies down to $\sim$ 0 eV. Furthermore, the data on electron attachment and electron scattering for this molecule discussed earlier in this section show that the lowest vertical attachment energy of CCl_2F_2 is 0.9 eV and its adiabatic electron affinity is $\sim$ 0.4 eV. The latter finding suggests that the potential energy surface of the lowest negative ion state of CCl_2F_2 has a minimum at about 0.4 eV below that of the neutral molecule. It is possible (see Fig. 6.59) that the potential energy surface of this state rises steeply in the Franck-Condon region to account for the vertical attachment energy at 0.9 eV. The preponderance of electron attachment reactions below about 1 eV (see Sec. 3.5.6 of Ch. 6) lead to dissociation of $CCl_2F_2^{-*}$ producing Cl^-. Since the CF_2Cl-Cl bond dissociation energy ((3.3 $\pm$ 0.2) eV [225]; 3.58 eV [235]; 3.1 eV [288]) is smaller than the electron affinity (3.61 eV [74]) of the Cl atom, the reaction

$$CCl_2F_2 + e \rightarrow CClF_2 + Cl^- \tag{6.19}$$

is exoergic by $\sim$ 0.28 eV.

The energy position of this (lowest) negative ion state would make the dissociative electron attachment Reaction (6.19) highly temperature dependent [289], and hence the rate constant would be expected to in-

Table 6.37. Recommended total electron attachment rate constant as a function of mean electron energy, $k_{a,t}(\langle\varepsilon\rangle)$, for CCl_2F_2.

$\langle\varepsilon\rangle$ (eV)	$k_{a,t}$ (10^{-9} cm^3 s^{-1})	$\langle\varepsilon\rangle$ (eV)	$k_{a,t}$ (10^{-9} cm^3 s^{-1})
0.05	1.77	1.5	0.86
0.06	1.75	1.6	0.79
0.07	1.73	1.7	0.77
0.08	1.71	1.8	0.76
0.09	1.68	2.0	0.79
0.10	1.66	2.2	0.85
0.20	1.44	2.4	0.87
0.30	1.39	2.6	0.86
0.40	1.53	2.8	0.85
0.50	1.72	3.0	0.83
0.60	1.86	3.2	0.80
0.65	1.90	3.4	0.77
0.70	1.90	3.6	0.75
0.75	1.89	3.8	0.74
0.80	1.85	4.0	0.77
0.90	1.75	4.2	0.76
1.0	1.59	4.4	0.68
1.1	1.42	4.6	0.67
1.2	1.26	4.8	0.62
1.3	1.10	5.0	0.50
1.4	0.97		

crease with increasing temperature. Indeed, this has been shown to be the case both for the thermal value, $(k_{a,t})_{th}$, of $k_{a,t}$ [134, 233, 287, 288, 290–292] and for the values of $k_{a,t}$ over a wider range of electron energies [233]. In Table 6.39 measurements are listed of $(k_{a,t})_{th}(T)$ at temperatures ranging from 74–777 K. In order to discern the temperature variation of $(k_{a,t})_{th}$, the $(k_{a,t})_{th}(T)$ data are normalized to the average value of $(k_{a,t})_{th}$ at $\sim$ 300 K and plotted in Fig. 6.60(a). Over this temperature range, $(k_{a,t})_{th}$ increases by more than a factor of 300. The temperature enhancement of $k_{a,t}(\langle\varepsilon\rangle)$ for $\langle\varepsilon\rangle \lesssim 1$ eV is shown in Fig. 6.60(b). These measurements are consistent with the results of a recent crossed beam study [293] (see also Ref. [294]) on the temperature dependence of the production of Cl^- from CCl_2F_2 shown in Fig. 6.61. The observed pronounced enhancement suggests that dissociative electron attachment to hot $CCl_2F_2^*$ molecules is an effective way to decompose the CCl_2F_2 molecules. In contrast to these generally accepted data,

Table 6.38. Thermal values, $(k_{a,t})_{th}$, of the total electron attachment rate constant[a] for CCl_2F_2.

$(k_{a,t})_{th}$ (10^{-10} cm^3 s^{-1})	T (K)	Method	Reference
16.6	298	Electron swarm	[233]
9.6	295?	Electron swarm	[280]
13	298	Microwave conductivity	[281]
8.3	300	Electron cyclotron resonance	[282]
7	298	Electron cyclotron resonance	[283]
18	293	Electron cyclotron resonance	[284]
19	298	Electron swarm	[285]
12.3	298	Electron swarm	[231, 286]
22	298	Electron swarm	[230]
32	300	Flowing afterglow	[134, 287]

[a]Average value $(15.7 \pm 7.1) \times 10^{-10}$ cm^3 s^{-1}.

one study [238] showed the rather peculiar behavior of the total electron attachment cross section at 563 K being lower than at 393 K. Finally, it should be noted that substantial increases in electron attachment have been observed [295] in ArF-excimer-laser irradiated CCl_2F_2 compared to non-irradiated CCl_2F_2. This photoenhanced electron attachment has been attributed [295] to electron attachment to highly excited electronic states of CCl_2F_2 produced by the laser light.

3.5.5 Total Electron Attachment Cross Section, $\sigma_{a,t}(\varepsilon)$, of CCl_2F_2

There are three sources of total electron attachment cross sections for CCl_2F_2: (i) swarm-unfolded total cross sections using electron attachment rate constants measured in mixtures of CCl_2F_2 with N_2 [230, 231, 233] and in mixtures with N_2 and with Ar (Ref. [232] using the data of Ref. [184]); (ii) electron beam measurements using monoenergetic electrons [234, 238, 244, 291]; and (iii) threshold electron attachment using very-low-energy electrons produced by photoionization [229]. The results of these methods are compared in Fig. 6.62 up to 5 eV. There is considerable variation in these data. With the exception of the data

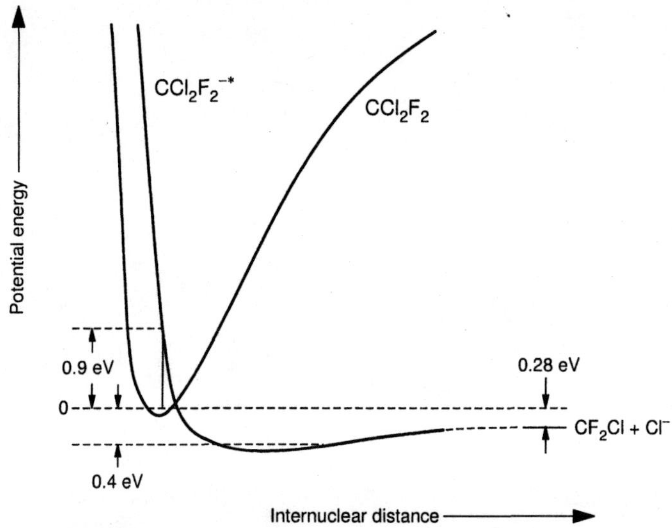

Figure 6.59. Schematic potential energy curves for CF_2Cl–Cl and the lowest negative ion state of $CCl_2F_2^{-*}$ consistent with the (adiabatic) electron affinity (= 0.4 eV) of CCl_2F_2, the vertical attachment energy at 0.9 eV of the lowest negative ion state of CCl_2F_2, and the observation (see next section) that the dissociative electron attachment cross section rises steeply as the electron energy decreases towards zero. The asymptotic limit $CF_2Cl + Cl^-$ lies 0.28 eV below the 0 eV level taken to be at the $\nu = 0$ level of the CF_2Cl–Cl symmetric stretch vibration ν_3, using a value of 3.33 eV for the CF_2Cl–Cl dissociation energy and 3.61 eV for the electron affinity of the Cl atom (see text).

of McCorkle *et al.* [231] and Illenberger *et al.* [235–237] which show a downward trend at the extreme low energy range, the rest of the measurements show a steep increase in the attachment cross section as the electron energy approaches zero, including the very-low-energy data of Chutjian and Alajajian [229]. Moreover, recent measurements on Cl^- production from CCl_2F_2 by Kiendler *et al.* [293] using a crossed beam experiment with a 60 meV energy resolution gave a cross section which rises steeply as the energy decreases in the extreme low-energy range, in agreement with the rest of the data. We thus believe that the cross section rises as the energy decreases toward zero. Additionally, all data show a cross section maximum near 0.9 eV, and the beam data of Pejčev *et al.* [234], Underwood-Lemons *et al.* [238], and Aflatooni and Burrow [244] also show a maximum near 3.5 eV. We can therefore conclude that the electron attachment data indicate three negative ion states of CCl_2F_2 below 4 eV at ≤ 0.0 eV, 0.9 eV, and 3.5 eV. This conclusion, as we discussed in Sec. 3.1 of Ch. 6 (see Fig. 6.39), is consistent with the

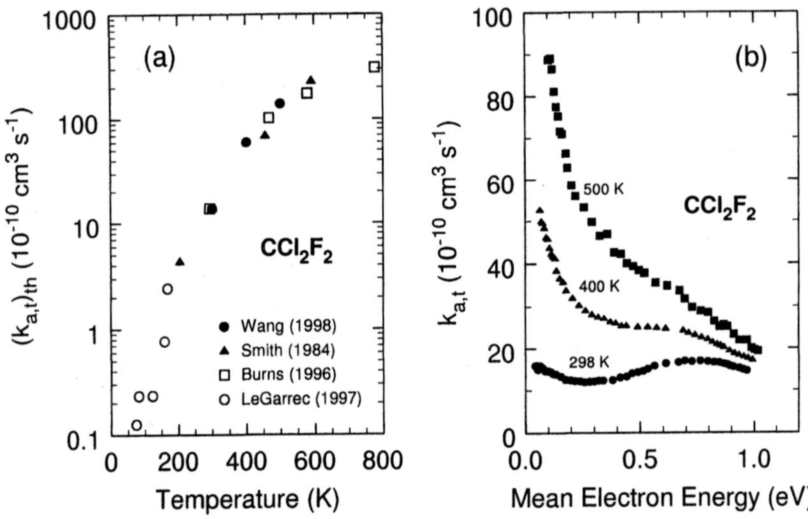

Figure 6.60. (a) Variation of the $(k_{a,t})_{th}$ of CCl_2F_2 with T: (●) Ref. [233]; (▲) Ref. [134]; (□) Ref. [290]; (○) Ref. [292]. (b) Variation of the $k_{a,t}(\langle\varepsilon\rangle)$ of CCl_2F_2 with T measured by Wang *et al.* [233] (measurements were made in N_2 buffer gas).

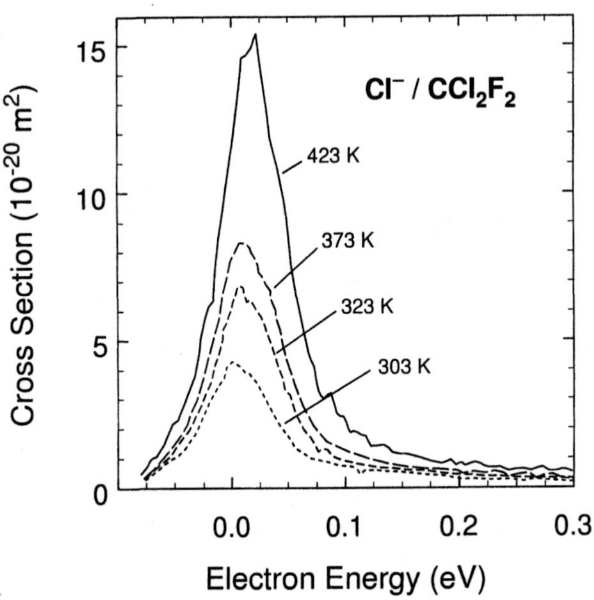

Figure 6.61. Temperature dependence of the cross section for the production of Cl^- from CCl_2F_2 measured in an electron beam experiment by Kiendler *et al.* [293].

Table 6.39. $(k_{a,t})_{th}$ of CCl_2F_2 as a function of gas temperature.

$(k_{a,t})_{th}$ (10^{-10} cm^3 s^{-1})	T (K)	Reference
< 10	205	[287, 134]
32	300	
160	455	
530	590	
16.6	298	[233]
60	400	
< 140	500	
19	293	[290]
140	467	
240	579	
420	777	
0.125	74	[292][a]
0.233	82	
0.234	123	
0.771	159	
2.44	168	

[a]The values of Le Garrec *et al.* were taken off of Fig. 9 in Ref. [292].

electron scattering data. Two swarm-unfolded total electron attachment cross sections [230, 231] and one electron beam total electron attachment study [234] indicate structure at about 0.25 eV. Since no negative ion state is expected in this energy range, this possible structure may arise from dissociative electron attachment to vibrationally excited CCl_2F_2 molecules.

We have attempted to deduce recommended values of the total electron attachment cross section by using a least squares fit to the various data in Fig. 6.62 in the range of energies where they are the most reliable.

Below 0.1 eV: In this energy range only the electron swarm [230, 231, 233] and the threshold electron attachment [229] data were used to obtain the average cross section because the electron beam measurements are known to be uncertain in this extreme low-energy range. In the averaging we excluded the lowest three points of the unfolded cross section given in Table V of McCorkle *et al.* [231], because in this energy range all other cross sections increase rather than decrease with decreasing electron energy. We also multiplied the cross section of Christophorou *et al.* [230] by 0.7 since their electron attachment rate constants had to

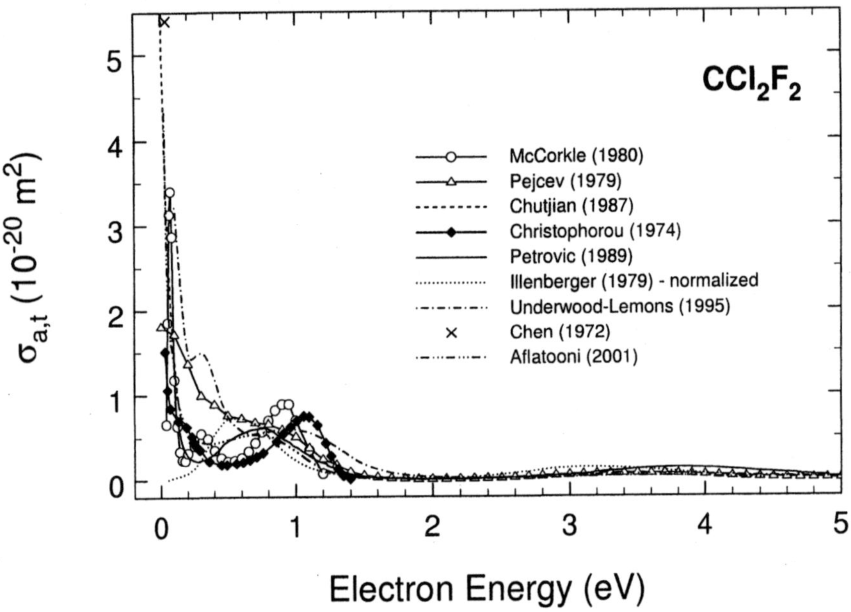

Figure 6.62. Total electron attachment cross section, $\sigma_{a,t}(\varepsilon)$, for CCl_2F_2 as determined by various methods. Swarm unfolded: (○) Ref. [231]; (♦) Ref. [230] (multiplied by 0.7, see text); (—) Ref. [232]. Electron beam: (△) Ref. [234]; (— · —) Ref. [238]; (— · · · —) Ref. [244]; (×) Ref. [291]. Threshold attachment: (- - -) Ref. [229]. In addition, the dotted line (· · ·) shows the sum of the relative yields of all observed fragment negative ions detected by Illenberger *et al.* [235–237] (see the text) normalized to the data of Ref. [234] at 0.7 eV.

be multiplied by this factor to be brought into agreement with other data of this type.

Between 0.1 eV and 1.2 eV: In this energy range all data were used except we excluded the normalized cross section of Illenberger *et al.* [235–237] on three grounds. First, the accuracy of the relative yields for the various negative ion fragments, especially at low electron energies, is uncertain as the authors themselves stated in their article. Second, below about 0.3 eV, the data of Illenberger *et al.* show the total electron attachment cross section decreasing precipitously with decreasing electron energy, in contrast to the most reliable beam and swarm data which show that the total electron attachment cross section rises steeply as the electron energy decreases toward zero. Third, the data are not absolute.

Energy > 1.2 eV: In this energy range the accuracy of the electron beam data is superior to the swarm-unfolded cross sections (the latter

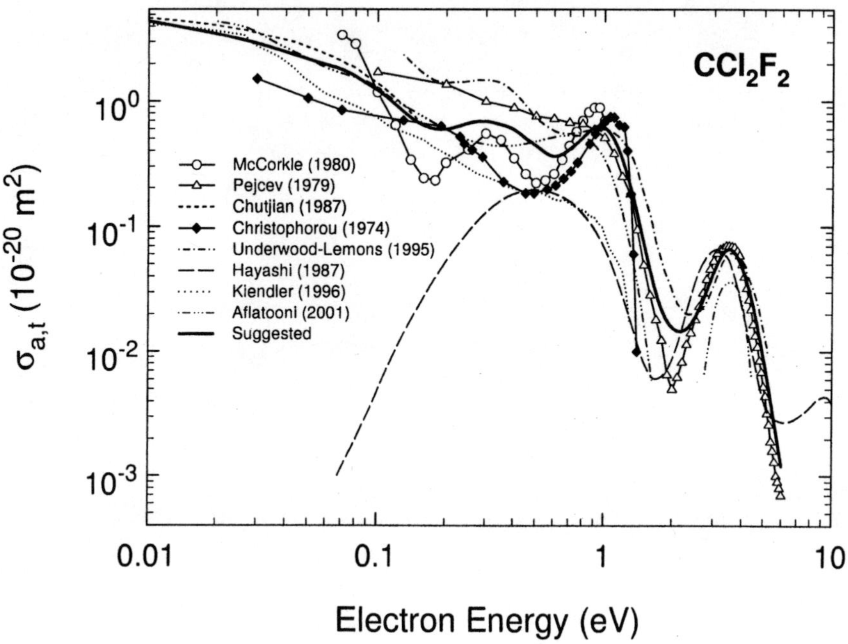

Figure 6.63. Suggested total electron attachment cross section $\sigma_{a,t}(\varepsilon)$ (—) for CCl_2F_2 based on an assessment of the various measurements below 0.1 eV, between 0.1 and 1.2 eV, and above 1.2 eV (see text). The rest of the $\sigma_{a,t}(\varepsilon)$ plotted are from the following sources: (○) Ref. [231]; (△) Ref. [234]; (- - -) Ref. [229]; (♦) Ref. [230] (multiplied by 0.7, see text); (– · –) Ref. [238]; (– –) Ref. [245]; (···) Ref. [293]; (– · · · –) Ref. [244].

[232] might also be influenced by the effect of the attaching gas on the distribution functions in pure argon used in the unfolding) and for this reason we averaged only the cross sections of the two electron beam studies, namely, those of Pejčev *et al.* [234] and Underwood-Lemons *et al.* [238].

The cross sections for the data used in each of the three energy regions, as discussed above, are plotted in Fig. 6.63 as a log-log plot, and the solid line in the figure is a least square fit to these measurements. Values taken from the fit are listed in Table 6.40 as our suggested values for the $\sigma_{a,t}(\varepsilon)$ of CCl_2F_2. The cross section deduced by Hayashi [245] is shown for reference and is clearly in error. In Fig. 6.63 is also plotted the cross section for Cl^- production from CCl_2F_2 measured by Kiendler *et al.* [293] using a crossed beam experiment. This cross section agrees with the rest of the data in the extreme low-energy range, but it progressively

Table 6.40. Suggested total electron attachment cross section, $\sigma_{a,t}(\varepsilon)$, for CCl_2F_2.

ε (eV)	$\sigma_{a,t}(\varepsilon)$ (10^{-20} m^2)	ε (eV)	$\sigma_{a,t}(\varepsilon)$ (10^{-20} m^2)	ε (eV)	$\sigma_{a,t}(\varepsilon)$ (10^{-20} m^2)
0.010	4.42	0.10	1.23	1.00	0.615
0.015	3.85	0.15	0.680	1.25	0.269
0.020	3.48	0.20	0.597	1.50	0.073
0.025	3.16	0.25	0.665	1.75	0.024
0.030	2.90	0.30	0.691	2.0	0.015
0.035	2.67	0.35	0.650	2.5	0.019
0.040	2.47	0.40	0.592	3.0	0.043
0.045	2.31	0.45	0.504	3.5	0.066
0.050	2.17	0.50	0.441	4.0	0.047
0.060	1.96	0.60	0.362	4.5	0.023
0.070	1.79	0.70	0.411	5.0	0.009
0.080	1.58	0.80	0.509	6.0	0.001
0.090	1.38	0.90	0.616		

falls below the rest of the data; at ~ 1 eV where all the data are virtually in agreement, the cross section of Kiendler *et al.* is much smaller.

3.5.6 Dissociative-Electron-Attachment Fragment Negative Ions of CCl_2F_2

A number of electron beam studies [235–237, 239, 291, 296, 297] reported relative yields of fragment negative ions by electron impact on CCl_2F_2 as a function of electron energy (see also Table 6.26). Rosenbaum and Neuert [296] detected Cl^- and F^- with maximum intensities at 1.7 and 3.7 eV, respectively, Hickam and Berg [297] observed Cl^- with an appearance onset of 0.5 eV, and Verhaart *et al.* [239] found the yields of Cl^-, F^-, and CCl_2F^- to maximize, respectively, at 0.7, 3.2, and 3.7 eV. In another electron beam study, Chen and Chantry [291] found that the yield of Cl^- from CCl_2F_2 peaked at "very-near-zero energy" with a cross section at this energy of $\sim 5.4 \times 10^{-20}$ m^2. They also found that at temperatures above 500 K the production of Cl_2^- from CCl_2F_2 exhibits a small zero-energy peak which increases rapidly with increasing temperature.

Illenberger and coworkers [235–237], carried out the most comprehensive investigation of the relative intensities of the various fragment negative ions generated by electron impact on CCl_2F_2 as a function of electron impact energy. They measured the energy dependence of the

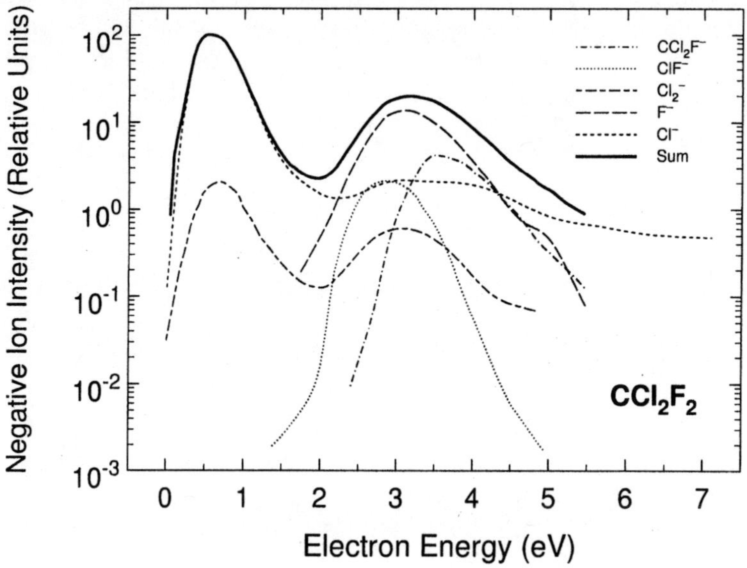

Figure 6.64. Relative negative ion intensity as a function of electron energy for the production of Cl^-, F^-, Cl_2^-, ClF^-, and CCl_2F^- by electron impact on CCl_2F_2 as reported by Illenberger *et al.* [235–237]. The data have been put on a "relative absolute scale" using the energy-integrated ion intensities given by Illenberger *et al.* [235]. The solid line represents the sum of the relative absolute intensities of all the ions in the figure (see discussion in text).

intensities of the fragment anions F^-, Cl^-, ClF^-, Cl_2^-, and CCl_2F^- and reported approximate values of their relative energy-integrated intensities. We have multiplied the relative intensities of the various fragment negative ions reported by Illenberger and co-workers by the corresponding values of the energy-integrated intensities of the negative ions as given by Illenberger *et al.* [235] and the resultant relative cross sections are shown in Fig. 6.64. Clearly the resonance below 1 eV is predominantly due to Cl^- with some contribution from the Cl_2^- fragment. Many fragment negative ions contribute to the second broad maximum near 3.5 eV, foremost F^-. These fragmentations are consistent with the symmetry of the negative ion states for the resonances at these energies as we have discussed earlier in this section.

We have taken the sum of the relative cross sections in Fig. 6.64 which is shown by the solid line in the figure. This sum also has been plotted in Fig. 6.62 after it was normalized to the cross section of Pejčev *et al.* [234] at 0.7 eV. The shape of Illenberger's total cross section is not consistent with the rest of the $\sigma_{a,t}(\varepsilon)$ data, especially below ~ 0.5 eV, but these data still provide useful information concerning the relative yields of negative ions. In this connection, it is noted that the relative yields of negative ions produced by electron impact on CCl_2F_2 have been found to depend on phase conditions (e.g., see Refs. [298–300]).

3.5.7 Negative Ions in CCl_2F_2 Discharges

A number of studies have been conducted which utilize negative ions produced by dissociative electron attachment to CCl_2F_2 to study radio frequency (rf) discharges of this gas. Many of these studies dealt with the measurement of negative ion densities in rf discharges using a combination of microwave resonance and photodetachment techniques. As expected, these studies have found that Cl^- is the dominant fragment negative ion [301]. Askaryan *et al.* [302] found that "the mechanism of dissociative electron attachment which is manifested in a cold decaying plasma of a pulsed microwave discharge is a principal mechanism causing dissociation of chlorofluorocarbons (CCl_2F_2)." In another study involving negative ions of CCl_2F_2, the role of negative ions in particle formation in low-pressure discharges of the CCl_2F_2/Ar/Si system has been investigated [303]. While in this study the density of the various negative ion species has been determined by detecting the extra electrons created by laser photodetachment, other studies [304] of this general type detected Cl atoms and chlorine-containing negative ions in rf plasmas using a two-photon laser-induced fluorescence technique. In this later method the negative ions were detected by laser photodetachment followed by two-photon excitation of the atomic chlorine, i.e., the Cl^- ions were detected by looking at the Cl atom rather than by looking at the released electron. The spatially resolved plasma concentration measurements of Selwyn *et al.* [304], under certain etching conditions, indicated an anomalously large signal spike at the plasma/sheath boundary which they attributed to an aggregation of chlorine-containing negative ions.

3.6 Electron Transport in CCl_2F_2

3.6.1 Electron Drift Velocity, $w(E/N)$, in CCl_2F_2

There is only one measurement [305] of the electron drift velocity, $w(E/N)$, in CCl_2F_2. The measurements of Naidu and Prasad [305] were conducted at 293 K with an estimated uncertainty of $\pm 5\%$. Data taken off of Fig. 2 of their paper (solid line) are plotted in Fig. 6.65 and are listed in Table 6.41.

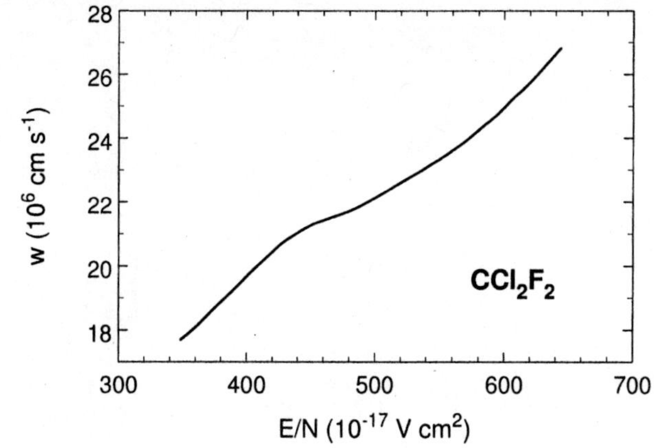

Figure 6.65. Electron drift velocity, $w\,(E/N)$, for CCl_2F_2 ($T = 293$ K) measured by Naidu and Prasad [305].

Table 6.41. Electron drift velocity, $w(E/N)$, in CCl_2F_2 measured by Naidu and Prasad [305].

E/N (10^{-17} V cm^2)	w (10^6 cm s^{-1})	E/N (10^{-17} V cm^2)	w (10^6 cm s^{-1})
350	17.8	525	22.7
375	18.7	550	23.4
400	19.7	575	24.1
425	20.6	600	25.0
450	21.3	625	26.0
475	21.6	640	26.6
500	22.1		

3.6.2 Ratio of Transverse Electron Diffusion Coefficient to Electron Mobility, $D_T/\mu\,(E/N)$, for CCl_2F_2

Two measurements [305, 306] have been made of $D_T/\mu\,(E/N)$, for CCl_2F_2. These were both made at 293 K and are plotted in Fig. 6.66. The data plotted were taken off the solid lines of Fig. 2 of Naidu and Prasad [305] and Maller and Naidu [306]. The uncertainty quoted for both measurements is ±5%. A fit to the two sets of measurements is shown in Fig. 6.66 and numerical values are listed in Table 6.42. Interestingly, D_T/μ increases rather slowly with increasing E/N. More measurements are needed over a wider E/N range.

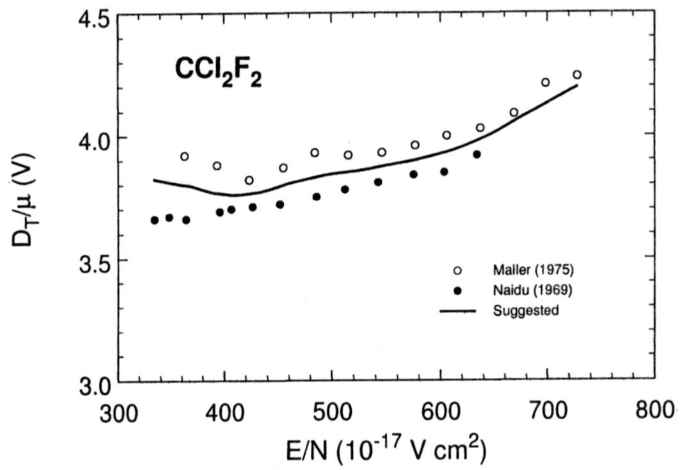

Figure 6.66. $D_T/\mu\,(E/N)$ for CCl_2F_2: (•) Ref. [305]; (∘) Ref. [306]; (—) suggested.

Table 6.42. Suggested values of $D_T/\mu\,(E/N)$ for CCl_2F_2 ($T = 293$ K).

E/N (10^{-17} V cm^2)	D_T/μ (V)	E/N (10^{-17} V cm^2)	D_T/μ (V)
335	3.82	550	3.88
350	3.81	575	3.90
400	3.76	600	3.93
425	3.77	625	3.96
450	3.79	650	4.01
475	3.82	675	4.07
500	3.84	700	4.13
525	3.86	725	4.19

Limited measurements of $D_T/\mu\,(E/N)$ in CCl_2F_2/N_2 mixtures have been made by Maller [307].

3.6.3 Density-Normalized Effective Ionization Coefficient, $(\alpha - \eta)/N\,(E/N)$, and $(E/N)_{lim}$ of CCl_2F_2

Shown in Fig. 6.67 are the values of the density-normalized effective ionization coefficient, $(\alpha - \eta)/N\,(E/N)$, of CCl_2F_2. This quantity was reported by Fréchette [272, 274]. We have also derived values of this quantity using our recommended values for $\alpha/N\,(E/N)$ (Fig. 6.52) and

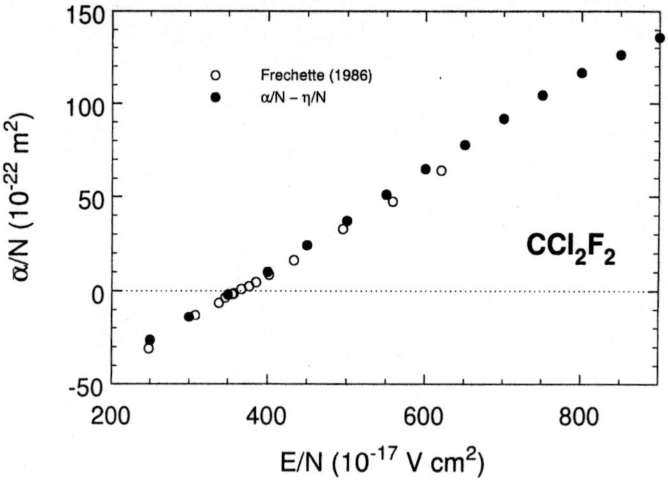

Figure 6.67. Density-normalized effective ionization coefficient, $(\alpha - \eta)/N\,(E/N)$, for CCl_2F_2: (○) Ref. [274]; (•) suggested values derived from the difference of the recommended values for α/N (Fig. 6.52) and η/N (Fig. 6.54).

Table 6.43. Suggested values of $(\alpha - \eta)/N\,(E/N)$ for CCl_2F_2.

E/N (10^{-17} V cm²)	$(\alpha - \eta)/N$ (10^{-22} m²)	E/N (10^{-17} V cm²)	$(\alpha - \eta)/N$ (10^{-22} m²)
250	−26.4	600	64.8
300	−14.0	650	77.8
350	−2.0	700	91.8
400	10.0	750	104.4
450	24.0	800	116.5
500	37.0	850	126.2
550	51.0	900	135.6

$\eta/N\,(E/N)$ (Fig. 6.54), which are shown as solid circles in Fig. 6.67. The latter data are given in Table 6.43 as our suggested values.

Measurements have also been made of $(\alpha - \eta)/N\,(E/N)$ in a number of CCl_2F_2 binary gas mixtures: CCl_2F_2/N_2 [269, 271, 273]; CCl_2F_2/SF_6 [274, 275]; and CCl_2F_2/CO_2 [276, 277].

In Table 6.44 are listed measurements [266–268, 270, 271, 276, 286, 308, 309] of $(E/N)_{\rm lim}$, the value of E/N at which $\alpha/N = \eta/N$, and also $(E/N)_{\rm Br}$, the E/N value at which gas breakdown occurs under uniform field conditions. The average of the $(E/N)_{\rm lim}$ values, determined from

Table 6.44. Values of $(E/N)_{\rm lim}$ for CCl_2F_2.

$(E/N)_{\rm lim}(10^{-17}$ V cm$^2)$	Reference
360[a]	[276]
372[a]	[271]
372[a]	[268]
373[a]	[266]
373[a]	[270]
375[a]	[267]
357	[309]
379	[308]
390	[286]

[a]These are values of $(E/N)_{\rm lim}$ determined from values of E/N at which $\alpha/N = \eta/N$; the remainder of the data are values of $(E/N)_{\rm Br}$ determined from uniform field breakdown measurements.

the α/N and η/N measurements listed in Table 6.44 is $(371 \pm 5) \times 10^{-17}$ V cm^2.

Measurements have also been reported on the values of $(E/N)_{\rm lim}$ for binary mixtures of CCl_2F_2 with air [267], N_2 [269–272, 307, 308, 310], CO_2 [276, 277, 308], SF_6 [310, 311], and in tertiary gas mixtures with SF_6/CO_2 [312] and SF_6/N_2 [309].

3.7 Optical Emission Under Electron Impact on CCl_2F_2

Allcock and McConkey [313] used time-of-flight mass spectroscopy to study electron-impact fragmentation of CCl_2F_2 via the detection of metastable fragments. The metastable fragments they observed included C, F, Cl Rydberg atoms, fluorine atoms in the $3s\,^4P_{5/2}$ state, and chlorine molecules in the metastable state $c\,^3\Sigma_{\rm iu}$ (excitation energy 7.2 eV). The kinetic energies of the fragments for many of the processes they studied were high, indicating steeply repulsive potential energy surfaces in the Franck-Condon region. The reaction producing the $F(^4P_{5/2})$ species has an asymptotic energy or dissociation limit (defined as the energy required to break a bond and separate two fragments to infinite separation plus the internal excitation energy of these fragments) of (27.3 ± 1.0) eV and was identified as

$$e + CCl_2F_2 \rightarrow CCl_2F^+ + F(^4P_{5/2}) + 2e. \qquad (6.20)$$

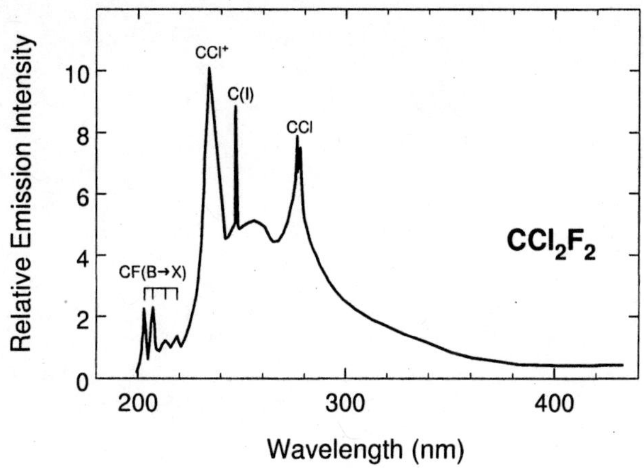

Figure 6.68. Electron-impact-induced emission spectrum of CCl_2F_2 in the wavelength range 200–440 nm from Van Sprang *et al.* [314].

They estimated a value of 14.6 eV for the appearance threshold of CCl_2F^+ based on the value of 27.3 eV for the asymptotic energy of Reaction (6.20) and the excitation energy of 12.7 eV for $F(^4P_{5/2})$.

While the study of Allcock and McConkey [313] detected metastable fragments, it did not do so by looking at light emission. An early such study on electron impact light-induced emission from CCl_2F_2 was conducted by Van Sprang *et al.* [314]. The emission spectrum they observed using 100 eV incident energy electrons exhibited emission lines of atomic fragments at long wavelengths and continuous emission with diatomic fragment emissions superimposed at the short wavelength side. Figure 6.68 shows the emission spectrum they observed in the wavelength range 200–440 nm. It contains emission from diatomic fragments, pronounced emission at 236.7 nm from the CCl^+ ionic species, and superimposed continuous radiation. The continuous emission has an estimated threshold of (15.7 ± 0.5) eV and was ascribed to the $CCl_2F_2^+$ $(E\,^2B_1 - A\,^2B_1)$ transition (see Van Sprang *et al.* for the energy dependence of the continuum emission cross section). In Table 6.45 are listed the emission cross sections, σ_{em} (100 eV), measured by Van Sprang *et al.* using electrons with 100 eV incident energy, for the various F and Cl atomic lines. The uncertainty of these measurements was quoted to be about ±10%.

More recently, Jabbour and Becker [315] analyzed the optical emissions in the wavelength region 200–800 nm produced by dissociative electron impact on CCl_2F_2. They determined photoemission cross sections for a variety of neutral and ionic fluorine and chlorine lines as well as

Table 6.45. Emission cross sections, σ_{em} (100 eV), for various atomic F and Cl lines, resulting from the impact of 100 eV electrons on CCl_2F_2.

Line	Wavelength[a] (Å)	σ_{em} (10^{-19} cm^2) Ref. [314]	Ref. [315]
$F(I)^2P_0 - {}^2P$	7037	0.9	0.9
	7127	1.0	0.5
	7202*	—	0.3
$F(I)^2D_0 - {}^2P$	7607	0.6	—
	7754	3.1	2.7
	7800	1.7	1.6
$F(I)^2S_0 - {}^2P$	7311	1.6	—
$F(I)^4P_0 - {}^4P$	7331	1.5	—
	7398	2.8	2.8
	7425	0.6	0.6
$F(I)^4D_0 - {}^4P$	6773	1.4	0.6
	6795*	—	0.1
	6834*	—	0.9
	6856	5.1	3.5
	6870*	—	0.7
	6902	3.5	2.0
	6910*	—	0.7
$F(I)^4S_0 - {}^4P$	6240*	—	0.9
	6349*	—	0.6
	6413*	—	0.4
$Cl(I)^4D_0 - {}^4P$	8212	9.0	
	8333	4.5	
	8375	29	
	8428	3.1	
	8586	12	
$Cl(I)^4S_0 - {}^4P$	7256	2.0	2.0
	7546	2.6	2.8
	7745*	—	0.6

[a]The wavelength numbers are from Van Sprang *et al.* [314] except those indicated by an asterisk which are from Jabbour and Becker [315].

for the strong CCl and CCl^+ bands at 277.8 and 236.8 nm, respectively. Photoemission cross sections at 100 eV energy for the most intense lines are given in Table 6.45 where they are compared with the data of Van Sprang *et al.* [314]. The overall agreement between the cross section values of the two groups is good, although the cross sections of Jabbour and Becker [315] are generally smaller by about 15%. The wavelength region between 400–600 nm was found to be dominated by ionic fluorine and chlorine emissions and by atomic chlorine lines corresponding to the $5p \rightarrow 4s$ manifold. From a comparison of the calculated minimum energies required for the formation of excited F* ($3p$) and Cl* ($4p$) fragments via several partial and total fragmentation channels and the measured appearance potentials, they concluded that the total fragmentation of the parent molecule is the most probable dissociation channel, viz.

$$CCl_2F_2 \rightarrow C + 2Cl + F + F^*(3p) \quad (31.2 \text{ eV}) \tag{6.21a}$$

$$\rightarrow C + 2F + Cl + Cl^*(4p) \quad (27.1 \text{ eV}). \tag{6.21b}$$

The wavelength region 200–400 nm showed several structured emissions superimposed on a continuum. The most prominent structured emissions were identified as arising from diatomic fragments, the $A\,^2\Delta \rightarrow X\,^2\Pi$ system of CCl centered at 277.8 nm and the CCl^+ emission at 236.8 nm assigned to the $A\,^1\Pi \rightarrow X\,^1\Sigma$ transition. On energetic grounds, Jabbour and Becker [315] concluded that the excited CCl^+ fragments are the result of a breakup of the parent CCl_2F_2 molecule with simultaneous formation of atomic ground state chlorine and fluorine atoms, viz.,

$$CCl_2F_2 \rightarrow CCl^+(A\,^1\Pi) + Cl + 2F + e \quad (31.2 \text{ eV}) \tag{6.22}$$

and the CCl ($A\,^2\Delta$) via the channel

$$CCl_2F_2 \rightarrow CCl(A\,^2\Delta) + Cl + 2F \quad (17.6 \text{ eV}). \tag{6.23}$$

With regard to the continuous emission, Jabbour and Becker [315] determined its onset to be (14.2 ± 1.0) eV which is close to the vertical ionization onset of the $\Delta\,^2B_2$ state of $CCl_2F_2^+$ parent ion of 14.4 eV. Consequently, they assigned the continuous emission to the optically allowed $CCl_2F_2^+$ ($\Delta\,^2B_2 \rightarrow X\,^2B_2$) transition. This assignment differs from the assignment of Van Sprang *et al.* [314] and both assignments differ from the assignment of Creasey *et al.* [316] who studied fluorescence processes in CCl_2F_2 following electron impact, He and Ne metastable impact, and VUV photons for excitation. According to Creasy *et al.* [316] the broad emission centered at 270 nm should be assigned to the CF_2 $A\,^1B_1 \rightarrow X\,^1A_1$ transition of the CF_2 radical and not to electronic transitions in the parent molecular ion. They did not observe parent

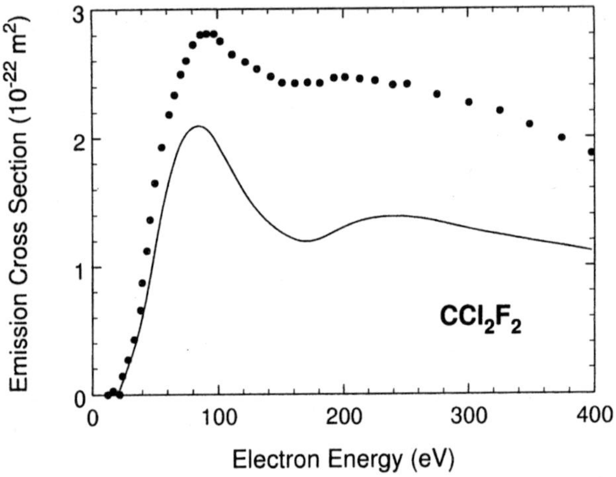

Figure 6.69. Emission cross section of the fluorine $^2P \rightarrow {}^2P$ multiplet at 95.5 nm as a function of electron energy, produced by electron impact dissociative excitation of CCl_2F_2. The data points are measured apparent cross sections and the solid line is the cascade-corrected contribution due to direct excitation (data of Roque *et al.* [317]).

ion emission. Creasey *et al.* reported that the emission spectrum they recorded from electron impact on supersonic molecular beam of CCl_2F_2 was similar to the room temperature electron impact spectrum of Van Sprang *et al.* and Jabbour and Becker.

Finally, Roque *et al.* [317] studied the emission of fluorine $(2p^4\,3s)\,{}^{2,4}P \rightarrow (2p^5)\,{}^2P$ resonance lines in the VUV following dissociative excitation of CCl_2F_2 by electron impact. In Fig. 6.69 is shown the emission cross section of the fluorine $^2P \rightarrow {}^2P$ multiplet at 95.5 nm as a function of the electron energy. The data points are the measured apparent cross section values and the solid line is the cascade-corrected contribution due to direct excitation. Roque *et al.* [317] determined the onset of this emission to be (21.5 ± 1.5) eV, which is lower than the minimum energy of 29.0 eV for total fragmentation of the CCl_2F_2 molecule, viz.,

$$CCl_2F_2 \rightarrow C + 2Cl + F + F^*(3s\,{}^2P). \tag{6.24}$$

This finding, coupled with the shape of the energy dependence of the emission cross section in Fig. 6.69 which indicates the opening up of another dissociation channel at energies above 35 eV, led Roque *et al.* to conclude that partial fragmentation channels (e.g., $CCl_2F_2 \rightarrow CF + Cl_2 + F^*$, threshold = 21.4 eV ; $CCl_2F_2 \rightarrow CCl_2 + F + F^*$, threshold =

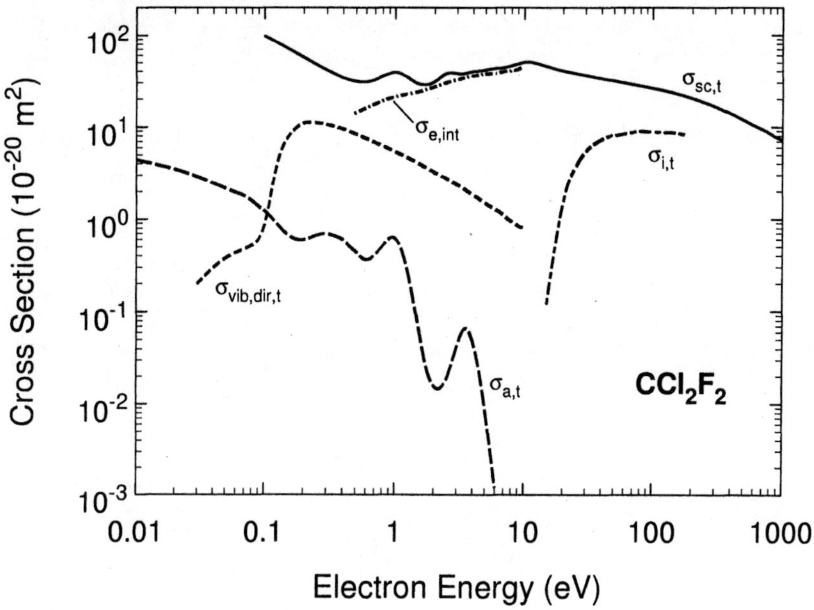

Figure 6.70. Recommended and suggested cross sections for CCl_2F_2 (see text).

22.1 eV) play an important role in the breakup of the CCl_2F_2 molecule, along with the total dissociation of the molecule via Process (6.24).

3.8 Recommended and Suggested Cross Sections and Coefficients for CCl_2F_2

In Fig. 6.70 are plotted the cross sections that have been designated as recommended or suggested for this molecule. These are:

- $\sigma_{\text{sc,t}}(\varepsilon)$ — Table 6.27, Fig. 6.40;
- $\sigma_{\text{e,int}}(\varepsilon)$ — Table 6.29, Fig. 6.43;
- $\sigma_{\text{vib,dir,t}}(\varepsilon)$ — Table 6.30, Fig. 6.45;
- $\sigma_{\text{i,t}}(\varepsilon)$ — Table 6.31, Fig.6.49; and
- $\sigma_{\text{a,t}}(\varepsilon)$ — Table 6.40, Fig. 6.63.

In addition to the cross sections presented in Fig. 6.70, the partial (Table 6.32, Fig. 6.50) and double (Table 6.34, Fig. 6.51) ionization cross sections are suggested in the absence of any other measurements.

Our recommended or suggested data for the electron attachment rate constant, electron attachment coefficient, ionization coefficient, effective ionization coefficient, and ratio of lateral diffusion coefficient to mobility are as follows:

- $k_{\mathrm{a,t}}(\langle\varepsilon\rangle)$ — Table 6.37, Fig. 6.58;
- $\eta/N\,(E/N)$ — Table 6.36, Fig. 6.54;
- $\alpha/N(E/N)$ — Table 6.35, Fig. 6.52;
- $(\alpha-\eta)/N\,(E/N)$ — Table 6.43, Fig. 6.67; and
- $D_{\mathrm{T}}/\mu\,(E/N)$ — Table 6.42, Fig. 6.66.

* * *

4. Electron Interactions with BCl_3

The boron trifluoride (BCl_3) molecule is a major source of reactive radicals generated via electron-impact dissociative processes, including electron attachment. For this reason, it is widely used as a plasma processing gas for the etching of metals (e.g., aluminum) and semiconductors (e.g., GaAs, Si) (see, for example, Refs. [318–324]). BCl_3 is used in a variety of gas mixtures (i.e., BCl_3 mixed with Ar, N_2, Cl_2, SF_6, CF_4, CHF_3, Ar + O_2, Ar + Cl_2, Ar + CCl_2F_2, Ar + SF_6, Ar + H_2 + N_2, $SiCl_4$ + H_2, and Ar + Cl_2 + N_2) for the etching of GaAs, GaSb, GaN, GaP, InAs, InP, AlGaAs, AlGaP, AlGaN, AlInAs, AlInP, InGaAs, InGaP, InAlP, InAlN, NiMnSb, and InGaAsP (e.g., see Refs. [324–330]).

In this section we review, assess, synthesize, discuss, and suggest cross sections for collisions of low-energy electrons with the BCl_3 molecule. The high reactivity and corrosiveness of the BCl_3 gas make experimental measurements difficult and partially account for the meager electron collision cross section data that are available for this gas and their large uncertainties. The only available experimental cross section data are for partial and total ionization and electron attachment, and the electron attachment cross sections are uncertain. Calculated values are available for the total elastic, elastic differential, and momentum transfer cross sections, and derived cross sections have been published for vibrational excitation and dissociation. Besides some rather uncertain data on electron attachment rate constants for BCl_3 (see Sec. 4.4 of Ch. 6) and some measurements of the electron drift velocity in BCl_3/Ar and

BCl_3/He mixtures [331], there are no measurements of the electron attachment, ionization, or transport coefficients for this gas. Analysis of the experimental data for the electron affinity, electron attachment, and electron scattering enabled identification of negative ion states of the BCl_3 molecule at about 1.0, 2.8, 5.2, 7.6, and 9.0 eV. Because the existing electron collision data are few and uncertain, we provide relevant data involving photon impact with the BCl_3 molecule. For calculated cross sections for some electron collision processes for the BCl and BCl_2 fragments see McKoy *et al.* [332].

4.1 Structural and Electronic Properties of BCl_3

Boron trichloride (BCl_3) is a nonpolar, plane symmetric (D_{3h} symmetry) molecule [333]. Its ground state outer valence shell molecular orbital configuration is [334, 335]: $(2a_1')^2\,(2e')^4\,(1a_2'')^2\,(3e')^4\,(1e'')^4\,(1a_2')^2\,X\,^1A_1'$. The outermost three occupied valence orbitals ($1a_2'$, $1e''$, and $3e'$) are nonbonding (Cl lone pair), the $1a_2''$ orbital is B–Cl π bonding, and the $2a_1'$ and $2e'$ orbitals are B–Cl σ bonding. These six orbitals have binding energies between 10 and 20 eV [334].

The lowest unoccupied molecular orbital (LUMO) of boron trichloride is $2a_2''$ (B2p–Cl3$p\,\pi^*$), and the next two higher-lying empty orbitals are $3a_1'$ and $4e'$ (B2p–Cl3$p\,\sigma^*$) [336, 337]. The presence of these unoccupied molecular orbitals helps explain the observed negative ion states of the BCl_3 molecule, and consequently the observed or calculated structure in the electron scattering and electron attachment cross sections. A detailed calculation by Baeck and Bartlett [336] optimizing bond lengths and bond angles within D_{3h} symmetry for BCl_3 and C_{3v} symmetry for BCl_3^- provided a number of structural and electronic parameters such as ionization energies, electronic vertical excitation energies, and vibrational frequencies for the BCl_3, BCl_3^-, BCl_3^+ species [336].

A number of experimental studies contribute to an understanding of the molecular and electronic structure of the BCl_3 molecule. These include electron-impact excitation and ionization [338–340], negative-ion formation [338–344], electron-impact dissociation into excited fragments [345–347], photoelectron spectra [335, 348, 349], photoionization and photodissociation [350–356], and photoabsorption and fluorescence measurements in the ultraviolet/visible (UV/VIS) region [351–359]. Some of the photophysical studies incorporated mass analysis and fluorescence measurements and established the fragmentation patterns of BCl_3 at specific photon wavelengths (see Sec. 4.3.3 of Ch. 6).

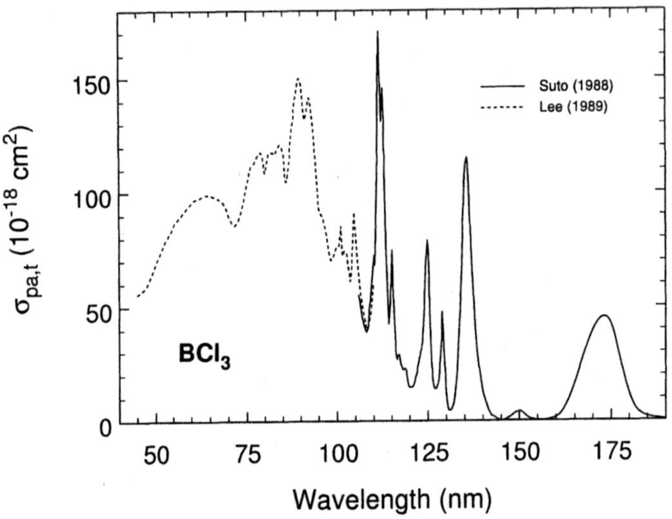

Figure 6.71. Total photoabsorption cross section, $\sigma_{pa,t}(\lambda)$, of BCl_3: (—) Ref. [351]; (- - -) Ref. [352].

4.1.1 Total Photoabsorption Cross Section, $\sigma_{pa,t}(\lambda)$, of BCl_3

The vacuum ultraviolet (VUV) absorption spectrum of BCl_3 (absorbance vs. wave number) has been measured by Planckaert *et al.* [358] from 200 to 120 nm. It has two well defined but broad bands with maxima near 57 970 cm^{-1} (7.187 eV, 173 nm) and 73 400 cm^{-1} (9.100 eV, 136 nm). Planckaert *et al.* assigned the lowest frequency band to a valence-shell transition of the $\pi^* \leftarrow \pi$ type, and described the second band as a Rydberg transition to a $3s$ orbital or as a valence-shell transition of the $\sigma^* \leftarrow \sigma$ type. Another measurement of the relative photoabsorption cross section of BCl_3 was made by Maria *et al.* [357] who reported a broad low-energy absorption band in the 200–220 nm region. Subsequently, Suto *et al.* [351] measured the total photoabsorption cross section, $\sigma_{pa,t}(\lambda)$, of BCl_3 as a function of the photon wavelength λ in the 106–190 nm range (Fig. 6.71). Suto *et al.* estimated the uncertainty in their measurements to be about ±15%. According to Suto *et al.*, the peak at 129 nm is likely to be due to the impurity HCl which could be produced by the reaction of BCl_3 with water adsorbed on the wall of their gas handling system. Another photoabsorption cross section measurement was made by Lee *et al.* [352] at shorter wavelengths

(45–106 nm). These measurements have an uncertainty of ±15% and also are shown in Fig. 6.71.

Measurements of the photoabsorption cross section of BCl_3 at still higher energies (190–280 eV) have been made by Ishiguro *et al.* [360] using synchrotron light. Synchrotron light was similarly employed by Ueda *et al.* [334] for the measurement of Auger electron spectra. The latter have also been investigated by Cini *et al.* [361] using 2 keV electrons.

In Table 6.46 are listed vertical and adiabatic ionization energies for the BCl_3 molecule, along with suggested corresponding assignments.

4.1.2 Electron Affinity and Negative Ion States of BCl_3

As indicated earlier in this section, the three lowest unoccupied molecular orbitals of BCl_3 are $2a_2''$, $3a_1'$, and $4e'$. These unoccupied molecular orbitals help explain the observed negative ion states of the BCl_3 molecule and, consequently, the observed structure in the electron scattering and electron attachment cross sections. Table 6.47 lists values of the electron affinity and negative ion states of the BCl_3 molecule and the vertical detachment energy of BCl_3^-, as determined by various methods. The experimental data in Table 6.47 (see also Fig. 6.72) identify negative ion states of BCl_3 at about 1.0, 2.8, 5.2, 7.6, and 9.0 eV. The negative ion state with vertical attachment energy of 1.0 eV and adiabatic electron affinity of 0.3 eV is associated with the lowest empty orbital $2a_2''$, while the negative ion state at 2.8 eV is associated with the $3a_1'$ empty orbital, and that at 5.2 eV with the $4e'$ empty orbital. The 7.6 and 9.0 eV values are associated with the positions of electron-excited Feshbach resonances in the 6–10 eV energy range.

4.1.3 Fundamental Vibrational Modes of BCl_3 and Other Data

Herzberg [373] lists for the BCl_3 vibrational frequencies ν_1 (symmetric stretching), ν_2 (out-of-plane bending), ν_3 (asymmetric stretching), and ν_4 (asymmetric bending), respectively, the values of 471 cm^{-1} (0.0584 eV), 462 cm^{-1} (0.0573 eV), 958 cm^{-1} (0.1188 eV), and 243 cm^{-1} (0.030 eV). Stull and Prophet [374] list for ν_1, ν_2, ν_3, and ν_4, respectively, the values of 0.0584, 0.0583, 0.1223, and 0.0301 eV. Knowledge of these frequencies can be helpful in efforts to use infrared laser measurements of the extinction coefficient of this gas in a discharge to determine the rotational and translational temperature and degree of dissociation of the gas, as well as plasma diagnostics (see, for example, Ref. [375]).

Table 6.46. Vertical and adiabatic[a] ionization energies[b] of BCl_3.

Ionization energy (eV)	Comment/ Orbital assignment	Reference	Method
(10.9 ± 0.2)	Threshold energy[c,d] for the production of ground state BCl_3^+	[362]	EI-MS[e]
11.82 (11.60)	Values for the production of BCl_3^+	[349]	PES[f]
(11.60 ± 0.01)	Adiabatic value for the production of BCl_3^+	[350]	PI-MS[g]
(11.16–11.44)[h]	Adiabatic values	[336]	C[i]
(11.48–11.74)[h]	Adiabatic values	[336]	C
11.49–11.83[h]	Vertical values	[336]	C
11.68–12.34[h]	Vertical values	[336]	C
11.60 ± 0.02	Threshold energy for the production of BCl_3^+	[350]	PI-MS
11.64	Threshold energy for the production of BCl_3^+ $(\tilde{X}\,^2A_2') + e$	[352]	PA[j]
11.67 ± 0.15	Vertical value	[363]	EI[k]
11.72	Vertical value, $1a_2'$ [l]	[335]	HeI EPS[m]
11.73 (11.64)	$1a_2'$; $2a_2'$ [n]	[348]	PES
11.73 (11.64)	$\tilde{X}\,^2A_2'$	[353]	PA-SR[o]
11.78	$2a_2'$	[365]	C[p]
11.82 (11.60)	$\tilde{X}\,^2A_2'$	[355, 349]	PES
11.97	e'	[366]	PES
12.03 ± 0.02	"Appearance potential" for BCl_3^+	[338]	EI-MS
12.0 ± 0.5	"Appearance potential" for BCl_3^+	[367]	EI-MS

[a]Data in parenthesis are values designated by the corresponding authors as adiabatic.
[b]See Table 6.49 for electron impact values.
[c]Koski *et al.* [362] measured the thresholds 11.0 ± 0.2 eV, 10.9 ± 0.2 eV, 10.9 ± 0.2 eV, and 10.6 ± 0.2 eV, respectively, for the production of $B^{10}Cl_2^{37}Cl^{35+}$, $B^{10}Cl_2^{35}Cl^{37+}$, $B^{11}Cl_3^{35+}$, and $B^{10}Cl_3^{35+}$.
[d]Koski *et al.* [362] measured mass spectrometrically the ionization threshold energies for the production of BCl_2^+ and BCl^+ by electron impact on BCl_3 and found them to be, 7.20 eV and $\sim$ 10.44 eV, respectively. Dibeler and Walker [350] estimated an ionization threshold for BCl_2 = 7.74 eV from photoionization studies.
[e]EI-MS = Electron Impact-Mass Spectrometry.
[f]PES = Photoelectron Spectroscopy.
[g]PI-MS = Photoionization-Mass Spectrometry.
[h]The range of values corresponds to the results of various types of calculation.
[i]C = Calculation.
[j]PA = Photoabsorption.
[k]EI = Electron Impact.
[l]Probable orbital.
[m]HeI EPS = HeI excited photoelectron spectrum.
[n]Assignments of Berger *et al.* [364] based on their *ab initio* calculations.
[o]PA-SR = Photoabsorption-Synchrotron Radiation.
[p]*Ab initio* calculation.

Table 6.46 (continued)
Vertical and adiabatic[a] ionization energies[b] of BCl_3.

Ionization energy (eV)	Comment/ Orbital assignment	Reference	Method
12.19	Threshold energy for the production of BCl_3^{+*} $(\tilde{A}\,^2E'') + e$	[352]	PA
12.214 ($\nu = 0$)	$1e''$ [l]	[335]	HeI EPS
12.28	$2e''$	[365]	C[p]
12.30	Threshold energy for the production of BCl_2^+ + Cl + e	[352]	PA
12.30 ± 0.02	Threshold energy for the production of BCl_2^+ + Cl + e	[350]	PI-MS
12.39 (12.19)	$1e''$; $2e''$ [n]	[348]	PES
12.39 (12.19)	$\tilde{A}\,^2E''$	[353]	PA-SR
12.43	a_2'	[366]	PES
12.65 (12.23)	$\tilde{A}\,^2E''$	[355, 349]	PES
12.41	$7e'$	[365]	C[p]
12.608 (line 5)	$3e'$ [l]	[335]	HeI EPS
12.66	Threshold energy for the production of BCl_3^+ $(\tilde{B}\,^2E') + e$	[352]	PA
12.66	$3e'$; $7e'$ [n]	[348]	PES
12.66	$\tilde{B}\,^2E'$	[353]	PA-SR
12.75	$\tilde{B}\,^2E'$	[355, 349]	PES
12.77	e''	[366]	PES
14.10	$2a_2''$	[365]	C[p]
14.22	Threshold energy for the production of BCl_3^{+*} $(\tilde{C}\,^2A_2'') + e$	[352]	PA
14.240 ($\nu_1^+ = 0$)	$1a_2''$ [l]	[335]	HeI EPS
14.40 (14.23)	$\tilde{C}\,^2A_2''$	[349]	PES
14.41 (14.30)	$\tilde{C}\,^2A_2''$	[355]	PES
14.42 (14.22)	$1a_2''$; $2a_2''$ [n]	[348]	PES
14.42 (14.22)	$\tilde{C}\,^2A_2''$	[353]	PA-SR
14.50	a_2''	[366]	PES

[a] Data in parenthesis are values designated by the corresponding authors as adiabatic.
[b] See Table 6.49 for electron impact values.
[l] Probable orbital.
[n] Assignments of Berger *et al.* [364] based on their *ab initio* calculations.
[p] *Ab initio* calculation.

The vibrational frequencies of BCl_3^- are expected to lie energetically lower than the corresponding ones of the neutral molecule BCl_3. Jacox *et al.* [376] in a matrix isolation study of the interaction of excited Ne atoms with BCl_3 assigned an absorption band they observed at 686 cm^{-1}

Table 6.46 (continued)
Vertical and adiabatic[a] ionization energies[b] of BCl_3.

Ionization energy (eV)	Comment/ Orbital assignment	Reference	Method
15.27	$6e'$	[365]	C[p]
15.32	Threshold energy for the production of BCl_3^{+*} $(\tilde{D}\,^2E') + e$	[352]	PA
15.320[q]	$2e'$	[335]	HeI EPS
15.51 (15.28)	$\tilde{D}\,^2E''$	[355, 349]	PES
15.54 (15.32)	$2e'$; $6e'$ [n]	[348]	PES
15.54 (15.32)	$\tilde{D}\,^2E''$	[353]	PA-SR
15.75	e'	[366]	PES
17.74 (17.74)	Threshold energy for the production of BCl_3^{+*} $(\tilde{E}\,^2A_1') + e$	[352, 353]	PA, PA-SR
17.74	$6a_1'$	[365]	C[p]
17.699 ($\nu_1^+ = 0$)	$2a_1'$ [l]	[335]	HeI EPS
17.70 (17.70)	$\tilde{E}\,^2A_1'$	[355, 349]	PES
— (17.74)	$2a_1'$; $6a_1'$ [n]	[348]	PES
17.79	a_1'	[366]	PES
18.37	Threshold energy for the production of $BCl^+ + 2Cl + e$	[352, 350]	PA, PI-MS

[a]Data in parenthesis are values designated by the corresponding authors as adiabatic.
[b]See Table 6.49 for electron impact values.
[l]Probable orbital.
[n]Assignments of Berger *et al.* [364] based on their *ab initio* calculations.
[p]*Ab initio* calculation.
[q]Progression 6, $\nu = 0$.

(0.085 eV) to the ν_3 of $^{11}BCl_3^-$, and Baeck and Bartlett [336] calculated values for all the vibrational frequencies of BCl_3^-. The averages of the two values they give for each frequency ν_1, ν_2, ν_3, and ν_4 of BCl_3^- in Table VII of their paper are, respectively, 551 cm^{-1} (0.068 eV), 320 cm^{-1} (0.0397 eV), 704 cm^{-1} (0.0873 eV), and 218 cm^{-1} (0.0270 eV).

In Table 6.48 are listed other data on BCl_3 and BCl_3^- which are relevant to the present discussion on the interactions of slow electrons with the BCl_3 molecule.

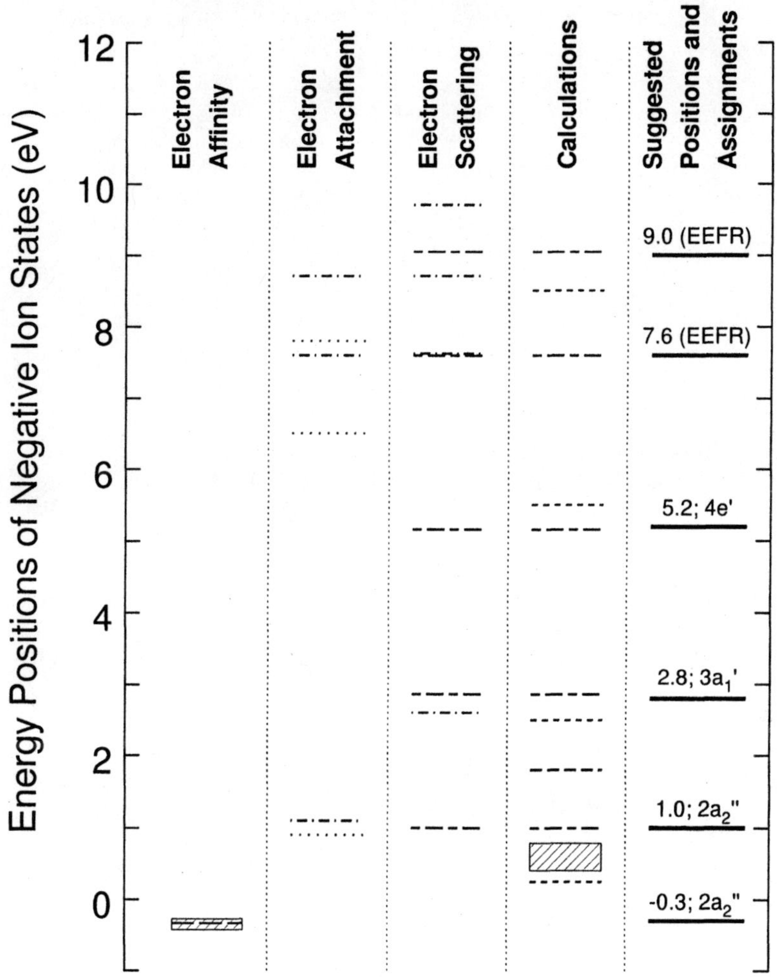

Figure 6.72. Negative ion states of the BCl_3 molecule as determined by electron affinity measurements, electron attachment measurements, electron scattering measurements, and calculations: Electron affinity: (– –) experimental data from Ref. [342]; (///) calculations from Ref. [336]. Dissociative electron attachment producing Cl^-: experimental data from (···) Ref. [344] and (–·–) Ref. [339]. Electron scattering: experimental data from (- – -) Refs. [337, 343] and (– · –) Ref. [339]. Electron scattering: calculated results from (- - -) Ref. [369], (///) Ref. [336], (– –) Ref. [370]; and (- – -) Ref. [337]. The last column gives the suggested positions and assignments. It should be noted that the adiabatic position of the $\sim$ 1.0 eV negative ion state is at about -0.3 eV [$-EA(BCl_3)$]. EEFR labels refer to assignments made from electron excited Feshbach resonance data.

Table 6.47. Electron affinity and negative ion states of BCl_3 and vertical detachment energy of BCl_3^-.

Value / Energy position (eV)	Method of determination	Comment/ Assignment	Ref.
Electron affinity			
0.33 ± 0.2	Measurement[a]	Adiabatic	[342]
0.27 to 0.42[b]	Calculation	Adiabatic	[336]
> 0.0	Swarm measurements	Formation of BCl_3^-	[368]
> 0.0	Beam measurements	Formation of BCl_3^-	[343, 344]
Negative ion states			
~ 0.9	Dissociative attachment producing Cl^-		[344]
1.1 ± 0.1	Dissociative attachment producing Cl^-		[339]
< 1.0	Electron transmission experiment/Calculation	a_2''	[343, 337]
0.25[c]	Calculation	B_2	[369]
0.41–0.79[b]	Calculation	Vertical attachment energy	[370]
0.00–1.30	Calculation		[371]
~ 1.8	Calculation[d]	B_2	[370]
~ 2.6	Threshold-electron excitation[e]		[339]
2.86	Electron transmission experiment/Calculation	a_1'	[343, 337]
2.5	Calculation[f]	A_1	[369]
5.16	Electron transmission experiment/Calculation	e'	[343, 337]
5.5	Calculation[f]	A_1; B_1	[369]
~ 6.5	Dissociative attachment producing Cl^-		[344]
~ 7.6	Threshold-electron excitation[e]		[339]
~ 7.6	Electron transmission experiment/Calculation	Core-excited shape resonance[g]	[343, 337]

[a]Determined using collisions of Cs atoms with BCl_3.
[b]The range of values corresponds to the results of various types of calculation.
[c]The position of this B_2 symmetry shape resonance moves from $\sim$ 2.0 eV in the static exchange calculation to 0.25 eV with the inclusion of polarization effects [369].
[d]Static exchange approximation calculation.
[e]SF_6^- scavenger technique.
[f]Polarized-SCF calculations.
[g]Observed in electron transmission and in dissociative electron attachment experiments.

Table 6.47 (continued)
Electron affinity and negative ion states of BCl_3 and vertical detachment energy of BCl_3^-.

Value / Energy position (eV)	Method of determination	Comment/ Assignment	Ref.
~ 7.8	Dissociative attachment producing Cl^-		[344]
8.7	SF_6^- scavenger technique		[339]
9.05	Electron transmission experiment/Calculation	Core-excited shape[g] resonance	[337, 343]
9.7	Threshold-electron excitation[e]		[339]
8.5	Calculation[d]	A_1, B_1	[369]
Vertical detachment energy			
1.77	Optogalvanic experiments		[372]
1.73–1.99[b]	Calculation		[336]

[a]Determined using collisions of Cs atoms with BCl_3.
[b]The range of values corresponds to the results of various types of calculation.
[c]The position of this B_2 symmetry shape resonance moves from ~ 2.0 eV in the static exchange calculation to 0.25 eV with the inclusion of polarization effects [369].
[d]Static exchange approximation calculation.
[e]SF_6^- scavenger technique.
[f]Polarized-SCF calculations.
[g]Observed in electron transmission and in dissociative electron attachment experiments.

4.2 Electron Scattering from BCl_3

4.2.1 Total Electron Scattering Cross Section, $\sigma_{sc,t}(\varepsilon)$, of BCl_3

To our knowledge, there are no measured or calculated values of the total electron scattering cross section, $\sigma_{sc,t}(\varepsilon)$, of BCl_3.

4.2.2 Total Elastic Electron Scattering Cross Section, $\sigma_{e,t}(\varepsilon)$, of BCl_3

There are no experimental data on the total elastic electron scattering cross section, $\sigma_{e,t}(\varepsilon)$, of BCl_3. There are, however, a number of calculated values of the total (integral) elastic electron scattering cross section. Figure 6.73 shows the $\sigma_{e,t}(\varepsilon)$ as calculated by a number of groups: Tossell *et al.* [337] using the multiple scattering $X\alpha$ method; Winstead and McKoy [332, 385] using the Schwinger multichannel variational method in the static exchange approximation with and without

Table 6.48. Other basic properties of BCl_3 and BCl_3^-.

Parameter	Value	Reference
Equilibrium distance, R_{B-Cl}	1.754 Å, 1.755 Å	[336]
	1.754 Å	[369]
	1.751 Å[a]	[377]
	1.75 Å	[374]
	1.745 Å[b]	[378]
	1.742 Å	[370]
	1.742 Å	[337]
Dissociation energy		
BCl_2–Cl	4.78 ± 0.02 eV	[350]
	~4.59 eV	[368]
	4.61 eV	[379]
BCl + Cl_2	5.65 eV	[379]
BCl–Cl	3.39 ± 0.02 eV	[350]
B–Cl	5.51 ± 0.04 eV	[350]
Polarizability	9.6×10^{-24} cm^3	[329]
	9.47×10^{-24} cm^3	[380]
	9.38×10^{-24} cm^3	[380]
	8.99×10^{-24} cm^3 [c]	[381]
	8.21×10^{-24} cm^3 [d]	[382]
Bond angle		
Cl–B–Cl	120°	[374]
BCl_3^-		
R (B–Cl)	1.876 Å, 1.879 Å[e]	[336]
	1.836 Å[f]	[383]
Bond angle[g]		
	107.3°, 107.6° [e]	[336]
	110.9° [f]	[383]

[a] Experimental value.
[b] CNDO/2 calculation.
[c] In plane electrical polarizability measured in dioxan.
[d] Mean polarizability.
[e] *Ab initio* calculations.
[f] Determined from studies of the molecular structure of donor-acceptor complexes of trimethyl-amine and boron trichloride.
[g] The BCl_2 radical is bent with a Cl–B–Cl angle of ~ 122° [384].

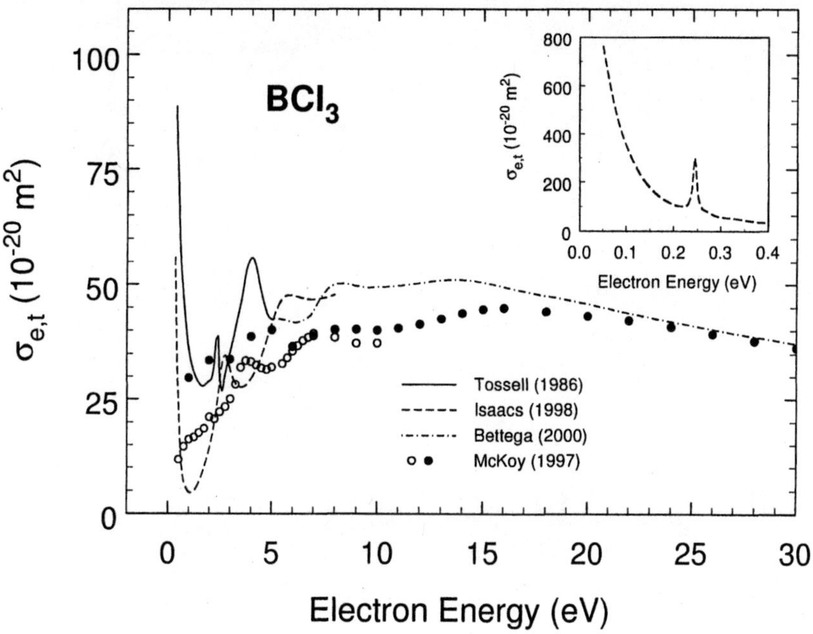

Figure 6.73. Calculated total (integral) elastic electron scattering cross section, $\sigma_{e,t}(\varepsilon)$, of BCl_3: (—) Ref. [337]; (•) Ref. [332] using static-exchange approximation without polarization; (∘) Ref. [332] using static-exchange approximation with polarization; (- - -) Ref. [369]; (— · —) Ref. [370]. Inset: Low energy data of Isaacs *et al.*

inclusion of polarization; Isaacs *et al.* [369] using the complex Kohn method, and Bettega [370] using the Schwinger multichannel method with pseudopotentials at the static exchange polarization approximation. The inset in Fig. 6.73 shows an expanded view of the calculation of Isaacs *et al.* at low energies.

The contributions of the lowest unoccupied electronic states of BCl_3 to the magnitude and structure of $\sigma_{e,t}(\varepsilon)$ shown in Fig. 6.73 can be seen from the calculated partial elastic electron scattering cross sections, $\sigma_{e,part}(\varepsilon)$, shown in Fig. 6.74. Figure 6.74(a) shows the $\sigma_{e,part}(\varepsilon)$ calculated by Tossell *et al.* [337] using the multiple scattering $X\alpha$ method. These results show the symmetry of the resonances which give rise to the peaks in the total elastic electron scattering cross section (see also Table 6.47). Figure 6.74(b) shows the polarized self-consistent field (SCF) results of Isaacs *et al.* [369] for the A_1, A_2, B_1, and B_2 symmetries. The A_1 cross section is very large as the electron energy approaches zero and shows two broad maxima at 2.5 and 5.5 eV. Although the polarized-

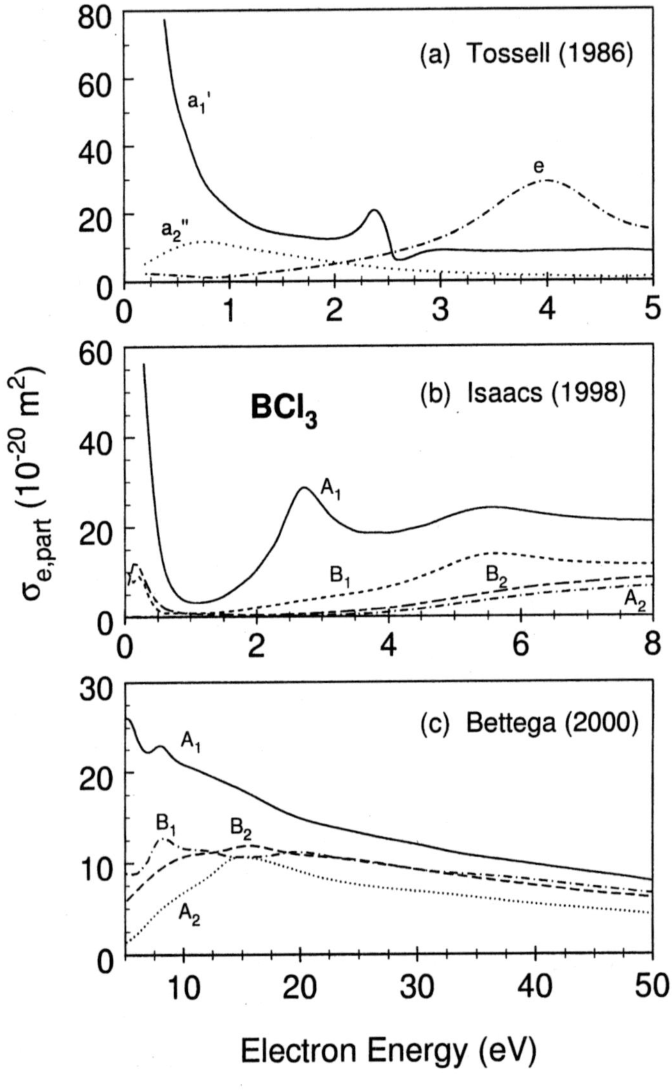

Figure 6.74. Calculated partial elastic electron scattering cross sections, $\sigma_{\mathrm{e,part}}(\varepsilon)$, for BCl_3 for the identified symmetries. (a) Ref. [337]; (b) Ref. [369]; (c) Ref. [370].

SCF B_2 cross section does not show a distinct peak due to the negative ion (shape) resonance near 2 eV, it does show this feature in the static exchange calculation results. Figure 6.74(c) shows the $\sigma_{\mathrm{e,part}}(\varepsilon)$ results of Bettega [370] to 50 eV.

4.2.3 Elastic Differential Electron Scattering Cross Section, $\sigma_{\mathrm{e,diff}}(\varepsilon)$, of BCl_3

Figure 4.75 shows the elastic differential electron scattering cross section, $\sigma_{\mathrm{e,diff}}(\varepsilon)$, of BCl_3 calculated by Isaacs *et al.* [369] and Bettega [370] at several electron energies. Unfortunately, the results of these two calculations can be compared at only one common energy (5.0 eV). There are no experimental measurements of $\sigma_{\mathrm{e,diff}}(\varepsilon)$ for this molecule.

4.2.4 Momentum Transfer (Elastic) Cross Section, $\sigma_{\mathrm{m}}(\varepsilon)$, of BCl_3

Figure 6.76 shows the elastic momentum transfer cross section, $\sigma_{\mathrm{m}}(\varepsilon)$, of BCl_3 as calculated by McKoy *et al.* [332, 385], Isaacs *et al.* [369], and Bettega [370]. The inset in Fig. 6.76 is an expansion of the results of Isaacs *et al.* [369] showing the B_2 resonance.

4.2.5 Inelastic Electron Scattering Cross Sections, $\sigma_{\mathrm{in}}(\varepsilon)$, of BCl_3

Vibrational Excitation Cross Sections for BCl_3. Nagpal and Garscadden [379] derived three vibrational excitation cross sections for BCl_3 in the energy range 0–30 eV using the conventional "two-term" solution of the time-dependent Boltzmann transport equation and the electron drift velocity data for BCl_3/Ar and BCl_3/He mixtures of Mosteller *et al.* [331]. These are shown in Fig. 6.77. The cross sections designated by $\sigma_{\nu 2,\nu 1}$, $\sigma_{\nu 4}$, and $\sigma_{\nu 3}$, are for the vibrational frequencies indicated by the subscripts. McKoy *et al.* [385] questioned these data on the basis that they are based on experimental measurements taken under conditions for which electron attachment is significant and was not considered in the analysis.

Electronic Excitation Cross Sections of BCl_3. Cross sections for some electronic transitions of BCl_3 (and also for the BCl and BCl_2 fragments) have been computed by McKoy *et al.* [332].

Cross Sections for Dissociation of BCl_3 into Neutral Fragments. Using the same experimental data and computational procedure as for the vibrational excitation cross sections (Fig. 6.77), Nagpal and Garscadden [379] derived two dissociation cross sections for BCl_3. These are shown in Fig. 6.77 and are designated by σ_{d1} and σ_{d2}. They refer, respectively, to the dissociation processes leading to BCl_2 + Cl

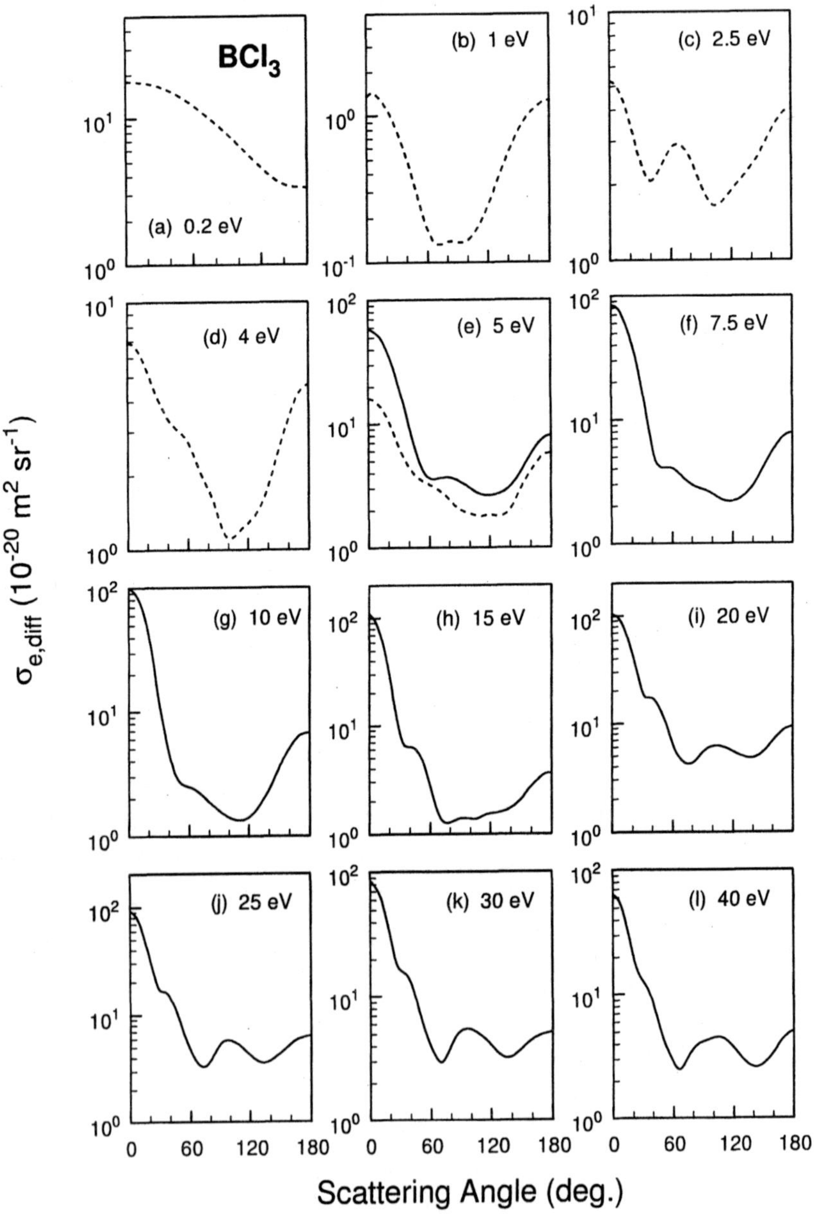

Figure 6.75. Calculated elastic differential electron scattering cross sections, $\sigma_{e,\text{diff}}$, for BCl_3 for a number of electron energies: (- - -) results of Isaacs *et al.* [369]; (—) results of Bettega [370].

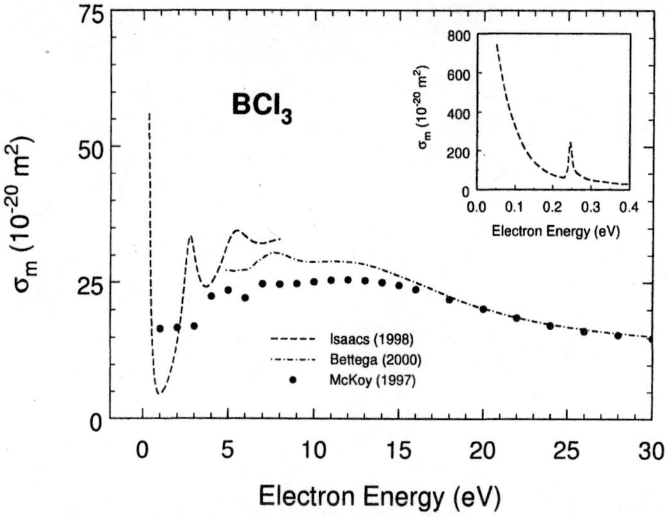

Figure 6.76. Calculated elastic momentum transfer cross section, $\sigma_m(\varepsilon)$, for BCl_3: (•) static-exchange approximation results from Ref. [332]; (- - -) Ref. [369]; (— · —) [370]. Inset: expanded results by Isaacs *et al.* [369] showing the B_2 resonance.

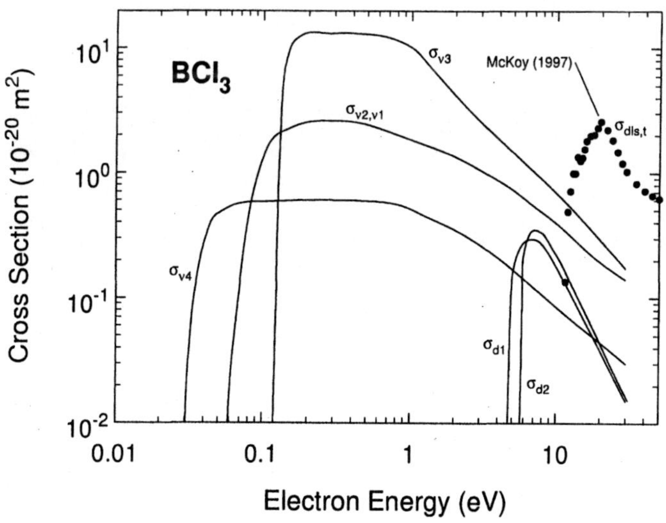

Figure 6.77. Derived collisional cross sections for BCl_3 by Nagpal and Garscadden [379]: (—) cross sections $\sigma_{\nu2,\nu1}(\varepsilon)$, $\sigma_{\nu4}(\varepsilon)$, and $\sigma_{\nu3}(\varepsilon)$ for the vibrational modes indicated by the subscripts, and cross sections $\sigma_{d1}(\varepsilon)$ and $\sigma_{d2}(\varepsilon)$ for the dissociation processes leading to BCl_2 + Cl and BCl + Cl_2, respectively. (•) Computed cross section, $\sigma_{dis,t}(\varepsilon)$, for dissociation of BCl_3 by electron-impact [332] (see text).

and BCl + Cl_2 (see values of the energy thresholds for these processes in Table 6.48). In Fig. 6.77 is also plotted an estimate of the electron-impact dissociation cross section as computed by McKoy *et al.* [332]. This latter cross section is essentially the sum of the electron-impact excitation cross sections they computed for electronic states presumed to be dissociative. In this regard, the estimated cross section by McKoy *et al.* for dissociation of BCl_3 into neutrals should at best be a lower limit value.

The cross sections of Nagpal and Garscadden [379] are observed to have a lower magnitude and to lie at lower energies than the cross section of McKoy *et al.* [332]. This has been attributed by McKoy *et al.* [332, 385] to the neglect of the effect of electron attachment in the calculation of Nagpal and Garscadden, and to the fact that the dissociation energies are larger than their threshold values used by Nagpal and Garscadden.

4.3 Electron-Impact Ionization of BCl_3

4.3.1 Partial Ionization Cross Sections, $\sigma_{i,part}(\varepsilon)$, of BCl_3

An early electron-impact ionization study of BCl_3 by Marriott and Craggs [338] showed that the most abundant positive ion is BCl_2^+, followed by BCl_3^+. Subsequently, in a mass spectrometric study of the photoionization of BCl_3, Diebeler and Walker [350] found that the relative abundances of the positive ions BCl_3^+, BCl_2^+, BCl^+, and B^+ at 58.4 nm (21.23 eV) are: 0.59, 1.0, 0.072, and ~ 0.001. A more recent study of the fragmentation of valence electronic states of BCl_3^+ using photoelectron spectroscopy by Biehl *et al.* [349] found that between 11 and 19 eV the BCl_2^+ ion is produced from the fragmentation of the excited states of BCl_3^+.

The most detailed study of electron-impact ionization of BCl_3 was conducted by Jiao *et al.* [340] using Fourier transform mass spectrometry. They observed the formation of the parent positive ion BCl_3^+ and the fragment positive ions BCl_2^+, BCl^+, and Cl^+. The positive-ion fragments BCl^+ and Cl^+ were found to react with neutral BCl_3 to generate BCl_2^+ with bimolecular rate constants equal to $(5.3 \pm 0.5) \times 10^{-10}$ cm^3 s^{-1} and $(6.2 \pm 0.5) \times 10^{-10}$ cm^3 s^{-1}, respectively. The BCl_2^+ ion was found not to react with BCl_3. Jiao *et al.* [340] did not observe B^+ or Cl_2^+, thus confirming the suspicion of Marriott and Craggs [338] that the observation of these two positive ions in their experiment was an artifact of pyrolysis on the filament. Table 6.49 lists the threshold energies ("appearance potentials") of the various positive ions produced by electron impact on BCl_3.

Table 6.49. Threshold energies[a] ("appearance potentials") of positive ions produced by electron impact on BCl_3.

Positive ion fragment	Threshold energy (eV)	Reference
$B^{10}Cl_2^{37}Cl^{35+}$	11.0 ± 0.2	[362]
$B^{10}Cl_2^{35}Cl^{37+}$	10.9 ± 0.2	[362]
$B^{11}Cl_3^{35+}$	10.9 ± 0.2	[362]
$B^{10}Cl_3^{35+}$	10.6 ± 0.2	[362]
BCl_3^+	12.0 ± 0.5	[367]
BCl_3^+	12.03 ± 0.02	[338]
$B^{10}Cl_2^{37+}$	12.0 ± 0.2	[362]
$B^{11}Cl_2^{35+}$	11.8 ± 0.2	[362]
BCl_2^+	13.2 ± 0.5	[367]
BCl_2^+	13.01 ± 0.02	[338]
$B^{10}Cl^{35+}$	20.0 ± 0.2	[362]
$B^{11}Cl^{35+}$	17.2 ± 0.2	[362]
BCl^+	19.2 ± 0.5	[367]
BCl^+	18.54 ± 0.07	[338]
B^{11+}	19.5 ± 0.2	[362]
B^{10+}	20.8 ± 0.2	[362]
B^+	19.5 ± 1.0	[386]
B^+	22.5 ± 0.5	[367]
B^+	13.6 ± 0.2	[338]
B^+	18.4 ± 0.2	[338]
B^+	22.35 ± 0.06	[338]
Cl^+	12.9 ± 0.15	[338]
Cl^+	17.1 ± 0.15	[338]
Cl_2^+	12.03 ± 0.05	[338]
BCl_2^{++}	33.77 ± 0.07	[338]

[a]See also Table 6.46.

Jiao *et al.* [340] measured the BCl_3 partial ionization cross sections, $\sigma_{i,part}(\varepsilon)$, relative to the ionization cross section of argon and calibrated their data using the measurements of Wetzel *et al.* [387] and Krishnakumar and Srivastava [388] for Ar. They gave no uncertainty for the magnitude of $\sigma_{i,part}(\varepsilon)$, but the uncertainty in their energy scale was quoted as ± 0.5 eV. Their data for $\sigma_{i,part}(\varepsilon)$ are given in Fig. 6.78 and are listed in Table 6.50. The cross section for BCl_3^+ is seen to be sub-

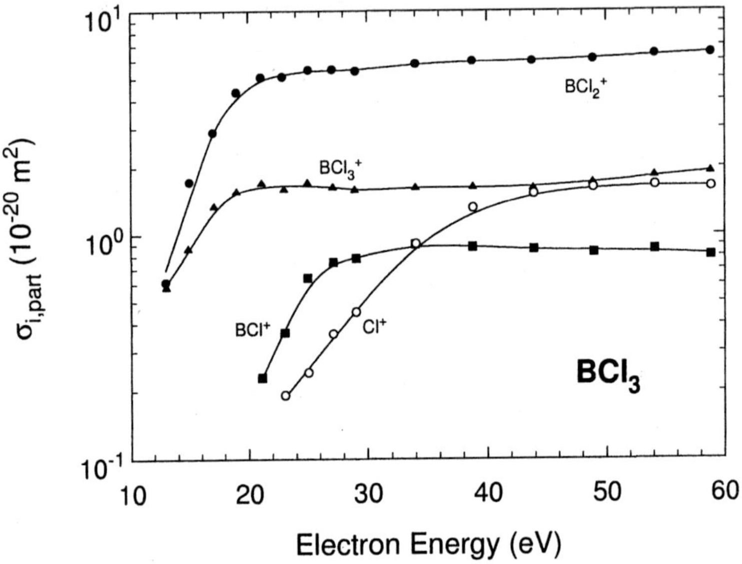

Figure 6.78. Partial electron-impact ionization cross sections, $\sigma_{i,part}(\varepsilon)$, for BCl_3 from Jiao *et al.* [340].

stantial. Interestingly, Overzet and Luo [389] in a study of ionization in rf plasmas of pure BCl_3 observed only the positive ions BCl_2^+, BCl^+, and B^+. The data of Jiao *et al.* show that from threshold to 60 eV the most abundant positive ion is BCl_2^+.

4.3.2 Total Ionization Cross Section, $\sigma_{i,t}(\varepsilon)$, of BCl_3

The sum of the partial ionization cross sections of Jiao *et al.* for BCl_3^+, BCl_2^+, BCl^+, and Cl^+ has been plotted in Fig. 6.79, and is taken to represent the total ionization cross section, $\sigma_{i,t}(\varepsilon)$, of BCl_3. These data are listed in Table 6.51, and represent our suggested values, but they should be considered a lower limit to $\sigma_{i,t}(\varepsilon)$. To our knowledge there are no other measurements of $\sigma_{i,t}(\varepsilon)$. According to Becker [393], the preliminary measurements by Tarnovsky and Becker quoted in Deutsch *et al.* [390] need further scrutiny.

In Fig. 6.79 the measurements of Jiao *et al.* are compared with the calculated values obtained using the modified additivity-rule (MAR) [390, 392], the Deutsch-Märk (DM) formalism [392], and the Binary Encounter-Bethe (BEB) theory [391]. All of the calculations exhibit

Table 6.50. Suggested partial ionization cross sections, $\sigma_{i,part}(\varepsilon)$, for BCl_3 from Jiao *et al.* [340].

	$\sigma_{i,part}(\varepsilon)$ $(10^{-20}\ m^2)$			
ε (eV)	BCl_2^+	BCl_3^+	BCl^+	Cl^+
13	0.6	0.6		
15	1.7	0.9		
17	2.9	1.3		
19	4.4	1.6		
21	5.1	1.7	0.2	
23	5.1	1.6	0.4	0.2
25	5.5	1.7	0.6	0.2
27	5.5	1.6	0.8	0.4
29	5.4	1.6	0.8	0.5

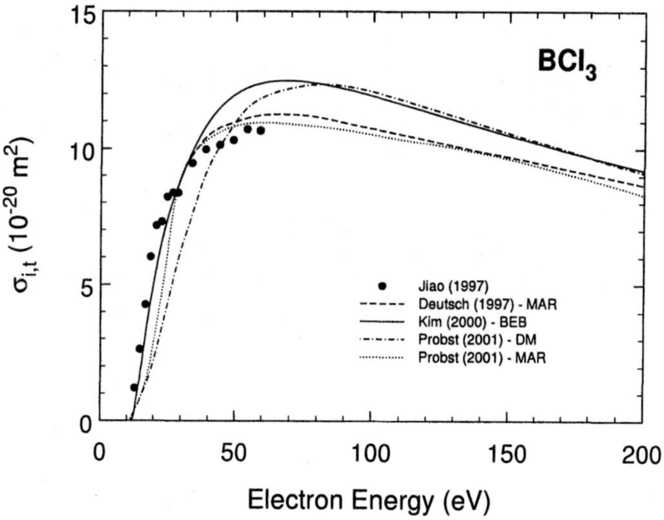

Figure 6.79. Total ionization cross section, $\sigma_{i,t}(\varepsilon)$, for BCl_3: (•) experimental values (sum of the partial ionization cross sections in Fig. 6.78, see Table 6.51) from Ref. [340]; (- - -) calculated values/MAR from Ref. [390]; (—) calculated values/BEB from Ref. [391]; (—·—) calculated values/DM from Ref. [392]; (···) calculated values/MAR from Ref. [392].

reasonable agreement with the measurements, but none agrees over the entire energy range from threshold to 70 eV.

Table 6.51. Suggested values of the total ionization cross section, $\sigma_{i,t}(\varepsilon)$, of BCl_3 for energies less than 30 eV from Jiao *et al.* [340].

ε (eV)	$\sigma_{i,t}(\varepsilon)$ (10^{-20} m^2)	ε (eV)	$\sigma_{i,t}(\varepsilon)$ (10^{-20} m^2))
13	1.2	23	7.3
15	2.6	24	8.0
17	4.4	27	8.3
19	6.0	29	8.3
21	7.0		

4.3.3 Dissociation of BCl_3 by Electron and Photon Impact

Besides the calculated results discussed in Sec. 4.2.5 of Ch. 6, there are no measurements of the cross section for electron-impact dissociation of BCl_3 into neutral fragments. However, there have been a number of studies on electron-impact-induced dissociation [345–347, 394] and photon-impact-induced dissociation (e.g., Refs. [351–356]) of BCl_3 and identification of the resultant (excited) radicals via their fluorescence emissions. The results of these electron-impact and photon-impact studies are briefly elaborated upon in this section, since they provide useful spectroscopic data for the identification of the species in plasma reactors containing BCl_3.

Jabbour *et al.* [345] measured the photoemission cross section of the BCl^* $A\,^1\Pi \rightarrow X\,^1\Sigma^+$ system around 272 nm produced by dissociative electron impact on BCl_3. This system is used for optical plasma diagnostics. The cross section was found to rise from the threshold at (14.0 ± 1.5) eV to a maximum near 25 eV and then to rise again to its overall maximum at near 40 eV. The maximum cross section value was measured to be 2.3×10^{-22} m^2 with an uncertainty of $\pm 20\%$. In a subsequent study by this group, Gilbert *et al.* [346] reported photoemission cross sections and appearance potentials for the most intense UV emissions produced by electron-impact dissociation of BCl_3. In particular, Gilbert *et al.* [346] studied the two atomic boron emissions $B^*\,2p^2\,^2D \rightarrow 2p\,^2P^0$ line at 208.9 nm and $B^*\,3s\,^2S \rightarrow 2p\,^2P^0$ line at 249.8 nm, and the BCl^* $A\,^1\Pi \rightarrow X\,^1\Sigma^+$ system around 272.4 nm. The 249.8 nm boron line was found to have the largest emission cross section (5.4×10^{-22} m^2) at 75 eV. The BCl $A\,^1\Pi \rightarrow X\,^1\Sigma^+$ emission cross section around 272.4 nm was found to be 3×10^{-22} m^2 at 40 eV, which is $\sim 20\%$ higher than their earlier measurement [345]. Figure 6.80 shows

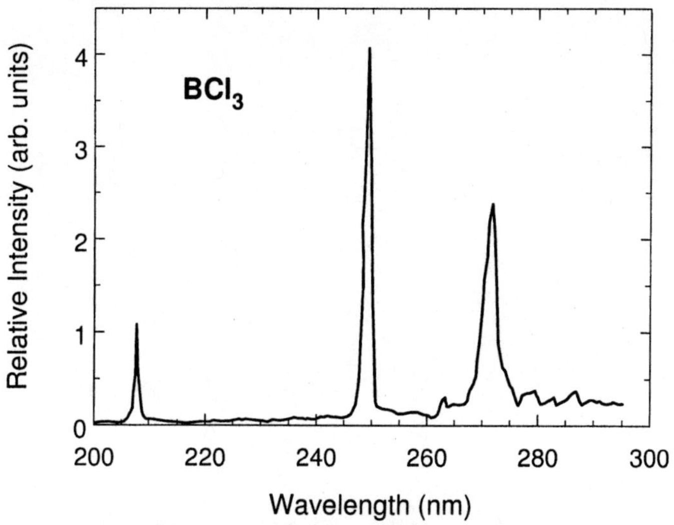

Figure 6.80. Emission spectrum between 200 and 300 nm induced by impact of 200 eV electrons on BCl_3 from Gilbert *et al.* [346].

the emission spectrum recorded by Gilbert *et al.* [346] in the wavelength range 200–300 nm, generated by impact of 200 eV electrons on BCl_3. The two B* lines at 208.9 nm and 249.8 nm and the BCl* $A\,^1\Pi \rightarrow X\,^1\Sigma^+$ emission band around 272.4 nm are clearly seen in the figure. The radiative lifetime of the BCl* A state [for the spectrally unresolved (0,0), (1,1) and (2,2) bands at 272 nm] was measured by Hesser [394] to be 19.1 ns.

Figure 6.81 gives the emission cross sections measured by Gilbert *et al.* [346] for the BCl* $A\,^1\Pi \rightarrow X\,^1\Sigma^+$ band and the B* $3s\,^2S \rightarrow 2p\,^2P^0$ line at 249.8 nm as a function of the electron energy from threshold to 100 eV. Based on the data in Tables 6.52 and 6.53, the two onsets at 14.0 and 29.0 eV [indicated by the arrows in Fig. 6.81(a)], have been ascribed [346], respectively, to the dissociation processes

$$\mathrm{BCl_3^*} \rightarrow \mathrm{BCl^*}(A\,^1\Pi) + 2\mathrm{Cl} \qquad (6.25)$$

and

$$\mathrm{BCl_3^*} \rightarrow \mathrm{BCl^*}(A\,^1\Pi) + \mathrm{Cl} + \mathrm{Cl^+}. \qquad (6.26)$$

Similarly, on the basis of the data in Tables 6.52 and 6.53, the two onsets at 22.0 and 34.0 eV [indicated by the arrows in Fig. 6.81(b)] have been ascribed [346], respectively, to the dissociation processes

$$\mathrm{BCl_3^*} \rightarrow \mathrm{B^*}(3s\,^2S) + 3\mathrm{Cl} \qquad (6.27)$$

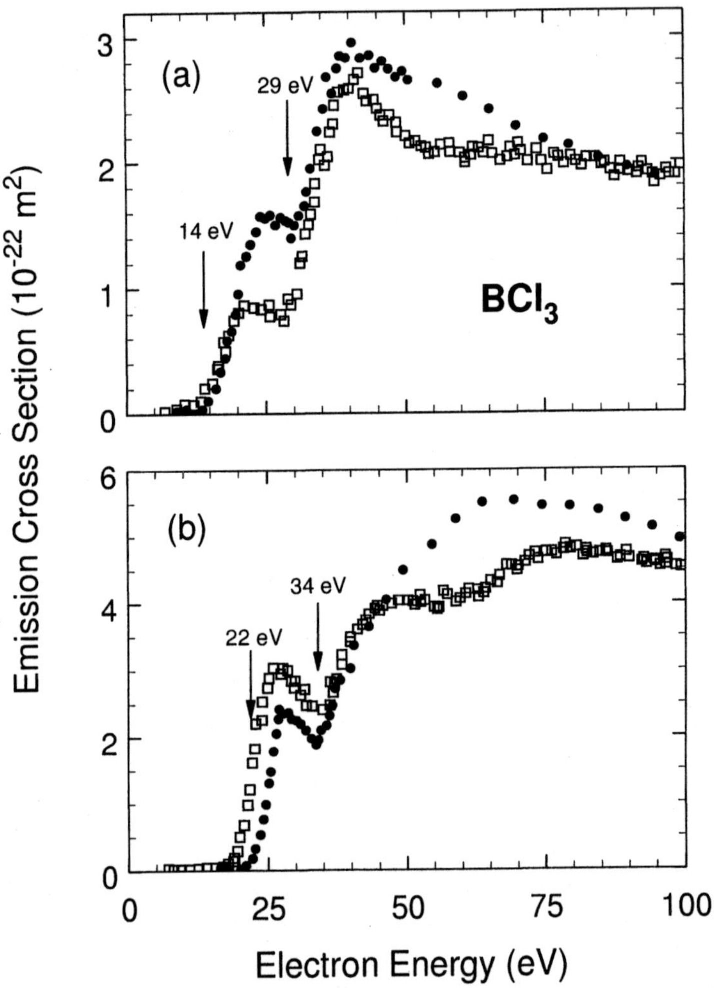

Figure 6.81. (a) Emission cross section for the BCl* $A\,^1\Pi \rightarrow X\,^1\Sigma^+$ band as a function of the electron energy: (•) Gilbert *et al.* [346]; (□) Tokue *et al.* [347]. (b) Emission cross section for the transition B* $2s\,^2S \rightarrow 2p\,^2P^0$ as a function of the electron energy: (•) Gilbert *et al.* [346]; (□) Tokue *et al.* [347].

and

$$BCl_3^* \rightarrow B^*(3s\,^2S) + 2Cl + Cl^+. \tag{6.28}$$

Tokue *et al.* [347] also investigated the emission spectra of excited fragments generated by electron impact on BCl_3 in the wavelength range 190–600 nm for electron energies up to 110 eV. Their cross section measurements for the B* $3s\,^2S \rightarrow 2p\,^2P^0$ transition and the BCl* $A\,^1\Pi \rightarrow$

Table 6.52. Energies and threshold wavelengths for the dissociation and ionization of BCl_3 as calculated by Lee *et al.* [352].

Process	Energy (eV)	Wavelength (nm)
Dissociation		
BCl_2 + Cl	4.61	
BCl + Cl_2	5.65	
BCl + 2Cl	8.16	
BCl* (A) + Cl_2	10.21	121.4
BCl* (A) + 2Cl	12.72	97.5
BCl_2 + Cl* $(4P'\,{}^2P^0)$	16.31	76.0
B + 3Cl	13.77	
B* $(3s\,{}^2S)$+ 3Cl	18.73	66.2
B* $(2p^2\,{}^2D)$ + 3Cl	19.70	62.9
Ionization		
$BCl_3^+\,(\tilde{X}\,{}^2A_2') + e$	11.64	106.5
$BCl_3^{+*}\,(\tilde{A}\,{}^2E'') + e$	12.19	101.7
BCl_2^+ + Cl	12.30	
$BCl_3^+\,(\tilde{B}\,{}^2E') + e$	12.66	97.9
$BCl_3^{+*}\,(\tilde{C}\,{}^2A_2'') + e$	14.22	87.2
$BCl_3^{+*}\,(\tilde{D}\,{}^2E') + e$	15.32	80.9
$BCl_3^{+*}\,(\tilde{E}\,{}^2A_1') + e$	17.74	69.9
BCl^+ + 2Cl + e	18.37	

$X\,{}^1\Sigma^+$ band are compared with the measurements of Gilbert *et al.*, respectively, in Figs. 6.81(a) and 6.81(b). The cross section values for these emissions – and also for the $B^*\,2p^2\,{}^2D \rightarrow 2p\,{}^2P^0$ transition – are compared in Table 6.54 for 100 eV electrons. Although the cross section value of Gilbert *et al.* for the transition $B^*\,2p^2\,{}^2D \rightarrow 2p\,{}^2P^0$ is about a factor of 2 lower than that of Tokue *et al.*, the agreement between the measurements of the two groups is generally good. In addition to the emissions from the three species just mentioned, Tokue *et al.* [347] observed two continuous emissions in the wavelength regions 230–380 nm and 400–580 nm which they attributed to BCl_2^*.

Figure 6.82(a) shows the fluorescence cross section, and Fig. 6.82(b) the fluorescence quantum yield, of BCl_3 in the wavelength range 106–190 nm as measured by Suto *et al.* [351] using synchrotron light. The fluorescence cross section measurements have a quoted uncertainly of ±30%. They indicate a fluorescence threshold of $\sim$ 173 nm (7.17 eV) and a number of emission bands in the visible and the UV regions attributed by Suto *et al.* to excited states of the BCl_2 photofragment. The

Table 6.53. Calculated and observed minimum energies for the electron-impact-induced fragmentation of BCl_3 into excited boron (B^*), boron chloride (BCl^*), and BCl_2^* fragments.

Fragments[a]	Minimum energy (eV)	Reference
B^* $(3s\,^2S)$ + Cl_2 + Cl	[16.2][b]	[346]
	[16.28]	[347]
B^* $(3s\,^2S)$ + 3Cl	[18.70]	[347]
	[18.8]	[346]
	22.0 ± 1.5[c]	[346]
B^* $(3s\,^2S)$ + Cl_2 + Cl^+	[29.1]	[346]
B^* $(3s\,^2S)$ + 2Cl + Cl^+	[31.7]	[346]
B^* $(2p^2\,^2D)$ + Cl_2 + Cl	[17.1]	[346]
	[17.25]	[347]
	17.7 ± 0.2	[347]
B^* $(2p^2\,^2D)$ + 3Cl	[19.67]	[347]
	[19.7]	[346]
	20.0 ± 0.5	[347]
B^* $(2p^2\,^2D)$ + Cl_2 + Cl^+	[30.0]	[346]
	[30.28]	[347]
B^* $(2p^2\,^2D)$ + 2Cl + Cl^+	[32.6]	[346]
	[32.70]	[347]
B^* $(3p\,^2P_0)$ + Cl_2 + Cl	[17.34]	[347]
	17.7 ± 0.4	[347]
B^* $(3p\,^2P_0)$ + 3Cl	[19.76]	[347]
	19.7 ± 0.3	[347]
B^* $(3p\,^2P_0)$ + Cl_2 + Cl^+	[30.37]	[347]
B^* $(3p\,^2P_0)$ + 2Cl + Cl^+	[32.79]	[347]
BCl^* $(A\,^1\Pi)$ + Cl_2	[10.50]	[346, 347]
BCl^* $(A\,^1\Pi)$ + 2Cl	[13.1]	[346]
	[12.92]	[347]
	13.9 ± 0.5	[347]
	14.0 ± 1.5	[346]
BCl^* $(A\,^1\Pi)$ + Cl + Cl^+	[26.0]	[346]
	[25.95]	[347]
	25–29	[347]
	29.0 ± 2.0	[346]
BCl_2^* (468 nm) + Cl	[6.75]	[347]
	7.0 ± 0.6	[347]
	7.17	[351]
BCl_2^* (330 nm) + Cl	[8.90]	[347]
	9.4 ± 0.4	[347]
	9.80	[351]

[a]Excited fragments are designated by the asterisk (*).
[b]Values in brackets are calculated by the respective authors using the enthalpies of formation for the parent molecule, the related fragments, and electronic energies of the excited states involved (see respective references for details).
[c]Values not in brackets are measurements.

Table 6.54. Absolute cross sections of boron and boron chloride fragment emissions formed by 100 eV electrons impacting on BCl_3, and associated wavelengths (λ), transitions, and measured onsets.

Fragment	λ (nm)	Transition	Cross section (10^{-22} m^2)	Onset energy (eV)	Ref.
Boron	208.9	$2p^2\,{}^2D \rightarrow 2p\,{}^2P^0$	1.8 ± 0.9^a	23.0 ± 1.5	[346]
				34.5 ± 2.0	[346]
	208.9	$2p^2\,{}^2D \rightarrow 2p\,{}^2P^0$	4.9 ± 1.0		[347]
Boron	249.8	$3s\,{}^2S \rightarrow 2p\,{}^2P^0$	4.9 ± 1.6	22.0 ± 1.5	[346]
				34.0 ± 2.0	[346]
	249.7	$3s\,{}^2S \rightarrow 2p\,{}^2P^0$	4.5 ± 0.7		[347]
Boron chloride	272.4^b	$A\,{}^1\Pi \rightarrow X\,{}^1\Sigma^+$	1.9 ± 0.4	14.0 ± 1.5	[346]
				29.0 ± 2.0	[346]
	271.5^c	$A\,{}^1\Pi \rightarrow X\,{}^1\Sigma^+$	1.9 ± 0.3		[347]

[a] Estimated uncertainty ±50%.
[b] The wavelength refers to the location of the maximum intensity of the $\Delta\nu = 0$ vibrational sequence of the $A\,{}^1\Pi \rightarrow X\,{}^1\Sigma^+$ band.
[c] Unresolved (0,0), (1,1), and (2,2) bands.

fluorescence quantum yield was calculated from the ratio of the fluorescence and absorption cross sections (see Figs. 6.82(a) and 6.71). It shows three broad bands with peaks near 118, 130, and 142 nm, which indicate that the dissociating states have, respectively, vertical excitation energies of 10.5, 9.5, and 8.7 eV.

Lee *et al.* [352] extended these room-temperature investigations to excitation wavelengths in the region 45–106 nm. Their fluorescence data are also shown in Fig. 6.82 and have a quoted uncertainly of ±30%. The observed emissions at excitation wavelengths longer than 96 nm were attributed to BCl_2^* and those at excitation wavelengths shorter than 97.5 nm to BCl* (A-X) [352]. Emissions at the thresholds of 88 and 81 nm were attributed to BCl_3^{+*} and emissions appearing in the 48–64 nm region to B* atoms.

Synchrotron light was also employed by Boyle *et al.* [356] to study the fluorescence processes in BCl_3 following VUV photoexcitation of BCl_3 in the energy range 9 eV (137.7 nm) to 22 eV (56.4 nm). Photoexcitation of BCl_3 in this energy range resulted in complicated fluorescence spectra in which the observed emissions were assigned to BCl_2^*, BCl*, BCl_3^{+*}, BCl_2^{+*}, and B*. As the authors noted, their experiments are sensitive to those Rydberg states of BCl_3 which photodissociate to an excited state of a neutral fragment that fluoresces, and to valence states of the parent or a fragment ion that fluoresce. These two processes can be represented

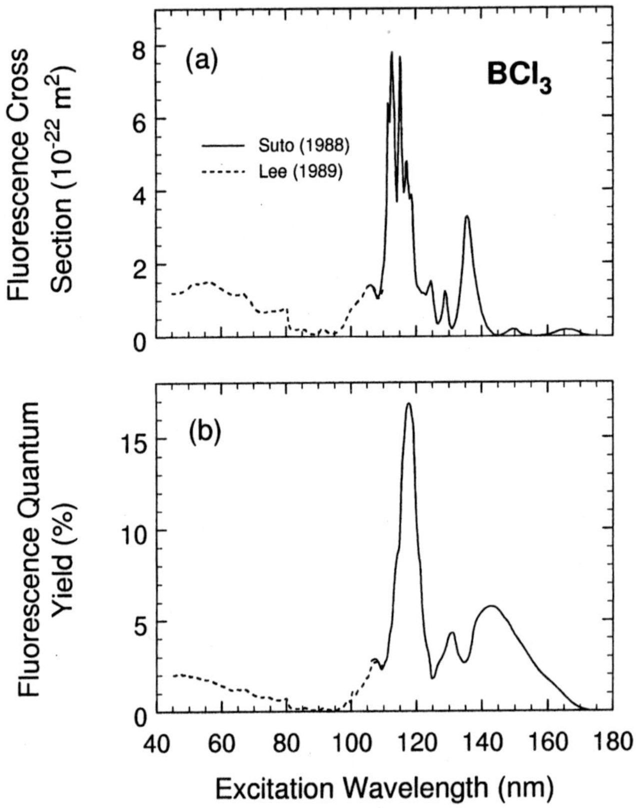

Figure 6.82. (a) Fluorescence cross section and (b) fluorescence quantum yield as a function of excitation wavelength for BCl_3: (—) Ref. [351]; (- - -) Ref. [352].

as [356]

$$BCl_3 + h\nu_1 \rightarrow BCl_3^*, \tag{6.29a}$$

$$BCl_3^* \rightarrow BCl_2^* + Cl, \tag{6.29b}$$

$$BCl_3^* \rightarrow BCl^* + (Cl_2 \text{ or } 2Cl), \tag{6.29c}$$

$$BCl_3^* \rightarrow B^* + (3Cl \text{ or } Cl_2 + Cl), \tag{6.29d}$$

$$BCl_2^* \text{ or } BCl^* \text{ or } B^* \rightarrow BCl_2 \text{ or } BCl \text{ or } B + h\nu_2, \tag{6.29e}$$

and

$$BCl_3 + h\nu_1 \rightarrow BCl_3^{+*} + e, \tag{6.30a}$$

$$BCl_3^{+*} \rightarrow BCl_3^+ + h\nu_2, \tag{6.30b}$$

$$BCl_3^{+*} \rightarrow BCl_2^{+*} + Cl \rightarrow BCl_2 + Cl + h\nu_2, \tag{6.30c}$$

where $h\nu_1$ and $h\nu_2$ represent, respectively, photons in the VUV (50–150 nm) and UV/VIS (200–700 nm).

Finally, a number of spectroscopic studies have been performed on discharges in gas mixtures containing BCl_3 (e.g., see Refs. [395–397] for mixtures of BCl_3 with N_2 or Ar), which provided additional information regarding the dissociation products of BCl_3 and the role of metastable states of the buffer gas employed. Thus, Breitbarth [395] concluded that in BCl_3/N_2 discharges, direct electron-impact excitation is not responsible for the BCl* and B* emissions, but rather an energy transfer from the N_2* (W) state to the BCl_x species present in his system. In the case of BCl_3/Ar discharges, Scheller *et al.* [396] argued that argon metastables indirectly enhance molecular dissociation by increasing the positive ion density via Penning ionization (that is, especially at concentrations below 5%, Penning ionization generates BCl_3^+ which subsequently undergoes dissociative recombination). Similarly, Breitbarth and Ducke [397] observed three broad emissions with maxima at 305, 350, and 480 nm in rf discharges using pure BCl_3. They attributed all three emissions to the electronically excited BCl_2^* fragment. In discharges with pure BCl_3, they showed the dominant process for formation of excited BCl*(A) fragments to be direct electron-impact excitation of BCl(X). For mixtures of BCl_3/Ar, however, the BCl*($A\,^1\Pi$) species are also produced via energy transfer from argon metastables.

It should perhaps be mentioned that a number of other photophysical studies have been made to probe the properties of BCl_3 gas discharges. Discharges containing BCl_3 provide a convenient source of BCl radicals whose Stark-mixed laser-induced fluorescence spectrum can be used to determine the magnitude of the sheath electric field with high spatial and temporal resolution [398]. For instance, Gottscho [399] used Stark-mixed laser-induced fluorescence to measure space-time-resolved maps of sheath electric fields in discharges through BCl_3 and in mixtures of BCl_3 with Ar as a function of power density (0.14–0.41 W cm^{-3}) and frequency (DC to 10 MHz).

4.4 Electron Attachment to BCl_3

Electron attachment to BCl_3 has been experimentally investigated using electron beam [338–341, 343, 344] and electron swarm [368, 400] methods. There have also been some studies of negative ion formation in BCl_3 gas discharges [372, 401, 402]. The results of these investigations are synthesized, assessed, and discussed in this section.

4.4.1 Electron Attachment Processes in BCl_3

Collectively, the results of a number of electron-beam mass spectrometric studies [338–340, 343, 344] on the relative cross sections for the production of fragment negative ions by low-energy electron impact on BCl_3, are consistent with the reactions

$$BCl_3 + e \rightarrow BCl_3^{-*} \rightarrow BCl_2 + Cl^-, \tag{6.31}$$

$$BCl_3 + e \rightarrow BCl_3^{-*} \rightarrow BCl + Cl_2^-, \tag{6.32}$$

and

$$BCl_3 + e \rightarrow BCl_3^{-*} \text{ (long lived)}. \tag{6.33}$$

They all show that dissociative electron attachment to BCl_3 produces primarily Cl^- via Reaction (6.31). These studies [339, 340, 343, 344] also showed the production of Cl_2^- via Reaction (6.32) albeit with much smaller probability (the amount of Cl_2^- ions may be affected by ion-molecule reactions [340]). The existence of long lived BCl_3^{-*} ions [Reaction (6.33)] at thermal electron energies has been indicated by electron beam [343, 344] and by electron swarm studies (see later in this section). These findings are discussed further below.

The dissociative electron attachment Reaction (6.31) is endoergic by approximately 1 eV. Olthoff [343] estimated the energy threshold for Cl^- to be ~ 0.7 eV from thermodynamic data, and Baeck and Bartlett [336] calculated the dissociation energy of BCl_3^- into $BCl_2 + Cl^-$ to be 1.59 eV. In accord with these values, Stockdale *et al.* [339] found the lowest peak in the relative cross section for the formation of Cl^- from BCl_3 to be located at (1.1 ± 0.1) eV, and Brüning [344] observed the most intense peak in the Cl^- production to be at ~ 0.9 eV (Fig. 6.83). However, in similar studies Olthoff [343] found the cross section for Cl^- to peak at an energy (see Fig. 6.83) well below the thermodynamic threshold of 0.7 eV. Although the production of Cl^- is expected to be temperature dependent [289] — and thus possible differences in the temperature of the various experimental studies may influence the magnitude and position of the low-energy Cl^- resonance — the temperature in these experiments is not expected to differ appreciably. Brüning, for instance, quotes the temperature of his experiment to be ~ 340 K, and the temperature in the experiment of Olthoff is expected to be similar. Nonetheless, a study of the effect of temperature on the cross section for the production of Cl^- ions is indicated for this molecule. The observation of Cl^- may also be attributed to impurities produced by reactions of the BCl_3 with the walls of the gas manifold.

Besides the Cl^- peak near 1 eV, the relative cross section for the production of Cl^- from BCl_3 was found [339, 343, 344] to have a number

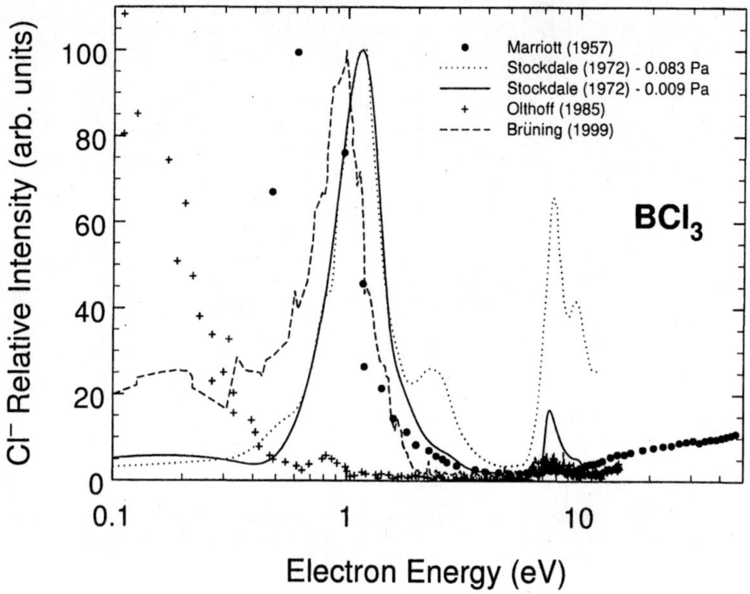

Figure 6.83. Relative cross section for the production of Cl^- by electron impact on BCl_3: (•) Ref. [338]; (···) Ref. [339] for two pressures (0.009 and 0.083 Pa) showing the production of Cl^- directly via Reaction (6.31) at 1.1 eV and indirectly via Reactions (6.34a) and (6.34b) at higher energies (see text); (+) Ref. [343] under single-collision conditions; (- - -) Ref. [344] at a pressure of ~ 0.001 Pa.

of weaker peaks between 6 and 10 eV (see Fig. 6.83 and Table 6.47) which are near the location of electron-excited Feshbach resonances observed in electron transmission studies [337]. Since the pressure in the experiments of Brüning [344] and Olthoff [343] was sufficiently low (of the order of 10^{-5} mbar) to ensure single-collision conditions, the Cl^- observed in these investigations between 6 and 10 eV is due to primary processes, that is, due to the decay of the negative ion states in this energy range via dissociative electron attachment. However, in the experiments of Stockdale *et al.* [339] the pressure was much higher [(9.2 to 82) $\times 10^{-5}$ mbar], and in this case, in addition to the 1.1 eV peak, Stockdale *et al.* observed Cl^- peaks around 2.5 and 8 eV. The intensity of these peaks varied as the square of the BCl_3 pressure. They were attributed [339] to the processes

$$e + BCl_3 \rightarrow BCl_3^* + e_{slow}, \qquad (6.34a)$$

$$e_{slow} + BCl_3 \rightarrow Cl^- + BCl_2, \qquad (6.34b)$$

that is, to the production of Cl^- by dissociative attachment to BCl_3 of electrons, e_{slow}, which have first been slowed down to near 1 eV via the negative ion states near 2.5 and 8 eV.

In addition, the measurements of Marriott and Craggs [338] shown in Fig. 6.83 indicate the production of Cl^- at energies $\gtrsim$ 9 eV, which may be due to the ion-pair process $BCl_3 + e \rightarrow BCl_2^+ + Cl^- + e$. Marriott and Craggs were careful to point out however that because BCl_3 hydrolyzes readily ($BCl_3 + 3H_2O \rightarrow H_3BO_3 + 3HCl$), the production of HCl may affect the measurements dealing with the products Cl^-, Cl^+, and Cl^{++} which may originate from electron impact on HCl.

The dissociative electron attachment Reaction (6.32) is also endoergic. Olthoff [343] estimated the threshold energy for Reaction (6.32) to be $\sim$ 3.3 eV, and Baeck and Bartlett [336] calculated the dissociation energy of BCl_3^- into BCl + Cl_2^- to be 4.01 eV. These values are incompatible with the observation by Stockdale *et al.* [339] that the relative cross section of Cl_2^- ions from BCl_3 peaks at (1.1±0.1) eV and the observations by Olthoff [343] and Brüning [344] that the relative cross section for the production of Cl_2^- ions from BCl_3 peaks near 0.0 eV. Indeed, because of this discrepancy, the observed Cl_2^- signals in these experiments may be due to impurities. Interestingly, Brüning observed BCl_4^-, which Jiao *et al.* [340] found to be an ion-molecule reaction product resulting from the rapid reaction of Cl_2^- with BCl_3. No BCl_2^- or BCl^- fragment negative ions were reported in any of these studies.

Reaction (6.33) occurs at thermal and near-thermal energies. The transient BCl_3^{-*} species, which is initially formed by electron capture, is metastable and thus, can decay by autodetachment. It is, however, long lived [343, 344, 372, 400]. Its autodetachment lifetime was inferred [343] to be at least 60 μs at thermal energies and temperatures somewhat higher than ambient (due to the heating of the gas from the hot filament in these experiments). This long lifetime may result from the fact that the equilibrium geometries of BCl_3 and BCl_3^- are different — the BCl_3 molecule is planar (D_{3h}) and the BCl_3^- ion is pyramidal (C_{3v}) (Sec. 4.1 of Ch. 6). Such geometrical changes are known to lead to long lifetimes for a number of transient anions [403].

The long-lived BCl_3^{-*} ion can rid itself of its extra energy by collision (or radiation) and lead to the formation of a stable parent negative ion BCl_3^- since the BCl_3 molecule has a positive electron affinity (Table 6.47). The electron attachment rate constants measured in swarm experiments (see below) at low E/N may thus be due to Reaction (6.33). They may be sensitive to the gas temperature (decrease with increasing temperature above ambient) since the electron affinity of BCl_3 is small (Table 6.47) and the probability of thermally induced electron detach-

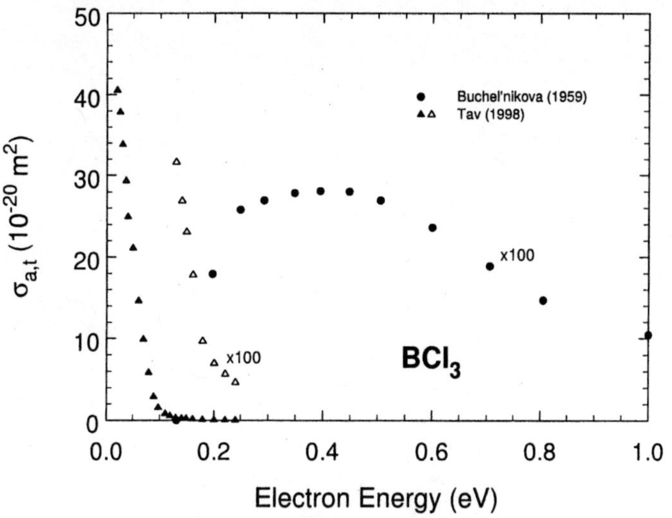

Figure 6.84. Total electron attachment cross section as a function of electron energy, $\sigma_{a,t}(\varepsilon)$, for BCl_3: (•) electron beam data from Ref. [341] (Note that these data have been multiplied by a factor of 100 for the convenience of display); (▲) electron swarm data from Ref. [368] (the data shown with open triangles are multiplied by 100 to facilitate comparison).

ment from the stabilized BCl_3^- (BCl_3^- + heat → BCl_3 + e) can be large (e.g., see Refs. [289, 404]).

A summary of the data on negative ions found in electron beam studies to be produced by low-energy electron impact on BCl_3 is given in Table 6.47 (see also Fig. 4.72).

4.4.2 Total Electron Attachment Cross Section, $\sigma_{a,t}(\varepsilon)$, for BCl_3

There has been only one measurement of the total electron attachment cross section, $\sigma_{a,t}(\varepsilon)$, of BCl_3 using the electron beam method, namely, that by Buchel'nikova [341]. This total electron attachment cross section is shown in Fig. 6.84. It has a maximum at near 0.4 eV and a cross section value at this energy of 2.8×10^{-21} m^2. While Buchel'nikova's technique yields total electron attachment values, the data in Fig. 6.84 were attributed to the production of Cl^- via Reaction (6.31). As has been found for other molecules that Buchel'nikova had investigated, these cross section data for BCl_3 are considered to be lower than their true values.

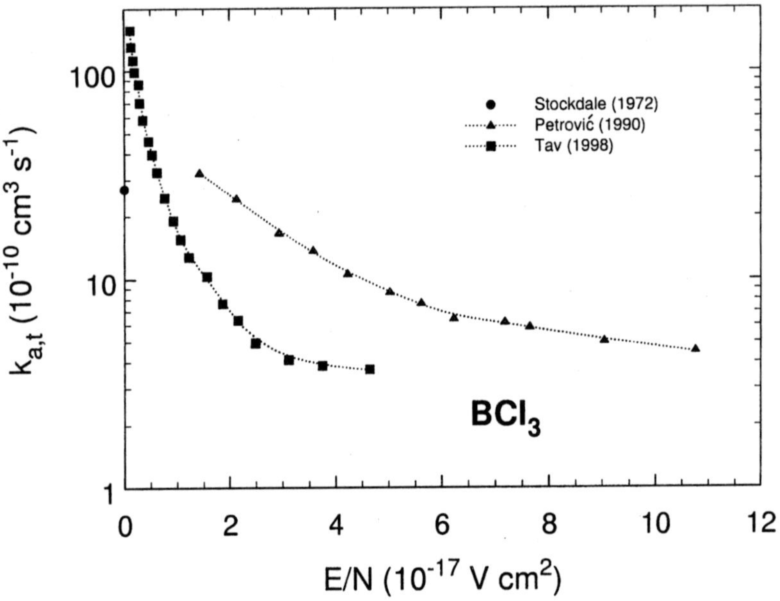

Figure 6.85. Total electron attachment rate constant $k_{a,t}(E/N)$, for BCl_3: (▲) Ref. [400]; (■) Ref. [368]; (•) thermal value of Ref. [339].

4.4.3 Total Electron Attachment Rate Constant, $k_{a,t}$, as a Function of E/N and $\langle \varepsilon \rangle$ for BCl_3

There have been two [368, 400] investigations of electron attachment to BCl_3 using the electron swarm method. Both of these studies employed mixtures of BCl_3 with N_2 and were conducted at room temperature ($\sim$ 295 K). In both studies the total electron attachment rate constant $k_{a,t}(E/N)$ was measured and is shown in Fig. 6.85. The measurements of Petrović *et al.* [400] have a quoted uncertainty of $\pm 30\%$. Although Tav *et al.* [368] gave no uncertainty for their measurements, the uncertainty in their measurements is expected to be at least as large as that of Petrović *et al.* [400] considering the large variation of their data with time following the introduction of BCl_3 into their reaction chamber. Also shown in Fig. 6.85 is an earlier measurement [339] of the thermal electron attachment rate constant. There are substantial differences between these data, stressing the need for new measurements.

Petrović *et al.* [400] argued that since the threshold for dissociative electron attachment producing either Cl^- or Cl_2^- is greater than ~ 1 eV and since for the E/N range they investigated the values of the mean electron energy are less than ~ 1 eV, the electron attachment rate constants they measured in their experiments (Fig. 6.85) are due to the formation of BCl_3^-. This interpretation would be consistent with the mass spectrometric observation [343, 344] of long lived BCl_3^{-*}, and with the conclusion of Gottscho and Gaebe [372] that BCl_3^- is the dominant negative ion formed in rf discharges using pure BCl_3. Although Petrović *et al.* observed only Cl^- in dc discharges of BCl_3, they argued that the difference between the results of the dc and rf discharge studies may be due to the difference in the electron energies in the 2 types of discharges. In dc discharges the mean electron energies may be too high to enable production of BCl_3^-, in contrast to the bulk of rf plasma where the electron energies are much lower allowing production of BCl_3^- (see also Ref. [372]).

Similarly, Tav *et al.* [368] attributed the total electron attachment rate constant they measured at low electron energies ($\lesssim 0.1$ eV) to the formation of BCl_3^-, and the weaker electron attachment at higher energies to Cl^- formation. However, neither in the study of Petrović *et al.* nor in the study of Tav *et al.* were the negative ions directly identified. It is emphasized that this indirect inference of the identity of the ions needs further scrutiny. The argument, for instance, that Cl^- cannot be formed at low energies for energetic reasons may not be correct, if there is a large effect of temperature on the dissociative electron attachment process generating Cl^-. Also, since the electron affinity of the BCl_3 molecule is small (Table 6.47), the electron may be detached thermally from the stabilized BCl_3^- ion at temperatures at and above ambient. Therefore, to fully characterize the electron attachment processes for this molecule, an investigation is needed of the temperature dependence of the formation of both Cl^- and BCl_3^-.

The $k_{a,t}\,(E/N)$ data in Fig. 6.85 are plotted in Fig. 6.86 as a function of the mean electron energy, $\langle\varepsilon\rangle$, using the $\langle\varepsilon\rangle\,(E/N)$ data for N_2 ($T =$ 300 K) given in Table A.II of Christophorou and Olthoff [289] (see also Table 6.82 of Ch. 6).

4.4.4 Thermal Electron Attachment Rate Constant, $(k_{a,t})_{th}$, of BCl_3

The thermal ($T \approx 295$ K) value, $(k_{a,t})_{th}$, of the total electron attachment rate constant of BCl_3 has been measured by Stockdale *et al.* [339] to be 27×10^{-10} cm^3 s^{-1}. However, there is potential disagreement with the other measurements [368, 400] of $k_{a,t}$ (see Figs. 6.85 and 6.86).

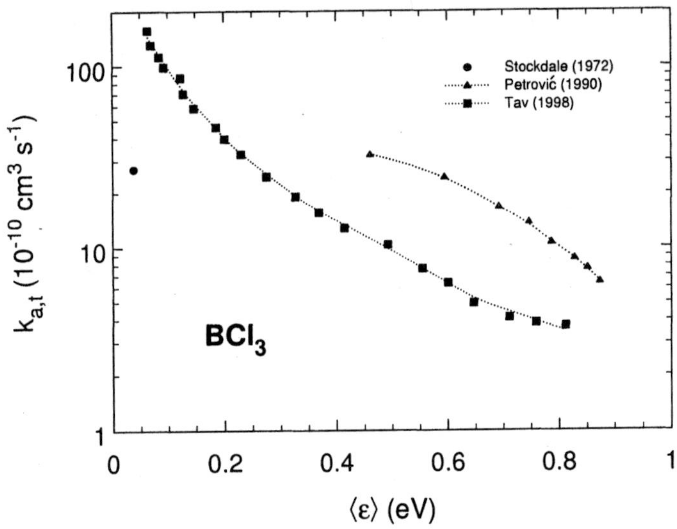

Figure 6.86. Total electron attachment rate constant $k_{a,t}(\langle\varepsilon\rangle)$, for BCl_3: (▲) Ref. [400]; (■) Ref. [368]; (•) thermal value of Ref. [339].

4.4.5 Swarm-Unfolded Total Electron Attachment Cross Section, $\sigma_{a,t}(\varepsilon)$, of BCl_3

Tav *et al.* [368] unfolded the total electron attachment cross section $\sigma_{a,t}(\varepsilon)$ shown in Fig. 6.84 using their data on $k_{a,t}\,(E/N)$ and the known electron energy distribution functions in the buffer gas N_2. They attributed the rise in the cross section below $\sim$ 0.1 eV to the formation of BCl_3^-. The smaller cross section at higher energies was attributed to Cl^- formation, but no direct identification of the negative ions was made. They indicated that their derived cross section has an uncertainty of $\pm$20%, but the uncertainty is probably much larger than indicated. The disagreement between the swarm-unfolded total electron attachment cross section [368] and the beam-determined values [341] is large. Clearly, there is a need for further measurements.

4.4.6 Negative Ion Photodetachment in BCl_3 Plasmas

Gaebe *et al.* [401] described a spectroscopic diagnostic technique that provides the response of a radio-frequency discharge to electrons photodetached from negative ions in BCl_3. In this technique the spatially and temporally resolved charges in the local electric field of a 50 kHz discharge through BCl_3 (changes in the local electric field which resulted from photodetachment of electrons from negative ions) are monitored using Stark-mixing spectroscopy (e.g., see Refs. [398, 405]).

Similarly, Fleddermann and Hebner [402] investigated the Cl^- concentration and the electron density in BCl_3-containing inductively coupled plasmas using laser light to photodetach electrons from Cl^- ions. They detected the photodetached electrons and the steady-state electron density using a microwave interferometer, which allowed measurement of negative ion densities as a function of gas mixture and reactor parameters. The 266 nm radiation from a frequency-quadrupled Nd:YAG laser was used for photodetachment. The plasma was probed at lower photon energy (355 nm) to photodetach electrons from other possible negative ions present in the plasma. To within the sensitivity of their measurement (1.3×10^8 cm^{-3}), no other negative ions (such as BCl_x^- or Cl_2^-) were detected.

4.5 Electron Transport Coefficients for BCl_3

There are no measurements of the electron transport coefficients in pure BCl_3. However, there have been some measurements of the electron drift velocity, $w(E/N)$, in BCl_3/Ar gas mixtures at concentrations of BCl_3 in Ar ranging from 2.5×10^{-5} to 5×10^{-3} which are representative of those used in plasma applications [331]. According to Mosteller *et al.* [331], these mixtures exhibit strong electron attachment at low E/N. The $w(E/N)$ measurements are shown in Fig. 6.87(a) and indicate regions of negative differential conductivity. Mosteller *et al.*, also measured the $w(E/N)$ in a mixture of 0.54% BCl_3 in He. These measurements are shown in Fig. 6.87(b), where they are compared with the $w(E/N)$ of Milloy and Crompton [406] and Nygaard *et al.* [407] for pure He.

4.6 Suggested and Needed Data for BCl_3

The only suggested electron collision data for BCl_3 are those for $\sigma_{i,t}(\varepsilon)$ (Fig. 6.79, Table 6.51). There is a need for further measurements of this cross section, and also for measurements of the cross sections of all other principal electron scattering processes for this molecule. Measurements are also needed of the electron transport, ionization, and attachment coefficients in pure BCl_3 and in mixtures.

* * *

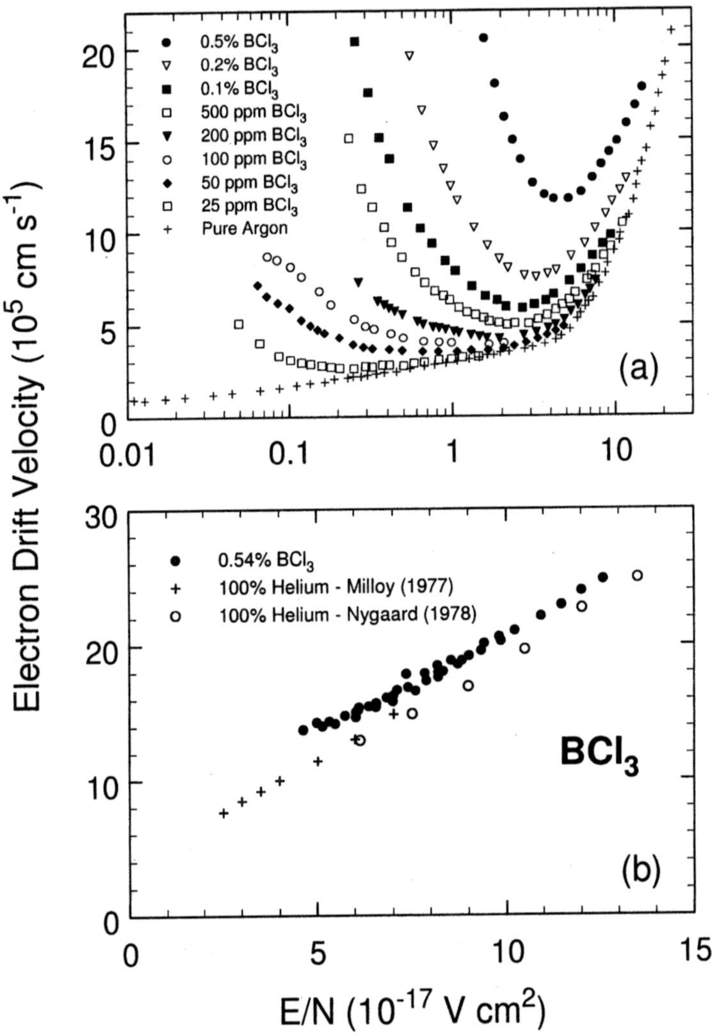

Figure 6.87. (a) Electron drift velocity $w(E/N)$ for mixtures of BCl_3 in argon as measured by Mosteller *et al.* [331]. (b) Electron drift velocity $w(E/N)$ for a 0.54 % BCl_3 mixture with He (•, Ref. [331]), in comparison with the $w(E/N)$ for pure He (+, Ref. [406]; ○, Ref. [407]).

5. Electron Interactions with SF_6

Sulfur hexafluoride (SF_6) is a man-made molecule used in many applied areas [5, 408]: as an insulating medium in high-voltage transmission and distribution equipment; a high-dielectric strength gas in ultraviolet

(UV) laser-triggered spark-gap closing switches; a fluorine donor in, for instance, rare gas-halide excimer lasers; a gas for plasma etching of silicon and GaAs-based semiconductors; a blanket gas for magnesium casting; a reactive gas in aluminum recycling to reduce porosity; a gas for thermal and sound insulation; and a gas for use in many other wide ranging areas including retinal detachment surgery, airplane tires, AWACS radar domes, and x-ray machines. Its general properties, especially as they relate to its use by the electric power industry, have been summarized and discussed by Christophorou *et al.* [5]. Sulfur hexafluoride is also of environmental concern in that it has been shown to be a potent greenhouse gas [409]. It is an efficient absorber of infrared radiation, and because it is very stable, is largely immune to chemical and photolytic degradation. The latter accounts for its long residence lifetime in the environment. The strong infrared absorption and the long (800–3200 year) residence lifetime [409–413] in the environment account for its high global warming potential, which for a 100 year period is estimated to be approximately 24 000 times greater than that of CO_2 [409]. Because of its unique structure and applied interest, the SF_6 molecule has been the subject of many experimental, theoretical, and computational studies. There also have been a number of reviews of its interactions with slow electrons [5, 414–420] and photons [64, 421–423].

In the present work we synthesize and critically evaluate existing knowledge on the interactions of low-energy (mostly below about 100 eV) electrons with the SF_6 molecule and recommend, suggest, or simply provide values for the various electron scattering cross sections, coefficients, and rate constants that are used to quantify these processes. Data are presented and discussed on the following cross sections: total electron scattering; elastic differential; elastic integral; elastic momentum; total vibrational; total and partial ionization; total dissociative and nondissociative electron attachment; and dissociation into neutrals. Coefficients for electron-impact ionization, effective ionization, electron attachment, electron drift, and electron diffusion are also reviewed and assessed. In addition, complementary information on the electronic and molecular structure of the SF_6 molecule and on electron detachment and ion transport in parent SF_6 gas is provided that allows a better understanding of the nature of the cross sections and swarm parameters. The assessed data are used to deduce cross sections and coefficients for which there exist no direct measurements at the present time.

The present work on electron interactions with the SF_6 molecule reveals a rather simple picture which can be summarized as follows: (1) Elastic electron scattering is the most significant electron scattering process over most of the electron energy range from ~ 0.01 to ~ 1000 eV.

(2) Below 15 eV the most distinct inelastic energy-loss process is vibrational excitation — direct dipole excitation involving the ν_3 mode and indirect vibrational excitation via negative ion states involving the ν_1 mode. (3) Below $\sim$ 0.1 eV electron attachment forming SF_6^- is the most dominant interaction (along with elastic scattering). Above this energy, the cross sections for dissociative electron attachment forming fragment anions [principally SF_x^- ($x = 3$, 4, and 5) and F^-] are appreciable, with the room temperature total electron attachment cross section dominated by the formation of SF_5^- between $\sim$ 0.3 and 1.5 eV and by the formation of F^- beyond $\sim$ 2.0 eV. (4) Above $\sim$ 16 eV dissociative ionization becomes significant, generating principally SF_x^+ (x = 1–5), F^+, and S^+ positive-ion fragments which, together with elastic electron scattering, makes up most of the total electron scattering cross section. (5) Electron-impact dissociation into neutral fragments SF_x (x = 1, 2, and 3) and F occurs above $\sim$ 15 eV, with cross section values potentially exceeding those for ionization for electron energies near 20 eV. (6) The total electron scattering cross section exhibits distinct structure due to negative-ion resonances near 0.0, 2.5, 7.0, and 11.9 eV.

No attempt is made here to evaluate the predictions of the many Boltzmann-code or Monte-Carlo-type analyses of electron transport in SF_6 [417, 424–442] that are available in the literature.

5.1 Electronic and Molecular Structure of SF_6

The electronic structure of this highly symmetric molecule has long been a subject of much interest both because of its unusual properties (e.g., unusual spectroscopy) and because of its many industrial uses. In the SF_6 molecule the sulfur atom is at the center of a regular octahedron, the corners of which are occupied by the six fluorine atoms. It has ten valence orbitals containing 48 electrons and belongs to the point group O_h [421]. Its ground-state valence electron configuration is [443–448]: $(\text{core})^{22}\,(4a_{1g})^2(3t_{1u})^6\,(2e_g)^4(5a_{1g})^2\,(4t_{1u})^6\,(1t_{2g})^6\,(3e_g)^4\,[(1t_{2u})^6\,(5t_{1u})^6]$ $(1t_{1g})^6\,A_{1g}$, where the first three orbitals constitute the inner-shell valence region. The four lowest empty (valence) orbitals [446, 449] are: $(6a_{1g})^0\,(6t_{1u})^0\,(2t_{2g})^0\,(4e_g)^0$. However, there are still some questions as to the energetic sequence of the various orbitals. For instance, the $(5t_{1u})$ and $(1t_{2u})$ orbitals are nearly degenerate and their ordering unresolved. Actually, their sequence has been interchanged by some authors (e.g., see Refs. [419, 449], and also discussions in Refs. [422, 423, 443, 444, 450–452]). As noted by Gianturco and Jain [453], the SF_6 molecule, being an assembly of atoms containing 70 bound electrons, is difficult to handle by *ab initio* computational methods. The rather large nuclear charges of the fluorine atoms away from the center-of-mass, makes the single-center approach unlikely to work. Thus, computational models are by necessity approximate models.

5.1.1 Total Photoabsorption Cross Section, $\sigma_{pa,t}(\lambda)$, of SF_6

The photoabsorption (and photoelectron) spectrum of SF_6 has been investigated extensively over broad energy ranges. It is unusual in that the most prominent features are due to intervalence transitions. Rydberg series, which often dominate molecular absorption cross sections at energies of a few electron volts above the ionization threshold, are strongly suppressed in SF_6. This anomalous intensity distribution has been interpreted by Nefedov [454] and Dehmer [455] in terms of a potential barrier experienced by the outgoing electron.

Total photoabsorption cross sections, $\sigma_{pa,t}(\lambda)$, have been discussed by a number of authors (e.g., Hitchcock and Van der Wiel [456], Berkowitz [422], Holland *et al.* [445]). Also, Gallagher *et al.* [64] reviewed the data on the photoabsorption, photoionization, and ionic photofragmentation cross sections for this molecule. The reader is referred to these studies and the sources cited therein for detailed data and discussion.

In Fig. 6.88(a) are shown the total photoabsorption cross sections for SF_6 of Codling [457], Blechschmidt *et al.* [458], Lee *et al.* [450], Sasanuma *et al.* [459], and Holland *et al.* [445]. Collectively these measurements cover the photon energy range between $\sim$ 10 and $\sim$ 50 eV. There are considerable differences in these measurements. For example, the most recent cross section measurements by Holland *et al.* [445] lie below those of Blechschmidt *et al.* [458] by nearly a factor of 2 in some regions of the spectrum. They are in closer overall agreement with the measurements of Lee *et al.* [450] and Sasanuma *et al.* [459]. All the data however show that the most intense feature in the entire absorption spectrum occurs around 23 eV and is due to the $5a_{1g} \rightarrow 6t_{1u}$ and $5t_{1u} \rightarrow 2t_{2g}$ intervalence transitions which are enhanced by shape resonances [445]. For a spectral assignment see Mitsuke *et al.* [462], Sze and Brion [449], and Holland *et al.* [445] (see also Table 6.56 in Sec. 5.1.3 of Ch. 6).

For comparison with the photoabsorption measurements shown in Fig. 6.88(a), excitation cross sections [456, 460, 461] determined from high-energy electron energy-loss spectra of SF_6 are shown in Fig. 6.88(b). These spectra are equivalent to the total photoabsorption cross section as a function of the photon energy. The electron energy-loss spectrum of Simpson *et al.* [460] was obtained at 0° scattering angle with 400 eV incident electrons, and was normalized to the photoabsorption results of Codling [457] at 23.0 eV. The electron energy-loss spectrum of Hitch-

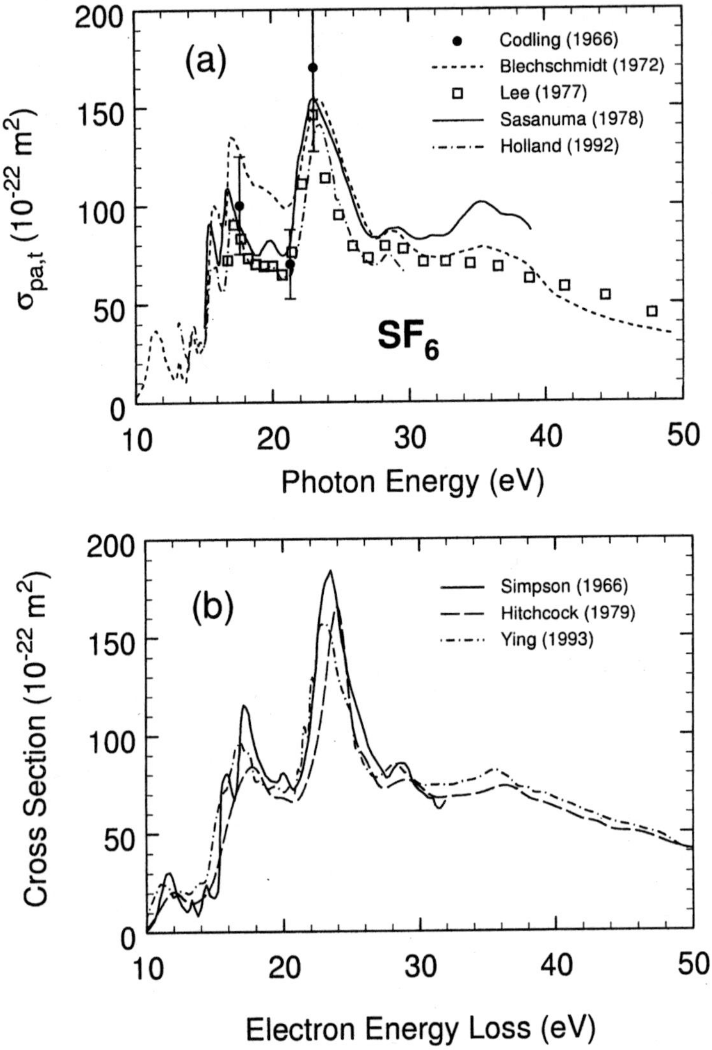

Figure 6.88. (a) Total photoabsorption cross section, $\sigma_{\mathrm{pa,t}}(\lambda)$, of SF_6 as a function of photon energy: (•) Ref. [457]; (- - -) Ref. [458]; (□) Ref. [450]; (—) Ref. [459]; (– · –) Ref. [445]. (b) Total excitation (photoabsorption) cross section obtained from high-energy electron energy-loss spectra (see text): (—) Ref. [460]; (– –) Ref. [456]; (– · –) Ref. [461].

cock and Van der Wiel [456] was also obtained at 0° scattering angle but with 8 keV incident electrons, and was normalized to the photoabsorption data of Blechschmidt *et al.* [458] at 27 eV. Ying *et al.* [461] used 0.5° scattering angle and 2.5 keV incident electrons. Their measurements of generalized oscillator strengths, df/dE were put on a cross section scale via the relation $\sigma_{\text{pa,t}}\,(10^{-18}\ \text{cm}^2) = 109.75 df/dE\,(\text{eV}^{-1})$.

Photoabsorption cross section measurements also have been extended into the extreme UV and soft x-ray regions using synchrotron radiation by a number of investigators [445, 447, 450, 451, 457, 458, 463–467].

5.1.2 Total Photoionization, $\sigma_{\text{pi,t}}(\lambda)$, Partial Photoionization, $\sigma_{\text{pi,part}}(\lambda)$, and Total Photodissociation, $\sigma_{\text{pdis,t}}(\lambda)$, Cross Sections of SF_6

Because the lowest ionic state of SF_6 is antibonding, the ionization process even at threshold is dissociative and results in the formation of the fragments SF_5^+ and F. No SF_6^+ ion has been observed in the preponderance of investigations, and its abundance is estimated to be less than 10^{-4} compared to that of SF_5^+. Thus, the photoionization of SF_6 is viewed as totally dissociative, and can be attributed to any of the following processes: direct, indirect, double, and Auger. Measurements of dissociative photoionization and dissociative and nondissociative double photoionization can be found in Gallagher *et al.* [64] and Hitchcock and Van der Wiel [456]. Holland *et al.* [445] used their measurements on the total photoabsorption cross section and the photoionization quantum efficiency as a function of the photon wavelength to calculate the total photoionization cross section $\sigma_{\text{pi,t}}(\lambda)$ and the total photodissociation cross section $\sigma_{\text{pdis,t}}(\lambda)$ for SF_6 shown in Fig. 6.89. In deducing the photodissociation cross section it has been implicitly assumed that an excited SF_6 molecule can decay only by dissociation into neutral fragments or by autoionization.

In Fig. 6.90 are shown the partial photoionization cross sections, $\sigma_{\text{pi,part}}(h\nu)$, of Hitchcock and Van der Wiel [456] deduced from dipole oscillator strength measurements over a range of energies up to an equivalent photon energy of 63 eV. In the same figure are also plotted the cross sections for dissociative and double photoionization of SF_6 in the energy range 75–125 eV obtained by Masuoka and Samson [468] using time-of-flight mass spectrometry and synchrotron radiation. The most abundant positive fragment ions are seen to be SF_5^+ and SF_3^+ (see also Creasey *et al.* [469] and Peterka *et al.* [470]). The relatively large cross section of the doubly charged positive ion SF_4^{++} was also observed under electron impact and was taken to indicate [471] the formation of SF_4^{++} by removal of two F^- ions.

Gustafsson [451] also measured partial photoionization cross sections of SF_6 between 20 and 54 eV using synchrotron light. In general, these

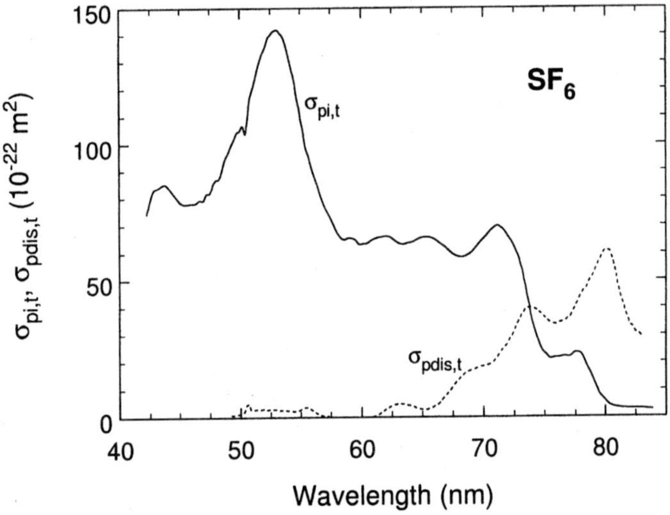

Figure 6.89. Total cross sections for photodissociation (- - -), $\sigma_{pdis,t}(\lambda)$, and photoionization (—), $\sigma_{pi,t}(\lambda)$, of SF_6 as a function of wavelength measured by Holland *et al.* [445].

cross sections (not shown in Fig. 6.90) are rich in structure showing distinct resonance effects (e.g., see Gustafsson [451] and Dehmer [455]).

Studies employing photoelectron spectroscopy allowed detailed investigations of the photoionization process, including the relative probabilities (partial cross sections) of exciting specific electronic, vibrational, or rotational states of the molecular ion. Photoelectron-photoion coincidence experiments also allowed determination of the subsequent fate of the excited molecular ions. In addition, photoion-photoion coincidence experiments (e.g., see Refs. [472, 473]) have allowed probing of double ionization processes. These can be either direct or via an Auger decay. In the first case, two highly correlated valence electrons are ejected. In the second case, an inner-shell electron is ejected, an Auger decay occurs, and the resulting doubly-charged positive molecular ion dissociates into two fragment ions (plus neutrals). Using such techniques in conjunction with synchrotron radiation, Frasinski *et al.* [473] detected the two photoions from the double photoionization of SF_6. The thresholds they determined for the production of the ion pairs $SF_5^+ + F^+$, $SF_3^+ + F^+$, $SF_2^+ + F^+$, $SF^+ + F_2^+$, and $SF_3^+ + F_2^+$ are listed in Table 6.55. According to Eland *et al.* [479] the major dissociation process of SF_6^{++} after HeII ionization is $SF_6^{++} \rightarrow SF_3^+ + F^+ + F_2$. It is possible, however, that SF_6^{++} dissociation may proceed via doubly-charged intermediates

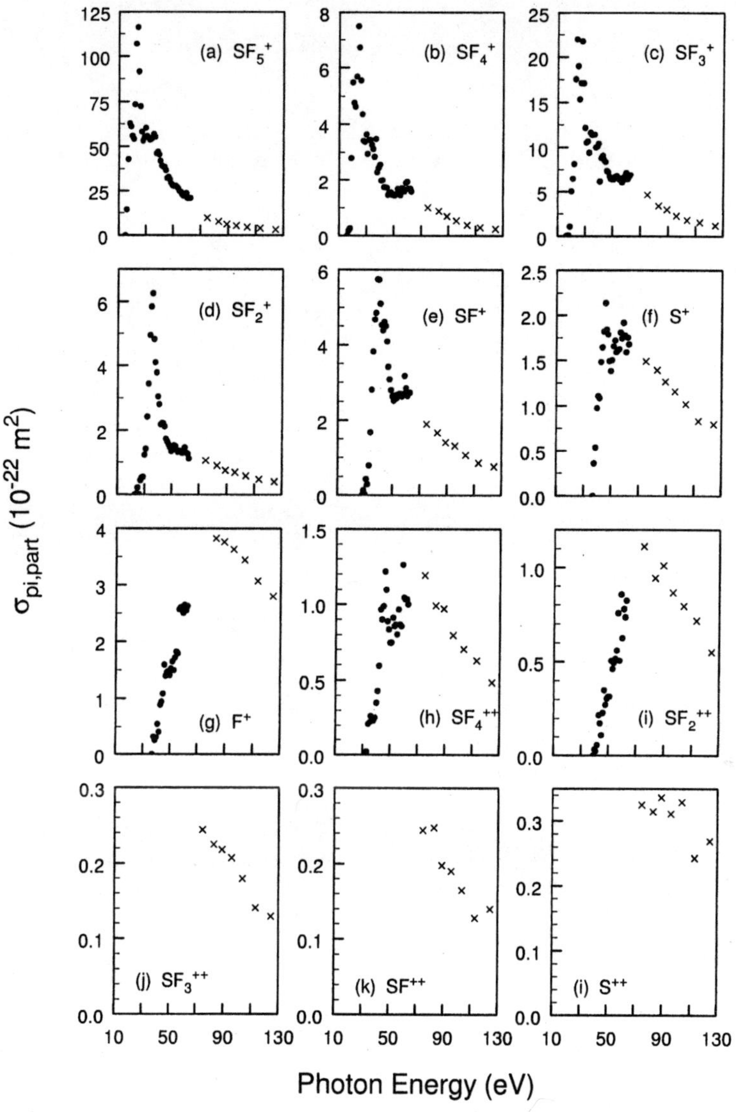

Figure 6.90. Partial photoionization cross sections, $\sigma_{\text{pi,part}}(h\nu)$, of SF_6 as a function of photon energy: (•) Hitchcock and Van der Wiel [456]; (×) Masuoka and Samson [468] (see text).

such as SF_4^{++} [479]. Table 6.55 lists the ionization energies of the various orbitals of the SF_6 molecule.

Ion-pair formation from photoexcitation of SF_6 is another interesting process which has revealed a great deal about the dissociation energetics

Table 6.55. Ionization energies of various orbitals of SF_6.

Type of orbital	Energy (eV)	Ref.	Comments
	15.29	[474][a]	Photoionization/Formation of SF_5^+
	15.3 ± 0.2	[475]	Photoionization/Formation of SF_5^+
	15.32	[462]	Photoexcitation/Formation of SF_5^+
	15.5	[476]	VES[b]
$1t_{1g}$	15.67	[476]	VES
t_{1u}	15.69	[477]	PES[c]
$1t_{1g}$	15.7	[452]	PES
$1t_{1g}$	15.7	[449]	Formation of SF_5^+
	15.9 ± 0.2	[471]	EI[d]; Formation of SF_5^+
	15.9 ± 0.2	[478]	EI; Formation of SF_5^+
	16.0	[456]	Formation of SF_5^+
	16.7	[476]	VES
$3t_{1u}$, $1t_{2u}$	16.93	[476]	VES
t_{1g}	16.96	[477]	PES
$1t_{2u}$, $5t_{1u}$	17.0	[452]	PES
$1t_{2u}$, $5t_{1u}$	17.0	[449]	EELS[e]
	18.0	[456]	Formation of SF_4^+
	18.0	[476]	VES
$2e_g$	18.3	[476]	VES
e_{1g}	18.40	[477]	PES
$3e_g$	18.6	[452]	PES
$3e_g$	18.6	[449]	EELS
	18.7 ± 0.2	[478]	EI; Formation of SF_4^+
t_{2u}	18.71	[477]	PES
	18.79 ± 0.14	[462]	Formation of SF_3^+
	18.9 ± 0.2	[471]	EI; Formation of SF_4^+
	19.1 ± 0.5	[475]	Photoionization/Formation of SF_4^+
	19.0	[456]	Formation of SF_3^+
$1t_{2g}$	19.245 (0-0)	[476]	VES
	19.4 ± 0.5	[475]	Photoionization/Formation of SF_3^+
t_{1u}	19.68	[477]	PES
$1t_{2g}$	19.8	[452]	PES
$1t_{2g}$	19.8	[449]	EELS
	20.1 ± 0.3	[471]	EI; Formation of SF_3^+
	20.3 ± 0.2	[478]	EI; Formation of SF_3^+

[a]The authors observed a weak SF_6^+ signal using the 21 eV resonance line of helium.
[b]VES = HeI induced valence electron spectra.
[c]PES = Photoelectron spectra.
[d]EI = Electron impact.
[e]EELS = Electron energy-loss spectra.
[f]DDP = Dissociative double photoionization.

Table 6.55 (continued)
Ionization energies of various orbitals of SF_6.

Type of orbital	Energy (eV)	Ref.	Comments
t_{2g}	22.5	[477]	PES
$4t_{1u}$	22.6	[452]	PES
$2t_{1u}$	22.7	[476]	VES
$4t_{1u}$	22.9	[449]	EELS
	26.0	[456]	Formation of SF_2^+
a_1	26.8	[477]	PES
	26.8 ± 0.3	[471]	Formation of SF_2^+
$5a_{1g}$	26.85	[452]	PES
$5a_{1g}$	27.0	[449]	EELS
$2a_{1g}$	27.0	[476]	VES
	31.0	[456]	Formation of SF^+
	31.3 ± 0.3	[471]	Formation of SF^+
	33	[456]	Formation of SF_4^{++}
	35.8 ± 1.0	[471]	Formation of F^+
	37.0	[456]	Formation of S^+
	37.3 ± 1.0	[471]	Formation of S^+
	38.0	[456]	Formation of F^+
$2e_g$	39.3	[449]	EELS
	40.0	[456]	Formation of SF_2^{++}
	40.6 ± 0.5	[471]	Formation of SF_4^{++}
$3t_{1u}$	41.2	[449]	EELS
	41.1 ± 1	[473]	DDP[f] producing $SF_5^+ + F^+$, $SF_3^+ + F^+$, or $SF_3^+ + F_2^+$
	45.1 ± 1	[473]	DDP producing $SF_2^+ + F^+$
	46.5 ± 0.5	[471]	Formation of SF_2^{++}
	52.8 ± 1	[473]	DDP producing $SF^+ + F_2^+$

[a]The authors reported observation of a weak SF_6^+ signal using the 21 eV resonance line of helium.
[b]VES = HeI induced valence electron spectra.
[c]PES = Photoelectron spectra.
[d]EI = Electron impact.
[e]EELS = Electron energy-loss spectra.
[f]DDP = Dissociative double photoionization.

and the structure of the SF_6 molecule. For instance, Mitsuke *et al.* [462] investigated the process

$$SF_6 + h\nu \rightarrow SF_6^{**} \rightarrow F^- + SF_x^+ + (5 - x)F \; (x \leq 5) \qquad (6.35)$$

employing negative-ion mass spectrometry and synchrotron radiation in the 11.27–31.0 eV photon energy range. Mitsuke *et al.* measured the photodissociation efficiency curve for F^- produced from SF_6 as a function of wavelength and compared the observed structure with that in the photoabsorption cross section data. The results of this study are compared with those from photoabsorption and electron energy-loss experiments in Table 6.56. Both the photon energy and/or energy-loss and the transition assignments are given in the table. For a fuller account and more detailed list of transitions see Refs. [449, 459, 462].

5.1.3 Electron Energy-Loss Spectra of SF_6

There have been a number of electron-impact energy-loss studies of SF_6 [449, 456, 460, 461, 480, 483–485]. The studies of Simpson *et al.* [460], Hitchcock and Van der Wiel [456], Sze and Brion [449], and Ying *et al.* [461] are of particular interest in that their high-energy electron-impact energy-loss spectra can be used to generate UV absorption cross sections. As discussed in the previous section, Simpson *et al.* obtained an energy-loss spectrum corresponding to optical electronic excitation of SF_6 using 400 eV incident energy and 0° scattering angle. Their high-energy electron-impact energy-loss spectrum was used to generate UV absorption cross sections in the 10–30 eV region by normalization to the photoabsorption results of Codling [457] at 23.0 eV (see comparison with photoabsorption data in Fig. 6.88). Hitchcock and Van der Wiel [456] also obtained electron-impact energy-loss spectra at 0° using electron-ion coincidence measurements and incident electrons with 8 keV kinetic energy. Their electron-energy loss data were also used to determine cross sections for dipole photoabsorption and photofragmentation of SF_6 in the equivalent photon energy range 5–63 eV by normalization to the photoabsorption measurements of Blechschmidt *et al.* [458] at 27 eV. Their data were compared earlier in Fig. 6.88(b) with the data of Simpson *et al.* [460].

Furthermore, Hitchcock and Van der Wiel [456] used their electron energy-loss measurements, along with the photoionization branching ratios they measured for the six lowest energy states of SF_6^+, to obtain partial ionization cross sections for the production of specific positive ions. These data were presented earlier in Fig. 6.90. Sze and Brion [449] studied inner-shell electron energy-loss spectra of SF_6 using 2.0–3.7 keV electrons and 0° scattering angle. Their findings for valence-shell excitation spectra have been discussed earlier (Table 6.55). Ying *et al.* [461] performed angle-resolved electron-energy loss measurements using a coplanar electron spectrometer, and determined generalized oscillator strengths of valence shell electronic transitions in the range 7–70 eV using electrons with initial energy of 2.5 keV. Their valence-shell

Table 6.56. Photon energy/energy loss and transition assignments for SF_6.[a]

Photon energy/ Energy loss (eV)	Assignment Ref. [462][b]	Assignment Ref. [459][c]	Assignment Ref. [449][d]	Assignment Ref. [480][d]
6.494 [449][e]				
7.93 [481][c,f]		$5t_{1u} \rightarrow 6a_{1g}$		
9.596 [449]			$1t_{1g} \rightarrow 6a_{1g}$ [g]	
9.8 [480][h]				$1t_{1g} \rightarrow 6a_{1g}$
10.982 - 12.266 [449]	$1t_{1g} \rightarrow 6t_{1u}/5t_{1u} \rightarrow 6a_{1g}$		$1t_{1g} \rightarrow 6t_{1u}/1t_{2u},\ 5t_{1u} \rightarrow 6a_{1g}$	
11.0 [480][h]				$1t_{2u} \rightarrow 6a_{1g}$
11.5; 11.6 [480][i]				$5t_{1u} \rightarrow 6a_{1g}$
11.54 [460][d,j]				
11.729 [481][c]				
11.76 [482][c]				
12.8 [480][h]				$1t_{1g} \rightarrow 7a_{1g}(4s)$
13.2 [462][k]	$1t_{1g} \rightarrow 4p$	$5t_{1u} \rightarrow 6a_{1g}$		
13.2 [480][h]				$1t_{1g} \rightarrow 6t_{1u}(4p)$
13.143-13.309 [449]			$1t_{1g} \rightarrow 4p$	
13.244 [482][c]				
13.27 [481][c]		$4t_{1u} \rightarrow 6a_{1g}$		
13.30 [460][d,j]		$4t_{1u} \rightarrow 6a_{1g}$		
13.3 [480][i]				$1t_{1g} \rightarrow 6t_{1u}(4p)$
14.1 [480]				$1t_{2g} \rightarrow 6a_{1g}$
14.218 [482][c]				
14.283 [481][c]				
14.3 [462][k]	$5t_{1u} \rightarrow 4s/3e_g \rightarrow 6t_{1u}$			
14.35 [460][d,j]				
13.970 – 14.415 [449]			$3e_g \rightarrow 6t_{1u}$	
14.3 ; 14.7 [480][i]				$5t_{1u} \rightarrow 7a_{1g}(4s)$
14.6 [462][k]	$1t_{1g} \rightarrow 5p$			
14.707 [449]			$1t_{2u},\ 5t_{1u} \rightarrow 4s$	

Table 6.56 (continued)
Photon energy/energy loss and transition assignments for SF_6.[a]

Photon energy/ Energy loss (eV)	Assignment Ref. [462][b]	Assignment Ref. [459][c]	Assignment Ref. [449][d]	Assignment Ref. [480][d]
14.90 [460][d]				
14.9 [480][i]				$1t_{1g} \rightarrow 7t_{1u}(5p)$
14.933 [482][c]				
15.163 [481][c]				
15.546 [449]			$3e_g \rightarrow 6t_{1u}$	
15.7 [462][k]	$5t_{1u} \rightarrow 5s/1t_{2g} \rightarrow 6t_{1u}$			
15.7 [459][c]		$1t_{2g} \rightarrow 6t_{1u}$		
16.0; 16.1 [480]				$3e_g \rightarrow 6t_{1u}(4p)$
17.00 [459][c]		$1t_{2g} \rightarrow 6t_{1u}$	$1t_{2g} \rightarrow 6t_{1u}$	
17.00 [462][k]	$4t_{1u} \rightarrow 6a_{1g}$			
17.021			$4t_{1u} \rightarrow 6a_{1g}$ [g]	
19.6 [462][k]	$4t_{1u} \rightarrow 4s$			
19.8 [449][j]		$5t_{1u} \rightarrow 2t_{2g}$	$4t_{1u} \rightarrow 4s$	
19.8 [459][c]		$5t_{1u} \rightarrow 2t_{2g}$		
20.9 [449][j]			$1t_{1g} \rightarrow 2t_{2g}/5a_{1g} \rightarrow 6a_{1g}$ [g]	
21.2 [462][k]	$4t_{1u} \rightarrow 5s$		$4t_{1u} \rightarrow 5s$	
22.1 [462][k]	$5t_{1u} \rightarrow 2t_{2g}/4t_{1u} \rightarrow ns\,(n \geq 6)$			
23.2 [449][l]		$5a_{1g} \rightarrow 6t_{1u}$	$5a_{1g} \rightarrow 6t_{1u}/1t_{2u},\ 5t_{1u} \rightarrow 2t_{2g}$	
23.2 [459][c]		$5a_{1g} \rightarrow 6t_{1u}$		
24.6 [462][k] and [449][l]	$5a_{1g} \rightarrow 4p$	$5a_{1g} \rightarrow 4p$		
25.7 [462][k]	$5a_{1g} \rightarrow 5p$			
25.8 [449][l]		$5a_{1g} \rightarrow 5p$		
26.2 [462][k]	$5a_{1g} \rightarrow 6p$			
28.3 [449][l]		$4t_{1u} \rightarrow 2t_{2g}$		

Table 6.56 (continued)

Photon energy/energy loss and transition assignments for SF_6.[a]

Photon energy/ Energy loss (eV)	Assignment Ref. [462][b]	Assignment Ref. [459][c]	Assignment Ref. [449][d]	Assignment Ref. [480][d]
28.5 [457][c]				
28.8 [459][c]		$1t_{2u}$, $5a_{1g} \rightarrow (6a_{1g})^2$		
31.8 [459][c]		$3t_{1u} \rightarrow 6a_{1g}$		
35.4 [459, 457][c]		$2e_g \rightarrow 6t_{1u}$		
37.6 [459][c]		$4t_{1u} \rightarrow 4e_g$		

[a]See Refs. [449, 459, 462] for more details and for information on other transitions.
[b]Data from studies on ion-pair formation from photoexcitation.
[c]Data from studies on photoabsorption spectra.
[d]Data from electron energy-loss experiments.
[e]Estimated uncertainty ±0.020 eV.
[f]The photoabsorption data show weak continuous absorption which begins at 7.93 eV [481].
[g]Forbidden transition.
[h]Triplet.
[i]Singlet.
[j]Uncertainty ±0.05 eV.
[k]Energy position of peak maximum in the F^- efficiency curve.
[l]Estimated uncertainty ±0.5 eV.

electron energy-loss spectrum of SF_6 measured at 0.5° (converted to photoabsorption cross section using the relationship $\sigma_{\text{pa,t}}\,(10^{-18}\,\text{cm}^2) = 109.75 df/dE\,(\text{eV}^{-1})$, where df/dE is the differential oscillator strength) is compared with other electron energy-loss spectra and photoabsorption data in Fig. 6.88(b). There is general agreement among the photoabsorption data and the electron energy-loss spectra.

Besides the above three studies, Chutjian *et al.* [483] and Trajmar and Chutjian [480] investigated the electron-impact excitation of SF_6 using 20 eV incident electrons in the energy-loss range of 7–14 eV. In their experiments the scattering angles ranged from 20° to 135°. From a comparison of their data at low energies and large scattering angles with data they obtained earlier using 400 eV and 0° scattering angle (i.e., under conditions favoring excitation of optically allowed transitions), they identified a number of spin and/or symmetry forbidden transitions in the energy range of ~ 10 to ~ 16 eV which we have listed in Table 6.56.

In addition to the electron energy-loss studies involving low-lying transitions, Francis *et al.* [485] recorded electron energy-loss spectra of SF_6 in the region of S 2*p*, S 2*s*, and F 1*s* excitation, using dipole and nondipole electron-scattering conditions. Electron-impact energies between 700 and 3200 eV and scattering angles between 0° and 30° were used. It was found that relative to dipole electron energy-loss or photoabsorption spectra, there are large intensity redistributions in both the S 2*p* and S 2*s* spectra under nondipole conditions. In contrast, the F 1*s* spectrum was found to be essentially the same in near-dipole and nondipole scattering regimes. Fomunung *et al.* [486] conducted a theoretical study of electron excitation of optically allowed transitions between 20 eV and 2.5 keV, and a number of investigations were conducted on the generalized oscillator of SF_6 (e.g., Refs. [485, 487]).

Energy-loss spectra involving vibrational excitation of SF_6 were investigated by Trajmar and Chutjian [480], Rohr [488], and Randell *et al.* [489]. These are discussed in Sec. 5.2.5 of Ch. 6.

5.1.4 The Electron Affinity and Negative Ion States of SF_6

The four lowest lying virtual orbitals of SF_6 are [446, 449]: $(6a_{1g})^0$ $(6t_{1u})^0\,(2t_{2g})^0(4e_g)^0$. Calculations for the octahedral symmetry (O_h) of SF_6^- (e.g., see Refs. [490–493]) have shown that the S–F antibonding character of the highest occupied, totally symmetric, nondegenerate $6a_{1g}$ molecular orbital of SF_6^- results in a substantial increase of the S–F bond length in SF_6^- compared to neutral SF_6. Although a number of authors (e.g., Refs. [494–496]) considered electron capture into

Table 6.57. Values of the electron affinity of the SF_6 molecule reported since 1983.

Electron affinity (eV)	Ref.	Comments
0.0–1.49 (0.76)[a]	[74]	Range of 20 values listed in Ref. [74] which were reported prior to 1983.
1.05 ± 0.1	[500]	Energetics of electron transfer reactions
1.15 ± 0.15	[501]	Thermal electron attachment studies
1.07 ± 0.07[b]	[502]	Thermal electron attachment studies
1.19	[492]	Calculation
1.3–1.4	[493]	Calculation
1.06	[503]	Calculation
3.44	[498]	Calculation
3.46	[497]	Calculation
1.06 eV		**Recommended value (see text)**

[a]Average of the values listed in Ref. [74], excluding the lowest and the highest listed values.
[b]This value is a refinement of their earlier result in Ref. [501].

degenerate molecular orbitals of lower symmetry than a_{1g}, calculations for a lower symmetry configuration (see Gutsev [497, 498] and Richman and Banerjee [499]) showed the O_h octahedral structure to be the most stable.

There has been widespread unanimity that SF_6 attaches thermal and near thermal electrons with a very large cross section forming SF_6^-, and that the SF_6 molecule must thus have a positive electron affinity. However, up until recently there has been no consensus regarding the size of its electron affinity. Prior to 1983 there were at least 20 reported values for this quantity [74] that varied from near 0.0 eV to $\sim$ 1.5 eV. Values reported since 1983 are listed in Table 6.57 along with the range of pre-1983 values reviewed in Ref. [74]. We take the average of the two most recent and widely accepted values determined in Refs. [500, 502] to be our presently recommended value of (1.06 ± 0.06) eV for the electron affinity of the SF_6 molecule.

In addition to the electron attachment resonance at $\sim$ 0.0 eV, electron attachment and electron scattering experiments and calculations have shown the existence of several negative ion states at energies below 30 eV. The energies of these states, as determined by the positions of the maxima in the measured electron attachment cross sections, measured electron scattering cross sections, and the calculated cross sections are listed in Table 6.58. Also listed in Table 6.58 are possible symmetry assignments.

Table 6.58. Negative ion states of SF_6.

Energy position (eV)	Type of study	Symmetry/ Orbital	Reference
~0.0	Many electron attachment studies		
0.38	SF_5^- from SF_6		[424, 504–506]
0.5 ± 0.1	SF_5^- from SF_6		[507]
~0.1	Excitation function of ν_1	A_{1g}	[489]
~0.1	Vibrational excitation by electron impact		[488]
~0.0	Multiple scattering calculation	a_{1g}	[508]
2.0	F_2^- from SF_6		[424]
2.2	F_2^- from SF_6		[509]
2.4	F_2^- from SF_6		[510][a]
2.6	F^- from SF_6		[424]
2.8	F^- from SF_6		[510]
2.8	F^- from SF_6		[509]
~2.9	F^- from SF_6		[511]
2.3	Total electron scattering cross section	a_{1g}	[512]
2.5	Total electron scattering cross section	A_{1g}	[513]
2.5	Total electron scattering cross section	a_{1g}	[514]
2.52 ± 0.15	Trochoidal derivative spectrum	a_{1g}	[515]
2.56 ± 0.15	Total electron scattering cross section	a_{1g}	[515]
2.7	Angular dependence of vibrationally elastic electron scattering	a_{1g}	[516]
2.1	Multiple scattering calculation	a_{1g}	[508]
3.30	Multichannel calculation with close-coupling	A_{1g}	[517]
4.4	F_2^- from SF_6		[424]
4.8	F_2^- from SF_6		[510]
4.8	F_2^- from SF_6		[509]
5.0	SF_4^- from SF_6		[509]
5.1	F^- from SF_6		[424]
5.2	F^- from SF_6		[510]
5.3	F^- from SF_6		[509]
5.4	SF_4^- from SF_6		[424]
5.4	SF_4^- from SF_6		[510]
~5.4[b]	F^-, F_2^-, and SF_4^- from SF_6		[511]
5.7 ± 0.1	F^- from SF_6		[507]
6.0 ± 0.1	SF_4^- from SF_6		[507]
6.7	Total electron scattering cross section	t_{1g}	[512]

[a]The values of Lehmann [510] given in this table have been deduced from his figures and were reduced by 0.6 eV as discussed in the text.
[b]Main resonance.

Table 6.58 (continued)
Negative ion states of SF_6.

Energy position (eV)	Type of study	Symmetry/ Orbital	Ref.
7.0	Total electron scattering cross section	t_{1u}	[514]
7	Angular dependence of vibrationally elastic electron scattering	t_{1u}	[516]
7	Total electron scattering cross section	t_{1u}	[513]
7.01 ± 0.16	Trochoidal derivative spectrum	t_{1u}	[515]
7.05 ± 0.10	Total electron scattering cross section	t_{1u}	[515]
7.2	Elastic electron scattering cross section	t_{1u}	[518]
7.2	Multiple scattering calculation	t_{1u}	[508]
8.8	F^- from SF_6		[424]
8.9	F^- from SF_6		[510]
9.3 ± 0.1	F^- from SF_6		[507]
9.4	F^- from SF_6		[509]
9.4	SF_3^- from SF_6		[424]
8.7; 9.1	Multiple scattering calculation	T_{1u}	[519]
9.85	Multichannel calculation with close-coupling	T_{1u}	[517]
11.2	F_2^- from SF_6		[424]
11.2	SF_3^- from SF_6		[510]
11.3	SF_3^- from SF_6		[509]
11.3	F^- from SF_6		[424]
11.5	F^- from SF_6		[510]
11.6	F^- from SF_6		[509]
11.6	F_2^- from SF_6		[509]
11.7	F_2^- from SF_6		[424]
11.8 ± 0.1	F^- from SF_6		[507]
12.0	SF_2^- from SF_6		[424]
12.3	SF_2^- from SF_6		[509]
13.0	SF_2^- from SF_6		[510]
11.87 ± 0.10	Total electron scattering cross section	t_{2g}	[515]
11.88 ± 0.07	Trochoidal derivative spectrum	t_{2g}	[515]
11.9	Total electron scattering cross section	t_{2g}	[514]
11.9	Total electron scattering cross section	t_{2g}	[512]
12	Total electron scattering cross section	t_{2g}	[513]
12	Elastic electron scattering cross section	t_{2g}	[518]
12	Elastic and vibrationally inelastic electron-impact excitation		[480]
11	Multiple scattering calculation	T_{1u}	[520]
11.0; 11.8; 13.7	Multiple scattering calculation	T_{2g}	[519]
12.7	Multiple scattering calculation	t_{2g}	[508]
13.10	Multichannel calculation with close coupling	T_{2g}	[517]

Table 6.58 (continued)
Negative ion states of SF_6.

Energy position (eV)	Type of study	Symmetry/ Orbital	Ref.
16.8	Total electron scattering cross section		[514]
17	Multiple scattering calculation of total cross section for elastic electron scattering	T_{2g}	[520]
23–24	Angular distribution of photoelectrons		[443]
24	Multiple scattering calculation	E_g	[520]
27.0	Multiple scattering calculation	e_g	[508]
27.0	Multichannel calculation with close coupling	e_g	[517]
27.1; 28.3; 28.6	Multiple scattering calculation	E_g	[519]
28.84	Multichannel calculation with close coupling	E_g	[517]
25–55[c]	Total electron scattering cross section	e_g	[515]

[c]Possibly one or more resonances in this energy range.

These experimental and the calculated values (below 15 eV) are shown in Fig. 6.91. The electron attachment data for each resonance exhibit a considerable spread partly because these data involve a number of different fragment negative ions. The energy range for each resonance as determined by the electron attachment studies, extends to lower energies than the respective energy range determined by the electron-scattering studies. This difference is especially large for the 7.0 eV resonance, and can be attributed to the competition between dissociation and autodetachment of SF_6^{-*} in the electron attachment studies. It is interesting to note that although electron attachment studies and theory indicate a resonance near 9 eV, none of the electron scattering investigations show any evidence of it.

The last column in Fig. 6.91 gives the recommended energy positions of the SF_6 resonances as determined by considering only the electron scattering data, along with their symmetry assignments. The energy positions of the lowest four negative ion states of the SF_6 molecule and their symmetries are: $\sim$ 0 eV (a_{1g}), 2.5 eV (a_{1g}), 7.0 eV (t_{1u}), and 11.9 eV (t_{2g}). It should be noted that the adiabatic position of the $\sim$ 0 eV negative ion state is at -1.06 eV [$= -$EA (SF_6)]. Other negative ion states are indicated near 17 eV and higher energies (Table 6.58). The resonances near 27 and 50 eV can be assigned as core-excited valence negative ion states involving the σ^* molecular orbitals.

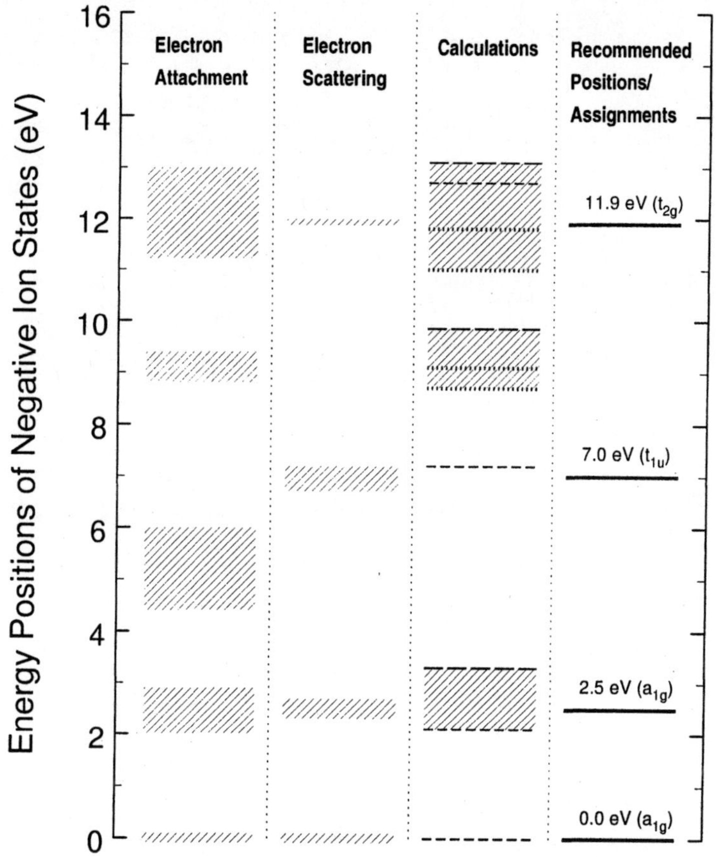

Figure 6.91. Negative ion states of the SF_6 molecule as determined by electron attachment measurements, electron scattering measurements, and calculations: (- - -) Dehmer *et al.* [508]; (– –) Gianturco *et al.* [517]; ($\cdots$) Gyemant *et al.* [519]. Shaded areas represent the range of measured resonant energies. The last column gives the recommended average positions and assignments (see text). It should be noted that the adiabatic position of the ~ 0.0 eV negative ion state is at -1.06 eV [$= -EA(SF_6)$].

5.1.5 Fundamental Vibrational Modes and Other Data for SF_6 and SF_6^-

The fundamental vibrational frequencies $\nu_1, \ldots, \nu_6$ of the SF_6 molecule have a high degree of degeneracy due to the high symmetry of the molecule. Table 6.59 lists the vibrational assignment, energy, symmetry, infrared activity, and degeneracy of the six fundamental vibrational modes [421, 521–526]. The low energies of these vibrations would result in vibration excitation of these modes at relatively low temperatures

Table 6.59. Vibrational assignment, energy, symmetry, infrared activity, and degeneracy of the fundamental frequencies of the SF_6 molecule [421, 521–526].[a]

Vibration	Energy (eV)	Ref.	Symmetry	Infrared Activity	Degeneracy
ν_1	0.0955	[523]	A_{1g}	No	1
	0.0959	[522]			
	0.0959	[525]			
	0.096	[526]			
ν_2	0.0793	[523]	E_g	No	2
	0.0796	[522]			
	0.0796	[525]			
	0.0798	[526]			
ν_3	0.1174	[521]	F_{1u}	Yes	3
	0.1175	[523]			
	0.1175	[525]			
	0.1175	[524]			
	0.1175	[526]			
ν_4	0.0761	[523]	F_{1u}	Yes	3
	0.0762	[526]			
	0.07625	[524]			
	0.0763	[525]			
ν_5	0.0647	[523]	F_{2g}	No	3
	0.0649	[526]			
	0.0651	[522]			
	0.0651	[525]			
ν_6	0.0430	[522]	F_{2u}	No	3
	0.0430	[525]			
	0.0432	[526]			
	0.0434	[523]			

[a]See text and Chase *et al.* [525] for the vibrational frequencies of SF_6^-.

above ambient. Even at room temperature the population of the $\nu = 1$ level of ν_6 is one half of the ground-state population.

The vibrational frequencies for SF_6^- are somewhat lower relative to the respective ones for the neutral SF_6 due to the weaker bonding in the negative ion. The values (in eV) of the six fundamental frequencies of SF_6^- and their degeneracies (values in parenthesis) as listed by Chase *et*

Table 6.60. Other data on SF_6 and SF_6^-.[a]

Physical quantity	Value	Reference
SF_5–F bond dissociation energy[b]	3.38 eV	[528][c]
	$\leq (3.39 \pm 0.15)$ eV	[529][c]
	(3.9 ± 0.15) eV	[530]
	(4.1 ± 0.13) eV	[531]
S–F bond length of SF_6	1.5568 Å	[532]
	(1.564 ± 0.01) Å	[525]
	1.565 Å[d]	[493]
	1.567 Å[d]	[491]
F–S–F bond angle of SF_6 and SF_6^-	90°	[525]
S–F bond length of SF_6^-	1.72 Å	[525]
	1.717 Å[d]	[498]
	1.710 Å[d]	[491]
	1.704 Å[d]	[493]

[a]See text for values of the vibrational frequencies of SF_6^-.
[b]Hubers and Los [527] obtained a value of (1.0 ± 0.1) eV for the dissociation energy of SF_5^-–F.
[c]According to Fenzlaff *et al.* [511] this value is too low due to the low value used by these authors for the production of F^- from SF_6.
[d]Octahedral geometry.

al. [525] are: 0.0868 (1), 0.0775 (2), 0.1147 (3), 0.0736 (3), 0.0620 (3), and 0.0403 (3).

In Table 6.60 are listed other data on SF_6 and SF_6^- which are relevant to the present discussion on electron interactions with the SF_6 molecule.

5.2 Electron Scattering by SF_6

5.2.1 Total Electron Scattering Cross Section, $\sigma_{sc,t}(\varepsilon)$, of SF_6

There have been eight sets [512–516, 533–535] of experimental determinations of the $\sigma_{sc,t}(\varepsilon)$ of SF_6. These are shown in Fig. 6.92. Rohr [516] determined the integrated vibrationally elastic ($\Delta\nu = 0$) electron scattering cross section for SF_6 below ~ 6 eV. He then obtained a value of $\sigma_{sc,t}(\varepsilon)$ by adding to this cross section the total vibrational excitation cross section he measured earlier [488]. His data on the total vibrational

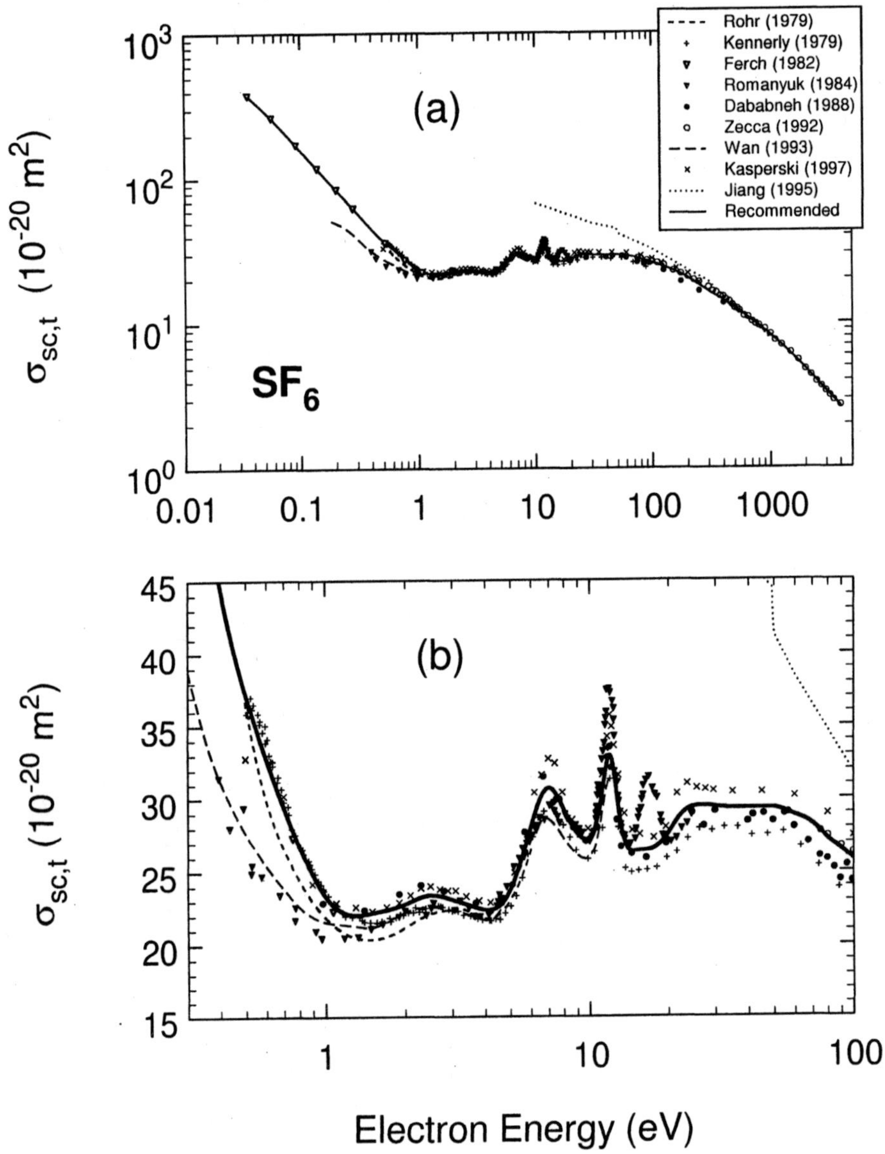

Figure 6.92. (a) Total electron scattering cross section, $\sigma_{sc,t}(\varepsilon)$, of SF_6 as a function of electron energy: (- - -) Ref. [516]; (+) Ref. [515]; ($\triangledown$) Ref. [533]; ($\blacktriangledown$) Ref. [514]; ($\bullet$) Ref. [512]; ($\circ$) Ref. [246]; (– –) Ref. [535]; ($\times$) Ref. [513]; ($\cdots$) Ref. [247]; (—) recommended values. (b) Expanded view.

excitation cross section, however, need to be corrected because they have originally been normalized to the data of Srivastava *et al.* [536] which were subsequently corrected by Trajmar *et al.* [416]. According to Trajmar *et al.* [416], the vibrational cross section of Rohr [488] needs to be multiplied by a factor of 1.28 at 5 eV to correspond to the renormalized data of Srivastava *et al.* The data identified in Fig. 6.92 as "Rohr (1979)" are the original values of Rohr without this correction. The $\sigma_{sc,t}(\varepsilon)$ of Rohr is not a direct measurement and for this reason it is not included in the data used to determine the recommended values of $\sigma_{sc,t}(\varepsilon)$ derived later in this section.

Another low-energy measurement of $\sigma_{sc,t}(\varepsilon)$ was made by Ferch *et al.* [533] in a transmission experiment using a time-of-flight spectrometer. It covered the electron energy range from 0.036 to 1.0 eV with a reported uncertainty of ±7% above 0.2 eV and ±9% below 0.2 eV. The results of Ferch *et al.* are in excellent agreement with the other available measurements of $\sigma_{sc,t}(\varepsilon)$ above 0.5 eV.

Measurements of $\sigma_{sc,t}(\varepsilon)$ extending to higher electron energies were also made using a variety of techniques. Wan *et al.* [535] measured the $\sigma_{sc,t}(\varepsilon)$ of SF_6 using an electron transmission spectrometer employing a trochoidal monochromator. The overall uncertainty of their measurements is within ±15% above 1 eV, but it may increase to as much as ±50% at energies below 0.2 eV. These data are in reasonable agreement with the other measurements above ~ 1.0 eV, but fall systematically below most of the other measurements for lower electron energies. Kennerly *et al.* [515] measured the $\sigma_{sc,t}(\varepsilon)$ of SF_6 for incident-electron energies between 0.5 and 100 eV. The uncertainty in their measurements was reported to be ±5% from 0.5 to 30 eV, ±10% from 30 to 50 eV, and ±15% from 50 to 100 eV. The principal source of error above 30 eV is the detection of electrons scattered inelastically in the forward direction which contribute to a higher transmission value and a lower cross section. Hence, the values above 30 eV are a lower bound [515]. Indeed, as can be seen from Fig. 6.92(b), the values of Kennerly *et al.* lie below those of the other groups at energies above ~ 10 eV.

Romanyuk *et al.* [514] measured the $\sigma_{sc,t}(\varepsilon)$ of SF_6 from ~ 0.4 to ~ 25 eV using an electron trap method. No uncertainty is given in their paper, but their values below ~ 1.0 eV are low. In addition, the measurements of Romanyuk *et al.* show a resonance peak at 16.8 eV, which is absent from all the other data. The measurements of Dababneh *et al.* [512] extend over a larger electron energy range stretching from 1 to 500 eV and were made using an electron transmission technique. The values of $\sigma_{sc,t}(\varepsilon)$ at the resonance energies 2.3, 6.7, and 11.9 eV are, respectively, reported to be $(24.1 \pm 0.2) \times 10^{-20}$ m^2, $(31.5 \pm 0.5) \times$

10^{-20} m^2, and $(33.5 \pm 0.2) \times 10^{-20}$ m^2. Measurements of the $\sigma_{\text{sc,t}}(\varepsilon)$ of SF_6 to higher energies (75–4000 eV) have been reported by Zecca *et al.* [246], and were made using a Ramsauer-type electron spectrometer. The total uncertainty in these measurements is $\sim \pm 6\%$. The data of Zecca *et al.* [246] are generally higher than those of Dababneh *et al.* [512] in the energy range from 75 to 700 eV where the two sets of data overlap. They are also higher than those of Kennerly *et al.* [515] in the range from 75 to 100 eV [see Fig. 6.92(b)].

Finally, a more recent measurement of the $\sigma_{\text{sc,t}}(\varepsilon)$ of SF_6 for electron energies between 0.6 and 250 eV has been made by Kasperski *et al.* [513] who employed the electron transmission method. The estimated uncertainty of these measurements is up to $\pm 7\%$ below 5 eV, less than $\pm 4\%$ between 10 and 100 eV, and approximately $\pm 5\%$ at the highest energies they investigated. These data are consistently higher than the other measurements between 20 and 100 eV, but agree well with the high-energy measurements of Zecca *et al.* [246]. In Fig. 6.92 are also plotted the values of $\sigma_{\text{sc,t}}(\varepsilon)$ for SF_6 calculated by Jiang *et al.* [247] who employed the additivity rule and the model complex optical potential. These calculated values of $\sigma_{\text{sc,t}}(\varepsilon)$ are not in good agreement with the measurements except at high energies. The additivity rule ignores molecular geometry and reduces molecular scattering to atomic scattering. Since the contribution from the interference occurring between the scattering amplitudes originating from the different constituent atoms of the molecule is not included in the additivity rule, the results of the total scattering cross section using the additivity rule show larger discrepancies in the low-energy range.

Besides the calculations by Jiang *et al.* [247] there have also been two other calculations, one by Dehmer *et al.* [508] and another by Benedict and Gyemant [520]. However, these authors calculated cross sections for total elastic electron scattering and for this reason their results are not shown in Fig. 6.92 (see Sec. 5.2.3 of Ch. 6).

The experimental data in Fig. 6.92 by-and-large lie within the quoted or expected experimental uncertainties for most of the electron energy range investigated. There are however systematic differences between the various sets of experimental measurements. For instance, there exist large uncertainties in the data below ~ 1 eV and, as mentioned earlier, the measurements of Romanyuk *et al.* [514] show a resonance peak at 16.8 eV which is not present in the other measurements. We thus deduced a recommended cross section $\sigma_{\text{sc,t}}(\varepsilon)$ for SF_6 by considering only the measurements of Kennerly *et al.* [515], Ferch *et al.* [533], Dababneh *et al.* [512], Zecca *et al.* [246], and Kasperski *et al.* [513]. The data of Rohr were not considered because they are not direct measurements, and

Table 6.61. Recommended values for the total electron scattering cross section, $\sigma_{sc,t}(\varepsilon)$, of SF_6.

ε (eV)	$\sigma_{sc,t}(\varepsilon)$ (10^{-20} m^2)	ε (eV)	$\sigma_{sc,t}(\varepsilon)$ (10^{-20} m^2)	ε (eV)	$\sigma_{sc,t}(\varepsilon)$ (10^{-20} m^2)
0.035	379.8	3.0	23.1	70	28.2
0.040	344.7	3.5	22.7	80	27.2
0.045	315.3	4.0	22.4	90	26.5
0.050	290.1	4.5	22.6	100	25.8
0.060	249.4	5.0	23.7	150	22.7
0.070	217.6	6.0	28.0	200	20.2
0.080	192.7	7.0	30.7	250	18.0
0.090	173.1	8.0	29.2	300	16.3
0.10	157.4	9.0	27.6	350	15.1
0.15	109.5	10.0	27.3	400	14.1
0.20	84.3	11.0	29.6	450	13.1
0.25	68.7	12.0	32.9	500	12.4
0.30	58.1	13.0	28.9	600	11.2
0.35	50.4	14.0	26.6	700	10.2
0.40	44.6	15.0	26.4	800	9.41
0.45	40.3	18.0	26.7	900	8.73
0.50	37.2	20.0	27.6	1000	8.15
0.60	32.8	22.0	28.7	1500	6.07
0.70	29.4	25.0	29.4	2000	4.80
0.80	26.8	30.0	29.4	2500	3.94
0.90	24.9	35.0	29.3	3000	3.36
1.0	23.4	40.0	29.3	3500	2.95
1.5	22.2	45.0	29.3	4000	2.64
2.0	22.8	50.0	29.3		
2.5	23.4	60.0	28.8		

the data of Wan *et al.* [535] and Romanyuk *et al.* [514] were not considered because of their differences with the rest of the measurements. The recommended cross section is shown by the solid line in Fig. 6.92 [Fig. 6.92(b) is an expanded view of part of Fig. 6.92(a) to more clearly show the individual measurements]. Values obtained from the solid line are listed in Table 6.61 as our recommended values for the $\sigma_{sc,t}(\varepsilon)$ of the SF_6 molecule. It should be noted that while these values are well defined for energies above 0.5 eV, below this energy the uncertainties are potentially larger since the data rely solely on one measurement, namely, that of Ferch *et al.* [533].

5.2.2 Elastic Differential Electron Scattering Cross Section, $\sigma_{e,diff}(\varepsilon)$, of SF_6

There have been a number of published measurements [416, 516, 518, 536–539] and calculations [517, 540] of the elastic differential electron scattering cross section, $\sigma_{e,diff}(\varepsilon)$, of SF_6 which are plotted in Fig. 6.93. The measurements of $\sigma_{e,diff}(\varepsilon)$ by Srivastava *et al.* [536] for incident electron energies between 5 and 75 eV were made relative to those for helium. It was subsequently realized [416, 541] that the data of Srivastava *et al.* [536] needed to be renormalized to more accurate helium cross section values than those used originally. The data identified in Fig. 6.93 by the squares are the data of Srivastava *et al.* [536] for $\sigma_{e,diff}(\varepsilon)$ as recalculated by Trajmar *et al.* [416] using the improved He cross sections of Register *et al.* [542]. Another measurement of the $\sigma_{e,diff}(\varepsilon)$ of SF_6 was made by Rohr [516]. He measured differential cross sections for vibrationally elastic ($\Delta\nu = 0$) electron scattering by SF_6 up to 10 eV. Absolute values of the cross section $\sigma_{e,diff}(\varepsilon)$ were obtained by comparison with data on elastic electron scattering by He and have an estimated uncertainty of about ±20%. Interestingly, Rohr [516] found that below 1 eV the cross section appears to be determined essentially by direct scattering, while between 2 and 3 eV and around 7 eV the scattering is dominated by resonances. In Fig. 6.93 are shown the angular dependencies measured by Rohr for 0.5, 2.7, and 7 eV. (Note that the cross section values at 2.7 eV as presented in Fig. 2 of Ref. [516] were too large by a factor of 2. The data have been corrected as presented here in Fig. 6.93.) The 0.5 eV data peak sharply in the forward direction and the scattering is dominated by direct processes. The 2.7 and 7 eV measurements indicate strong resonant contributions to the scattering process. These data compare favorably with other data obtained at similar electron energies.

The measurements of the $\sigma_{e,diff}(\varepsilon)$ of SF_6 by Johnstone and Newell [518] were made over a wider electron energy range (5–75 eV) using a hemispherical electron spectrometer with an angular range of 10°–120°. The magnitude of $\sigma_{e,diff}(\varepsilon)$ was determined with reference to cross sections for He (calculated values by Nesbet [543] below 20 eV and measurements by Register *et al.* [542] for energies greater than 20 eV), and they have a quoted uncertainty of ±12% to ±14%. While these data are in qualitative agreement with the earlier measurements, the degree of agreement varies with the electron energy and the scattering angle.

The measurements of Sakae *et al.* [537] were made using a crossed-beam method and extend up to 700 eV with an angular range of 5°–135°. They were put on absolute scale by normalization to the differential electron scattering cross section for He (data of Jansen *et al.* [544] for

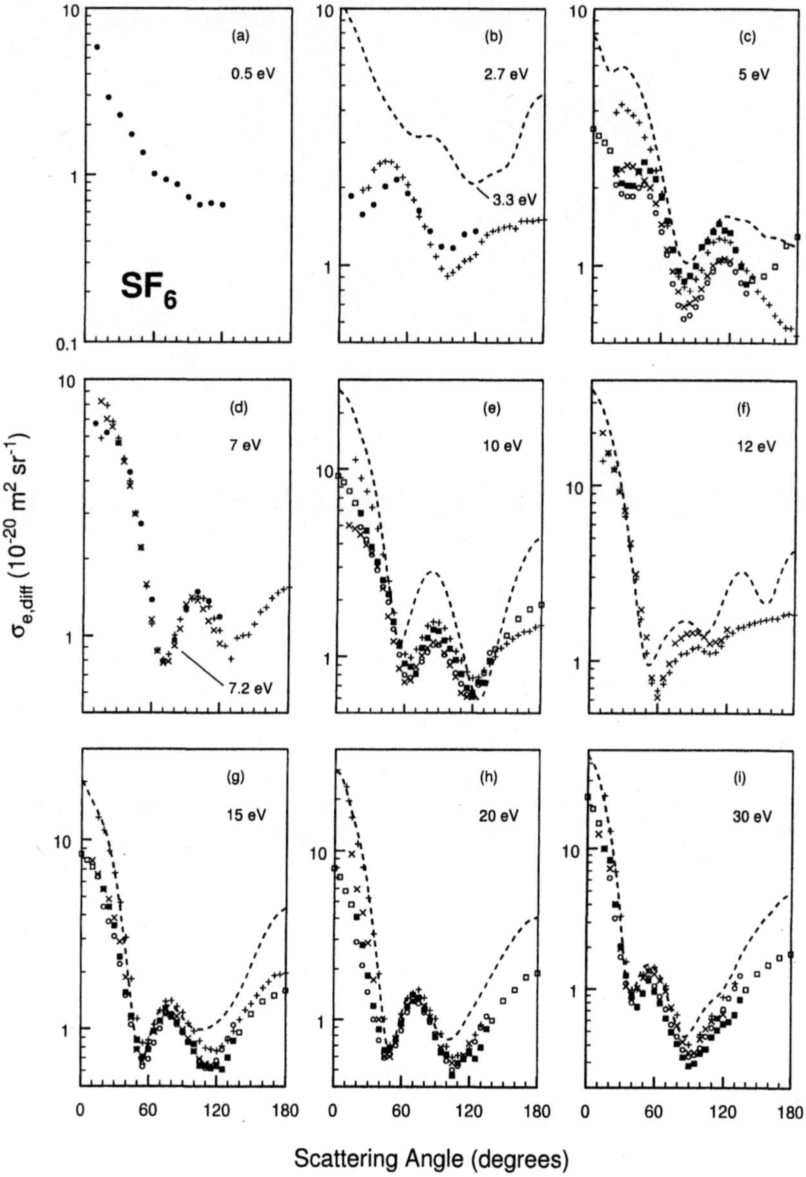

Figure 6.93. Elastic differential electron scattering cross section, $\sigma_{e,\text{diff}}(\varepsilon)$, of SF_6 as a function of electron energy ε: (○) data of Srivastava *et al.* [536]; (■) experimental measurements from Ref. [536] as corrected by Trajmar *et al.* [416]; (□) extrapolated values of Trajmar *et al.* [416]; (•) data of Rohr [516]; (♦) data of Sakae *et al.* [537]; (×) data of Johnstone and Newell [518]; (+) data of Cho *et al.* [538, 539]; (- - -) calculations of Gianturco *et al.* [517]; (– –) calculations of Jiang *et al.* [540].

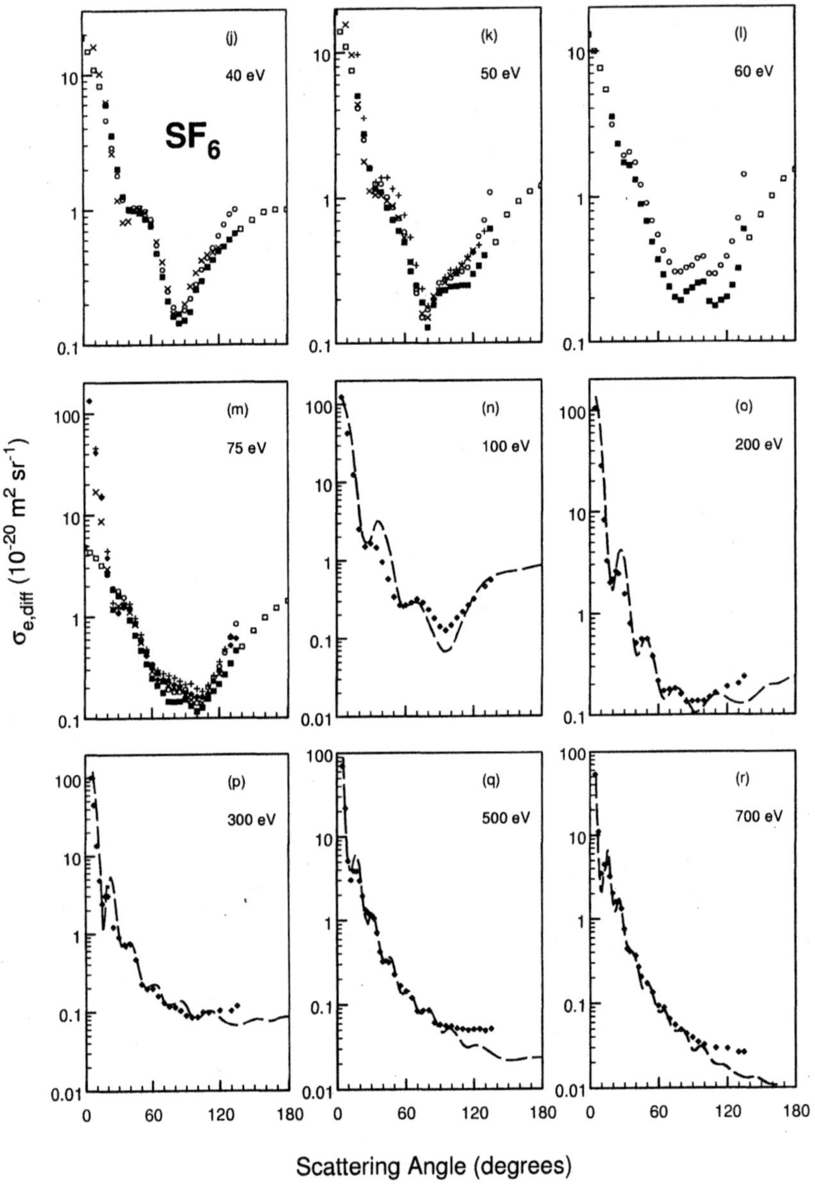

Figure 6.93 (continued)
Elastic differential electron scattering cross section, $\sigma_{e,\text{diff}}(\varepsilon)$, of SF_6 as a function of electron energy ε: (○) data of Srivastava *et al.* [536]; (■) experimental measurements from Ref. [536] as corrected by Trajmar *et al.* [416]; (□) extrapolated values of Trajmar *et al.* [416]; (•) data of Rohr [516]; (♦) data of Sakae *et al.* [537]; (×) data of Johnstone and Newell [518]; (+) data of Cho *et al.* [538, 539]; (- - -) calculations of Gianturco *et al.* [517]; (– –) calculations of Jiang *et al.* [540].

energies between 100 and 700 eV and data of Register *et al.* [542] for 75 eV), and have an overall uncertainty of about ±10%.

The recent measurements of the elastic differential electron scattering cross section for SF_6 at energies ranging from 2.5 to 75 eV by Cho *et al.* [538, 539] were made in response to the discrepancies in the resultant determinations of the elastic integral electron scattering cross section, $\sigma_{e,int}(\varepsilon)$, by Trajmar *et al.* [416], Johnstone and Newell [518], and Sakae *et al.* [537] (see Sec. 5.2.3 of Ch. 6). The data of Cho *et al.* [538, 539] are shown in Fig. 6.93 and have a typical uncertainty of $< \pm 10\%$. They are in good agreement with the data of Sakae *et al.* [537] at 75 eV and with the data of Rohr [516] at 2.7 and 7 eV. Cho *et al.* [538, 539] directly measured the cross sections at large angles using a new magnetic deflection technique that allows cross section measurements at scattering angles up to 180° (see Refs. [538, 545, 546] and Ch. 2). This reduces extrapolation uncertainties when using these data to calculate the elastic integral and momentum transfer cross sections. At most electron energies, the measured values of $\sigma_{e,diff}(\varepsilon)$ by Cho *et al.* are higher than the corresponding measurements of Trajmar *et al.* [416] and Johnstone and Newell [518].

In Fig. 6.93 are also shown the results of two calculations of $\sigma_{e,diff}(\varepsilon)$; those of Gianturco *et al.* [517] who used an *ab initio* approach and solved the multichannel scattering problem within the close-coupling methodology, and those of Jiang *et al.* [540] who calculated elastic differential electron scattering cross sections at six electron energies between 100 and 700 eV using the independent-atom model with partial waves. There is qualitative agreement between these calculated results and the experimental data, particularly at higher energies. The values calculated by Gianturco *et al.* [517] are generally larger than the measurements, especially at small and large scattering angles.

In Tables 6.62, 6.63, and 6.64 are listed, respectively, the $\sigma_{e,diff}(\varepsilon)$ values of Rohr [516] for low electron energies, the values of Cho *et al.* [538, 539] at intermediate electron energies, and the values of Sakae *et al.* [537] at higher energies. Collectively these data provide values of the $\sigma_{e,diff}(\varepsilon)$ of SF_6 from 0.5 to 700 eV. The tabulated values of $\sigma_{e,diff}(\varepsilon)$ of Trajmar *et al.* and Johnstone and Newell are presented in their original publications [416, 518].

5.2.3 Elastic Integral Electron Scattering Cross Section, $\sigma_{e,int}(\varepsilon)$, of SF_6

In Fig. 6.94 are compared the five sets of values of the elastic integral electron scattering cross section, $\sigma_{e,int}(\varepsilon)$, for SF_6 derived from the

Table 6.62. Elastic differential electron scattering cross section, $\sigma_{e,diff}(\varepsilon)$, for SF_6 in units of 10^{-20} m^2 sr^{-1}. The data are those of Rohr [516] and were obtained by digitizing the curves in Fig. 2 of Ref. [516].

	Electron energy (eV)		
Angle	0.5	2.7[a]	7.0
10°	5.82	1.85	6.75
20°	2.91	1.57	6.21
30°	2.28	1.72	5.63
40°	1.75	2.02	4.34
50°	1.36	2.15	2.73
60°	1.01	1.90	1.38
70°	0.93	1.63	0.80
80°	0.87	1.35	0.96
90°	0.74	1.18	1.26
100°	0.66	1.16	1.49
110°	0.68	1.32	1.37
120°	0.66	1.35	1.19

[a]The data shown in Fig. 2 of Ref. [516] for an electron energy of 2.7 eV were larger by a factor of 2 than the correct values. The values listed here are those presented in the original reference divided by two.

elastic differential cross sections of Srivastava *et al.* [536] (as corrected by Trajmar *et al.* [416]), Rohr [516], Sakae *et al.* [537], Johnstone and Newell [518], and Cho *et al.* [538, 539]. The uncertainties of $\sigma_{e,int}(\varepsilon)$ are somewhat larger than those quoted in the preceding section for $\sigma_{e,diff}(\varepsilon)$, largely because of errors arising from the extrapolation of the elastic differential electron scattering cross section measurements to 0° and 180° scattering angles. The quoted estimated uncertainties for the $\sigma_{e,int}(\varepsilon)$ data are as follows: Srivastava *et al.* [536] (±18%); Rohr [516] (∼ ±20%); Johnstone and Newell [518] (±17%); Sakae *et al.* [537] (∼ ±10%); and Cho *et al.* [538, 539] (±15% to ±20%). The errors are generally larger when the differential cross sections rise steeply at small scattering angles.

The $\sigma_{e,int}(\varepsilon)$ data of Sakae *et al.* [537], Cho *et al.* [538, 539], and Rohr [516] agree well within the overlapping energy ranges. The data of Trajmar *et al.* [416] and Johnstone and Newell [518] are seen to fall significantly below the other data, particularly at higher electron energies. In view of the agreement between the data of Sakae *et al.* [537], Cho *et al.* [538, 539] and Rohr [516], we have chosen to fit these data in order to obtain our suggested values for this cross section. This fit is shown by

Table 6.63. Elastic differential electron scattering cross section, $\sigma_{e,diff}(\varepsilon)$, for SF_6 in units of 10^{-20} m^2 sr^{-1}, as measured by Cho *et al.* [538, 539].

	Electron energy (eV)						
Angle	2.7	5	7	8.5	10	12	15
10°						13.79	
15°			5.90	8.50	11.17	15.22	12.99
20°	1.95	3.94	7.90	8.82	8.89	12.41	11.13
25°	1.99	4.22	6.84	7.72	7.56	9.31	8.70
30°	2.34	4.00	5.90	6.40	6.24	6.65	6.65
35°	2.47	3.83	4.89	5.20	4.81	4.51	4.65
40°	2.52	3.60	3.99	3.89	3.51	2.91	3.05
45°	2.51	3.14	3.00	2.81	2.52	1.72	1.84
50°	2.40	2.82	2.20	2.07	1.71	1.10	1.13
55°	2.18	2.33	1.55	1.44	1.19	0.75	0.84
60°	2.05	1.89	1.11	1.09	1.03	0.68	0.87
65°	1.79	1.42	0.87	0.95	0.98	0.74	1.05
70°	1.54	1.15	0.80	0.94	1.09	0.86	1.25
75°	1.41	0.91	0.84	1.07	1.26	0.99	1.39
80°	1.20	0.82	1.00	1.26	1.46	1.09	1.40
85°	1.08	0.80	1.15	1.32	1.53	1.10	1.31
90°	0.97	0.88	1.29	1.39	1.52	1.17	1.22
95°	0.91	0.96	1.39	1.39	1.41	1.19	1.07
100°	0.94	1.12	1.42	1.35	1.24	1.13	0.97
105°	0.98	1.23	1.40	1.22	1.07	1.10	0.86
110°	1.04	1.27	1.30	1.07	0.95	1.12	0.79
115°	1.06	1.25	1.17	0.94	0.83	1.21	0.78
120°	1.10	1.23	1.05	0.82	0.77	1.38	0.77
125°	1.24	1.15	0.91	0.75	0.78	1.45	0.81
130°	1.31	1.02	0.81	0.83	0.85	1.54	0.87
135°	1.35	0.96	0.98	0.96	0.92	1.57	0.97
140°	1.37	0.91	1.00	1.00	1.00	1.62	1.12
145°	1.40	0.79	1.01	1.07	1.11	1.65	1.26
150°	1.42	0.74	1.11	1.23	1.17	1.69	1.35
155°	1.39	0.70	1.24	1.30	1.25	1.71	1.50
160°	1.49	0.64	1.29	1.47	1.31	1.76	1.65
165°	1.49	0.61	1.40	1.57	1.36	1.75	1.77
170°	1.49	0.58	1.47	1.62	1.37	1.83	1.92
175°	1.50	0.57	1.51	1.71	1.44	1.86	1.94
180°	1.50	0.54	1.54	1.77	1.47	1.84	1.97

Table 6.63 (continued)
Elastic differential electron scattering cross section, $\sigma_{e,diff}(\varepsilon)$, for SF_6 in units of 10^{-20} m^2 sr^{-1}, as measured by Cho *et al.* [538, 539].

	Electron energy (eV)			
Angle	20	30	50	75
10°	23.71	55.57	49.46	45.48
12°	19.54			
15°	15.66	23.33	24.68	15.42
20°	10.98	13.26	9.67	4.41
25°	7.99	6.92	3.51	1.37
30°	5.25	3.31	1.61	1.23
35°	3.23	1.57	1.29	1.38
40°	1.87	0.99	1.38	1.33
45°	1.00	0.99	1.38	0.96
50°	0.70	1.27	1.16	0.67
55°	0.74	1.37	1.04	0.48
60°	0.98	1.43	0.76	0.34
65°	1.33		0.53	0.30
70°	1.43	1.05	0.34	0.27
75°	1.50		0.24	0.26
80°	1.34	0.66	0.18	0.25
85°	1.13		0.20	0.23
90°	1.01	0.40	0.23	0.21
95°	0.82		0.28	0.22
100°	0.69	0.47	0.32	0.19
105°	0.60		0.32	0.18
110°	0.61	0.64	0.35	0.21
115°	0.60		0.39	0.26
120°	0.67	0.87	0.42	0.36
125°	0.81		0.47	0.47
130°	0.93	1.10	0.58	0.63

the solid line in Fig. 6.94, and values obtained from the fit are listed in Table 6.65 as our suggested values for the $\sigma_{e,int}(\varepsilon)$ of the SF_6 molecule.

The cross section labeled "Randell (1992)/Present" in Fig. 6.94 was deduced by us from the relative cross section for total elastic electron scattering measured by Randell *et al.* [489]. This cross section was put on an absolute scale by normalizing its value at 1 eV to the data of Rohr [516] at 1 eV. The total elastic electron scattering cross section determined this way is shown in Fig. 6.94 by the crosses and is seen to be in excellent agreement with the measurements of Rohr between about 0.5 and 1 eV. Below $\sim$ 0.5 eV the deduced values are significantly lower

Table 6.64. Elastic differential electron scattering cross section, $\sigma_{e,\text{diff}}(\varepsilon)$, for SF_6 in units of 10^{-20} m^2 sr^{-1}. The data are those of Sakae *et al.* [537] as supplied by Professor M. Hayashi (Private communication, 2000). Stated uncertainties are approximately ±8% to ±10%.

	Electron energy (eV)						
Angle	75	100	150	200	300	500	700
5°	133	125	117	105	102	70.3	53.2
7.5°			68.5		45.3	21.9	11.1
10°	41.3	43.2	35.7	28.4	13.6	5.11	3.48
12.5°			16.8	8.34	4.86	3.03	4.46
15°	15.2	12.4	6.18	3.28	2.43	3.88	4.60
17.5°			2.40	2.00	3.05	3.86	3.21
20°	3.80	2.53	1.64	2.17	3.04	2.95	2.03
22.5°			1.93	2.57		1.94	1.60
25°	1.17	1.51	1.96	2.44	1.22	1.34	1.60
27.5°			1.82			1.22	1.33
30°	1.09	1.64	1.70	1.55	0.906	1.16	0.758
32.5°						1.05	0.442
35°	1.23	1.46	1.09	0.795	0.726	0.712	0.409
40°	1.16	0.953	0.592	0.509	0.738	0.322	0.365
42.5°						0.329	0.268
45°	0.878	0.573	0.390	0.557	0.468	0.317	0.205
50°	0.575	0.338	0.372	0.559	0.227	0.227	0.171
55°	0.416	0.265	0.385	0.375	0.196	0.166	0.134
60°	0.316	0.267	0.350	0.216	0.199	0.143	0.093
65°	0.241	0.286	0.243	0.172	0.159	0.119	0.088
70°	0.241	0.316	0.164	0.178	0.131	0.083	0.065
75°	0.236	0.289	0.134	0.182	0.117	0.085	0.056
80°	0.214	0.234	0.139	0.159	0.114	0.085	0.049
85°	0.204	0.180	0.139	0.134	0.104	0.062	0.044
90°	0.177	0.142	0.144	0.135	0.090	0.058	0.039
95°	0.172	0.127	0.140	0.136	0.086	0.056	0.035
100°	0.162	0.148	0.130	0.136	0.086	0.056	0.032
105°	0.161	0.182	0.127	0.149	0.098	0.052	
110°	0.174	0.218	0.146	0.164	0.097	0.051	0.030
115°	0.230	0.264				0.045	
120°	0.295	0.318	0.222	0.189	0.104	0.051	0.029
130°	0.514	0.456	0.345	0.202	0.103	0.049	0.026
135°	0.606	0.548	0.406	0.238	0.102	0.052	0.026

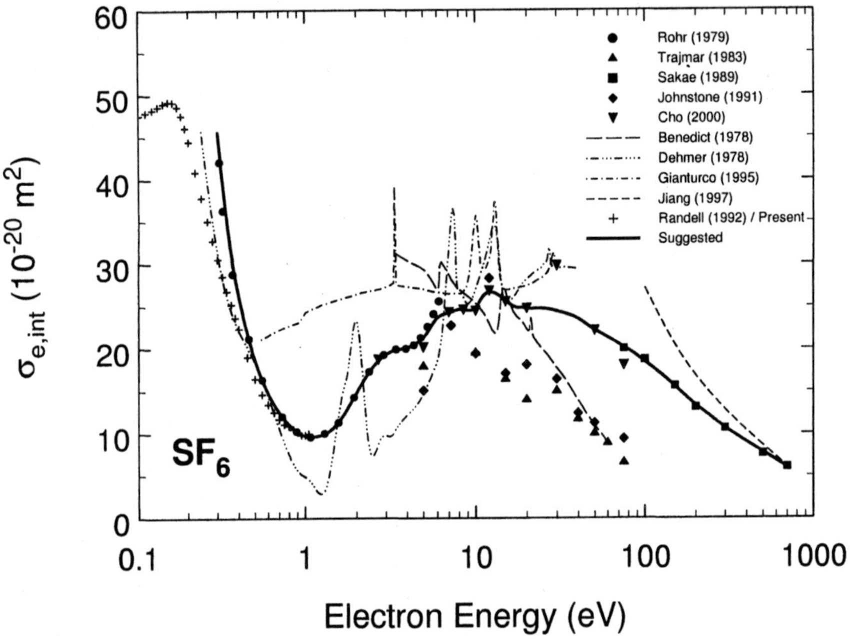

Figure 6.94. Elastic integral electron scattering cross section, $\sigma_{e,int}(\varepsilon)$, of SF_6 as a function of electron energy: (●) Ref. [516]; (▲) Ref. [536] as corrected by Trajmar *et al.* [416]; (■) Ref. [537], (♦) Ref. [518]; (▼) Ref. [538, 539]; (– –) Ref. [520]; (– · · · –) Ref. [508]; (– · –) Ref. [517]; (- - -) Ref. [540]; (+) deduced in present work from data in Randell *et al.* [489]; (—) suggested values.

than the values measured by Rohr, perhaps due to the fact that the measurements of Randell *et al.* were made at only a single scattering angle (90°).

Also shown in Fig. 6.94 are the results of four calculations of the total elastic electron scattering cross section, $\sigma_{e,t}(\varepsilon)$, by Benedict and Gyemant [520], Dehmer *et al.* [508], Gianturco *et al.* [517], and Jiang *et al.* [540]. The results of the first three calculations show distinct structure due to negative ion resonances, but they agree with the experimental data only in limited energy regions. The values of $\sigma_{e,t}(\varepsilon)$ calculated by Jiang *et al.* using the independent-atom model with partial waves are higher than the experimental data of Sakae *et al.* between 100 and 500 eV, and only merge with the experimental data at electron energies approaching 1000 eV.

Table 6.65. Suggested values for the elastic integral electron scattering cross section, $\sigma_{e,int}(\varepsilon)$, of SF_6.

ε (eV)	$\sigma_{e,int}(\varepsilon)$ (10^{-20} m^2)	ε (eV)	$\sigma_{e,int}(\varepsilon)$ (10^{-20} m^2)	ε (eV)	$\sigma_{e,int}(\varepsilon)$ (10^{-20} m^2)
0.30	45.6	7	24.2	50	22.2
0.35	33.0	8	24.6	60	21.3
0.40	26.2	9	24.5	70	20.5
0.45	21.8	10	24.8	75	20.2
0.50	18.7	11	26.1	80	19.8
0.60	14.8	12	26.6	90	19.1
0.70	12.5	13	26.5	100	18.4
0.80	11.1	14	26.1	125	16.9
0.90	10.2	15	25.6	150	15.5
1.0	9.72	16	25.2	200	13.3
1.2	9.73	17	24.9	250	11.8
1.5	10.9	18	24.8	300	10.6
2.0	14.8	19	24.8	350	9.71
2.5	17.8	20	24.7	400	8.95
3.0	19.3	22	24.7	450	8.31
3.5	19.9	25	24.7	500	7.74
4.0	20.1	30	24.4	600	6.76
4.5	20.6	35	24.0	700	5.94
5.0	21.3	40	23.5		
6.0	23.6	45	22.8		

5.2.4 Momentum Transfer (Elastic) Electron Scattering Cross Section, $\sigma_m(\varepsilon)$, of SF_6

The momentum transfer (elastic) electron scattering cross section, $\sigma_m(\varepsilon)$, of SF_6 has been determined by Srivastava *et al.* [536], Sakae *et al.* [537], Johnstone and Newell [518], and Cho *et al.* [538, 539] from their respective elastic differential electron scattering cross section measurements extrapolated to 0° and 180° scattering angles. These determinations of $\sigma_m(\varepsilon)$ are shown in Fig. 6.95. The data of Srivastava *et al.* [536] shown in Fig. 6.95 are those that have been renormalized by Trajmar *et al.* [416] (see Sec. 5.2.2 of Ch. 6). The quoted uncertainty in the data is ±20% for Srivastava *et al.* [536], ∼ ±9% for Sakae *et al.* [537], ±20% for Johnstone and Newell [518], and ±15% to ±20% for Cho *et al.* [538, 539]. The uncertainty is generally larger at the higher energies due to the difficulty in accurately measuring the steeply rising differential cross section at small scattering angles.

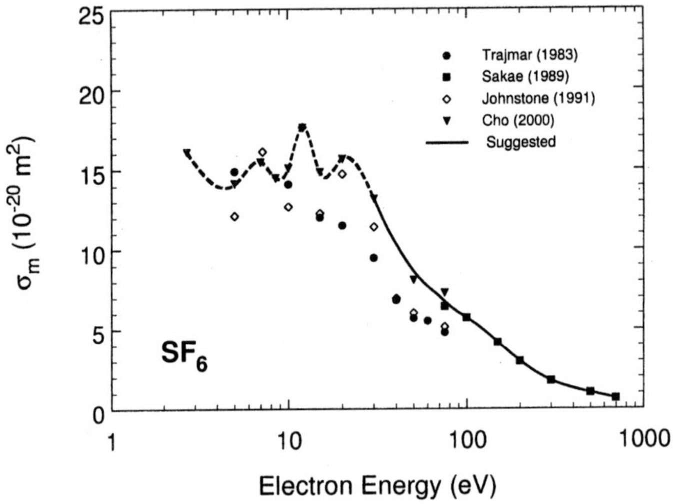

Figure 6.95. Momentum transfer (elastic) cross section, $\sigma_m(\varepsilon)$, of SF_6 as a function of electron energy: (•) Ref. [536] as corrected by Trajmar *et al.* [416]; (■) Ref. [537]; (◊) Ref. [518]; (▼) Ref. [538, 539]; (- - -, —) suggested values. The broken line indicates an uncertainty in the significance of the structure observed at energies between 2.7 and 30 eV.

At 75 eV the data of Trajmar *et al.* [416] and Johnstone and Newell [518] lie somewhat below those of Sakae *et al.* [537], while the data of Cho *et al.* [538, 539] are in reasonable agreement with Sakae *et al.* Based upon this agreement, we have made a modified spline fit to the data of Cho *et al.* and of Sakae *et al.* which is shown by the line in Fig. 6.95. Values extracted from this fit are listed in Table 6.66 as our suggested data for the $\sigma_m(\varepsilon)$ of the SF_6 molecule. The data of Cho *et al.* [538, 539] and Johnstone and Newell [518] indicate structure which can be associated with the negative ion states of the SF_6 molecule (Table 6.58). However, it should be noted that the variations in the measured values of $\sigma_m(\varepsilon)$ in the energy region of 2.7–30 eV are smaller than the stated uncertainties of the measurements. Thus the line designating our suggested values for $\sigma_m(\varepsilon)$ in this energy range is broken in order to highlight the potential uncertainty of the indicated structure.

5.2.5 Inelastic Electron Scattering Cross Section, $\sigma_{inel}(\varepsilon)$, for SF_6

Vibrational Excitation of SF_6. Two low-energy electron-impact vibrational-excitation studies [488, 489] of SF_6 have shown that at elec-

Table 6.66. Suggested values for the momentum transfer (elastic) electron scattering cross section, $\sigma_m(\varepsilon)$, of SF_6.

ε (eV)	$\sigma_m(\varepsilon)$ (10^{-20} m^2)	ε (eV)	$\sigma_m(\varepsilon)$ (10^{-20} m^2)	ε (eV)	$\sigma_m(\varepsilon)$ (10^{-20} m^2)
2.75	16.0	15	14.9	75	6.74
3.0	15.4	16	14.5	80	6.46
3.5	14.5	17	14.7	90	6.03
4.0	14.0	18	15.0	100	5.70
4.5	13.9	19	15.4	125	4.92
5.0	14.1	20	15.7	150	4.16
6.0	15.1	22	15.7	200	2.98
7.0	15.5	25	15.0	250	2.23
8.0	14.8	30	13.2	300	1.76
9.0	14.4	35	11.5	350	1.47
10.0	15.1	40	10.3	400	1.28
11.0	16.7	45	9.37	450	1.13
12.0	17.6	50	8.65	500	1.02
13.0	17.1	60	7.69	600	0.82
14.0	15.8	70	7.06	700	0.66

tron energies below ~ 3 eV the energy-loss spectra of SF_6 are characterized by strong vibrational excitation and vibrationally elastic electron scattering. In this energy range these studies have also shown that the vibrationally inelastic processes are dominated by excitation of the ν_1 and ν_3 modes. This is seen from the energy-loss spectrum (collision energy of 1 eV and scattering angle of 10°) in Fig. 6.96(a) that is characterized by the large magnitude of the excitation of the ν_3 fundamental mode which exceeds the elastic peak by about a factor of 3. Rohr [488] estimates that at 1 eV the integral total vibrational excitation cross section is $\sim 10^{-19}$ m^2. The ν_3 (0.1175 eV, Table 6.59) vibrational mode is infrared (IR) active and is associated with vertical displacement of the apical F atoms in the octahedral structure for normal modes of vibration of SF_6, and the ν_1 (0.0956 eV; Table 6.59) vibrational mode is the Raman active totally symmetric breathing mode.

Figure 6.96(b) shows the energy-loss spectrum again at a collision energy of 1 eV, but at a scattering angle of 80°. The excitation is now seen to be dominated by the ν_1 mode and its higher harmonics. No other modes or intercombination lines are seen. Rohr observed the same behavior at all angles greater than 60° and all the energies he employed.

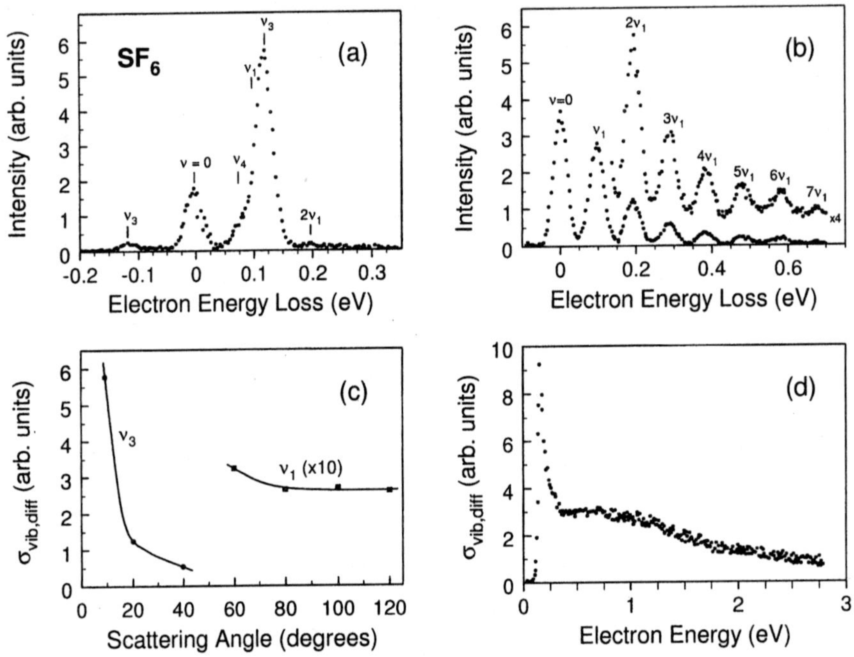

Figure 6.96. Low-energy electron energy-loss spectra for SF_6 from Rohr [488]. (a) Energy-loss spectrum of SF_6 for a collision energy of 1 eV and a scattering angle of 10°. (b) Energy-loss spectrum of SF_6 at 1 eV and a scattering angle of 80°. (c) Angular dependencies of the differential cross sections for the ν_1 and ν_3 vibrational excitations (the cross section scales reflect true intensity ratios). (d) Differential cross section for excitation of the ν_1 vibrational mode of SF_6 at a scattering angle of 80°.

In Fig. 6.96(c) are shown the angular dependencies of the differential cross sections, $\sigma_{\text{vib,diff}}$, for the ν_1 and ν_3 vibrational excitations in the angular range 10°–120° for a collision energy of 1 eV (note that the cross section scales reflect true intensity ratios and that the values of the cross section for ν_1 have been multiplied by a factor of 10). The differential cross section of the IR active ν_3 vibrational mode is peaked in the forward direction, indicating that a direct process via the dipole interaction is involved. For the Raman active totally symmetric vibration ν_1 the dipole mechanism is excluded and the angular distribution is nearly isotropic.

Figure 6.96(d) shows the differential cross section $\sigma_{\text{vib,diff}}(\varepsilon)$ for excitation of the ν_1 vibration at a scattering angle of 80°. The sharp peak at 0.060 eV was attributed [488] to the formation of SF_6^- negative ions and the broad maximum near 0.7 eV to vibrational excitation via a negative ion state.

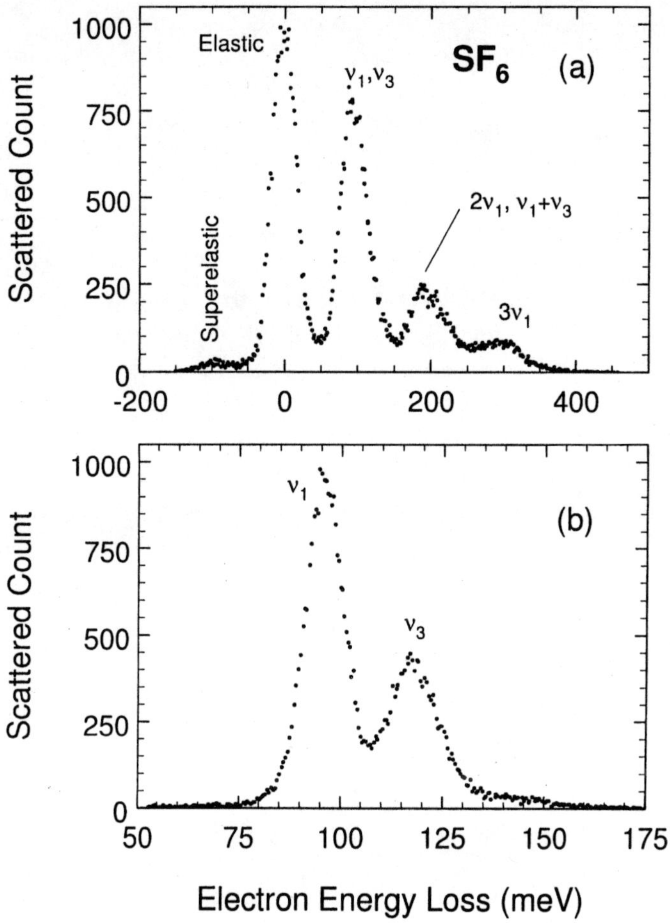

Figure 6.97. Low-energy electron energy-loss spectra of SF_6 from Randell *et al.* [489]. (a) Energy-loss spectrum of SF_6 for a collision energy of 0.4 eV and a scattering angle of 90°. (b) Energy-loss spectrum of SF_6 for an electron collision energy of 0.18 eV and a scattering angle of 90° showing resolution of the ν_1 and ν_3 vibrational modes.

A similar crossed-beam study of SF_6 was conducted by Randell *et al.* [489] using 0.05–1 eV incident-energy electrons scattered at a fixed angle of 90°. Figure 6.97(a) shows the electron-energy loss function measured at an incident electron energy of 0.4 eV due to excitation of the ν_1, ν_3, $2\nu_1$, $\nu_1 + \nu_3$, and $3\nu_1$ vibrations. The contribution to the total electron scattering cross section of inelastic scattering is roughly equal to the elastic scattering at this energy. The weak peak at a negative energy loss was attributed by Randell *et al.* to superelastic scattering from the thermal population of $\nu_3 = 1$. In agreement with Rohr [488], Randell *et al.* have attributed the excitation of ν_3 to a direct dipole mechanism and the excitation of ν_1 to a long-lived negative ion resonance, i.e., to the excitation of the symmetric breathing mode, ν_1, via autoionization of SF_6^{-*}. The latter interpretation is supported by the high-resolution electron attachment studies of Klar *et al.* [547, 548] which show that the attachment cross section has a pronounced cusp at 0.096 eV associated with ν_1 excitation. It is also consistent with the work of Trajmar and Chutjian [480] whose electron-energy-loss spectra using 12 eV incident electrons showed strong enhancement of vibrational excitation attributed mostly to the ν_1 mode. Figure 6.97(b) shows an energy-loss spectrum obtained by Randell *et al.* [489] for an electron-impact energy of 0.18 eV and a scattering angle of 90° showing resolution of electrons scattered inelastically through energies corresponding to the ν_1 and ν_3 vibrational modes. Within experimental error, the thresholds for ν_1 and ν_3 are at 0.095 eV and 0.117 eV, respectively, and excitation is thus at the energy threshold.

Figures 6.98(a), 6.98(b), 6.98(c), and 6.98(d) show the excitation functions, respectively, for ν_1, ν_3, $2\nu_1$, and $\nu_1+\nu_3$ as reported by Randell *et al.* [489]. The scales in Figs. 6.98 are not absolute, but the vertical axes give the correct relative values of these quantities. Clearly, the excitation cross section for ν_1 is the largest, and the excitation functions for ν_1 and ν_3 are markedly different, reflecting the different mechanisms by which these vibrations are excited. The measured functional dependence of the cross section for ν_3 was found by Randell *et al.* to be in agreement with that predicted by the Born point-dipole approximation.

The relative cross sections shown in Figs. 6.98(a), 6.98(b), 6.98(c), and 6.98(d) have been used by us in conjunction with other assessed cross sections in this work to deduce absolute values for the total vibrational excitation cross section, $\sigma_{\text{vib,t}}(\varepsilon)$, of SF_6 for electron energies less than 1.1 eV. This was done as follows: Since the vibrational energy-loss measurements by both Rohr [488] and Randell *et al.* [489] show that the major contribution to vibrational excitation by electron impact below $\sim$ 3 eV comes from ν_1 (indirect excitation via the SF_6^{-*} resonances) and ν_3 (direct excitation via the dipole interaction) vibrations and their combinational excitations, we have summed the relative vibrational excitation cross sections shown in Figs. 6.98(a), 6.98(b), 6.98(c), and 6.98(d), and plotted the result in Fig. 6.98(e). In Fig. 6.98(e) is also shown the relative total elastic electron scattering cross section measured by Randell *et al.* To put the deduced total vibrational cross section of Randell *et al.* on an absolute scale, the relative vibrational values in Fig. 6.98(e) were multiplied by the same factor used to normalize the relative $\sigma_{\text{e,int}}(\varepsilon)$ values of Randell *et al.* that are shown in Fig. 6.94 and

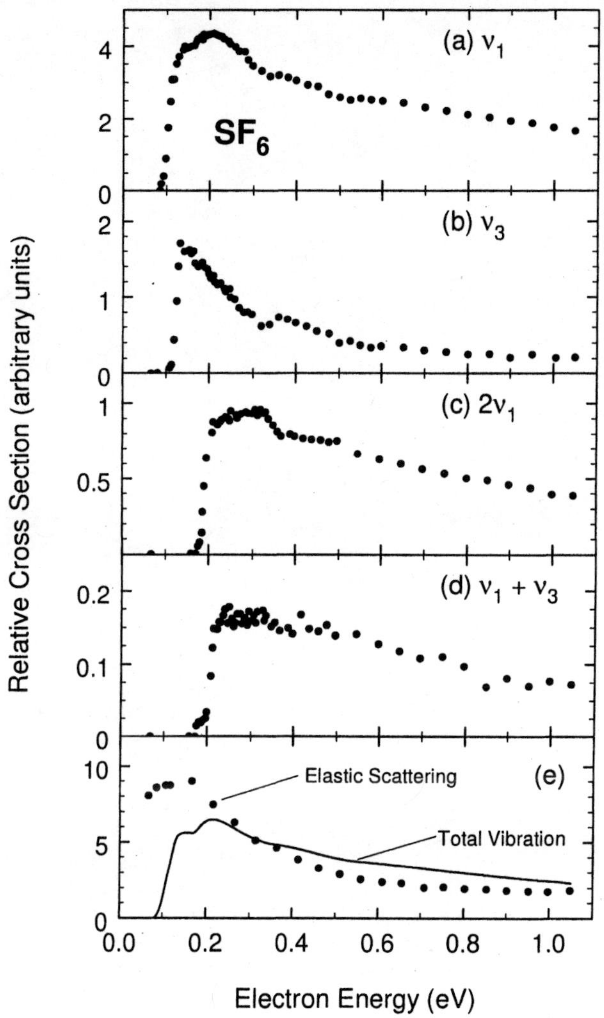

Figure 6.98. Low-energy electron energy-loss spectra for SF_6 from Randell *et al.* [489]. Excitation functions for (a) ν_1, (b) ν_3, (c) $2\nu_1$, (d) $\nu_1 + \nu_3$, and (e) elastic electron scattering. The solid line labelled "Total Vibration" in (e) is the sum of the relative excitation cross sections for ν_1, ν_3 , $2\nu_1$, $\nu_1 + \nu_3$ shown in (a)–(d).

discussed in Sec. 5.2.3. of Ch. 6. The values of $\sigma_{\mathrm{vib,t}}(\varepsilon)$ determined this way are shown in Fig. 6.99 (closed circles).

Confirmation of these deduced values of $\sigma_{\mathrm{vib,t}}(\varepsilon)$ below 1.1 eV, and extension of $\sigma_{\mathrm{vib,t}}(\varepsilon)$ to higher electron energies has been obtained by the following analysis. At energies above ~ 1 eV and below the region of significant electronic excitation (< 16 eV), the total electron attach-

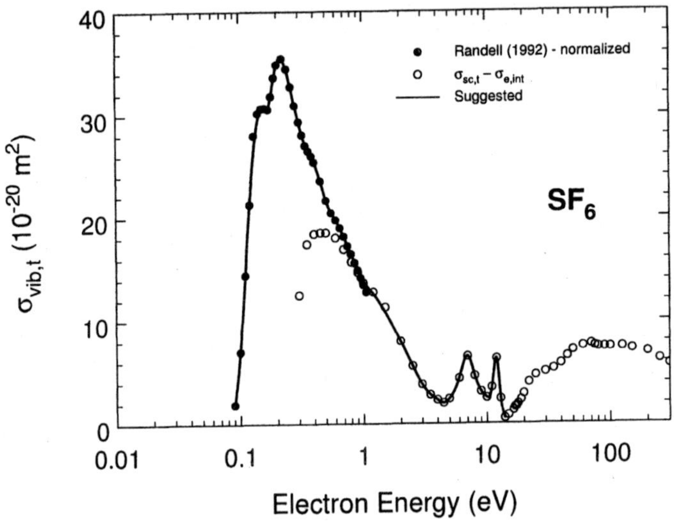

Figure 6.99. Deduced total vibrational excitation cross section, $\sigma_{\text{vib,t}}(\varepsilon)$, for SF_6: (•) Ref. [489] normalized as discussed in text; (○) difference, $\sigma_{\text{sc,t}}(\varepsilon) - \sigma_{\text{e,int}}(\varepsilon)$, of the recommended values for $\sigma_{\text{sc,t}}(\varepsilon)$ and $\sigma_{\text{e,int}}(\varepsilon)$ [see text]; (—) suggested values.

ment cross section of SF_6 is very much smaller than the total electron scattering cross section (see Fig. 6.135 in Sec. 5.9 of Ch. 6), so the cross section $\sigma_{\text{vib,t}}(\varepsilon)$ can be estimated by the difference $\sigma_{\text{sc,t}}(\varepsilon) - \sigma_{\text{e,int}}(\varepsilon)$, viz.,

$$\sigma_{\text{vib,t}}(\varepsilon) = \sigma_{\text{sc,t}}(\varepsilon) - \sigma_{\text{e,int}}(\varepsilon) - \sigma_{\text{a,t}}(\varepsilon) \approx \sigma_{\text{sc,t}}(\varepsilon) - \sigma_{\text{e,int}}(\varepsilon). \qquad (6.36)$$

Using the assessed values of $\sigma_{\text{sc,t}}(\varepsilon)$ and $\sigma_{\text{e,int}}(\varepsilon)$ determined in this work, we obtained the difference $\sigma_{\text{sc,t}}(\varepsilon) - \sigma_{\text{e,int}}(\varepsilon)$, and have shown these values in Fig. 6.99 by the open symbols. It is encouraging to see the agreement of the two determinations of $\sigma_{\text{vib,t}}(\varepsilon)$ in the energy range from ~ 0.5 to 1 eV, thus confirming the normalization of the data of Randell *et al.* [489]. However, the two sets of values differ significantly at energies below ~ 0.5 eV, indicating potential errors in the various cross sections employed. Nonetheless, in the absence of any direct measurements of $\sigma_{\text{vib,t}}(\varepsilon)$, we take the difference $\sigma_{\text{sc,t}}(\varepsilon) - \sigma_{\text{e,int}}(\varepsilon)$, between 1 and 14 eV, and the normalized data of Randell *et al.* [489] below 1 eV, to be a good approximation of $\sigma_{\text{vib,t}}(\varepsilon)$. The solid line in Fig. 6.99 represents our suggested values for the total vibrational cross section, $\sigma_{\text{vib,t}}(\varepsilon)$, for SF_6 using the cross section values deduced as just described. Values of $\sigma_{\text{vib,t}}(\varepsilon)$ extracted from the solid line in Fig. 6.99 are listed in Table 6.67. In the absence of direct measurements, the values represented by the

Table 6.67. Deduced values of the total vibrational excitation cross section, $\sigma_{vib,t}(\varepsilon)$, of SF_6.

ε (eV)	$\sigma_{vib,t}(\varepsilon)$ (10^{-20} m^2)	ε (eV)	$\sigma_{vib,t}(\varepsilon)$ (10^{-20} m^2)	ε (eV)	$\sigma_{vib,t}(\varepsilon)$ (10^{-20} m^2)
0.09	1.9	0.5	21.6	4.5	2.0
0.10	7.0	0.6	19.7	5.0	2.4
0.12	21.3	0.7	18.1	6.0	4.4
0.15	30.6	0.8	16.4	7.0	6.5
0.17	30.6	0.9	15.1	8.0	4.6
0.20	34.9	1.0	13.9	9.0	3.1
0.22	35.5	1.2	12.6	10.0	2.5
0.25	33.6	1.5	10.8	11.0	3.5
0.28	30.9	2.0	8.0	12.0	6.3
0.30	29.4	2.5	5.6	13.0	2.4
0.35	26.8	3.0	3.8	14.0	0.5
0.40	25.4	3.5	2.8		
0.45	23.5	4.0	2.3		

solid line in Fig. 6.99 can be used in combination with the relative data in Figs. 6.98(a), 6.98(b), 6.98(c), and 6.98(d) to deduce cross sections for vibrational excitation of ν_1, ν_3 , $2\nu_1$, $\nu_1 + \nu_3$ up to $\sim$ 4 eV. An estimate of the cross section for indirect vibrational excitation via the negative ion states near 7 and 12 eV can also be obtained from Fig. 6.99 and Table 6.67.

Above 15 eV, where $\sigma_{vib,t}(\varepsilon)$ is small, the difference $\sigma_{sc,t}(\varepsilon) - \sigma_{e,int}(\varepsilon)$ can be attributed to other inelastic processes such as ionization and dissociation, and may be used to deduce a value of the total cross section for dissociation of SF_6 into neutral fragments. This will be further discussed in Sec. 5.4 of Ch. 6.

Electronic Excitation of SF_6. Excitation of the SF_6 molecule by electron impact to any electronic state is generally assumed to lead to dissociation, i.e., all excited electronic states of the SF_6 molecule are antibonding. This assumption is consistent with measurements of optical emission following electron-impact excitation of SF_6 (e.g., Refs. [549, 550]). Electron-impact electronic excitation of SF_6 is thus an important process in controlling the rates of decomposition of this gas in electrical discharges and also indirectly the rates of electron-impact ionization and electron attachment. In spite of this, to our knowledge, there are no cross sections for electron-impact excitation functions for

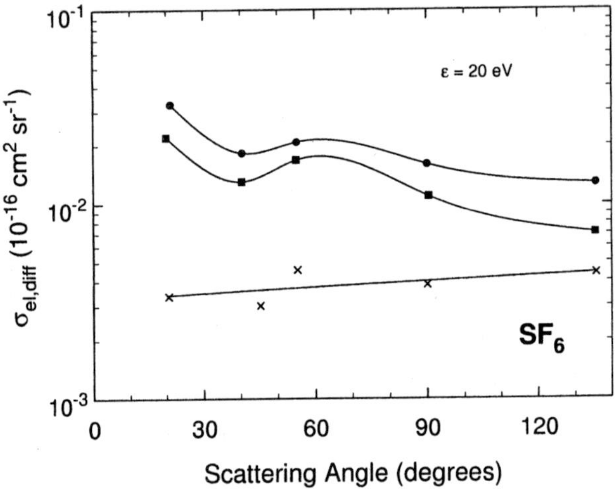

Figure 6.100. Differential electronic excitation cross section, $\sigma_{el,diff}$, for excited electronic states of SF_6 at (×) 9.8 eV; (■) 11.0, 11.6 and 12.8 eV; and (•) 13.3 eV. The incident electron energy was 20 eV (data of Trajmar and Chutjian [480]).

specific electronic transitions in SF_6. We can however direct attention to the following investigations: the electron-impact energy-loss spectra and the generalized oscillator-strength results presented in Sec. 5.1 of Ch. 6; the differential inelastic electron scattering cross sections, $\sigma_{el,diff}(\varepsilon = 20 \text{ eV})$, for five electronic states located at 9.8, 11.0, 11.6, 12.8 and 13.3 eV obtained by Trajmar and Chutjian [480] at one (20 eV) incident-electron energy (Fig. 6.100); and the Boltzmann-type calculations (e.g., Yoshizawa *et al.* [425] and Itoh *et al.* [426]) which determined an overall electronic excitation cross section function consistent with the measured ionization and attachment coefficients.

5.3 Electron-Impact Ionization of SF_6

5.3.1 Total Ionization Cross Section, $\sigma_{i,t}(\varepsilon)$, of SF_6

There have been six measurements [116, 478, 551–554] of the total ionization cross section, $\sigma_{i,t}(\varepsilon)$, of SF_6, which are compared in Fig. 6.101. Two [116, 478] of these were direct measurements (measurements of total charge production) of the total ionization cross section. The uncertainty in the data of Rapp and Englander-Golden [116] is probably ±7%, while that of Asundi and Craggs [478] is much larger. Asundi and Craggs observed structure in their data at 14.3, 15.9, 17.5, 18.7, 20.3, and 22.2 eV. The processes at 15.9, 18.7, and 20.3 eV were attributed to the forma-

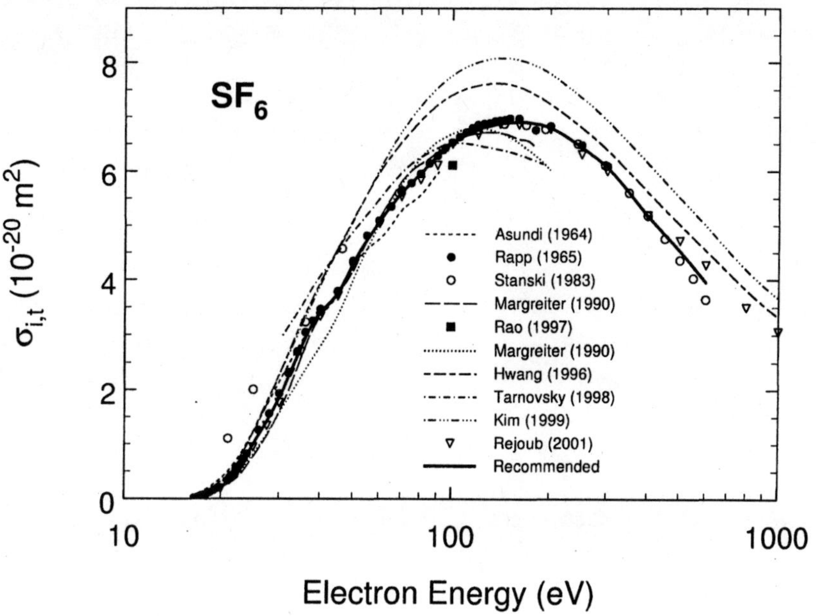

Figure 6.101. Total electron-impact ionization cross section, $\sigma_{i,t}(\varepsilon)$, for SF_6 as a function of electron energy: (- - -) Ref. [478]; (•) Ref. [116]; (○) Ref. [551]; (— —) Ref. [552]; (■) Ref. [553]; (▽) Ref. [554]; (···) Ref. [555]; (– – –) Ref. [556]; (– · –) Ref. [558]; (– ··· –) Ref. [559]; (—) recommended.

tion of SF_5^+, SF_4^+, and SF_3^+, respectively, and agree within experimental error (±0.2 eV) with the values of Dibeler and Mohler [471] which were similarly attributed to these positive ions (Table 6.55).

The other four experimental determinations [551–554] of $\sigma_{i,t}(\varepsilon)$ were made by summation of the respective partial ionization cross sections, $\sigma_{i,part}(\varepsilon)$, presented in Sec. 5.3.2 of Ch. 6. Stanski and Adamczyk [551] measured electron-impact dissociative ionization of SF_6 as a function of electron energy to 600 eV, and obtained partial ionization cross sections for the positive ion fragments SF_5^+, SF_4^+, SF_3^+, SF_2^+, SF^+, S^+, F^+, SF_4^{++}, and SF_2^{++} by normalizing their sum to the total ionization cross section of Rapp and Englander-Golden [116]. Margreiter *et al.* [552] measured electron-impact ionization of SF_6 from threshold to 180 eV using a Nier-type ion source in combination with a double-focusing sector field mass spectrometer. They determined partial ionization cross sections for the production of SF_5^+, SF_4^+, SF_3^+, SF_2^+, SF^+, S^+, F^+, SF_4^{++}, and SF_2^{++} using an improved experimental method based on ion trajectory calculations. They determined $\sigma_{i,part}(\varepsilon)$ using various procedures which gave cross section values at 100 eV that differed by 7%–40% for the larger ion fragments, and by as much as a factor of 3 for F^+. The partial cross section values used to determine the values of $\sigma_{i,t}(\varepsilon)$ shown in Fig. 6.101 for Margreiter *et al.* [552] are for their "preferred method of analysis" (case A).

Rao and Srivastava [553] reported only one value (6.11×10^{-20} m^2) of the total ionization cross section at 100 eV (the sum of the partial ionization cross sections they measured at this energy). The stated uncertainty in their measurements is ±10%.

Rejoub *et al.* [554] reported values of $\sigma_{i,t}(\varepsilon)$ for SF_6 from threshold to 1000 eV, which they obtained by summing the partial ionization cross sections $\sigma_{i,part}(\varepsilon)$ they measured for SF_6. Their $\sigma_{i,part}(\varepsilon)$ measurements have a quoted uncertainty of ±5% for all singly charged ions, except S^+ for which the quoted uncertainty is ±8%. The data of Rejoub *et al.* for the $\sigma_{i,t}(\varepsilon)$ of SF_6 are plotted in Fig. 6.101 and are seen to be in excellent agreement with the data of Rapp and Englander-Golden [116].

No SF_6^+ was detected by Dibeler and Mohler [471], Rao and Srivastava [553], Al-Nasir *et al.* [560], or Pullen and Stockdale [561]. It is believed that the SF_6^+ ion is unstable both in its ground state and in its excited electronic states in the symmetrical configuration. Nevertheless, some authors have reported weak SF_6^+ signals. For example, Stanski and Adamczyk [551] indicated that at 100 eV the cross section for SF_6^+ is about 5×10^{-23} m^2, and Shibata *et al.* [562] reported observation of SF_6^+ in frozen-gas-target experiments where a high-energy positive-ion beam collided with SF_6 frozen on a thin Al foil held at ~ 15 K.

There have been a number of calculations [555–559] of $\sigma_{i,t}(\varepsilon)$ for SF_6. Hwang *et al.* [556] and Kim and Rudd [559] calculated the total electron-impact ionization cross section using the Binary-Encounter-Bethe (BEB) theory, the former without and the latter with contributions from multiple ionization. Margreiter *et al.* [555] and Tarnovsky *et al.* [557, 558] calculated the total single ionization cross section using the additivity rule. The results of these calculations are compared with the experimental data in Fig. 6.101. (See also Deutsch *et al.* [253] for a calculation of electron-impact ionization cross sections "utilizing empirically modified collision theories"). Surprisingly, the calculated $\sigma_{i,t}(\varepsilon)$ by Kim and collaborators for this molecule are higher than the experimental measurements, although the results of similar calculations for other molecules (e.g., CF_4 [563, 564], C_2F_6 [564, 565], and C_3F_8 [566]) are in good agreement with experiment.

The experimental data shown in Fig 6.101 are in reasonable agreement with each other except in some cases at low electron energies. The data of Rapp and Englander-Golden [116] and Rejoub *et al.* [554] are considered

Table 6.68. Recommended values for the total ionization cross section, $\sigma_{i,t}(\varepsilon)$, of SF_6.

ε (eV)	$\sigma_{i,t}(\varepsilon)$ (10^{-20} m^2)	ε (eV)	$\sigma_{i,t}(\varepsilon)$ (10^{-20} m^2)	ε (eV)	$\sigma_{i,t}(\varepsilon)$ (10^{-20} m^2)
16.5	0.02	35	2.78	100	6.51
17	0.04	40	3.40	150	6.89
18	0.07	45	3.75	200	6.79
19	0.14	50	4.29	250	6.43
20	0.22	60	5.06	300	6.06
22	0.44	70	5.59	400	5.18
25	0.99	80	5.95	500	4.53
30	1.85	90	6.28	600	3.95

to be the most reliable, and they form the basis for our recommended data below 300 eV. The solid line in Fig. 6.101 is a fit to the data of Rapp and Englander-Golden and Rejoub *et al.* below 300 eV, and Rejoub *et al.* and Stanski and Adamczyk [551] above 300 eV. Values derived from this fit are listed in Table 6.68 as our recommended values of the $\sigma_{i,t}(\varepsilon)$ of the SF_6 molecule.

The high ionization thresholds and the repulsiveness of the dissociating excited electronic states of SF_6 account for the large values of W (the energy required by energetic charged particles to produce an electron-positive ion pair in SF_6). Hilal and Christophorou [567] measured the mean energy required to produce an electron-positive ion pair in SF_6 by α particles (initial energy $\sim$ 5.1 MeV) and found it to be 35.45 eV per ion pair. This value is larger than that measured by Lopes *et al.* [568] using ^{60}Co γ rays [(34.0 $\pm$ 0.4) eV per ion pair]. The lower value for γ rays is consistent with other measurements (e.g., see Christophorou [569]) which show that for molecular gases, the energy to produce an electron-positive ion pair for α particles exceeds that for γ rays and β particles.

5.3.2 Partial Ionization Cross Sections, $\sigma_{i,part}(\varepsilon)$, of SF_6

It was indicated in the preceding section that all electron-impact ionization processes in SF_6 are dissociative and that a number of investigators [551–554] measured the cross sections for the production of the various positive-ion fragments following electron impact on SF_6. Thus, Stanski and Adamczyk [551] measured electron-impact dissociative ionization cross sections, $\sigma_{\text{i,part}}(\varepsilon)$, for SF_6 as a function of electron energy to 600 eV. They reported $\sigma_{\text{i,part}}(\varepsilon)$ for the production of the positive-ion fragments SF_5^+, SF_4^+, SF_3^+, SF_2^+, SF^+, S^+, F^+, SF_4^{++}, and $SF_2^{++}$from SF_6. They obtained these cross sections by normalizing the sum of the relative intensities of the fragment ions to the total ionization cross section of Rapp and Englander-Golden [116] for SF_6. Their partial ionization cross sections, $\sigma_{\text{i,part}}(\varepsilon)$, are shown in Fig. 6.102. They generally maximize at or above 100 eV.

Similarly, Margreiter *et al.* [552] investigated electron-impact dissociative ionization of SF_6 and determined (see previous section for details) the $\sigma_{\text{i,part}}(\varepsilon)$ for the same positive ion fragments as Stanski and Adamczyk [551]. The $\sigma_{\text{i,part}}(\varepsilon)$ data for their preferred method (case A in their paper) are also plotted in Fig. 6.102. There is general overall agreement between most of the data of Stanski and Adamczyk [551] and Margreiter *et al.* [552], although for some fragments the two sets of measurements differ considerably. The observed differences may be due to the uncertainty in the detection efficiency of energetic fragments, a problem which was reportedly addressed in the experiments of Margreiter *et al.* [552].

The data of Rao and Srivastava [553] are for only one incident electron energy (100 eV) and were obtained by normalization to the electron-impact ionization cross sections of the rare gases. They have a quoted uncertainty of $\pm 10\%$.

The most recent measurements of the $\sigma_{\text{i,part}}(\varepsilon)$ of SF_6 are those of Rejoub *et al.* [554]. In the experiments of Rejoub *et al.*, the positive ion fragments were mass analyzed using a time-of-flight mass spectrometer and were detected with a position sensitive detector having the same detection efficiency irrespective of the ion kinetic energy. The $\sigma_{\text{i,part}}(\varepsilon)$ of Rejoub *et al.* for SF_5^+, SF_4^+, SF_3^+, SF_2^+, F^+, S^+, and SF_2^{++} from SF_6 are shown in Fig. 6.102. The overall uncertainty in the partial ionization cross sections of the singly charged ions is $\pm 5\%$, except for S^+ which is $\pm 8\%$. The uncertainty of the partial ionization cross sections for the doubly charged ions (SF_3^{++}, SF_2^{++}, and SF^{++}) was quoted by Rejoub *et al.* to be $\pm 12\%$ to $\pm 15\%$. The apparatus used by Rejoub *et al.* had insufficient mass resolution to separate SF^+ and SF_4^{++} ions, so they reported the combined cross section for production of these two ions (not shown in Fig. 6.102).

The agreement between the four sets of data varies from fragment to fragment. For some positive ion fragments (such as the energetic fragment F^+) the agreement between the four sets of measurements is poor. The fact that the cross section values of Rejoub *et al.* [554] for the production of the lighter ions are larger than those of the other groups would indicate experimental problems in the latter due to the incomplete

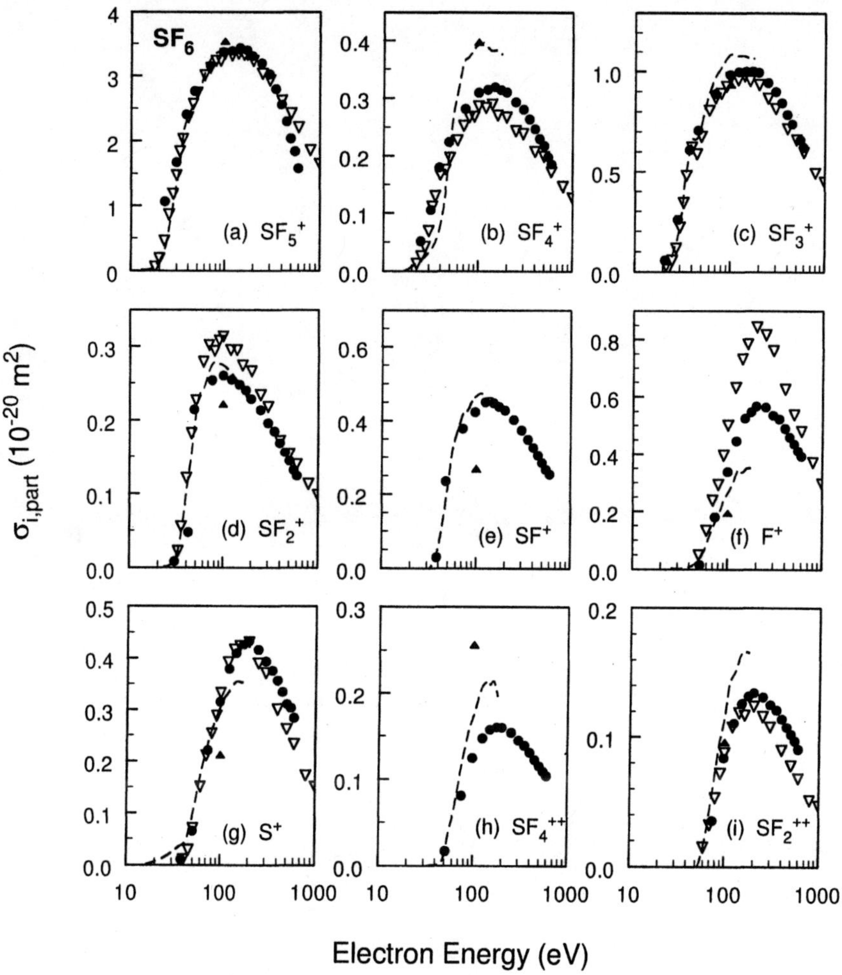

Figure 6.102. Partial ionization cross sections, $\sigma_{i,part}(\varepsilon)$, as a function of electron energy for SF_6: (•) Stanski and Adamczyk [551]; (- - -) Margreiter *et al.* [552] [these data are normalized to their "case A" values at 100 eV]; (▲) Rao and Srivastava [553]; (▽) Rejoub *et al.* [554].

collection of energetic ions. We have chosen the data of Rejoub *et al.* as the best available data. This is based upon the agreement between the $\sigma_{i,part}(\varepsilon)$ data of Rejoub *et al.* and the recommended values of $\sigma_{i,t}(\varepsilon)$ derived in Sec. 5.3.1, the ability of the time-of-flight apparatus to detect ion fragments of all kinetic energies, and the extended energy range of the data of Rejoub *et al.* These data are presented in Table 6.69 as our suggested values of $\sigma_{i,part}(\varepsilon)$ for SF_6.

Table 6.69. Partial ionization cross sections, $\sigma_{i,part}(\varepsilon)$, for SF_6 measured by Rejoub *et al.* [554] in units of 10^{-20} m^2.

ε (eV)	SF_5^+	SF_4^+	SF_3^+	SF_2^+	$(SF^+ + SF_4^{++})$
18	0.06				
20	0.19				
22.5	0.46	0.013	0.012		
25	0.86	0.026	0.056		
27.5	1.18	0.042	0.117		
30	1.47	0.069	0.221		
32.5	1.84	0.111	0.343	0.022	
35	2.03	0.130	0.481	0.055	
40	2.36	0.167	0.623	0.121	0.052
45	2.57	0.176	0.589	0.182	0.159
50	2.74	0.197	0.676	0.227	0.275
60	3.01	0.227	0.808	0.279	0.403
70	3.12	0.252	0.869	0.302	0.489
80	3.23	0.267	0.892	0.294	0.530
90	3.23	0.268	0.926	0.308	0.594
100	3.34	0.286	0.971	0.314	0.634
120	3.31	0.283	0.959	0.295	0.650
140	3.34	0.289	0.979	0.295	0.649
160	3.32	0.270	0.955	0.274	0.661
200	3.23	0.267	0.935	0.266	0.627
250	3.03	0.243	0.868	0.234	0.569
300	2.93	0.238	0.817	0.218	0.517
400	2.62	0.208	0.717	0.172	0.425
500	2.42	0.199	0.660	0.154	0.383
600	2.21	0.172	0.598	0.140	0.342
800	1.86	0.146	0.493	0.113	0.267
1000	1.65	0.127	0.449	0.098	0.221

Rejoub *et al.* [554] also reported measurement of the cross section, $\sigma_{pip}(\varepsilon)$, for positive ion pair production by electron-impact ionization of SF_6, for a number of ion pairs (SF_n^+, F^+). These are shown in Fig. 103. Over most of the electron energy range the uncertainty of these cross section data is ±30% for (SF_3^+, F^+), ±25% for (SF_2^+, F^+), ±20% for (SF^+, F^+), and ±10% for (S^+, F^+) positive ion pairs. The magnitude of the cross section for positive ion pair production by electron-impact ionization of SF_6 is generally smaller than the magnitude of the cross section for singly charged ions from electron impact on SF_6.

Finally, Al-Nasir *et al.* [560] employed an ejected-electron/produced-ion coincidence technique and measured relative partial doubly differen-

Table 6.69 (continued)
Partial ionization cross sections, $\sigma_{i,part}(\varepsilon)$, for SF_6 measured by Rejoub *et al.* [554] in units of 10^{-20} m^2.

ε (eV)	S^+	F^+	SF_3^{++}	SF_2^{++}	SF^{++}
40	0.011				
45	0.029				
50	0.070	0.049			
60	0.151	0.134	0.0032	0.015	0.0035
70	0.211	0.239	0.0073	0.032	0.0052
80	0.254	0.294	0.0089	0.053	0.0111
90	0.288	0.396	0.0119	0.072	0.0141
100	0.332	0.501	0.0133	0.088	0.0175
120	0.392	0.633	0.0154	0.108	0.0229
140	0.417	0.732	0.0172	0.119	0.0271
160	0.424	0.785	0.0167	0.117	0.0287
200	0.431	0.845	0.0154	0.124	0.0287
250	0.390	0.817	0.0141	0.116	0.0245
300	0.371	0.762	0.0139	0.108	0.0200
400	0.300	0.627	0.0117	0.090	0.0172
500	0.263	0.538	0.0104	0.078	0.0139
600	0.233	0.481	0.0098	0.068	0.0116
800	0.173	0.372	0.0072	0.051	0.0102
1000	0.151	0.297	0.0059	0.047	0.0082

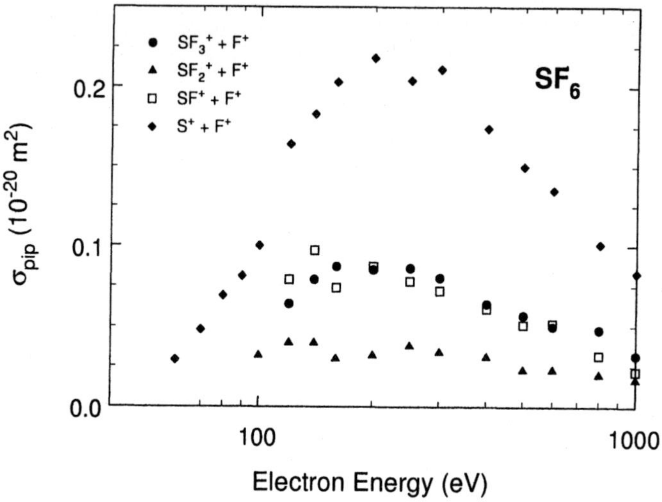

Figure 6.103. Cross sections, $\sigma_{pip}(\varepsilon)$, for positive-ion pair production by electron impact on SF_6 measured by Rejoub *et al.* [554].

tial cross sections $d^2\sigma^{(n)}/dEd\Omega$ (partial doubly differential cross section for the production of ions in the charge state n) for the ions SF_5^+, SF_4^+, SF_3^+, SF_2^+, (SF^+, SF_4^{++}), S^+, SF_5^{++}, SF_3^{++}, and SF_2^{++} resulting from electron-impact dissociative ionization of the SF_6 molecule. They made such measurements at incident electron energies of 100 and 200 eV, ejected electron energies of 30, 50, 65, 100, and 150 eV, and ejected-electron angular range of 10° to 120°. Their results confirm previous observations [456, 507] that the probability of production of ions of the form SF_x^+ with x odd is higher than that for similar ions with even values of x, and that the probability of production of ions of the form SF_x^{++} with x even is higher than that of similar ions with x odd.

All four experimental studies show that SF_5^+ is by far the most abundant positive ion fragment and that the cross sections of doubly charged positive ion fragments are generally small (Fig. 6.102, Table 6.69). Besides the doubly charged species SF_4^{++}, SF_3^{++}, SF_2^{++}, and SF^{++}, Al-Nasir *et al.* [560] found a measurable signal due to SF_5^{++}, and Harland and Thynne [507] detected S^{++} from SF_6 under impact of 70 eV electrons.

5.3.3 Electron-Impact Ionization of SF_6 Neutral Fragments

Cross sections for electron-impact ionization and dissociative ionization of the SF_5 and SF_3 radicals have been measured by Tarnovsky *et al.* [558] from threshold to 200 eV using the fast-neutral-beam technique. It was found that dissociative ionization is important for SF_5, but not for SF_3. For the SF_5 radical the cross sections for formation of the parent SF_5^+ positive ion and the fragment SF_4^+ positive ion were found to be about the same, while electron-impact ionization of the SF_3 radical predominantly generates SF_3^+ parent ions. The reported uncertainties of these cross section measurements are ±15% for the parent ionization cross sections and ±18% for the dissociative ionization cross sections. Figure 6.104(a) shows the cross sections for the formation of the SF_5^+ parent ion and the SF_4^+ fragment ion from the SF_5 free radical. The two cross section curves are similar except for the shift in the appearance energy of the SF_4^+ cross section to higher energy. The sum of the two cross sections is also plotted in the figure and represents the total single ionization cross section for this radical, since Tarnovsky *et al.* put an upper limit on the contribution from all other singly charged positive ions of 0.25×10^{-20} m^2 at 70 eV.

The cross section for the formation of the SF_3^+ parent positive ion from the SF_3 radical is shown in Fig. 6.104(b). Based on the very weak signals Tarnovsky *et al.* observed for ions other than SF_3^+ from SF_3, they

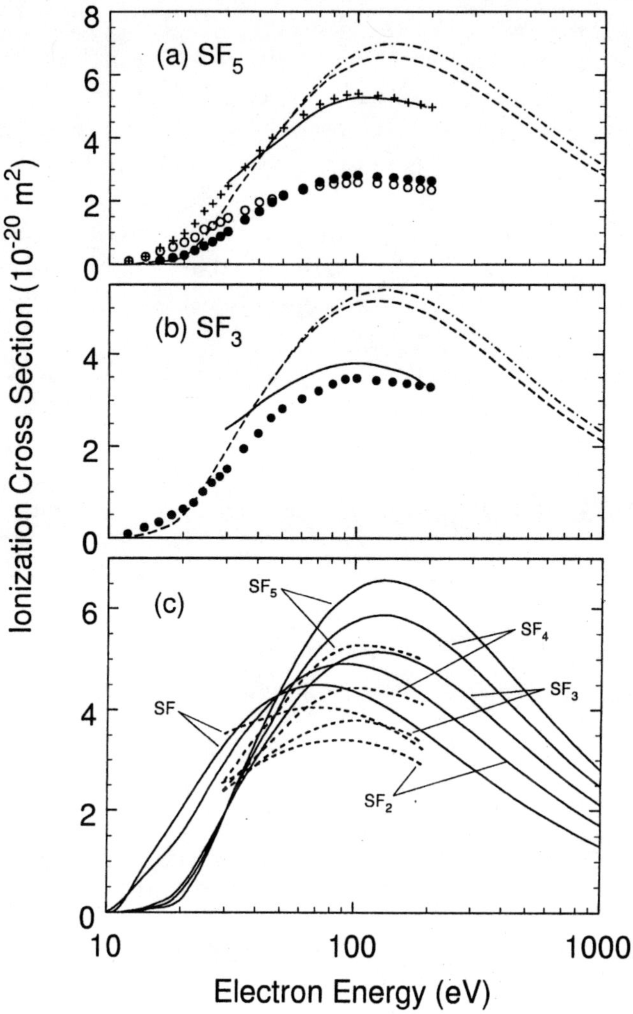

Figure 6.104. Electron-impact ionization of SF_6 radicals: (a) Cross section for the formation of SF_5^+ and SF_4^+ by electron-impact on the radical SF_5. Measurements of Tarnovsky *et al.* [558]: (o) SF_5^+; (•) SF_4^+; (+) sum of the measured cross sections for SF_5^+ and SF_4^+ generated by electron-impact on SF_5. Calculations of total electron-impact single ionization cross section of the radical SF_5: (—) data of Tarnovsky *et al.* [558]; (- - -) data of Ali *et al.* [570] (without double ionization); (— · —) data of Ali *et al.* [570] (with double ionization). (b) Cross section for the formation of SF_3^+ by electron-impact on the radical SF_3: (•) measurements of Tarnovsky *et al.* [558]; (—) calculation of Tarnovsky *et al.* [558]; (- - -) calculation of Ali *et al.* [570] (without double ionization); (— · —) calculation of Ali *et al.* [570] (with double ionization). (c) Calculated electron-impact total single ionization cross sections as a function of the electron energy for the radicals SF_5, SF_4, SF_3, SF_2, and SF: (- - -) calculations of Tarnovsky *et al.* [558]; (—) calculations of Ali *et al.* [570] (without double ionization).

put an upper limit on the possible contributions from all other singly charged ions of 0.2×10^{-20} m^2 at 70 eV.

Besides the measurements in Figs. 6.104(a) and 6.104(b), there have been two calculations of the ionization cross sections of these two radicals by electron impact. The modified additivity rule calculation of the total electron-impact single ionization cross section of Tarnovsky *et al.* [557, 558], and the BEB calculation of the total electron-impact ionization cross section with and without double ionization of Ali *et al.* [570]. In comparing the results of the two BEB calculations with the experimental measurements one should consider the calculated values without double ionization. The overall agreement between the experimental and calculated values is reasonable although, because the BEB cross sections are higher than the experimental data for most of the electron energy range, further scrutiny of both the theoretical and the experimental results is indicated.

Besides the total electron-impact single ionization cross sections for the SF_5 and SF_3, Tarnovsky *et al.* [557, 558] and Ali et al. [570] made similar calculations for the radicals SF_4, SF_2, and SF. The calculated total electron-impact single ionization cross sections for all five radicals SF_5, SF_4, SF_3, SF_2, and SF are presented in Fig. 6.104(c). In all cases the BEB cross sections lie generally higher than the modified additivity rule results.

Table 6.70 lists values of the threshold energies for the appearance of positive ions from SF_6 radicals.

5.3.4 Density-Normalized Electron-Impact Ionization Coefficient, $\alpha/N\,(E/N)$, of SF_6

There have been a number of measurements of this coefficient using the steady-state Townsend method [265, 424, 573–581], and the pulsed-Townsend method [582–586]. These measurements were made at room temperature (293–296 K) and collectively cover the E/N range from 212×10^{-17} V cm^2 to 8482×10^{-17} V cm^2. Not all of these studies quote uncertainties for the data. Bhalla and Craggs [574] and de Urquijo *et al.* [586] quote uncertainties varying from about ±5% to about ±15%, while Hayashi and Wang [579] estimate the uncertainty in their measurements to be about ±5%. These may be considered typical of the uncertainties in the data of the other groups.

Figure 6.105 shows the available data on $\alpha/N\,(E/N)$ for SF_6. The solid line represents a least squares fit to all the data plotted in the figure up to 2000×10^{-17} V cm^2, except those of Maller and Naidu [577] which clearly differ from the rest in their functional dependence on E/N. Values derived from this curve are listed in Table 6.71 as our

Table 6.70. Threshold energies for appearance of positive ions from SF_6 radicals.

Ion	Radical	Threshold energy (eV)	Reference
SF_5^+	SF_5	11.2 ± 1.0	[558]
		11.7^a	[571]
		10.5 ± 0.1^b	[572]
SF_4^+	SF_5	14.5 ± 1.0	[558]
	SF_5	15.5^a	[571]
	SF_4	13.0^a	[571]
	SF_4	12.03 ± 0.05^b	[572]
SF_3^+	SF_5	$17.0^{a,c}$, $16.2^{a,d}$	[571]
	SF_4	14.5^a	[571]
	SF_3	10.6^a, 11.4^e	[571]
	SF_3	11.0 ± 1.0	[558]
SF_2^+	SF_5	$21.8^{a,f}$, $21.0^{a,g}$	[571]
	SF_4	$19.3^{a,c}$, $18.5^{a,d}$	[571]
	SF_3	15.4^a	[571]
	SF_2	12.8^a, 11.8^e	[571]
	SF_2	10.08^b	[572]
SF^+	SF_5	$27.8^{a,h}$, $27.0^{a,i}$, $26.2^{a,j}$	[571]
	SF_4	$25.3^{a,f}$, $24.5^{a,g}$	[571]
	SF_3	$21.4^{a,c}$, $20.6^{a,d}$	[571]
	SF_2	18.8^a	[571]
	SF	14.7^a, 15.0^e	[571]
	SF	10.09^b	[572]

[a] Estimated.
[b] Listed value of the ionization threshold energy.
[c] Estimated using reactions producing two F.
[d] Estimated using reactions producing F_2.
[e] Measured.
[f] Estimated using reactions producing three F.
[g] Estimated using reactions producing F_2 + F.
[h] Estimated using reactions producing four F.
[i] Estimated using reactions producing F_2 + 2F.
[j] Estimated using reactions producing two F.

recommended data for the α/N (E/N) of SF_6. Above 2000×10^{-17} V cm^2 the data taken using the pulsed Townsend method [582, 585] lie higher than those taken using the steady state Townsend technique [579]. The steady state data of Hayashi and Wang [579, 580] for α/N above 2000×10^{-17} V cm^2 agree well with the high E/N values of $(\alpha - \eta)/N$ of

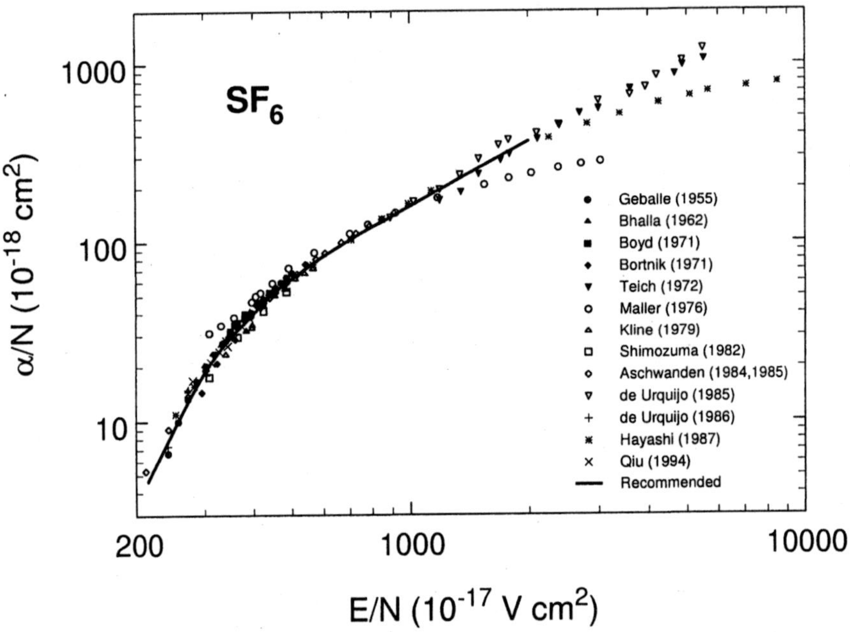

Figure 6.105. Density-normalized electron-impact ionization coefficient, $\alpha/N\,(E/N)$, for SF_6 as a function of E/N (T = 293–296 K): (●) Ref. [573]; (▲) Ref. [574]; (■) Ref. [575]; (♦) Ref. [576]; (▼) Ref. [582]; (○) Ref. [577]; (△) Ref. [424]; (□) Ref. [578]; (◊) Refs. [583, 584]; (▽) Ref. [585]; (+) Ref. [586]; (*) Ref. [579]; (×) Ref. [581]; (—) recommended.

Table 6.71. Recommended values of the density-normalized electron-impact ionization coefficient, $\alpha/N\,(E/N)$, for SF_6.

E/N (10^{-17} V cm^2)	α/N (10^{-18} cm^2)	E/N (10^{-17} V cm^2)	α/N (10^{-18} cm^2)
215	4.66	700	102.9
250	8.90	750	112.4
300	18.5	800	121.5
350	29.5	850	130.1
400	40.1	900	138.9
450	52.3	950	148.6
500	62.7	1000	158.2
550	73.2	1250	208.5
600	83.7	1500	259.8
650	93.3	2000	366.3

Hasegawa *et al.* [587] (also obtained using the steady state Townsend method, see Sec. 5.5.6 of Ch. 6). However, it is possible that the values obtained by the steady state method at high E/N may be adversely affected by the low pressures and/or small drift distances used in such high E/N measurements.

Besides the above measurements, there have been a number of Monte Carlo [429] and Boltzmann [417, 424–426, 428, 430–432, 436, 438, 441, 442] code calculations of $\alpha/N\,(E/N)$ of SF_6. The results of these calculations depend on the details of the calculation, especially the cross sections used as inputs and the associated energy losses (for such data and details see the individual papers).

5.4 Cross Sections, $\sigma_{\rm dis,neut}(\varepsilon)$, for Electron-Impact Dissociation of SF_6 into Neutral Fragments

Relatively little work has been done on the process of electron-impact dissociation of the SF_6 molecule into neutral fragments in spite of the practical significance of this process in plasma processing and gaseous dielectrics. There have been no measurements of the cross sections for total or partial dissociation of SF_6 into neutral fragments. To our knowledge the only study to date is that of Ito *et al.* [571] who used appearance mass spectrometry in a dual electron beam device and measured the energy dependence of the partial cross sections of electron-impact dissociation of SF_6 into the neutral radicals SF, SF_2, and SF_3 from threshold to 200 eV. The dissociation products SF_4 and SF_5 were below the detection limit of their system. The relative cross sections for electron-impact dissociation of SF_6 into SF, SF_2, and SF_3 reported by Ito *et al.* are shown in Fig. 6.106(a). The quoted relative uncertainty of these data is ±20%. Ito *et al.* [571] were unable to determine the absolute cross section scale in Fig. 6.106(a) because of a lack of ionization cross sections for the SF_x radicals required to normalize their relative measurements. However, such measurements have become available recently (Fig. 6.104), and could be used to determine absolute cross sections from the relative measurements of Ito *et al.*, assuming that the other experimental variables involved are accurately known. It should be noted that the relative cross sections for the three radicals in Fig. 6.106(a) are absolute with respect to each other, and that in Table 4 of the paper by Iio *et al.* [571] the cross sections for the three fragments are listed as being in units of 10^{-20} m^2. As for the case of the CF_4 molecule [563, 564], these cross sections are expected to be much smaller than the cross sections $\sigma_{\rm i,t}(\varepsilon)$ and $\sigma_{\rm e,int}(\varepsilon)$ of SF_6, at electron energies above $\sim$ 50 eV.

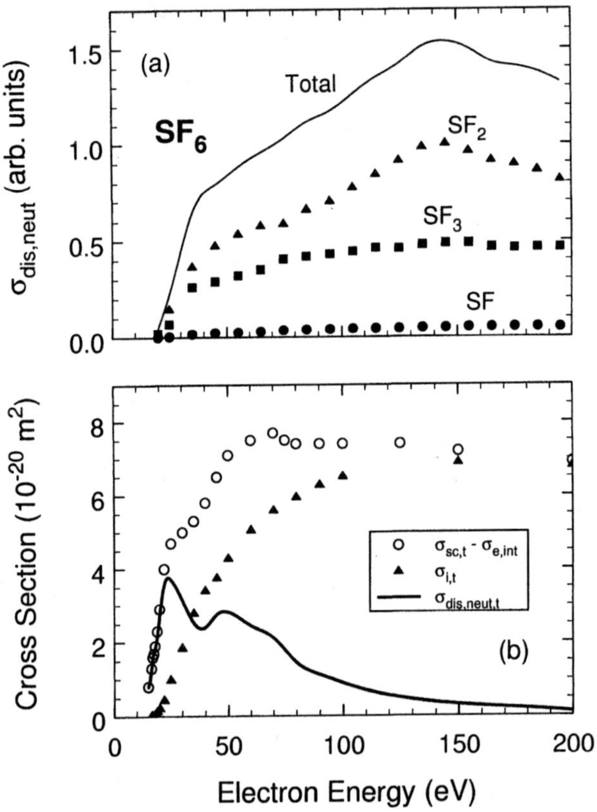

Figure 6.106. (a) Relative cross sections, $\sigma_{\text{dis,neut}}(\varepsilon)$, for electron-impact dissociation of SF_6 into SF_3 (■), SF_2 (▲), and SF (•) neutral radicals from the measurements of Ito *et al.* [571]. The line (—) represents the relative sum of the cross sections for the three radicals. (b) Deduced total cross section for dissociation into neutrals, $\sigma_{\text{dis,neut,t}}(\varepsilon)$, for SF_6 from other recommended cross sections (see text).

As we have discussed in Sec. 5.2.5 of Ch. 6, at energies above ~ 15 eV where the cross section $\sigma_{\text{vib,t}}(\varepsilon)$ is small, the difference $\sigma_{\text{sc,t}}(\varepsilon) - \sigma_{\text{e,int}}(\varepsilon)$ can be used to deduce a value for the total cross section, $\sigma_{\text{dis,neut,t}}(\varepsilon)$, for dissociation of SF_6 into neutral fragments via the relation

$$\sigma_{\text{dis,neut,t}}(\varepsilon) \cong [\sigma_{\text{sc,t}}(\varepsilon) - \sigma_{\text{e,int}}(\varepsilon)] - \sigma_{\text{i,t}}(\varepsilon). \tag{6.37}$$

In Eq. (6.37) the quantity $[\sigma_{\text{sc,t}}(\varepsilon) - \sigma_{\text{e,int}}(\varepsilon)]$ is as determined in Sec. 5.2.5 of Ch. 6 (open symbols in Fig. 6.99 above 15 eV), and $\sigma_{\text{i,t}}(\varepsilon)$ is our recommended cross section for the total ionization cross section (Table 6.68). These two quantities, along with their difference, $\sigma_{\text{dis,neut,t}}(\varepsilon)$, are shown in Fig. 6.106(b). The cross section $\sigma_{\text{dis,neut,t}}(\varepsilon)$ deduced this

Table 6.72. Deduced values of the total cross section, $\sigma_{\mathrm{dis,neut,t}}(\varepsilon)$, for electron-impact dissociation of SF_6 into neutral fragments.

ε (eV)	$\sigma_{\mathrm{dis,neut,t}}(\varepsilon)$ (10^{-20} m^2)	ε (eV)	$\sigma_{\mathrm{dis,neut,t}}(\varepsilon)$ (10^{-20} m^2)	ε (eV)	$\sigma_{\mathrm{dis,neut,t}}(\varepsilon)$ (10^{-20} m^2)
15	0.8	25	3.7	70	2.11
16.5	1.3	30	3.1	80	1.45
17	1.6	35	2.5	90	1.12
18	1.8	40	2.4	100	0.89
19	2.2	45	2.8	150	0.31
20	2.7	50	2.8	200	0.11
22	3.6	60	2.4		

way is seen to have a different energy dependence than the sum of the relative cross sections for dissociation of SF_6 into the neutral fragments SF_3, SF_2, and SF reported by Ito *et al.* [Fig. 6.106(a)]. This discrepancy is consistent with previous observations [564] of possible inaccuracies in magnitude and energy dependence in the cross sections for dissociation into neutral fragments measured by Sugai and co-workers for CF_4 and CHF_3. Ito *et al.* obtained the relative cross sections in Fig. 6.106(a) under the assumption that the total ionization cross section for each of the SF_6 radicals is similar in shape to that of SF_6 [588].

In the absence of any direct measurement of the $\sigma_{\mathrm{dis,neut,t}}(\varepsilon)$ of SF_6, we list in Table 6.72 the values of $\sigma_{\mathrm{dis,neut,t}}(\varepsilon)$ deduced as described above as our presently suggested values. It should be realized that these values are approximate and require experimental confirmation.

Table 6.73 lists the expected or measured threshold energies for dissociation of SF_6 into SF_5, SF_4, SF_3, SF_2, SF, F_2, and F as given by Ito *et al.* [571].

There have been a few other studies [549, 550, 589, 590] dealing with electron-impact dissociation of SF_6 into excited neutral fragments. These provide complementary information on the dissociation mechanisms of this molecule. Thus, Corr *et al.* [589] studied the dissociation of the SF_6 molecule into neutral metastable fragments as a function of the incident electron energy from threshold to 300 eV. They identified a number of dissociation channels involving both partial and total fragmentation of the SF_6 molecule by electron impact. In Table 6.74 are listed the threshold energies they determined for the production of the various excited fragments. The work of Corr *et al.* [589] shows that dissociation leading to total fragmentation of SF_6 is responsible for the forma-

Table 6.73. Expected or measured threshold energies given by Ito *et al.* [571] for electron-impact dissociation of SF_6 into neutrals.

Reaction	Expected threshold energy (eV)	Measured^a threshold energy (eV)
$SF_6 + e \rightarrow SF_5 + F + e$	9.6	–
$SF_6 + e \rightarrow SF_4 + 2F + e$	12.1	–
$SF_6 + e \rightarrow SF_4 + F_2 + e$	11.3	–
$SF_6 + e \rightarrow SF_3 + 3F + e$	16.0	16.0^b
$SF_6 + e \rightarrow SF_3 + F + F_2 + e$	15.2	
$SF_6 + e \rightarrow SF_2 + 4F + e$	18.6	19.5^b
$SF_6 + e \rightarrow SF_2 + 2F + F_2 + e$	17.8	
$SF_6 + e \rightarrow SF_2 + 2F_2 + e$	17.0	
$SF_6 + e \rightarrow SF + 5F + e$	22.7	22.0^b
$SF_6 + e \rightarrow SF + 3F + F_2 + e$	21.9	
$SF_6 + e \rightarrow SF + F + 2F_2 + e$	21.1	

^{a}Estimated uncertainty: ±0.5 eV.
^{b}The experimental value is affected by contributions from all respective processes given in the first column of the table.

tion of the S fragments and also the dominant process for the formation of F atoms. The finding that the F atoms are predominantly generated by complete dissociation of the SF_6 molecule is also consistent with the results of three other studies by Becker and collaborators [549, 550, 590] on the optical emission from excited atomic fragments generated by electron-impact dissociation of SF_6. Thus, Blanks *et al.* [550] using a crossed electron beam-gas beam apparatus, measured emission cross sections and appearance potentials for the most intense $3p \rightarrow 3s$ atomic fluorine lines in the visible range of the optical spectrum between 620 and 780 nm. According to Blanks *et al.* these lines are emitted by atomic fluorine fragments that are formed in the various excited states associated with the $1s^2\,2s^2\,2p^4\,3p$ electron configuration. Figure 6.107 shows the emission cross section, $\sigma_{\rm em}(\varepsilon)$, of the integrated FI $3p\,^4D^0 \rightarrow 3s\,^4P$ multiplet as a function of the incident electron energy, and Table 6.75 lists the emission cross sections, $\sigma_{\rm em}$ (100 eV), they measured for various visible FI emission lines for an incident-electron energy of 100 eV.

The threshold energies for excited neutral fragments produced by electron impact on SF_6 as obtained by these workers are also listed in Table

Table 6.74. Threshold energies for the production of excited neutral fragments upon electron impact on SF_6.

Fragment	Threshold energy (eV)	Reference	Species detected
F_2^R	18.7 ± 1.0	[589]	Rydbergs
	17.9 ± 1.0	[589]	Photons/total
	18.0 ± 1.0	[561]	F_2^+
Metastable fragments	27	[589]	Low-energy fragments
	30.0 ± 1.0	[589]	Neutrals/total
	26.8 ± 0.3	[471]	SF_2^+
	27.0 ± 0.3	[561]	SF_2^+
F^R	36.0 ± 1.0	[589]	Rydbergs
	35.0 ± 1.0	[589]	Photons/total
	35.8 ± 1.0	[589]	Neutrals/total
	35.8 ± 1.0	[471]	F^+
	37.5 ± 1.0	[561]	F^+
	35.8 ± 1.5	[549]	F^*
F^R, S^R	44.3 ± 1.0	[589]	Rydbergs
	44.5 ± 1.0	[589]	Photons/total
	44.9 ± 1.0	[589]	Neutrals/total
	42.0 ± 1.5	[549]	F^*
	41.6 ± 1.5	[550]	F^*
	40 ± 2	[590]	F^*
	43.6 ± 1.5	[549]	S^*

6.74. Blanks *et al.* [550] determined a value of (41.6 ± 1.5) eV for the threshold energy of the strongest line emission at 685.6 nm. This relatively high onset indicates that total fragmentation of the molecule plays a leading role in the electron-impact-induced dissociation of the SF_6.

Becker and collaborators [549, 590] also investigated light emission resulting from electron impact on SF_6 in other (shorter) wavelength regions: Blanks and Becker [590] investigated the wavelength range 200 to 600 nm, and Forand *et al.* [549] studied the wavelength range 45–185 nm using 200 eV incident electrons. Blanks and Becker [590] found that the $^{2,4}D \rightarrow {}^4P$ multiplet at 366.8 nm has the largest emission cross section with a cross section of $(4.4 \pm 1.2) \times 10^{-20}$ cm^2 at 100 eV. Fig. 6.108 shows the emissions produced by 200 eV electrons on SF_6. The spectrum consists of a broad structure from 200 to 320 nm and several line emissions. The strongest line emissions originate from excited fluorine fragments in the $^{2,4}S^0$, $^{2,4}P^0$, and $^{2,4}D^0$ states associated with the $2p^4\,4p$ electron

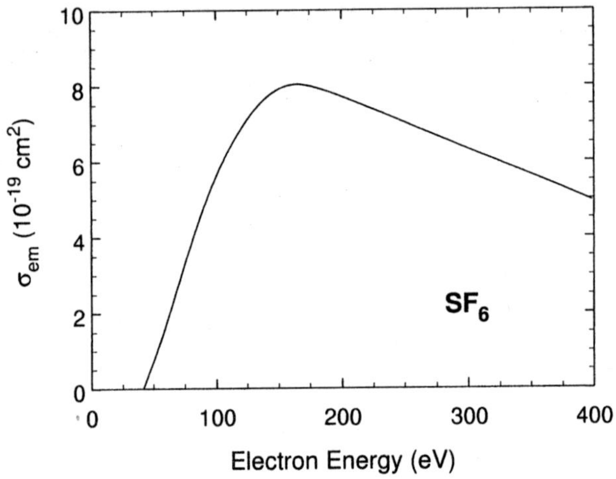

Figure 6.107. Emission cross section for the integrated FI $3p\,^4D^0 \rightarrow 3s\,^4P$ multiplet as a function of the energy of the incident electrons impacting on SF_6 (data of Blanks *et al.* [550]).

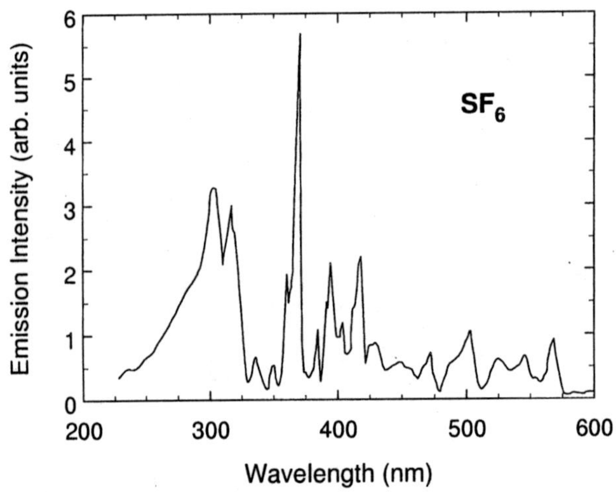

Figure 6.108. Emission spectrum of SF_6 between about 200 and 600 nm generated by collisions with 200 eV incident electrons (data uncorrected for the wavelength dependence of the detector efficiency) from Blanks and Becker [590].

configuration. These states lie approximately 16 eV above the ground state atom and decay into the $2p^4\,3s\,^{2,4}P$ states emitting several multiplets in the wavelength region 350 and 400 nm. The threshold energy of the strongest line was found by Blanks and Becker to be (40 ± 2) eV

Table 6.75. Emission cross sections, σ_{em} (100 eV), for various visible FI emission lines for 100 eV incident electrons impacting on SF_6 (data of Blanks *et al.* [550].)

Transition	Line (nm)	σ_{em} (100 eV)[a] (10^{-19} cm^2)
$^4S^0 \rightarrow {}^4P$	624.0	0.6
	634.9	0.4
	641.3	0.3
$^4D^0 \rightarrow {}^4P$	677.4	0.4
	679.5	<0.1
	683.4	0.6
	685.6	2.5
	687.0	0.4
	690.2	1.4
	691.0	0.4
$^2P^0 \rightarrow {}^2P$	703.7	0.6
	712.8	0.3
	720.2	0.2
$^2S^0 \rightarrow {}^2P$	731.1	1.4
$^4P^0 \rightarrow {}^4P$	733.2	
	739.9	2.1
	746.5	0.3
	755.2	
	757.3	1.1
$^2D^0 \rightarrow {}^2P$	760.7	
	775.5	2.0
	788.0	1.1

[a]The uncertainty is ±20% for σ_{em} (100 eV) values larger than 10^{-19} cm^2, ±25% for σ_{em} (100 eV) values smaller than 10^{-19} cm^2, and about ±40% for the σ_{em} (100 eV) of emissions at wavelengths longer than 750.0 nm.

(Table 6.74). Interestingly, no emission from ionic fluorine was observed by Becker and collaborators. A later study by Field and Eland [591] using wavelengths of 58.4, 30.4, and 25.6 nm to excite SF_6 is consistent with the electron-impact result in that no detectable ionic emission from either parent or fragment ions was observed. However, according to Forand *et al.* [549], at wavelengths in the range 45–185 nm, the observed lines following electron impact of 200 eV electrons on SF_6 could

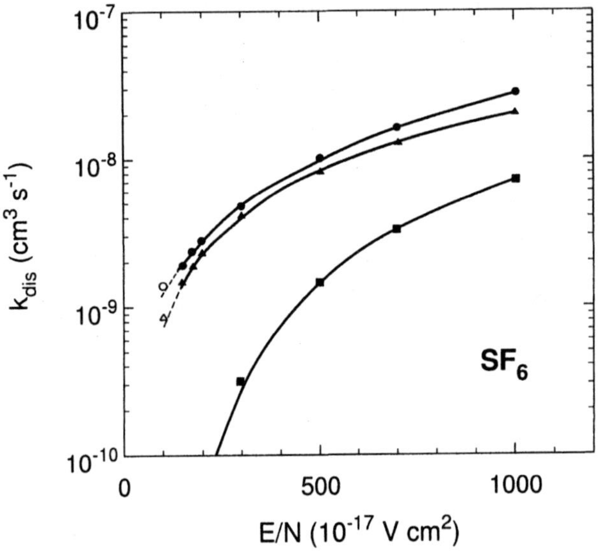

Figure 6.109. Calculated electron-impact dissociation rate, $k_{dis}(E/N)$, as a function of E/N for (•) total dissociation, (▲) total electronic excitation, and (■) total dissociative ionization (results of Phelps and Van Brunt [438]). The open points (○, △) are calculated results by Masek *et al.* [593] at $E/N = 100 \times 10^{-17}$ V cm^2. The dashed lines are extrapolations by Phelps and Van Brunt of their calculations to lower E/N.

be identified with singly ionized S atoms, besides neutral S and F atomic transitions.

From the work of Becker and collaborators [549, 550, 590], it can be concluded that electron impact on SF_6 results in the formation of neutral and ionized excited atomic fragments with significant probabilities, and that light observed following electron impact on SF_6 is due to de-excitation of these excited fragments. These results would indicate that the weak light emission observed by some investigators from the motion of electron swarms in SF_6 at high E/N, or from SF_6 discharges may be due to such excited fragments, although another possible source of these weak emissions may be impurities. It has, for instance, been shown that small amounts of N_2 in SF_6 greatly enhance light emission as the nitrogen molecule has a large excitation coefficient for the second positive group radiation ($C\,^3\Pi_u \rightarrow B\,^3\Pi_g$) and quenching of the C states of N_2 by SF_6 is not very strong [592].

Finally, in the absence of any measurements of the electron-impact dissociation rate for SF_6, we show in Fig. 6.109 the results [438] of a Boltzmann calculation of the electron-impact induced dissociation rate,

$k_{dis}(E/N)$, of SF_6 (also see Ref. [594]). It should be realized that these data are model-specific.

5.5 Electron Attachment to SF_6

5.5.1 Cross Sections for the Formation of SF_6^-, SF_5^-, SF_4^-, SF_3^-, SF_2^-, F_2^-, and F^- by electron attachment to SF_6

It has long been established that SF_6 is an electronegative gas. Electron attachment to SF_6 leads to the formation of both parent (SF_6^-) and fragment (SF_5^-, SF_4^-, SF_3^-, SF_2^-, F_2^-, and F^-) negative ions. The experimental data on the cross sections for the formation of each of these negative ions are presented and discussed in this section. The measured cross sections for the various negative ions formed by low-energy electron impact on SF_6 show that the parent negative ion SF_6^- is formed only at extremely low electron energies (below $\sim$ 0.2 eV), and that the fragment negative ions are formed via a number of negative ion resonances located between 0 and $\sim$ 15 eV.

SF_6^-. The stable parent negative ion SF_6^- produced at zero and near-zero energy is initially formed as a metastable SF_6^{-*} ion which subsequent to its formation is stabilized by collision or radiation. Its autodetachment lifetime, τ_a, under collision-free conditions has been found to be between 10 and 68 μs [403, 595–599] using time-of-flight (TOF) mass spectrometric measurements. Similar measurements using ion cyclotron resonance (ICR) techniques found the lifetime of SF_6^{-*} to be much longer, in the ms range [600–602]. This is not surprising, since a number of studies [403, 599, 603] have shown that the autodetachment lifetime of a number of polyatomic long-lived transient negative ions depends on the experimental conditions (kinetic energy of the captured electron, internal energy of the electron capturing molecule, and hence the gas temperature), and the electron energies in ICR experiments are normally much lower than in TOF experiments. While no effect of the attaching electron's energy on the τ_a of SF_6^{-*} has been reported, such energy dependence has been found for other transient anions [599, 604–608]. Furthermore, Delmore and Appelhans [609] observed an increase of $\sim$ 33% in the τ_a of the metastable SF_6^{-*} ion when the temperature of the ion source was decreased from $\sim$ 475 to $\sim$ 375 K, which they attributed to the contribution of various excited states to the autodetachment process. It has also been shown [601] that the τ_a of SF_6^{-*} in the ICR spectrometer varies with the observation time of the experiment, thus accounting for radiative and collisional cooling of SF_6^{-*} [601, 610].

Consistent with these findings are the recent measurements [611] of the lifetime of SF_6^{-*} in a free-jet expansion experiment, which showed that the lifetime of SF_6^{-*} at low temperatures ($T < 10$ K) is 19.1 ± 2.7 μs and independent of the electron energy in the range 0.0 to 0.1 eV.

Besides the formation of SF_6^- by direct attachment of low-energy electrons, SF_6^- can also be formed indirectly by relatively higher energy electrons when these electrons lose "all" of their energy to inelastic electron scattering and are then attached efficiently as "zero-energy" electrons to SF_6 molecules. Indeed, this is the basis of the so-called "SF_6^- threshold-electron-excitation technique" used to locate negative-ion states and excited electronic states, especially optically-forbidden states, of neutral molecules (e.g., Refs. [612–615]).

Many early studies (e.g., Refs. [341, 478, 510, 596, 612, 613, 616–619]) showed that SF_6^- is formed at near 0.0 eV with a very large cross section. These studies also indicated that the shape of the measured cross section for the formation of SF_6^- is instrumental, i.e., the SF_6^- resonance is narrower than the energy spread of the electron pulse or beam used to measure it. This narrowness of the SF_6^- resonance is largely responsible for the large uncertainty and spread in the early published values of the electron attachment cross section, $\sigma_{a,SF_6^-}(\varepsilon)$, for formation of SF_6^- at thermal and near-thermal energies. There have been a number of subsequent (room temperature) studies using both conventional electron beam [424, 535, 620] and electron swarm [231, 504, 505, 621–624] techniques. Additionally, newer very-high resolution and very-low energy electron beam techniques have been used to provide accurate values of the cross section for the attachment of free electrons to SF_6 to form SF_6^- from ~ 1 eV down to the μeV range [547, 548, 625–632]. Besides these methods, information on $\sigma_{a,SF_6^-}(\varepsilon)$ has been provided in the thermal and subthermal electron energy range by another new technique using "bound-electron" capture in collisions of SF_6 with high-Rydberg atoms [630, 633–649]. The results obtained by these methods are complementary to those obtained by direct "free" electron-impact methods.

The cross section data for total electron attachment determined by these methods are plotted in Fig. 6.110 for energies between 0.0001 and ~ 0.2 eV, and are briefly discussed below.

Data obtained by conventional electron swarm and electron beam techniques — The swarm data [231, 504, 505, 621–624] plotted in Fig. 6.110 have been unfolded from the total electron attachment rate-constant measurements made in mixtures of SF_6 with N_2 buffer gas. They are therefore total electron attachment cross sections. However, since SF_6^-

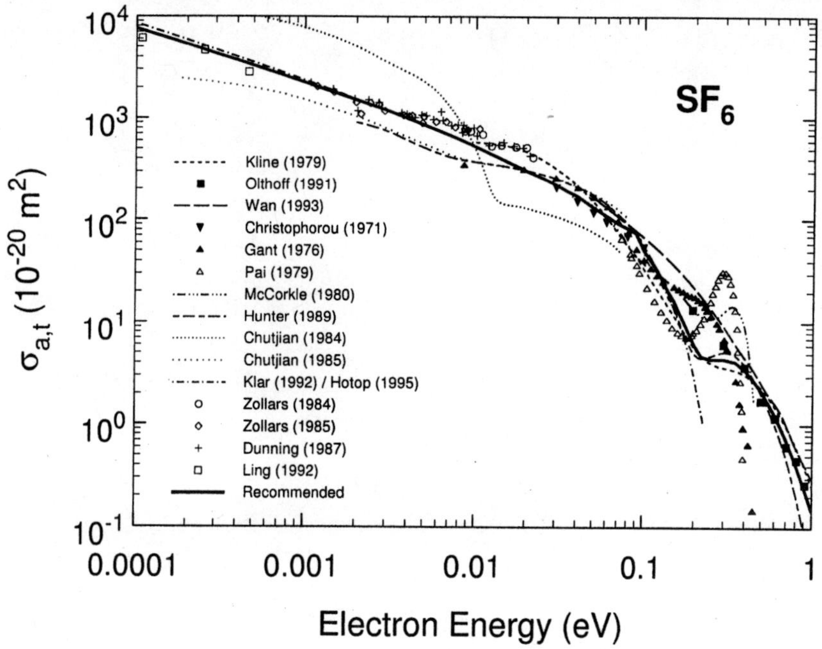

Figure 6.110. Measured total electron attachment cross section, $\sigma_{\mathrm{a,t}}(\varepsilon)$, for SF_6 as a function of the electron energy, for free electrons and for electrons bound in high-Rydberg states of atoms (see text). Below ~ 0.1 eV this cross section is about equal to $\sigma_{\mathrm{a,SF}_6^-}(\varepsilon)$ for the production of SF_6^-. *Conventional electron-beam data:* (- - -) Ref. [424]; (■) Ref. [620]; (– –) Ref. [535]. *Swarm-unfolded data:* (▼) Ref. [504]; (▲) Ref. [622]; (△) Ref. [623]; (– · · · –) Ref. [231]; (— - —) Ref. [624]. *TPSA technique:* (· · ·) Data from Ref. [628] normalized to the thermal electron attachment rate constant value of Crompton and Haddad in Ref. [650]; (· · ·) data of Ref. [629] normalized to the swarm-unfolded cross section of McCorkle *et al.* in Ref. [231]. *LPA technique:* (– · –) Refs. [547, 651, 652]. *High-Rydberg data:* (○) Xe*(nf), Ref. [636]; (◊) Rb*(ns, nd), Ref. [637]; (+) Rb*(ns, nd), K*(nd), Xe*(nf), Ref. [639]; (□) K*(np), Ref. [644]. (—) Recommended values.

is the predominant negative ion in the energy region below ~ 0.1 eV, these cross sections give the value of $\sigma_{\mathrm{a,SF}_6^-}(\varepsilon)$. It is difficult to assess the uncertainty of these cross section data. Although the rate constant measurements are expected to have an uncertainty around ±10%, the cross section values have a larger uncertainly. Similarly, the electron beam data have a large uncertainty at low energies. Kline *et al.* [424] used the retarding potential difference method in their experiments and their energy resolution was typically between 80 and 100 meV. Their cross section below ~ 0.1 eV is therefore instrumental. The measurements of Olthoff *et al.* [620] and Wan *et al.* [535] are for the total electron attachment cross section and were made using a trochoidal monochromator ($T \approx 328$ K). They have a quoted uncertainty of ±15% above 1 eV and a larger uncertainty below this energy, which perhaps increases to ±50% at 0.2 eV.

Data obtained from very-high resolution, very-low energy electron-beam techniques — Two such techniques have been developed, one by Chutjian and collaborators [625, 626, 628, 629] and the other by Hotop and collaborators [547, 548, 630–632]. The two techniques are similar in principle (see also Ch. 2). The technique of the former group uses VUV photoionization of rare-gas atoms mixed with SF_6 to generate *in situ* photoelectrons of well defined and variable energy. It is referred to as the threshold-photoelectron-spectroscopy-by-electron-attachment (TPSA) technique and provides relative cross sections for electron energies in the range from a few meV to 150 meV with electron-energy resolution between 6 to 8 meV (full-width-at-half-maximum) [628]. The energy dependence of the cross section is obtained using convolution techniques and the absolute value is determined by normalization to electron swarm data. The data of Chutjian *et al.* [628] were normalized to the thermal electron attachment rate constant (2.27×10^{-7} cm^3 s^{-1}) of Crompton and Haddad [650], while the data of Chutjian and Alajajian [629] were normalized to the electron swarm-unfolded cross section data of McCorkle *et al.* [231].

The technique of Hotop and collaborators is referred to as the laser-photoelectron-attachment (LPA) method and is based on the controlled production of variable-energy photoelectrons using lasers to photoionize a well-collimated beam of metastable $Ar^*(4s\,^3P_2)$ atoms "from a differentially-pumped dc discharge source in a field-fee region." The resulting negative ions are then detected by application of pulsed electric fields [547]. This technique has permitted the energy dependence of the electron attachment cross section to be measured directly with sub-meV resolution. The values measured by Hotop and collaborators [547, 651] for the $\sigma_{a,SF_6^-}(\varepsilon)$ in the range 0.0001–0.250 eV are shown in Fig. 6.110. These cross-section values were obtained by normalization to the thermal ($T = 300$ K) rate constant measurement of Crompton *et al.* [650, 653]. It should be emphasized that the cross section of Hotop *et al.* in Fig. 6.110 is only for the formation of SF_6^-, while the rest of the data for electron energies above 0.1 eV are total electron attachment cross sections.

Data obtained from high-Rydberg atom bound-electron capture methods — A number of investigators, foremost the group at Rice University, have used beams of atoms excited in high-Rydberg states, $A^{**}(nl)$, to measure the rate constant for the reaction

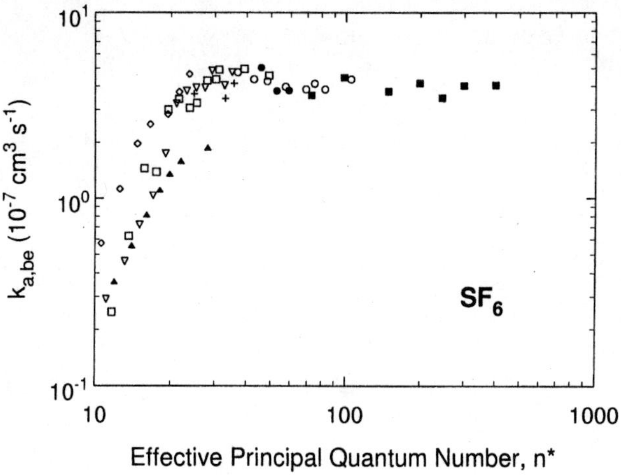

Figure 6.111. Rate constant, $k_{\mathrm{a,be}}(n^*)$, for bound-electron capture as a function of the effective principal quantum number n^* in collisions of Rydberg atoms with SF_6 from Ling *et al.* [644]. (■) K*(np), Ref. [644]; (○) Rb*(nd), Ref. [637]; (•) Rb*(ns), Ref. [637]; (□) K*(nd), Ref. [639]; (▽) Na*(np), Ref. [640]; (◇) Ne*(ns), Ref. [630]; (▲) Ne*(nd), Ref. [630].

$$A^{**}(nl) + \mathrm{SF}_6 \rightarrow A^+ + \mathrm{SF}_6^- \tag{6.38}$$

down to $\sim 4\ \mu$eV [649]. In almost all such investigations, $A^{**}(nl)$ is a rare-gas or an alkali-metal atom in a Rydberg state with principal quantum number n and angular momentum quantum number l. In these studies the rate constant $k_{\mathrm{a,be}}$ for the formation of SF_6^- is measured as a function of the principal quantum number n, or as a function of the effective principal quantum number n^* ($= n - \delta_l$, where δ_l is the l-dependent quantum defect). An example of the $k_{\mathrm{a,be}}(n^*)$ measurements is shown in Fig. 6.111. From data such as in Fig. 6.111, the cross section for bound-electron capture, $\sigma_{\mathrm{a,be}}(v)$, is determined via the relation

$$\sigma_{\mathrm{a,be}}(v) = k_{\mathrm{a,be}}(n^*)/v_{\mathrm{rms}}, \tag{6.39}$$

where v_{rms} is the root-mean-square velocity of the Rydberg electron. While a number of authors (e.g., Refs. [647, 649]) have shown Eq. (6.39) to be inappropriate for Rydberg states with l small compared to n, and while postattachment electrostatic attraction between the products (A^+, SF_6^-) may have significant effects on the size of the measured cross section (e.g., Ref. [649]), a number of studies (e.g., Refs. [637, 644, 648, 649]) have shown that the cross sections for bound-electron capture are consistent with the "free" electron model for sufficiently large values (> 30)

of the principal quantum number n. In such cases, then, the excited electron and the core can be considered independent particles, the electron in the Rydberg atom essentially being equivalent to a free electron having the same kinetic energy as the binding energy of the Rydberg electron. These studies, as well as those of Chutjian and collaborators and Hotop and collaborators (referred to earlier in this section) have shown that at very low energies ($\leq$ 1 meV, Refs. [547, 632]) the attachment cross section for the formation of SF_6^- varies inversely with the electron velocity in accordance with the Wigner threshold law for s-wave electron attachment [654].

Reaction (6.38) has been studied using a number of Rydberg atoms: He*(n = 14, ^1P) [646]; Ne* (ns, nd) [630]; Xe*(nf) [633, 634, 636, 639]; K*(nd) [638, 639, 641]; K*(np) [644]; Na*(np) [640]; Rb*(ns, nd) [637, 639]; and Cs*(ns, np, nd) [642]. The results of these high-Rydberg-atom studies on $\sigma_{\mathrm{a,be}}(v)$ are plotted in Fig. 6.110.

The wide spread in the data shown in Fig. 6.110 and the many factors that can affect them (many of which are often not clearly discerned) makes the deduction of a recommended total electron attachment cross section, $\sigma_{\mathrm{a,t}}(\varepsilon)$, for SF_6 difficult. However, since at energies below $\sim$ 1.5 eV, SF_6^- and SF_5^- are the only negative ions produced by electron attachment to SF_6 with significant intensity, we determined the recommended values of $\sigma_{\mathrm{a,t}}(\varepsilon)$ for SF_6 shown by the solid line in Fig. 6.110, by summing our recommended cross sections for SF_6^- and SF_5^- formation, which are determined separately below.

The determination of the recommended cross section, $\sigma_{\mathrm{a,SF_6^-}}(\varepsilon)$, for the formation of SF_6^- was made by considering the data in Fig. 6.112(a). The dot-dash line (– · –) represents the measurements of Hotop and collaborators [547, 651] obtained using the LPA technique (much of this curve from $\sim$ 0.003 to 0.2 eV is overlaid by the solid recommended line discussed later). The broken lines (– - –) and (- - -) are the cross sections for the production of only SF_6^- as obtained, respectively, from the electron-swarm-based measurements by Hunter *et al.* [624] and from electron-beam-based measurements by Kline *et al.* [424]. The symbols shown in the figure are the bound-electron attachment cross sections of Refs. [636, 637, 639, 644]. At the lowest energies ($<$ 0.005 eV), the cross section values of Hotop and collaborators are in excellent agreement with the bound-electron attachment measurements for $n > 30$. At higher energies, the cross section of Hotop and collaborators drops off abruptly at the onsets of the ν_1, ν_3, and $2\nu_1$ vibrational excitations (see discussions in Refs. [547, 651]), and is in overall agreement with the data of Hunter *et al.* and Kline *et al.* considering the precipitous decline in the magnitude of the cross section with increasing electron energy in this energy range.

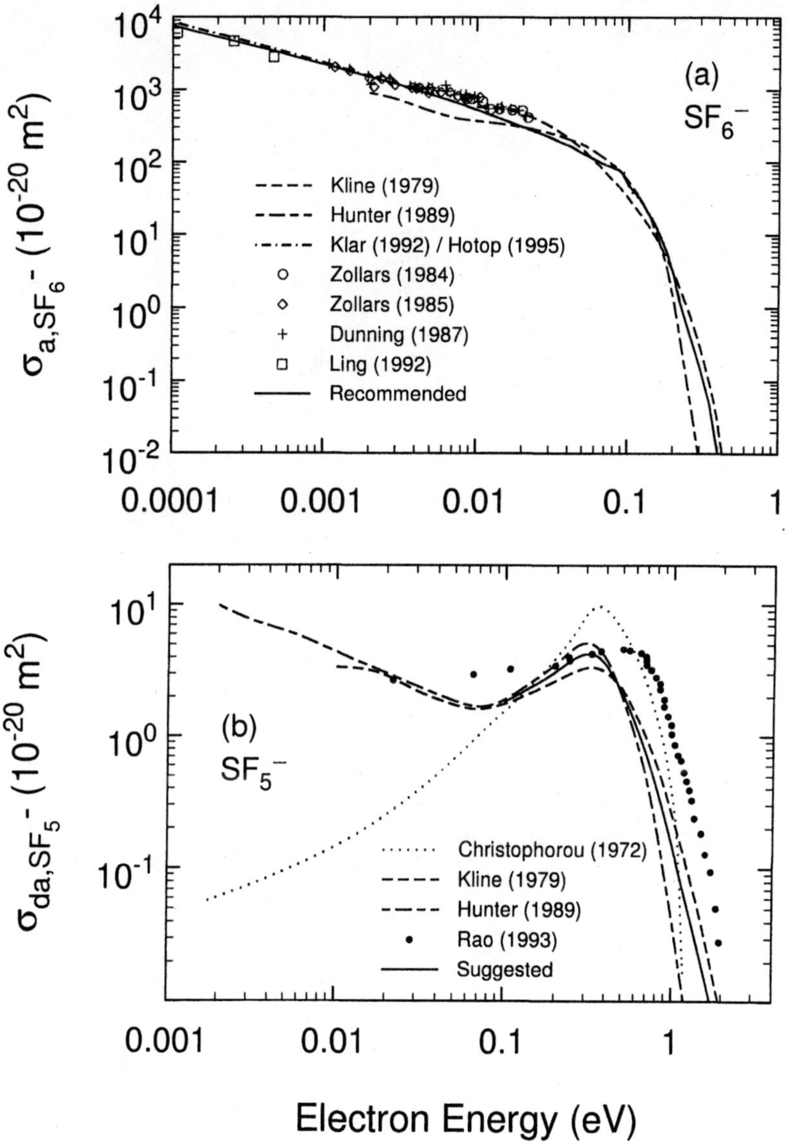

Figure 6.112. (a) Selected measurements of the electron attachment cross section, $\sigma_{a,SF_6^-}(\varepsilon)$, for the formation of SF_6^- from SF_6 as a function of the electron energy: (- - -) Ref. [424]; (– – –) Ref. [624]; (–·–) Refs. [547, 651, 652]; (○) Ref. [636]; (◇) Ref. [637]; (+) Ref. [639]; (□) Ref. [644]; (—) recommended. (b) Selected measurements on the dissociative electron attachment cross section, $\sigma_{da,SF_5^-}(\varepsilon)$, for the formation of SF_5^- from SF_6 as a function of the electron energy: (···) Refs. [504, 505]; (- - -) Ref. [424]; (– – –) Ref. [624]; (•) Ref. [509]; (—) suggested.

Table 6.76. Recommended values of the electron attachment cross section, $\sigma_{a,SF_6^-}(\varepsilon)$, for the formation of SF_6^- by electron impact on SF_6 in the electron energy range 0.0001 eV to 0.40 eV.

ε (eV)	$\sigma_{a,SF_6^-}(\varepsilon)$ (10^{-20} m^2)	ε (eV)	$\sigma_{a,SF_6^-}(\varepsilon)$ (10^{-20} m^2)
0.0001	7617	0.030	221
0.0002	5283	0.035	190
0.0003	4284	0.040	171
0.0004	3692	0.045	149
0.0005	3280	0.050	132
0.0006	2968	0.060	109
0.0007	2724	0.070	92.7
0.0008	2529	0.080	82.9
0.0009	2369	0.090	74.3
0.001	2237	0.10	49.5
0.002	1511	0.12	30.8
0.003	1202	0.14	17.8
0.004	993	0.15	14.2
0.005	859	0.16	10.5
0.006	760	0.18	5.85
0.007	683	0.20	2.86
0.008	621	0.22	1.24
0.009	569	0.25	0.52
0.010	526	0.28	0.25
0.015	383	0.30	0.16
0.020	304	0.35	0.05
0.025	257	0.40	0.01

The measurements of Hotop and collaborators [547, 651] from 0.0001 to 0.23 eV, the bound-electron attachment data below 0.005 eV, and the data of Hunter *et al.* and Kline *et al.* above 0.2 eV were fit to yield the solid line in Fig. 6.112(a). This line represents our recommended values for $\sigma_{a,SF_6^-}(\varepsilon)$ in the energy range 0.0001–0.4 eV, and values extracted from this line are listed in Table 6.76.

The other free-electron data for $\sigma_{a,SF_6^-}(\varepsilon)$ shown in Fig. 6.112(a) were not considered for various reasons: (1) The data of Chutjian and collaborators are indirect measurements both in absolute magnitude and energy dependence. Also, according to Chutjian [655] part of the disagreement between the data they obtained using the TPSA technique and the data obtained using the LPA and bound-electron attachment methods are due to the effects of stray fields on the TPSA measurements. (2) The electron beam data of Olthoff *et al.* [620] and Wan *et al.* [535] do not show the enhancement at ~ 0.4 eV due to SF_5^- borne out by all other electron beam and electron swarm studies, and no means exists to differentiate between the SF_6^- and SF_5^- contributions. (3) From the various swarm-based cross sections we considered only the more recent data of Hunter *et al.* [624] since these measurements were more accurate and more extensive than the previous ones, and since more recent cross sections were employed in the determination of the electron energy distribution functions used to obtain the swarm-based cross sections from the measured rate constants.

SF_5^-. A number of electron-beam studies [424, 504–507, 509, 511, 656], have shown that at energies below ~ 2 eV, SF_5^- is produced via the dissociative electron attachment reaction

$$e + SF_6 \rightarrow SF_5^- + F \tag{6.40}$$

with maxima at ~ 0.0 and 0.38 eV. It has been shown by Chen and Chantry [506] that the first resonance at ~ 0.0 eV is sensitive to gas temperature (Sec. 5.5.4 of Ch. 6) and that the second peak at 0.38 eV is rather independent of temperature. Interestingly, Matejcik *et al.* [656] have recently found that the first peak at ~ 0.0 eV disappears at temperatures below ambient.

In Fig. 6.112(b) are compared the published values of the cross section, $\sigma_{da,SF_5^-}(\varepsilon)$, for the formation of SF_5^- by low-energy electron-impact on SF_6 as a function of electron energy for swarm-based data [504, 505, 624] and electron-beam data [424, 509]. None of these studies quoted uncertainties. As can be seen from Fig. 6.112(b) there are large differences in the cross section data for the production of SF_5^- from SF_6. The large discrepancies below 0.1 eV may be partly due to the varied effects of temperature on the formation of SF_5^- at these low energies (e.g., see Refs. [547, 657]). For this reason, we suggest values for the cross section, $\sigma_{da,SF_5^-}(\varepsilon)$, for the formation of SF_5^- by electron attachment to SF_6 down to only 0.1 eV. These suggested values are shown in Fig. 6.112(b) by the solid line which was obtained by a least squares fit to the measurements of Kline *et al.* [424] and Hunter *et al.* [624]. Values derived from the line are listed in Table 6.77. The data of Rao and Srivastava [509] were excluded from the fit since they exhibit a larger energy uncertainty due to the difficulty of obtaining accurate values from the sharp SF_5^- peak shown in the figure of their paper. It should be noted, that at room temperature the contribution of SF_5^- to the total electron attachment cross section below ~ 0.1 eV is small compared to SF_6^- production.

Table 6.77. Suggested values for the dissociative electron attachment cross section, $\sigma_{da,SF_5^-}(\varepsilon)$, for the formation of SF_5^- by electron attachment to SF_6 as a function of the electron energy.

ε (eV)	$\sigma_{da,SF_5^-}(\varepsilon)$ (10^{-20} m^2)	ε (eV)	$\sigma_{da,SF_5^-}(\varepsilon)$ (10^{-20} m^2)	ε (eV)	$\sigma_{da,SF_5^-}(\varepsilon)$ (10^{-20} m^2)
0.10	1.85	0.25	3.86	0.7	0.722
0.12	2.09	0.28	4.15	0.8	0.416
0.14	2.36	0.30	4.23	0.9	0.245
0.15	2.48	0.35	4.06	1.0	0.146
0.16	2.61	0.40	3.44	1.2	0.060
0.18	2.87	0.45	2.75	1.5	0.020
0.20	3.15	0.50	2.15	1.9	0.005
0.22	3.45	0.60	1.25		

SF_4^-, SF_3^-, and SF_2^-. The smaller SF_x^- ($x = 2, 3, 4$) ions are generated mostly through negative-ion states located at progressively higher energies (see Table 6.78). The SF_4^- fragment is generated from a broad resonance between 4 and 8 eV. Harland and Thynne [507] attributed the production of this ion to the reaction

$$SF_6 + e \rightarrow SF_4^- + 2F \tag{6.41}$$

and reported the cross section maximum for this ion to be at (6.0 ± 0.1) eV. In contrast, Fenzlaff *et al.* [511] although not excluding Reaction (6.41) for the formation of SF_4^-, attributed its formation "at least on the high energy side of the resonance," to the reaction

$$e + SF_6 \rightarrow SF_6^{-*}\,(5.4\text{ eV}) \rightarrow SF_4^- + F_2 \tag{6.42}$$

with a threshold at (2.8 ± 0.1) eV and a peak at 5.4 eV.

Cross sections for all three fragment negative ions were measured at room temperature by Lehmann [510], Kline *et al.* [424], and Rao and Srivastava [509], and are shown in Fig. 6.113. Kline *et al.* established the magnitudes of the cross sections by measuring the various positive-ion cross sections and calibrating their sum with the total ionization cross section measurements of Rapp and Englander Golden [116]. Their effective electron energy resolution was between 0.08 and 0.1 eV with an estimated uncertainty of the electron energy scale of ±0.1 eV. Rao and Srivastava [509] measured peak cross section values for SF_4^-, SF_3^-, and SF_2^-, respectively, equal to 4.0×10^{-23}, 8.8×10^{-24}, and 1.3×10^{-24} m^2. The data of Rao and Srivastava shown in Fig. 6.113 were taken off their

Table 6.78. Energy positions of cross section maxima in the formation of negative ions by low-energy electron impact on SF_6.

Negative ion	Energy position of maximum (eV)	Ref.	Suggested reaction
SF_6^-	~0.0	see text	$e + SF_6 \rightarrow SF_6^{-*}$
SF_5^-	0.50 ± 0.1	[507]	$e + SF_6 \rightarrow SF_5^- + F$
	0.38	[504, 505]	
	~0.0; 0.38	[506, 424]	
	~0.0[a]; 0.38	[511]	
	~0.0	[509][b]	
SF_4^-	6.0 ± 0.1	[507]	$e + SF_6 \rightarrow SF_4^- + 2F$
	5.4	[510][b,c]	
	5.4	[424][b]	
	5.4	[511]	$e + SF_6 \rightarrow SF_4^- + F_2$ and ($SF_4^- + 2F$)
	5.0	[509][b]	
SF_3^-	11.2	[510][b,c]	
	9.4; 10.9	[424][b]	
	11.3	[509][b]	
SF_2^-	13.0	[510][b,c]	
	12.0	[424][b]	
	12.3	[509][b]	
F_2^-	2.4; 4.8; 11.7	[510][b,c]	
	2.0; 4.4; 11.2	[424][b]	
	5.4	[511]	$e + SF_6 \rightarrow SF_6^{-*}$ (5.4 eV) $\rightarrow SF_4 + F_2^-$
	2.2; 4.8; 11.6	[509][b]	
F^-	5.7 ± 0.1	[507]	$e + SF_6 \rightarrow SF_6^{-*} \rightarrow SF_4 + F + F^-$
	9.3 ± 0.1	[507]	$e + SF_6 \rightarrow SF_6^{-*} \rightarrow SF_3 + 2F + F^-$
	11.8 ± 0.1	[507]	$e + SF_6 \rightarrow SF_6^{-*} \rightarrow SF_2 + 3F + F^-$
	2.8; 5.2; 8.9; 11.5	[510][b,c]	
	2.6; 5.1; 8.8; 11.3	[424][b]	
	~2.9	[511]	$e + SF_6 \rightarrow SF_6^{-*} \rightarrow SF_5 + F^-$
	5.4	[511]	$e + SF_6 \rightarrow SF_6^{-*} \rightarrow SF_4 + F + F^-$
	2.8; 5.3; 9.4; 11.6	[509][b]	

[a]The zero-energy peak was found to "disappear" at temperatures below ambient [656].
[b]Determined from the respective figures of each of these references.
[c]All energy values determined from the figures given in Ref. [510] were reduced by 0.6 eV (see text).

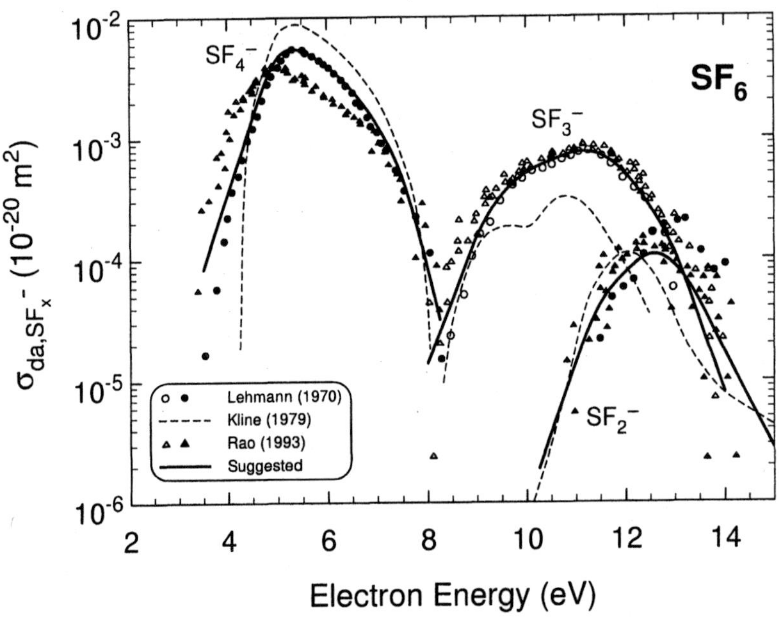

Figure 6.113. Cross sections for the formation of SF_4^-, SF_3^-, and SF_2^- by dissociative electron attachment to SF_6 as a function of the electron energy. (•, ∘) Ref. [510]; (- - -) Ref. [424]; (△, ▲) Ref. [509]; (—) suggested values.

plots [509] and were put on an absolute scale using these peak cross section values. In Fig. 6.113 the data of Lehmann were shifted to lower energies by 0.6 eV, as suggested by Kline *et al.* [424]. With the applied shift in the data of Lehmann [510], the agreement in the magnitude and energy dependence of the cross sections for these three ions as measured by Lehmann [510] and by Rao and Srivastava [509] is reasonable. The energy position of the maxima in the cross sections for each of these ions as determined from the data reported by Lehmann [510], Kline *et al.* [424], and Rao and Srivastava [509] are listed in Table 6.78, along with similar data derived from the relative cross section measurements of Fenzlaff *et al.* [511].

The solid lines shown in Fig. 6.113 for each negative ion fragment represent our respective suggested cross sections. They were obtained by a least squares fit to all three sets of measurements for each ion, except for the case of the SF_3^- ion for which only the data of Refs. [509, 510] were considered. In Table 6.79 are listed suggested values for the cross sections of these three fragment negative ions derived from the respective solid lines in Fig. 6.113.

Table 6.79. Suggested cross sections for the formation of SF_4^-, SF_3^-, SF_2^-, F_2^-, and F^- by dissociative electron attachment to SF_6 as a function of the electron energy.

	Cross section (10^{-22} m^2)				
ε (eV)	σ_{da,SF_4^-}	σ_{da,SF_3^-}	σ_{da,SF_2^-}	σ_{da,F_2^-}	σ_{da,F^-}
1.50	–	–	–	0.0030	–
2.00	–	–	–	0.0080	0.022
2.25	–	–	–	0.0075	0.075
2.50	–	–	–	0.0060	0.149
2.75	–	–	–	0.0042	0.162
3.0	–	–	–	0.0026	0.098
3.5	0.0084	–	–	0.0026	0.171
4.0	0.0350	–	–	0.0265	0.856
4.5	0.144	–	–	0.0707	2.69
5.0	0.457	–	–	0.0489	4.39
5.5	0.528	–	–	0.0155	4.63
6.0	0.394	–	–	0.0038	2.78
6.5	0.251	–	–	0.0019	1.37
7.0	0.1300	–	–	0.0015	0.749
7.5	0.0460	–	–	0.0015	0.615
8.0	0.0084	0.0014	–	0.0018	0.977
8.5	0.0032	0.0046	–	0.0031	1.42
9.0	–	0.0150	–	0.0065	1.57
9.5	–	0.0330	–	0.0104	1.39
10.0	–	0.0510	–	0.0200	1.13
10.25	–	0.0577	0.00019	0.0308	1.22
10.5	–	0.0640	0.00037	0.0503	1.31
11.0	–	0.0750	0.00140	0.0910	2.10
11.5	–	0.0710	0.00420	0.0954	2.35
12.0	–	0.0490	0.00760	0.0615	1.95
12.5	–	0.0260	0.01060	0.0274	1.22
13.0	–	0.0110	0.00870	0.0111	0.629
13.5	–	0.0031	0.00440	0.0050	0.400
14.0	–	0.0008	0.00170	0.0032	0.340
14.5	–	–	0.00068	0.0018	0.310
15.0	–	–	0.00028	0.0012	0.300

F_2^-. Figure 6.114 compares the cross section measurements at room temperature of Lehmann [510], Kline *et al.* [424], and Rao and Srivastava [509] for the production of F_2^-. The F_2^- ion is produced in the energy range from ~ 1 to ~ 14 eV, and the cross sections exhibit maxima near 2.2, 4.7, and 11.5 eV (Table 6.78). The data of Lehmann shown here

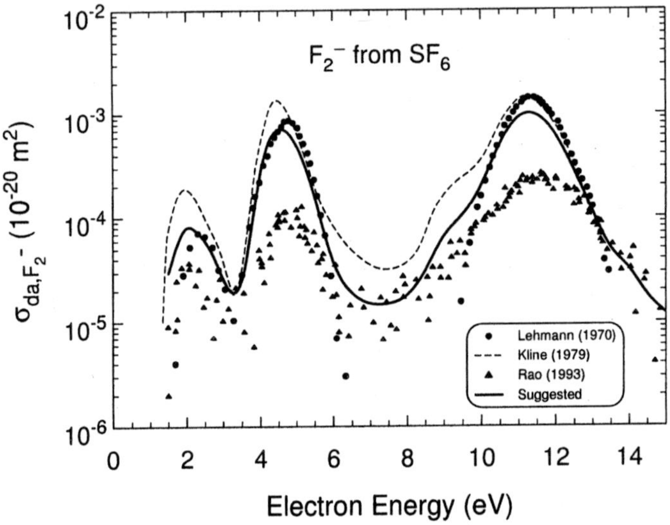

Figure 6.114. Cross section for the formation of F_2^- by dissociative electron attachment to SF_6 as a function of the electron energy: (•) Ref. [510]; (- - -) Ref. [424]; (▲) Ref. [509]; (—) suggested data.

were again shifted to lower energy by 0.6 eV as discussed previously. Rao and Srivastava [509] gave a value of 2.6×10^{-24} m^2 for the cross section of this ion at the maximum of the third peak at 11.5 eV, which is lower than the other two sets of measurements by nearly a factor of 10. The solid line in Fig. 6.114 is our suggested cross section, $\sigma_{da,F_2^-}(\varepsilon)$, for this ion that was obtained by considering all three sets of experimental data. Values derived from this curve are listed in Table 6.79.

F^-. The F^- ion is the predominant fragment negative ion produced by dissociative electron attachment to SF_6 at ambient temperature at electron energies above $\sim$ 3 eV. Cross sections for the formation of this ion are shown in Fig. 6.115, and these exhibit maxima near 2.8, 5.2, 9.1, and 11.5 eV. Above $\sim$ 15 eV the data of Rao and Srivastava [509] show formation of F^- possibly via ion-pair processes. The energy positions of the maxima in the F^- cross section function as determined from the various sources and possible reaction mechanisms of formation are also given in Table 6.78. Not included in Table 6.78 is an early observation by Curran [529] of F^- formation at $\sim$ 0.0 eV (presumably via the reaction $e + SF_6 \rightarrow F^- + SF_5$). None of the subsequent room temperature investigations identified this resonance, and the observation by Curran [529] has been questioned by Fenzlaff *et al.* [511].

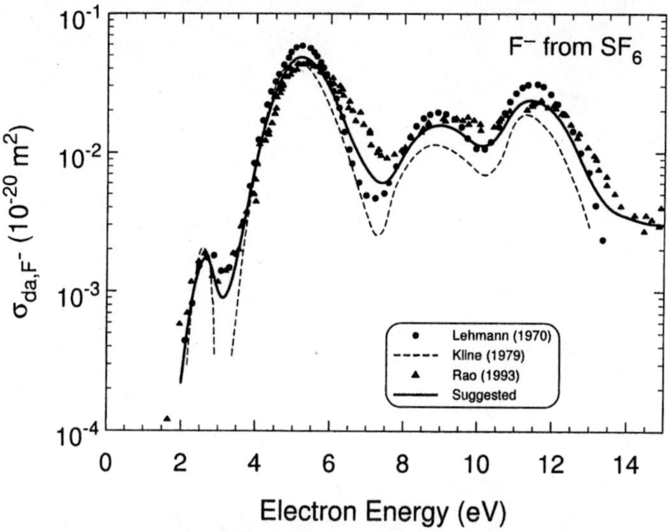

Figure 6.115. Cross section for the formation of F^- by dissociative electron attachment to SF_6 as a function of the electron energy: (•) Ref. [510]; (- - -) Ref. [424]; (▲) data of Rao and Srivastava [509] adjusted upwards by a factor of 10 as discussed in the text; (—) suggested data.

The original data of Rao and Srivastava [509] for this ion were mistakenly reported at values that were low by a factor of 10 [658]. The data plotted in Fig. 6.115 are the original values adjusted upward by a factor of 10. The various cross section measurements along with a least squares fit to the three sets of data (solid line) are shown in Fig. 6.115. Values obtained from the solid line in Fig. 6.115 are listed in Table 6.79 as our suggested data for the $\sigma_{da,F^-}(\varepsilon)$.

5.5.2 Total Electron Attachment Cross Section, $\sigma_{a,t}(\varepsilon)$, and Total Dissociative Electron Attachment Cross Section, $\sigma_{da,t}(\varepsilon)$, of SF_6

In Fig. 6.116 are plotted the recommended or suggested cross sections for the various negative ions as determined in the preceding sections, namely, for SF_6^- [Fig. 6.112(a)], SF_5^- [Fig. 6.112(b)], SF_4^-, SF_3^-, and SF_2^- (Fig. 6.113), F_2^- (Fig. 6.114), and F^- (Fig. 6.115). The total electron attachment cross section, $\sigma_{a,t}(\varepsilon)$, as represented by the sum of all these recommended or suggested cross sections is shown in Fig. 6.116 by the dotted line, which overlaps with the solid line for SF_5^- below $\sim$ 1.5 eV and the solid line for SF_6^- below $\sim$ 0.2 eV. Data derived from the

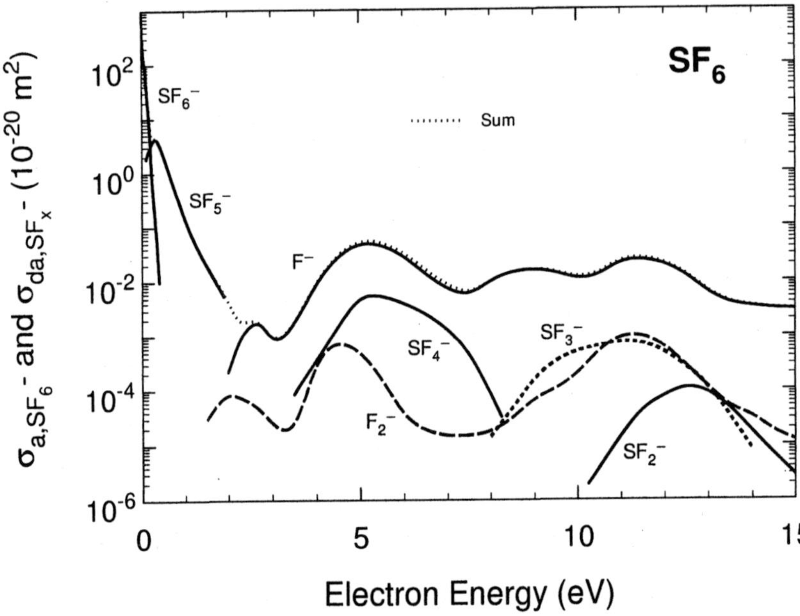

Figure 6.116. Recommended or suggested cross sections for SF_6^- [Fig. 6.112(a)], SF_5^- [Fig. 6.112(b)], SF_4^-, SF_3^-, and SF_2^- (Fig. 6.113), F_2^- (Fig. 6.114), and F^- (Fig. 6.115) by electron attachment or dissociative electron attachment. ($\cdots$) Suggested $\sigma_{a,t}(\varepsilon)$ (sum of all cross sections in the figure).

dotted line are listed in Table 6.80 as our recommended values for the $\sigma_{a,t}(\varepsilon)$ of the SF_6 molecule in the energy range 0.0001–15 eV. The total dissociative electron attachment cross section $\sigma_{da,t}(\varepsilon)$, as represented by the sum of the suggested dissociative electron attachment cross sections for all fragment negative ions, is also listed in Table 6.80. The values listed represent our suggested data for the $\sigma_{da,t}(\varepsilon)$ of the SF_6 molecule. It is seen from Fig. 6.116 that beyond $\sim$ 0.3 eV, $\sigma_{da,t}(\varepsilon) = \sigma_{a,t}(\varepsilon)$. It is also seen that $\sigma_{da,t}(\varepsilon)$ is dominated by the formation of SF_5^- below $\sim$ 1.5 eV and by the formation of F^- above this energy.

5.5.3 Total Electron Attachment Rate Constant, $k_{a,t}$, as a Function of E/N and $\langle\varepsilon\rangle$ for SF_6

$k_{a,t}(E/N)$ in Ar, Xe, and N_2. Hunter *et al.* [624] measured the total electron attachment rate constant $k_{a,t}(E/N)$ near 300 K in mixtures of SF_6 with N_2, Ar, and Xe buffer gases. The E/N range they covered for the three buffer gases corresponds to the mean electron energy range

Table 6.80. Suggested room temperature values of the total electron attachment cross section, $\sigma_{a,t}(\varepsilon)$, and the total dissociative electron attachment, $\sigma_{da,t}(\varepsilon)$, for SF_6.

ε (eV)	$\sigma_{a,t}(\varepsilon)$ (10^{-20} m^2)	$\sigma_{da,t}(\varepsilon)$ (10^{-20} m^2)	ε (eV)	$\sigma_{a,t}(\varepsilon)$ (10^{-20} m^2)	$\sigma_{da,t}(\varepsilon)$ (10^{-20} m^2)
0.0001	7617	–	0.30	4.40	4.24
0.0002	5283	–	0.35	4.12	4.07
0.0003	4284	–	0.40	3.46	3.45
0.0004	3692	–	0.45	2.75	2.75
0.0005	3280	–	0.50	2.15	2.15
0.0006	2968	–	0.60	1.25	1.25
0.0007	2724	–	0.70	0.722	0.722
0.0008	2529	–	0.80	0.416	0.416
0.0009	2369	–	0.90	0.245	0.245
0.001	2237	–	1.00	0.147	0.147
0.002	1511	–	1.20	0.060	0.060
0.003	1202	–	1.50	0.020	0.020
0.004	993	–	2.00	0.0043	0.0043
0.005	859	–	2.25	0.0020	0.0020
0.006	760	–	2.50	0.0019	0.0019
0.007	683	–	2.75	0.0017	0.0017
0.008	621	–	3.0	0.0010	0.0010
0.009	569	–	3.5	0.0018	0.0018
0.010	526	–	4.0	0.0092	0.0092
0.015	383	–	4.5	0.0290	0.0290
0.020	304	–	5.0	0.0514	0.0514
0.025	257	–	5.5	0.0493	0.0493
0.030	221	–	6.0	0.0317	0.0317
0.035	190	–	6.5	0.0162	0.0162
0.040	171	–	7.0	0.0088	0.0088
0.045	149	–	7.5	0.0066	0.0066
0.050	132	–	8.0	0.0099	0.0099
0.060	109	–	8.5	0.0143	0.0143
0.070	92.7	–	9.0	0.0159	0.0159
0.080	82.9	–	9.5	0.0144	0.0144
0.090	74.3	–	10.0	0.0120	0.0120
0.10	51.4	1.85	10.5	0.0142	0.0142
0.12	32.9	2.09	11.0	0.0227	0.0227
0.14	20.2	2.36	11.5	0.0252	0.0252
0.15	16.7	2.48	12.0	0.0206	0.0206
0.16	13.1	2.61	12.5	0.0128	0.0128
0.18	8.72	2.87	13.0	0.0066	0.0066
0.20	6.01	3.15	13.5	0.0041	0.0041
0.22	4.69	3.45	14.0	0.0035	0.0035
0.25	4.38	3.86	14.5	0.0031	0.0031
0.28	4.40	4.15	15.0	0.0030	0.0030

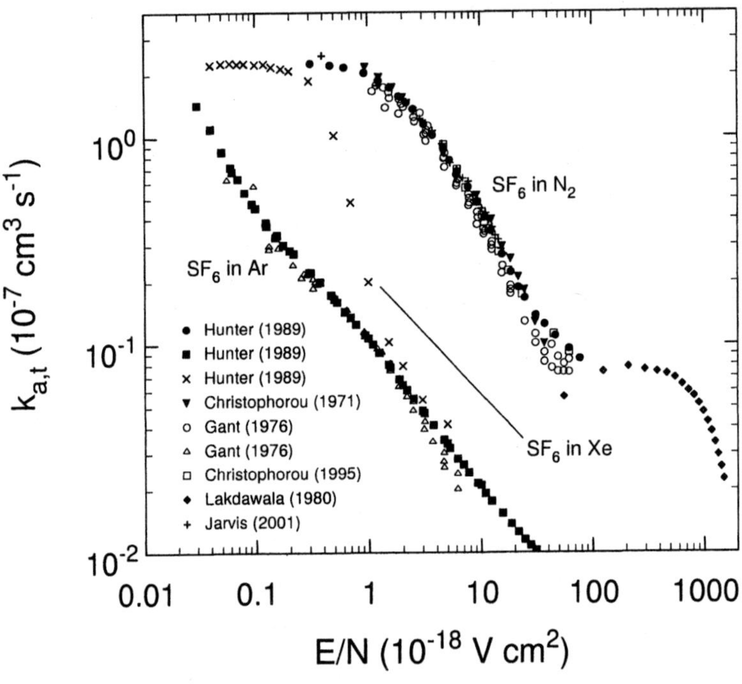

Figure 6.117. Total electron attachment rate constant, $k_{a,t}(E/N)$, as a function of E/N for SF_6 measured in mixtures with various buffer gases: N_2 – (•) Ref. [624]; (▼) Ref. [504]; (○) Ref. [622]; (◆) Ref. [660]; (□) Ref. [404]; (+) Ref. [659]. Ar – (■) Ref. [624]; (△) Ref. [622]. Xe – (×) Ref. [624].

from ~ 0.04 to ~ 4.3 eV. The uncertainty in their measurements varies from ±5% to ±15% depending on the value of E/N. Hunter *et al.* obtained values of the rate constants uninfluenced by changes in the electron energy distribution function of the pure buffer gases by using extremely low concentrations of SF_6 (the SF_6 fractional concentrations were as low as 1.3×10^{-9} for N_2, 1.5×10^{-9} for Ar, and 2.2×10^{-10} for Xe). This was achieved by extrapolating the measured electron attachment coefficient $\eta/N_a(E/N)$ at each value of E/N to zero SF_6 concentration and multiplying the extrapolated values of $\eta/N_a(E/N)$ by the electron drift velocity for the pure gas. The data of Hunter *et al.* are shown in Fig. 6.117. In Table 6.81 are listed the values of $k_{a,t}(E/N)$ and $\eta/N_a(E/N)$ for mixtures of SF_6 with Ar and Xe, along with the values of the electron drift velocity $w(E/N)$ and the mean electron energy $\langle\varepsilon\rangle(E/N)$ for the Ar and Xe buffer gases given by Hunter *et al.* [624]. Similar data are given in Table 6.82 for the N_2 buffer gas.

Table 6.81. Values of $k_{a,t}(E/N)$, and $\eta/N_a(E/N)$ for SF_6 measured in the buffer gases Ar and Xe by Hunter *et al.* [624], and values of $w(E/N)$, and $\langle\varepsilon\rangle(E/N)$ for these buffer gases as reported by Hunter *et al.* [624].

	SF_6/Ar			
E/N (10^{-17} V cm^2)	$\langle\varepsilon\rangle$ (eV)	η/N_a (10^{-13} cm^2)	w (10^5 cm s^{-1})	$k_{a,t}$ (10^{-7} cm^3 s^{-1})
0.003	0.130	26.8	0.533	1.43
0.004	0.178	16.5	0.668	1.10
0.005	0.213	11.4	0.750	0.854
0.006	0.240	8.92	0.803	0.716
0.007	0.260	7.49	0.840	0.629
0.008	0.277	6.19	0.869	0.538
0.010	0.304	4.94	0.915	0.452
0.0125	0.331	3.87	0.962	0.372
0.015	0.354	3.28	1.00	0.329
0.0175	0.375	2.91	1.04	0.302
0.02	0.394	2.66	1.07	0.285
0.03	0.461	1.89	1.19	0.224
0.05	0.570	1.21	1.36	0.165
0.07	0.663	0.904	1.49	0.135
0.10	0.784	0.648	1.65	0.107
0.15	0.956	0.433	1.85	0.080
0.20	1.10	0.319	1.99	0.0637
0.30	1.34	0.217	2.21	0.0480
0.50	1.72	0.133	2.51	0.0335
0.70	2.02	0.097	2.72	0.0265
1.00	2.39	0.071	2.96	0.0210
1.09[a]	2.49	0.0656	3.02	0.0192
1.24	2.65	0.0581	3.12	0.0176
1.55	2.95	0.0473	3.29	0.0154
1.86	3.22	0.0403	3.45	0.0137
2.17	3.48	0.0354	3.59	0.0125
2.48	3.72	0.0317	3.72	0.0115
2.79	3.94	0.0290	3.83	0.0108
3.11	4.15	0.0263	3.96	0.0102
3.42	4.33	0.0244	4.10	0.0096

[a]Data beyond this E/N value were taken with another (higher pressure) electron swarm apparatus. The rest of the measurements were made using a Pulsed Townsend Technique.

Also plotted in Fig. 6.117 are the earlier measurements of $k_{a,t}(E/N)$ of SF_6 conducted by Gant [622] in mixtures of SF_6 with N_2 and Ar buffer gases, and by Christophorou *et al.* [504], Jarvis *et al.* [659], Christophorou

Table 6.81 (continued)
Values of $k_{a,t}\,(E/N)$, and $\eta/N_a\,(E/N)$ for SF_6 measured in the buffer gases Ar and Xe by Hunter *et al.* [624], and values of $w\,(E/N)$, and $\langle\varepsilon\rangle\,(E/N)$ for these buffer gases as reported by Hunter *et al.* [624].

	SF_6/Xe			
E/N (10^{-17} V cm^2)	$\langle\varepsilon\rangle$ (eV)	η/N_a (10^{-13} cm^2)	w (10^5 cm s^{-1})	$k_{a,t}$ (10^{-7} cm^3 s^{-1})
0.004	0.0391	145.8	0.0153	2.23
0.005	0.0392	118.3	0.0191	2.26
0.006	0.0393	99.1	0.0230	2.28
0.007	0.0394	83.9	0.0268	2.25
0.008	0.0396	73.4	0.0308	2.26
0.010	0.0401	57.9	0.0387	2.24
0.012	0.0406	47.9	0.0469	2.25
0.014	0.0414	39.4	0.0554	2.18
0.017	0.0432	31.0	0.0691	2.14
0.02	0.0463	24.6	0.0849	2.09
0.03	0.0801	11.1	0.1690	1.87
0.05	0.343	2.00	0.5110	1.02
0.07	0.596	0.65	0.7420	0.48
0.10	0.777	0.23	0.8610	0.20
0.15	0.929	0.111	0.9240	0.102
0.20	1.03	0.0802	0.9820	0.0787
0.30	1.18	0.0517	1.0500	0.0543
0.50	1.40	0.0359	1.1500	0.0413

and Datskos [404], and Lakdawala and Moruzzi [660] in mixtures of SF_6 with N_2. Christophorou and Datskos [404] reported $k_{a,t}$ only as a function of the mean electron energy $\langle\varepsilon\rangle$. For presentation in Fig. 6.117, we converted their $k_{a,t}(\langle\varepsilon\rangle)$ data to $k_{a,t}(E/N)$ by using the $\langle\varepsilon\rangle\,(E/N)$ in Table 6.82. We deduced the $k_{a,t}(E/N)$ of Lakdawala and Moruzzi [660] from their measurements of $\eta/N_a\,(E/N)$ in mixtures of SF_6 with N_2. The data in Fig. 2 of their paper represent measurements for a number of SF_6 concentrations in N_2 ranging from 0.03% to 0.6%. We performed a least squares fit to all their $\eta/N_a\,(E/N)$ data points and values from this fit were converted to $k_{a,t}(E/N)$ using the electron drift velocities in N_2 given by Hunter and Christophorou [661] and Wedding *et al.* [662]. The resultant values of $k_{a,t}(E/N)$ are plotted in Fig. 6.117. It is seen that although the data of Lakdawala and Moruzzi considerably expand the E/N range of the measurements, they are incompatible with the rest of the data in this buffer gas.

Table 6.82. Values of $k_{a,t}\,(E/N)$ and $\eta/N_a\,(E/N)$ for SF_6 measured in the buffer gas N_2 by Hunter *et al.* [624]), and values of $w\,(E/N)$, and $\langle\varepsilon\rangle\,(E/N)$ for this buffer gas as reported by Hunter *et al.* [624].

	SF_6/N_2			
E/N (10^{-17} V cm^2)	$\langle\varepsilon\rangle$ (eV)	η/N_a (10^{-13} cm^2)	w (10^5 cm s^{-1})	$k_{a,t}$ (10^{-7} cm^3 s^{-1})
0.0311	0.0403	21.2	1.07	2.27
0.0466	0.0433	14.9	1.49	2.22
0.0621	0.0466	11.9	1.83	2.17
0.0932	0.0548	8.88	2.30	2.04
0.124	0.0644	7.27	2.57	1.87
0.155	0.0751	6.34	2.73	1.73
0.186	0.0864	5.55	2.83	1.57
0.217	0.0977	5.02	2.91	1.46
0.249	0.109	4.58	2.97	1.36
0.311	0.130	3.70	3.10	1.15
0.373	0.150	3.16	3.23	1.02
0.466	0.179	2.54	3.45	0.878
0.528	0.199	2.14	3.59	0.769
0.621	0.229	1.79	3.80	0.679
0.777	0.282	1.39	4.09	0.569
0.932	0.335	1.10	4.35	0.479
1.09	0.388	0.885	4.60	0.407
1.24	0.436	0.725	4.82	0.350
1.55	0.514	0.512	5.29	0.271
1.86	0.578	0.387	5.77	0.223
2.17	0.628	0.300	6.26	0.188
2.48	0.668	0.247	6.76	0.167
3.11	0.729	0.177	7.78	0.138
3.73	0.772	0.142	8.79	0.125
4.66	0.817	0.107	10.3	0.110
5.28	0.840	0.0903	11.3	0.102
6.21	0.870	0.0748	12.6	0.095
7.76	0.901	0.0567	15.0	0.085

For an indirect determination of the total electron attachment rate constant $k_{a,t}(E/N)$ in pure SF_6 see Table 6.89 in Sec. 5.6.1 of Ch. 6.

$k_{a,t}(\langle\varepsilon\rangle)$. Since the electron energy distribution can be calculated as a function of E/N for Ar, Xe, and N_2, one can determine the total electron attachment rate constant as a function of the mean electron energy, $k_{a,t}(\langle\varepsilon\rangle)$, from measurements of $k_{a,t}(E/N)$ for SF_6 in low-concentration

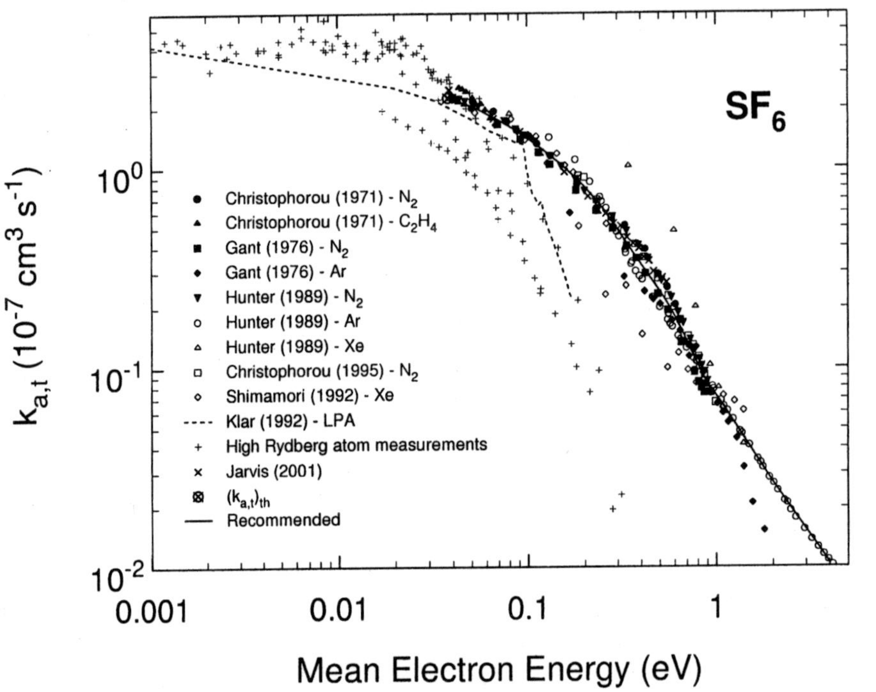

Figure 6.118. Total electron attachment rate constant, $k_{a,t}(\langle\varepsilon\rangle)$, for SF_6 as function of the mean electron energy $\langle\varepsilon\rangle$. For the electron swarm data, measurements were made in the buffer gases indicated in the legend. *Electron swarm measurements:* (●, ▲) Ref. [504]; (■, ◆) Ref. [622]; (▼, ○, △) Ref. [624]; (□) Ref. [404]; (◇) Ref. [663]; (×) Ref. [659]; (⊗) Recommended value of $(k_{a,t})_{th}$. *Laser photoelectron attachment measurements:* (- - -) Ref. [547]. *High-Rydberg-atom measurements:* (+) this same symbol was used for data from all of the following sources: West *et al.* [633] (Xe*(nf)); Foltz *et al.* [634] (Xe*(nf)); Zollars *et al.* [637] (Rb*(ns, nd)); Zollars *et al.* [638] (K*(nd)); Beterov *et al.* [640] (Na*(np)); Zheng *et al.* [641] (K*(nd)); Harth *et al.* [630] (Ne*(ns, nd)); and Ling *et al.* [644] (K*(np)). The recommended values of $k_{a,t}(\langle\varepsilon\rangle)$ are shown by the solid line.

mixtures of SF_6 with each of these buffer gases. This facilitates a comparison of the measurements made in different buffer gases. In Fig. 6.118 are plotted the $k_{a,t}(\langle\varepsilon\rangle)$ for SF_6 based on the measurements of $k_{a,t}(E/N)$ made at room temperature in the buffer gases N_2 [404, 504, 622, 624], Ar [622, 624], Xe [624], and also C_2H_4 [504]. The uncertainty in these data is likely to vary from ±5% to ±15% depending on E/N. Differences in the $k_{a,t}(\langle\varepsilon\rangle)$ values measured in the various buffer gases may reflect differences in the electron energy distribution functions $f(\varepsilon, E/N)$ and hence the mean electron energies. This may be especially true in the case of the Xe data since the cross sections for this buffer gas are not as well established (see below). The $\langle\varepsilon\rangle\,(E/N)$ values for ethylene (C_2H_4) buffer gas are those determined from $\langle\varepsilon\rangle\,(E/N) = 3/2(eD_T/\mu)(E/N)$, where $(D_T/\mu)(E/N)$ is the ratio of the lateral electron diffusion coefficient to electron mobility as a function of E/N for this gas.

Also plotted in Fig. 6.118 are the $k_{a,t}(\langle\varepsilon\rangle)$ for SF_6 measured at room temperature in a Xe buffer gas by Shimamori *et al.* [663] using a pulse-radiolysis microwave cavity technique. The calibration of the mean electron energy in this study was made by analyzing the time profile of the microwave conductivity signals for "thermalized" electrons produced by pulsed x-rays in gaseous Xe and assuming a Maxwellian form for $f(\varepsilon, E/N)$. The data of Shimamori *et al.* obtained from measurements in a Xe buffer gas disagree with the data of Hunter *et al.* [624] obtained using the same buffer gas, and are not in good overall agreement with the other electron swarm data. This may indicate that the electron energy distribution functions in Xe as assumed by Shimamori *et al.* and as determined by Hunter *et al.* are different, and both are less accurate than those for the N_2 and Ar buffer gases.

Also plotted in Fig. 6.118 is the recommended value (2.25×10^{-7} cm^3 s^{-1}) of the thermal ($T \sim 300$ K) electron attachment rate constant, $(k_{a,t})_{th}$ (see Table 6.84 and discussion later in this section).

Besides the measurements of $k_{a,t}(\langle\varepsilon\rangle)$ made using the electron swarm method, there have been a number of other measurements of this quantity using the laser-photoelectron-attachment and the high-Rydberg-atom methods which have extended the values of $k_{a,t}(\langle\varepsilon\rangle)$ for SF_6 down to the μeV range. The results of these methods dealing with the electron attachment cross section for the formation of SF_6^- have been discussed earlier in Sec. 5.5.1 of Ch. 6. Most of the rate-constant measurements involving high-Rydberg atoms are normally given in terms of the dependence of the rate constant on the principal quantum number n or the effective principal quantum number n^* of the Rydberg atom involved (see Fig. 6.111). However, the bound-electron attachment rate constant, $k_{a,be}(\langle\varepsilon\rangle)$, can be expressed as a function of the average kinetic energy of the Rydberg electron which is given by R/n^2 or R/n^{*2} where R is the Rydberg constant (13.6 eV). In Fig. 6.118 are plotted the $k_{a,t}(\langle\varepsilon\rangle)$ data of Klar *et al.* [547] using the laser-photoelectron-attachment technique and the data of a number of investigators using the high-Rydberg-atom method. All data on the bound-electron attachment rate constants, $k_{a,be}(\langle\varepsilon\rangle)$, are shown in Fig. 6.118 by the same symbol (+), although they involve a number of studies and various Rydberg atoms: Xe*(nf) [633, 634], Rb*(ns, nd) [637], K*(nd) [638, 641], K*(np) [644], Na*(np) [640], and Ne*(ns, nd) [630]. They refer to the production of SF_6^- and

Table 6.83. Recommended values for the total electron attachment rate constant, $k_{a,t}(\langle\varepsilon\rangle)$, of SF_6 as a function of the mean electron energy $\langle\varepsilon\rangle$.

$\langle\varepsilon\rangle$ (eV)	$k_{a,t}$ (10^{-7} cm^3 s^{-1})	$\langle\varepsilon\rangle$ (eV)	$k_{a,t}$ (10^{-7} cm^3 s^{-1})
0.038	2.25	0.5	0.241
0.04	2.20	0.6	0.181
0.05	2.05	0.7	0.135
0.06	1.88	0.8	0.103
0.07	1.74	0.9	0.0833
0.08	1.62	1.0	0.0721
0.09	1.51	1.5	0.0406
0.10	1.41	2.0	0.0268
0.15	1.04	2.5	0.0197
0.20	0.801	3.0	0.0154
0.25	0.622	3.5	0.0125
0.30	0.487	4.0	0.0104
0.40	0.327		

are consistent with the electron swarm measurements only for large n values (small values of $\langle\varepsilon\rangle$). The $k_{a,be}(\langle\varepsilon\rangle)$ becomes progressively smaller than $k_{a,t}(\langle\varepsilon\rangle)$ as the value of n decreases, that is, as the energy increases for the reasons discussed in Sec. 5.5.1 of Ch. 6.

We performed a least squares fit to only the swarm data in Fig. 6.118, excluding the measurements in the Xe buffer and also those in C_2H_4, and required the fit to agree with the recommended thermal value (2.25×10^{-7} cm^3 s^{-1}) of $k_{a,t}(\langle\varepsilon\rangle)$. This fit is shown by the solid line in Fig. 6.118, and recommended values of $k_{a,t}(\langle\varepsilon\rangle)$ in the energy range from 0.038 to 4.0 eV obtained from this fit are listed in Table 6.83.

Thermal Value, $(k_{a,t})_{th}$, of the Total Electron Attachment Rate Constant of SF_6. Many measurements have been made of this quantity which vary significantly. They are summarized in Table 6.84. The most reliable data lie between 2.2×10^{-7} and 2.3×10^{-7} cm^3 s^{-1}. The average of the ten values in this range is 2.25×10^{-7} cm^3 s^{-1} which is our recommended value.

5.5.4 Effect of Temperature on Electron Attachment to SF_6

Effect of Temperature on the Production of SF_5^- from SF_6. In Fig. 6.119 are shown the relative cross section measurements of Chen

Table 6.84. Thermal (T = 295 to 300 K) values, $(k_{a,t})_{th}$, of the total electron attachment rate constant for SF_6.

$(k_{a,t})_{th}$ (10^{-7} cm^3 s^{-1})	Reference
0.385	[596]
2.06 ± 0.08	[664]
2.13 ± 0.04	[665]
2.2	[281]
2.2 ± 0.1	[666]
2.20	[667]
2.27 ± 0.07	[650]
2.24 ± 0.15	[653]
2.27 ± 0.09	[653]
2.3	[668]
2.3 ± 0.2	[663]
2.25 ± 0.2	[624]
2.30 ± 0.1	[624]
2.41	[669]
2.5	[670]
2.6	[671]
2.70	[504]
2.78	[504]
2.77	[292]
2.8 ± 0.3	[672]
3.1	[673]
3.1 ± 0.47	[134]
2.25×10^{-7} cm^3 s^{-1}	**Recommended Value**

and Chantry [506] for the production of SF_5^- from SF_6 at temperatures ranging from 300 to 880 K. Two peaks are seen in the energy dependence of the production of SF_5^-, one at $\sim$ 0.0 eV and the other near 0.38 eV. The near-zero-energy peak is very sensitive to temperature, has an activation energy of $\sim$ 0.2 eV [506], and is due to dissociative electron attachment to "hot" SF_6 molecules, i.e., due to attachment of near-zero-energy electrons to vibrational/rotational states of SF_6 lying at energies $\sim$ 0.2 eV above the molecular ground state,

$$SF_6^*(\nu, r \gtrsim 0.2\ \text{eV}) + e(\sim 0\ \text{eV}) \rightarrow SF_5^- + F. \tag{6.43}$$

The 0.38-eV peak is broad, insensitive to temperature variation, and results from the capture of $\sim$ 0.38 eV electrons by unexcited SF_6 molecules via the lowest negative ion state of SF_6^-, that is, due to the reaction

$$SF_6\,(\nu = 0) + e(\sim 0.38\ \text{eV}) \rightarrow SF_5^- + F \tag{6.44}$$

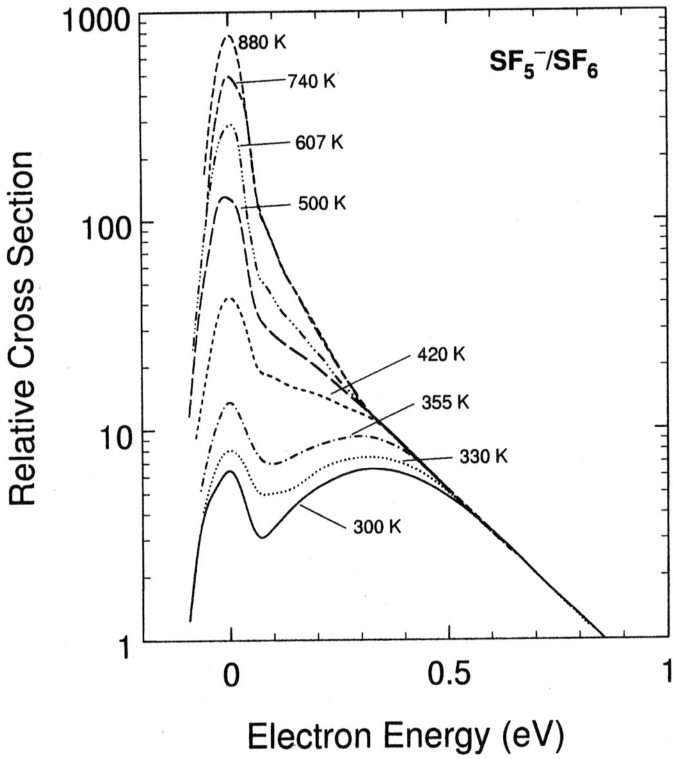

Figure 6.119. Temperature dependence of the relative cross section for the formation of SF_5^- by electron attachment to SF_6 (data of Chen and Chantry [506]).

(see, however, Refs. [656, 674]). It is interesting to observe that in spite of this large increase in the formation of SF_5^- from SF_6 with increasing T, the total electron attachment rate constant and cross section at near-zero-electron energy is independent of T. This may imply that near 0.0 eV, the formation of SF_6^- in electron beam experiments where the SF_6^{-*} is not stabilized by collision or radiation prior to its detection, decreases with T. Indeed, such a decrease has been observed [617] and would be consistent with the decrease in the autodetachment lifetime of SF_6^{-*} with increasing internal energy of the transient anion [599]. The data in Fig. 6.119 can be put on an absolute scale by considering the magnitude of the cross section for the production of SF_5^- at 0.38 eV at room temperature. A value of $\sim 3.8 \times 10^{-20}$ m^2 is suggested from the data in Fig. 6.112(b) and Table 6.77.

The strong increase of the cross section for the production of SF_5^- from SF_6 with increasing vibrational excitation energy of the SF_6 molecule has led Chen and Chantry [506] to anticipate that the zero-energy peak in Fig. 6.119 could be photoenhanced. They, thus, used specific lines from a CO_2 laser to produce (via n-photon absorption processes, $n \geq 1$) vibrationally excited SF_6 molecules, SF_6^*, and observed the production of SF_5^- from electron impact on SF_6^* as a function of the electron energy, viz.

$$n(h\nu)_{\text{laser}} + SF_6 \rightarrow (SF_6^*)_{\text{laser}} \tag{6.45a}$$

followed by

$$(SF_6^*)_{\text{laser}} + e \rightarrow (SF_5)^-_{\text{laser}} + F. \tag{6.45b}$$

Chen and Chantry found that the production of SF_5^- from electron impact on SF_6^* is enhanced and that this infrared enhancement of SF_5^- production is different for the ^{32}S and ^{34}S isotopes. Thus, the formation of SF_5^- was found to be radiation-wavelength dependent. With the 936.856 cm^{-1} CO_2 line the $^{32}SF_5^-$ enhancement was optimized and the $^{34}SF_5^-$ was unaffected. With the 920.810 cm^{-1} CO_2 laser line the $^{34}SF_5^-$ was optimized and the $^{32}SF_5^-$ was unaffected. This isotopic specificity is an illustration of possible isotope separation processes using knowledge on low-energy electron-molecule reactions. It is consistent with the earlier observations of Beterov *et al.* [675] and Beterov and Fateyev [676] of optogalvanic effects when they used a CO_2 laser to vibrationally excite the SF_6 molecules (see also Avrillier and Schermann [677]).

Effect of Temperature on the Total Electron Attachment Rate Constant, $k_{a,t}(\langle\varepsilon\rangle, T)$, of SF_6. In Table 6.85 are listed the thermal values, $(k_{a,t})_{th}$, of the total electron attachment rate constant for SF_6 as a function of gas temperature [134, 292, 504, 596, 653, 667, 668]. Although the data listed in Table 6.85 show $(k_{a,t})_{th}$ to be virtually constant for T between about 300 and 600 K, the more recent measurements of Le Garrec *et al.* [292] below room temperature indicate a small increase with increasing T in the low-temperature range they investigated [see Table 6.85 and Fig. 6.120(a)].

Consistent with the finding that the thermal electron attachment rate constant for SF_6 is independent of T for temperatures of a few hundred degrees above ambient, are the results of a beam study [619] which showed the energy-integrated total electron attachment cross section at thermal (and near-thermal) energies to be independent of T in the range from 300 to 1200 K. Interestingly, the temperature independence of the total rate of electron attachment to SF_6 is evident for mean electron

Table 6.85. Variation with temperature of the thermal electron attachment rate constant, $(k_{a,t})_{th}$, of SF_6.

$(k_{a,t})_{th}$ (10^{-7} cm^3 s^{-1})	Temperature (K)	Reference
2.70	298–418	[504, 596]
2.20	293–523	[667]
3.10	205	[134]
3.10	300	
4.50	455	
4.00	590	
2.27	294	[653]
2.20	500	
2.30	300	[668]
2.60	329	
2.80	362	
3.10	411	
2.80	449	
2.70	498	
2.20	545	
1.41	49	[292].
1.44	85	
1.42	126	
1.65	162	
1.73	174	
2.77	304	

energies up to at least 1 eV as can be seen from the data in Fig. 6.120(b) [404, 678].

5.5.5 Density-Normalized Electron Attachment Coefficient, $\eta/N\,(E/N)$, of SF_6

There have been at least 17 room-temperature (293–298 K) measurements of $\eta/N\,(E/N)$ in pure SF_6 [274, 424, 573–575, 577–579, 581–584, 586, 587, 679–682] over a period of 40 years, normally at high E/N to avoid the difficulty arising from the strong electron attachment at low E/N. They mostly employed the steady-state Townsend or the pulsed-

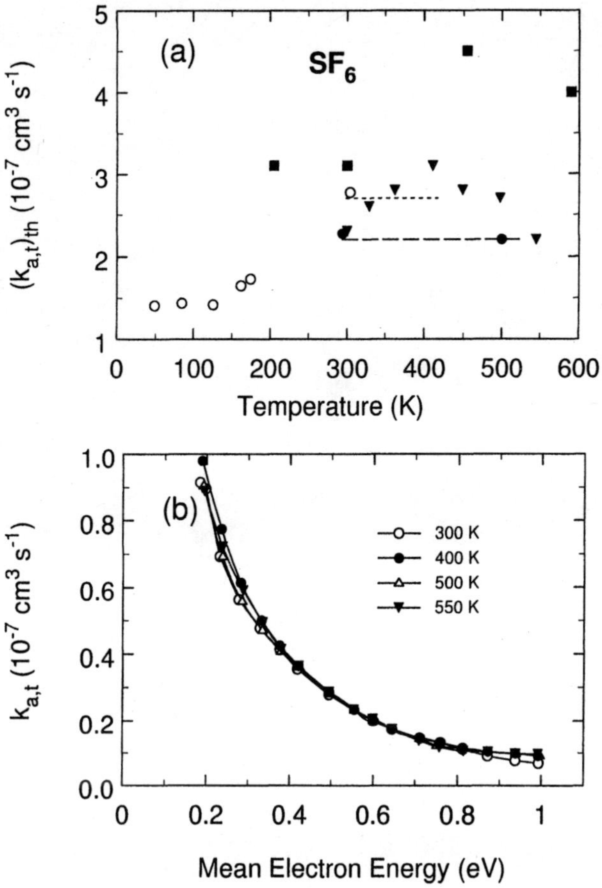

Figure 6.120. (a) Variation of the rate constant for electron attachment to SF_6 at thermal electron energies with gas temperature (see Table 6.85): (- - -) Christophorou *et al.* [504, 596]; (– –) Petrović and Crompton [653]; (▼) Miller *et al.* [668]; (○) Le Garrec *et al.* [292]; (■) Smith *et al.* [134]; (•) Fehsenfeld [667]. (b) Variation of the rate constant for electron attachment to SF_6 as a function of mean electron energy for a range of temperatures. The data are those of Christophorou and Datskos [404] taken in mixtures of SF_6 in N_2 buffer gas.

Townsend methods. The more recent time-resolved measuring techniques provided an improvement over the traditional avalanche-shape method. All determinations except that of Kline *et al.* [424] assumed that electron detachment is negligible in these measurements and this is a reasonable assumption [424, 577, 683] (see also Sec. 5.7 of Ch. 6).

The experimental results obtained by these methods are shown in Fig 6.121. Their quoted uncertainties range from ±5% to ±15%, but clearly

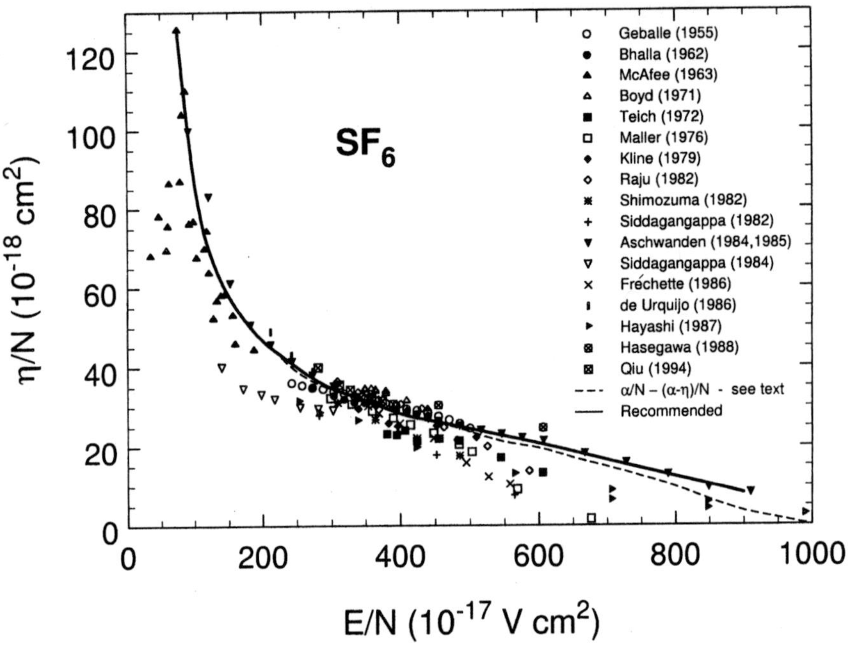

Figure 6.121. Measured density-normalized electron attachment coefficient, $\eta/N\,(E/N)$, for SF_6: (○) Ref. [573]; (●) Ref. [574]; (▲) Ref. [679]; (△) Ref. [575]; (■) Ref. [582]; (□) Ref. [577]; (♦) Ref. [424]; (◇) Ref. [680]; (∗) Ref. [578]; (+) Ref. [681]; (▼) Refs. [583, 584]; (▽) Ref. [682]; (×) Ref. [274]; (|) Ref. [586]; (▶) Ref. [579]; (⊗) Ref. [587]; (⊠) Ref. [581]; (- - -) $(\alpha/N) - (\alpha - \eta)/N$ (see text); (—) recommended.

some of the data diverge considerably more than this. For instance, this is the case for the lowest five data points of McAffee and Edelson [679] [their E/N dependence is inconsistent with the well-established steep rise of the electron attachment rate constant as the E/N (or $\langle\varepsilon\rangle$) decreases toward zero (Figs. 6.117 and 6.118)], and for the measurements of Siddagangappa *et al.* [681, 682]. The solid line in Fig. 6.121 up to $E/N = 450 \times 10^{-17}$ V cm^2 is a least squares fit to all the data except those just mentioned. Aschwanden [584] reported two values of η/N at each value of E/N for which he made measurements below 243×10^{-17} V cm^2. In Fig. 6.121 are only plotted the lower set of values since these are consistent with the other available measurements. For E/N values above 450×10^{-17} V cm^2, we prefer the data of Aschwanden [583, 584] as opposed to the other diverging data [274, 577–579, 582, 587]. Aschwanden's data are more consistent with the values (broken line in Fig. 6.121) of $\eta/N\,(E/N)$ we obtained from $\alpha/N\,(E/N) - [(\alpha - \eta)/N\,(E/N)]$ using our recommended values for $\alpha/N\,(E/N)$ (Fig. 6.105)

Table 6.86. Recommended values of the density-normalized electron attachment coefficient, $\eta/N\,(E/N)$, for SF_6 as a function of E/N.

E/N (10^{-17} V cm^2)	η/N (10^{-18} cm^2)	E/N (10^{-17} V cm^2)	η/N (10^{-18} cm^2)
75	126	450	26.3
100	87	500	24.5
125	68	550	22.6
150	58.0	600	20.7
175	51.7	650	18.6
200	46.8	700	16.4
225	43.3	750	14.3
250	40.1	800	12.2
300	34.8	850	10.2
350	31.6	900	8.06
400	28.3		

and $(\alpha - \eta)/N\,(E/N)$ (Fig. 6.122). Values obtained from the solid line in Fig. 6.121 are listed in Table 6.86 as our recommended values for the $\eta/N\,(E/N)$ of SF_6.

It should be noted that there have been a number of calculations of the $\eta/N\,(E/N)$ of SF_6 using Boltzmann and Monte Carlo codes (e.g., see Refs. [417, 424–426, 428–432, 436, 438, 439, 442]). These, however, are not discussed here.

Finally, it should be pointed out that one can determine values of $\eta/N_a\,(E/N)$ [where N_a refers to the SF_6 gas number density in a mixture with a buffer gas of density N ($N \gg N_a$)] for SF_6 in various buffer gases (N_2, Ar, Xe) from the electron-attachment-rate constant data presented in Sec. 5.5.3 of Ch. 6 by making use of the $w\,(E/N)$ values in the corresponding buffer gases (Tables 6.81 and 6.82). Furthermore, if the $\langle\varepsilon\rangle(E/N)$ is known (as in the measurements of $\eta/N_a\,(E/N)$ for SF_6 in N_2, Ar, Xe buffer gases by Hunter *et al.* [624]), values of $\eta/N_a\,(\langle\varepsilon\rangle)$ can be obtained.

5.5.6 Density-Normalized Effective Ionization Coefficient, $(\alpha - \eta)/N\,(E/N)$, of SF_6

The data from direct experimental measurements of $(\alpha-\eta)/N\,(E/N)$ [274, 424, 573–575, 578, 581, 583, 584, 587, 680, 684–687] are compared in Fig. 6.122 [Fig. 6.122(b) is an expanded view of a portion of Fig. 6.122(a)]. The solid line in Fig. 6.122 represents our recommended

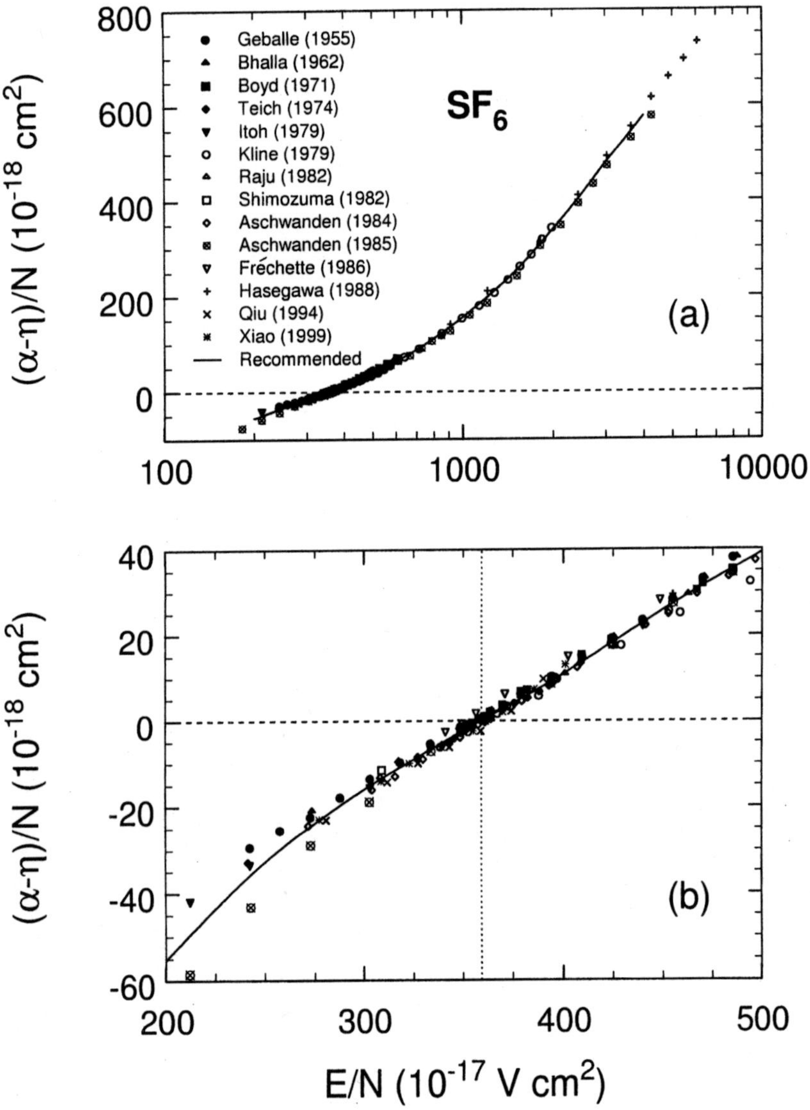

Figure 6.122. (a) Measured density-normalized effective ionization coefficient, $(\alpha - \eta)/N\,(E/N)$, for SF_6, (b) an expanded portion of Fig. 6.122(a): (•) Ref. [573]; (▲) Ref. [574]; (■) Ref. [575]; (♦) Ref. [687]; (▼) Ref. [684]; (○) Ref. [424]; (△) Ref. [680]; (□) Ref. [578]; (◊) Ref. [583]; (⊗) Ref. [584]; (▽) Ref. [274]; (+) Ref. [587]; (×) Ref. [581]; (∗) Ref. [685]; (—) recommended. The dotted vertical line in Fig. 6.122(b) shows $(E/N)_{\rm lim} = (359.3 \pm 3) \times 10^{-17}$ V cm^2.

Table 6.87. Recommended values of the density-normalized effective ionization coefficient, $(\alpha - \eta)/N\,(E/N)$, for SF_6 as a function of E/N.

E/N (10^{-17} V cm^2)	$(\alpha - \eta)/N$ (10^{-18} cm^2)	E/N (10^{-17} V cm^2)	$(\alpha - \eta)/N$ (10^{-18} cm^2)
200	−55.3	800	110
250	−32.8	850	122
300	−16.1	900	132
350	−2.43	950	143
400	10.9	1000	154
450	25.8	1250	204
500	39.3	1500	250
550	51.9	2000	338
600	63.8	2500	413
650	75.2	3000	478
700	87.0	3500	531
750	98.8	4000	578

data for $(\alpha - \eta)/N\,(E/N)$ up to an $E/N = 4000 \times 10^{-17}$ V cm^2, based upon a fit to all of the available data. Values obtained from this line are listed in Table 6.87. Analysis of the data in Fig. 6.122 in the region near where $(\alpha - \eta)/N\,(E/N) = 0$ gives a value for $(E/N)_{\text{lim}} = (359.3 \pm 3) \times 10^{-17}$ V cm^2.

For $E/N > 4000 \times 10^{-17}$ V cm^2, there is only one set of measurements, namely, that of Hasegawa *et al.* [587] made using the steady-state Townsend method. As discussed previously, the values of Hasegawa *et al.* [587] of $(\alpha - \eta)/N\,(E/N)$ for $E/N > 4000 \times 10^{-17}$ V cm^2 agree with the $\alpha/N\,(E/N)$ values of Hayashi and Wang [579, 580].

Finally, it should be noted that measurements of $(\alpha - \eta)/N\,(E/N)$ in mixtures of SF_6 with Xe have been made by Hilmert *et al.* [688] and Xiao *et al.* [689, 690]. Similar measurements have also been made by Qiu and Xiao [581] in mixtures of SF_6 with Ar and Kr, by Xiao *et al.* [685] in mixtures of SF_6 with Ne, by Urquijo *et al.* [691] in mixtures of SF_6 with He, by Xiao *et al.* [692] in mixtures of SF_6 with CO_2, and by Hernández-Ávila *et al.* [693] in mixtures of SF_6 with CHF_3. Similar measurements have been made by de Urquijo *et al.* [694] for mixtures of SF_6 with He, Ar, CO_2, and CH_4. The references cited in this paragraph also contain data for other electron swarm parameters for the respective mixtures.

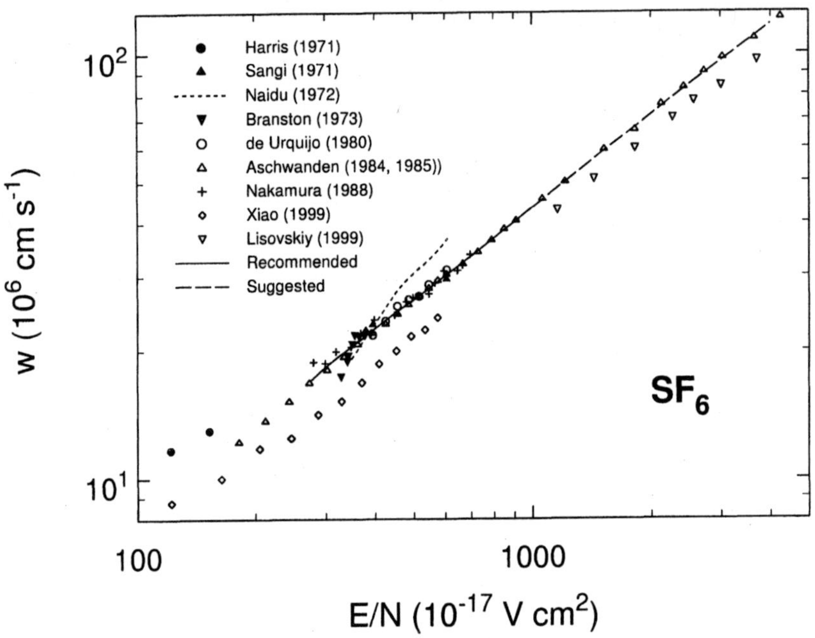

Figure 6.123. Electron drift velocity, $w(E/N)$, in SF_6: (•) Ref. [695]; (▲) Ref. [696]; (- - -) Ref. [697]; (▼) Ref. [698]; (○) Ref. [699]; (△) Refs. [583] and [584]; (+) Ref. [700]; (◇) Ref. [685]; (▽) Ref. [701]; (—) recommended values; (– –) suggested values.

5.6 Electron Transport in SF_6

5.6.1 Electron Drift Velocity, $w(E/N)$, for SF_6

The experimental data [583, 584, 685, 686, 695–701] on $w(E/N)$ for SF_6 are plotted in Fig. 6.123. These measurements were taken at "room temperature" ($\sim$ 293–300 K) and generally at high values of E/N. The reported uncertainties range from $< \pm 1\%$ to $\pm 15\%$, as evidenced by the uncertainty of $\pm 15\%$ reported by Harris and Jones [695], $\pm 5\%$ claimed by Naidu and Prasad [697], and the very low uncertainty of $\pm 0.5\%$ quoted by Aschwanden [583]. In general, the lower the E/N, the higher the expected uncertainty.

In the E/N range between $\sim 250 \times 10^{-17}$ and $\sim 600 \times 10^{-17}$ V cm^2, the data are more consistent than in the low E/N region. The solid line in Fig. 6.123 represents a fit to all measurements in the E/N range of 250×10^{-17} to 1000×10^{-17} V cm^2, except those of Branston [698], Naidu and Prasad [697], and Xiao *et al.* [685, 686]. The reason for excluding these three sets of measurements is because the first two do not show

Table 6.88. Recommended, suggested, and deduced values of the electron drift velocity, $w\,(E/N)$, for SF_6 (see text).

E/N (10^{-17} V cm^2)	w (10^6 cm s^{-1})	E/N (10^{-17} V cm^2)	w (10^6 cm s^{-1})
0	[0.0][a]	650	31.7
25	[4.1]	700	33.4
50	[6.8]	750	35.1
100	[10.2]	800	36.8
150	[12.1]	850	38.5
200	[13.5]	900	40.1
250	[15.6]	950	41.8
275	17.0[b]	1000	43.4
300	18.3	1500	(58.3)[c]
350	20.5	2000	(71.7)
400	22.6	2500	(83.9)
450	24.6	3000	(95.4)
500	26.4	3500	(106.3)
550	28.2	4000	(116.8)
600	30.0		

[a]Values in square brackets are deduced from available data.
[b]Recommended values.
[c]Values in parenthesis are suggested data based upon the measurements of Aschwanden [584].

the correct dependence of the electron drift velocity on E/N, and the third is consistently $\sim$ 20% lower in magnitude than the rest of the data. Values from the solid line in Fig. 6.123 are given in Table 6.88 as our recommended data for the $w\,(E/N)$ in SF_6.

Above $E/N = 1000 \times 10^{-17}$ V cm^2, there are only two sets of measurements, namely, those of Aschwanden [584] and those of Lisovskiy and Yegorenkov [701]. The latter lie systematically lower than the former, but the two data sets agree within combined uncertainties. Since the measurements of Aschwanden agree well with the other data at lower E/N, we choose his data as our suggested values of $w\,(E/N)$ for $E/N > 1000 \times 10^{-17}$ V cm^2. These suggested values are shown in Fig. 6.123 by the broken line and are listed in parenthesis in Table 6.88.

There are no reliable experimental measurements for the electron drift velocity $w\,(E/N)$ in SF_6 at low E/N. The data of Harris and Jones [695] are not direct measurements, but rather values based on an empirical formula they deduced from their measurements. The values of w at low E/N are difficult to measure because the strong electron at-

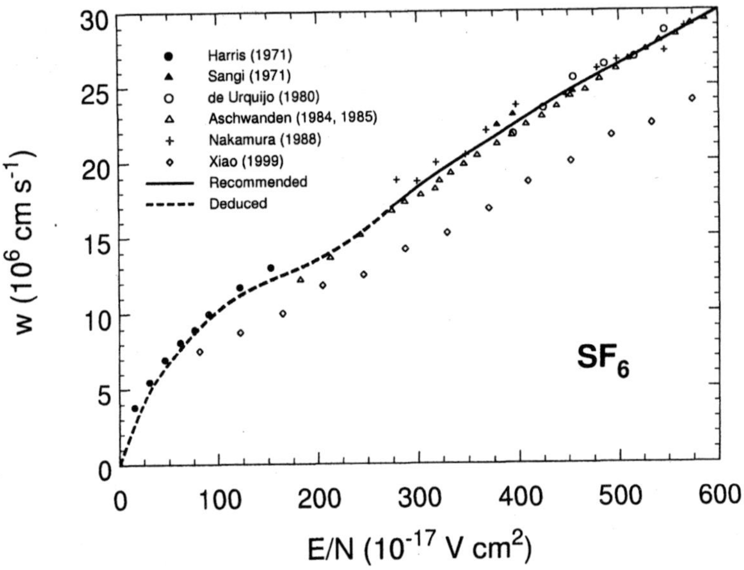

Figure 6.124. Electron drift velocity, $w\,(E/N)$, in SF_6 at low E/N(see text): (•) Ref. [695]; (▲) Ref. [696]; (○) Ref. [699]; (△) Refs. [583] and [584]; (+) Ref. [700]; (◊) Ref. [685]; (—) recommended values from Fig. 6.123; (- - -) deduced values for low E/N.

tachment at low E/N depletes the electrons and results in very weak electron currents. This may also prevent the electron swarm from reaching a steady-state condition with regard to electron attachment [702]. Unfortunately, there are no calculated values of $w\,(E/N)$ at low E/N either. Calculations using the Boltzmann transport equation or the Monte Carlo method (e.g., Refs. [417, 424, 425, 428–430, 432, 436–439, 442, 703]) do not extend to low E/N. However, in an effort to provide an estimate of the $w\,(E/N)$ in SF_6 at low E/N, we have fit a curve to the limited data below 300×10^{-17} V cm^2, requiring that the fitted curve goes through the origin since $w\,(E/N) = 0$ at $E/N = 0$. This fit is shown by the broken line in Fig. 6.124, and values from the broken curve are represented (in brackets) in Table 6.88 as our deduced values for $w\,(E/N)$ at $E/N \leq 250 \times 10^{-17}$ V cm^2 in lieu of direct measurements. It is emphasized that for electronegative gases, such as SF_6 for which the electron attachment cross section is very large at thermal and near thermal energies, the definition of electron drift velocity becomes questionable in the E/N region where the majority of the electrons are attached prior to reaching a steady-state electron energy distribution.

Table 6.89. Assessed values of $\eta/N\,(E/N)$ and $w\,(E/N)$, and their product – total electron attachment rate constant, $k_{a,t}\,(E/N)$ – for pure SF_6 (see text).

E/N (10^{-17} V cm^2)	w (10^6 cm s^{-1})	η/N (10^{-18} cm^2)	$k_{a,t}$ (10^{-12} cm^3 s^{-1})
100	10.2	87.0	887
150	12.1	58.0	702
200	13.5	46.8	632
250	15.6	40.1	626
300	18.3	34.8	637
350	20.5	31.6	648
400	22.6	28.3	640
450	24.6	26.3	647
500	26.4	24.5	647
550	28.2	22.6	637
600	30.0	20.7	621
650	31.7	18.6	590

Using the independent values of $\eta/N\,(E/N)$ and $w\,(E/N)$ determined here, we can calculate their product, $\eta/N \times w$, which represents the total electron attachment rate constant, $k_{a,t}(E/N)$, for SF_6 as a function of E/N in pure SF_6. Values of this quantity are listed in Table 6.89. Since, moreover, plots of the electron attachment rate constant as a function of the mean electron energy, $k_{a,t}(\langle\varepsilon\rangle)$, in various buffer gases is a unique function of the mean electron energy (see Fig. 6.118), one may infer that the rate constant $k_{a,t}$ measured in a buffer gas A has the same magnitude as measured in a buffer gas B, when the E/N values in A and B are such that the mean electron energies in the two buffer gases are the same. From such considerations, we infer from the data in Table 6.89 and Fig. 6.118, that the mean electron energy in pure SF_6 at $E/N = 100 \times 10^{-17}$ V cm^2 is $\sim$ 4.5 eV. Unfortunately, the data in Table 6.89 do not extend to sufficiently low E/N values to allow determination of $\langle\varepsilon\rangle$ for SF_6 at $E/N < 100 \times 10^{-17}$ V cm^2. An estimate of the mean electron energies at 350×10^{-17} V cm^2 $< E/N < 700 \times 10^{-17}$ V cm^2 can be made by considering the values of the characteristic electron energy $[3/2(eD_T/\mu)]$ inferred from the $D_T/\mu\,(E/N)$ data in Sec. 5.6.2 of Ch. 6 (Fig. 6.125). From these data, the characteristic energy in SF_6 is estimated to be 7.4 eV at $E/N = 400 \times 10^{-17}$ V cm^2 and 8.6 eV at $E/N = 650 \times 10^{-17}$ V cm^2.

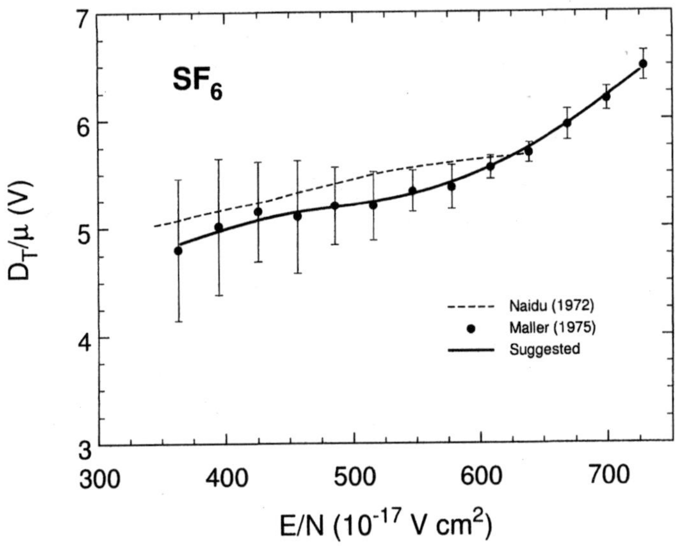

Figure 6.125. Transverse electron diffusion coefficient to electron mobility ratio as a function of E/N, $D_{\mathrm{T}}/\mu\,(E/N)$, for SF_6: (- - -) Ref. [697]; (•) Ref. [270]; (—) suggested.

5.6.2 Transverse Electron Diffusion Coefficient to Electron Mobility Ratio, $D_{\mathrm{T}}/\mu\,(E/N)$, for SF_6

In Fig. 6.125 are plotted the values of $D_{\mathrm{T}}/\mu\,(E/N)$ as measured by Naidu and Prasad [697] at 293 K. These are the mean values of a number of measurements made at a pressure of 0.27 kPa. Also plotted in Fig. 6.125 are similar subsequent measurements made by Maller and Naidu [270] along with their indicated errors. The two sets of measurements are consistent with each other. A fit to the most recent data is shown by the solid line in the figure and is listed in Table 6.90 as our suggested values of the $D_{\mathrm{T}}/\mu\,(E/N)$ for SF_6. They can be used to determine the characteristic energy, $3/2(eD_{\mathrm{T}}/\mu)$, of electrons in SF_6 gas as a function of E/N. Calculated values of $D_{\mathrm{T}}/\mu\,(E/N)$ for SF_6 using a Boltzmann code analysis or the Monte Carlo method can be found in Refs. [417, 424–426, 428–430, 432, 437–439, 441, 442].

5.6.3 Product of Gas Number Density and Longitudinal Electron Diffusion Coefficient, $ND_{\mathrm{L}}\,(E/N)$, for SF_6

There are three sets of measurements [583, 584, 685] of this quantity which are plotted in Fig. 6.126. Two of these are by Aschwanden [583,

Table 6.90. Suggested values of the transverse electron diffusion coefficient to electron mobility ratio, $D_T/\mu\,(E/N)$, for SF_6.

E/N (10^{-17} V cm^2)	D_T/μ (V)	E/N (10^{-17} V cm^2)	D_T/μ (V)
365	4.86	600	5.52
400	4.99	650	5.82
450	5.13	700	6.22
500	5.21	725	6.44
550	5.32		

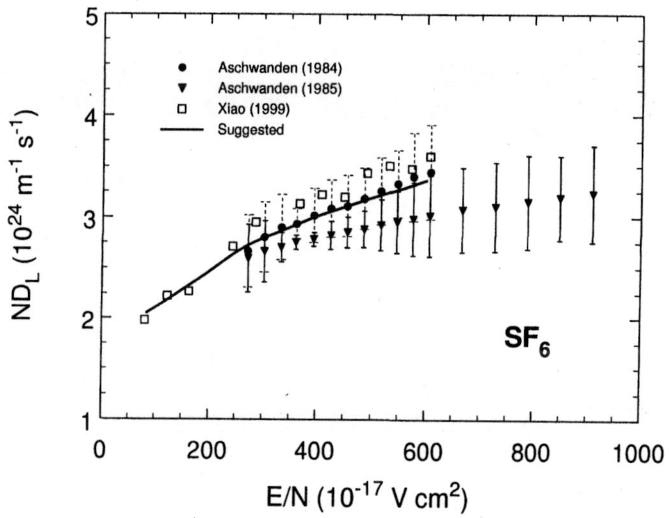

Figure 6.126. Measured values of the product, $ND_L\,(E/N)$, of the gas number density N and the longitudinal electron diffusion coefficient D_L as a function of E/N for SF_6: (•) Ref. [583]; (▼) Ref. [584]; (□) Ref. [685]; (—) suggested.

584] and have uncertainties as shown by the error bars in the figure. The other set of measurements is by Xiao *et al.* [685] who state no uncertainties. The solid line in Fig. 6.126 represents a fit to the three data sets, and values from this fit are listed in Table 6.91 as our suggested values for the $ND_L\,(E/N)$ of SF_6. Calculated values of the longitudinal electron diffusion coefficient D_L (or ND_L or D_L/μ) using a Boltzmann code analysis or the Monte Carlo method can be found in Refs. [425, 437–439, 441, 442].

Table 6.91. Suggested values of the product of the gas number density and the longitudinal electron diffusion coefficient, $ND_L\,(E/N)$, for SF_6.

E/N (10^{-17} V cm^2)	ND_L (10^{24} m^{-1} s^{-1})	E/N (10^{-17} V cm^2)	ND_L (10^{24} m^{-1} s^{-1})
85	2.05	350	2.90
100	2.10	400	3.01
150	2.28	450	3.10
200	2.46	500	3.19
250	2.65	550	3.27
300	2.79	600	3.36

5.7 Autodetachment, Thermally Induced Detachment, Photodetachment, and Collisional Detachment of SF_6^-

Electron detachment from SF_6^- ions is an important process of basic and applied interest. It can be induced via a number of ways, for instance, automatically from unstable SF_6^{-*}, or from SF_6^{-*} or SF_6^- by an applied electric field, or by heat, or by collision with photons, electrons, or neutral species. In this section we briefly elaborate on the autodetachment of SF_6^{-*} and summarize pertinent findings on thermally induced detachment, photodetachment, and collisional detachment of SF_6^- (see, also, Refs. [179, 180, 403, 420]).

5.7.1 Autodetachment

As shown in the previous section, low-energy free electrons attach very efficiently to the SF_6 molecule forming SF_6^{-*} parent anions. The initially produced isolated SF_6^{-*} parent ion is long lived, but metastable. Its mean lifetime toward autodetachment, that is, for the process

$$SF_6^{-*} \rightarrow SF_6^{(*)} + e \tag{6.46}$$

is $> 1\,\mu$s. Process (6.46) is a function of the internal energy of the unstable SF_6^{-*} [403]. Alternatively, the unstable SF_6^{-*} species can be stabilized by collision

$$SF_6^{-*} + \text{collision} \rightarrow SF_6^- + \text{energy} \tag{6.47}$$

and/or emission of infrared radiation $h\nu$

$$SF_6^{-*} \rightarrow SF_6^- + h\nu. \tag{6.48}$$

5.7.2 Thermally Induced Detachment

The possible effect of temperature on electron detachment from stabilized SF_6^- ions has recently been investigated by Christophorou and collaborators [404, 678]. These studies have shown that when the gas temperature was raised from 300 to $\sim$ 600 K, thermally enhanced electron detachment is observed for such negative ions as $C_6F_6^-$ and c-$C_4F_6^-$, but not for SF_6^-, that is, not for the reaction

$$SF_6^- + \text{heat} \rightarrow SF_6 + e. \tag{6.49}$$

This may be related to the relatively larger electron affinity of the SF_6 molecule (1.06 eV, Table 4.57) compared to those of the two other molecules (< 0.5 eV, Refs. [404, 678, 704]). Also, the energetically stabilized SF_6^- ion does not detach in collisions with SF_6 for collision energies less than $\sim$ 90 eV (see Refs. [705, 706] and Sec. 5.7.4 of Ch. 6).

5.7.3 Photodetachment

The first attempt to investigate the photodetachment process in SF_6^- was made by Drzaic and Brauman [495, 496] who reported observation of electron detachment from SF_6^- induced by IR multiphoton absorption. However, they did not observe photodetachment of SF_6^- with "visible light." According to Hay [491], Freiser and Beauchamp [707] observed a monotonically increasing photodetachment signal at wavelengths shorter than 350 nm (3.55 eV), but were unable to detect photodetachment from SF_6^- below 370–380 nm (3.25–3.35 eV). Following these initial investigations, new photodetachment methods [708–712] enabled a quantitative measurement of the photodetachment cross section and energetics of the SF_6^- ion. Thus, Mock and Grimsrud [708] observed photodetachment from SF_6^- (in a mixture of SF_6 with nitrogen at 1 atm and 423 K) over a portion of the wavelength range between 300 to 450 nm and estimated a photodetachment onset of 387 nm (3.2 eV). No photodetachment was detectable in the spectral range from 450 to 1200 nm. Mock and Grimsrud [708] were able to determine absolute values of the photodetachment cross section $\sigma_{\text{pd}}(\lambda)$ in the wavelength range between 300 to 400 nm, which are shown in Fig. 6.127 [the data plotted were taken from Fig. 1(a) of their paper].

A more detailed set of measurements was made by Datskos *et al.* [712] using a more sensitive method [709, 710], which allowed accurate determination of the photodetachment cross section, $\sigma_{\text{pd}}(h\nu)$, of SF_6^- as

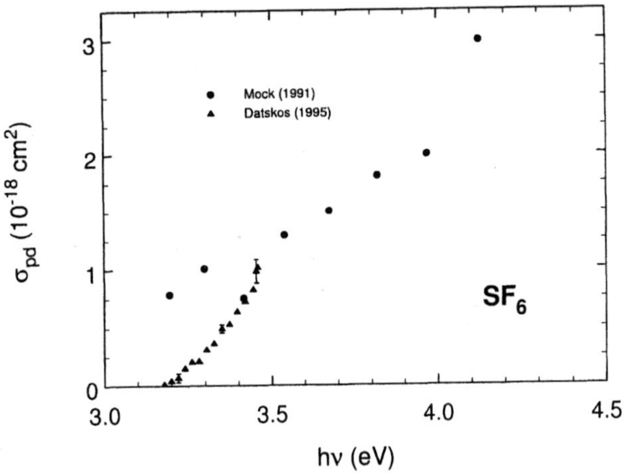

Figure 6.127. Photodetachment cross section, $\sigma_{\mathrm{pd}}(h\nu)$, for SF_6^- as a function of the photon energy $h\nu$: (•) Ref. [708]; (▲) Ref. [712].

a function of photon energy $h\nu$. In this method the photodetachment cross section and the photodetachment threshold E_{th} are determined using a two-laser arrangement (see Refs. [709, 710] and Sec. 3.4.3 of Ch. 2). Photoelectrons generated by one laser pulse are attached to SF_6 forming SF_6^-. Another tunable laser pulse, delayed with respect to the first, photodetaches the electrons from the SF_6^- ions when $h\nu > E_{\mathrm{th}}$. The photodetached electrons, being more mobile than the SF_6^- ions, induce a detectable transient voltage signal while drifting as free electrons prior to being again attached to SF_6. The size of this transient signal due to the photodetached electrons is measured and used to determine the cross section $\sigma_{\mathrm{pd}}(h\nu)$ (see Refs. [709, 710, 712] and Sec. 3.4.3 of Ch. 2 for details).

In Fig. 6.127 are shown the data for the photodetachment cross section, $\sigma_{\mathrm{pd}}(h\nu)$, for the reaction

$$SF_6^- + h\nu \rightarrow SF_6 + e, \tag{6.50}$$

obtained by Datskos *et al.* [712] using this technique, along with their associated errors. They show that the photodetachment cross section has a threshold (i.e., a vertical detachment energy) at 3.16 eV which is about three times the size of the electron affinity of the SF_6 molecule. The cross section increases from a zero value at the threshold to $\sim 1.0 \times 10^{-18}$ cm^2 at a photon energy of 3.46 eV. The small size of the measured cross section $\sigma_{\mathrm{pd}}(h\nu)$ has been attributed to the large relaxation in the equilibrium internuclear positions of SF_6^- compared to SF_6 [491, 627,

Table 6.92. Suggested values of the photodetachment cross section, $\sigma_{\rm pd}(h\nu)$, of SF_6^- as a function of photon energy measured by Datskos *et al.* [712].

Photon energy (eV)	$\sigma_{\rm pd}(h\nu)$ (10^{-18} cm^2)	Photon energy (eV)	$\sigma_{\rm pd}(h\nu)$ (10^{-18} cm^2)
3.18	0.013	3.35	0.492
3.20	0.042	3.37	0.530
3.22	0.070	3.40	0.638
3.24	0.152	3.42	0.725
3.26	0.205	3.44	0.825
3.28	0.211	3.45	0.983
3.30	0.313	3.46	1.015
3.33	0.366		

712]. With the exception of the lowest two data points of Mock and Grimsrud [708] (which are considered uncertain due to the low sensitivity of the experiment at these wavelengths), the measurements of Mock and Grimsrud are in agreement with those of Datskos *et al.* in the small wavelength range they overlap. Since the measurements of $\sigma_{\rm pd}(h\nu)$ for SF_6^- by Datskos *et al.* [712] are more extensive and their uncertainty indicated, they are listed in Table 6.92 as our suggested data.

Ingólfsson *et al.* [711] employed a somewhat similar technique to that of Christophorou and co-workers [709, 710, 712], but used only a fixed wavelength (337 nm) from a pulsed nitrogen laser. At this wavelength (which corresponds to a photon energy of 3.68 eV) they observed photodetachment from SF_6^- with an estimated cross section of $\sim 10^{-18}$ cm^2. This value is not inconsistent with the data in Fig. 6.127.

The cross section data in Fig. 6.127 should be valuable to investigations attempting to measure the SF_6^- ion densities in various types of gas discharges (e.g., see Ishikawa *et al.* [713] and Kono *et al.* [714]). The small measured cross section for photodetachment of SF_6^- would also indicate that this electron detachment mechanism is of minor significance compared to collisional detachment in gas discharges and gas-insulated high-voltage equipment using SF_6 gas [715–718].

Finally, it should be noted that no photodetachment cross sections appear to have been published for the other negative ions produced by dissociative electron attachment to SF_6, except for F^- [180].

5.7.4 Collisional Detachment

Collisional detachment studies involving the three major negative ions (SF_6^-, SF_5^- and F^-) formed by electron attachment to SF_6 have been performed by Champion and collaborators [705, 706, 719]. Collisional detachment from SF_6^- and SF_5^- in SF_6 was found to have anomalously high energy thresholds [see Figs. 6.128(a) and 6.128(b)]. The thresholds for electron release in the reactions

$$SF_6^- + SF_6 \rightarrow SF_6 + SF_6 + e \tag{6.51}$$

and

$$SF_5^- + SF_6 \rightarrow SF_5 + SF_6 + e \tag{6.52}$$

are $\sim$ 90 eV in the center-of-mass (cm) system. These values are much larger than the respective electron affinities of SF_6 (1.06 eV, Table 6.57) and SF_5 (2.7–3.7 eV, Ref. [74]). The anomalously high energy threshold for the collisional detachment reaction (6.51) has been attributed [705, 706, 719] to the competing ion-conversion and charge-transfer reactions at center-of-mass energies below $\sim$ 100 eV, viz.

$$SF_6^- + SF_6 \rightarrow F^- + SF_5 + SF_6, \tag{6.53}$$

$$SF_6^- + SF_6 \rightarrow SF_5^- + F + SF_6, \tag{6.54}$$

and

$$SF_6^- + SF_6 \rightarrow \text{charge transfer products} + SF_6. \tag{6.55}$$

As can be seen from Fig. 6.128(a), the cross sections, $\sigma_{ic,F^-}(\mathcal{E}_{cm})$, $\sigma_{ic,SF_5^-}(\mathcal{E}_{cm})$, and $\sigma_{ct}(\mathcal{E}_{cm})$, for reactions (6.53), (6.54), and (6.55) respectively, are much larger than the cross section, $\sigma_{cd}(\mathcal{E}_{cm})$, for the collisional detachment reaction (6.51), and their thresholds lie at much lower collision energies $\mathcal{E}_{cm}$ (collision energy in the center-of-mass system). The energetic thresholds for collisionally decomposing ground-state SF_6^- via the reactions (6.51), (6.53), and (6.54) are, respectively, 1.06 eV [= EA (SF_6), Table 6.57], 1.61 eV [= D (SF_5–F) + EA (SF_6) − EA (F), with D (SF_5–F) = 3.95 eV (Table 6.60), EA (SF_6) = 1.06 eV, and EA (F) = 3.40 eV [74]], and 1.34 eV [= D (SF_5–F) + EA (SF_6) − EA (SF_5), with D (SF_5–F) = 3.95 eV (Table 6.60), EA (SF_6) = 1.06 eV, and EA (SF_5) = 3.66 eV (average of the two highest values listed in Ref. [74])]. Clearly, the data in Fig. 6.128(a) show that electron detachment from SF_6^- by SF_6 has a very small probability of occurrence for collision energies $\mathcal{E}_{cm} < 60$ eV, and that the dominant collisional decomposition mechanism of SF_6^- is dissociation. It is interesting to note also that studies of reactions of the form $A^- + B \rightarrow A + B^-$, with A^- being SF_6^-,

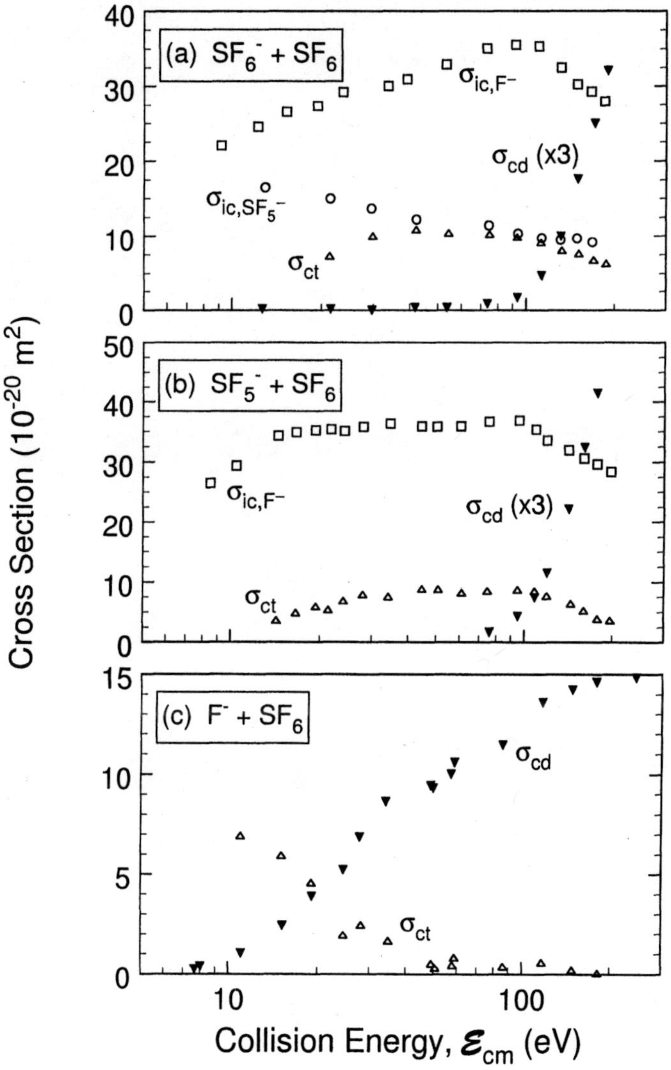

Figure 6.128. Measured cross sections for collisions of (a) SF_6^-, (b) SF_5^-, and (c) F^- on SF_6 for the following processes: (▼) collisional detachment, $\sigma_{cd}(\mathcal{E}_{cm})$; (□) collisional production of F^-, $\sigma_{ic,F^-}(\mathcal{E}_{cm})$; (○) collisional production of SF_5^-, $\sigma_{ic,SF_5^-}(\mathcal{E}_{cm})$; and (△) charge transfer processes, $\sigma_{ct}(\mathcal{E}_{cm})$. All data in this figure are those of Wang *et al.* [719]. Note that in Fig. 6.128(a) and Fig. 6.128(b) the cross section values for $\sigma_{cd}(\mathcal{E}_{cm})$ have been multiplied by a factor of 3 for convenience of display.

have shown [720] that the large geometrical change between SF_6^- and SF_6 introduces an internal barrier in the reaction coordinates and hence such reactions proceed with a low reaction rate constant.

Similarly, the measurements of Wang *et al.* [719] shown in Fig. 6.128(b) for SF_5^-, indicate that the cross sections for the reactions

$$SF_5^- + SF_6 \rightarrow F^- + SF_4 + SF_6 \tag{6.56}$$

and

$$SF_5^- + SF_6 \rightarrow \text{charge transfer products} + SF_6 \tag{6.57}$$

have much lower energy thresholds than that for the electron detachment reaction (6.52). The cross sections for reactions (6.56) and (6.57) far exceed that for reaction (6.52) below $\sim$ 100 eV.

In contrast to the above findings on the collisional detachment in SF_6 of the SF_6^- and SF_5^- ions, the measurements of Wang *et al.* shown in Fig. 6.128(c) for the collisional detachment of F^- in SF_6, viz.,

$$F^- + SF_6 \rightarrow F + SF_6 + e \tag{6.58}$$

indicate a rather low energy threshold ($\sim$ 8.0 eV). Additionally, the cross-section values for reaction (6.58) progressively exceed those of the charge-transfer reactions as the reaction energy increases above $\sim$ 20 eV. It is clear then that charge transfer and collision-induced dissociation processes are the dominant inelastic channels in SF_6 involving the destruction of SF_6^-, SF_5^-, and F^- at low collision energies. These findings are consistent with the results of Wang *et al.* [719] on the destruction of SF_6^-, SF_5^-, and F^- in rare gases . They are also consistent with the measurements of Champion *et al.* [721] of the total cross section for electron detachment and collision induced dissociation for collisions of SF_6^- with N_2 as a function of collision energy.

Interestingly, evidence has been obtained [705, 719] that long-lived energetically unstable states of SF_6^- can contribute to collisional detachment. Also, experimental evidence has been provided [722–725] that negative-ion clusters such as $(SF_6^-)SF_6$ (and also clusters involving F^-) can form in parent SF_6 gas and can contribute to the pressure dependence [724] of the measured density-normalized electron detachment coefficient δ/N in SF_6 (the gas pressure in these experiments was 7.4 and 11.1 kPa).

Shown in Fig. 6.129 are the data for the density-normalized electron detachment coefficient, $\delta/N\,(E/N)$, measured in SF_6 by O'Neill and Craggs [683], Hansen *et al.* [724], and Hilmert and Schmidt [725]. The density dependence of $\delta/N\,(E/N)$ observed by O'Neill and Craggs (at pressures between 0.33 and 1.35 kPa) is seen to continue up to the higher

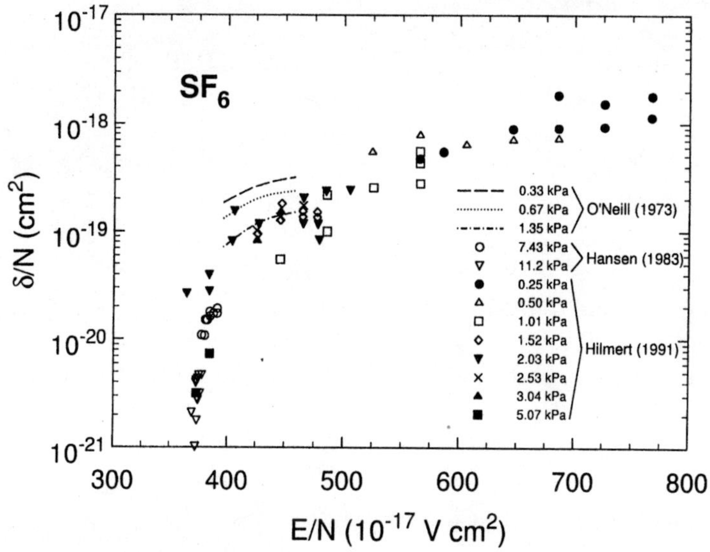

Figure 6.129. Density-normalized electron detachment coefficient, $\delta/N\,(E/N)$, as a function of E/N for SF_6 (T = 293–296 K) at the pressures indicated in the legend: (– –,· · ·,– · –) Ref. [683]; (○, ▽) Ref. [724]; (●, △, □, ◇, ▼, ×, ▲, ■) Ref. [725].

pressures (7.43 and 11.2 kPa) employed by Hansen *et al.* [724]. The coefficient $\delta/N\,(E/N)$ decreases with N. They attributed this decrease to a three-body process $SF_6^- + SF_6 + SF_6 \rightarrow (SF_6^-)\,SF_6 + SF_6$, which leads to the formation of a stable complex $(SF_6^-)SF_6$ in which the electron is more tightly bound than in SF_6^-. This process then competes with the two-body electron detachment reaction which is pressure-independent. The data shown in Fig.6.129 were obtained using swarm techniques. Data obtained by time-lag to breakdown experiments [717, 718, 726] are not plotted since they gave very different and less accurate values because of the way the coefficient was evaluated in these experiments (see discussion in Hilmert and Schmidt [725]).

The high-energy collisional detachment thresholds of the SF_6 negative ions in SF_6 gas accounts for the observed small values of the collisional detachment coefficient compared to the values of the coefficients of other negative ion-molecule reactions in SF_6 [705]. As a consequence, collisional detachment in SF_6, and electrical properties of SF_6 such as discharge initiation probabilities that depend on electron detachment, are sensitive to the presence of impurities, such as water and air, that may form clusters with SF_6-originated negative ions for which collisional detachment may be more probable [727–729].

5.8 Ion Transport in SF_6

The motion of ions through SF_6 gas is an important aspect of modeling many systems utilizing SF_6. The available ion transport data are briefly summarized here.

A number of studies have attempted to quantify the motion of negative ions [582, 700, 722, 730–735] and positive ions [731, 734, 736–740] formed by electron impact on SF_6 in parent SF_6 gas. These studies include measurements of ion mobilities [582, 700, 722, 731–740] and ion diffusion [700, 732, 735, 737, 738]. They are not only limited in number, but the interpretation of the measurements is often hindered by the absence of mass spectrometric identification of the ions present in the system and also by the absence of reliable information on the pertinent ion-transformation processes under varied experimental conditions. In fact, only the work of Patterson [722] and de Urquijo *et al.* [735] on negative ions and the work of de Urquijo *et al.* [737] on positive ions included mass analysis. The rest of the studies inferred the nature of the ions by reference to the mass spectrometric data of these three studies. The early data on the mobilities of positive ions, negative ions, and ion clusters in the parent SF_6 gas have been summarized and discussed by Brand and Jungblut [723] and by Morrow [418].

In Fig. 6.130 are plotted the reduced mobilities [that is, mobilities referred to standard pressure (101.325 kPa) and temperature (273.16 K)], $\mu_0^-(E/N)$, for the negative ions SF_5^-, SF_6^-, $SF_6^-(SF_6)$, and $SF_6^-(SF_6)_2$ in SF_6. As mentioned in the preceding paragraph, only Patterson [722] and de Urquijo *et al.* [735] identified the ions by mass spectrometry. The identification of the ionic mobilities with the anionic species by other researchers was made primarily by virtue of their agreement with the $\mu_0^-(E/N)$ data of Patterson for the corresponding negative-ion species. According to Brand and Jungblut [723], all measured negative ion mobilities can be assigned to one or more of the following ions: SF_5^-, SF_6^-, $SF_6^-(SF_6)$, and $SF_6^-(SF_6)_2$.

In Fig. 6.131 are shown the $\mu_0^-(E/N)$ measurements of Nakamura [700] for the fastest negative ion observed in parent SF_6 gas in his study, which he presumed to be F^-.

Similar measurements of the reduced mobilities, $\mu_0^+(E/N)$, of positive ions in parent SF_6 gas are shown in Fig. 6.132. Along with the early data [582, 731, 734, 736] plotted in Fig. 6.132, are shown the more recent mass-identified measurements of de Urquijo *et al.* [737, 738]) and Talib and Saporoschenko [739, 740].

In Fig. 6.133(a) are shown the $ND_L^-(E/N)$ measurements of Nakamura [700] for negative ions in SF_6 (closed symbols). Although in the

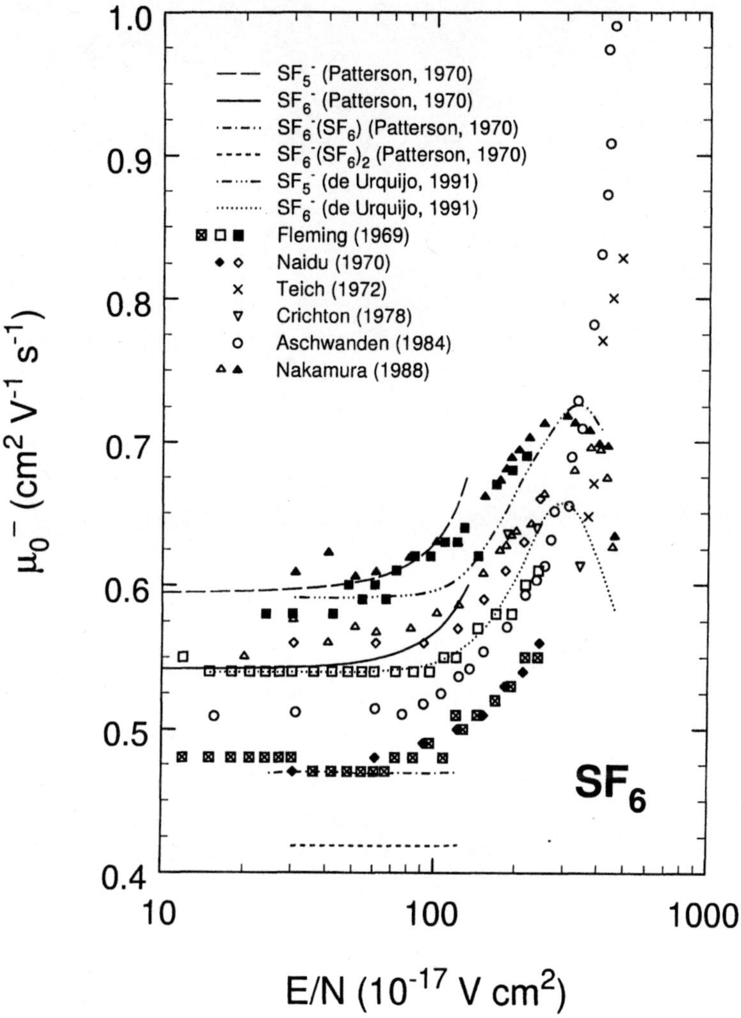

Figure 6.130. Measured reduced mobilities, $\mu_0^-(E/N)$, for negative ions in SF_6. Mass identified data of Patterson [722]: (– –) SF_5^-; (—) SF_6^-; (– · –) $SF_6^-(SF_6)$; (- - -) $SF_6^-(SF_6)_2$. Mass identified data of de Urquijo *et al.* [735]: (–· · · ·–) SF_5^-; (· · ·), SF_6^-. Measurements without mass analysis: (⊠, □, ■) Ref. [731]; (◇, ♦) Ref. [732]; (×) Ref. [582]; (▽) Ref. [733]; (○) Ref. [734]; (△, ▲) Ref. [700].

experiments of Nakamura there was no mass analysis of the negative ions present in the system, he ascribed his mobility measurements to the SF_6^-, SF_5^-, and F^- ions as shown in the figure. More recent mass-identified measurements of $ND_L^-(E/N)$ for SF_6^- and SF_5^- in SF_6 by de Urquijo *et al.* [735] are also shown in Fig. 6.133(a) [open symbols]. De Urquijo *et*

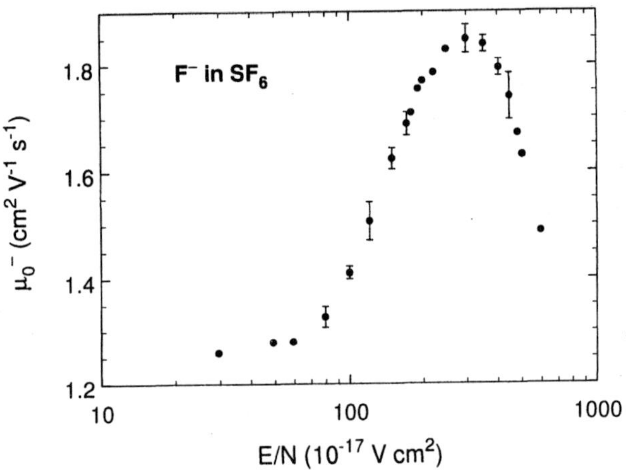

Figure 6.131. Measured reduced mobilities, $\mu_0^-(E/N)$, for the fastest ion in the study of Nakamura [700], presumed to be F^- in SF_6.

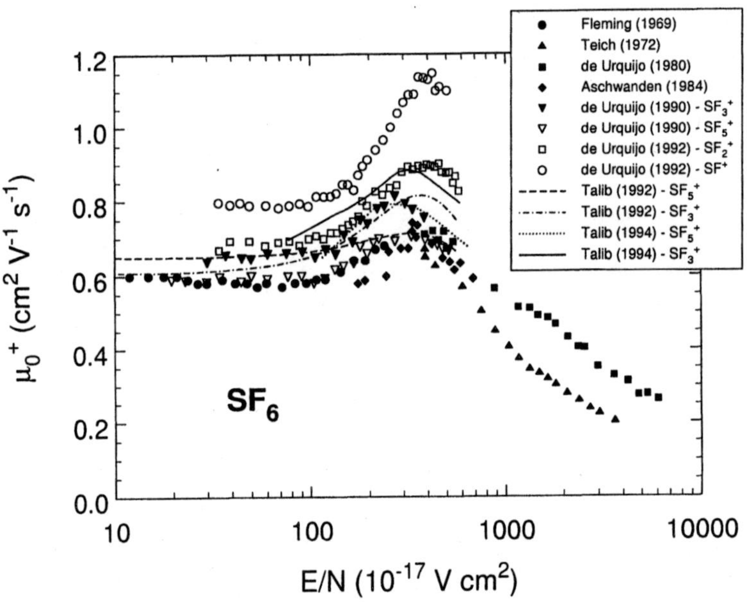

Figure 6.132. Measured reduced mobilities, $\mu_0^+(E/N)$, for SF_5^+ and SF_3^+ in parent SF_6 gas ($T = 290$–300 K): (•) Ref. [731]; (▲) Ref. [582]; (■) Ref. [736]; (♦) Ref. [734]; (▼) SF_3^+ from Ref. [737]; (▽) SF_5^+ from Ref. [737]; (- - -) SF_5^+ from Ref. [739]; (— · —) SF_3^+ from Ref. [739]; (—) SF_3^+ from Ref. [740]; (· · ·) SF_5^+ from Ref. [740]. Also shown are the reduced mobilities of (○) SF^+ and (□) SF_2^+ from Ref. [738].

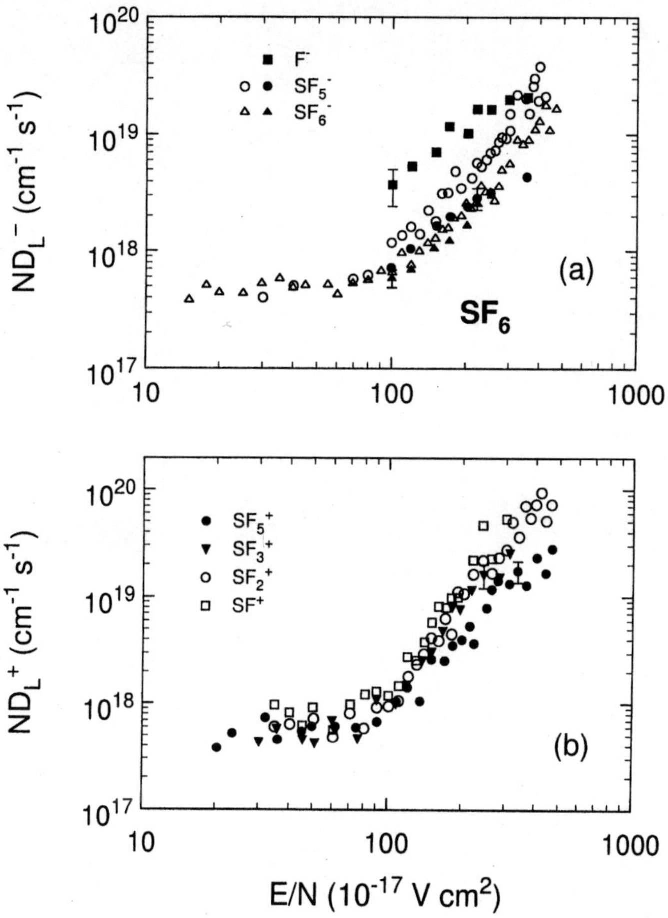

Figure 6.133. (a) $ND_L^-(E/N)$ for the negative ions SF_6^-, SF_5^-, and F^- in SF_6 (closed symbols, data of Nakamura [700]; open symbols, data of de Urquijo *et al.* [735]); (b) $ND_L^+(E/N)$ for the positive ions SF_5^+, SF_3^+, SF_2^+, and SF^+ in SF_6 (data of de Urquijo *et al.* [737, 738]).

al. [737, 738] also measured $ND_L^+(E/N)$ for the positive ions SF_5^+, SF_3^+, SF_2^+, and SF^+ in SF_6, and these results are shown in Fig. 6.133(b).

There seem to be no measurements of $D_T^-/\mu\,(E/N)$ for negative ions in SF_6. However, Naidu and Prasad [732] measured the so-called Townsend energy factor k_T (ratio of the mean ion energy to the mean energy of the gas molecules) as a function E/N at $T = 293$ K for pressures equal to 0.16 and 0.32 kPa. Table 6.93 lists the measurements of Naidu and Prasad [732] for $k_T^-\,(E/N)$ and the $D_T^-/\mu\,(E/N)$ values we calculated from these using the equation $k_T^-(E/N) = 39.6 D_T^-/\mu\,(E/N)$ (for $T =$

Table 6.93. $k_T^-(E/N)$ and $D_T^-/\mu(E/N)$ for negative ions in SF_6 ($T = 293$ K) measured by Naidu and Prasad [732].

E/N (10^{-17} V cm^2)	$k_T^-(E/N)$	$D_T^-/\mu(E/N)$ (V)
30.3	1.0	0.025
60.6	1.2	0.030
91.0	1.7	0.043
121.3	2.6	0.066
151.6	3.5	0.088
181.9	4.6	0.116
212.2	5.9	0.149

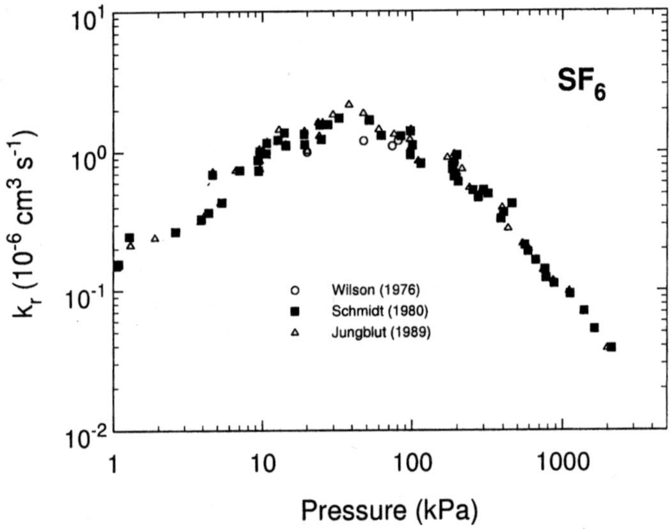

Figure 6.134. Volume (ion-ion) recombination rate coefficient, k_r, in SF_6 as a function of gas pressure: (○) Ref. [741] ($T = 298$ K); (■) Ref. [742] ($T = 296$ K); (△) Ref. [743] ($T = 296$ K).

293 K and a Maxwell distribution function for the ion kinetic energies). The quantity $D_T^-/\mu(E/N)$ is seen to increase from a value of 0.025 V at $E/N = 30.3 \times 10^{-17}$ V cm^2 to a value of 0.149 V at $E/N = 212.2 \times 10^{-17}$ V cm^2.

Finally, in Fig. 6.134 are shown the measurements of Wilson *et al.* [741] and Schmidt and collaborators [742, 743] of the volume (ion-ion)

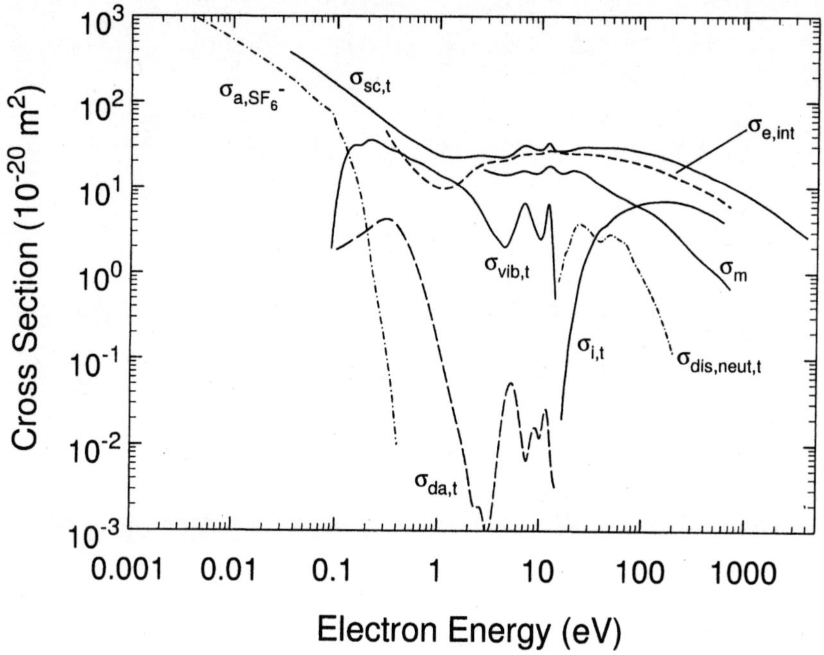

Figure 6.135. Recommended and suggested electron collision cross sections for SF_6.

recombination coefficient k_r for SF_6 as a function of gas pressure (T = 296–298 K).

5.9 Recommended and Suggested Electron Collision Cross Sections and Electron Transport Coefficients for SF_6

In Fig. 6.135 are plotted the recommended or suggested cross sections for SF_6 discussed in Sec. 5 of Ch. 6:

- $\sigma_{sc,t}(\varepsilon)$ — Table 6.61, Fig. 6.92;
- $\sigma_{e,int}(\varepsilon)$ — Table 6.65, Fig. 6.94;
- $\sigma_m(\varepsilon)$ — Table 6.66, Fig. 6.95;
- $\sigma_{vib,t}(\varepsilon)$ — Table 6.67, Fig. 6.99;
- $\sigma_{i,t}(\varepsilon)$ — Table 6.68, Fig. 6.101;
- $\sigma_{dis,neut,t}(\varepsilon)$ — Table 6.72; Fig. 6.106;

- $\sigma_{a,SF_6^-}(\varepsilon)$ — Table 6.76, Fig. 6.112(a) and Fig. 6.116; and
- $\sigma_{da,t}(\varepsilon)$ — Table 6.80, Fig. 6.116.

A number of additional electron collision cross sections have been discussed in the paper and data on these can be found as follows:

- $\sigma_{e,diff}(\varepsilon)$ — Table 6.62, Table 6.63, and Table 6.64, Fig. 6.93;
- $\sigma_{i,part}(\varepsilon)$ — Table 6.69, Fig. 6.102;
- $\sigma_{a,t}(\varepsilon)$ — Table 6.80, Fig. 6.110 and Fig. 6.116;
- $\sigma_{a,SF_5^-}(\varepsilon)$ — Table 6.77, Fig. 6.112(b);
- $\sigma_{a,SF_4^-}(\varepsilon)$, $\sigma_{a,SF_3^-}(\varepsilon)$, and $\sigma_{a,SF_2^-}(\varepsilon)$ — Table 6.79, Fig. 6.113 and Fig. 6.116;
- $\sigma_{a,F_2^-}(\varepsilon)$ — Table 6.79, Fig. 6.114 and Fig. 6.116; and
- $\sigma_{a,F^-}(\varepsilon)$ — Table 6.79, Fig. 6.115 and Fig. 6.116.

Based on the discussions in Sec. 5 of Ch. 6, recommended or suggested data for the electron transport coefficients, density-normalized electron attachment and ionization coefficients, and electron attachment rate constants have been presented as follows:

- $\alpha/N\,(E/N)$ — Table 6.71, Fig. 6.105;
- $\eta/N\,(E/N)$ — Table 6.86, Fig. 6.121;
- $\eta/N_a\,(\langle\varepsilon\rangle)$ — Table 6.81 and Table 6.82;
- $(\alpha-\eta)/N\,(E/N)$ — Table 6.87, Fig. 6.122;
- $k_{a,t}(E/N)$ in Ar, Xe, or N_2 — Table 6.81 and Table 6.82, Fig. 6.117;
- $k_{a,t}(\langle\varepsilon\rangle)$ — Table 6.83, Fig. 6.118;
- $w\,(E/N)$ — Table 6.88, Fig. 6.123 and Fig. 6.124;
- $D_T/\mu\,(E/N)$ — Table 6.90, Fig. 6.125; and
- $ND_L\,(E/N)$, — Table 6.91, Fig. 6.126.

In spite of the many studies on electron interactions with the SF_6 molecule there still exist distinct gaps in the available data. The most significant data need is for direct measurements of cross sections for vibrational excitation and for electron-impact dissociation into neutral fragments. Additionally, a determination of the momentum transfer cross section at energies below 2.5 eV is needed. These data are necessary to extend the recommended and suggested data in the present work, and hence provide a reliable source of electron-SF_6 interaction data for use in computer modeling codes.

6. References

[1] L. G. Christophorou and J. K. Olthoff, J. Phys. Chem. Ref. Data **28**, 131 (1999).

[2] L. G. Christophorou and J. K. Olthoff, J. Phys. Chem. Ref. Data **26**, 1205 (1997).

[3] L. G. Christophorou and J. K. Olthoff, J. Phys. Chem. Ref. Data **31**, 971 (2002).

[4] L. G. Christophorou and J. K. Olthoff, J. Phys. Chem. Ref. Data **29**, 267 (2000).

[5] L. G. Christophorou, J. K. Olthoff, and D. S. Green, *Gases for Electrical Insulation and Arc Interruption: Possible Present and Future Alternatives to Pure SF_6*, NIST Technical Note 1425, November 1997.

[6] G. L. Rogoff, J. M. Kramer, R. B. Piejak, IEEE Trans. Plasma. Sci. **PS-14**, 103 (1986).

[7] A. D. Richards and H. H. Sawin, J. Appl. Phys. **62**, 799 (1987).

[8] D. S. Fischl and D. W. Hess, J. Vac. Sci. Technol. B **6**, 1577 (1988).

[9] R. d'Agostino, F. Cramarossa, F. Fracassi, F. Illuzzi, and M. N. Armenise, J. Vac. Sci. Technol. B **6**, 1584 (1988).

[10] D. M. Manos and D. L. Flamm (Eds.), *Plasma Etching*, Academic, Boston, 1989.

[11] L. E. Kline and M. J. Kushner, Crit. Rev. Solid State Mater. Sci. **16**, 1 (1989).

[12] T. Matsuura, H. Uetake, T. Ohmi, J. Murota, K. Fukuda, N. Mikoshiba, T. Kawashima, and Y. Yamashita, Appl. Phys. Lett. **56**, 1339 (1990).

[13] V. V. Boiko, A. T. Rakhimov, and N. V. Suetin, Sov. Phys. Tech. Phys. **35**, 1268 (1990).

[14] L. Peters, Semicond. Int., May, 66 (1992).

[15] E. S. Aydil and D. J. Economou, J. Electrochem. Soc. **139**, 1396 (1992).

[16] N. L. Bassett and D. J. Economou, J. Appl. Phys. **75**, 1931 (1994).

[17] S. C. Deshmukh and D. J. Economou, J. Appl. Phys. **72**, 4597 (1992).

[18] P. L. G. Ventzek, M. Grapperhaus, and M. J. Kushner, J. Vac. Sci. Technol. B **12**, 3118 (1994).

[19] C. C. Cheng, K. V. Guinn, V. M. Donnelly, and I. P. Herman, J. Vac. Sc. Technol. A **12**, 2630 (1994).

[20] D. P. Lymberopoulos and D. J. Economou, IEEE Trans. Plasma Sci. **23**, 573 (1995).

[21] C. Lee and M. A. Lieberman, J. Vac. Sci. Technol. A **13**, 368 (1995).

[22] M. J. Kushner, J. Appl. Phys. **82**, 5312 (1997).

[23] Y. T. Lee, M. A. Lieberman, A. J. Lichtenberg, F. Bose, H. Baltes, and R. Patrick, J. Vac. Sci. Technol. A **15**, 113 (1997).

[24] G. I. Font and I. D. Boyd, J. Vac. Sci. Technol. A **15**, 313 (1997).

[25] J. P. Chang, A. P. Mahorowala, and H. H. Sawin, J. Vac. Sci. Technol. A **16**, 217 (1998).

[26] H. Doshita, K. Ohtani, and A. Namiki, J. Vac. Sci. Technol. A **16**, 265 (1998).

[27] G. P. Kota, J. W. Coburn, and D. B. Graves, J. Vac. Sci. Technol. A **16**, 270 (1998).

[28] M. V. Malyshev, V. M. Donnelly, A. Kornblit, and N. A. Ciampa, J. Appl. Phys. **84**, 137 (1998).

[29] M. V. Malyshev, V. M. Donnelly, and S. Samukawa, J. Appl. Phys. **84**, 1222 (1998).

[30] A. K. Hays, Opt. Commun. **28**, 209 (1979).

[31] M. C. Castex, J. Le Calvé, D. Haaks, B. Jordan, and G. Zimmerer, Chem. Phys. Lett. **70**, 106 (1980).

[32] M. Rokni and J. H. Jacob, in *Applied Atomic Collision Physics, Vol. 3: Gas Lasers*, H. S. W. Massey, E. W. McDaniel, and B. Bederson (Eds.), Academic, New York, 1982, p. 273.

[33] W. L. Nighan, in *Applied Atomic Collision Physics, Vol. 3: Gas Lasers*, H. S. W. Massey, E. W. McDaniel, and B. Bederson (Eds.), Academic, New York, 1982, p. 319.

[34] M. R. Flannery, in *Applied Atomic Collision Physics, Vol. 3: Gas Lasers*, H. S. W. Massey, E. W. McDaniel, and B. Bederson (Eds.), Academic, New York, 1982, p. 141.

[35] D. R. Bates, Adv. At. Mol. Phys. **20**, 1 (1985).

[36] W. L. Morgan, Plasma Chem. Plasma Process. **12**, 449 (1992).

[37] T. E. Graedel and P. J. Crutzen, *Chemie der Atmosphäre*, Spektrum Akademischer, Heidelberg, 1994.

[38] D. Maric, J. P. Burrows, R. Meller, and G. K. Moortgat, J. Photochem. Photobiol. A: Chem. **70**, 205 (1993).

[39] R. Nagpal and A. Garscaddden, Contrib. Plasma Phys. **35**, 301 (1995).

[40] N. Pinhão and A. Chouki, in *Proceedings XXII International Conference on Phenomena in Ionized Gases*, K. H. Becker, W. E. Carr, and E. E. Kunhardt (Eds.), Hoboken, NJ, July/August 1995, Contributed Papers 2, p. 5.

[41] A. M. Efremov, V. I. Svettsov, and V. P. Mikhalkin, High Energy Chem. **29**, 433 (1995).

[42] G. Herzberg, *Molecular Spectra and Molecular Structure I. Spectra of Diatomic Molecules*, 2nd ed., Van Nostrand Reinhold, New York, 1950.

[43] J. Jureta, S. Cvejanović, M. Kurepa, and D. Cvejanović, Z. Phys. A **304**, 143 (1982).

[44] D. Spence, R. H. Huebner, H. Tanaka, M. A. Dillon, and R.-G. Wang, J. Chem. Phys. **80**, 2989 (1984).

[45] K. P. Huber and G. Herzberg, *Molecular Spectra and Molecular Structure. IV. Constants of Diatomic Molecules*, Van Nostrand Reinhold, New York, 1979.

[46] S. D. Peyerimhoff and R. J. Buenker, Chem. Phys. **57**, 279 (1981).

[47] H. von Halban and K. Siedentopf, Z. Phys. Chem. **103**, 71 (1922).

[48] H. von Halban and K. Siedentopf, Z. Electrochem. **28**, 496 (1922).

[49] G. E. Gibson and N. S. Bayliss, Phys. Rev. **44**, 188 (1933).

[50] F. W. Jones and W. Spooner, Trans. Faraday Soc. **31**, 811 (1935).

[51] W. C. Fergusson, L. Slotkin, and W. G. Style, Trans. Faraday Soc. **32**, 956 (1936).

[52] R. G. Aickin and N. S. Bayliss, Trans. Faraday Soc. **33**, 1333 (1937).

[53] K. Watanabe, J. Chem. Phys. **26**, 542 (1957).

[54] J. Lee and A. D. Walsh, Trans. Faraday Soc. **55**, 1281 (1959).

[55] R. P. Iczkowski, J. L. Margrave, and J. W. Green, J. Chem. Phys. **33**, 1261 (1960).

[56] A. E. Douglas, C. K. Moeller, and B. P. Stoicheff, Can. J. Phys. **41**, 1174 (1963).

[57] D. J. Seery and D. Britton, J. Phys. Chem. **68**, 2263 (1964).

[58] H. Okabe, *Photochemistry of Small Molecules*, Wiley-Interscience, New York, 1978.

[59] C. Roxlo and A. Mandl, J. Appl. Phys. **51**, 2969 (1980).

[60] J. A. Coxon, J. Mol. Spectrosc. **82**, 264 (1980).

[61] A. E. Douglas, Can. J. Phys. **59**, 835 (1981).

[62] J. B. Burkholder and E. J. Bair, J. Phys. Chem. **87**, 1859 (1983).

[63] J. A. R. Samson and G. C. Angel, J. Chem. Phys. **86**, 1814 (1987).

[64] J. W. Gallagher, C. E. Brion, J. A. R. Samson, and P. W. Langhoff, J. Phys. Chem. Ref. Data **17**, 9 (1988).

[65] J. A. Ganske, H. N. Berko, and B. J. Finlayson-Pitts, J. Geophys. Res. **97**, 7651 (1992).

[66] S. Hubinger and J. B. Nee, J. Photochem. Photobiol. A **86**, 1 (1995).

[67] D. C. Frost, C. A. McDowell, and D. A. Vroom, J. Chem. Phys. **46**, 4255 (1967).

[68] A. W. Potts and W. C. Price, J. Chem. Soc. Faraday Trans. 2 **67**, 1242 (1971).

[69] A. B. Cornford, D. C. Frost, C. A. McDowell, J. L. Ragle, and I. A. Stenhouse, J. Chem. Phys. **54**, 2651 (1971).

[70] J. H. D. Eland, in *Electron Spectroscopy: Theory, Techniques, and Applications*, C. R. Brundle and A. D. Baker (Eds.), Academic, New York, 1979, Vol. 3, Ch. 5.

[71] T. A. Carlson, M. O. Krause, F. A. Grimm, and T. A. Whitley, J. Chem. Phys. **78**, 638 (1983).

[72] H. Van Lonkhuyzen and C. A. De Lange, Chem. Phys. **89**, 313 (1984).

[73] A. G. McConkey, G. Dawber, L. Avaldi, M. A. MacDonald, G. C. King, and R. I. Hall, J. Phys. B **27**, 271 (1994).

[74] A. A. Christodoulides, D. L. McCorkle, and L. G. Christophorou, in *Electron-Molecule Interactions and Their Applications*, L. G. Christophorou (Ed.), Academic, New York, Vol. 2, Ch. 6.

[75] S. G. Lias, J. E. Bartmess, J. F. Liebman, J. L. Holmes, R. D. Levin, and W. G. Mallard, J. Phys. Chem. Ref. Data **17**, 592 (1988).

[76] D. C. Frost and C. A. McDowell, Can. J. Chem. **38**, 407 (1960).

[77] L. Frost, A. M. Grisogono, I. E. McCarthy, E. Weigold, C. E. Brion, A. O. Bawagan, P. K. Mukherjee, W. Von Niessen, M. Rosi, and A. Sgamellotti, Chem. Phys. **113**, 1 (1987).

[78] J. D. Morrison and A. J. C. Nicholson, J. Chem. Phys. **20**, 1021 (1952).

[79] R. Thorburn, Proc. Phys. Soc. (London) **123**, 122 (1959).

[80] M. V. Kurepa and D. S. Belić, Chem. Phys. Lett. **49**, 608 (1977).

[81] V. E. Bondybey and C. Fletcher, J. Chem. Phys. **64**, 3615 (1976).

[82] T. Moeller, B. Jordan, P. Gürtler, G. Zimmerer, D. Haaks, J. Le Calvé, and M.-C. Castex, Chem. Phys. **76**, 295 (1983).

[83] R. G. McLoughlin, J. D. Morrison, and D. L. Smith, Int. J. Mass Spectrom. Ion Processes **58**, 201 (1984).

[84] R. J. Stubbs, T. A. York, and J. Comer, J. Phys. B **18**, 3229 (1985).

[85] T. L. Gilbert and A. C. Wahl, J. Chem. Phys. **55**, 5247 (1971).

[86] P. W. Tasker, G. G. Balint-Kurti, and R. N. Dixon, Mol. Phys. **32**, 1651 (1976).

[87] W.-C. Tam and S. F. Wong, J. Chem. Phys. **68**, 5626 (1978).

[88] M. Rokni, J. H. Jacob, and J. A. Mangano, Appl. Phys. Lett. **34**, 187 (1979).

[89] L. C. Lee, G. P. Smith, J. T. Moseley, P. C. Cosby, and J. A. Guest, J. Chem. Phys. **70**, 3237 (1979).

[90] J. G. Dojahn, E. C. M. Chen, and W. E. Wentworth, J. Phys. Chem. **100**, 9649 (1996).

[91] T. Leininger and F. X. Gadéa, J. Phys. B **33**, 735 (2000).

[92] J. B. Fisk, Phys. Rev. **51**, 25 (1937).

[93] R. J. Gulley, T. A. Field, W. A. Steer, N. J. Mason, S. L. Lunt, J.-P. Ziesel, and D. Field, J. Phys. B **31**, 2971 (1998).

[94] G. D. Cooper, J. E. Sanabia, J. H. Moore, J. K. Olthoff, and L. G. Christophorou, J. Chem. Phys. **110**, 682 (1999).

[95] M. Gote and H. Ehrhardt, J. Phys. B **28**, 3957 (1995).

[96] M. V. Kurepa and D. S. Belić , J. Phys. B **11**, 3719 (1978).

[97] R. Azria, R. Abouaf, and D. Teillet-Billy, J. Phys. B **15**, L569 (1982).

[98] D. Spence, Phys. Rev. A **10**, 1045 (1974).

[99] H. Kutz and H.-D. Meyer, Phys. Rev. A **51**, 3819 (1995).

[100] T. N. Rescigno, Phys. Rev. A **50**, 1382 (1994).

[101] V. A. Bailey and R. H. Healey, Philos. Mag. **19**, 725 (1935).

[102] S. E. Božin and C. C. Goodyear, Br. J. Appl. Phys. **18**, 49 (1967).

[103] A. Ernesti, M. Gote, and H. J. Kotsch, Phys. Rev. A **52**, 1266 (1995).

[104] M.-C. Bordage, P. Ségur, and A. Chouki, J. Appl. Phys. **80**, 1325 (1996).

[105] M.-C. Bordage, P. Ségur, L. G. Christophorou, and J. K. Olthoff, J. Appl. Phys. **86**, 3558 (1999).

[106] L. G. Christophorou, *Atomic and Molecular Radiation Physics*, Wiley-Interscience, New York, 1971, p. 328.

[107] H. Ehrhardt, L. Langhans, F. Linder, and H. S. Taylor, Phys. Rev. **173**, 222 (1968).

[108] R. E. Center and A. Mandl, J. Chem. Phys. **57**, 4104 (1972).

[109] F. A. Stevie and M. J. Vasile, J. Chem. Phys. **74**, 5106 (1981).

[110] S. K. Srivastava and R. Boivin, Bull. Am. Phys. Soc. **42**, 1738 (1997).

[111] S. K. Srivastava, private communication, January 1998.

[112] Y.-K. Kim, private communication, January 1998.

[113] H. Deutsch, K. Becker, and T. D. Märk, private communication, February 1998.

[114] K. N. Joshipura and C. G. Limbachiya, Int. J. Mass Spectrom. **216**, 239 (2002).

[115] P. Calandra, C. S. S. O'Connor, and S. D. Price, J. Chem. Phys. **112**, 10821 (2000).

[116] D. Rapp and P. Englander-Golden, J. Chem. Phys. **43**, 1464 (1965).

[117] P. C. Cosby, Bull. Am. Phys. Soc. **35**, 1822 (1990).

[118] P. C. Cosby and H. Helm, Wright Laboratory Report No. WL-TR-93-2004, Wright Patterson AFB, OH 45433-7650, 1992.

[119] P. C. Cosby, private communication, June 1998.

[120] W. C. Wells and E. C. Zipf, J. Chem. Phys. **66**, 5828 (1977).

[121] J. B. A. Mitchell, Phys. Rep. **186**, 216 (1990).

[122] D. L. McCorkle, A. A. Christodoulides, and L. G. Christophorou, Chem. Phys. Lett. **109**, 276 (1984).

[123] R. Azria, L. Parenteau, and L. Sanche, J. Chem. Phys. **87**, 2292 (1987).

[124] L. G. Christophorou, in *Linking the Gaseous and the Condensed Phases of Matter: the Behavior of Slow Electrons*, L. G. Christophorou, E. Illenberger, and W. F. Schmidt (Eds.), Plenum, New York, 1994, p. 3.

[125] D. Muigg, G. Denifl, A. Stamatović , E. Illenberger, I. Walker, and T. D. Märk, Chem. Phys. Lett. (Submitted).

[126] T. D. Märk, private communication, October 1998.

[127] W. A. Chupka, J. Berkowitz, and D. Gutman, J. Chem. Phys. **55**, 2724 (1971).

[128] B. M. Hughes, C. Lifshitz, and T. O. Tiernan, J. Chem. Phys. **59**, 3162 (1973).

[129] L. G. Christophorou, Z. Phys. Chem. **195**, 195 (1996).

[130] I. I. Fabrikant, T. Leininger, and F. X. Gadéa, J. Phys. B **33**, 4575 (2000).

[131] S. Barsotti, M.-W. Ruf, and H. Hotop, Phys. Rev. Lett. **89**, 083201 (2002).

[132] R. C. Sze, A. E. Greene, and C. A. Brau, J. Appl. Phys. **53**, 1312 (1982).

[133] N. E. Bradbury, J. Chem. Phys. **2**, 827 (1934).

[134] D. Smith, N. G. Adams, and E. Alge, J. Phys. B **17**, 461 (1984).

[135] P. J. Chantry, in *Applied Atomic Collision Physics, Vol 3: Gas Lasers*, H. S. W. Massey, E. W. McDaniel, and B. Bederson (Eds.), Academic, New York, 1982, p. 35.

[136] M. V. Kurepa, D. S. Babić, and D. S. Balić, Chem. Phys. **59**, 125 (1981).

[137] J. A. Ayala, W. E. Wentrworth, and E. C. M. Chen, J. Phys. Chem. **85**, 768 (1981).

[138] A. A. Christodoulides, R. Schumacher, and R. N. Schindler, J. Phys. Chem. **79**, 1904 (1975).

[139] E. Schultes, A. A. Christodoulides, and R. N. Schindler, Chem. Phys. **8**, 354 (1975).

[140] G. D. Sides, T. O. Tiernan, and R. J. Hanrahan, J. Chem. Phys. **65**, 1966 (1976).

[141] J. C. Han, M. Suto, J. C. Lee, and Z. Lj. Petrović, J. Appl. Phys. **68**, 2649 (1990).

[142] G. A. Hebner, J. Vac. Sci. Technol. A **14**, 2158 (1996).

[143] Y. Ota and Y. Uchida, Jpn. J. Phys. **5**, 53 (1928).

[144] A. Elliott and W. H. B. Cameron, Proc. Roy. Soc. (London) A **158**, 681 (1937).

[145] A. Elliott and W. H. B. Cameron, Proc. Roy. Soc. (London) A **164**, 531 (1938).

[146] H. G. Howell, Proc. Phys. Soc. (London) A **66**, 759 (1953).

[147] V. V. Rao and P. T. Rao, Can. J. Phys. **36**, 1557 (1958).

[148] P. B. V. Haranath and P. T. Rao, Indian J. Phys. **32**, 401 (1958).

[149] F. P. Huberman, J. Mol. Spectrosc. **20**, 29 (1966).

[150] R. S. Berry and C. W. Reimann, J. Chem. Phys. **38**, 1540 (1963).

[151] G. Pietsch and L. Rehder, Z. Naturforsch. **22a**, 2127 (1967).

[152] G. Mück and H.-P. Popp, Z. Naturforsch. **23a**, 1213 (1968).

[153] D. M. de Leeuw, R. Mooyman, and C. A. de Lange, Chem. Phys. Lett. **54**, 231 (1978).

[154] K. Kimura, T. Yamazaki and Y. Achiba, Chem. Phys. Lett. **58**, 104 (1978).

[155] R. E. Huffman, J. C. Larrabee, and Y. Tanaka, J. Chem. Phys. **47**, 856 (1967).

[156] R. E. Huffman, J. C. Larrabee, and Y. Tanaka, J. Chem. Phys. **48**, 3835 (1968).

[157] C. A. de Lange, P. Van der Meulen, and W. J. Van der Meer, J. Mol. Struct. **173**, 215 (1988).

[158] J. A. R. Samson, Y. Shefer, and G. C. Angel, Phys. Rev. Lett. **56**, 2020, 1986.

[159] D. C. Griffin, M. S. Pindzola, T. W. Gorczyca, and N. R. Badnell, Phys. Rev. A **51**, 2265 (1995).

[160] H. P. Saha, Phys. Rev. A **53**, 1553 (1996).

[161] E. J. Robinson and S. Geltman, Phys. Rev. **153**, 4 (1967).

[162] I. I. Fabrikant, J. Phys. B **27**, 4545 (1994).

[163] P. S. Ganas, J. Appl. Phys. **63**, 277 (1988).

[164] T. R. Hayes, R. C. Wetzel, and R. S. Freund, Phys. Rev. A **35**, 578 (1987).

[165] W. Lotz, Z. Phys. **216**, 241 (1968).

[166] Gopaljee, S. N. Chatterjee, and B. N. Roy, J. Phys. (India) **36**, 325 (1991).

[167] M. A. Lennon, K. L. Bell, H. B. Gilbody, J. G. Hughes, A. E. Kingston, M. J. Murray, and F. J. Smith, J. Phys. Chem. Ref. Data **17**, 1285 (1988).

[168] D. E. Rothe, Phys. Rev. **177**, 93 (1969).

[169] G. Pietsch, Z. Naturforsch. **27a**, 989 (1972).

[170] H.-P. Popp, Phys. Rep. **16**, 169 (1975).

[171] L. G. Christophorou, *Atomic and Molecular Radiation Physics*, Wiley-Interscience, New York, 1971, Ch. 7.

[172] I. Yamada, A. Danjo, T. Hirayama, A. Matsumoto, S. Ohtani, H. Suzuki, T. Takayanaki, H. Tawara, K. Wakiya, and M. Yoshino, J. Phys. Soc. Jpn. **58**, 3151 (1989).

[173] W. Shi, D. Fang, F. Lu, H. Gao, J. Gu, S. Wu, W. Wu, J. Tang, and F. Yang, Chin. Phys. Lett. **11**, 73 (1994).

[174] N. Djurić, E. W. Bell, E. Daniel, and G. H. Dunn, Phys. Rev. A **46**, 270 (1992).

[175] D. W. Mueller, T. J. Morgan, G. H. Dunn, D. C. Gregory, and D. H. Crandall, Phys. Rev. **31**, 2905 (1985).

[176] R. Rackwitz, D. Feldmann, E. Heinicke, and H. J. Kaiser, Z. Naturforsch. **29a**, 1797 (1974).

[177] O. I. Asubiojo, H. L. McPeters, W. N. Olmstead, and J. I. Brauman, Chem. Phys. Lett. **48**, 127 (1977).

[178] S. A. Sullivan, B. S. Freiser, and J. L. Beauchamp, Chem. Phys. Lett. **48**, 294 (1977).

[179] R. L. Champion and L. D. Doverspike, in *Electron-Molecule Interactions and Their Applications*, L. G. Christophorou (Ed.), Academic, New York, Vol. 1, 1984, Ch. 7.

[180] L. G. Christophorou, Contrib. Plasma Phys. **27**, 237 (1987).

[181] R. S. Berry, C. W. Reimann, and G. N. Spokes, J. Chem. Phys. **37**, 2278 (1962).

[182] A. Mandl, Phys. Rev. A **14**, 345 (1976).

[183] W. C. Wang and L. C. Lee, J. Phys. D **21**, 675 (1988).

[184] W. C. Wang and L. C. Lee, IEEE Trans. Plasma Sci. **PS-15**, 460 (1987).

[185] J. W. Cooper and J. B. Martin, Phys. Rev. **126**, 1482 (1962).

[186] Yu. V. Moskvin, High Temp. **3**, 765 (1965).

[187] V. Radojević, H. P. Kelly, and W. R. Johnson, Phys. Rev. A **35**, 2117 (1987).

[188] C. J. Howard, F. C. Fehsenfeld, and M. McFarland, J. Chem. Phys. **60**, 5086 (1974).

[189] F. C. Fehsenfeld, C. J. Howard, and E. E. Ferguson, J. Chem. Phys. **58**, 5841 (1973).

[190] F. C. Fehsenfeld, J. Chem. Phys. **54**, 438 (1971).

[191] F. C. Fehsenfeld, E. E. Ferguson, and A. L. Schmeltekopf, J. Chem. Phys. **45**, 1844 (1966).

[192] L. M. Babcock and G. E. Streit, J. Chem. Phys. **76**, 2407 (1982).

[193] L. D. Doverspike, B. T. Smith, and R. L. Champion, Phys. Rev. A **22**, 393 (1980).

[194] Iu. F. Bydin and V. M. Dukel'skii, Sov. Phys.-JETP **4**, 474 (1957).

[195] M. S. Huq, D. Scott, N. R. White, R. L. Champion, and L. D. Doverspike, J. Chem. Phys. **80**, 3651 (1984).

[196] J. B. Hasted and R. A. Smith, Proc. Roy. Soc. (London) A **235**, 349 (1956).

[197] M. J. Church and D. Smith, J. Phys. D **11**, 2199 (1978).

[198] R. K. Boyd and G. Burns, J. Phys. Chem. **83**, 88 (1979).

[199] Y.-T. Chang, C.-J. Hwang, and T.-M. Su, Chem. Phys. Lett. **114**, 92 (1985).

[200] C.-J. Hwang, R.-C. Jiang, and T.-M. Su, J. Chem. Phys. **84**, 5095 (1986).

[201] F. L. Eisele, M. G. Thackston, W. M. Pope, H. W. Ellis, and E. W. McDaniel, J. Chem. Phys. **70**, 5918 (1979).

[202] L. A. Vieland and E. A. Mason, J. Chem. Phys. **63**, 2913 (1975).

[203] H. R. Skullerud, J. Phys. B **9**, 535 (1976).

[204] E. A. Mason and E. W. McDaniel, *Transport Properties of Ions in Gases*, John Wiley & Sons, New York, 1988.

[205] M. G. Thackston, M. S. Byers, F. B. Holleman, R. D. Chelf, J. R. Twist, and E. W. McDaniel, J. Chem. Phys. **78**, 4781 (1983).

[206] G. Smolinsky, R. P. Chang, and T. M. Mayer, J. Vac. Sci. Technol. **18**, 12 (1981).

[207] F. S. Rowland and M. J. Molina, Rev. Geophys. Space Phys. **13**, 1 (1975).

[208] F. S. Rowland, Environ. Conservation **15**, 101 (1988).

[209] *The 1994 Report of the Scientific Assessment Working Group of IPCC*, Intergovernmental Panel on Climate Change, p. 28.

[210] W. Zhang, G. Cooper, T. Ibuki, and C. E. Brion, Chem. Phys. **151**, 357 (1991).

[211] T. Cvitǎs, H. Güsten, and L. Klasinc, J. Chem. Phys. **67**, 2687 (1977).

[212] A. W. Potts, I. Novak, F. Quinn, G. V. Marr, B. Dobson, I. H. Hillier, and J. B. West, J. Phys. B **18**, 3177 (1985).

[213] A. L. McClellan, *Tables of Experimental Dipole Moments*, W. H. Freeman, San Francisco, CA, 1963, p. 35.

[214] J. A. Beran and L. Kevan, J. Phys. Chem. **73**, 3860 (1969).

[215] H. W. Jochims, W. Lohr, and H. Baumgärtel, Ber. Bunsenges. Phys. Chem. **80**, 130 (1976).

[216] J. Doucet, P. Sauvageau, and C. Sandorfy, J. Chem. Phys. **58**, 3708 (1973).

[217] C. Y. R. Wu, L. C. Lee, and D. L. Judge, J. Chem. Phys. **71**, 5221 (1979).

[218] R. Gilbert, P. Sauvageau, and C. Sandorfy, J. Chem. Phys. **60**, 4820 (1974).

[219] G. C. King and J. W. McConkey, J. Phys. B **11**, 1861 (1978).

[220] W. Zhang, T. Ibuki, and C. E. Brion, Chem. Phys. **160**, 435 (1992).

[221] R. H. Huebner, D. L. Bushnell, Jr., R. J. Celotta, S. R. Mielczarek, and C. E. Kuyatt, Nature **257**, 376 (1975).

[222] J. C. Person, D. E. Fowler, and P. P Nicole, Argonne National Laboratory Report ANL-75-60 Part 1, 1975 (see Ref. [220]).

[223] A. Mann and F. Linder, J. Phys. B **25**, 1633 (1992).

[224] T. Shimanouchi, J. Phys. Chem. Ref. Data **3**, 269 (1974).

[225] H. Dispert and K. Lacmann, Int. J. Mass Spectrom. Ion Phys. **28**, 49 (1978).

[226] H. J. T. Preston and J. J. Kaufman, Chem. Phys. Lett. **50**, 157 (1977).

[227] J. J. Kaufman, H. E. Popkie, and H. J. T. Preston, Int. J. Quantum Chem. **XI**, 1005 (1977).

[228] J. A. Tossell, private communication, October 1996.

[229] A. Chutjian and S. H. Alajajian, J. Phys. B **20**, 839 (1987).

[230] L. G. Christophorou, D. L. McCorkle, and D. Pittman, J. Chem. Phys. **60**, 1183 (1974).

[231] D. L. McCorkle, A. A. Christodoulides, L. G. Christophorou, and I. Szamrej, J. Chem. Phys. **72**, 4049 (1980).

[232] Z. Lj. Petrović, W. C. Wang, and L. C. Lee, J. Chem. Phys. **90**, 3145 (1989).

[233] Y. Wang, L. G. Christophorou, and J. K. Verbrugge, J. Chem. Phys. **109**, 8304 (1998).

[234] V. M. Pejčev, M. V. Kurepa, and I. M. Čadež, Chem. Phys. Lett. **63**, 301 (1979).

[235] E. Illenberger, H.-U. Scheunemann, and H. Baumgärtel, Chem. Phys. **37**, 21 (1979).

[236] E. Illenberger, Ber. Bunsenges. Phys. Chem. **86**, 252 (1982).

[237] T. Oster, A. Kühn, and E. Illenberger, Int. J. Mass Spectrom. Ion Processes **89**, 1 (1989).

[238] T. Underwood-Lemons, T. J. Gergel, and J. H. Moore, J. Chem. Phys. **102**, 119 (1995).

[239] G. J. Verhaart, W. J. van der Hart, and H. H. Brongersma, J. Chem. Phys. **34**, 161 (1978).

[240] R. K. Jones, J. Chem. Phys. **84**, 813 (1986).

[241] P. D. Burrow, A. Modelli, N. S. Chiu, and K. D. Jordan, J. Chem. Phys. **77**, 2699 (1982).

[242] T. Underwood-Lemons, D. C. Winkler, J. A. Tossell, and J. H. Moore, J. Chem. Phys. **10**, 9117 (1994).

[243] D. Field, N. C. Jones, S. L. Lunt, J.-P. Ziesel, R. J. Gulley, J. Chem. Phys. **115**, 3045 (2001).

[244] K. Aflatooni and P. D. Burrow, Int. J. Mass Spectrom. **205**, 149 (2001).

[245] M. Hayashi, in *Swarm Studies and Inelastic Electron-Molecule Collisions*, L. C. Pitchford, B. V. McKoy, A. Chutjian and S. Trajmar (Eds.), Springer, New York, 1987, p. 167.

[246] A. Zecca, G. P. Karwasz, and R. S. Brusa, Phys. Rev. A **46**, 3877 (1992).

[247] Y. Jiang, J. Sun, and L. Wan, Phys. Rev. A **52**, 398 (1995).

[248] J. P. Novak and M. F. Fréchette, J. Appl. Phys. **57**, 4368 (1985).

[249] K. Rohr, in *Proceedings of the 11th International Conference on Physics of Electronic and Atomic Collisions, Abstracts of Papers*, North Holland, Amsterdam, 1979, p. 322. Also, unpublished results quoted in Ref. [223].

[250] J. Randell, J.-P. Ziesel, S. L. Lunt, G. Mrotzek, and D. Field, J. Phys. B. **26**, 3423 (1993).

[251] J. A. Beran and L. Kevan, J. Phys. Chem. **73**, 3866 (1969).

[252] K. Leiter, P. Scheier, G. Walder, and T. D. Märk, Int. J. Mass Spectrom. Ion Processes **87**, 209 (1989).

[253] H. Deutsch, P. Scheier, and T. D. Märk, Int. J. Mass Spectrom. Ion Processes **74**, 81 (1986).

[254] W. Lotz, Z. Phys. **206**, 205 (1967).

[255] M. Gryzinski, Phys. Rev. A **138**, 305 (1965).

[256] L. Vriens, Phys. Rev. **141**, 88 (1966).

[257] G. Elwert, Z. Naturforsch. **7a**, 432 (1952).

[258] T. D. Märk, private communication, February 1997.

[259] J. M. Ajello, W. T. Huntress, Jr., and P. Rayermann, J. Chem. Phys. **64**, 4746 (1976).

[260] F. C.-Y. Wang and G. E. Leroi, Ann. Isr. Phys. Soc. **6**, 210 (1984).

[261] R. Jadrny, L. Karlsson, L.Mattsson, and K. Siegbahn, Phys. Scr. **16**, 235 (1977).

[262] W. Kischlat and H. Morgner , J. Electron Spectrosc. Relat. Phenom. **35**, 273 (1985).

[263] K. Watanabe, T. Nakayama, and J. Mottl, J. Quant. Spectrosc. Radiat. Transfer **2**, 369 (1962).

[264] H. Schenk, H. Oertel, and H. Baumgärtel, Ber. Bunsenges. Phys. Chem. **83**, 683 (1979).

[265] M. A. Harrison and R. Gebarlle, Phys. Rev. **91**, 1 (1953).

[266] J. L. Moruzzi, Br. J. Appl. Phys. **14**, 938 (1963).

[267] C. R. Rao and G. R. G. Raju, Int. J. Electron. **35**, 49 (1973).

[268] H. A. Boyd, G. C. Crichton, and T. M. Nielsen, IEE Conference Publication No. 70, 1970, p. 426.

[269] V. N. Maller and M. S. Naidu, in *Third IEE Conference on Gas Discharges*, London, 1974 (IEE Torbridge, 1974), p. 409.

[270] V. N. Maller and M. S. Naidu, IEEE Trans. Plasma Sci. **PS-3**, 49 (1975).

[271] M. C. Siddagangappa, C. S. Lakshminarasimha, and M. N. Naidu, J. Phys. D **16**, 763 (1983).

[272] M. F. Fréchette, J. Appl. Phys. **61**, 5254 (1987).

[273] Y. Qiu, X. Ren, and Z. Y. Liu, J. Phys. D **23**, 751 (1990).

[274] M. F. Fréchette, J. Appl. Phys. **59**, 3684 (1986).

[275] Y. Qiu and X. Weng, J. Phys. D **20**, 1203 (1987).

[276] M. S. Dincer and G. R. G. Raju, IEEE Trans. Electr. Insul. **EI-20**, 595 (1985).

[277] Y. Qiu, X. Ren, Z. Y. Liu, and M. C. Zhang, J. Phys. D **22**, 1553 (1989).

[278] L. G. Christophorou, *Atomic and Molecular Radiation Physics*, Wiley-Interscience, New York, 1971, p. 38.

[279] P. Popp, J. Leonhardt, and J. I. Baumbach, Radiat. Phys. Chem. **26**, 567 (1985).

[280] I. Szamrej, W. Tchórzewska, H. Kość, and M. Foryś, Radiat. Phys. Chem. **47**, 269 (1996).

[281] K. M. Bansal and R. W. Fessenden, J. Chem. Phys. **59**, 1760 (1973).

[282] K. G. Mothes, E. Schultes, and R. N. Schindler, J. Phys. Chem. **76**, 3758 (1972).

[283] R. Schumacher, H.-R. Sprünken, A. A. Christodoulides, and R. N. Schindler, J. Phys. Chem. **82**, 2248 (1778).

[284] C. J. Marotta, C. Tsai, and D. L. McFadden, J. Chem. Phys. **91**, 2194 (1989).

[285] L. G. Christophorou, Chem. Rev. **76**, 409 (1976).

[286] L. G. Christophorou, R. A. Mathis, D. R. James, and D. L. McCorkle, J. Phys. D **14**, 1889 (1981).

[287] D. Smith and P. Španěl, Adv. At. Mol. Opt. Phys. **32**, 307 (1994).

[288] W. E. Wentworth, R. George, and H. Keith, J. Chem. Phys. **51**, 1791 (1969).

[289] L. G. Christophorou and J. K. Olthoff, Adv. At. Mol. Opt. Phys. **44**, 155 (2000).

[290] S. J. Burns, J. M. Matthews, and D. L. McFadden, J. Phys. Chem. **100**, 19436 (1996).

[291] C. L. Chen and P. J. Chantry, Bull. Am. Phys. Soc. **17**, 1133 (1972).

[292] J. L. Le Garrec, O. Sidko, J. L. Queffelec, S. Hamon, J. B. A. Mitchell, and B. R. Rowe, J. Chem. Phys. **107**, 54 (1997).

[293] A. Kiendler, S. Matejcik, J. D. Skalny, A. Stamatovic, and T. D. Märk, J. Phys. B **29**, 6217 (1996).

[294] I. Hahndorf and E. Illenberger, Int. J. Mass Spectrom. Ion Processes **167/168**, 87 (1997).

[295] L. A. Pinnaduwage, P. G. Datskos, and C. Tav, J. Appl. Phys. **84**, 3442 (1998).

[296] Von O. Rosenbaum and H. Neuert, Z. Naturforsch. **9a**, 990 (1954).

[297] W. M. Hickam and D. Berg, Adv. Mass Spectrometry **1**, 458 (1959).

[298] J. Langer, S. Matt, M. Meinke, P. Tegeder, A. Stamatovic, and E. Illenberger, J. Chem. Phys. **113**, 11063 (2000).

[299] Q.-B. Lu and T. E. Madey, Surface Science **451**, 238 (2000).

[300] Q.-B. Lu, T. E. Madey, L. Parenteau, F. Weik, and L. Sanche, Chem. Phys. Lett. **342**, 1 (2001).

[301] E. Stoffels, W. W. Stoffels, D. Vender, G. M. W. Kroesen, and F. J. de Hoog, J. Vac. Sci. Technol. A **13**, 2051 (1995).

[302] G. A. Askaryan, G. M. Batanov, A. E. Barkhudarov, S. I. Gritsinin, E. G. Korchagina, I. A. Kossyi, V. P. Silakov, and N. M. Tarasova, J. Phys. D **27**, 1311 (1994).

[303] G. M. W. Kroesen, W. W. Stoffels, E. Stoffels, M. Haverlag, J. H. W. G. den Boer, and F. J. de Hoog, Plasma Sources Sci. Technol. **3**, 246 (1994).

[304] G. S. Selwyn, L. D. Baston, and H. H. Sawin, Appl. Phys. Lett. **51**, 898 (1987).

[305] M. S. Naidu and A. N. Prasad, Brit. J. Appl. Phys. **2**, 1431 (1969).

[306] V. N. Maller and M. S. Naidu, IEEE Trans. Plasma Sci. **PS-3**, 205 (1975).

[307] V. N. Maller, IEEE-IAS Annual Meeting, New York, 1978, pp. 243-246.

[308] R. S. Nema, S. V. Kulkarni, and E. Husain, IEEE Trans. Electr. Insul. **EI-17**, 434 (1982).

[309] R. E. Wootton, S. J. Dale, and N. J. Zimmerman, in *Gaseous Dielectrics II*, L. G. Christophorou (Ed.), Pergamon, New York 1980, p.137.

[310] R. J. Van Brunt, J. Appl. Phys. **61**, 1773 (1987).

[311] G. R. G. Raju and R. Hackam, J. Appl. Phys. **53**, 5557 (1982).

[312] D. Raghavender and M. S. Naidu, in *Fifth International Symposium on High Voltage Engineering*, Braunschweig, Germany, August 1987, paper 15.14.

[313] G. Allcock and J. W. McConkey, J. Phys. B **11**, 741 (1978).

[314] H. A. Van Sprang, H. H. Brongersma, and F. J. De Heer, Chem. Phys. **35**, 51 (1978).

[315] Z. J. Jabbour and K. Becker, J. Chem. Phys. **90**, 4819 (1989).

[316] J. C. Creasey, I. R. Lambert, R. P. Tuckett, and A. Hopkirk, Mol. Phys. **71**, 1367 (1990).

[317] M. B. Roque, R. B. Siegel, K. E. Martus, V. Tarnovsky, and K. Becker, J. Chem. Phys. **94**, 341 (1991).

[318] D. L. Flamm and V. M. Donnelly, Plasma Chem. Plasma Process. **1**, 317 (1981).

[319] R. J. Saia and B. Gorowitz, Solid State Technol., April 1983, p. 247.

[320] G. J. Sonek and J. M. Ballantyryne, J. Vac. Sci. Technol. B **2**, 653 (1984).

[321] R. A. Morgan, *Plasma Etching in Semiconductor Fabrication*, Elsevier, Amsterdam, 1985.

[322] T. Sugano (Ed.), *Application of Plasma Processes to VLSI Technology*, Wiley-Interscience, New York, 1985.

[323] S. S. Cooperman, H. K. Choi, H. H. Sawin, and D. F. Kolesar, J. Vac. Sci. Technol. B **7**, 41 (1989).

[324] J. W. Lee, J. Hong, E. S. Lambers, C. R. Abernathy, S. J. Pearton, W. S. Hobson, and F. Ren, Plasma Chem. Plasma Process. **17**, 155 (1997).

[325] J. W. Coburn, Plasma Chem. Plasma Process. **2**, 1 (1982).

[326] S. J. Pearton, W. S. Hobson, C. R. Abernathy, F. Ren, T. R. Fullowan, A. Katz, and A. P. Perley, Plasma Chem. Plasma Process. **13**, 311 (1993).

[327] J. W. Lee, J. Hong, E. S. Lambers, C. R. Abernathy, S. J. Pearton, W. S. Hobson, and F. Ren, Plasma Chem. Plasma Process. **17**, 169 (1997).

[328] J. Hong, H. Cho, T. Maeda, C. R. Abernathy, S. J. Pearton, R. J. Shul, and W. S. Hobson, J. Vac. Sci. Technol. B **16**, 2690 (1998).

[329] E. Meeks, P. Ho, A. Ting, and R. J. Buss, J. Vac. Sci. Technol. A **16**, 2227 (1998).

[330] J. Hong, J. A. Caballero, E. S. Lambers, J. R. Childress, and S. J. Pearton, J. Vac. Sci. Technol. B **16**, 3349 (1998).

[331] D. L. Mosteller, Jr., M. L. Andrews, J. D. Clark, and A. Garscadden, J. Appl. Phys. **74**, 2247 (1993).

[332] V. McKoy, C. Winstead, and W. L. Morgan, *Data Compilation for Plasma Chemistries*, Technology Transfer No. 97043274A-TR, SEMATECH, August 22, 1997.

[333] G. Herzberg, *Molecular Spectra and Molecular Structure II: Infrared and Raman Spectra of Polyatomic Molecules*, Van Nostrand Reinhold Co., New York, 1945, p. 302.

[334] K. Ueda, H. Chiba, Y. Sato, T. Hayashi, E. Shigemasa, and A. Yagishita, J. Chem. Phys. **101**, 7320 (1994).

[335] L. G. Shpinkova, D. M. P. Holland, and D. A. Shaw, Molec. Phys. **96**, 323 (1999).

[336] K. K. Baeck and R. J. Bartlett, J. Chem. Phys. **106**, 4604 (1997).

[337] J. A. Tossell, J. H. Moore, and J. K. Olthoff, Int. J. Quantum Chem. **XXIX**, 1117 (1986).

[338] J. Marriott and J. D. Craggs, J. Electron. Control **3**, 194 (1957).

[339] J. A. Stockdale, D. R. Nelson, F. J. Davis, and R. N. Compton, J. Chem. Phys. **56**, 3336 (1972).

[340] C. Q. Jiao, R. Nagpal, and P. Haaland, Chem. Phys. Lett. **265**, 239 (1997).

[341] I. S. Buchel'nikova, Sov. Phys. JETP **35**, 783 (1959).

[342] E. W. Rothe, B. P. Mathur, and G. P. Reck, Inorg. Chem. **19**, 829 (1980).

[343] J. K. Olthoff, Ph.D. dissertation, University of Maryland, 1985.

[344] F. Brüning, Ph.D. dissertation, Freie Universität Berlin, 1999.

[345] Z. J. Jabbour, K. E. Martus, and K. Becker, Z. Phys. D **9**, 263 (1988).

[346] P. G. Gilbert, R. B. Siegel, and K. Becker, Phys. Rev. A **41**, 5594 (1990).

[347] I. Tokue, M. Kudo, M. Kusakabe, T. Honda, and Y. Ito, J. Chem. Phys. **96**, 8889 (1992).

[348] P. J. Bassett and D. R. Lloyd, J. Chem. Soc. A, 1551 (1971).

[349] H. Biehl, K. J. Boyle, D. M. Smith, R. P. Tuckett, K. R. Yoxall, K. Codling, P. A. Hatherly, and M. Stankiewicz, J. Chem. Soc. Faraday Trans. **92**, 185 (1996).

[350] V. H. Dibeler and J. A. Walker, Inorg. Chem. **8**, 50 (1969).

[351] M. Suto, C. Ye, J. C. Han, and L. C. Lee, J. Chem. Phys. **89**, 6653 (1988).

[352] L. C. Lee, J. C. Han, and M. Suto, J. Chem. Phys. **91**, 2036 (1989).

[353] J. C. Creasey, P. A. Hatherly, I. R. Lambert, and R. P. Tuckett, Mol. Phys. **79**, 413 (1993).

[354] S. Georgiou, E. Raptakis, X. Xing, E. Hontzopoulos, and Y. P. Vlahoyannis, J. Chem. Soc. Faraday Trans. **90**, 3633 (1994).

[355] H. Biehl, J. C. Creasey, D. M. Smith, R. P. Tuckett, K. R. Yoxall, H. Baumgärtel, H. W. Jochims, and U. Rockland, J. Chem. Soc. Faraday Trans. **91**, 3073 (1995).

[356] K. J. Boyle, D. P. Seccombe, R. P. Tuckett, H. Baumgärtel, and H. W. Jochims, J. Phys. B **32**, 2569 (1999).

[357] H. J. Maria, J. R. McDonald, and S. P. McGlynn, J. Am. Chem. Soc. **95**, 1050 (1973).

[358] A. A. Planckaert, P. Sauvageau, and C. Sandorfy, Chem. Phys. Lett. **20**, 170 (1973).

[359] A. Slaoui, F. Foulon, C. Fuchs, E. Fogarassy, and P. Siffert, Appl. Phys. A **50**, 317 (1990).

[360] E. Ishiguro, S. Iwata, Y. Suzuki, A. Mikuni, and T. Sasaki, J. Phys. B **15**, 1841 (1982).

[361] M. Cini, F. Maracci, and R. Platania, J. Electron Spectrosc. Related Phenomena **41**, 37 (1986).

[362] W. S. Koski, J. J. Kaufman, and C. F. Pachucki, J. Am. Chem. Soc. **81**, 1326 (1959).

[363] M. F. Lappert, M. R. Litzow, J. B. Pedley, P. N. K. Riley, T. R. Spalding, and A. Tweedale, J. Chem. Soc. A, 2320 (1970).

[364] H.-O. Berger, J. Kroner, and H. Nöth, Chem. Ber. **109**, 2266 (1976).

[365] D. Goutier and L. A. Burnelle, Chem. Phys. Lett. **18**, 460 (1973).

[366] R. J. Boyd and D. C. Frost, Chem. Phys. Lett. **1**, 649 (1968).

[367] O. Osberghaus, Z. Phys. **128**, 366 (1950).

[368] C. Tav, P. G. Datskos, and L. A. Pinnaduwage, J. Appl. Phys. **84**, 5805 (1998).

[369] W. A. Isaacs, C. W. McCurdy, and T. N. Rescigno, Phys. Rev. A **58**, 2881 (1998).

[370] M. H. F. Bettega, Phys. Rev. A **61**, 042703 (2000).

[371] D. R. Armstrong and P. G. Perkins, J. Chem. Soc. A, 1218 (1967).

[372] R. A. Gottscho and C. E. Gaebe, IEEE Trans. Plasma Sci. **PS-14**, 92 (1986).

[373] G. Herzberg, *Molecular Spectra and Molecular Structure II: Infrared and Raman Spectra of Polyatomic Molecules*, Van Nostrand Reinhold Co., New York, 1945, p. 178.

[374] D. R. Stull and H. Prophet, *JANAF Thermochemical Tables*, 3rd ed., NSRDS-NBS 37, June 1971.

[375] L. A. Farrow, J. Chem. Phys. **82**, 3625 (1985).

[376] M. E. Jacox, K. K. Irikura, and W. E. Thompson, J. Chem. Phys. **104**, 8871 (1996).

[377] J. H. Callomon, E. Horota, K. Kuchitsu, W. J. Lafferty, A. G. Maki, and C. S. Pole, in *Landolt-Bornstein, Numerical Data and Functional Relationships in Science and Technology, New Series, Group II Atomic and Molecular Physics*, K. H. Hellwege and A. M. Hellwege (Ed.), Springer, New York, 1976, Vol. 7.

[378] C. Leibovici, J. Mol. Struct. **14**, 459 (1972).

[379] R. Nagpal and A. Garscadden, Appl. Phys. Lett. **64**, 1626 (1994).

[380] R. C. Weast, D. R. Lide, M. J. Astle, and W. H. Beyer (Eds.), *CRC Handbook of Chemistry and Physics*, 70th edition, CRC Press Inc., Boca Raton, FL, 1990, p. E-74.

[381] R. S. Armstrong, M. J. Aroney, A. Hector, and R. J. W. Le Févre, J. Chem. Soc. B, 1203 (1968).

[382] R. I. Keir and G. L. D. Ritchie, Chem. Phys. Lett. **290**, 409 (1998).

[383] K. Iijima and S. Shibata, Bull. Chem. Soc. Jpn. **53**, 1908 (1980).

[384] R. Franzi, M. Geoffroy, E. A. Lucken, and N. Leray, J. Chem. Phys. **78**, 708 (1983).

[385] V. McKoy, C. Winstead, W. L. Morgan, and P. D. Haaland, *Data Compilation for Plasma Chemistries No. 2*, Technology Transfer No. 98063515A-TR, SEMATECH, June 30, 1998.

[386] R. W. Law and J. L. Margrave, J. Chem. Phys. **25**, 1086 (1956).

[387] R. C. Wetzel, F. A. Baiocchi, T. R. Hayes, and R. C. Freund, Phys. Rev. A **35**, 559 (1987).

[388] E. Krishnakumar and S. K. Srivastava, J. Phys. B **21**, 1055 (1988).

[389] L. J. Overzet and L. Luo, Appl. Phys. Lett. **59**, 161 (1991).

[390] H. Deutsch, K. Becker, and T. D. Märk, Int. J. Mass Spectrom. Ion Processes **167/168**, 503 (1997).

[391] Y.-K. Kim and K. K. Irikura, *Atomic and Molecular Data and Their Applications*, K. A. Berrington and K. L. Bell (Eds.), American Institute of Physics 1-56396-971-8, CP543, 2000, p. 220.

[392] M. Probst, H. Deutsch, K. Becker, and T. D. Märk, Int. J. Mass Spectrom. Ion Processes **206**, 13 (2001).

[393] K. Becker, private communication, 2000.

[394] J. E. Hesser, J. Chem. Phys. **48**, 2518 (1968).

[395] F.-W. Breitbarth, Plasma Chem. Plasma Process. **12**, 261 (1992).

[396] G. R. Scheller, R. A. Gottscho, T. Intrator, and D. B. Graves, J. Appl. Phys. **64**, 4384 (1988).

[397] F.-W. Breitbarth and E. Ducke, Contrib. Plasma Phys. **30**, 691 (1990).

[398] M. L. Mandich, C. E. Gaebe, and R. A. Gottscho, J. Chem. Phys. **83**, 3349 (1985).

[399] R. A. Gottscho, Phys. Rev. A **36**, 2233 (1987).

[400] Z. Lj. Petrović, W. C. Wang, M. Suto, J. C. Han, and L. C. Lee, J. Appl. Phys. **67**, 675 (1990).

[401] C. E. Gaebe, T. R. Hayes, and R. A. Gottscho, Phys. Rev. A **35**, 2993 (1987).

[402] C. B. Fleddermann and G. A. Hebner, J. Vac. Sci. Tech. A **15**, 1955 (1997).

[403] L. G. Christophorou, D. L. McCorkle, and A. A. Christodoulides, in *Electron-Molecule Interactions and Their Applications*, Academic, New York, 1984, Vol. 1, Ch. 6.

[404] L. G. Christophorou and P. G. Datskos, Int. J. Mass Spectrom. Ion Processes **149/150**, 59 (1995).

[405] C. A. Moore, G. P. Davis, and R. A. Gottscho, Phys. Rev. Lett. **52**, 538 (1984)

[406] H. B. Milloy and R. W. Crompton, Phys. Rev. A **15**, 1847 (1977).

[407] R. J. Nygaard, J. Fletcher, S. R. Hunter, and S. R. Foltyn, Appl. Phys. Lett. **32**, 612 (1978).

[408] L. G. Christophorou and S. R. Hunter, in *Electron-Molecule Interactions and Their Applications*, L. G. Christophorou (Ed.), Academic, New York, 1984, Vol. 2, Ch. 5.

[409] Intergovernmental Panel on Climate Change (IPCC), *Climate Change 1995*, J. T. Houghton, L. G. M. Filho, B. A. Callander, N. Harris, A. Kattenberg, and K. Maskell (Eds.), Cambridge University Press, Cambridge, New York, 1996, p. 22.

[410] M. K. W. Ko, N. D. Sze, W.-C. Wang, G. Shia, A. Goldman, F. J. Murcray, D. G. Murcray, and C. P. Rins, J. Geophys. Res. **98**, 10 499 (1993).

[411] R. A. Morris, T. M. Miller, A. A. Viggiano, J. F. Paulson, S. Solomon, and G. Reid, J. Geophys. Res. **100**, 1287 (1995).

[412] D. J. Wuebbles and A. K. Jain, in *Electrical Transmission and Distribution Systems, Sulfur Hexafluoride, and Atmospheric Effects of Greenhouse Gas Emissions Conference*, U. S. Environmental Protection Agency Conference Proceedings Report, Washington, DC, August 1995, p. 8.

[413] L. S. Geller, J. W. Elkins, J. M. Lobert, A. D. Clarke, D. F. Hurst, J. H. Butler, and R. C. Myers, Geophys. Res. Lett. **24**, 675 (1997).

[414] T. H. Teich, in *Electron and Ion Swarms*, L. G. Christophorou (Ed.), Pergamon, New York, 1981, p. 241.

[415] J. W. Gallagher, E. C. Beaty, J. Dutton, and L. C. Pitchford, J. Phys. Chem. Ref. Data **12**, 109 (1983).

[416] S. Trajmar, D. F. Register, and A. Chutjian, Phys. Rep. **97**, 219 (1983).

[417] J. P. Novak and M. F. Fréchette, J. Appl. Phys. **55**, 107 (1984).

[418] R. Morrow, IEEE Trans. Plasma Sci. **PS-14**, 234 (1986).

[419] E. Weigold, Y. Zheng, and W. von Niessen, Chem. Phys. **150**, 405 (1991).

[420] L. G. Christophorou and R. J. Van Brunt, IEEE Trans. Dielect. Electr. Insul. **2**, 952 (1995).

[421] G. Herzberg, *Molecular Spectra and Molecular Structure III. Electronic Spectra and Electronic Structure of Polyatomic Molecules*, D. Van Nostrand Company, Inc., Princeton, NJ, 1966.

[422] J. Berkowitz, *Photoabsorption, Photoionization, and Photoelectron Spectroscopy*, Academic, New York, 1979.

[423] M. B. Robin, *Higher Excited States of Polyatomic Molecules, Academic*, Orlando, FL, Vol. III, 1985.

[424] L. E. Kline, D. K. Davies, C. L. Chen, and P. J. Chantry, J. Appl. Phys. **50**, 6789 (1979).

[425] T. Yoshizawa, Y. Saki, H. Tagashira, and S. Sakamoto, J. Phys. D **12**, 1839 (1979).

[426] H. Itoh, M. Shimozuma, and H. Tagashira, J. Phys. D **13**, 1201 (1980).

[427] G. R. Govinda Raju and R. Hackam, J. Appl. Phys. **52**, 3912 (1981).

[428] J. P. Novak and M. F. Fréchette, J. Phys. D **15**, L105 (1982).

[429] M. S. Dincer and G. R. Govinda Raju, J. Appl. Phys. **54**, 6311 (1983).

[430] J. P. Novak and M. F. Fréchette, in *Gaseous Dielectrics IV*, L. G. Christophorou and M. O. Pace (Eds.), Pergamon, New York, 1984, p. 34.

[431] M. Hayashi and T. Nimura, J. Phys. D **17**, 2215 (1984).

[432] I. M. Bortnik, A. N. Kushko, and A. N. Lobanov, Sov. Phys. Tech. Phys. **30**, 1015 (1985).

[433] M. F. Fréchette and J. P. Novak, J. Phys. D **20**, 438 (1987).

[434] M. Yousfi and A. Chatwiti, J. Phys. D **20**, 1457 (1987).

[435] A. V. Phelps, in *Gaseous Dielectrics V*, L. G. Christophorou and D. W. Bouldin (Eds.), Pergamon, New York, 1987, p. 1.

[436] H. Itoh, Y. Miura, N. Ikuta, Y. Nakao, and H. Tagashira, J. Appl. Phys. **21**, 922 (1988).

[437] K. Satoh, H. Itoh, Y. Nakao, and H. Tagashira, J. Phys. D **21**, 931 (1988).

[438] A. V. Phelps and R. J. Van Brunt, J. Appl. Phys. **64**, 4269 (1988).

[439] H. Itoh, M. Kawaguchi, K. Satoh, Y. Miura, Y. Nakao, and H. Tagashira, J. Phys. D **23**, 299 (1990).

[440] M. Hayashi and S. Hara, in *Proceedings of Joint Symposium on Electron and Ion Swarms and Low-Energy Electron Scattering*, Gold Coast, Australia, July 1991, p. 109.

[441] H. Itoh, T. Miyachi, M. Kawaguchi, Y. Nakao, and H. Tagashira, J. Phys. D **24**, 277 (1991).

[442] H. Itoh, T. Matsumura, K. Satoh, H. Date, Y. Nakao, and H. Tagashira, J. Phys. D **26**, 1975 (1993).

[443] J. L. Dehmer, A. C. Parr, S. Wallace, and D. Dill, Phys. Rev. A **26**, 3283 (1982).

[444] B. M. Addison-Jones, K. H. Tan, B. W. Yates, J. N. Cutler, G. M. Bancroft, and J. S. Tse, J. Electron Spectrosc. Relat. Phenom. **48**, 155 (1989).

[445] D. M. P. Holland, D. A. Shaw, A. Hopkirk, M. A. MacDonald, and S. M. McSweeney, J. Phys. B **25**, 4823 (1992).

[446] D. M. P. Holland, M. A. MacDonald, P. Baltzer, L. Karlsson, M. Lundqvist, B. Wannberg, and W. von Niessen, Chem. Phys. **192**, 333 (1995).

[447] A. J. Yencha, D. B. Thompson, A. J. Cormack, D. R. Cooper, M. Zubek, P. Bolognesi, and G. C. King, Chem. Phys. **216**, 227 (1997).

[448] E. Fainelli, F. Maracci, R. Platania, and L. Avaldi, J. Electron Spectrosc. Relat. Phenom. **87**, 169 (1998).

[449] K. H. Sze and C. E. Brion, Chem. Phys. **140**, 439 (1990).

[450] L. C. Lee, E. Phillips, and D. L. Judge, J. Chem. Phys. **67**, 1237 (1977).

[451] J. Gustafsson, Phys. Rev. A **18**, 1481 (1978).

[452] G. Bieri, L. Åsbrink, and W. von Niessen, J. Electron Spectrosc. Relat. Phenom. **27**, 129 (1982).

[453] F. A. Gianturco and A. Jain, Phys. Rep. **143**, 347 (1986).

[454] V. I. Nefedov, J. Struct. Chem. **11**, 272 (1970).

[455] J. L. Dehmer, J. Chem. Phys. **56**, 4496 (1972).

[456] A. P. Hitchcock and M. J. Van der Wiel, J. Phys. B **12**, 2153 (1979).

[457] K. Codling, J. Chem Phys. **44**, 4401 (1966).

[458] D. Blechschmidt, R. Haensel, E. E. Koch, U. Nielsen, and T. Sagawa, Chem. Phys. Lett. **14**, 33 (1972).

[459] M. Sasanuma, E. Ishiguro, H. Masuko, Y. Morioka, and M. Nakamura, J. Phys. B **11**, 3655 (1978).

[460] J. A. Simpson, C. E. Kuyatt, and S. R. Mielczarek, J. Chem. Phys. **44**, 4403 (1966).

[461] J. F. Ying, T. A. Daniels, C. P. Mathers, H. Zhu, and K. T. Leung, J. Chem. Phys. **99**, 3390 (1993).

[462] K. Mitsuke, S. Suzuki, T. Imamura, and I. Koyano, J. Chem. Phys. **93**, 8717 (1990).

[463] T. M. Zimkina and V. A. Fomichev, Sov. Phys.-Dokl. **11**, 726 (1967).

[464] A. S. Vinogradov and T. M. Zimkina, Opt. Spectrosc. **32**, 17 (1972).

[465] E. S. Gluskin, A. A. Krasnoperova, and L. N. Mazalov, J. Struct. Chem. **18**, 156 (1977).

[466] Y. Sato, K. Ueda, H. Chiba, E. Shigemasa, and A. Yagishita, Chem. Phys. Lett. **196**, 475 (1991).

[467] K. Mochiji, K. Lee, C.-I. Ma, D. Y. Kim, M. Mahalingam, and D. M. Hanson, J. Vac. Sci. Technol. A **12**, 216 (1994).

[468] T. Masuoka and J. A. R. Samson, J. Chem. Phys. **75**, 4946 (1981).

[469] J. C. Creasey, I. R. Lambert, R. P. Tuckett, K. Codling, L. J. Frasinski, P. A. Hatherly, and M. Stankiewicz, J. Chem. Soc. Faraday Trans. **87**, 1287 (1991).

[470] D. S. Peterka, M. Ahmed, C.-Y. Ng, and A. G. Suits, Chem. Phys. Lett. **312**, 108 (1999).

[471] V. H. Dibeler and F. L. Mohler, J. Res. Natl. Bur. Stand. **40**, 25 (1948).

[472] D. M. Curtis and J. H. D. Eland, Int. J. Mass Spectrom. Ion Phys. **63**, 241 (1985).

[473] L. J. Frasinski, M. Stankiewicz, K. J. Randall, P. A. Hatherly, and K. Codling, J. Phys. B **19**, L819 (1986).

[474] V. H. Dibeler and J. A. Walker, J. Chem. Phys. **44**, 4405 (1966).

[475] M. Sasanuma, E. Ishiguro, T. Hayaisha, H. Masuko, Y. Morioka, T. Nakajima, and M. Nakamura, J. Phys. B **12**, 4057 (1979).

[476] L. Karlsson, L. Mattsson, R. Jadrny, T. Bergmark, and K. Siegbahn, Phys. Scr. **14**, 230 (1976).

[477] A. W. Potts, H. J. Lempka, D. G. Streets, and W. C. Price, Philos. Trans. Roy. Soc. London A **268**, 59 (1970).

[478] R. K. Asundi and J. D. Craggs, Proc. Phys. Soc. **83**, 611 (1964).

[479] J. H. D. Eland, F. S. Wort, P. Lablanquie, and I. Nenner, Z. Phys. D **4**, 31 (1986).

[480] S. Trajmar and A. Chutjian, J. Phys. B **10**, 2943 (1977).

[481] E. D. Nostrand and A. B. F. Duncan, J. Am. Chem. Soc. **76**, 3377 (1954).

[482] T.-K. Liu, G. Moe, and A. B. F. Duncan, J. Chem. Phys. **19**, 71 (1951).

[483] A. Chutjian, S. K. Srivastava, and S. Trajmar, in *Ninth International Conference on the Physics of Electronic and Atomic Collisions*, Seattle, WA, 1975, Vol. 1, p. 137.

[484] C. C. Turci, J. T. Francis, T. Tyliszczak, G. G. B. de Souza, and A. P. Hitchcock, Phys. Rev. A **52**, 4678 (1995).

[485] J. T. Francis, C. C. Turci, T. Tyliszczak, G. G. B.de Souza, N. Kosugi, and A. P. Hitchcock, Phys. Rev. A **52**, 4665 (1995).

[486] I. W. Fomunung, Z. Chen, and A. Z. Msezane, Phys. Rev. A **53**, 806 (1996).

[487] Z. Felfli, I. Fomunung, D. Bessis, and A. Z. Msezane, Chem. Phys. **211**, 325 (1996).

[488] K. Rohr, J. Phys. B **10**, 1175 (1977).

[489] J. Randell, D. Field, S. L. Lunt, G. Mrotzek, and J. P. Ziesel, J. Phys. B **25**, 2899 (1992).

[490] P. J. Hay, J. Am. Chem. Soc. **99**, 1013 (1977).

[491] P. J. Hay, J. Chem. Phys. **76**, 502 (1982).

[492] R. Tang and J. Callaway, J. Chem. Phys. **84**, 6854 (1986).

[493] M. Klobukowski, Z. Barandiaran, L. Seijo, and S. Huzinaga, J. Chem. Phys. **86**, 1637 (1987).

[494] J. A. D. Stockdale, R. N. Compton, and H. C. Schweinler, J. Chem. Phys. **53**, 1502 (1969).

[495] P. S. Drzaic and J. I. Brauman, Chem. Phys. Lett. **83**, 508 (1981).

[496] P. S. Drzaic and J. I. Brauman, J. Am. Chem. Soc. **104**, 13 (1982).

[497] G. L. Gutsev, Plenum, New York, 1992, p. 504 [translated from Izv. Ak. Nauk, Se. Khimi. No. 3, 641 (1992)].

[498] G. L. Gutsev, Int. J. Mass Spectrom. Ion Processes **115**, 185 (1992).

[499] K. W. Richman and A. Banerjee, Int. J. Quantum Chem.: Quantum Chem. Symp. **27**, 759 (1993).

[500] E. P. Grimsrud, S. Chowdhury, and P. Kebarle, J. Chem. Phys. **83**, 1059 (1985).

[501] E. C. M. Chen, L.-R. Shuie, E. D. D'sa, C. F. Batten, and W. E. Wentworth, J. Chem. Phys. **88**, 4711 (1988).

[502] E. C. M. Chen, J. R. Wiley, C. F. Batten, and W. E. Wentworth, J. Phys. Chem. **98**, 88 (1994).

[503] E. Miyoshi, Y. Sakai, and S. Miyoshi, J. Chem. Phys. **88**, 1470 (1988).

[504] L. G. Christophorou, D. L. McCorkle, and J. G. Carter, J. Chem. Phys. **54**, 253 (1971).

[505] L. G. Christophorou, D. L. McCorkle, and J. G. Carter, J. Chem. Phys. **57**, 2228E (1972).

[506] C. L. Chen and P. J. Chantry, J. Chem. Phys. **71**, 3897 (1979).

[507] P. W. Harland and J. C. J. Thynne, J. Phys. Chem. **73**, 4031 (1969).

[508] J. L. Dehmer, J. Siegel, and D. Dill, J. Chem. Phys. **69**, 5205 (1978).

[509] M. V. V. S. Rao and S. K. Srivastava, in *Proceedings of the XVIIIth International Conference on the Physics of Electronic and Atomic Collisions, Abstracts of Contributed Papers*, T. Andersen, B. Fastrup, F. Folkmann, and H. Knudsen (Eds.), Aarhus, Denmark, July 1993; p. 345.

[510] B. Lehmann, Z. Naturforschg. **25A**, 1755 (1970).

[511] M. Fenzlaff, R. Gerhard, and E. Illenberger, J. Chem. Phys. **88**, 149 (1988).

[512] M. S. Dababneh, Y.-F. Hsieh, W. E. Kauppila, C. K. Kwan, S. J. Smith, T. S. Stein, and M. N. Uddin, Phys. Rev. A **38**, 1207 (1988).

[513] G. Kasperski, P. Mozejko, and C. Szmytkowski, Z. Phys. D **42**, 187 (1997).

[514] N. I. Romanyuk, I. V. Chernyshova, and O. B. Shpenik, Sov. Phys. Tech. Phys. **29**, 1204 (1984).

[515] R. E. Kennerly, R. A. Bonham, and M. McMillan, J. Chem. Phys. **70**, 2039 (1979).

[516] K. Rohr, J. Phys. B **12**, L185 (1979).

[517] F. A. Gianturco, R. R. Lucchese, and N. Sanna, J. Chem. Phys. **102**, 5743 (1995).

[518] W. M. Johnstone and W. R. Newell, J. Phys. B **24**, 473 (1991).

[519] I. Gyemant, Zs. Varga, and M. G. Benedict, Int. J. Quantum Chem. **XVII**, 255 (1980).

[520] M. G. Benedict and I. Gyemant, Int. J. Quantum Chem. **XIII**, 597 (1978).

[521] D. Edelson and K. B. McAffe, J. Chem. Phys. **19**, 1311 (1951).

[522] H. H. Claassen, G. L. Goodman, J. H. Holloway, and H. Selig, J. Chem. Phys. **53**, 341 (1970).

[523] H. Brunet, IEEE J. Quantum Electron. **QE-6**, 678 (1970).

[524] K. Kim, R. S. McDowell, and W. T. King, J. Chem. Phys. **73**, 36 (1980).

[525] M. W. Chase, Jr., C. A. Davis, J. R. Downey, Jr., D. J. Frurip, R. A. McDonald, and A. N. Syverud, J. Phys. Chem. Ref. Data **14**, 1163 (1985).

[526] C. Chapados and G. Birnbaum, J. Mol. Spectrosc. **132**, 323 (1988).

[527] M. M. Hubers and J. Los, Chem. Phys. **10**, 235 (1975).

[528] D. L. Hildenbrand, J. Phys. Chem. **77**, 897 (1973).

[529] R. K. Curran, J. Chem. Phys. **34**, 1069 (1961).

[530] T. Kiang, R. C. Estler, and R. N. Zare, J. Chem. Phys. **70**, 5925 (1979).

[531] L. M. Babcock and G. E. Streit, J. Chem. Phys. **74**, 5700 (1981).

[532] A. A. Ischenko, J. D. Ewbank, and L. Schäfer, J. Chem. Phys. **98**, 4287 (1994).

[533] J. Ferch, W. Raith, and K. Schröder, J. Phys. B **15**, L175 (1982).

[534] A. Zecca, G. Karwasz, and R. S. Brusa, Chem. Phys. Lett. **199**, 423 (1992).

[535] H.-X. Wan, J. H. Moore, J. K. Olthoff, and R. J. Van Brunt, Plasma Chem. Plasma Process. **13**, 1 (1993).

[536] S. K. Srivastava, S. Trajmar, A. Chutjian, and W. Williams, J. Chem. Phys. **64**, 2767 (1976).

[537] T. Sakae, S. Sumiyoshi, E. Murakami, Y. Matsumoto, K. Ishibashi, and A. Katase, J. Phys. B **22**, 1385 (1989).

[538] H. Cho, R. J. Gulley, and S. J. Buckman, J. Phys. B **33**, L309 (200).

[539] H. Cho, R. J. Gulley, K. W. Trantham, L. J. Uhlmann, C. J. Dedman, and S. J. Buckman, J. Phys. B **33**, 3531 (2000).

[540] Y. Jiang, J. Sun, and L. Wan, Phys. Lett. A **231**, 231 (1997).

[541] J. A. Rees, in *Gaseous Dielectrics III* , L. G. Christophorou (Ed.), Pergamon, New York, 1982, p. 1.

[542] D. F. Register, S. Trajmar, and S. K. Srivastava, Phys. Rev. A **21**, 1134 (1980).

[543] R. K. Nesbet, Phys. Rev. A **20**, 58 (1979).

[544] R. H. J. Jansen, F. J. de Heer, H. J. Luyken, B. van Wingerden, H. J. Blaauw, J. Phys. B **9**, 185 (1976).

[545] J. C. Gibson, M. A. Green, K. W. Trantham, S. J. Buckman, P. J. O. Teubner, and M. J. Brugner, J. Phys. B **32**, 213 (1999).

[546] F. H. Read and J. M. Channing, Rev. Sci. Instrum. **67**, 2372 (1996).

[547] D. Klar, M.-W. Ruf, and H. Hotop, Aust. J. Phys. **45**, 263 (1992).

[548] D. Klar, M.-W. Ruf, and H. Hotop, Meas. Sci. Technol. **5**, 1248 (1994).

[549] J. L. Forand, K. Becker, and J. W. McConkey, Can. J. Phys. **64**, 269 (1986).

[550] K. A. Blanks, A. E. Tabor, and K. Becker, J. Chem. Phys. **86**, 4871 (1987).

[551] T. Stanski and B. Adamczyk, Int. J. Mass Spectrom. Ion Phys. **46**, 31 (1983).

[552] D. Margreiter, G. Walder, H. Deutsch, H. U. Poll, C. Winkler, K. Stephan, and T. D. Märk, Int. J. Mass Spectrom. Ion Processes **100**, 143 (1990).

[553] M. V. V. S. Rao and S. K. Srivastava, in *Proceedings of the XXth International Conference on the Physics of Electronic and Atomic Collisions, Scientific Program and Abstracts of Contributed Papers*, Vienna, Austria, July 1997, Vol. II, Paper MO 151.

[554] R. Rejoub, D. R. Sieglaff, B. G. Lindsay, and R. F. Stebbings, J. Phys. B **34**, 1289 (2001).

[555] D. Margreiter, H. Deutsch, M. Schmidt, and T. D. Märk, Int. J. Mass Spectrom. Ion Processes **100**, 157 (1990).

[556] W. Hwang, Y.-K. Kim, and M. E. Rudd, J. Chem. Phys. **104**, 2956 (1996).

[557] V. Tarnovsky, H. Deutsch, S. Matt, T. D. Märk, R. Basner, M. Schmidt, and K. Becker, in *Gaseous Dielectrics VIII*, L. G. Christophorou and J. K. Olthoff (Eds.), Plenum, New York, 1998, p. 3.

[558] V. Tarnovsky, H. Deutsch, K. E. Martus, and K. Becker, J. Chem. Phys. **109**, 6596 (1998).

[559] Y.-K. Kim and M. E. Rudd, Comments At. Mol. Phys. **34**, 309 (1999).

[560] A. H. Al-Nasir, M. A. Chaudhry, A. J. Duncan, R. Hippler, D. M. Campbell, and H. Kleinpoppen, J. Phys. B **29**, 1849 (1996).

[561] B. P. Pullen and J. A. D. Stockdale, Int. J. Mass Spectrom. Ion Phys. **19**, 35 (1976).

[562] H. Shibata, T. Tonuma, T. Matsuo, H. Kumagai, and H. Tawara, in *VIth International Conference on the Physics of Highly Charged Ions*, P. Richard,

M. Stöckli, C. L. Cocke, and C. D. Lin (Eds.), American Institute of Physics (AIP) Conf. Proc. 274, AIP, New York, 1993, p. 335.

[563] L. G. Christophorou, J. K. Olthoff, and M. V. V. S. Rao, J. Phys. Chem. Ref. Data **25**, 1341 (1996).

[564] L. G. Christophorou and J. K. Olthoff, J. Phys. Chem. Ref. Data **28**, 967 (1999).

[565] L. G. Christophorou and J. K. Olthoff, J. Phys. Chem. Ref. Data **27**, 1 (1998).

[566] L. G. Christophorou and J. K. Olthoff, J. Phys. Chem. Ref. Data **27**, 889 (1998).

[567] Y. H. Hilal and L. G. Christophorou, J. Phys. D **20**, 975 (1987).

[568] I. Lopes, H. Hilmert, and W. F. Schmidt, J. Phys. D **19**, L107 (1986).

[569] L. G. Christophorou, *Atomic and Molecular Radiation Physics*, Wiley-Interscience, New York, 1971, Ch. 2.

[570] M. A. Ali, K. K. Irikura, and Y.-K. Kim, Int. J. Mass Spectrom. **201**, 187 (2000).

[571] M. Iio, M. Goto, H. Toyoda, and H. Sugai, Contrib. Plasma Phys. **35**, 405 (1995).

[572] S. G. Lias, J. E. Bartmess, J. F. Liebman, J. L. Holmes, R. D. Levin, and W. G. Mallard, J. Phys. Chem. Ref. Data **17**, 1 (1988).

[573] R. Geballe and M. A. Harrison as reported by L. B. Loeb, *Basic Processes of Gaseous Electronics* (University of California Press, Berkeley, Los Angeles, 1955), Ch. 5, p. 415.

[574] M. S. Bhalla and J. D. Craggs, Proc. Phys. Soc. (London), **80**, 151 (1962).

[575] H. A. Boyd and G. C. Crichton, Proc. IEE **118**, 1872 (1971).

[576] I. M. Bortnik and A. A. Panov, Sov. Phys.-Tech. Phys. **16**, 571 (1971).

[577] V. N. Maller and M. S. Naidu, Proc. IEE **123**, 107 (1976).

[578] M. Shimozuma, H. Itoh, and H. Tagashira, J. Phys. D **15**, 2443 (1982).

[579] M. Hayashi and G. Wang, unpublished results, 1987, as quoted by Ref. [580].

[580] M. Hayashi, private communication, 1999.

[581] Y. Qiu and D. M. Xiao, J. Phys. D **27**, 2663 (1994).

[582] T. H. Teich and R. Sangi, in *Proceedings of the First International Symposium on High Voltage Technology*, F. Heidbromer (Ed.), Munich 1972, Vol. 1, p. 391.

[583] Th. Aschwanden, in *Gaseous Dielectrics IV*, L. G. Christophorou and M. O. Pace (Eds.), Pergamon, New York, 1984, p. 24.

[584] Th. Aschwanden, Ph.D. dissertation, ETH, Zürich, 1985.

[585] J. de Urquijo-Carmona, C. Cisneros, and I. Alvarez, J. Phys. D **18**, 2017 (1985).

[586] J. de Urquijo-Carmona, I. Alvarez, and C. Cisneros, J. Phys. D **19**, L207 (1986).

[587] H. Hasegawa, A. Taneda, K. Murai, M. Shimozuma, and H. Tagashira, J. Phys. D **21**, 1745 (1988).

[588] H. Toyoda, private communication, 2000.

[589] J. J. Corr, M. A. Khakoo, and J. W. McConkey, J. Phys. B **20**, 2597 (1987).

[590] K. A. Blanks and K. Becker, J. Phys. B **20**, 6157 (1987).

[591] T. Field and J. H. D. Eland, Chem. Phys. Lett. **197**, 542 (1992).

[592] T. H. Teich and R. Bräunlich, in *Gaseous Dielectrics IV*, L. G. Christophorou and M. O. Pace (Eds.), Pergamon, New York, 1984, p. 71.

[593] K. Masek, L. Laska, V. Perina, and J. Krasa, Acta Phys. Slovaca **33**, 145 (1983).

[594] R. J. Van Brunt and J. T. Herron, IEEE Trans. Electr. Insul. **25**, 55 (1990).

[595] D. Edelson, J. E. Griffiths, and K. B. McAfee, Jr., J. Chem. Phys. **37**, 917 (1962).

[596] R. N. Compton, L. G. Christophorou, G. S. Hurst, and P. W. Reinhardt, J. Chem. Phys. **45**, 4634 (1966).

[597] P. W. Harland and J. C. J. Thynne, J. Phys. Chem. **75**, 3517 (1971).

[598] L. G. Christophorou, *Atomic and Molecular Radiation Physics*, Wiley-Interscience, New York, 1971, Ch. 6.

[599] L. G. Christophorou, Adv. Electron. Electron Phys. **46**, 55 (1978).

[600] J. M. S. Henis and C. A. Mabie, J. Chem. Phys. **53**, 2999 (1970).

[601] R. W. Odom, D. L. Smith, and J. H. Futrell, J. Phys. B **8**, 1349 (1975).

[602] M. S. Foster and J. L. Beauchamp, Chem. Phys. Lett. **31**, 482 (1975).

[603] C. E. Klots, J. Chem. Phys. **46**, 1197 (1967).

[604] P. M. Collins, L. G. Christophorou, E. L. Chaney, and J. G. Carter, Chem. Phys. Lett. **4**, 646 (1970).

[605] L. G. Christophorou, A. Hadjiantoniou, and J. G. Carter, J. Chem. Soc., Faraday Trans. **2** 69, 1713 (1975).

[606] C. D. Cooper and R. N. Compton, J. Chem. Phys. **59**, 3550 (1973).

[607] J. P. Johnson, D. L. McCorkle, L. G. Christophorou, and J. G. Carter, J. Chem. Soc., Faraday Trans. **2**, 71, 1742 (1975).

[608] S. M. Spyrou, I. Sauers, and L. G. Christophorou, J. Chem. Phys. **78**, 7200 (1983).

[609] J. E. Delmore and A. D. Appelhans, J. Chem. Phys. **84**, 6238 (1986).

[610] S. P. Heneghan and S. W. Benson, Int. J. Chem. Kinet. **15**, 109 (1983).

[611] J.-L. Le Garrec, D. A. Steinhurst, and M. A. Smith, J. Chem. Phys. **114**, 8831 (2001).

[612] R. K. Curran, J. Chem. Phys. **38**, 780 (1963).

[613] R. N. Compton, L. G. Christophorou, and R. H. Huebner, Phys. Lett. **23**, 656 (1966).

[614] R. N. Compton, R. H. Huebner, P. W. Reinhardt, and L. G. Christophorou, J. Chem. Phys. **48**, 901 (1968).

[615] L. G. Christophorou, D. L. McCorkle, and J. G. Carter, J. Chem. Phys. **60**, 3779 (1974).

[616] W. M. Hickam and R. E. Fox, J. Chem. Phys. **25**, 642 (1956).

[617] W. M. Hickam and D. Berg, J. Chem. Phys. **29**, 517 (1958).

[618] D. Rapp and D. D. Briglia, J. Chem. Phys. **43**, 1480 (1965).

[619] D. Spence and G. J. Schulz, J. Chem. Phys. **58**, 1800 (1973).

[620] J. K. Olthoff, R. J. Van Brunt, H.-X. Wan, J. H. Moore, and J. A. Tossell, in *Gaseous Dielectrics VI*, L. G. Christophorou and I. Sauers (Eds.), Plenum, New York, 1991, p. 19.

[621] K. S. Gant and L. G. Christophorou, J. Chem. Phys. **65**, 2977 (1976).

[622] K. S. Gant, Ph.D. dissertation, University of Tennessee, 1976.

[623] R. Y. Pai, L. G. Christophorou, and A. A. Christodoulides, J. Chem. Phys. **70**, 1169 (1979).

[624] S. R. Hunter, J. G. Carter, and L. G. Christophorou, J. Chem. Phys. **90**, 4879 (1989).

[625] J. M. Ajello and A. Chutjian, J. Chem. Phys. **65**, 5524 (1976).

[626] A. Chutjian, Phys. Rev. Lett. **46**, 1511 (1981); **48**, 289E (1982).

[627] A. J. Chutjian, J. Phys. Chem. **86**, 3518 (1982).

[628] A. Chutjian, S. H. Alajajian, and O. J. Orient, in *Gaseous Dielectrics IV*, L. G. Christophorou and M. O. Pace (Eds.), Pergamon, New York, 1984, p. 42.

[629] A. Chutjian and S. H. Alajajian, Phys. Rev. A **31**, 2885 (1985).

[630] K. Harth, M.-W. Ruf, and H. Hotop, Z. Phys. D **14**, 149 (1989).

[631] D. Klar, M.-W. Ruf, and H. Hotop, Chem. Phys. Lett. **189**, 448 (1992).

[632] A. Schramm, J. M. Weber, J. Kreil, D. Klar, M.-W. Ruf, and H. Hotop, Phys. Rev. Lett. **81**, 778 (1998).

[633] W. P. West, G. W. Foltz, F. B. Dunning, C. J. Latimer, and R. F. Stebbings, Phys. Rev. Lett. **36**, 854 (1976).

[634] G. W. Foltz, C. J. Latimer, G. F. Hildebrandt, F. G. Kellert, K. A. Smith, W. P. West, F. B. Dunning, and R. F. Stebbings, J. Chem. Phys. **67**, 1352 (1977).

[635] J. P. Astruc, R. Barbé, and J. P. Schermann, J. Phys. B **12**, L377 (1979).

[636] B. G. Zollars, K. A. Smith, and F. B. Dunning, J. Chem. Phys. **81**, 3158 (1984).

[637] B. G. Zollars, C. Higgs, F. Lu, C. W. Walter, L. G. Gray, K. A. Smith, F. B. Dunning, and R. F. Stebbings, Phys. Rev. A **32**, 3330 (1985).

[638] B. G. Zollars, C. W. Walter, F. Lu, C. B. Johnson, K. A. Smith, and F. B. Dunning, J. Chem. Phys. **84**, 5589 (1986).

[639] F. B. Dunning, J. Phys. Chem. **91**, 2244 (1987).

[640] I. M. Beterov, G. L. Vasilenko, I. I. Riabtsev, B. M. Smirnov, and N. V. Fatayev, Z. Phys. D **6**, 55 (1987).

[641] Z. Zheng, K. A. Smith, and F. B. Dunning, J. Chem. Phys. **89**, 6295 (1988).

[642] H. S. Carman, Jr., C. E. Klots, and R. N. Compton, J. Chem. Phys. **90**, 2580 (1989).

[643] A. Pesnelle, C. Ronge, M. Perdrix, and G. Watel, Phys. Rev. A **42**, 273 (1990).

[644] X. Ling, B. G. Lindsay, K. A. Smith, and F. B. Dunning, Phys. Rev. A **45**, 242 (1992).

[645] R. A. Popple, M. A. Durham, R. W. Marawar, B. G. Lindsay, K. A. Smith, and F. B. Dunning, Phys. Rev. A **45**, 247 (1992).

[646] A. Pesnelle, M. Perdrix, and G. Watel, J. Chem. Phys. **96**, 4303 (1992).

[647] D. Klar, B. Mirbach, H. J. Korsch, M.-W. Ruf, and H. Hotop, Z. Phys. D **31**, 235 (1994).

[648] M. T. Frey, S. B. Hill, K. A. Smith, F. B. Dunning, and I. I. Fabrikant, Phys. Rev. Lett. **75**, 810 (1995).

[649] F. B. Dunning, J. Phys. B **28**, 1645 (1995).

[650] R. W. Crompton and G. N. Haddad, Aust. J. Phys. **36**, 15 (1983).

[651] H. Hotop, D. Klar, J. Kreil, M.-W. Ruf, A. Schramm, and J. M. Weber, in *The physics of Electronic and Atomic Collisions*, L. J. Dubé, J. B. A. Mitchell, J. W. McConkey, and C. E. Brion (Eds.), AIP Press, Woodbury, NY [AIP Conf. Proc. 360, 267 (1995)].

[652] H. Hotop, private communication, March 2000.

[653] Z. Lj. Petrović and R. W. Crompton, J. Phys. B **18**, 2777 (1985).

[654] E. P. Wigner, Phys. Rev. **73**, 1002 (1948).

[655] A. Chutjian, private communication, 1999.

[656] S. Matejcik, P. Eichberger, B. Plunger, A. Kiendler, A. Stamatovic, and T. D. Märk, Int. J. Mass Spectrom. Ion Processes **144**, L13 (1995).

[657] A. Schramm, Ph.D. dissertation, University of Kaiserslautern, 1998.

[658] M. V. V. S. Rao, private communication, 1999.

[659] G. K. Jarvis, R. A. Kennedy, and C. A. Mayhew, Int. J. Mass Spectrom. **205**, 253 (2001).

[660] V. K. Lakdawala and J. L. Moruzzi, J. Phys. D **13**, 1439 (1980).

[661] S. R. Hunter and L. G. Christophorou, in *Electron-Molecule Interactions and Their Applications*, L. G. Christophorou (Ed.), Academic, New York, 1984, Vol. 2, Ch. 3.

[662] A. B. Wedding, H. A. Blevin, and J. Fletcher, J. Phys. D **18**, 2361 (1985).

[663] H. Shimamori, Y. Tatsumi, Y. Ogawa, and T. Sunagawa, J. Chem. Phys. **97**, 6335 (1992).

[664] J. Sherwell, R. Cooper, D. C. Nguyen, and S. P. Mezyk, Austr. J. Chem. **41**, 1491 (1988).

[665] F. J. Davis and D. R. Nelson, Chem. Phys. Lett. **6**, 277 (1970).

[666] H. Shimamori and R. W. Fessenden, J. Chem. Phys. **71**, 3009 (1979).

[667] F. C. Fehsenfeld, J. Chem. Phys. **53**, 2000 (1970).

[668] T. M. Miller, A. E. S. Miller, J. F. Paulson, and X. Liu, J. Chem. Phys. **100**, 8841 (1994).

[669] E. C. M. Chen, R. D. George, and W. E. Wentworth, J. Chem. Phys. **49**, 1973 (1968).

[670] I. Szamrej and M. Forys, in *Gaseous Dielectrics VI*, L. G. Christophorou and I. Sauers (Eds.), Plenum, New York, 1991, p. 43.

[671] K. G. Mothes and R. N. Schindler, Ber. Bunsen-Ges. **75**, 938 (1971).

[672] J. A. Ayala, W. E. Wentworth, and E. C. M. Chen, J. Phys. Chem. **85**, 3989 (1981).

[673] B. H. Mahan and C. E. Young, J. Chem. Phys. **44**, 2192 (1966).

[674] D. Smith, P. Španěl, S. Matejcik, A. Stamatovic, T. D. Märk, T. Jaffke, and E. Illenberger, Chem. Phys. Lett. **240**, 481 (1995).

[675] I. M. Beterov, V. P. Chebotaev, N. V. Fateev, and D. V. Yakovin, Sov. J. Quantum Electron. **8**, 533 (1978).

[676] I. M. Beterov and N. V. Fateyev, J. Phys. Colloq. **44**, C7-447 (1983).

[677] S. Avrillier and J.-P. Schermann, Opt. Commun. **19**, 87 (1976).

[678] P. G. Datskos, L. G. Christophorou, and J. G. Carter, J. Chem. Phys. **99**, 8607 (1993).

[679] K. B. McAfee, Jr., and D. Edelson, Proc. Phys. Soc. (London) **81**, 382 (1963).

[680] G. R. G. Raju and M. S. Dincer, J. Appl. Phys. **53**, 8562 (1982).

[681] M. C. Siddagangappa, C. S. Lakshminarasimha, and M. S. Naidu, J. Phys. D **15**, L83 (1982).

[682] M. C. Siddagangappa, C. S. Lakshminarasimha, and M. S. Naidu, in *Gaseous Dielectrics IV*, L. G. Christophorou and M. O. Pace (Eds.), Pergamon, New York, 1984, p. 49.

[683] B. C. O'Neill and J. D. Craggs. J. Phys. B **6**, 2634 (1973).

[684] H. Itoh, M. Shimozuma, H. Tagashira, and S. Sakamoto, J. Phys. D **12**, 2167 (1979).

[685] D. M. Xiao, H. L. Liu, and Y. Z. Chen, J. Appl. Phys. **86**, 6611 (1999).

[686] D. M. Xiao, L. L. Zhu, and Y. Z. Chen, J. Phys. D **32**, L18 (1999).

[687] T. H. Teich and D. W. Branston, in *Third International Conference on Gas Discharges*, IEE Conf. Publ. **118**, 109 (1974).

[688] H. Hilmert, W. F. Schmidt, Y. Sakai, and S. Sawada, Jpn. J. Appl. Phys. **30**, 1487 (1991).

[689] D. M. Xiao, H. L. Liu, and L. Qin, Jpn. J. Appl. Phys. **38**, L875 (1999).

[690] D. M. Xiao, L. L. Zhu, and X. G. Li, J. Phys. D **33**, L145 (2000).

[691] J. de Urquijo, E. Basurto, and J. L. Hernández-Ávila, J. Phys D **34**, 2151 (2001).

[692] D. M. Xiao, X. G. Li, and X. Xu, J. Phys. D **34**, L133 (2001).

[693] J. L. Hernández-Ávila, E. Basurto, and J. de Urquijo, in *Proceedings of the XXVth International Conference on Phenomena in Ionized Gases, Contributed Papers*, T. Goto (Ed.), Nagoya, Japan, July 2001, Vol. 3, p. 215.

[694] J. de Urquijo, C. Cisneros, I. Alvarez, and E. Basurto, in *Proceedings of the XXIVth International Conference on Phenomena in Ionized Gases, Contributed Papers*, P. Pisarczyk, T. Pisarczyk, and J. Wołowski (Eds.), Warsaw, Poland, July 1999, Vol. I, p. 67.

[695] F. M. Harris and G. J. Jones, J. Phys. B **4**, 1536 (1971).

[696] B. Sangi, Ph.D. thesis, University of Manchester, 1971, as quoted in Ref. [414].

[697] M. S. Naidu and A. N. Prasad, J. Phys. D **5**, 1090 (1972).

[698] D. W. Branston, Ph.D. thesis, University of Manchester, 1973, as quoted in Ref. [414].

[699] J. de Urquijo-Carmona, Ph.D. thesis, University of Manchester, 1980, as quoted in Ref. [414].

[700] Y. Nakamura, J. Phys. D **21**, 67 (1988).

[701] V. A. Lisovskiy and V. D. Yegorenkov, in *Proceedings of the International Symposium on Electron-Molecule Collisions and Swarms*, Y. Hatano, H. Tanaka and N. Kouchi (Eds.), July 1999, Tokyo, Japan, p. 156; J. Phys. D **32**, 2645 (1999).

[702] B. Shizgal, J. Phys. B **21**, 1699 (1988).

[703] M. S. Dincer, J. Phys. D **23**, 622 (1990).

[704] S. Chowdhury, E. P. Grimsrud, T. Heinis, and P. Kebarle, J. Am. Chem. Soc. **108**, 3630 (1986).

[705] J. K. Olthoff, R. J. Van Brunt, Y. Wang, R. L. Champion, and L. D. Doverspike, J. Chem. Phys. **91**, 2261 (1989).

[706] R. L. Champion, in *Gaseous Dielectrics VI*, L. G. Christophorou and I. Sauers (Eds.) Plenum, New York, 1991, p.1.

[707] B. Freiser and J. L. Beauchamp, quoted in Ref. [491].

[708] R. S. Mock and E. P. Grimsrud, Chem. Phys. Lett. **184**, 99 (1991).

[709] H. Faidas, L. G. Christophorou, and D. L. McCorkle, Chem. Phys. Lett. **193**, 487 (1992).

[710] L. G. Christophorou, P.G. Datskos, and H. Faidas, J. Chem. Phys. **101**, 6728 (1994).

[711] O. Ingólfsson, E. Illenberger, and W. F. Schmidt, Int. J. Mass Spectrom. Ion Processes **139**, 103 (1994).

[712] P. G. Datskos, J. G. Carter, and L. G. Christophorou, Chem. Phys. Lett. **239**, 38 (1995).

[713] I. Ishikawa, K. Koike, T. Akitsu, S. Suganomata, and H. Matsuzawa, Jpn. J. Appl. Phys. **29**, 767 (1990).

[714] A. Kono, M. Endo, K. Ohata, S. Kishimoto, and T. Goto, J. Appl. Phys. **76**, 7221 (1994).

[715] W. F. Schmidt and R. J. Van Brunt, in *Gaseous Dielectrics III*, L. G. Christophorou (Ed.), Pergamon, New York, 1982, p. 561.

[716] R. J. Van Brunt and M. Misakian, J. Appl. Phys. **54**, 3074 (1983).

[717] N. Wiegart, IEEE Trans. Electr. Insul. **EI-20**, 587 (1985).

[718] J. Kindersberger, in *Proceedings of the VIIIth International Conference on Gas Discharges and Their Applications*, September 1985, Oxford, U.K., University of Leeds Press, 1985, p. 263.

[719] Y. Wang, R. L. Champion, L. D. Doverspike, J. K. Olthoff, and R. J. Van Brunt, J. Chem. Phys. **91**, 2254 (1989).

[720] S. Chowdhury and P. Kebarle, J. Chem. Phys. **85**, 4989 (1986).

[721] R. L. Champion, I. V. Dyakov, B. L. Peko, and Y. Wang, J. Chem. Phys. **115**, 1765 (2001).

[722] P. L. Patterson, J. Chem. Phys. **53**, 696 (1970).

[723] K. P. Brand and H. Jungblut, J. Chem. Phys. **78**, 1999 (1983).

[724] D. Hansen, H. Jungblut and W. F. Schmidt, J. Phys. D **16**, 1623 (1983).

[725] H. Hilmert and W. F. Schmidt, J. Phys. D **24**, 915 (1991).

[726] N. Wiegart, L. Niemeyer, F. Pinnekamp, W. Boeck, J. Kindersberger, R. Morrow, W. Zaengl, M. Zwicky, I. Galimberti, and S. A. Boggs, IEEE Trans. Power Del. **3**, 923 (1988); **3**, 931 (1988); **3**, 939 (1988).

[727] R. J. Van Brunt, J. Appl. Phys. **59**, 2314 (1986).

[728] G. Berger and B. Senouci, J. Phys. D **19**, 2337 (1986).

[729] A. Gilbert, J. Dupuy, P. Domens, G. Riquel, and B. Hutzler, J. Phys. D **26**, 773 (1993).

[730] H. Schlumbohm, Z. Phys. **166**, 192 (1962).

[731] I. A. Fleming and J. A. Rees, J. Phys. B **2**, 777 (1969).

[732] M. S. Naidu and A. N. Prasad, J. Phys. D **3**, 951 (1970).

[733] B. H. Crichton and D.-i. Lee, in *Proceedings of the 5th International Conference on Gas Discharges*, Liverpool, U.K., 1978, p. 254.

[734] Th. Aschwanden (private communication, 1984), as quoted in Morrow [418].

[735] J. de Urquijo-Carmona, I. Alvarez, H. Martinez, and C. Cisneros, J. Phys. D **24**, 664 (1991).

[736] J. de Urquijo-Carmona, Ph.D. thesis, University of Manchester, 1980, as quoted in Morrow [418].

[737] J. de Urquijo-Carmona, I. Alvarez, C. Cisneros, and H. Martinez, J. Phys. D **23**, 778 (1990).

[738] J. de Urquijo-Carmona, C. Cisneros, H. Martinez, and I. Alvarez, J. Phys. D **25**, 1277 (1992).

[739] Z. A. Talib and M. Saporoschenko, Int. J. Mass Spectrom. Ion Processes **116**, 1 (1992).

[740] Z. A. Talib and M. Saporoschenko, J. Phys. D **27**, 2307 (1994).

[741] D. E. Wilson, W. J. Quiring, and D. A. Armstrong, J. Appl. Phys. **47**, 1194 (1976).

[742] W. F. Schmidt, H. Jungblut, D. Hansen, and H. Tagashira, in *Gaseous Dielectrics II*, L. G. Christophorou (Ed.), Pergamon, 1980, p. 1.

[743] H. Jungblut, D. Hansen, and W. F. Schmidt, IEEE Trans. Electr. Insul. **24**, 343 (1989).

Index

About the Authors

LOUCAS G. CHRISTOPHOROU was with the Oak Ridge National Laboratory and the University of Tennessee from 1963 to 1994. At Oak Ridge National Laboratory he served as Head of the Atomic, Molecular, and High Voltage Physics Group, and was appointed Corporate Fellow in 1981 and Senior Corporate Fellow in 1991. At the University of Tennessee he has been a Ford Foundation Professor from 1969 to 1997. From 1995 to 2002 he has been with the National Institute of Standards and Technology in Gaithersburg, Maryland. He has conducted extensive and broad research in atomic, molecular, radiation, chemical, electron, and ion physics and on materials for radiation detectors, pulsed power, gaseous dielectrics, and plasma processing. He is the author /editor of 17 books and the author/co-author of about 400 research papers in these areas. He is a Fellow of the American Physical Society, a Fellow of the American Association for the Advancement of Science, a Permanent Member of the Academy of Athens (Greece), and a recipient of the Alexander von Humboldt Senior Scientist Award (Germany) and the Phoenix Medal (Greece).

JAMES K. OLTHOFF obtained his Ph.D. in physics in 1985 from the University of Maryland, and then served as a postdoctoral fellow at The Johns Hopkins School of Medicine. He has been at the National Institute of Standards and Technology in Gaithersburg, Maryland since 1987. While at NIST, he has worked as a research physicist and as the leader of the Electrical Systems Group. He is now the Chief of the Electricity Division of the Electronics and Electrical Engineering Laboratory. He has conducted extensive research related to electron-molecule interactions, mass spectrometry, gaseous dielectrics, low temperature plasma physics, and plasma processing. He is the author/coauthor of over 130 research papers in these areas, and is the coeditor of two books.

Series Publications

Below is a chronological listing of all the published volumes in the *Physics of Atoms and Molecules* series.

ELECTRON AND PHOTON INTERACTIONS WITH ATOMS
Edited by H. Kleinpoppen and M. R. C. McDowell

ATOM–MOLECULE COLLISION THEORY: A Guide for the Experimentalist
Edited by Richard B. Bernstein

COHERENCE AND CORRELATION IN ATOMIC COLLISIONS
Edited by H. Kleinpoppen and J. F. Williams

VARIATIONAL METHODS IN ELECTRON–ATOM SCATTERING THEORY
R. K. Nesbet

DENSITY MATRIX THEORY AND APPLICATIONS
Karl Blum

INNER-SHELL AND X-RAYS PHYSICS OF ATOMS AND SOLIDS
Edited by Derek J. Fabian, Hans Kleinpoppen, and Lewis M. Watson

INTRODUCTION TO THE THEORY OF LASER–ATOM INTERACTIONS
Marvin H. Mittleman

ATOMS IN ASTROPHYSICS
Edited by P. G. Burke, W. B. Eissner, D. G. Hummer, and I. C. Percival

ELECTRON–ATOM AND ELECTRON–MOLECULE COLLISIONS
Edited by Juergen Hinze

ELECTRON–MOLECULE COLLISIONS
Edited by Isao Shimamura and Kazuo Takayanagi

ISOTOPE SHIFTS IN ATOMIC SPECTRA
W. H. King

AUTOIONIZATION: Recent Developments and Applications
Edited by Aaron Temkin

ATOMIC INNER-SHELL PHYSICS
Edited by Bernd Crasemann

COLLISIONS OF ELECTRONS WITH ATOMS AND MOLECULES
G. F. Drukarev

THEORY OF MULTIPHOTON PROCESSES
Farhad H. M. Faisal

PROGRESS IN ATOMIC SPECTROSCOPY, Parts A, B, C, and D
Edited by W. Hanle, H. Kleinpoppen, and H. J. Beyer

RECENT STUDIES IN ATOMIC AND MOLECULAR PROCESSES
Edited by Arthur E. Kingston

QUANTUM MECHANICS VERSUS LOCAL REALISM: The Einstein-Podolsky-Rosen Paradox
Edited by Franco Selleri

ZERO-RANGE POTENTIALS AND THEIR APPLICATIONS IN ATOMIC PHYSICS
Yu. N. Demkov and V. N. Ostrovskii

COHERENCE IN ATOMIC COLLISION PHYSICS
Edited by H. J. Beyer, K. Blum, and R. Hippler

ELECTRON–MOLECULE SCATTERING AND PHOTOIONIZATION
Edited by P. G. Burke and J. B. West

ATOMIC SPECTRA AND COLLISIONS IN EXTERNAL FIELDS
Edited by K. T. Taylor, M. H. Nayfeh, and C. W. Clark

ATOMIC PHOTOEFFECT
M. Ya. Amusia

MOLECULAR PROCESSES IN SPACE
Edited by Tsutomu Watanabe, Isao Shimamura, Mikio Shimizu, and Yukikazu Itikawa

THE HANLE EFFECT AND LEVEL CROSSING SPECTROSCOPY
Edited by Giovanni Moruzzi and Franco Strumia

ATOMS AND LIGHT: INTERACTIONS
John N. Dodd

POLARIZATION BREMSSTRAHLUNG
Edited by V. N. Tsytovich and I. M. Ojringel

INTRODUCTION TO THE THEORY OF LASER–ATOM INTERACTIONS (Second Edition)
Marvin H. Mittleman

ELECTRON COLLISIONS WITH MOLECULES, CLUSTERS, AND SURFACES
Edited by H. Ehrhardt and L. A. Morgan

THEORY OF ELECTRON–ATOM COLLISIONS, Part 1: Potential Scattering
Philip G. Burke and Charles J. Joachain

POLARIZED ELECTRON/POLARIZED PHOTON PHYSICS
Edited by H. Kleinpoppen and W. R. Newell

INTRODUCTION TO THE THEORY OF X-RAY AND ELECTRONIC SPECTRA OF FREE ATOMS
Romas Karazija

VUV AND SOFT X-RAY PHOTOIONIZATION
Edited by Uwe Becker and David A. Shirley

DENSITY MATRIX THEORY AND APPLICATIONS (Second Edition)
Karl Blum

SELECTED TOPICS ON ELECTRON PHYSICS
Edited by D. Murray Campbell and Hans Kleinpoppen

PHOTON AND ELECTRON COLLISIONS WITH ATOMS AND MOLECULES
Edited by Philip G. Burke and Charles J. Joachain

COINCIDENCE STUDIES OF ELECTRON AND PHOTON IMPACT IONIZATION
Edited by Colm T. Whelan and H. R. J. Walters

PRACTICAL SPECTROSCOPY OF HIGH-FREQUENCY DISCHARGES
Sergei A. Kazantsev, Vyacheslav I. Khutorshchikov, Günter H. Guthöhrlein, and Laurentius Windholz

IMPACT SPECTROPOLARIMETRIC SENSING
S. A. Kazantsev, A. G. Petrashen, and N. M. Firstova

NEW DIRECTIONS IN ATOMIC PHYSICS
Edited by Colm T. Whelan, R. M. Dreizler, J. H. Macek, and H. R. J. Walters

COMPLETE SCATTERING EXPERIMENTS
Edited by Uwe Becker and Albert Crowe

ELECTRON MOMENTUM SPECTROSCOPY
Erich Weigold and Ian McCarthy

POLARIZATION AND CORRELATION PHENOMENA IN ATOMIC COLLISIONS: A Practical Theory Course
Vsevolod V. Balashov, Alexei N. Grum-Grzhimailo, and Nikolai M. Kabachnik

RELATIVISTIC HEAVY-PARTICLE COLLISION THEORY
Derrick S. F. Crothers

INTRODUCTION TO THE THEORY OF COLLISIONS OF ELECTRONS WITH ATOMS AND MOLECULES
S. P. Khare

FUNDAMENTAL ELECTRON INTERACTIONS WITH PLASMA PROCESSING GASES
Loucas G. Christophorou and James K. Olthoff